THE SCIENCE AND ENGINEERING OF MATERIALS

THE SCIENCE AND ENGINEERING OF MATERIALS

SI Edition

DONALD R. ASKELAND

University of Missouri—Rolla

SI Edition prepared by
JANE RANDALL and MAURICE DENTON

VNR
International

First published in 1984 in the USA by
PWS Publishers, 20 Park Plaza, Boston, Massachusetts 02116

SI edition first published in UK 1988 by
Van Nostrand Reinhold (International) Co. Ltd
11 New Fetter Lane, London EC4P 4EE

© 1984 Wadsworth Inc., Belmont, California 94002
© 1988 SI edition Van Nostrand Reinhold (International) Co. Ltd

Printed in Hong Kong

ISBN 0 278 00057 6

British Library Cataloguing in Publication Data

Askeland, Donald R.
   The science and engineering of materials.
   - SI ed.
   1. Materials
   I. Title
   620.1'1

ISBN 0-278-00057-6

PREFACE

Our understanding of the relationship between structure and properties provides the basis for our development of new materials. For instance, the electronic and atomic structure of materials has been our model for a myriad miniaturized electronic components. By manipulating molecular structure, we have produced a vast spectrum of polymers; by controlling microstructure, we have developed many new metal alloys and ceramics. And we've played wizard, juggling composite materials that have unique properties.

In engineering materials, we use such scientific understanding to shape materials into useful products. Our materials processing both depends on and influences the structure and properties we are using. For instance, we can obtain directional properties while casting or deformation-processing metals; the directional properties thus obtained in turn influence the subsequent processing and behavior of the metal. So we must link science and processing in order to understand and select materials for engineering applications.

The environments present during processing and use also affect the characteristics of a material. By melting and pouring aluminum alloys in air, for example, we can produce gas pores in the finished casting. High-strength alloys may (catastrophically) lose their properties when exposed to high temperatures. And the properties of polymers may change dramatically when the material is exposed to radiation.

This book presents the three-way relationship between structure, properties, and processing. The text can serve, first, those engineering students who are formally introduced to materials in only one course, who do not continue in this field. Such students need a basic understanding of material behavior, available materials, and the processing of materials so that they can help select materials. Second, this text can introduce the science of materials, the types of materials available, and the application and processing of materials to the materials-oriented engineering student. Such students will later go on to study the details in more advanced courses.

Part I of the text introduces atomic and crystal structure, the foundation for understanding the mechanical and physical behavior of materials. Part II explains how we control the structure and mechanical properties of metals, with an emphasis on strengthening mechanisms. This part also explores structure-property relationship by considering processing techniques—solidification, deformation, and heat treatment—as well as alloying. Part III describes the common alloys, ceramics, polymers, and composite materials, showing how to control their mechanical properties. For each, the structure-property relationship is developed, then individual materials are discussed, including processing and applications. Part IV presents physical properties via a reexamination of electronic and atomic structure. Changes in structure and processing affect engineering applications of materials. Finally, Part V describes the way materials perform during service, with an emphasis on preventing and analyzing corrosion and mechanical failure.

To use this book, students need some background in chemistry, physics, and higher mathematics; sophomore standing is recommended. By selecting topics, an instructor can emphasize metals, mechanical behavior, physical properties, or introductory materials science. The book offers many examples and practice problems to help students understand the principles of materials science and engineering.

Thanks are due to many people who have helped me prepare this text, especially Robert Wolf, Fred Kisslinger, Mike Norberg, Greg Lynch, and Darren Washausen. I am particularly indebted to my wife Mary and son Per.

NOTE TO SI EDITION

The adoption in Europe of the International System of units (SI) has prompted the editing of this textbook. An opportunity has been taken to change to SI units throughout the text with the exception of certain units which have been deliberately left in metric form because of the frequency of their use.

With the exception of very few minor alterations the text remains unchanged.

Certain British Standard Specifications have been included in Chapter 13 — Ferrous Alloys.

Appendix C has been introduced to include certain SI nomenclature and a limited number of conversion factors.

CONTENTS

Chapter 1 INTRODUCTION TO MATERIALS   1

1-1 Introduction   *1*
1-2 Types of Materials   *1*
1-3 Structure-Property-Processing Relationship   *5*
1-4 Environmental Effects on Material Behavior   *11*
1-5 Strength-to-Weight Ratio   *13*

Part I ATOMIC STRUCTURE, ARRANGEMENT, AND MOVEMENT   17

Chapter 2 ATOMIC STRUCTURE   19

2-1 Introduction   *19*
2-2 The Structure of the Atom   *19*
2-3 The Electronic Structure of the Atom   *20*
2-4 The Periodic Table   *27*
2-5 Atomic Bonding   *31*
2-6 Binding Energy and Interatomic Spacing   *37*

Chapter 3 ATOMIC ARRANGEMENT   41

3-1 Introduction   *41*
3-2 Short-Range Order versus Long-Range Order   *41*
3-3 Unit Cells   *43*
3-4 Points, Directions, and Planes in the Unit Cell   *49*
3-5 Allotropic Transformations   *60*
3-6 Complex Crystal Structures   *61*

Chapter 4   IMPERFECTIONS IN THE ATOMIC ARRANGEMENT   72

4-1 Introduction   72
4-2 Dislocations   72
4-3 Significance of Dislocations   74
4-4 Schmid's Law   76
4-5 Influence of Crystal Structure   78
4-6 Control of the Slip Process   81
4-7 Dislocation Interactions   82
4-8 Point Defects   82
4-9 Surface Defects   85

Chapter 5   ATOM MOVEMENT IN MATERIALS   93

5-1 Introduction   93
5-2 Self-diffusion   93
5-3 Diffusion in Alloys   93
5-4 Diffusion Mechanisms   94
5-5 Activation Energy for Diffusion   95
5-6 Rate of Diffusion (Fick's First Law)   97
5-7 Composition Profile (Fick's Second Law)   103
5-8 Interdiffusion and the Kirkendall Effect   105
5-9 Types of Diffusion   108
5-10 Grain Growth and Diffusion   110
5-11 Diffusion Bonding   111
5-12 Sintering and Powder Metallurgy   112
5-13 Diffusion in Ionic Compounds   113
5-14 Diffusion in Polymers   113

Part II   CONTROLLING THE MICROSTRUCTURE AND MECHANICAL PROPERTIES OF MATERIALS   119

Chapter 6   MECHANICAL TESTING AND PROPERTIES   121

6-1 Introduction   121
     Tensile Test   121
6-2 The Stress-Strain Diagram   121
6-3 Elastic versus Plastic Deformation   126
6-4 Yield Strength   126
6-5 Tensile Strength   128
6-6 True Stress-True Strain   128
6-7 Brittle Behavior   129
6-8 Modulus of Elasticity   130
6-9 Ductility   131
6-10 Temperature Effects   132

Impact Test   132

6-11 Nature of the Impact Test   *132*

6-12 Temperature Effects of the Impact Test   *133*

6-13 Notch Sensitivity   *134*

6-14 Relation of Impact Energy to True Stress-True Strain   *135*

6-15 Use and Precautions of Impact Properties   *135*

Fatigue Test   136

6-16 Nature of the Fatigue Test   *136*

6-17 Results of the Fatigue Test   *136*

6-18 Factors Affecting Fatigue Properties   *138*

Creep Test   138

6-19 Nature of the Creep Test   *138*

6-20 Use of Creep Data   *141*

Hardness Test   143

6-21 Nature of the Hardness Test   *143*

Chapter 7   SOLIDIFICATION AND GRAIN SIZE STRENGTHENING   151

7-1 Introduction   *151*

7-2 Nucleation   *151*

7-3 Growth   *156*

7-4 Solidification Time   *158*

7-5 Cooling Curves   *161*

7-6 Casting or Ingot Structure   *162*

7-7 Solidification Defects   *164*

7-8 Control of Casting Structure   *171*

7-9 Solidification and Metals Joining   *173*

Chapter 8   SOLIDIFICATION AND SOLID SOLUTION STRENGTHENING   181

8-1 Introduction   *181*

8-2 Phases, Solutions, and Solubility   *181*

8-3 Conditions for Unlimited Solid Solubility in Metals   *183*

8-4 Solid Solution Strengthening   *184*

8-5 Isomorphous Phase Diagrams   *187*

8-6 Relationship between Strength and the Phase Diagram   *193*

8-7 Solidification of a Solid Solution Alloy   *193*

8-8 Nonequilibrium Solidification of Solid Solution Alloys   *194*

8-9 Segregation   *197*

8-10 Castability of Alloys with a Freezing Range   *200*

8-11 Ceramic and Polymer Systems   *201*

Chapter 9   DEFORMATION, STRAIN HARDENING, AND ANNEALING   207

9-1 Introduction   *207*

Cold Working   **207**

9-2 Relationship to the Stress-Strain Curve   *207*

9-3 Dislocation Multiplication   *209*

9-4 Properties versus Percent Cold Work   *209*

9-5 Microstructure of Cold-Worked Metals   *213*

9-6 Residual Stresses   *216*

9-7 Characteristics of Cold Working   *217*

Annealing   **219**

9-8 Three Stages of Annealing   *219*

9-9 Control of Annealing   *222*

9-10 Annealing Textures   *224*

9-11 Control of Properties by Combining Cold Working and Annealing   *224*

9-12 Implications of Annealing on High-Temperature Properties   *225*

Hot Working   **225**

9-13 Characteristics of the Hot-Working Process   *225*

9-14 Deformation Processing by Hot Working   *228*

9-15 Deformation Bonding Processes   *228*

9-16 Superplastic Forming   *231*

Chapter 10   SOLIDIFICATION AND DISPERSION STRENGTHENING   239

10-1 Introduction   *239*

10-2 Principles of Dispersion Strengthening   *239*

10-3 Intermetallic Compounds   *240*

10-4 Phase Diagrams Containing Three-Phase Reactions   *242*

10-5 The Eutectic Phase Diagram   *245*

10-6 Strength of Eutectic Alloys   *252*

10-7 Nonequilibrium Freezing in the Eutectic System   *258*

10-8 The Peritectic Reaction   *259*

10-9 The Monotectic Reaction   *260*

10-10 Ternary Phase Diagrams   *261*

Chapter 11   DISPERSION STRENGTHENING BY PHASE TRANSFORMATION AND HEAT TREATMENT   275

11-1 Introduction   *275*

11-2 Nucleation and Growth in Solid-State Reactions   *275*

11-3 Alloys Strengthened by Exceeding the Solubility Limit   *276*

11-4 Age Hardening or Precipitation Hardening   *280*

11-5 Effects of Aging Temperature and Time   *283*

11-6   Requirements for Age Hardening   *284*
11-7   Use of Age Hardenable Alloys at High Temperatures   *285*
11-8   Residual Stresses During Quenching   *287*
11-9   The Eutectoid Reaction   *288*
11-10 Controlling the Eutectoid Reaction   *294*
11-11 The Martensitic Reaction   *300*
11-12 Tempering of Martensite   *304*

Part III   ENGINEERING MATERIALS   311

Chapter 12   NONFERROUS ALLOYS   313

12-1   Introduction   *313*
12-2   Aluminum Alloys   *313*
12-3   Magnesium Alloys   *322*
12-4   Beryllium   *324*
12-5   Copper Alloys   *326*
12-6   Nickel and Cobalt   *334*
12-7   Zinc Alloys   *337*
12-8   Titanium Alloys   *338*
12-9   Zirconium   *345*
12-10 Refractory Metals   *345*

Chapter 13   FERROUS ALLOYS   351

13-1   Introduction   *351*
13-2   Review of the Fe-Fe₃C Phase Diagram   *351*
13-3   Designation and Typical Structures of Steels   *352*
13-4   Simple Heat Treatments   *355*
13-5   Isothermal Heat Treatments and Dispersion Strengthening   *357*
13-6   Quench and Temper Heat Treatments   *362*
13-7   Purpose of Alloying Elements in Steels   *367*
13-8   Effect of Alloying Elements on the IT and CCT Diagrams   *367*
13-9   Hardenability Curves   *372*
13-10 Tool Steels   *377*
13-11 Special Steels   *377*
13-12 Surface Treatments   *378*
13-13 Weldability of Steel   *380*
13-14 Stainless Steels   *383*

Cast Irons   390

13-15 Solidification of Cast Irons   *390*
13-16 The Matrix Structure in Cast Irons   *396*
13-17 Characteristics and Production of the Cast Irons   *397*

Chapter **14**   CERAMIC MATERIALS   412

14-1 Introduction   *412*
14-2 Short-Range Order in Crystalline Ceramic Materials   *412*
14-3 Long-Range Order in Crystalline Ceramic Materials   *416*
14-4 Silicate Structures   *419*
14-5 Imperfections in Crystalline Ceramic Structures   *421*
14-6 Noncrystalline Ceramic Materials   *425*
14-7 Deformation and Failure   *428*
14-8 Phase Diagrams in Ceramic Materials   *430*
14-9 Processing of Ceramics   *438*
14-10 Applications and Properties of Ceramics   *447*

Chapter **15**   POLYMERS   455

15-1 Introduction   *455*
15-2 Fitting Polymers into Categories   *455*
15-3 Representing the Structure of Polymers   *456*
15-4 Chain Formation by the Addition Mechanism   *456*
15-5 Chain Formation by the Condensation Mechanism   *465*
15-6 Degree of Polymerization   *467*
15-7 Deformation of Thermoplastic Polymers   *470*
15-8 Effect of Temperature on Behavior of Thermoplastics   *472*
15-9 Controlling the Structure and Properties of Thermoplastics   *477*
15-10 Elastomers (Rubbers)   *487*
15-11 Thermosetting Polymers   *492*
15-12 Additives to Polymers   *498*
15-13 Forming of Polymers   *499*

Chapter **16**   COMPOSITE MATERIALS   507

16-1 Introduction   *507*
16-2 Particulate-Reinforced Composite Materials   *507*
16-3 Dispersion-Strengthened Composites   *507*
16-4 True Particulate Composites   *511*
16-5 Applications for Particulate Composites   *512*
16-6 Fiber-Reinforced Composites   *518*
16-7 Predicting Properties of Fiber-Reinforced Composites   *518*
16-8 Characteristics of Fiber-Reinforced Composites   *523*
16-9 Manufacturing Fibers and Composites   *528*
16-10 Fiber-Reinforced Systems   *530*
16-11 Laminar Composite Materials   *534*
16-12 Examples and Applications of Laminar Composites   *536*

16-13 Manufacturing Laminar Composites   539
16-14 Wood   540
16-15 Concrete and Asphalt   543
16-16 Sandwich Structures   545

Part IV PHYSICAL PROPERTIES OF ENGINEERING MATERIALS   549

Chapter 17 ELECTRICAL CONDUCTIVITY   551

17-1 Introduction   551
17-2 Relationship Between Ohm's Law and Electrical Conductivity   551
17-3 Band Theory   554
17-4 Band Structure of Alkali Metals   555
17-5 Band Structure of Other Metals   558
17-6 Controlling the Conductivity of Metals   560
17-7 Thermocouples   566
17-8 Superconductivity   568
17-9 Energy Gaps—Insulators and Semiconductors   570
17-10 Intrinsic Semiconductors   571
17-11 Extrinsic Semiconductors   575
17-12 Applications of Semiconductors to Electrical Devices   579
17-13 Manufacture and Fabrication of Semiconductor Devices   585
17-14 Conductivity of Ionic Materials   586

Chapter 18 DIELECTRIC AND MAGNETIC PROPERTIES   591

18-1 Introduction   591
18-2 Dipoles   591
18-3 Polarization in an Electric Field   591
18-4 Dielectric Properties and Capacitors   594
18-5 Controlling Dielectric Properties   598
18-6 Dielectric Properties and Electrical Insulators   603
18-7 Piezoelectricity and Electrostriction   603
18-8 Ferroelectricity   605
18-9 Magnetization versus Polarization   607
18-10 Magnetic Dipoles and Magnetic Moments   607
18-11 Magnetization, Permeability, and the Magnetic Field   609
18-12 Interactions Between Magnetic Dipoles and the Magnetic Field   611
18-13 Domain Structure   613
18-14 Application of the Magnetization-Field Curve   615
18-15 Temperature Effects   618
18-16 Magnetic Materials   619
18-17 Eddy Current Losses   624

Chapter 19   OPTICAL, THERMAL, AND ELASTIC PROPERTIES   629

19-1 Introduction   629
Optical Properties   629
19-2 Emission of Continuous and Characteristic Radiation   629
19-3 Examples of Emission Phenomena   633
19-4 Interaction of Photons with a Material   639
Thermal Properties   648
19-5 Heat Capacity   648
19-6 Thermal Expansion   650
19-7 Thermal Conductivity   653
Elastic Properties   657
19-8 Elastic Behavior   657
19-9 Anelastic and Thermoelastic Behavior   661

Part V   PROTECTION AGAINST DETERIORATION AND FAILURE OF MATERIALS   667

Chapter 20   CORROSION AND WEAR   669

20-1 Introduction   669
20-2 Chemical Corrosion   669
20-3 The Electrochemical Cell   672
20-4 The Electrode Potential in Electrochemical Cells   673
20-5 The Corrosion Current in the Electrochemical Cell   677
20-6 Sources of Polarization   684
20-7 Types of Electrochemical Corrosion   686
20-8 Protection Against Electrochemical Corrosion   691
20-9 Oxidation and Other Gas Reactions   699
20-10 Radiation Damage   703
20-11 Wear and Erosion   703

Chapter 21   FAILURE—ORIGIN, DETECTION, AND PREVENTION   711

21-1 Introduction   711
21-2 Determining the Fracture Mechanism in Metal Failures   711
21-3 Fracture in Nonmetallic Materials   720
21-4 Source and Prevention of Failures in Metals   721
21-5 Detection of Potentially Defective Materials   726
21-6 Fracture Mechanics   739

Appendix A   SELECTED PHYSICAL PROPERTIES OF METALS   746

Appendix B   THE ATOMIC AND IONIC RADII OF SELECTED
ELEMENTS   748

Appendix C   NOMENCLATURE AND CONVERSION FACTORS
FOR SI SYSTEM   749

ANSWERS TO SELECTED ODD-NUMBERED PRACTICE
PROBLEMS   A1

INDEX   A5

CHAPTER **1**

Introduction to Materials

1-1 Introduction

All engineers are involved with materials on a daily basis. We manufacture and process materials, design and construct components or structures using materials, select materials, analyze failures of materials, or simply hope the materials we are using perform adequately.

As responsible engineers, we are interested in improving the performance of the product we are designing or manufacturing. Electrical engineers want integrated circuits to perform properly, switches in computers to react instantly, and insulators to withstand high voltages even under the most adverse conditions. Civil, structural, and architectural engineers wish to construct strong, reliable structures that are aesthetic and resistant to corrosion. Petroleum and chemical engineers require drill bits or piping that survive in abrasive or corrosive conditions. Automotive engineers desire lightweight yet strong and durable materials. Aerospace engineers demand lightweight materials that perform well both at high temperatures and in the cold vacuum of outer space. Metallurgical, ceramic, and polymer engineers wish to produce and shape materials that are more economical and possess improved properties.

The intent of this text is to permit the student to become aware of the types of materials available, to understand their general behavior and capabilities, and to recognize the effects of the environment and service conditions on the material's performance.

1-2 Types of Materials

We will classify materials into four groups—metals, ceramics, polymers, and composite materials (Table 1-1).

Metals. Metals and alloys, which include steel, aluminum, magnesium, zinc, cast iron, titanium, copper, nickel, and many others, have the general characteristics of good electrical and thermal conductivity, relatively high strength, high stiffness, ductility or formability, and shock resistance (Figure 1-1). They are particularly useful for structural or load-bearing applications. Although pure metals are occasionally used, combinations of metals called *alloys* are normally designed to provide improvement in a particular desirable property or permit better combinations of properties.

1

TABLE 1-1   Representative examples, applications, and properties for each category of materials

	Applications	Properties
Metals		
Copper	Electrical conductor wire	High electrical conductivity, good formability
Gray cast iron	Automobile engine blocks	Castability, machinability, vibration damping
Fe—3% Si	Motors and generators	Excellent ferromagnetic properties
Alloy steels	Wrenches	Good strengthening by heat treatment
Ceramics		
SiO_2–Na_2O–CaO	Window glass	Good optical properties and thermal insulation
Al_2O_3, MgO, SiO_2	Refractories for containing molten metal	Thermal insulation, high melting temperature, relatively inert to molten metal
Barium titanate	Transducers for stereo record players	Piezoelectric behavior converting sound to electricity
Polymers		
Polyethylene	Food packaging	Easily formed into thin flexible airtight film
Epoxy	Encapsulation of integrated circuits	Good electrical insulation and moisture resistance
Phenolics	Adhesives to join plies in plywood for marine use	Strength and moisture resistance
Composites		
Graphite-epoxy	Aircraft components	High strength-to-weight ratio
Tungsten carbide-cobalt	Carbide cutting tools for machining	High hardness yet good shock resistance
Titanium-clad steel	Reactor vessels	Low cost and high strength of steel with good corrosion resistance of titanium

Ceramics.   Ceramics, such as brick, glass, tableware, insulators, and abrasives, have poor electrical and thermal conductivity. Although ceramics may have good strength and hardness, their ductility, formability, and shock resistance are poor. Consequently, ceramics are less often used for structural or load-bearing applications than are metals. However, many ceramics have excellent resistance to high temperatures and certain corrosive media and have a number of unusual and desirable optical, electrical, and thermal properties.

Polymers.   Polymers include rubber, plastics, and many types of adhesives. They are produced by creating large molecular structures from organic molecules, obtained from petroleum or agricultural products, in a process known as *polymerization* (Figure 1-2). Polymers have low electrical and thermal conductiv-

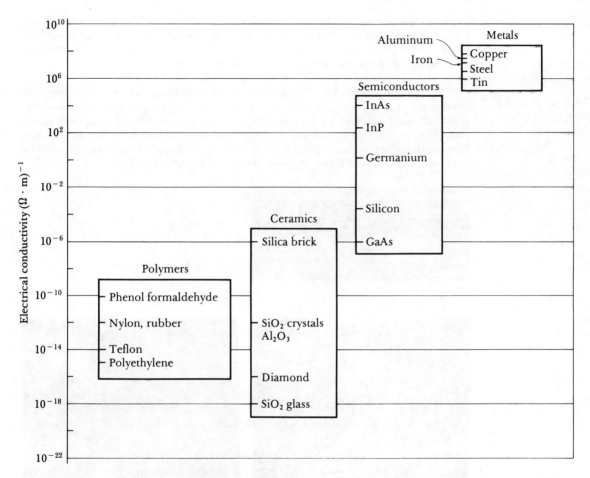

FIG. 1-1   Extremely large differences in electrical conductivity are observed between the different categories of materials.

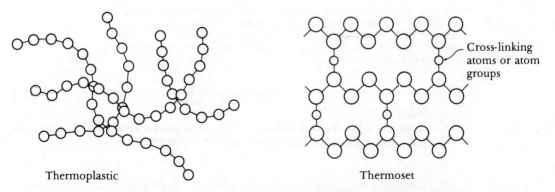

FIG. 1-2   Polymerization occurs when small molecules combine to produce larger molecules, or polymers. The polymer molecules can have a chainlike structure (thermoplastics) or can form three-dimensional networks (thermosets).

ity, have low strengths, and are not suitable for use at high temperatures. Some polymers (*thermoplastics*) have excellent ductility, formability, and shock resistance while others (*thermosets*) have the opposite properties. Polymers are lightweight and frequently have excellent resistance to corrosion.

Composite materials.   Composites are formed from two or more materials, producing properties that cannot be obtained by any single material. Concrete, plywood, and fiberglass are typical, although crude, examples of composite materials (Figure 1-3). With composites we can produce lightweight, strong, ductile, high temperature-resistant materials that are otherwise unobtainable, or produce hard yet shock-resistant cutting tools that would otherwise shatter.

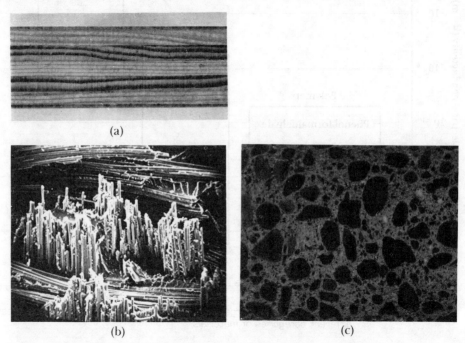

(a)

(b)   (c)

FIG. 1-3   Some examples of composite materials. (a) Plywood is a laminar composite composed of layers of wood veneer. (b) Fiberglass is a fiber-reinforced composite containing stiff, strong glass fibers in a softer polymer matrix. (c) Concrete is a particulate composite containing coarse sand or gravel in a cement matrix.

EXAMPLE 1-1

You wish to select the materials needed to carry a current between the components inside an electrical "black box." What materials would you select?

Answer:

The material that actually carries the current must have a high electrical conductivity. Thus, we need to select a *metal* wire. Copper, aluminum, gold, or silver might all serve. However, the metal wire must be insulated from the rest of the "black box" to prevent short circuits or arcing. Although a ceramic coating would be an excellent insulator, ceramics are brittle; the wire could not be bent without the ceramic coating breaking off. Instead we would select a *polymer* or plastic coating with good insulating characteristics yet good ductility.

EXAMPLE 1-2

What materials are used to make coffee cups? What particular property makes these materials suitable?

Answer:

Coffee cups are normally made of ceramic or plastic materials. Both ceramics and polymers have excellent thermal insulation due to their low thermal conductivity. Disposable expanded polystyrene cups are particularly effective, since they contain many gas bubbles which further improve insulation. Actually, we could consider the disposable cups to be made from a composite of a polymer and gas!

Metal cups, however, are seldom used because the high thermal conductivity permits the heat to be transferred, burning our hands.

1-3 Structure-Property-Processing Relationship

We are interested in producing a component that has the proper shape and properties, permitting the component to perform its task for its expected lifetime. The materials engineer meets this requirement by taking advantage of a complex three-part relationship between the internal structure of the material, the processing of the material, and the final properties of the material (Figure 1-4). When the materials engineer changes one of these three aspects of the relationship, either or both of the others also change. We must therefore determine how the three aspects interrelate in order to finally produce the required product.

Properties. We can consider the properties of a material in two categories—mechanical and physical (Table 1-2).

The mechanical properties describe how the material responds to an applied force or stress. *Stress* is defined as the force divided by the cross-sectional area on which the force acts. The most common mechanical properties are the strength, ductility, and stiffness of the material. However, we are often interested in how the material behaves when it is exposed to a sudden intense blow (impact), continually cycled through an alternating force (fatigue), exposed to high temperatures (creep), or subjected to abrasive conditions (wear). The mechanical properties not only determine how well the material performs in service, but also determine the ease with which the material can be formed into a useful shape. A metal part formed by forging must withstand the rapid application of a force without breaking and have a high enough ductility to deform to the proper shape. Often small changes in the structure have a profound effect on the mechanical properties of the material.

Physical properties include electrical, magnetic, optical, thermal, elastic, and chemical behavior. The physical properties depend both on structure and processing of the material. Even tiny changes in the composition cause a profound change in the electrical conductivity of many semiconducting metals and ceramics.

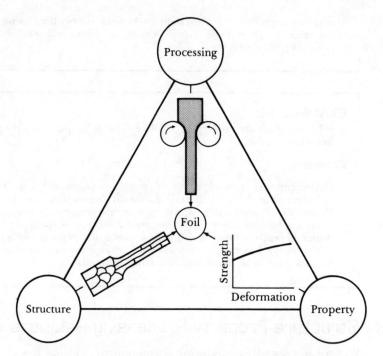

FIG. 1-4   The three-part relationship between structure, properties, and processing method. When aluminum is rolled into foil, the rolling process changes the metal's structure and increases its strength.

High firing temperatures may greatly reduce the thermal insulation characteristics of ceramic brick. Small amounts of impurities change the color of a glass or polymer.

EXAMPLE 1-3

Describe some of the key mechanical and physical properties we would consider in selecting a material for an airplane wing.

Answer:

First, let's look at mechanical properties. We obviously want the material to have a high strength to support the forces acting on the wing. We must also recognize that the wing is exposed to cyclical application of a force as well as vibration—this suggests that fatigue properties are important. Ordinarily, shock loading (impact), high temperatures (creep), and abrasive conditions (wear) are not encountered.

Important physical properties are density and corrosion resistance. The wing should be as light as possible, so the material should have a low density.

EXAMPLE 1-4

What properties are required for a solar cell used to generate electricity for a satellite?

Answer:

> The optical properties are most important in this case! The materials for solar cells, such as silicon, must interact with radiation, or light, to change the electron configuration of the atom. This interaction and change in structure in turn produce the electrical current that is desired.

Structure. The structure of a material can be considered on several levels, all of which influence the final behavior of the product (Figure 1-5). At the finest level is the structure of the individual atoms that compose the material. The arrangement of the electrons surrounding the nucleus of the atom significantly affects electrical, magnetic, thermal, and optical behavior and may also influence corrosion resistance. Furthermore, the electronic arrangement influences how the atoms are bonded to one another and helps determine the type of material—metal, ceramic, or polymer.

At the next level, the arrangement of the atoms in space is considered. Metals, many ceramics, and some polymers have a very regular atomic arrangement, or crystal structure. The crystal structure influences the mechanical proper-

TABLE 1-2 Typical examples of the properties of materials

Mechanical Properties	Physical Properties
Creep	Chemical
Creep rate	Corrosion
Stress-rupture properties	Refining
Ductility	Density
% Elongation	Electrical
% Reduction in area	Conductivity
Fatigue	Dielectric (insulation)
Endurance limit	Ferroelectricity
Fatigue life	Piezoelectricity
Hardness	Magnetic
Scratch resistance	Ferrimagnetic
Wear rates	Ferromagnetic
Impact	Paramagnetic
Absorbed energy	Optical
Toughness	Absorption
Transition temperature	Color
Strength	Diffraction
Modulus of elasticity	Lasing action
Tensile strength	Photoconduction
Yield strength	Reflection
	Refraction
	Transmission
	Thermal
	Heat capacity
	Thermal conductivity
	Thermal expansion

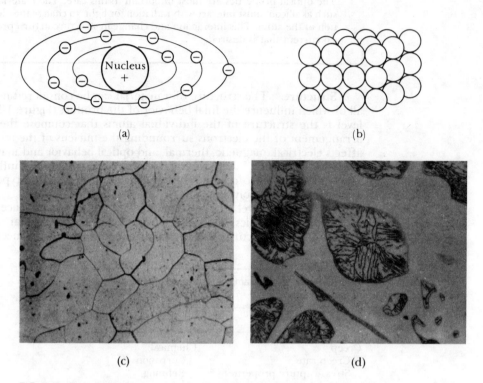

(a)

(b)

(c)

(d)

FIG. 1-5 Four levels of structure in a material. (a) Atomic structure, (b) atomic arrangement, (c) grain structure in iron, (d) multiple-phase structure in white cast iron.

ties of metals such as ductility, strength, and shock resistance. Other ceramic materials and most polymers have no orderly atomic arrangement—these amorphous or glassy materials behave much differently from crystalline materials. For instance, glassy polyethylene is transparent while crystalline polyethylene is translucent. Defects in this atomic arrangement exist and may be controlled to produce profound changes in properties.

A grain structure is found in most metals, some ceramics, and occasionally in polymers. Between the grains, the atomic arrangement changes its orientation and thus influences properties. The size and shape of the grains play a key role at this level.

Finally, in most materials, more than one phase is present, with each phase having its unique atomic arrangement and properties. Control of the type, size, distribution, and amount of these phases within the main body of the material provides an additional way to control properties.

Processing. Materials processing produces the desired shape of a component from the initial formless material (Table 1-3). Metals can be processed by pouring liquid metal into a mold (casting), joining individual pieces of metal (welding, brazing, soldering, adhesive bonding), forming the solid metal into

TABLE 1-3 Typical materials processing techniques

Metals

Casting: sand, die cast, permanent mold, investment, continuous casting	Liquid metal is poured or injected into a solid mold to produce a desired shape.
Forming: forging, wire drawing, deep drawing, bending	Solid metal is deformed by high pressure, often while hot, into useful shapes.
Joining: gas welding, resistance welding, brazing, arc welding, soldering, friction welding, diffusion bonding	Several pieces of metal are joined together, using liquid metal, deformation, or high pressures and temperatures to provide bonding.
Machining: turning, drilling, milling, cutting	Metal is removed by a cutting operation, leaving a finished shape.
Powder metallurgy	Metal powders are compacted at high pressures into a useful shape, then heated at high temperatures to permit the particles to join together.

Ceramics

Casting, including slip casting	Liquid or slurries of liquid plus solid ceramics are poured into a desirable shape.
Compaction: extrusion, pressing, isostatic forming	Solid or viscous slurries of liquid and solid ceramics are compacted into a useful shape.
Sintering	Compacted solids are heated at high temperatures to cause the solids to bond together.

Polymers

Molding: injection molding, transfer molding	Hot or even liquid polymer is forced into a mold; this resembles the casting process.
Forming: spinning, extrusion, vacuum forming	Heated polymer is forced through a die opening or around a pattern to produce a shape.

Composites

Casting, including infiltration	A liquid surrounds one of the constituents to produce the completed composite.
Forming	A soft constituent is forced by pressure to deform around a second constituent of the composite.
Joining: adhesive bonding, explosive bonding, diffusion bonding	The two constituents are joined together by gluing, deformation, or high-temperature processes.
Compaction and sintering	Powdered constituents are pressed into shapes, then heated to cause the powders to join.

useful shapes using high pressures (forging, drawing, extrusion, rolling, bending), compacting tiny metal powder particles into a solid mass (powder metallurgy), or removing excess material (machining). Similarly, ceramic materials can be formed into shapes by related processes such as casting, forming, extrusion, or compaction, often while wet, and heat treatment at high temperatures to drive off the fluids and to bond the individual constituents together. Polymers are produced by injection of softened plastic into molds (much like casting), drawing, and forming. Often a material is heat treated at some temperature below its melting temperature to effect a desired change in structure. The type of processing we use depends, at least partly, on the properties, and thus the structure, of the material.

EXAMPLE 1-5

The first step in the manufacture of tungsten filaments for light bulbs is by powder metallurgy rather than casting. Explain.

Answer:

One of the physical properties of tungsten is its high melting temperature, 3410°C. In order to make a casting, the tungsten must be heated to an exceptionally high temperature. Powder metallurgy processing, by which powdered tungsten is compacted into a solid mass, is done at much lower temperatures.

Structure-property-processing interaction.   The processing of a material affects the structure. The structure of a copper bar is very different if it is produced by casting rather than forming (Figure 1-6). The shape, size, and orientation of the grains may be different, the cast structure may contain voids due to shrinkage or gas bubbles, and nonmetallic particles (inclusions) may be trapped within the structure. The formed material may contain elongated nonmetallic particles and internal defects in the atomic arrangement. Thus, the structure and consequently the final properties of the casting are very different from those of the formed product.

On the other hand, the original structure and properties determine how we can process the material to produce a desired shape. A casting containing large

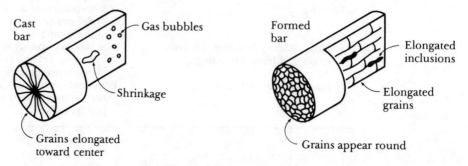

FIG. 1-6   The difference in structure between a copper bar produced by casting and a copper bar produced by forming.

shrinkage voids may crack during a subsequent processing step. Alloys that have been strengthened by introducing imperfections in the structure also become brittle and fail during forming. Elongated grains in a metal may lead to nonuniform shapes when subsequently formed. Thermosetting polymers cannot be formed, while thermoplastic polymers are easily formed.

1-4 Environmental Effects on Material Behavior

The structure-property-processing relationship is also influenced by the surroundings to which the material is subjected.

Loading. The type of force, or load, acting on the material may dramatically change its behavior. Normally, the yield strength, above which the material suffers a permanent change in its dimensions, is the most critical property and is usually the most important consideration in the design of a material component. However, a material that displays a high yield strength may fail rather easily at lower loads if the loading is cyclical (fatigue) or applied suddenly (impact). The engineer must recognize the type of loading to which the material is exposed.

Temperature. Changes in temperature dramatically alter the properties of materials. The strength of most materials decreases as the temperature increases. Furthermore, sudden catastrophic changes may occur when heating above critical temperatures. Metals that have been strengthened by certain heat treatments or forming techniques may suddenly lose their strength when heated (Figure 1-7). Very low temperatures may cause a metal to fail in a brittle manner even though

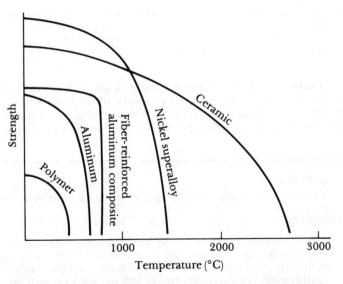

FIG. 1-7 Increasing temperature normally reduces the strength of a material. Polymers are suitable only at low temperatures. Some composites, special alloys, and ceramics have excellent properties at high temperatures.

the applied loads are low. High temperatures can also change the structure of ceramics or cause polymers to melt or char.

Atmosphere.   Most metals and polymers react with oxygen or other gases, particularly at elevated temperatures. Metals and ceramics may catastrophically disintegrate (Figure 1-8) or be chemically attacked, while others may be protected. Polymers often are hardened or may even depolymerize, char, or burn. Steels may react with hydrogen and become brittle.

Corrosion.   Metals are attacked by a variety of corrosive liquids. The metal may be uniformly or selectively consumed or may develop cracks or pits leading to premature failure (Figure 1-9). Ceramics can be attacked by other liquid ceramics, while solvents can dissolve polymers.

   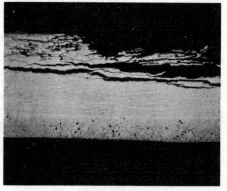

FIG. 1-8   When hydrogen dissolves in tough pitch copper, steam is produced at the grain boundaries, thus creating thin voids. The metal is then weak and brittle and fails easily.

FIG. 1-9   Attack of an aluminum fuel tank by bacteria in contaminated jet fuel causes severe corrosion, pitting, and eventual failure.

Radiation.   High-energy radiation, such as neutrons produced in nuclear reactors, can affect the internal structure of all materials, producing a loss of strength, embrittlement, or critical alteration of physical properties. External dimensions may also change, causing swelling or even cracking.

EXAMPLE 1-6

What precautions might have to be taken when joining titanium by a welding process?

Answer:

During welding, the titanium is heated to a high temperature. The high temperature may cause detrimental changes in the structure of the titanium, even eliminating some of the strengthening mechanisms by which the properties of the metal were obtained. Furthermore, titanium reacts rapidly with oxygen, hydrogen, and other gases at high temperatures. A welding process must supply a minimum of heat while protecting the metal from the surrounding atmosphere. Special gases, such as argon, or even a vacuum may be needed.

1-5   Strength-to-Weight Ratio

Material cost is normally based on dollars per pound or a similar figure. We must consider the density of the material, or its weight per unit volume, in our design and selection (Table 1-4). Aluminum may be more expensive per pound than steel, but it is only one-third the weight of steel. A component made from aluminum may be less expensive than one made from steel due to the weight difference. Yet the aluminum may be capable of carrying the same load as the steel. Strength is based on cross-sectional area, not weight.

TABLE 1-4   Strength-to-weight ratio of various materials

Material	Strength MPa	Density Mg m^{-3}	Strength-to-weight ratio
Polyethylene	6.9	0.83	7.6
Pure aluminum	44.8	2.7	16.6
Pure copper	207	8.9	23.3
Zinc-aluminum alloy	276	7.2	38.3
Low-carbon steel	393	7.8	50.4
Pure titanium	241	4.4	53.6
Nylon	76	1.11	66.1
Epoxy	103	1.4	73.6
High-carbon steel	614	7.8	78.7
Heat-treated Cu-Be alloy	1310	8.9	147.2
Heat-treated magnesium alloy	276	1.7	162.4
Heat-treated alloy steel	1655	7.8	212.2
Heat-treated aluminum alloy	593	2.7	219.6
Heat-treated titanium alloy	1172	4.4	266.4
Kevlar-epoxy composite	448	1.4	320

EXAMPLE 1-7

Suppose we wish to make a metal cable that can support a force of 9810 N. The strength of aluminum is 196 MPa while that of steel is 294 MPa. The densities of the two metals are 2.7 Mg m^{-3} for aluminum and 7.8 Mg m^{-3} for steel. The costs of the metals are £1650 Mg^{-1} for aluminum and £990 Mg^{-1} for steel. Compare the volume, weight, and cost of the cable per meter for each material.

Answer:

Let's find the cross-sectional area of the cable needed for each material. (1 MPa = 1 MN m^{-2} = 1 N mm^{-2})

$$\text{Area of Al cable} = \frac{9810 \text{ N}}{196 \text{ N mm}^{-2}} = 50 \text{ mm}^2 \, (50.10^{-6} \text{ m}^2)$$

$$\text{Area of steel cable} = \frac{9810 \text{ N}}{294 \text{ N mm}^{-2}} = 33 \text{ mm}^2 \, (33.10^{-6} \text{ m}^2)$$

The volume of metal per meter of cable is

$$\text{Volume of Al cable m}^{-1} = 50.10^{-6} \text{ m}^3 \text{ m}^{-1}$$

$$\text{Volume of steel cable m}^{-1} = 33.10^{-6} \text{ m}^3 \text{ m}^{-1}$$

The weight of metal per meter of cable is

$$\text{Weight of Al cable m}^{-1} = (50.10^{-6} \text{ m}^3 \text{ m}^{-1})(2.7 \text{ Mg m}^{-3})$$
$$= 135.10^{-6} \text{ Mg m}^{-1}$$

$$\text{Weight of steel cable m}^{-1} = (33.10^{-6} \text{ m}^3 \text{ m}^{-1})(7.8 \text{ Mg m}^{-3})$$
$$= 258.10^{-6} \text{ Mg m}^{-1}$$

The cost of cable per meter is

$$\text{Cost of Al cable m}^{-1} = (135.10^{-6})(\pounds 1650) = \pounds 0.22 \text{ m}^{-1}$$

$$\text{Cost of steel cable m}^{-1} = (258.10^{-6})(\pounds 900) = \pounds 0.26 \text{ m}^{-1}$$

The aluminium cable, although larger in volume and diameter, is lighter and very slightly less expensive than the steel cable.

SUMMARY

The procedure of selecting a material, processing the material into a useful shape, and obtaining the needed properties is a complicated process involving knowledge of the structure-property-processing relationship. The remainder of this text is intended to introduce the student to the wide variety of materials available. As we do so, we will come to understand the fundamentals of the structure of materials, how the structure affects the behavior of the material, and the role that processing and the environment play in shaping the relationship between structure and properties.

If you are ready, let's begin.

GLOSSARY

Alloys   Combinations of metals which enhance the general characteristics of metals.

Ceramics   A group of materials characterized by good strength and high melting temperatures, but poor ductility and electrical conductivity. Ceramic raw materials are typically compounds of metallic and nonmetallic elements.

Composites   A group of materials formed from combinations of metals, ceramics, or polymers in such a manner that unusual combinations of properties are obtained.

Metals   A group of materials having the general characteristics of good ductility, strength, and electrical conductivity.

Polymerization   The process by which organic molecules are joined into giant molecules, or polymers.

Polymers   A group of materials normally obtained by joining organic molecules into giant molecular chains or networks. Polymers are characterized by low strengths, low melting temperatures, and poor electrical conductivity.

Thermoplastics   A special group of polymers that are easily formed into useful shapes. Normally, these polymers have a chainlike structure.

Thermosets   A special group of polymers that are normally quite brittle. These polymers typically have a three-dimensional network structure.

PRACTICE PROBLEMS

1 Solid Al_2O_3 is strong, hard, and wear-resistant. Why isn't Al_2O_3 used to make sledge hammers?

2 Polyethylene is an inexpensive, easily formed material. Why isn't polyethylene used to make paper clips?

3 Classify the following materials as metals, ceramics, polymers, or composite materials.

brass	reinforced	rubber
sodium	concrete	asphalt
chloride	lead-tin solder	silicon
epoxy	magnesium alloys	carbide
concrete	fiberglass	graphite

4 What mechanical and physical properties are of particular importance when selecting materials for the following applications?

crankshaft for an automobile

piping to transport hot gases and fluids in a refinery

disposable beverage cans

furnace linings to contain liquid steel during refining

axle for an automobile

filament in a light bulb

windshield of an automobile

pair of scissors

screen of a television set

5 Describe the type of materials processing techniques that might be used to produce the following products.

automobile engine block

brick

paper clip

wrench

plywood

plastic toy

plastic water pipe

steel transmission gear

6 Would casting be a good way to form the following materials and products? Explain if casting is not a good choice.

aluminum	alumina (Al_2O_3)
beryllium	silicon carbide (SiC)
tungsten	glass
titanium	aluminum foil

PART I

ATOMIC STRUCTURE, ARRANGEMENT, AND MOVEMENT

The electronic structure of the atom determines the nature of the atomic bonding, which in turn imparts certain general properties to metals, ceramics, and polymers. Furthermore, the electronic structure plays a predominant role in determining the physical properties, such as electrical conductivity, dielectric and magnetic behavior, and optical, thermal and elastic characteristics, of the material.

The arrangement of atoms into a crystalline or amorphous structure influences the physical properties and in particular the mechanical behavior of materials. Imperfections in the normal atomic arrangement play a critical role in understanding the deformation and mechanical properties of materials.

Finally, the movement of atoms, known as diffusion, is important for many heat treatments and manufacturing processes, and both physical and mechanical properties of materials.

In the next four chapters, we will examine atomic structure, atomic arrangement, imperfections in the atomic arrangement, and atom movement. This examination will lay the groundwork needed to understand the structure and behavior of materials which we will discuss later.

CHAPTER 2

Atomic Structure

2-1 Introduction

The structure of a material may be divided into four levels—atomic structure, atomic arrangement, microstructure, and macrostructure. Although the main thrust of this text is to understand and control the microstructure and macro-structure of various materials, we must first understand the atomic and crystal structures.

Atomic structure influences how the atoms are bonded together, which in turn helps us to categorize materials as metals, ceramics, and polymers and permits us to draw some general conclusions concerning the mechanical properties and physical behavior of these three classes of materials.

2-2 The Structure of the Atom

Most of us know that an atom is composed of a nucleus surrounded by electrons. The nucleus contains neutrons and positively charged protons and thus carries a net positive charge. The negatively charged electrons are held to the nucleus by an electrostatic attraction. The electrical charge q carried by each electron and proton is 1.60×10^{-19} C. Because the numbers of electrons and protons in the atom are equal, the atom as a whole is electrically neutral.

The *atomic number* of an element is equal to the number of electrons or protons in each atom. Thus, an iron atom, which contains 26 electrons and 26 protons, has an atomic number of 26.

Most of the mass of the atom is contained within the nucleus. The mass of each proton and neutron is about 1.67×10^{-24} g, but the mass of each electron is only 9.11×10^{-28} g. The *atomic mass M*, which is equal to the average number of protons and neutrons in the atom, is the mass of the *Avogadro number N_A* of atoms.[1] $N_A = 6.02 \times 10^{23}$ mol^{-1} is the number of atoms or molecules in a g · mole. Therefore, the atomic mass has units of g/g · mole. An alternate unit for atomic mass is amu, which is $\frac{1}{12}$ the mass of carbon 12.

Atoms of the same element that contain a different number of neutrons in the nucleus are called *isotopes* and thus have a different atomic mass. The atomic mass used for such an element is an average value of those of the different isotopes and thus the atomic mass may not be a whole number.

---

[1] Often the atomic mass is called the *atomic weight*.

EXAMPLE 2-1

In a collection of nickel atoms, 70% of the atoms contain 30 neutrons and 30% of the atoms contain 32 neutrons. The atomic number for nickel is 28. Calculate the approximate average atomic mass of nickel.

Answer:

The atomic mass of the atoms containing 30 neutrons is M = number of protons + number of neutrons = 28 + 30 = 58 g/g · mole. These atoms are the Ni^{58} isotope.

The atomic mass of the atoms containing 32 neutrons is M = 28 + 32 = 60 g/g · mole, or the atomic mass of Ni^{60}.

The average atomic mass of nickel is

$$M \text{ of Ni} = (0.7)(M \text{ of } Ni^{58}) + (0.3)(M \text{ of } Ni^{60})$$

$$= (0.7)(58) + (0.3)(60) = 58.6 \text{ g/g} \cdot \text{mole}$$

2-3   The Electronic Structure of the Atom

Electrons occupy discrete energy levels within the atom. Each electron possesses a particular energy, with no more than two electrons in each atom having the same energy. This also implies that there is a definite energy difference between each electron.

Quantum numbers.   The energy level to which each electron belongs is determined by four quantum numbers. The number of possible energy levels is determined by the first three quantum numbers.

1. The *principal quantum number n* is assigned integral values 1, 2, 3, 4, 5, . . . , that refer to the quantum shell to which the electron belongs (Figure 2-1).

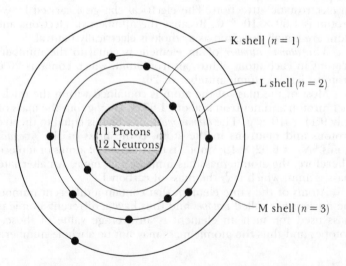

FIG. 2-1   The atomic structure of sodium, atomic number 11, showing the electrons in the K, L, and M quantum shells.

Often the quantum shells are assigned a letter rather than a number; the shell for $n = 1$ is designated K, for $n = 2$ is L, for $n = 3$ is M, and so on.

2. The number of energy levels in each quantum shell is determined by the *azimuthal quantum number* l and the *magnetic quantum number* m_l. The azimuthal quantum numbers may also be assigned numbers: $l = 0, 1, 2, \ldots, n - 1$. If $n = 2$, then there are also two azimuthal quantum numbers, $l = 0$ and $l = 1$. The azimuthal quantum numbers are often designated by lowercase letters,

s for $l = 0$ d for $l = 2$

p for $l = 1$ f for $l = 3$

The magnetic quantum number m_l gives the number of energy levels, or orbitals, for each azimuthal quantum number. The total number of magnetic quantum numbers for each l is $2l + 1$. The values for m_l are given by whole numbers between $-l$ to $+l$. For $l = 2$, there are $2(2) + 1 = 5$ magnetic quantum numbers, with values $-2, -1, 0, +1$, and $+2$.

EXAMPLE 2-2

Calculate the number of possible orbitals in the L shell, where $n = 2$.

Answer:

If $n = 2$, then $l = 0, 1$.

For $l = 0$, there are $2(0) + 1 = 1$ magnetic quantum numbers, so $m_l = 0$.

For $l = 1$, there are $2(1) + 1 = 3$ magnetic quantum numbers, so $m_l = -1, 0, +1$.

Consequently, there are a total of four orbitals possible in the L shell.

---

3. The *Pauli exclusion principle* specifies that no more than two electrons, each with opposing electronic spins, may be present in each orbital. The *spin quantum number* m_s is assigned values $+\frac{1}{2}$ and $-\frac{1}{2}$ to reflect the different spins. Figure 2-2 shows the quantum numbers and energy levels for each electron in a sodium atom.

EXAMPLE 2-3

Determine the maximum number of electrons in the M shell of an atom.

Answer:

The principal quantum number for the M shell is $n = 3$. If $n = 3$, then $l = 0, 1, 2$.

s level, $l = 0$, $m_l = \quad 0$, $m_s = +\frac{1}{2}, -\frac{1}{2}$ 2 electrons

p level, $l = 1$, $m_l = -1$, $m_s = +\frac{1}{2}, -\frac{1}{2}$

$\qquad\qquad\qquad = \quad 0$, $m_s = +\frac{1}{2}, -\frac{1}{2}$ 6 electrons

$\qquad\qquad\qquad = +1$, $m_s = +\frac{1}{2}, -\frac{1}{2}$

$$d \text{ level, } l = 2, \, m_l = -2, \, m_s = +\tfrac{1}{2}, -\tfrac{1}{2}$$
$$= -1, \, m_s = +\tfrac{1}{2}, -\tfrac{1}{2}$$
$$= 0, \, m_s = +\tfrac{1}{2}, -\tfrac{1}{2} \quad \right\} \quad 10 \text{ electrons}$$
$$= +1, \, m_s = +\tfrac{1}{2}, -\tfrac{1}{2}$$
$$= +2, \, m_s = +\tfrac{1}{2}, -\tfrac{1}{2}$$

Thus, a total of 18 electrons may be present in the M shell.

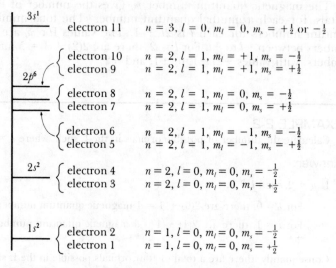

$3s^1$	electron 11	$n = 3, \, l = 0, \, m_l = 0, \, m_s = +\tfrac{1}{2} \text{ or } -\tfrac{1}{2}$
$2p^6$	electron 10	$n = 2, \, l = 1, \, m_l = +1, \, m_s = -\tfrac{1}{2}$
	electron 9	$n = 2, \, l = 1, \, m_l = +1, \, m_s = +\tfrac{1}{2}$
	electron 8	$n = 2, \, l = 1, \, m_l = 0, \, m_s = -\tfrac{1}{2}$
	electron 7	$n = 2, \, l = 1, \, m_l = 0, \, m_s = +\tfrac{1}{2}$
	electron 6	$n = 2, \, l = 1, \, m_l = -1, \, m_s = -\tfrac{1}{2}$
	electron 5	$n = 2, \, l = 1, \, m_l = -1, \, m_s = +\tfrac{1}{2}$
$2s^2$	electron 4	$n = 2, \, l = 0, \, m_l = 0, \, m_s = -\tfrac{1}{2}$
	electron 3	$n = 2, \, l = 0, \, m_l = 0, \, m_s = +\tfrac{1}{2}$
$1s^2$	electron 2	$n = 1, \, l = 0, \, m_l = 0, \, m_s = -\tfrac{1}{2}$
	electron 1	$n = 1, \, l = 0, \, m_l = 0, \, m_s = +\tfrac{1}{2}$

FIG. 2-2 The complete set of quantum numbers for each of the 11 electrons in sodium.

If we continue the exercises in Examples 2-2 and 2-3, we could represent the maximum number of electrons in each energy shell by the pattern in Table 2-1.

TABLE 2-1 The pattern used to assign electrons to energy levels

	$l = 0$ (s)	$l = 1$ (p)	$l = 2$ (d)	$l = 3$ (f)	$l = 4$ (g)	$l = 5$ (h)
$n = 1$ (K)	2					
$n = 2$ (L)	2	6				
$n = 3$ (M)	2	6	10			
$n = 4$ (N)	2	6	10	14		
$n = 5$ (O)	2	6	10	14	18	
$n = 6$ (P)	2	6	10	14	18	22

Note: 2, 6, 10, 14, . . . , refer to the number of electrons in the energy level.

EXAMPLE 2-4

Germanium has an atomic number of 32. What are the expected principal and azi-muthal quantum numbers for the outermost electron in germanium?

Answer:

With an atomic number of 32, germanium must have 32 electrons. From Table 2-1, germanium must have 2 electrons in the K shell, 8 electrons in the L shell, 18 electrons in the M shell, and $32 - 2 - 8 - 18 = 4$ electrons in the N shell. In the N shell, 2 electrons are in the s level and 2 are in the p level. The quantum numbers of the last electron are $n = 4$ and $l = 1$.

The shorthand notation frequently used to denote the electronic structure of an atom combines the numerical value of the principal quantum number, the lowercase letter notation for the azimuthal quantum number, and a superscript showing the number of electrons in each orbital. Thus, the shorthand notation for the electronic structure of germanium is

$$1s^2 2s^2 2p^6 3s^2 3p^6 3d^{10} 4s^2 4p^2$$

The electronic configurations for the elements are summarized in Table 2-2.

Deviations from expected electronic structures. The orderly building up of the electronic structure is not always followed, particularly when the atomic number is large and the d and f levels begin to fill. For example, we would expect the electronic structure of iron, atomic number 26, to be

$$1s^2 2s^2 2p^6 3s^2 3p^6 \boxed{3d^8}$$

However, from Table 2-2, we find that the actual structure is

$$1s^2 2s^2 2p^6 3s^2 3p^6 \boxed{3d^6 4s^2}$$

The unfilled $3d$ level causes the magnetic behavior of iron, as we will see in a later chapter. Many other examples of this behavior can be found in Table 2-2.

Valence. The *valence* of an atom is related to the ability of the atom to enter into chemical combination with other elements and is often determined by the number of electrons in the outermost combined sp level. Examples of the valence are

Mg: $1s^2 2s^2 2p^6 \boxed{3s^2}$ valence = 2

Al: $1s^2 2s^2 2p^6 \boxed{3s^2 3p^1}$ valence = 3

Ge: $1s^2 2s^2 2p^6 3s^2 3p^6 3d^{10} \boxed{4s^2 4p^2}$ valence = 4

The valence also depends on the nature of the chemical reaction. The electronic structure of phosphorous is

$$1s^2 2s^2 2p^6 \boxed{3s^2 3p^3}$$

Phosphorous has the expected valence of five when it combines with oxygen. But

TABLE 2-2 The electronic configuration for each of the elements

Atomic Number	Element	K 1s	L 2s	L 2p	M 3s	M 3p	M 3d	N 4s	N 4p	N 4d	N 4f	O 5s	O 5p	O 5d	P 6s	P 6p
1	Hydrogen	1														
2	Helium	2														
3	Lithium	2	1													
4	Beryllium	2	2													
5	Boron	2	2	1												
6	Carbon	2	2	2												
7	Nitrogen	2	2	3												
8	Oxygen	2	2	4												
9	Fluorine	2	2	5												
10	Neon	2	2	6												
11	Sodium	2	2	6	1											
12	Magnesium	2	2	6	2											
13	Aluminum	2	2	6	2	1										
14	Silicon	2	2	6	2	2										
15	Phosphorus	2	2	6	2	3										
16	Sulfur	2	2	6	2	4										
17	Chlorine	2	2	6	2	5										
18	Argon	2	2	6	2	6										
19	Potassium	2	2	6	2	6		1								
20	Calcium	2	2	6	2	6		2								
21	Scandium	2	2	6	2	6	1	2								
22	Titanium	2	2	6	2	6	2	2								
23	Vanadium	2	2	6	2	6	3	2								
24	Chromium	2	2	6	2	6	5	1								
25	Manganese	2	2	6	2	6	5	2								
26	Iron	2	2	6	2	6	6	2								
27	Cobalt	2	2	6	2	6	7	2								
28	Nickel	2	2	6	2	6	8	2								
29	Copper	2	2	6	2	6	10	1								
30	Zinc	2	2	6	2	6	10	2								
31	Gallium	2	2	6	2	6	10	2	1							
32	Germanium	2	2	6	2	6	10	2	2							
33	Arsenic	2	2	6	2	6	10	2	3							
34	Selenium	2	2	6	2	6	10	2	4							
35	Bromine	2	2	6	2	6	10	2	5							
36	Krypton	2	2	6	2	6	10	2	6							
37	Rubidium	2	2	6	2	6	10	2	6			1				
38	Strontium	2	2	6	2	6	10	2	6			2				

the valence of phosphorous is only three—the electrons in the $3p$ level—when it reacts with hydrogen.

Atomic stability. If an atom has a valence of zero, no electrons enter into chemical reactions and the element is inert. An example is argon, which has the electronic structure

$$1s^2 2s^2 2p^6\boxed{3s^2 3p^6}$$

Other atoms also prefer to behave as if their outer sp levels are either completely full, with eight electrons, or completely empty. Aluminum with the electronic structure

TABLE 2-2 The electronic configuration for each of the elements (continued)

Atomic Number	Element	K 1s	L 2s	L 2p	M 3s	M 3p	M 3d	N 4s	N 4p	N 4d	N 4f	O 5s	O 5p	O 5d	P 6s	P 6p
39	Yttrium	2	2	6	2	6	10	2	6	1		2				
40	Zirconium	2	2	6	2	6	10	2	6	2		2				
41	Niobium	2	2	6	2	6	10	2	6	4		1				
42	Molybdenum	2	2	6	2	6	10	2	6	5		1				
43	Technetium	2	2	6	2	6	10	2	6	6		1				
44	Ruthenium	2	2	6	2	6	10	2	6	7		1				
45	Rhodium	2	2	6	2	6	10	2	6	8		1				
46	Palladium	2	2	6	2	6	10	2	6	10						
47	Silver	2	2	6	2	6	10	2	6	10		1				
48	Cadmium	2	2	6	2	6	10	2	6	10		2				
49	Indium	2	2	6	2	6	10	2	6	10		2	1			
50	Tin	2	2	6	2	6	10	2	6	10		2	2			
51	Antimony	2	2	6	2	6	10	2	6	10		2	3			
52	Tellurium	2	2	6	2	6	10	2	6	10		2	4			
53	Iodine	2	2	6	2	6	10	2	6	10		2	5			
54	Xenon	2	2	6	2	6	10	2	6	10		2	6			
55	Cesium	2	2	6	2	6	10	2	6	10		2	6		1	
56	Barium	2	2	6	2	6	10	2	6	10		2	6		2	
57	Lanthanum	2	2	6	2	6	10	2	6	10	1	2	6		2	
⋮		⋮	⋮	⋮	⋮	⋮	⋮	⋮	⋮	⋮	⋮	⋮	⋮		⋮	
71	Lutetium	2	2	6	2	6	10	2	6	10	14	2	6	1	2	
72	Hafnium	2	2	6	2	6	10	2	6	10	14	2	6	2	2	
73	Tantalum	2	2	6	2	6	10	2	6	10	14	2	6	3	2	
74	Tungsten	2	2	6	2	6	10	2	6	10	14	2	6	4	2	
75	Rhenium	2	2	6	2	6	10	2	6	10	14	2	6	5		
76	Osmium	2	2	6	2	6	10	2	6	10	14	2	6	6		
77	Iridium	2	2	6	2	6	10	2	6	10	14	2	6	9		
78	Platinum	2	2	6	2	6	10	2	6	10	14	2	6	9	1	
79	Gold	2	2	6	2	6	10	2	6	10	14	2	6	10	1	
80	Mercury	2	2	6	2	6	10	2	6	10	14	2	6	10	2	
81	Thallium	2	2	6	2	6	10	2	6	10	14	2	6	10	2	1
82	Lead	2	2	6	2	6	10	2	6	10	14	2	6	10	2	2
83	Bismuth	2	2	6	2	6	10	2	6	10	14	2	6	10	2	3
84	Polonium	2	2	6	2	6	10	2	6	10	14	2	6	10	2	4
85	Astatine	2	2	6	2	6	10	2	6	10	14	2	6	10	2	5
86	Radon	2	2	6	2	6	10	2	6	10	14	2	6	10	2	6

$$1s^2 2s^2 2p^6 \boxed{3s^2 3p^1}$$

has three electrons in its outer sp level. An aluminum atom readily gives up its outer three electrons to empty the $3sp$ level. The nature of the atomic bonding and the chemical behavior of aluminum is determined by the mechanism through which these three electrons interact with surrounding atoms.

On the other hand, chlorine with an electronic structure

$$1s^2 2s^2 2p^6 \boxed{3s^2 3p^5}$$

contains seven electrons in the outer $3sp$ level. The reactivity of chlorine is caused by its desire to fill its outer energy level by accepting an electron.

Electronegativity.   *Electronegativity* describes the tendency of an atom to gain an electron. Atoms with almost completely filled outer energy levels, like chlorine, are strongly electronegative and readily accept electrons. However, atoms with nearly empty outer levels, such as sodium

$$1s^2 2s^2 2p^6 \boxed{3s^1}$$

readily give up electrons and are strongly *electropositive*. High atomic number elements also have a low electronegativity; because the outer electrons are at a greater distance from the positive nucleus, electrons are not as strongly attracted to the atom. Electronegativities for some elements are shown in Figure 2-3.

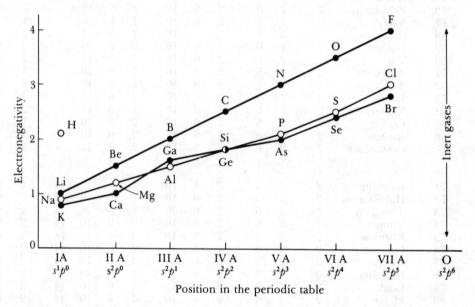

FIG. 2-3 The electronegativities of selected elements versus the position of the element in the periodic table.

EXAMPLE 2-5

Using the electronic structures, compare the electronegativities of calcium and bromine.

Answer:

The electronic structures, obtained from Table 2-2, are

Ca: $1s^2 2s^2 2p^6 3s^2 3p^6 \boxed{4s^2}$

Br: $1s^2 2s^2 2p^6 3s^2 3p^6 3d^{10} \boxed{4s^2 4p^5}$

Calcium has two electrons in its outer $4s$ orbital and bromine has seven electrons in its outer $4s4p$ orbital. Calcium tends to give up electrons and is strongly electropositive but bromine tends to accept electrons and is strongly electronegative.

2-4 The Periodic Table

The construction of the familiar periodic table (Figure 2-4) is based on the electronic configuration of the elements. Consequently, the periodic table can give us some clues to the behavior of the elements.

The IA to VIIA elements. The rows in the periodic table correspond to quantum shells, or principal quantum numbers [Figure 2-5(a)]. For example, the elements lithium through neon contain electrons in the L shell ($n = 2$), whereas sodium through argon contain electrons in the M shell ($n = 3$).

The columns refer to the number of electrons present in the outermost sp energy level and correspond to the most common valence [Figure 2-5(b)]. Thus lithium, sodium, and potassium in column IA all have a valence of one, whereas fluorine, chlorine, and bromine in column VIIA have a valence of seven. The column 0 on the far right represents the inert gases, which have the outer sp level full. Normally, the elements in each column have similar properties and behavior.

The IIIB to VIIIB elements. In each of these rows, an inner energy level is progressively filled [Figure 2-5(c)]. The elements in the fourth row, scandium through zinc, are the transition elements and contain valence electrons in the N shell. However, the inner $3d$ level in the M shell is not full. The electronic structures of the transition elements are given in Table 2-3.

Similar situations arise in subsequent rows; the $4d$ level is filled in the yttrium series, the $5d$ level in the lanthanide series, and the $5f$ level in the actinide series. Elements within each of these series tend to have similar behavior.

The IB and IIB elements. These elements, which include copper, silver, and gold, have complete inner shells and one or two valence electrons. We might compare copper to potassium, in Group IA.

$$\text{K:} \quad 1s^2 2s^2 2p^6 3s^2 3p^6 \boxed{4s^1}$$

$$\text{Cu:} \quad 1s^2 2s^2 2p^6 3s^2 3p^6 \boxed{3d^{10} 4s^1}$$

The filled $3d$ level in copper helps keep the valence electrons tightly held to the inner core; copper (as well as silver and gold) are consequently very stable and unreactive. In potassium, however, the valence electron is not tightly held by the inner $3s3p$ shell and is very reactive.

TABLE 2-3 The electronic configuration of the transition elements

Group	Element	Electronic Configuration
IIIB	Sc	$\ldots 3s^2 3p^6 \boxed{3d^1} 4s^2$
IVB	Ti	$\ldots 3s^2 3p^6 \boxed{3d^2} 4s^2$
VB	V	$\ldots 3s^2 3p^6 \boxed{3d^3} 4s^2$
VIB	Cr	$\ldots 3s^2 3p^6 \boxed{3d^5} 4s^1$
VIIB	Mn	$\ldots 3s^2 3p^6 \boxed{3d^5} 4s^2$
VIIIB	Fe	$\ldots 3s^2 3p^6 \boxed{3d^6} 4s^2$
VIIIB	Co	$\ldots 3s^2 3p^6 \boxed{3d^7} 4s^2$
VIIIB	Ni	$\ldots 3s^2 3p^6 \boxed{3d^8} 4s^2$

IA	IIA	IIIB	IVB	VB	VIB	VIIB	VIIIB			IB	IIB	IIIA	IVA	VA	VIA	VIIA	O
1 **H** 1.00797																	2 **He** 4.003
3 **Li** 6.939	4 **Be** 9.012											5 **B** 10.81	6 **C** 12.011	7 **N** 14.007	8 **O** 15.9994	9 **F** 19.00	10 **Ne** 20.183
11 **Na** 22.99	12 **Mg** 24.31											13 **Al** 26.98	14 **Si** 28.09	15 **P** 30.974	16 **S** 32.064	17 **Cl** 35.453	18 **Ar** 39.948
19 **K** 39.102	20 **Ca** 40.08	21 **Sc** 44.96	22 **Ti** 47.90	23 **V** 50.94	24 **Cr** 52.00	25 **Mn** 54.94	26 **Fe** 55.85	27 **Co** 58.93	28 **Ni** 58.71	29 **Cu** 63.54	30 **Zn** 65.37	31 **Ga** 69.72	32 **Ge** 72.59	33 **As** 74.92	34 **Se** 78.96	35 **Br** 79.909	36 **Kr** 83.80
37 **Rb** 85.47	38 **Sr** 87.62	39 **Y** 88.905	40 **Zr** 91.22	41 **Nb** 92.91	42 **Mo** 95.94	43 **Tc** 98	44 **Ru** 101.1	45 **Rh** 102.90	46 **Pd** 106.4	47 **Ag** 107.87	48 **Cd** 112.4	49 **In** 114.82	50 **Sn** 118.69	51 **Sb** 121.75	52 **Te** 127.60	53 **I** 126.90	54 **Xe** 131.30
55 **Cs** 132.905	56 **Ba** 137.34	57 **La** 138.91	72 **Hf** 178.49	73 **Ta** 180.95	74 **W** 183.85	75 **Re** 186.2	76 **Os** 190.2	77 **Ir** 192.2	78 **Pt** 195.09	79 **Au** 196.97	80 **Hg** 200.59	81 **Tl** 204.37	82 **Pb** 207.19	83 **Bi** 208.98	84 **Po** 210	85 **At** 210	86 **Rn** 222
87 **Fr** 223	88 **Ra** 226	89 **Ac** 227															

← Lanthanide series

58 **Ce** 140.12	59 **Pr** 140.91	60 **Nd** 144.24	61 **Rm** 147	62 **Sm** 150.35	63 **Eu** 152	64 **Gd** 157.25	65 **Tb** 158.92	66 **Dy** 162.50	67 **Ho** 164.93	68 **Er** 167.26	69 **Tm** 168.93	70 **Yb** 173.04	71 **Lu** 174.97

← Actinide series

90 **Th** 232.04	91 **Pa** 231	92 **U** 238.03	93 **Np** 237	94 **Pu** 242	95 **Am** 243	96 **Cm** 247	97 **Bk** 247	98 **Cf** 251	99 **Es** 254	100 **Fm** 253	101 **Md** 256	102 **No** 254	103 **Lw** 257

FIG. 2-4 The periodic table of the elements.

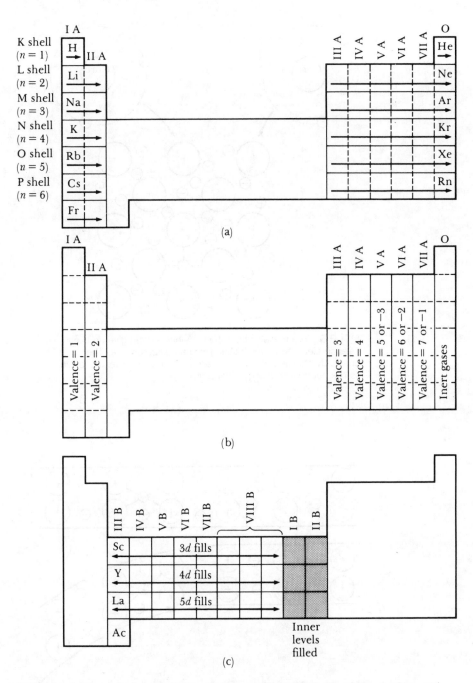

FIG. 2-5 In the periodic table, (a) rows correspond to principal quantum numbers and show the change in electronegativity, (b) columns correspond to similar electronegativity, and (c) transition elements have inner energy levels filling.

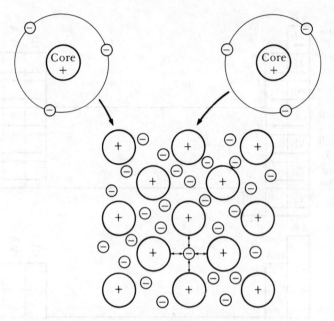

FIG. 2-6   The metallic bond forms when atoms give up their valence electrons, which then form an electron sea. The positively charged atom cores are bonded by mutual attraction to the negatively charged electrons.

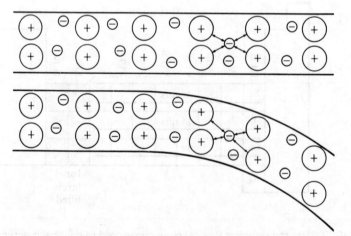

FIG. 2-7   Atoms joined by the metallic bond can shift their relative positions when the metal is deformed, permitting metals to have good ductility.

2-5 Atomic Bonding

There are four mechanisms by which atoms are bonded together. In three of the four mechanisms, bonding is achieved when the atoms fill their outer s and p levels.

The metallic bond. The metallic elements, which have a low valence, give up their valence electrons to form a "sea" of electrons surrounding the atoms (Figure 2-6). Aluminum, for example, gives up its three valence electrons, leaving behind a core consisting of the nucleus and inner electrons. Since three negatively charged electrons are missing from this core, the core becomes an ion with a positive charge of three. The valence electrons, which are no longer associated with any particular atom, move freely within the electron sea and become associated with several atom cores. The positively charged atom cores are held together by mutual attraction to the electron, thus producing the strong metallic bond.

Metallic bonds are *nondirectional*; the electrons holding the atoms together are not fixed in one position. When a metal is bent and the atoms attempt to change their relationship to one another, the direction of the bond merely shifts, rather than the bond breaking (Figure 2-7). This permits metals to have good ductility and to be deformed into useful shapes.

The metallic bond also allows metals to be good electrical conductors. Under the influence of an applied voltage, the valence electrons move (Figure 2-8) causing a current to flow if the circuit is complete. Other bonding mechanisms require much higher voltages to free the electrons from the bond.

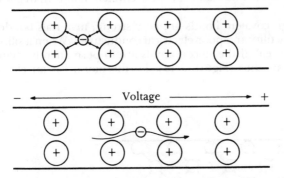

FIG. 2-8 When a voltage is applied to a metal, the electrons in the electron sea can easily move and carry a current.

The covalent bond. Covalently bonded materials share electrons between two or more atoms. For example, a silicon atom, which has a valence of four, obtains eight electrons in its outer energy shell by sharing its electrons with four surrounding silicon atoms (Figure 2-9). Each instance of sharing represents one covalent bond; thus each silicon atom is bonded to four neighboring atoms by four covalent bonds.

In order for the covalent bonds to be formed the silicon atoms must be arranged so the bonds have a fixed *directional* relationship with one another. In the case of silicon, this arrangement produces a *tetrahedron*, with angles of about

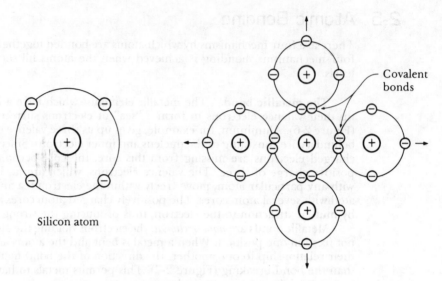

Silicon atom

FIG. 2-9 Covalent bonding requires that electrons be shared between atoms in such a way that each atom has its outer *sp* orbital filled. In silicon, with a valence of four, four covalent bonds must be formed.

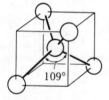

FIG. 2-10 Covalent bonds are directional. In silicon, a tetrahedral structure is formed, with angles of about 109° required between each covalent bond.

109° between the covalent bonds (Figure 2-10). There is a much higher probability that electrons are located near these covalent bonds than elsewhere around the atom core.

Although covalent bonds are very strong, materials bonded in this manner have poor ductility and poor electrical conductivity. When a silicon rod is bent, the bonds must break if the silicon atoms are to permanently change their relationship to one another (Figure 2-11). Furthermore, for an electron to move and

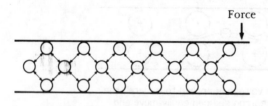

Force

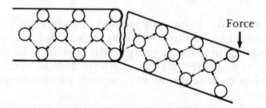

Force

FIG. 2-11 When a force is applied to a covalently bonded material, the bonds may break. Covalently bonded materials are brittle.

carry a current, the covalent bond must be broken, requiring high temperatures or voltages.

Thus, covalently bonded materials are brittle rather than ductile and behave as electrical insulators instead of conductors. Many ceramic and polymer materials are fully or partly bonded by covalent bonds, explaining why glass shatters when dropped and why bricks are good insulating materials.

EXAMPLE 2-6

Describe how covalent bonding joins oxygen and silicon atoms in silica (SiO_2).

Answer:

Silicon has a valence of four and shares electrons with four oxygen atoms, thus giving a total of eight electrons for each silicon atom. However, oxygen has a valence of six and shares electrons with two silicon atoms, giving oxygen a total of eight electrons.

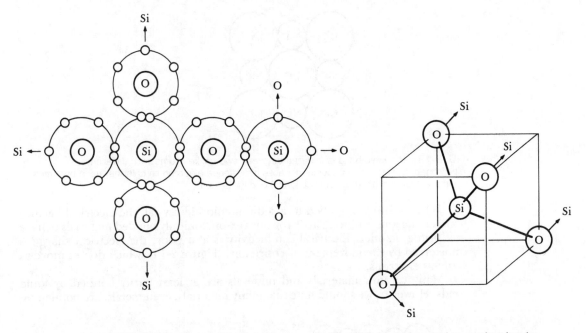

FIG. 2-12   The tetrahedral structure of silica (SiO_2), which contains covalent bonds between silicon and oxygen atoms.

Figure 2-12 shows one of the possible structures. Like silicon, a tetrahedral structure is produced.

The ionic bond.   When more than one type of atom is present in a material, one atom may donate its valence electrons to a different atom, filling the outer energy shell of the second atom. Both atoms now have filled (or empty) outer energy levels but both have acquired an electrical charge and behave as ions. The atom that contributes the electrons is left with a net positive charge and is a *cation*, while the atom that accepts the electrons acquires a net negative charge and is an *anion*. The oppositely charged ions are then attracted to one another and produce the ionic bond. For example, attraction between sodium and chloride ions (Figure 2-13) produces sodium chloride, or table salt, NaCl.

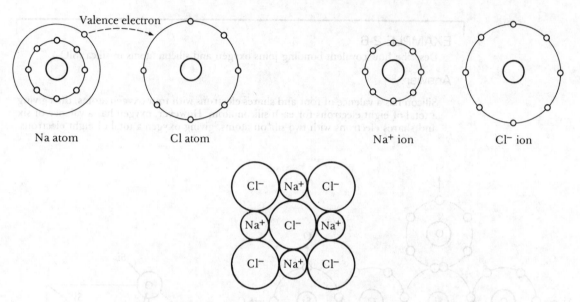

FIG. 2-13   The ionic bond is created between two unlike atoms with different electronegativities. When sodium donates its valence electron to chlorine, each becomes an ion, attraction occurs, and the ionic bond is formed.

When a force is applied to a sodium chloride crystal, the electrical balance between the ions is upset. Partly for this reason, ionically bonded materials behave in a brittle manner. Electrical conductivity is also poor; the electrical charge is transferred by the movement of entire ions (Figure 2-14), which do not move as easily as electrons.

Many ceramic materials and minerals are at least partly bonded by ionic bonds. However, we should note that many materials, even metals, are bonded by

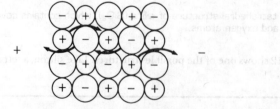

FIG. 2-14   When a voltage is applied to an ionic material, entire ions must move to cause a current to flow. Ion movement is slow and the electrical conductivity is poor.

a complex mixture of the three primary bonding mechanisms. Although iron is primarily bonded by the metallic bond, covalent bonding also makes a contribution. Silicon, which is predominantly bonded by covalent bonds, also contains some elements of the metallic bond.

EXAMPLE 2-7

Describe the ionic bonding between magnesium and chlorine.

Answer:

The electronic structures and valences are

$$Mg: \quad 1s^2 2s^2 2p^6 \boxed{3s^2} \qquad valence = 2$$
$$Cl: \quad 1s^2 2s^2 2p^6 \boxed{3s^2 3p^5} \qquad valence = 7$$

Each magnesium atom gives up its two valence electrons, becoming a Mg^{2+} ion. Each chlorine atom accepts one electron, becoming a Cl^- ion. To satisfy the ionic bonding, there must be twice as many chlorine atoms as magnesium atoms present and a compound $MgCl_2$ is formed.

EXAMPLE 2-8

As a general rule, would you expect anions or cations to have the larger ionic radius?

Answer:

Because the cation gives up its electron while the anion accepts an electron, we find that most anions are larger than cations.

Van der Waals bonding.   Van der Waals bonds join molecules or groups of atoms by weak electrostatic attractions. Many plastics, ceramics, water, and other molecules are permanently *polarized*; that is, some portions of the molecule tend to

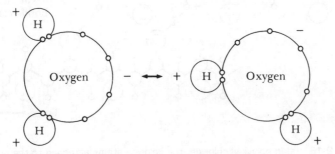

FIG. 2-15   The Van der Waals bond is formed due to polarization of molecules or groups of atoms. In water, electrons in the oxygen tend to concentrate away from the hydrogen. The resulting charge difference permits the molecule to be weakly bonded to other water molecules.

be positively charged, while other portions are negatively charged. The electrostatic attraction between the positively charged regions of one molecule and the negatively charged regions of a second molecule weakly bond the two molecules together (Figure 2-15).

Van der Waals bonding is a secondary bond, but the atoms within the molecule or group of atoms are joined by strong covalent or ionic bonds. Heating water to the boiling point breaks the Van der Waals bonds and changes water to steam, but much higher temperatures are required to break the covalent bonds joining oxygen and hydrogen atoms together.

Van der Waals bonds can dramatically change the properties of materials. Since polymers normally have covalent bonds, we would expect polyvinyl chloride (PVC plastic) to be very brittle. However, polyvinyl chloride contains many long chainlike molecules (Figure 2-16). Within each chain, bonding is covalent, but individual chains are bonded to one another by Van der Waals bonds. Polyvinyl chloride can be deformed significantly by breaking only the Van der Waals bonds as the chains slide past one another.

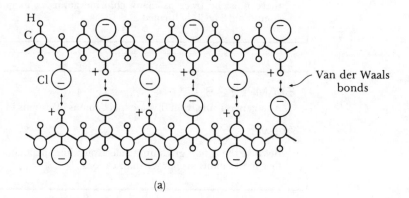

(a)

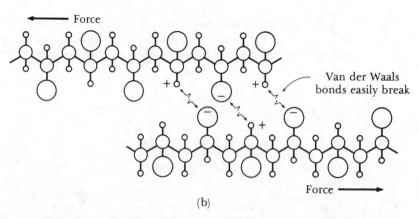

(b)

FIG. 2-16   (a) In polyvinyl chloride, the chlorine atoms attached to the polymer chain have a negative charge and the hydrogen atoms are positively charged. The chains are weakly bonded by Van der Waals bonds. (b) When a force is applied to the polymer, the Van der Waals bonds are broken and the chains slide past one another.

2-6 Binding Energy and Interatomic Spacing

The *interatomic spacing* is the equilibrium distance between the atoms and is caused by a balance between repulsive and attractive forces. In the metallic bond, for example, the attraction\between the electrons and the atom core is balanced by the repulsion between the atom cores. The equilibrium separation occurs when the total energy of the pair of atoms is at a minimum (Figure 2-17). The interatomic

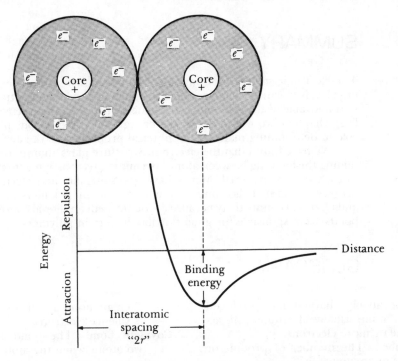

FIG. 2-17 Atoms or ions are separated by an equilibrium spacing that corresponds to the minimum energy of the atoms or ions.

spacing in a solid metal is equal to the atomic diameter, or twice the atomic radius r. We cannot use this approach for ionically bonded materials, however, since the interatomic spacing is the sum of the two different ionic radii. Atomic and ionic radii are listed for the elements in Appendix B and will be used in the next chapter.

The minimum energy in Figure 2-17 is the *binding energy*, or the energy required to create or break the bond. Ordinarily, we expect ionically bonded materials to have a particularly large binding energy due to the large difference in electronegativities between the ions, and metals to have lower binding energies since the electronegativities of the atoms are similar (Table 2-4).

TABLE 2-4   Binding energies for the
four bonding mechanisms

Bond	Binding Energy (kcal/mol)
Ionic	150–370
Covalent	125–300
Metallic	25–200
Van der Waals	<10

SUMMARY

The electronic structure of an atom may be characterized by examining the energy levels to which each electron is assigned by the four quantum numbers. The periodic table of the elements is constructed based on the electronic structure. In later chapters, we will find that the energies of the electrons play an important role in determining many of the physical properties of a material.

   We have found that the electronic structure plays an important role in determining the bonding between atoms, permitting us to assign general characteristics to each type of material. Metals have good ductility and electrical conductivity because of the metallic bond. Ceramics and many polymers have poor ductility and electrical conductivity because of the covalent and ionic bonds. Van der Waals bonds are responsible for good ductility in certain polymers.

GLOSSARY

Anion   A negatively charged ion produced when an atom, usually of a nonmetal, accepts one or more electrons.

Atomic number   The number of protons or electrons in an atom.

Atomic mass   The mass of the Avogadro number of atoms, g/g · mole. Normally, this is the average number of protons and neutrons in the atom. Also called the atomic weight.

Avogadro number   The number of atoms or molecules in a g · mole. The Avogadro number is 6.02×10^{23} atoms/g · mole or molecules/g · mole.

Binding energy   The energy required to separate two atoms from their equilibrium spacing to an infinite distance apart. Alternately, the binding energy is the strength of the bond between two atoms.

Cation   A positively charged ion produced when an atom, usually of a metal, gives up its valence electrons.

Covalent bond   The bond formed between two atoms when the atoms share their valence electrons.

Electronegativity   The relative tendency of an atom to accept an electron and become an anion. Strongly electronegative atoms readily accept electrons.

Interatomic spacing   The equilibrium spacing between the centers of two atoms. In solid elements, the interatomic spacing equals the apparent diameter of the atom.

Ionic bond   The bond formed between two different atom species when one atom (the cation) donates its valence electrons to the second atom (the anion). An electrostatic attraction binds the ions together.

Isotopes   Isotopes of an element have the same atomic number but a different number of neutrons, thus giving a different atomic mass.

Metallic bond   The electrostatic attraction between the valence electrons and the positively charged cores of the atoms.

Pauli exclusion principle   No more than two electrons in a material can have the same energy. The two electrons have opposite magnetic spins.

Polarized molecule   A molecule whose structure causes portions of the molecule to have a negative charge while other portions have a positive charge, leading to electrostatic attraction between the molecules.

Quantum numbers   The numbers that assign electrons in an atom to discrete energy levels. The four quantum numbers are the principal quantum number n, the azimuthal quantum number l, the magnetic quantum number m_l, and the spin quantum number m_s.

Valence   The number of electrons in an atom that participate in bonding or chemical reactions. Usually, the valence is the number of electrons in the outer sp energy level.

Van der Waals bond   A weak electrostatic attraction between polar molecules. The polar molecules have concentrations of positive and negative charges at different locations.

PRACTICE PROBLEMS

1   Magnesium, which has an atomic number of 12, contains three isotopes: 78.70% of Mg atoms contain 12 neutrons, 10.13% contain 13 neutrons, and 11.17% contain 14 neutrons. Calculate the atomic mass of magnesium.

2   Chromium, which has an atomic number of 24, has four isotopes: 4.31% of Cr atoms contain 26 neutrons, 83.76% contain 28 neutrons, 9.55% contain 29 neutrons, and 2.38% contain 30 neutrons. Calculate the atomic mass of chromium.

3   Gallium, which has an atomic number of 31 and an atomic mass of 69.72 g/g · mole, contains two isotopes—Ga^{69} and Ga^{71}. Determine the percentage of each isotope of gallium.

4   Copper, which has an atomic number of 29 and an atomic mass of 63.54 g/g · mole, contains two isotopes—Cu^{63} and Cu^{65}. Determine the percentage of each isotope of copper.

5   Calculate the maximum number of electrons in the N shell of an atom. Determine the atomic number if all energy levels in the K, L, M, and N shells are filled.

6   Calculate the maximum number of electrons in the O shell of an atom. Determine the atomic number if all energy levels in the K, L, M, N, and O shells are filled.

7   Indium, with an atomic number of 49, has all of its inner energy levels filled except the $4f$ level is empty. From the electronic structure, determine the expected valence of indium.

8   Platinum, atomic number 78, has only nine electrons in the $5d$ level and none in the $5f$ level. How many electrons are in the $6s$ level?

9   Without consulting the periodic table, use the electronic structure to determine whether an element with an atomic number of 54 should be strongly electronegative, electropositive, or inert. No electrons are present in the $4f$ level.

10   Repeat Problem 9 for an element with an atomic number of 35. All inner energy levels are filled.

11   Repeat Problem 9 for an element with an atomic number of 20 which has no electrons in the $3d$ level.

12   Would you expect molybdenum to be more or less stable than rubidium? Explain.

13   The electronic structure of silver and gold are similar. Which is expected to be more stable? Explain.

14   Calculate the number of atoms in 100 g of aluminum. If all of the valence electrons can carry an electrical current, calculate the number of these charge carriers in 100 g of aluminum.

15   Suppose there are 5×10^{10} electrons/100 g of silicon which are free to move. (a) What fraction of the total valence electrons are free to move? (b) What fraction of the covalent bonds must be broken? (On average there is one covalent bond per silicon atom and two electrons in each covalent bond.)

16   Draw a sketch, similar to Figure 2-9 or 2-12, showing the atomic arrangement and bonding you would expect to find in carbon dioxide (CO_2). (It is possible that atoms can be joined by more than one covalent bond.)

17 Draw a sketch, similar to Figure 2-9 or 2-12, showing the atomic arrangement and bonding you would expect to find in ethylene (C_2H_4).

18 Draw a sketch, similar to Figure 2-9 or 2-12, showing the atomic arrangement and bonding you would expect to find in methane (CH_4).

19 Calculate the exact angle between the covalent bonds in the silicon tetrahedron.

20 What elements, in addition to silicon, might you expect to display the same type of tetrahedral structure?

21 What is the formula of the ionic compound produced when magnesium and sulfur are brought together?

22 What is the formula of the ionic compound formed when beryllium is oxidized?

23 What is the formula of the ionic compound formed when magnesium reacts with fluorine?

24 What is the formula of the ionic compound formed when zinc oxidizes?

25 In the ionic compound NaGl, would you expect

the Na^+ ions or the Cl^- ions to move more easily through the crystal when a voltage is applied? Explain.

26 Sometimes sulfur behaves as if it has a valence of 6 whereas at other times it behaves as if it has a valence of 4. Explain this behavior.

27 Figure 2-18 shows the energy-separation curves for three materials—a metal, an ionic crystal, and a Van der Waals bonded material. Match the material to the curve.

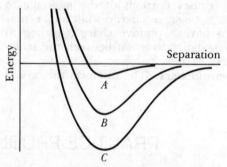

FIG. 2-18   Energy-separation curves for Problem 2-27.

3

Atomic Arrangement

3-1 Introduction

Atomic arrangement plays an important role in determining the microstructure and behavior of a solid material. In metals, some arrangements permit exceptional ductility, whereas others permit exceptional strength. Certain physical properties of ceramics rely on the atomic arrangement; transducers used to produce the electrical signal in a stereo record player rely on an atomic arrangement that produces a permanent displacement of electrical charge within the material. The diverse behavior of polymers, such as rubber, polyethylene, and epoxy, is caused by differences in the atomic arrangement.

In this chapter, we will describe typical atomic arrangements in perfect solid materials and develop the nomenclature used to characterize this arrangement. We will then be prepared to see how imperfections in the atomic arrangement permit us to understand both deformation and strengthening of many solid materials.

3-2 Short-Range Order versus Long-Range Order

If we neglect imperfections in materials, there are three levels of atomic arrangement (Figure 3-1).

No order. In gases such as argon, the atoms have no order; argon atoms randomly fill up the space to which the gas is confined.

Short-range order. A material displays short-range order if the special arrangement of the atoms extends only to the atom's nearest neighbors. Each water molecule in steam has a short-range order due to the covalent bonds between the hydrogen and oxygen atoms; that is, each oxygen atom is joined to two hydrogen atoms, forming an angle of about 104.5° between the bonds. However, the water molecules have no special arrangement but instead randomly fill up the space available to them.

A similar situation exists in ceramic glasses. In Example 2-6, we described the tetrahedral structure in silica, which satisfies the requirement that four oxygen atoms be covalently bonded to each silicon atom. Because the oxygen atoms must

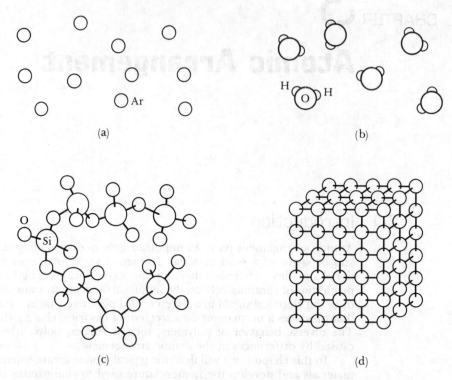

FIG. 3-1   The levels of atomic arrangement in materials. (a) Inert gases have no regular ordering of atoms. (b) and (c) Some materials, including steam and glass, have ordering only over a short distance. (d) Metals and many other solids have a regular ordering of atoms that extends throughout the material.

form angles of about 109° to satisfy the directionality requirements of the covalent bonds, a short-range order results. However, the tetrahedral units may be joined together in a random manner. Most polymers display similar short-range atomic arrangements.

   Long-range order.   In metals, many ceramics, and even some polymers, the atoms display long-range order; that is, the special atomic arrangement extends throughout the entire material. The atoms form a regular, repetitive gridlike pattern, or lattice. The *lattice* is a collection of points, called *lattice points*, which are arranged in a periodic pattern so that the surroundings of each point in the lattice are identical. One or more atoms are associated with each lattice point. Consequently, each atom has both a short-range order, since the surroundings of each lattice point are identical, and a long-range order, since the lattice extends periodically throughout the entire material.

   The lattice differs from material to material in both shape and size, depending on the size of the atoms and the type of bonding between the atoms. The *crystal structure* of a material refers to the size, shape, and atomic arrangement within the lattice.

3-3 Unit Cells

The *unit cell* is a subdivision of the lattice that still retains the overall characteristics of the entire lattice. A unit cell is shown in the lattice in Figure 3-2 by the heavy lines. By stacking identical unit cells, the entire lattice can be constructed.

We identify 14 types of unit cells, or *Bravais lattices*, grouped in seven crystal structures (Figure 3-3 and Table 3-1). Lattice points are located at the corners of the unit cells and, in some cases, at either faces or the center of the unit cell. Let's look at some of the characteristics of a lattice or unit cell.

Lattice parameter. The lattice parameters, which describe the size and shape of the unit cell, include the dimensions of the sides of the unit cell and the angles between the sides (Figure 3-4). In a cubic crystal system, only the length of one of the sides of the cube is necessary to completely describe the cell (angles of 90° are assumed unless otherwise specified). This length, measured at room temperature, is the *lattice parameter* a_0. The length is often given in angstrom units, still widely used, though strictly non-SI.

1 angstrom (Å) $= 10^{-1}$ nm $= 10^{-10}$ m

Several lattice parameters are required to define the size and shape of complex unit cells. For an orthorhombic unit cell, we must specify the dimensions of all three sides of the cell, a_0, b_0, and c_0. Hexagonal unit cells require two dimensions, a_0 and c_0, and the angle of 120° between the a_0 axes. The most complicated cell, the triclinic cell, is described by three lengths and three angles.

Number of atoms per unit cell. A specific number of lattice points define each of the unit cells. For example, the corners of the cells are easily identified, as are body- and face-centered positions (Figure 3-3). When counting the number of lattice points belonging to each unit cell, we must recognize that lattice points may be shared by more than one unit cell. A lattice point at a corner of one unit cell is shared by seven adjacent unit cells; only $\frac{1}{8}$ of each corner belongs to one particular cell. Thus, the number of lattice points from the corner positions of one unit cell is

$$\left(\frac{1}{8}\frac{\text{lattice point}}{\text{corner}}\right)\left(8\frac{\text{corners}}{\text{cell}}\right) = 1\frac{\text{lattice point}}{\text{unit cell}}$$

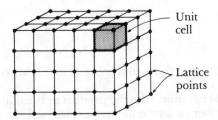

FIG. 3-2 A lattice is a periodic array of points that define space. The unit cell (heavy outline) is a subdivision of the lattice that still retains the characteristics of the lattice.

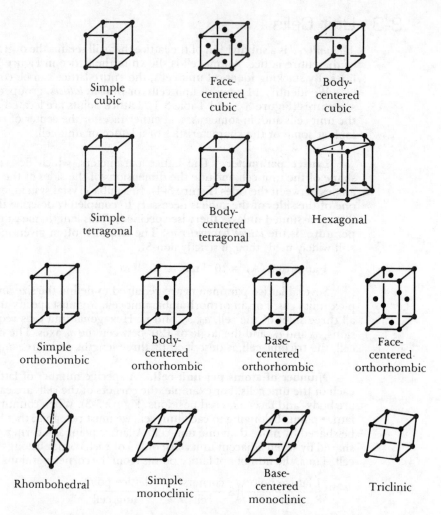

FIG. 3-3 The seven crystal systems and 14 Bravais lattices. Characteristics of the crystal systems are summarized in Table 3-1.

Corners contribute $\frac{1}{8}$ of a point, faces contribute $\frac{1}{2}$, and body-centered positions contribute a whole point.

The number of atoms per unit cell is the product of the number of atoms per lattice point and the number of lattice points per unit cell. In most metals, one atom is located at each lattice point, so the number of atoms is equal to the number of lattice points. The structures of simple cubic (SC), face-centered cubic (FCC), and body-centered cubic (BCC) unit cells, with one atom per lattice point, are shown in Figure 3-5. In more complicated structures, particularly compounds and ceramic materials, several or even hundreds of atoms may be associated with each lattice point, forming very complex unit cells.

TABLE 3-1   Characteristics of the seven crystal systems

Structure	Axes	Angles between Axes
Cubic	$a_1 = a_2 = a_3$	All angles equal 90°
Tetragonal	$a_1 = a_2 \neq c$	All angles equal 90°
Orthorhombic	$a \neq b \neq c$	All angles equal 90°
Hexagonal	$a_1 = a_2 \neq c$	Two angles equal 90° One angle equals 120°
Rhombohedral	$a_1 = a_2 = a_3$	All angles are equal and none equal 90°
Monoclinic	$a \neq b \neq c$	Two angles equal 90° One angle not equal to 90°
Triclinic	$a \neq b \neq c$	All angles are different and none equals 90°

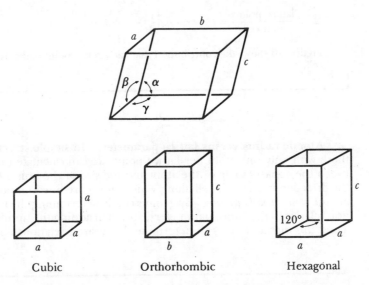

FIG. 3-4   Definition of the lattice parameters and their use in three crystal systems.

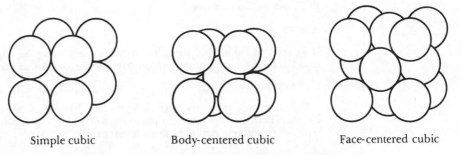

Simple cubic   Body-centered cubic   Face-centered cubic

FIG. 3-5   The models for simple cubic (SC), body-centered cubic (BCC), and face-centered cubic (FCC) unit cells assuming only one atom per lattice point.

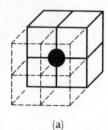

(a)

(b)

FIG. 3-6
(a) Corner atoms
are shared by
eight unit cells. (b)
Face-centered
atoms are shared
by two unit cells.

EXAMPLE 3-1

Determine the number of lattice points per cell in the cubic crystal systems.

Answer:

In the SC unit cell, lattice points are located only at the corners of the cube:

$$\frac{\text{lattice points}}{\text{unit cell}} = (8 \text{ corners})\left(\frac{1}{8}\right) = 1$$

In BCC unit cells, lattice points are located at the corners and the center of the cube:

$$\frac{\text{lattice points}}{\text{unit cell}} = (8 \text{ corners})\left(\frac{1}{8}\right) + (1 \text{ center})(1) = 2$$

In FCC unit cells, lattice points are located at the corners and faces of the cube:

$$\frac{\text{lattice points}}{\text{unit cell}} = (8 \text{ corners})\left(\frac{1}{8}\right) + (6 \text{ faces})\left(\frac{1}{2}\right) = 4$$

Figure 3-6 shows the contribution that each lattice point makes to the individual unit cell.

Atomic radius versus lattice parameter.   In simple structures, particularly those with only one atom per lattice point, we can calculate the relationship between the apparent size of the atom and the size of the unit cell. We must locate the direction in the unit cell along which atoms are in continuous contact. These are the *close-packed directions*. By geometrically determining the length of the direction relative to the lattice parameters, and then adding the number of atomic radii along this direction, we can determine the desired relationship.

EXAMPLE 3-2

Determine the relationship between the atomic radius and the lattice parameter in SC, BCC, and FCC structures.

Answer:

If we refer to Figure 3-7, we find that atoms touch along the edge of the cube in SC structures. The corner atoms are centered on the corners of the cube, so

$$a_0 = 2r \tag{3-1}$$

In FCC structures, atoms touch along the face diagonal of the cube, which is $\sqrt{2}a_0$ in length. There are four atomic radii along this length—two radii from the face-centered atom and one radius from each corner, so

$$a_0 = \frac{4r}{\sqrt{2}} \tag{3-2}$$

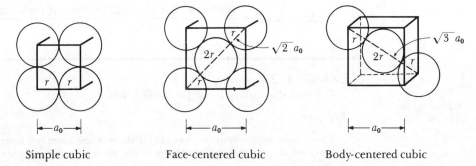

Simple cubic   Face-centered cubic   Body-centered cubic

FIG. 3-7 The relationship between the atomic radius and the lattice parameter in cubic systems. (See Example 3-2.)

In BCC structures, atoms touch along the body diagonal, which is $\sqrt{3}a_0$ in length. There are two atomic radii from the center atom and one atomic radius from each of the corner atoms on the body diagonal, so

$$a_0 = \frac{4r}{\sqrt{3}} \tag{3-3}$$

EXAMPLE 3-3

The atomic radii of BCC iron and FCC iron are 1.241 Å and 1.269 Å respectively. Calculate the lattice parameter of each form of iron.

Answer:

For BCC iron,

$$a_0 = \frac{4r}{\sqrt{3}} = \frac{(4)(1.241)}{\sqrt{3}} = 2.87 \text{ Å (0.287 nm)}$$

For FCC iron,

$$a_0 = \frac{4r}{\sqrt{2}} = \frac{(4)(1.269)}{\sqrt{2}} = 3.59 \text{ Å (0.359 nm)}$$

Coordination number. The number of atoms touching a particular atom, or the number of nearest neighbors, is the *coordination number* and is one indication of how tightly and efficiently atoms are packed together. In simple crystal structures containing only one atom per lattice point, we find that the atoms have a coordination number related to the lattice structure. By inspecting the unit cells in Figure 3-5, we see that each atom in the SC lattice has a coordination number of six, while each atom in the BCC lattice has eight nearest neighbors. We will show later that each atom in the FCC lattice has a coordination number of 12, which is the maximum.

Packing factor. The *packing factor* is the fraction of space occupied by atoms, assuming that atoms are hard spheres. The general expression for the packing factor is

$$\text{Packing factor} = \frac{(\text{number of atoms/cell})(\text{volume of each atom})}{\text{volume of unit cell}} \qquad (3\text{-}4)$$

EXAMPLE 3-4

Calculate the packing factor for the FCC cell.

Answer:

There are four lattice points per cell; if there is one atom per lattice point, there are also four atoms per cell. The volume of one atom is $4\pi r^3/3$ and the volume of the unit cell is a_0^3.

$$\text{Packing factor} = \frac{(4 \text{ atoms/cell})(\frac{4}{3}\pi r^3)}{a_0^3}$$

Since for FCC unit cells, $a_0 = 4r/\sqrt{2}$,

$$\text{Packing factor} = \frac{(4)(\frac{4}{3}\pi r^3)}{(4r/\sqrt{2})^3} = 0.74$$

In metals, the packing factor of 0.74 in the FCC unit cell is the most efficient packing possible. BCC cells have a packing factor of 0.68 and SC cells have a packing factor of 0.52. Materials may have low packing factors as a consequence of atomic bonding. Metals with only the metallic bond are packed as efficiently as possible. Metals with mixed bonding may have unit cells with less than the maximum packing factor. No common engineering metals have the SC structure.

Density.   The theoretical density of a metal can also be calculated using the properties of the crystal structure. The general formula is

$$\text{Density } \rho = \frac{(\text{atoms/cell})(\text{atomic mass of each atom})}{(\text{volume of unit cell})(\text{Avogadro number})} \qquad (3\text{-}5)$$

EXAMPLE 3-5

Determine the density of BCC iron, which has a lattice parameter of 2.866 Å. (2.866 × 10^{-10} m).

Answer:

Atoms/cell = 2

Atomic mass = 55.85 g g · mol^{-1}

Volume of unit cell = a_0^3 = $(2.866 \times 10^{-10})^3$ = 23.54×10^{-30} m^3/cell

Avogadro number N_A = 6.02×10^{23} atoms/g · mole

$$\rho = \frac{(2)(55.85)}{(23.54 \times 10^{-30})(6.023 \times 10^{23})(10^6)} = 7.882 \text{ Mg m}^{-3}$$

The measured density is 7.870 Mg m^{-3}. We will explain the slight discrepancy between the theoretical and measured density in the next chapter when we discuss vacancies.

The hexagonal close-packed structure. A special form of the hexagonal lattice, the hexagonal close-packed structure (HCP), is shown in Figure 3-8. The unit cell is the skewed prism outlined in the hexagonal lattice. The HCP cell has two lattice points per cell—one from the eight corners and one from within the cell. In ideal HCP metals the a_0 and c_0 axes are related by the ratio $c/a = 1.633$. Most HCP metals, however, have c/a ratios that differ slightly from the ideal value due to imperfect metallic bonding. Because the HCP structure, like the FCC structure, has a very efficient packing factor of 0.74, a number of materials possess this structure. Table 3-2 summarizes the characteristics of the most important crystal structures in metals. In each case, only one atom is located at each lattice point.

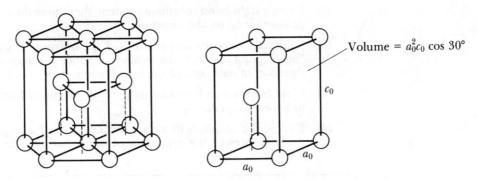

$$\text{Volume} = a_0^2 c_0 \cos 30°$$

FIG. 3-8 The hexagonal close-packed (HCP) lattice and its unit cell.

TABLE 3-2 Characteristics of common metallic crystals

Structure	a_0 versus r	Coordination Number	Packing Factor	Typical Metals
Simple cubic (SC)	$a_0 = 2r$	6	0.52	None
Body-centered cubic (BCC)	$a_0 = 4r/\sqrt{3}$	8	0.68	Fe, Ti, W, Mo, Nb, Ta, K, Na, V, Cr, Zr
Face-centered cubic (FCC)	$a_0 = 4r/\sqrt{2}$	12	0.74	Fe, Cu, Al, Au, Ag, Pb, Ni, Pt
Hexagonal close-packed (HCP)	$a_0 = 2r$ $c_0 = 1.633a_0$	12	0.74	Ti, Mg, Zn, Be, Co, Zr, Cd

3-4 Points, Directions, and Planes in the Unit Cell

Coordinates of points. We can locate certain points, such as atom positions, in the lattice or unit cell by constructing the right-handed coordinate system in

Figure 3-9. Distance is measured in terms of the number of lattice parameters we must move in each of the x, y, and z coordinates to get from the origin to the point in question. The coordinates are written as the three distances, with commas separating the numbers, as shown in Figure 3-9.

Directions in the unit cell. Certain directions in the unit cell are of particular importance. Metals deform, for example, in directions along which atoms are in closest contact. Properties of a material may depend on the direction in the crystal along which the property is measured. *Miller indices* for directions are the shorthand notation used to describe these directions. The procedure for finding the Miller indices for directions is as follows:

(a) Using a right-hand coordinate system, determine the coordinates of two points that lie on the direction.

(b) Subtract the coordinates of the "tail" point from the coordinates of the "head" point to obtain the number of lattice parameters traveled in the direction of each axis of the coordinate system.

(c) Clear fractions and/or reduce the results obtained from the subtraction to lowest integers.

(d) Enclose the numbers in square brackets [ ]. If a negative sign is produced, represent the negative sign with a bar over the number.

EXAMPLE 3-6

Determine the Miller indices of directions A, B, and C in Figure 3-10.

Answer:

Direction A

(a) Two points are 1, 0, 0, and 0, 0, 0

(b) 1, 0, 0 − 0, 0, 0 = 1, 0, 0

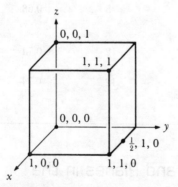

FIG. 3-9 Coordinates of points in the unit cell. The numbers refer to numbers of lattice parameters.

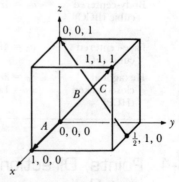

FIG. 3-10 Crystallographic directions and coordinates required for Example 3-6.

(c) No fractions to clear or integers to reduce

(d) [100]

Direction B

(a) Two points are 1, 1, 1 and 0, 0, 0

(b) 1, 1, 1 − 0, 0, 0 = 1, 1, 1

(c) No fractions to clear or integers to reduce

(d) [111]

Direction C

(a) Two points are 0, 0, 1 and $\frac{1}{2}$, 1, 0

(b) 0, 0, 1 − $\frac{1}{2}$, 1, 0 = −$\frac{1}{2}$, −1, 1

(c) 2(−$\frac{1}{2}$, −1, 1) = −1, −2, 2

(d) [$\bar{1}\bar{2}2$]

There are several points that should be noted about the use of Miller indices for directions.

1. A direction and its negative are not identical; [100] is not equal to [$\bar{1}$00]. They represent the same line but opposite directions. Going east isn't the same as going west!

2. A direction and its multiple are identical; [100] is the same direction as [200]. We just forgot to reduce to lowest integers.

3. Certain groups of directions are equivalent; they have their particular indices primarily because of the way we construct the coordinates. For example, a [100] direction is a [010] direction if we redefine the coordinate system as shown in Figure 3-11. We may refer to groups of equivalent directions as *directions of a form*. The special brackets ⟨ ⟩ are used to indicate this collection of directions. All of the directions of the form ⟨110⟩ are shown in Table 3-3.

Planes in the unit cell. Certain planes of atoms in a crystal are also significant; for example, metals deform along planes of atoms that are most tightly

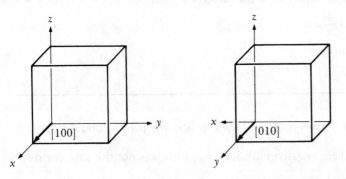

FIG. 3-11 Equivalency of crystallographic directions of a form.

TABLE 3-3 Directions of the form ⟨110⟩

$$\langle 110 \rangle = \begin{cases} [110] & [\bar{1}\bar{1}0] \\ [101] & [\bar{1}0\bar{1}] \\ [011] & [0\bar{1}\bar{1}] \\ [1\bar{1}0] & [\bar{1}10] \\ [10\bar{1}] & [\bar{1}01] \\ [01\bar{1}] & [0\bar{1}1] \end{cases}$$

packed together. Miller indices can be used as a shorthand notation to identify these important planes, as described in the following procedure.

(a) Identify the points at which the plane intercepts the x, y, and z coordinates in terms of the number of lattice parameters. If the plane passes through the origin, the origin of the coordinate system must be moved!

(b) Take reciprocals of these intercepts.

(c) Clear fractions but do not reduce to lowest integers.

(d) Enclose the resulting numbers in parentheses ( ). Again, negative numbers should be written with a bar over the number.

EXAMPLE 3-7

Determine the Miller indices of planes A, B, and C in Figure 3-12.

Answer:

Plane A

(a) $x = 1, y = 1, z = 1$

(b) $\dfrac{1}{x} = 1, \dfrac{1}{y} = 1, \dfrac{1}{z} = 1$

(c) No fractions to clear

(d) (111)

Plane B

(a) The plane never intercepts the x and z axes, so $x = \infty$, $y = \frac{1}{2}$, and $z = \infty$.

(b) $\dfrac{1}{x} = 0, \dfrac{1}{y} = 2, \dfrac{1}{z} = 0$

(c) No fractions to clear

(d) (020)

Plane C

(a) We must move the origin since the plane passes through 0, 0, 0. Let's move the origin one lattice parameter in the y-direction. Then, $x = \infty$, $y = -1$, and $z = \infty$.

(b) $\dfrac{1}{x} = 0, \dfrac{1}{y} = -1, \dfrac{1}{z} = 0$

(c) No fractions to clear

(d) $(0\bar{1}0)$

Several important aspects of the Miller indices for planes should be noted.

1. Planes and their negatives are identical (this was not the case for directions).

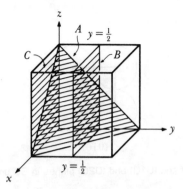

FIG. 3-12 Crystallographic planes and intercepts for Example 3-7.

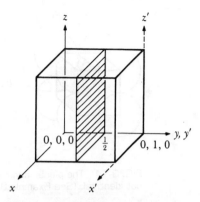

FIG. 3-13 A plane and its negative are identical. (See Example 3-8.)

EXAMPLE 3-8

Determine the Miller indices of the plane in Figure 3-13.

Answer:

If we locate our coordinate system at 0, 0, 0, then

(a) $x = \infty$, $y = \frac{1}{2}$, $z = \infty$

(b) $\dfrac{1}{x} = 0, \dfrac{1}{y} = 2, \dfrac{1}{z} = 0$

(c) No fractions to clear

(d) (020)

However, if we locate our coordinate system at 0, 1, 0, then

(a) $x' = \infty$, $y' = -\frac{1}{2}$, $z' = \infty$

(b) $\dfrac{1}{x'} = 0, \dfrac{1}{y'} = -2, \dfrac{1}{z'} = 0$

(c) No fractions to clear

(d) $(0\bar{2}0)$

But we are considering exactly the same plane! Therefore, $(020) = (0\bar{2}0)$.

---

2. Planes and their multiples are not identical (again this is the opposite of what we found for directions).

EXAMPLE 3-9

Show that the (010) and (020) planes in Figure 3-14 are not identical in a SC unit cell.

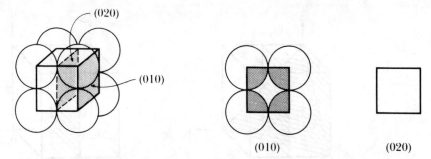

FIG. 3-14   The planar densities of the (010) and (020) planes in SC unit cells are not identical. (See Example 3-9.)

Answer:

Let's compare the *planar density*, or the fraction of a plane that is occupied by atoms. The (010) plane cuts through the center of the corner atoms in the SC cell, whereas the (020) plane does not intersect any of the atoms. Thus,

$$\text{Planar density for (010)} = \frac{(\frac{1}{4}\text{ atom/corner})(4\text{ corners})(\pi r^2)}{a_0^2}$$

$$= \frac{(\frac{1}{4})(4)(\pi r^2)}{(2r)^2} = 0.79$$

$$\text{Planar density (020)} = 0$$

Since the planar densities are not equal, the planes are not identical.

---

3. In each unit cell, *planes of a form* represent groups of equivalent planes that have their particular indices because of the orientation of the coordinates. We represent these groups of similar planes with the notation { }. The planes of the form {110} are shown in Table 3-4.

4. In cubic systems, a direction that has the same indices as a plane is perpendicular to that plane. Figure 3-15 shows a unit cell containing a (100) plane and a

TABLE 3-4
Planes of the form {110}

$$\{110\}\begin{cases} (110) \\ (101) \\ (011) \\ (1\bar{1}0) \\ (10\bar{1}) \\ (01\bar{1}) \end{cases}$$

The negatives of the planes are not unique planes.

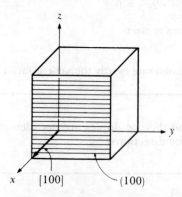

FIG. 3-15   A direction in a cubic unit cell is perpendicular to a plane with the same indices.

[100] direction and clearly indicates this property. This is not always true for noncubic cells.

EXAMPLE 3-10

A tetragonal unit cell with $a_0 = 3.0$ Å and $c_0 = 5.0$ Å is shown in Figure 3-16. Is the [001] direction perpendicular to the (001) plane? Is the [101] direction perpendicular to the (101) plane?

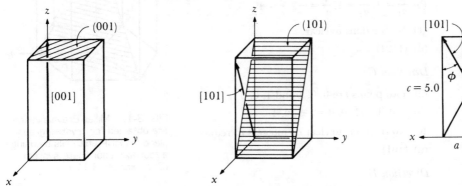
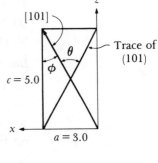

FIG. 3-16 A direction is not always perpendicular to a plane with the same indices in noncubic unit cells, such as the tetragonal cell. (See Example 3-10.)

Answer:

From inspection of Figure 3-16, we see that the [001] direction is obviously perpendicular to (001). However, when we examine the tetragonal cell from the y-axis, we see that [101] is not perpendicular to (101). The angle θ between the direction and the plane is

$$\theta = 2\phi = 2 \arctan \left(\tfrac{3}{5}\right) = 2(30.96°) = 61.92°$$

Miller indices of planes for hexagonal unit cells. A special set of *Miller-Bravais indices* has been devised for hexagonal unit cells because of the unique symmetry of the system (Figure 3-17). The coordinate system uses four axes instead of three axes, with one of the axes being redundant. The procedure for finding the indices is exactly the same as before, but four intercepts are required. Miller indices for directions, however, still use a three-axis coordinate system.

EXAMPLE 3-11

Determine the Miller-Bravais indices for planes A and B and directions C and D in Figure 3-17.

Answer:

Plane A

(a) $a_1 = a_2 = a_3 = \infty, c = 1$

(b) $\dfrac{1}{a_1} = \dfrac{1}{a_2} = \dfrac{1}{a_3} = 0, \dfrac{1}{c} = 1$

(c) No fractions to clear

(d) (0001)

Plane B

(a) $a_1 = 1, a_2 = 1, a_3 = -\tfrac{1}{2}, c = 1$

(b) $\dfrac{1}{a_1} = 1, \dfrac{1}{a_2} = 1, \dfrac{1}{a_3} = -2, \dfrac{1}{c} = 1$

(c) No fractions to clear

(d) $(11\bar{2}1)$

Direction C

(a) Two points are 0, 0, 1 and 1, 0, 0

(b) 0, 0, 1 − 1, 0, 0 = −1, 0, 1

(c) No fractions to clear or integers to reduce

(d) $[\bar{1}01]$

Direction D

(a) Two points are 0, 1, 0 and 1, 0, 0

(b) 0, 1, 0 − 1, 0, 0 = −1, 1, 0

(c) No fractions to clear or integers to reduce

(d) $[\bar{1}10]$

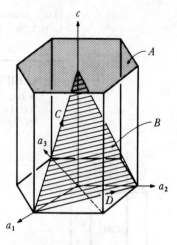

FIG. 3-17   Miller-Bravais indices are obtained for crystallographic planes in HCP unit cells by using a four-axis coordinate system. (See Example 3-11.)

Close-packed planes and directions.   In examining the relationship between atomic radius and lattice parameter, we looked for close-packed directions, where atoms are in continuous contact. We can now assign Miller indices to these close-packed directions, as shown in Table 3-5.

We can also examine FCC and HCP unit cells more closely and discover that there is at least one set of close-packed planes in each. A close-packed plane is shown in Figure 3-18. Notice that a hexagonal arrangement of atoms is produced in two dimensions. The close-packed planes are easy to find in the HCP unit cell;

TABLE 3-5   Close-packed planes and directions

Structure	Directions	Planes
SC	⟨100⟩	None
BCC	⟨111⟩	None
FCC	⟨110⟩	{111}
HCP	⟨100⟩, ⟨110⟩	(0001), (0002)

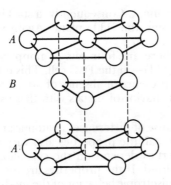

FIG. 3-18 The *ABABAB* stacking sequence of close-packed planes produces the HCP structure.

they are the (0001) and (0002) planes of the HCP structure and are given the special name *basal planes*. In fact, we can build up an HCP unit cell by stacking together close-packed planes in an . . . *ABABAB* . . . *stacking sequence* (Figure 3-18). Atoms on plane *B*, the (0002) plane, fit into the valleys between atoms on plane *A*, the bottom (0001) plane. If another plane identical in orientation to plane *A* is placed in the valleys of plane *B*, the HCP structure is created. Notice that all of the possible close-packed planes are parallel to one another. Only the basal planes—(0001) and (0002)—are close packed.

We can easily determine the coordination number of the atoms in the HCP structure from Figure 3-18. We find that the center atom in a basal plane is touched by six other atoms in the same plane. Three atoms in a lower plane and three atoms in an upper plane also touch the same atom. The coordination number is 12.

In the FCC structure, close-packed planes are of the form {111} (Figure 3-19). When stacking parallel (111) planes, atoms in plane *B* fit over valleys in plane *A* and atoms in plane *C* fit over valleys in both planes *A* and *B*. The fourth

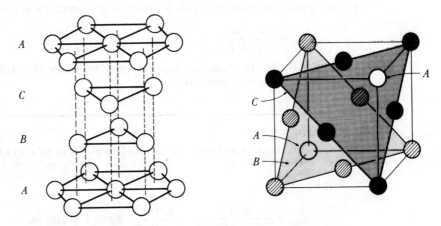

FIG. 3-19 The *ABCABCABC* stacking sequence of close-packed planes produces the FCC structure.

plane fits directly over atoms in plane A. Consequently, a stacking sequence . . . ABCABCABC . . . is produced using the (111) plane. Again we find that each atom has a coordination number of 12.

Unlike the HCP unit cell, there are four sets of nonparallel close-packed planes—(111), (11$\bar{1}$), (1$\bar{1}$1), and ($\bar{1}$11)—in the FCC cell. This difference between the FCC and HCP unit cells—the presence or absence of intersecting close-packed planes—significantly affects the behavior of metals with these structures.

Anisotropic behavior.   Because of differences in atomic arrangement in the planes and directions within a crystal, the properties also vary with direction. A material is *anisotropic* if its properties depend on the crystallographic direction along which the property is measured. If the properties are identical in all directions, the crystal is *isotropic*. The anisotropic behavior of the modulus of elasticity is illustrated in Table 3-6 for several materials.

TABLE 3-6   Variation of modulus of elasticity GPa with crystal direction

Material	[100]	[111]	Random
Al	63	76	69
Cu	67	192	125
Fe	132	279	207
Nb	152	81	103
W	408	408	408
MgO	245	336	310
NaCl	43	32	37

Interplanar spacing.   The distance between two adjacent parallel planes of atoms with the same Miller indices is called the *interplanar spacing d_{hkl}*. The interplanar spacing in cubic materials is given by the general equation

$$d_{hkl} = \frac{a_0}{\sqrt{h^2 + k^2 + l^2}} \tag{3-6}$$

where a_0 is the lattice parameter and h, k, and l represent the Miller indices of the adjacent planes being considered.

EXAMPLE 3-12

Calculate the distance between adjacent (111) planes in gold, which has a lattice parameter of 4.0786 Å.

Answer:

$$d_{111} = \frac{4.0786}{\sqrt{1^2 + 1^2 + 1^2}} = \frac{4.0786}{\sqrt{3}} = 2.355 \text{ Å } (0.2355 \text{ nm})$$

The interplanar spacings are important in X-ray diffraction. We can use the interaction between X-ray radiation and planes of atoms in a crystal to determine the lattice parameter of a material. X-ray diffraction can also be used to identify unknown materials or to determine the orientation of a crystal.

The *Debye-Scherrer X-ray diffraction* technique is used to determine interplanar spacings. X rays having a known wavelength strike a powdered specimen. The X-ray beam is diffracted at a specific angle from the incident beam according to Bragg's law,

$$\lambda = 2d_{hkl} \sin \theta \qquad (3\text{-}7)$$

where λ is the wavelength of the radiation, θ is the diffraction angle, and d_{hkl} is the interplanar spacing. If the wavelength is fixed, then atoms on any particular plane cause the X rays to be diffracted at a particular angle. The diffracted X-ray beam intersects and exposes a film surrounding the specimen (Figure 3-20). Only cones of radiation from planes whose interplanar spacings satisfy Bragg's law intersect the film. By analyzing the film and calculating the 2θ angle, we can determine the interplanar spacing.

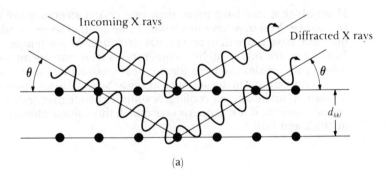

(a)

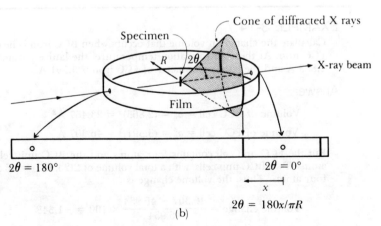

(b)

FIG. 3-20 (a) When X rays are diffracted from crystallographic planes at the Bragg angle, they produce a reinforced exit beam of radiation. (b) The Debye-Scherrer X-ray diffraction technique uses a film that intercepts a cone of X rays diffracted by the powdered specimen.

EXAMPLE 3-13

Suppose chromium radiation of wavelength 2.291 Å produced a diffracted line on a film at $2\theta = 58.2°$. What is the interplanar spacing of the crystallographic plane that produced the diffracted line?

Answer:

$$d_{hkl} = \frac{\lambda}{2 \sin \theta} = \frac{2.291}{2 \sin 29.1} = 2.355 \text{ Å}$$

If we were able to determine that the line represents a (111) plane, we might identify the specimen as gold (see Example 3-12).

3-5 Allotropic Transformations

Materials that can have more than one crystal structure are called *allotropic* or *polymorphic*. You may have noticed in Table 3-2 that some metals, such as iron and titanium, have more than one crystal structure. At low temperatures iron has the BCC structure but at higher temperatures iron transforms to a FCC structure. Allotropy provides the basis for the heat treatment of steel and titanium.

Many ceramic materials, such as silica (SiO_2), also go through allotropic transformations on heating or cooling. A volume change may accompany the allotropic transformation; if not properly controlled, this volume change causes the material to crack and fail.

EXAMPLE 3-14

Calculate the change in volume that occurs when BCC iron is heated and changes to FCC iron. At the transformation temperature, the lattice parameter of BCC iron is 2.863 Å and the lattice parameter of FCC iron is 3.591 Å.

Answer:

Volume of BCC cell = $a_0^3 = (2.863)^3 = 23.467 \text{ Å}^3$

Volume of FCC cell = $a_0^3 = (3.591)^3 = 46.307 \text{ Å}^3$

But the FCC unit cell contains four atoms and the BCC unit cell contains only two atoms. Two BCC unit cells, with a total volume of $2(23.467) = 46.934 \text{ Å}^3$, will contain four atoms. Thus, the volume change is

$$\text{Volume change} = \frac{46.307 - 46.934}{46.934} \times 100 = -1.34\%$$

This indicates that iron contracts on heating.

3-6 Complex Crystal Structures

Because of the nondirectional nature of the metallic bond, metals generally have simple crystal structures with only one atom per lattice point. However, covalently bonded materials, ionically bonded materials, and metallic compounds frequently must have more complex structures in order to satisfy the restraints imposed by bonding, ion size differences, and valence. We will examine a few of these structures in this section. In later chapters on ceramics and polymers, we will take a closer look at some of these complex crystal structures.

Diamond cubic structure. Elements such as silicon, germanium, tin, and carbon in its diamond form are bonded by four covalent bonds and produce a *tetrahedron* [Figure 3-21(a)]. Note that the coordination number for each silicon atom is four.

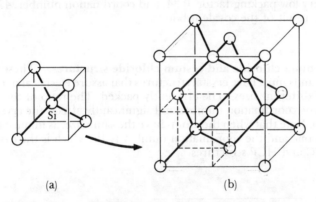

(a)   (b)

FIG. 3-21   (a) Tetrahedra and (b) the diamond cubic (DC) unit cell. This open structure is produced because of the requirements of covalent bonding.

As these tetrahedral groups are combined, a large cube can be constructed [Figure 3-21(b)]. This large cube contains eight smaller cubes that are the size of the tetrahedral cube; however, only four of the cubes contain tetrahedra. The large cube is the *diamond cubic,* or DC, unit cell. The lattice is a special FCC structure. The atoms on the corners of the tetrahedral cubes provide atoms at each of the regular FCC lattice points. However, four additional atoms are present within the DC unit cell from the atoms in the center of the tetrahedral cubes. We can describe the DC lattice as a FCC lattice with two atoms associated with each lattice point. Therefore there must be eight atoms per unit cell.

EXAMPLE 3-15

Determine the packing factor for DC silicon.

Answer:

We find that atoms touch along the body diagonal of the cell. Although atoms are not present at all locations along the body diagonal, there are voids that have the same diameter as atoms. Consequently,

$$\sqrt{3}a_0 = 8r$$

$$\text{Packing factor} = \frac{(8 \text{ atoms/cell})(\frac{4}{3}\pi r^3)}{a_0^3}$$

$$= \frac{(8)(\frac{4}{3}\pi r^3)}{(8r/\sqrt{3})^3}$$

$$= 0.34$$

(3-8)

The very low packing factor, 0.34, and coordination number, 4, of DC structures are the result of the covalent bonds.

Sodium chloride and cesium chloride structures.   These ionically bonded compounds must have crystal structures that assure electrical neutrality yet permit ions of different sizes to be efficiently packed. The ratio of the ionic radii determines the coordination number and significantly affects the crystal structure (Table 3-7). Note that when the atoms have the same size, as in pure metals, the radius ratio is one and the coordination number is 12, which is the case for metals with the FCC and HCP structures.

TABLE 3-7   Relationship between radius ratio and coordination number

Radius Ratio	Coordination Number
0–0.155	2
0.155–0.225	3
0.225–0.414	4
0.414–0.732	6
0.732–1.0	8
1.0	12

Sodium ions have a radius of 0.97 Å and chloride ions have a radius of 1.81 Å. Their radius ratio, $r_{Na}/r_{Cl} = 0.536$, indicates that the ions in sodium chloride should have a coordination number of six. The crystal structure of sodium chloride must permit each sodium ion to be surrounded by six chloride ions and vice versa. Figure 3-22 shows the sodium chloride structure, which can be viewed as a FCC structure with two ions—one sodium and one chloride—at each lattice point. The chloride ions are located at the normal FCC lattice points and the sodium ions are located at the edges of the cube and at the center of the cube.

EXAMPLE 3-16

Determine the packing factor and density of sodium chloride.

Answer:

The ionic radii are $r_{Na} = 0.97$ Å and $r_{Cl} = 1.81$ Å. Since the ions touch along the edge of the cube,

$$a_0 = 2r_{Na} + 2r_{Cl} = (2)(0.97) + (2)(1.81) = 5.56 \text{ Å } (5.56 \times 10^{-10} \text{ m})$$

$$\text{Packing factor} = \frac{(\text{Na atoms/cell})(\frac{4}{3}\pi r_{Na}^3) + (\text{Cl atoms/cell})(\frac{4}{3}\pi r_{Cl}^3)}{a_0^3}$$

$$= \frac{(4)(\frac{4}{3}\pi)(0.97)^3 + (4)(\frac{4}{3}\pi)(1.81)^3}{(5.56)^3} = 0.667$$

The atomic mass of sodium is 22.99 g g $\cdot$ mol^{-1} and the atomic mass of chlorine is 35.45 g g $\cdot$ mol^{-1}. The density is

$$\rho = \frac{(\text{Na atoms/cell})(M_{Na}) + (\text{Cl atoms/cell})(M_{Cl})}{(a_0)^3 \text{ (Avogadro number)}}$$

$$= \frac{(4)(22.99) + (4)(35.45)}{(5.56 \times 10^{-10})^3 (6.023 \times 10^{23})(10^6)} = 2.26 \text{ Mg m}^{-3}$$

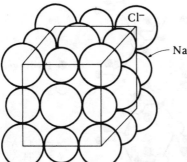

FIG. 3-22 The sodium chloride structure, an FCC unit cell with two ions per lattice point.

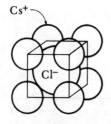

FIG. 3-23
The cesium chloride structure, a SC unit cell with two ions per lattice point.

Cesium ions have a radius of 1.67 Å. The radius ratio of cesium ions to chloride ions is $r_{Cs}/r_{Cl} = 0.922$ and indicates that a coordination number of eight is required. The crystal structure of cesium chloride in Figure 3-23, which is SC with two ions—one cesium and one chloride—associated with each lattice point, satisfies this requirement.

Crystalline silica. In a number of its forms, silica, or SiO_2, has a crystalline ceramic structure which is partly covalent and partly ionic. The ionic radii of silicon and oxygen are 0.42 Å and 1.32 Å, respectively, so the radius ratio is $r_{Si}/r_O = 0.318$ and the coordination number is four. Figure 3-24 shows the crystal structure of one of the forms of silica, β-cristobalite, which is a complicated FCC structure.

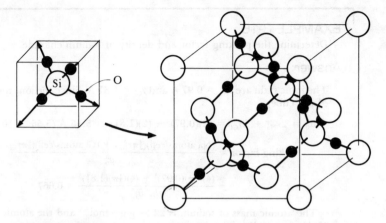

FIG. 3-24   The silicon-oxygen tetrahedra and their combination to form the β-cristobalite form of silica.

Crystalline polymers.   A number of polymers may form a crystalline structure. Figure 3-25 shows the unit cell for polyethylene. The chains composed of carbon and hydrogen atoms pack together to produce an orthorhombic unit cell. Some polymers, including nylon, can have several allotropic forms.

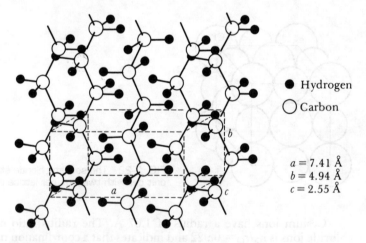

● Hydrogen

○ Carbon

$a = 7.41$ Å
$b = 4.94$ Å
$c = 2.55$ Å

FIG. 3-25   The unit cell of crystalline polyethylene.

EXAMPLE 3-17

How many carbon and hydrogen atoms are in each unit cell of crystalline polyethylene? There are twice as many hydrogen atoms as carbon atoms in the chain. The density of polyethylene is about 0.95 Mg m^{-3}.

Answer:

If we let x be the number of carbon atoms, then $2x$ is the number of hydrogen atoms.

$$\rho = \frac{(x)(12 \text{ g g} \cdot \text{mol}^{-1}) + (2x)(1 \text{ g g} \cdot \text{mol}^{-1})}{(7.41 \times 10^{-10})(4.94 \times 10^{-10})(2.55 \times 10^{-10})(6.02 \times 10^{23})}$$

$$0.95 = \frac{14x}{(56.2 \times 10^{-7})(10^6)}$$

$$x = 3.8 \approx 4 \text{ carbon atoms per cell}$$

$$2x = 8 \text{ hydrogen atoms per cell}$$

SUMMARY

Atoms are arranged in solid materials with either a short-range or long-range order. Amorphous materials, such as glasses and many polymers, have only a short-range order determined primarily by the restrictions of covalent bonding. Crystalline materials, including metals and many ceramics, have both a long- and short-range order. The long-range periodicity in these materials is described by the crystal structure.

Particularly in metals, the crystal structure is closely related to the general mechanical properties and behavior of the material. As we shall see in subsequent chapters, metals with FCC structures are normally soft and ductile, metals with BCC structures are much stronger, and metals with HCP structures tend to be relatively brittle. We will be able to explain the differences in behavior of metals by examining the role of crystal structure in deformation.

GLOSSARY

Allotropy The characteristic of being able to exist in more than one crystal structure, depending on temperature and pressure.

Anisotropy Having different properties in different directions.

Basal plane The special name given to the close-packed plane in hexagonal close-packed unit cells.

Bragg's law The relationship describing the angle at which a beam of X-rays of a particular wavelength diffracts from crystallographic planes of a given interplanar spacing.

Coordination number The number of nearest neighbors to an atom in its atomic arrangement.

Crystal structure The arrangement of the atoms in a material into a regular repeatable lattice.

Debye-Scherrer X-ray diffraction A technique using the interaction between X-ray radiation and planes of atoms in a crystal to obtain information concerning the identity and characteristics of a material.

Density Mass per unit volume of a material, usually in units of Mg m^{-3}.

Diamond cubic A special type of face-centered cubic crystal structure found in carbon, silicon, and other covalently bonded materials.

Directions of a form Crystallographic directions that all have the same characteristics, although their "sense" is different.

Glass A solid noncrystalline material that has only short-range order between the atoms.

Interplanar spacing Distance between two adjacent parallel planes with the same Miller indices.

Isotropy   Having the same properties in all directions.

Lattice   A collection of points that divide space into smaller equally sized segments.

Lattice parameters   The lengths of the sides of the unit cell and the angles between those sides. The lattice parameters describe the size and shape of the unit cell.

Lattice points   Points that make up the lattice. The surroundings of each lattice point are identical anywhere in the material.

Long-range order   A regular repetitive arrangement of the atoms in a solid which extends over a very large distance.

Miller indices   A shorthand notation to describe certain crystallographic directions and planes in a material.

Miller-Bravais indices   A special shorthand notation to describe the crystallographic planes in hexagonal close-packed unit cells.

Packing factor   The fraction of space occupied by atoms.

Planar density   The fraction of a plane in a lattice that intersects atoms.

Planes of a form   Crystallographic planes that all have the same characteristics, although their orientations are different.

Polymorphism   Allotropy, or having more than one crystal structure.

Short-range order   The arrangement of the atoms is regular and predictable only over a short distance, usually one or two atom spacings.

Stacking sequence   The sequence in which close-packed planes are stacked. If the sequence is *ABABAB* a hexagonal close-packed unit cell is produced; if the sequence is *ABCABCABC* a face-centered cubic structure is produced.

Tetrahedron   The structure produced when atoms are packed together with a fourfold coordination.

Unit cell   A subdivision of the lattice that still retains the overall characteristics of the entire lattice.

PRACTICE PROBLEMS

1   The atomic radius of FCC nickel is 1.243 Å. Calculate (a) the lattice parameter and (b) the density of nickel.

2   The atomic radius of BCC tungsten is 1.371 Å. Calculate (a) the lattice parameter and (b) the density of tungsten.

3   The lattice parameter of BCC molybdenum is 3.1468 Å. Calculate the atomic radius of the molybdenum atom.

4   The lattice parameter of FCC gold is 4.0786 Å. Calculate the atomic radius of the gold atom.

5   Chromium has a lattice parameter of 2.8844 Å and a density of 7.19 Mg m^{-3}. Determine by suitable calculations whether chromium is SC, BCC, or FCC.

6   Palladium has a lattice parameter of 3.8902 Å and a density of 12.02 Mg m^{-3}. Determine by suitable calculations whether palladium is SC, BCC, or FCC.

7   Gallium, which is face-centered orthorhombic, has two atoms per lattice point. The lattice parameters are $a_0 = 4.526$ Å, $b_0 = 4.520$ Å, and $c_0 = 7.660$ Å.

The atomic radius is 1.218 Å. Calculate (a) the packing factor and (b) the density of gallium.

8   Calculate the packing factors for SC and BCC unit cells, assuming one atom per lattice point.

9   Indium has a tetragonal structure. The atomic mass is 114.82 g g·mol^{-1}, the atomic radius is 1.625 Å, the lattice parameters are $a_0 = 3.2517$ Å and $c_0 = 4.9459$ Å, and the density is 7.286 Mg m^{-3}. (a) How many atoms are present in each unit cell? (b) What is the packing factor in indium?

10   Determine the number of lattice points per unit cell in a base-centered orthorhombic unit cell.

11   Germanium is DC with an atomic radius of 1.225 Å. Calculate (a) the lattice parameter and (b) the density of germanium.

12   One of the forms of manganese has a cubic structure with a lattice parameter of 6.326 Å. The density is 7.26 Mg m^{-3} and the atomic radius is 1.12 Å. (a) How many atoms are present in each unit cell? (b) Calculate the packing factor.

13 BCC niobium has a density of 8.57 Mg m^{-3} and one atom per lattice point. Calculate (a) the lattice parameter and (b) the atomic radius of niobium.

14 The body-centered tetragonal form of tin has lattice parameters $a_0 = 5.831$ Å and $c_0 = 3.182$ Å with a density of 7.298 Mg m^{-3}. Calculate the number of atoms per lattice point.

15 Uranium has an orthorhombic structure, with lattice parameters of $a_0 = 2.854$ Å, $b_0 = 5.869$ Å, and $c_0 = 4.955$ Å. Its atomic radius is 1.38 Å and density is 19.05 Mg m^{-3}. Determine (a) the number of atoms per unit cell and (b) the packing factor.

16 One of the forms of boron has 50 atoms in a tetragonal unit cell with $a_0 = 8.75$ Å and $c_0 = 5.06$ Å. Determine the density of this form of boron.

17 Lanthanum has an unusual HCP structure with $a_0 = 3.774$ Å and $c_0 = 12.17$ Å. The density of lanthanum is 6.146 Mg m^{-3}. Calculate the number of atoms per lattice point.

18 One of the forms of manganese has an atomic radius of 1.12 Å, a lattice parameter of 8.931 Å, and a packing factor of 0.479. How many atoms are in the unit cell?

19 Strontium is FCC with a lattice parameter of 6.0849 Å, but transforms to BCC with a lattice parameter of 4.85 Å on heating. (a) Calculate the percent volume change during the allotropic transformation. (b) Does strontium expand or contract on heating?

20 Under certain conditions, FCC iron transforms to a body-centered tetragonal (BCT) unit cell on cooling. The lattice parameter of FCC iron is 3.5656 Å and those of BCT iron are $a_0 = 2.8558$ Å and $c_0 = 2.9074$ Å. Calculate the volume change during this transformation (known as the martensitic reaction).

21 BCC titanium, with a lattice parameter of 3.32 Å, transforms to an HCP structure on cooling. The HCP structure has lattice parameters $a_0 = 2.978$ Å and $c_0 = 4.735$ Å. (a) Calculate the volume change during cooling. (b) Does titanium expand or contract on cooling?

22 Prove that the c/a ratio in HCP metals is 1.633.

23 Calculate the volume of a unit cell of zinc, which is HCP with $a_0 = 2.6648$ Å and $c_0 = 4.9470$ Å.

24 Prove that the packing factor in HCP cells is 0.74.

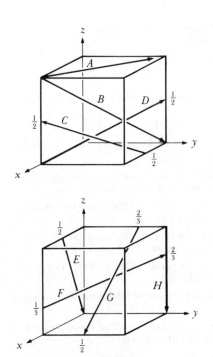

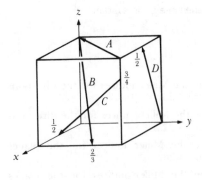

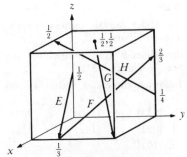

FIG. 3-26 Directions in a cubic system for Problem 3-25.

FIG. 3-27 Directions in a cubic system for Problem 3-26.

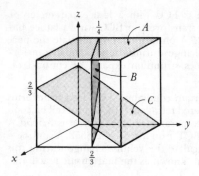

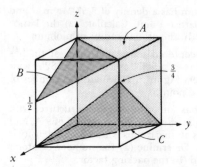

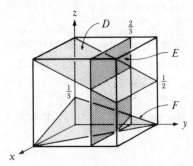

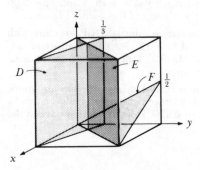

FIG. 3-28   Planes in a cubic system
for Problem 3-27.

FIG. 3-29   Planes in a cubic system
for Problem 3-28.

25   Determine the Miller indices for the directions shown in Figure 3-26.

26   Determine the Miller indices for the directions shown in Figure 3-27.

27   Determine the Miller indices for the planes shown in Figure 3-28.

28   Determine the Miller indices for the planes shown in Figure 3-29.

29   Determine the Miller-Bravais indices for the planes and directions in Figure 3-30.

30   Determine the Miller-Bravais indices for the planes and directions in Figure 3-31.

31   Sketch the following directions within a cubic unit cell.

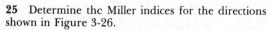

    a. [121]   b. [2$\bar{1}$3]   c. [$\bar{1}$10]   d. [$\bar{2}$01]

    e. [1$\bar{3}\bar{3}$]   f. [02$\bar{1}$]   g. [$\bar{1}$4$\bar{2}$]   h. [3$\bar{1}\bar{2}$]

    i. [1$\bar{1}$1]   j. [10$\bar{1}$]

32   Sketch the following planes within a cubic unit cell. You may have to move the origin of the coordi-

nate system to assure that the intercepts are within the cell.

    a. (121)   b. (2$\bar{1}$3)   c. ($\bar{1}$10)   d. ($\bar{2}$01)

    e. (1$\bar{3}\bar{3}$)   f. (02$\bar{1}$)   g. ($\bar{1}$4$\bar{2}$)   h. (3$\bar{1}\bar{2}$)

    i. (1$\bar{1}$1)   j. (10$\bar{1}$)

33   Determine the six directions of the form $\langle 110 \rangle$ that are contained in the (111) plane.

34   Determine the planes of the form $\{110\}$ that contain the [111] direction.

35   Determine the four directions of the form $\langle 111 \rangle$ that are contained in the ($\bar{1}$10) plane.

36   Determine the planes of the form $\{111\}$ that contain the [110] direction.

37   What are the indices of all the planes, including the negatives, of the form $\{111\}$ in cubic systems?

38   What are the indices of all of the directions of the form $\langle 112 \rangle$ in cubic systems?

39   What are the indices of all of the directions of the form $\langle 221 \rangle$ in cubic systems?

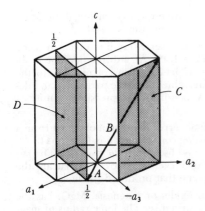

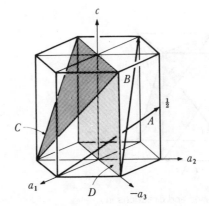

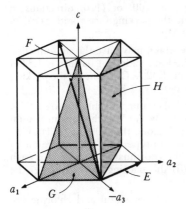

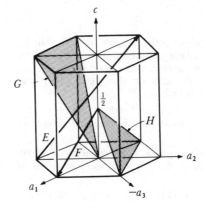

FIG. 3-30 Planes and directions in the hexagonal crystal structure for Problem 3-29.

FIG. 3-31 Planes and directions in the hexagonal crystal structure for Problem 3-30.

40 What are the indices of all of the planes of the form {110} in an orthorhombic system?

41 (a) Is the [210] direction perpendicular to the (210) plane in a tetragonal unit cell? (b) Is the [201] direction perpendicular to the (201) plane in a tetragonal unit cell?

42 Does a [$1\bar{2}1$] direction lie in the (111) plane?

43 Does a [221] direction lie in the (111) plane?

44 Determine the Miller indices for the planes and directions in the orthorhombic unit cell shown in Figure 3-32.

45 What are the indices of the plane that passes through the coordinates $\frac{1}{2}$, $\frac{1}{2}$, 0; 1, 0, 0; and 0, 0, 1?

46 Determine the indices of the direction that passes through the coordinates $\frac{1}{2}$, 0, $\frac{1}{2}$ and 0, 1, 0.

47 Determine the indices of the direction that passes through the coordinates $\frac{1}{2}$, $\frac{1}{4}$, $\frac{1}{2}$ and $\frac{1}{4}$, 1, 3.

48 Determine the indices of the plane that passes through the coordinates $\frac{1}{2}$, $\frac{1}{2}$, 0; 1, 0, 1; and 0, 1, 1.

49 Compare the planar densities on the (100), (200), and (110) planes of a BCC unit cell.

50 Compare the planar densities on the (100), (200), and (111) planes in FCC unit cells.

51 Determine the planar density of the (100) and (001) planes in a base-centered tetragonal unit cell with $c/a = 1.3$. (a) Are the (100) and (001) planes equivalent? (b) What are the planes of the form {100}? (c) What are the planes of the form {001}?

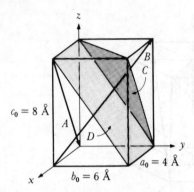

FIG. 3-32   Planes and directions in
the orthorhombic unit cell for
Problems 3-44 and 3-52.

52 Suppose that the orthorhombic structure shown in Figure 3-32 is face centered and that the atomic radius of the atoms at each lattice point is 1.8 Å. Compare the planar densities on the (100), (010), and (001) planes of the cell.

53 Determine the distance between adjacent (121) planes in copper, which has a lattice parameter of 3.615 Å.

54 What is the interplanar spacing between the following planes in iron, with a lattice parameter of 2.886 Å?

a. (111)   b. (212)   c. (423)   d. (201)

55 Suppose we obtained a Debye-Scherrer X-ray film from a copper powder, $a_0 = 3.615$ Å, and determined from the 2θ measurement that a particular line has an interplanar spacing of 1.278 Å. What are the Miller indices of the planes that caused the diffracted line?

56 In a Debye-Scherrer X-ray film obtained using copper radiation ($\lambda = 1.5418$ Å), a diffraction line is observed at $2\theta = 38.51°$. The plane causing the diffracted line is a (111) plane. Calculate the lattice parameter and determine from Appendix A the identity of the specimen.

57 An X-ray film with a 50 mm radius is used in an X-ray camera. A diffracted line 40 mm from the exit port of the beam is produced. Determine the interplanar spacing of the plane that produced the diffracted line if copper radiation ($\lambda = 1.5418$ Å) is used.

58 In a Debye-Scherrer X-ray film obtained from a powdered chromium specimen with copper radiation ($\lambda = 1.5418$ Å), the first three diffraction lines are located at $2\theta = 44.4°, 64.6°,$ and $81.8°$. Determine the

Miller indices corresponding to each line, if the lattice parameter of BCC chromium is 2.8845 Å.

59 In a Debye-Scherrer X-ray film obtained using iron radiation ($\lambda = 1.937$ Å), a diffraction line is observed at $2\theta = 153.44°$, which corresponds to the (310) plane. Determine the lattice parameter and determine from Appendix A the probable identity of the specimen.

60 In a Debye-Scherrer X-ray film obtained from a nickel powder using copper radiation ($\lambda = 1.5418$ Å), a diffraction line is observed at $2\theta = 76.64°$. The lattice parameter of nickel is 3.5167 Å. What are the indices of the plane that produced the line?

61 Magnesium oxide, or magnesia, MgO, has the sodium chloride structure. The ionic radius of magnesium is 0.78 Å and that of oxygen is 1.32 Å. (a) Will ions touch along the $\langle 100 \rangle$ or $\langle 110 \rangle$ directions in MgO? (b) Calculate the packing factor and (c) calculate the density of MgO.

62 Barium oxide (BaO) may have the sodium chloride structure. The ionic radius of barium is 1.34 Å and that of oxygen is 1.32 Å. Calculate (a) the packing factor and (b) the density of BaO.

63 The ionic radii of cesium and chlorine are 1.67 Å and 1.81 Å, respectively. (a) Will the ions in cesium chloride (CsCl) touch along the $\langle 100 \rangle$ or $\langle 111 \rangle$ direction? (b) How many atoms per unit cell are present in CsCl? (c) What is the coordination number for each atom? Calculate (d) the packing factor and (e) the density of CsCl.

64 The ionic radii of potassium and chlorine are 1.33 Å and 1.81 Å, respectively. (a) Show that potassium chloride (KCl) can have the CsCl structure. Calculate (b) the packing factor and (c) the density of KCl.

65 Determine the diameter of the largest atom that could fit into the center of a SC unit cell that has a lattice parameter of 4 Å.

66 A hole in the BCC unit cell occurs at the coordinates $0, \frac{1}{2}, \frac{1}{4}$. This is known as the *tetrahedral site*. Calculate the diameter of the largest atom that could fit in this site if the lattice parameter is 4 Å.

67 A hole in the FCC unit cell occurs at the coordinates $\frac{1}{2}, \frac{1}{2}, \frac{1}{2}$. This is known as the *octahedral site*. Calculate the diameter of the largest atom that could fit in this site if the lattice parameter is 4 Å.

68 (a) Calculate the lattice parameter of β-cristobalite from Figure 3-24. The silicon and oxygen ions touch along the body diagonal. (b) The density of β-cristobalite is 2.32 Mg m^{-3}. Determine the number of ions per unit cell. (There should be twice as many oxygen ions as silicon ions.)

69 Determine the coordination number for aluminum oxide (Al_2O_3). The ionic radius of aluminum is 0.51 Å and for oxygen is 1.32 Å.

70 Iron carbide (Fe_3C) is an intermetallic compound that has an orthorhombic structure with $a_0 = 4.514$ Å, $b_0 = 5.080$ Å, and $c_0 = 6.734$ Å. Its density is 7.66 Mg m^{-3}. Determine the number of iron and carbon atoms in each unit cell.

71 Polypropylene is a polymer made up of propylene molecules with the formula C_3H_6. The monoclinic unit cell for crystalline polypropylene has the lattice parameters $a_0 = 6.65$ Å, $b_0 = 20.96$ Å, $c_0 = 6.5$ Å, and $\beta = 99.3°$. The density of the polymer is about 0.91 Mg m^{-3}. Calculate the number of carbon and hydrogen atoms present in the unit cell. The volume of a monoclinic unit cell is given by $V = a_0b_0c_0 \sin \beta$.

CHAPTER **4**

Imperfections in the Atomic Arrangement

4-1 Introduction

All materials contain imperfections in the arrangement of the atoms which have a profound effect on the behavior of the material. By controlling lattice imperfections we create stronger metals and alloys, more powerful magnets, improved transistors and solar cells, glassware of striking colors, and many other materials of practical importance.

In this chapter we will introduce the three basic types of lattice imperfections—point defects, line defects (or dislocations), and surface defects. We must remember, however, that these imperfections represent only defects in the perfect atomic arrangement and should not suggest that the material itself is defective. Indeed, these "defects" are intentionally added to produce a desired set of mechanical and physical properties. In later chapters we will show how we control these defects through alloying, heat treatment, or processing techniques to produce improved engineering materials.

4-2 Dislocations

Dislocations are line imperfections in an otherwise perfect lattice. We can identify two types of dislocations—the screw dislocation and the edge dislocation. The *screw dislocation* (Figure 4-1) can be illustrated by cutting partway through a perfect crystal, then skewing the crystal one atom spacing. If we were to follow a crystallographic plane one revolution around the axis on which the crystal was skewed, traveling equal atom spacings in each direction, we would finish one atom spacing below our starting point. The vector required to complete the loop and return us to our starting point is the *Burgers vector* **b**. If we continued our rotation, we would trace out a spiral path. The axis, or line, around which we trace out this path is the screw dislocation. We see that the Burgers vector is parallel to the screw dislocation.

An *edge dislocation* (Figure 4-2) can be illustrated by slicing partway through a perfect crystal, spreading the crystal apart, and partly filling the cut with an extra plane of atoms. The bottom edge of this inserted plane represents the edge dislocation. If we describe a clockwise loop around the edge dislocation by going an equal number of atom spacings in each direction, we would finish one atom spacing from our starting point. The vector that is required to complete the loop is again the Burgers vector. In this case, the Burgers vector is perpendicular to the edge dislocation.

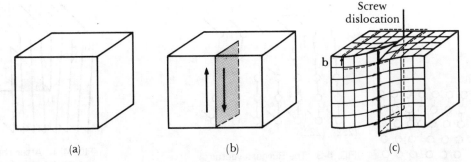

FIG. 4-1 The perfect crystal (a) is cut and sheared one atom spacing, (b) and (c). The line along which shearing occurs is a screw dislocation. A Burgers vector **b** is required to close a loop of equal atom spacings around the screw dislocation.

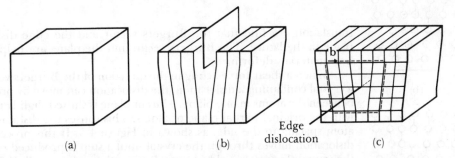

FIG. 4-2 The perfect crystal in (a) is cut and an extra plane of atoms is inserted (b). The bottom edge of the extra plane is an edge dislocation (c). A Burgers vector **b** is required to close a loop of equal atom spacings around the edge dislocation.

EXAMPLE 4-1

Suppose we have a BCC structure with a lattice parameter of 4.0 Å that contains the dislocation shown in Figure 4-3. Determine the direction and length of the Burgers vector.

Answer:

The clockwise loop around the dislocation is closed by the vector **b**. Because **b** is perpendicular to the (222) planes, the Miller indices of direction **b** must be [222] or, reducing to lowest integers, [111]. The length of **b** is the distance between two adjacent (222) planes. From Equation (3-6),

$$d_{222} = \frac{a_0}{\sqrt{h^2 + k^2 + l^2}} = \frac{4}{\sqrt{2^2 + 2^2 + 2^2}} = 1.155 \text{ Å}$$

The Burgers vector is a [111] direction that is 1.155 Å in length.

Because vectors can be moved about in space, we could translate the Burgers vector from the loop to the edge dislocation, as shown in Figure 4-4. After this

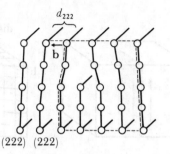

FIG. 4-3   The Burgers vector for Example 4-1 is perpendicular to the (222) planes and has a length equal to the interplanar spacing between (222) planes.

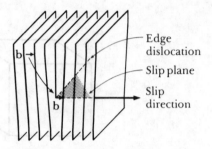

FIG. 4-4   After the Burgers vector is translated from the loop to the dislocation line, a plane is defined.

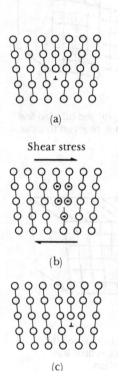

FIG. 4-5
When a shear force is applied to the dislocation in (a), the atoms are displaced (b), until the dislocation moves one Burgers vector in the slip direction (c).

translation, we find that the Burgers vector and the edge dislocation define a plane in the lattice. The Burgers vector and the plane are helpful in explaining how materials deform.

When a shear force acting in the direction of the Burgers vector is applied to a crystal containing a dislocation, the dislocation can move by breaking the bonds between the atoms in one plane. The cut plane is shifted slightly to establish bonds with the original partial plane of atoms. This causes the dislocation to move one atom spacing to the side, as shown in Figure 4-5. If this process continues, the dislocation moves through the crystal until a step is produced on the exterior of the crystal (Figure 4-6). The crystal has now been deformed. If dislocations could be continually introduced into one side of the crystal and moved along the same path through the crystal, the crystal would eventually be cut in half.

The process by which a dislocation moves and causes a metal to deform is called *slip*. The direction in which the dislocation line moves, the *slip direction*, is the direction of the Burgers vector for edge dislocations. During slip the edge dislocation sweeps out the plane formed by the Burgers vector and the dislocation; this plane is called the *slip plane*. The combination of slip direction and slip plane is the *slip system*. A screw dislocation produces the same result; the dislocation moves in a direction perpendicular to the Burgers vector, although the crystal deforms in a direction parallel to the Burgers vector. Whenever possible, the slip direction is a close-packed direction and the slip plane is a close-packed plane. The most common slip systems in metals are summarized in Table 4-1.

4-3   Significance of Dislocations

Although slip can occur in some ceramics and polymers, the slip process is particularly helpful to us in understanding the mechanical behavior of metals. First, slip explains why the strength of metals is much lower than the value predicted from the metallic bond. If we had to break an iron bar by breaking all of the metallic bonds across the cross section, as shown in Figure 4-7, we would have to exert a force of several million pounds per square inch. Instead we could deform the bar by causing slip, during which only a tiny fraction of all the metallic bonds need be broken at any one time. Perhaps a force of only 70 MPa would be required to deform the iron bar by slip.

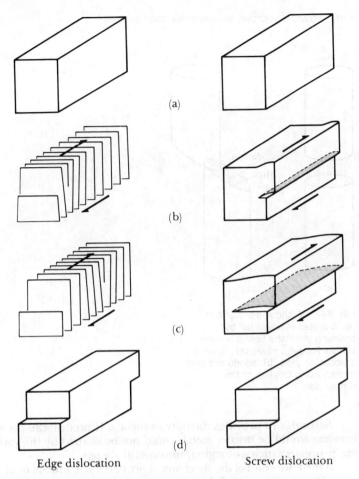

(a)

(b)

(c)

(d)

Edge dislocation   Screw dislocation

FIG. 4-6   A shear force acting on a dislocation introduced into a
perfect crystal (a) causes the dislocation to move through the
crystal until a step is created (d). The crystal is now deformed.

TABLE 4-1   Slip planes and directions in
metallic structures

Crystal Structure	Slip Plane	Slip Direction
BCC	{110} {112} {123}	⟨111⟩
FCC	{111}	⟨110⟩
HCP	(0001) {11$\bar{2}$0} {10$\bar{1}$0} {10$\bar{1}$1} } See note	⟨100⟩ ⟨110⟩

Note: These planes are active in some metals, alloys, or at
elevated temperatures.

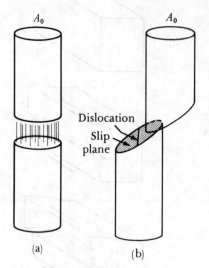

FIG. 4-7   Without dislocations
(a), a material would fail by
breaking all of the bonds across
the surface A_0. However, when a
dislocation slips (b), bonds are only
broken along the line of the
dislocation.

Second, slip provides ductility in metals. If no dislocations were present, the iron bar would be brittle; metals could not be shaped by the various metal-working processes, such as forging, into useful shapes.

Third, we control the mechanical properties of a metal or alloy by interfering with the movement of dislocations. An obstacle introduced into the crystal prevents a dislocation from slipping unless we apply higher forces. If we must apply a higher force, then the metal must be stronger!

4-4   Schmid's Law

We can understand the differences in behavior of metals that have different crystal structures by examining the slip process. Suppose we apply a unidirectional force F to a cylinder of metal which is one single crystal (Figure 4-8). We can orient the slip plane and slip direction to the applied force by defining the angles λ and ϕ. λ is the angle between the slip direction and the applied force and ϕ is the angle between the normal to the slip plane and the applied force.

In order for the dislocation to move in its slip system, a shear force acting in the slip direction must be produced by the applied force. This resolved shear force F_r is given by

$$F_r = F \cos \lambda$$

If we divide the equation by the area of the slip plane, $A = A_0/\cos \phi$, we obtain *Schmid's law*

$$\tau_r = \sigma \cos \phi \cos \lambda \tag{4-1}$$

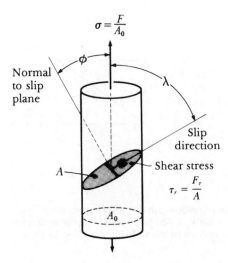

FIG. 4-8   A resolved shear stress τ may be produced on a slip system, causing the dislocation to move on the slip plane in the slip direction.

where

$$\tau_r = \frac{F_r}{A} = \text{resolved shear } \textit{stress} \text{ in the slip direction}$$

$$\sigma = \frac{F}{A_0} = \text{unidirectional } \textit{stress} \text{ applied to the cylinder}$$

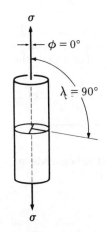

FIG. 4-9
When the slip plane is perpendicular to the applied stress σ, the angle λ is 90° and no shear stress is resolved.

EXAMPLE 4-2

Suppose the slip plane is perpendicular to the applied stress σ, as in Figure 4-9. Then, $\phi = 0°$, $\lambda = 90°$, $\cos \lambda = 0$, and therefore $\tau_r = 0$. Even if the applied stress σ is enormous, no resolved shear stress develops along the slip direction and the dislocation cannot move. (You could perform a simple experiment to demonstrate this with a deck of cards. If you push on the deck at an angle, the cards slide over one another, like the slip process. If you push perpendicular to the deck, however, the cards do not slide.) Slip cannot occur if the slip system is oriented so that either λ or ϕ is 90°.

EXAMPLE 4-3

Calculate the resolved shear stress on the (111) $[\bar{1}01]$ slip system if a stress of 70 MPa is applied in the $[001]$ direction of a FCC unit cell. (See Figure 4-10.)

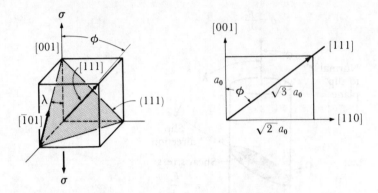

FIG. 4-10   A normal stress σ is applied in the [001] direction of the unit cell. This produces an angle λ of 45° to the [$\bar{1}$01] slip direction and an angle ϕ of 54.76° to the normal to the (111) plane. (See Example 4-3.)

Answer:

By inspection, $\lambda = 45°$ and $\cos \lambda = 0.707$. The normal to the (111) plane must be the [111] direction. We can then calculate that

$$\cos \phi = \frac{1}{\sqrt{3}} = 0.577 \quad \text{or} \quad \phi = 54.76°$$

$$\tau_r = \sigma \cos \lambda \cos \phi = (70 \text{ MPa})(0.707)(0.577) = 28.56 \text{ MPa}$$

The *critical resolved shear stress* τ_{crss} is the shear stress required to break enough metallic bonds in order for slip to occur. Thus slip occurs, causing the metal to deform, when the applied stress produces a resolved shear stress that equals the critical resolved shear stress.

$$\tau_r = \tau_{crss} \tag{4-2}$$

EXAMPLE 4-4

Consider a slip system in which $\lambda = 70°$ and $\phi = 30°$. Slip is found to begin when a stress of 35 MPa is applied. Calculate the critical resolved shear stress.

Answer:

$\tau_{crss} = \tau_r$ when slip begins

$$\tau_{crss} = \sigma \cos \lambda \cos \phi = (35 \text{ MPa})(\cos 70)(\cos 30) = 10.37 \text{ MPa}$$

4-5 Influence of Crystal Structure

We can use Schmid's law to compare the properties of metals having BCC, FCC, and HCP crystal structures. Table 4-2 lists three important factors that we can

TABLE 4-2 Summary of factors affecting slip in metallic structures

Factor	FCC	BCC	HCP
Critical resolved shear stress (MPa)	0.34–0.69	35–70	0.34–0.69
Number of slip systems	12	48	3^a
Cross-slip	Can occur	Can occur	Cannot occur^a
Summary of properties	Ductile	Strong	Relatively brittle

[a] By alloying or heating to elevated temperatures, additional slip systems are active in HCP metals, permitting cross-slip to occur and thereby improving ductility.

examine. We must be careful to note, however, that this discussion describes the behavior of nearly perfect single crystals. Real engineering materials are seldom single crystals and never quite perfect.

Critical resolved shear stress. If the critical resolved shear stress in a metal is very high, then the applied stress σ must also be high in order for τ, to equal τ_{crss}. If σ is large, then the metal must have a high strength! In FCC and HCP metals, which have close-packed planes, the critical resolved shear stress is low, about 50 to 100 psi in a perfect crystal; metals having either of these crystal structures are expected to have low strengths. On the other hand, BCC crystal structures contain no close-packed planes. Dislocations must move on nonclose-packed planes, such as {110}, {112}, and {123} planes. Now we must exceed a higher critical resolved shear stress, in the order of 10,000 psi in perfect crystals, before slip occurs and therefore BCC metals tend to have high strengths. We should note, however, that engineering metals have much higher strengths than those calculated from τ_{crss} because of other imperfections in the crystal.

EXAMPLE 4-5

Compare the planar densities for the close-packed (111) plane of FCC metals and the (110) plane of BCC metals.

Answer:

Figure 4-11 shows the arrangement of the atoms in each plane. Half of the area of the atoms at the sides of the triangular (111) plane lies within the triangle and one-sixth of the area of the atoms at the corners lies within the triangle. The planar density for the (111) slip plane in FCC metals is

$$\frac{A_{atoms}}{A_{plane}} = \frac{(3 \text{ corners})(\frac{1}{6})(\pi r^2) + (3 \text{ sides})(\frac{1}{2})(\pi r^2)}{\frac{1}{2}(\sqrt{2}a_0)(\sqrt{2}a_0 \cos 30)}$$

$$= \frac{2\pi r^2}{a_0^2 \cos 30} = \frac{2\pi r^2}{(4r/\sqrt{2})^2(0.866)} = 0.907$$

On the (110) plane of BCC metals, one quarter of each corner atom is in the

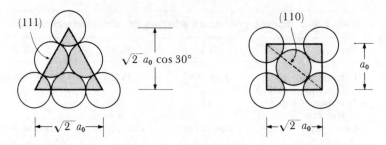

FIG. 4-11   The planar densities for the (111) plane in FCC metals and the (110) plane in BCC metals. (See Example 4-5.)

rectangle and all of the area of the center atom is within the rectangle. The planar density of the (110) slip plane in BCC metals is

$$\frac{A_{\text{atoms}}}{A_{\text{plane}}} = \frac{(4 \text{ corners})(\tfrac{1}{4})(\pi r^2) + (1 \text{ center})(\pi r^2)}{\sqrt{2}\, a_0^2} = \frac{2\pi r^2}{\sqrt{2}(4r\sqrt{3})^2} = 0.833$$

The slip planes in BCC metals are not nearly as close packed as in FCC metals.

Number of slip systems.   If at least one slip system is oriented to give the angles λ and ϕ near 45°, then τ_r equals τ_{crss} at low applied stresses. HCP metals possess only one set of parallel close-packed planes, the (0001) planes, and three close-packed directions, giving three slip systems (Figure 4-12). Consequently, the

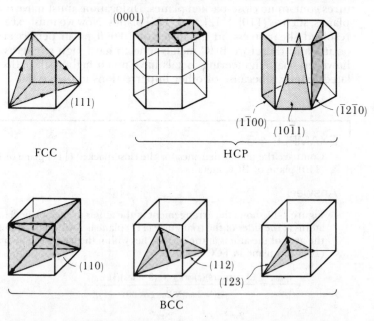

FIG. 4-12   Typical slip planes and directions in the common metallic crystal structures.

probability of the close-packed planes and directions being oriented with λ and ϕ near 45° is very low. The HCP crystal may fail in a brittle manner without a significant amount of slip.

On the other hand, FCC metals contain four nonparallel close-packed planes of the form {111} and three close-packed directions of the form ⟨110⟩ within each plane, giving a total of 12 slip systems (Figure 4-12). At least one slip system is favorably oriented for slip to occur at low applied stresses, causing FCC metals to have low strengths but high ductilities.

Finally, BCC metals have as many as 48 slip systems that are nearly close packed. Several slip systems are always properly oriented for slip to occur—in fact, there are too many possible slip systems. Dislocations moving on one slip plane may interfere with movement of dislocations on other active slip planes. This interference leads to high strengths in BCC metals yet still permits at least some ductility as well.

Cross-slip.   Suppose a dislocation moving on one slip plane encounters an obstacle and is blocked from further movement. The dislocation can shift to a second intersecting slip system, also properly oriented, and continue to move. This is called *cross-slip*. In many HCP metals, no cross-slip can occur because the slip planes are parallel, not intersecting. Therefore, the HCP metals tend to re-main brittle. Fortunately, additional slip systems become active when HCP metals are alloyed or heated, thus improving ductility. Cross-slip is possible in both FCC and BCC metals because a number of intersecting slip systems is present (Figure 4-13). Cross-slip consequently helps maintain ductility in these metals.

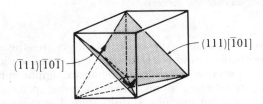

FIG. 4-13   Cross-slip of a dislocation from a $(\overline{1}11)$ $[\overline{1}0\overline{1}]$ system to a (111) $[\overline{1}01]$ system in a FCC crystal.

In summary, the most significant characteristics of the three important crys-tal structures in single crystals of metals are: FCC metals are ductile, BCC metals are strong, and HCP metals are relatively brittle. These general conclusions still tend to be followed even in real engineering metals and alloys which are not single crystals.

4-6   Control of the Slip Process

In a perfect crystal, the fixed repeated arrangement of the atoms gives the lowest possible energy within the crystal. Any imperfection in the lattice raises the inter-nal energy at the location of the imperfection. The local energy is increased because near the imperfection, the atoms either are squeezed too closely together (compression) or are forced too far apart (tension).

One dislocation in an otherwise perfect lattice can move easily through the crystal if the resolved shear stress equals the critical resolved shear stress. However, if the dislocation encounters a region where the atoms are displaced from their usual positions, a higher stress is required to force the dislocation past the region of high local energy. Thus, the metal is stronger.

4-7   Dislocation Interactions

Dislocations disrupt the perfection of the lattice. In Figure 4-14, the atoms below the dislocation line at point B are compressed, since an extra plane of atoms has

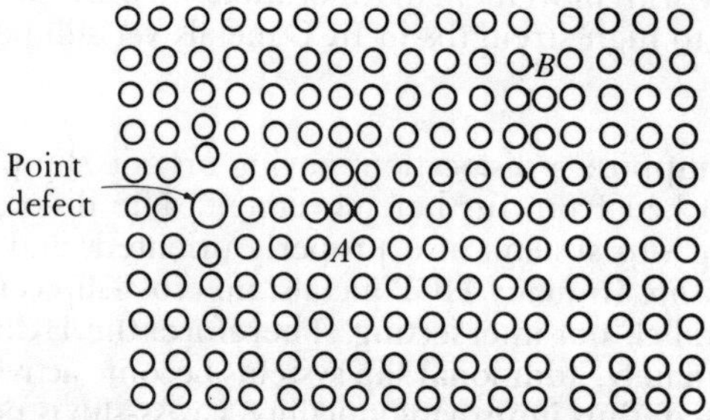

FIG. 4-14   If the dislocation at point A moves to the left, it is blocked by the point defect. If the dislocation moves to the right, it interacts with the disturbed lattice near the second dislocation at point B.

been inserted into the structure. Above the dislocation line, the atoms are further apart than they normally would be, leading to an area of tension. This effect is very widespread—atoms near the dislocation line are significantly out of position. They disrupt the surrounding atoms, which then disrupt other atoms further away, for a number of atom spacings. If dislocation A moves and passes near the dislocation at point B, dislocation A encounters a region where the atoms are not properly arranged. Higher stresses are required to keep the second dislocation moving and consequently the metal must be stronger. The more dislocations that are present in the lattice, the more likely it is that any one dislocation will be blocked. Therefore increasing the number of dislocations increases the strength of the metal. We will discuss this effect formally in Chapter 9 as *strain hardening*.

4-8   Point Defects

Point defects are localized disruptions of the lattice involving one or possibly several atoms (Figure 4-15). A *vacancy* is produced when an atom is missing from a normal lattice point. Vacancies are introduced into the crystal structure during solidification, at high temperatures, or as a consequence of radiation damage. Normally, fewer than one lattice site out of a million contains a vacancy at room temperature.

An *interstitial defect* is formed when an extra atom is inserted into the lattice

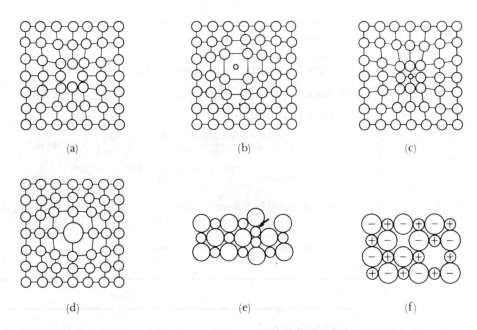

FIG. 4-15 Point defects: (a) vacancy, (b) interstitial atom, (c) small substitutional atom, (d) large substitutional atom, (e) Frenkel defect, and (f) Schottky defect. All of these defects disrupt the perfect arrangement of the surrounding atoms.

structure at a site which is not a normal lattice point. A *substitutional defect* is introduced when an atom is replaced by a different type of atom. The substitutional atom remains at the original normal lattice point. Both interstitial and substitutional defects are present in materials as impurities and may also be intentionally introduced as alloying elements. The number of these defects is usually independent of temperature.

A *Frenkel defect* is a vacancy-interstitial pair formed when an ion jumps from a normal lattice point to an interstitial site, leaving behind a vacancy. The *Schottky defect* is a pair of vacancies in an ionically bonded material. In order to maintain an equal charge in the crystal, both an anion and a cation must be missing from the lattice.

Point defects disturb the perfect arrangement of the surrounding atoms. When a vacancy or a small substitutional atom is present, the surrounding atoms collapse towards the point defect, stretching the bonds between the surrounding atoms and producing a tensile stress field. An interstitial or large substitutional atom pushes the surrounding atoms together, producing a compressive stress field. In either case the effect is widespread. A dislocation moving through the general vicinity of the point defect, as in Figure 4-14, encounters a lattice in which the atoms are not at their equilibrium positions. This disruption requires that a higher stress be applied to force the dislocation past the defect, therefore increasing the strength of the metal.

Intentional addition of interstitial and substitutional atoms into the structure of a material forms the basis for *solid solution strengthening* of materials, which is discussed in Chapter 8.

EXAMPLE 4-6

Iron has a measured density of 7.87 Mg m^{-3}. The lattice parameter of BCC iron is 2.866 Å. Calculate the percentage of vacancies in pure iron.

Answer:

From Equation (3-5),

$$\rho = \frac{(\text{atoms/cell})(55.85 \text{ g/g·mole})}{(2.866 \times 10^{-8})^3(6.02 \times 10^{23})} = 7.87 \text{ Mg m}^{-3}$$

$$\text{Atoms/cell} = \frac{(7.87)(2.866 \times 10^{-8})^3(6.02 \times 10^{23})}{55.85} = 1.998$$

There should be 2 atoms/cell in a perfect BCC iron crystal. Thus, the difference must be due to the presence of vacancies.

$$\text{Vacancies} = \frac{2 - 1.998}{2} \times 100 = 0.1\%$$

EXAMPLE 4-7

In FCC iron, carbon atoms are located at interstitial sites with coordinates $\frac{1}{4}, \frac{1}{4}, \frac{1}{4}$, whereas carbon atoms enter interstitial sites at $0, \frac{1}{2}, \frac{1}{4}$ in BCC iron. The lattice parameter is 3.571 Å for FCC iron and 2.866 Å for BCC iron. Carbon atoms have a radius of 0.71 Å. Would we expect a greater distortion of the lattice by an interstitial carbon atom for FCC or BCC iron?

Answer:

In the FCC crystal structure, the $\frac{1}{4}, \frac{1}{4}, \frac{1}{4}$ site is surrounded by four iron atoms in a tetrahedral arrangement. From Table 3-7, we know that the minimum radius ratio for four-fold coordination is

$$\frac{r \text{ (interstitial)}}{r \text{ (iron atom)}} = 0.225$$

$$r \text{ (interstitial)} = (0.225)\left(\frac{\sqrt{2}a_0}{4}\right) = \frac{(0.225)(\sqrt{2})(3.571)}{4}$$

$$= 0.284 \text{ Å}$$

In the BCC iron, the $0, \frac{1}{2}, \frac{1}{4}$ site is shown in Figure 4-16.

$$R \text{ (iron atom)} = \frac{\sqrt{3}a_0}{4} = \frac{(\sqrt{3})(2.866)}{4} = 1.241 \text{ Å}$$

$$(\tfrac{1}{2}a_0)^2 + (\tfrac{1}{4}a_0)^2 = (r + R)^2$$

$$(r + R)^2 = 2.566$$

$$r \text{ (interstitial)} = 1.602 - 1.241 = 0.361 \text{ Å}$$

The carbon atom, 0.71 Å, is larger than either interstitial site. A compressive field is developed, which in this case will be larger in the FCC lattice than in the BCC lattice. Carbon atoms will likely disrupt slip more in the FCC iron.

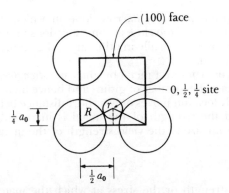

FIG. 4-16 The location of the 0, $\frac{1}{2}$, $\frac{1}{4}$ interstitial site in BCC metals, showing the arrangement of the normal atoms and the interstitial atom.

4-9 Surface Defects

Surface defects are the boundaries that separate a material into regions, each region having the same crystal structure but different orientations.

Grain boundaries.   The microstructure of metals and many other solid materials consists of many grains. A *grain* is a portion of the material within which the arrangement of the atoms is identical. However, the orientation of the atom arrangement, or crystal structure, is different for each adjoining grain. Three grains are shown schematically in Figure 4-17; the lattice in each grain is identical but the lattices are oriented differently. A *grain boundary* is the surface that sepa-

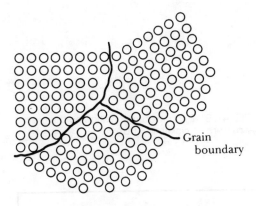

FIG. 4-17 The atoms near the boundaries of the three grains do not have an equilibrium spacing or arrangement.

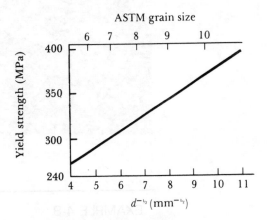

FIG. 4-18 The effects of grain size on the yield strength of a steel at room temperature.

rates the individual grains and is a narrow zone in which the atoms are not properly spaced. The atoms are too close at some locations in the grain boundary, causing a region of compression, while in other areas the atoms are too far apart, causing a region of tension.

We control the properties of a metal by *grain size strengthening*. By reducing the grain size, we increase the number of grains and hence increase the amount of grain boundary. Any dislocation moves only a short distance before encountering a grain boundary and the strength of the metal is increased. The Hall-Petch equation relates the grain size to the yield strength of the metal,

$$\sigma_y = \sigma_0 + Kd^{-1/2} \tag{4-3}$$

where σ_y is the yield strength or the stress at which the material permanently deforms, d is the average diameter of the grains, and σ_0 and K are constants for the metal. Figure 4-18 shows this relationship in steel. We will describe in later chapters how the grain size can be controlled through solidification, alloying, and heat treatment.

One technique by which grain size is specified is the ASTM (American Society for Testing & Materials) grain size number. The number of grains per square meter is determined from a photograph of the metal taken at magnification $\times$ 100 (Figure 4-19). The number of grains per square meter N is entered into Equation (4-4) and the ASTM grain size number n is calculated.

$$N = 2^{n+3} \tag{4-4}$$

A large ASTM number indicates many grains, or a fine grain size, and correlates with high strengths.

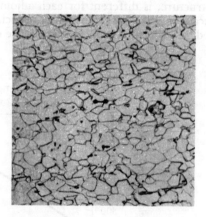

FIG. 4-19   Photomicrograph showing the grain structure of nickel. (×100)

EXAMPLE 4-8

Suppose we count 256 grains per square meter at magnification $\times$ 100 in a photomicrograph of a metal. Determine the ASTM grain size number.

Answer:

$$N = 256 = 2^{n+3}$$
$$\log 256 = (n + 3)\log 2$$
$$2.408 = (n + 3)(0.301)$$
$$n = 5$$

EXAMPLE 4-9

Suppose we count 256 grains per square meter in a photomicrograph taken at magnification $\times 250$. What is the ASTM grain size number?

Answer:

If we count 256 grains per square meter at magnification $\times 250$, we must have at magnification $\times 100$:

$$N = \left(\frac{250}{100}\right)^2 (256) = 1600 = 2^{n+3}$$
$$\log 1600 = (n + 3)\log 2$$
$$3.2 = (n + 3)(0.301)$$
$$n = 3.64$$

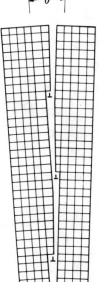

FIG. 4-20
The small angle grain boundary is produced by an array of dislocations, causing an angular mismatch θ between the lattices on either side of the boundary.

Small angle grain boundaries.   A *small angle grain boundary* is an array of dislocations that produces a small misorientation between the adjoining lattices (Figure 4-20). Because the energy of the surface is less than that at a regular grain boundary, the small angle grain boundaries are not as effective in blocking slip. Small angle boundaries formed by edge dislocations are called *tilt boundaries* and those caused by screw dislocations are called *twist boundaries*.

EXAMPLE 4-10

Determine the angle θ across a small angle grain boundary in copper when the dislocations in the boundary are 1000 Å apart.

Answer:

The grains are tilted one Burgers vector in each direction every 1000 Å. The Burgers vector in FCC copper is [110], so the length of the Burgers vector is d_{110}.

$$d_{110} = \frac{a_0}{\sqrt{h^2 + k^2 + l^2}} = \frac{3.615}{\sqrt{2}} = 2.557 \text{ Å}$$

$$\sin \frac{\theta}{2} = \frac{2.557}{1000} = 0.002557$$

$$\theta = 0.293°$$

Stacking faults.   *Stacking faults* occur in FCC metals and represent an error in the stacking sequence of close-packed planes. Normally, a stacking sequence of *ABCABCABC* is produced in a perfect FCC lattice. But suppose the following sequence is produced:

ABCABABCABC

In the portion of the sequence indicated, a type *A* plane is shown where a type *C* plane would normally be located. This small region, which has an HCP stacking sequence instead of the FCC stacking sequence, represents a stacking fault. Stacking faults interfere with the slip process.

Twin boundaries.   A *twin boundary* is a plane across which there is a special mirror image misorientation of the lattice structure (Figure 4-21). Twins can be produced when a shear force, acting along the twin boundary, causes the atoms to shift out of position. Twinning occurs during deformation or heat treatment of certain metals. The twin boundaries interfere with the slip process and increase the strength of the metal. Movement of twin boundaries can also cause a metal to

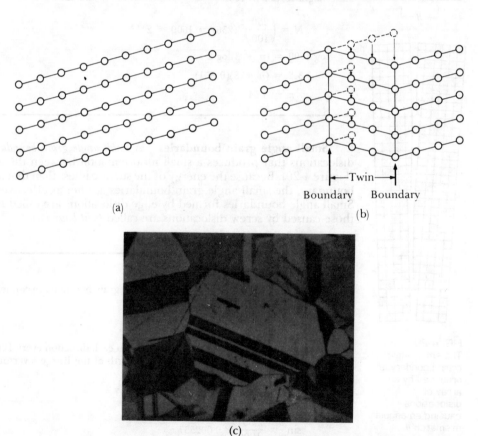

FIG. 4-21   Application of a stress to the perfect crystal (a) may cause a displacement of the atoms (b), causing the formation of a twin. Note that the crystal has deformed as a result of twinning. (c) A photomicrograph of a twin within a grain of brass.

deform. Figure 4-21 shows that the formation of a twin has changed the shape of the metal.

The effectiveness of the surface defects in interfering with the slip process can be judged from the surface energies (Table 4-3). The high-energy grain boundaries are much more effective in blocking dislocations than either stacking faults or twin boundaries.

TABLE 4-3 Energies of surface imperfections in selected metals[a]

Surface Imperfection (mJ m^{-2})	Al	Cu	Pt	Fe
Stacking fault energy	200	75	95	—[b]
Twin boundary energy	120	45	195	190
Grain boundary energy	625	645	1000	780

[a] After R.E. Reed-Hill, *Physical Metallurgy Principles*, 2nd Ed. D. Van Nostrand, 1973.

[b] Stacking faults do not occur in BCC metals such as iron.

SUMMARY

We have found in this chapter that the deformation of a metal can be explained in terms of the movement of certain lattice imperfections, namely dislocations, by a mechanism known as slip. The ease with which slip occurs depends first on the type of crystal structure and second on the number of other imperfections in the lattice. Based on this analysis, FCC metals should be ductile but weak, BCC metals should be strong with moderate ductility, and HCP metals should be relatively brittle with moderate strength. We can strengthen any of these structures by introducing larger numbers of lattice imperfections.

GLOSSARY

ASTM grain size number A measure of the size of the grains in a crystalline material obtained by counting the number of grains per square inch at magnification ×100.

Burgers vector The direction and distance that a dislocation moves in each step.

Critical resolved shear stress The shear stress required to cause a dislocation to move and cause slip.

Cross-slip A change in the slip system of a dislocation.

Dislocation A line imperfection in the lattice of a crystalline material. Movement of dislocations helps explain how materials deform. Interference with the movement of dislocations helps explain how materials are strengthened.

Edge dislocation A dislocation introduced

into the lattice by adding an "extra half plane" of atoms.

Frenkel defect A pair of point defects produced when an ion moves to create an interstitial site, leaving behind a vacancy.

Grain A portion of a solid material within which the lattice is identical and oriented in only a single direction.

Grain boundary A surface defect representing the boundary between two grains. The lattice has a different orientation on either side of the grain boundary.

Interstitial defect A point defect produced when an atom is placed into the lattice at a site that is normally not a lattice point.

Schottky defect A pair of point defects in ionically bonded materials. In order to maintain a neutral charge, both a cation and an anion vacancy must form.

Screw dislocation A dislocation produced by skewing a crystal so that one atomic plane produces a spiral ramp about the dislocation.

Slip Deformation of a material by the movement of dislocations through the lattice.

Slip direction The direction in the lattice in which the dislocation moves. The slip direction is the same as the direction of the Burgers vector.

Slip plane The plane swept out by the dislocation line during slip. Normally, the slip plane is a close-packed plane, if one exists in the crystal structure.

Slip system The combination of the slip plane and the slip direction.

Small angle grain boundary An array of dislocations causing a small misorientation of the lattice across the surface of the imperfection.

Stacking fault A surface defect in FCC metals caused by the improper stacking sequence of close-packed planes.

Substitutional defect A point defect produced when an atom is removed from a regular lattice point and replaced with a different atom, usually of a different size.

Tilt boundary A small angle grain boundary composed of an array of edge dislocations.

Twin boundary A surface defect across which there is a mirror image misorientation of the lattice. Twin boundaries can also move and cause deformation of the material.

Twist boundary A small angle grain boundary composed of an array of screw dislocations.

Vacancy A vacancy is created when an atom is missing from a lattice point.

PRACTICE PROBLEMS

1 Determine the Miller indices and the length of the Burgers vector for the edge dislocation shown in Figure 4-22.

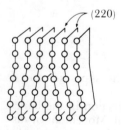

(220)

FIG. 4-22 An edge dislocation in a FCC metal having a lattice parameter of 3.5 Å. (See Problem 4-1.)

2 Determine the Miller indices and the length of the Burgers vector for the edge dislocation shown in Figure 4-23.

3 Determine the Miller indices and the length of the Burgers vector for the screw dislocation shown in Figure 4-24.

4 What are the indices of each slip direction that lies in the (111) slip plane in FCC metals?

5 What are the indices of each slip direction that lies in the $(\bar{1}1\bar{1})$ slip plane in FCC metals?

6 Show, using an appropriate sketch, whether the $[1\bar{1}1]$ slip direction lies in the (112) slip plane in BCC metals.

7 What are the indices of each slip direction that lies in the $(1\bar{1}0)$ slip plane in BCC metals?

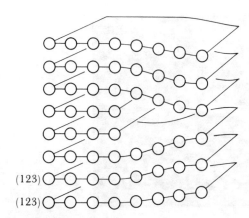

(123)
(123)

FIG. 4-23 An edge dislocation in a BCC metal having a lattice parameter of 4 Å. (See Problem 4-2.)

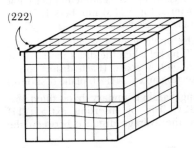

(222)

FIG. 4-24 A screw dislocation in a BCC metal having a lattice parameter of 3.5 Å. (See Problem 4-3.)

8 What slip directions lie in the (123) slip plane in BCC metals?

9 What close-packed directions lie in the $(01\bar{1}0)$ slip plane in HCP metals?

10 What close-packed directions lie in the $(10\bar{1}1)$ slip plane in HCP metals?

11 A FCC crystal has a stress of 68.97 MPa acting in the [001] direction. Calculate (a) the resolved shear stress acting on the $(111)[10\bar{1}]$ slip system and (b) the resolved shear stress acting on the $(111)[\bar{1}10]$ slip system.

12 A stress of 70 MPa is applied in the [001] direction of a BCC crystal. Calculate (a) the resolved shear stress acting on the $(101)[\bar{1}11]$ system and (b) the resolved shear stress acting on the $(110)[\bar{1}11]$ system.

13 A dislocation begins to slip in a $(111)[10\bar{1}]$ slip system when a stress of 1.4 MPa is applied in the [001] direction of a FCC crystal. Calculate the critical resolved shear stress.

14 A dislocation begins to slip in a $(101)[\bar{1}11]$ slip system when a stress of 55.2 MPa is applied in the [001] direction of a BCC crystal. (a) Calculate the critical resolved shear stress. (b) Will a dislocation in the $(\bar{2}11)[111]$ slip system move?

15 The critical resolved shear stress in the (111) $[\bar{1}0\bar{1}]$ slip system of a FCC crystal is 3.3 MPa. What stress must be applied (a) in the [001] direction to produce slip and (b) in the [010] direction to produce slip?

16 The critical resolved shear stress in the $(1\bar{1}0)[111]$ slip system of a BCC crystal is 58.6 MPa. What stress must be applied (a) in the [001] direction to produce slip and (b) in the [010] direction to produce slip?

17 A FCC crystal has a stress of 3.4 MPa applied in the [001] direction. Calculate the resolved shear stress acting on each of the 12 possible slip systems.

18 Nickel substitutional atoms are added to a copper lattice until the final FCC alloy has a lattice parameter ot 3.5806 Å and a density of 8.9478 Mg m^{-3}. Calculate the fraction of the total atoms that are nickel substitutional atoms.

19 Gold substitutional atoms are added to a silver lattice until the final FCC alloy has a lattice parameter of 4.0843 Å and a density of 12.686 Mg m^{-3}. Calculate the fraction of the total atoms that are gold substitutional atoms.

20 BCC niobium has a density of 8.57 Mg m^{-3} and a lattice parameter of 3.294 Å. Calculate the number of vacancies per 10^6 Nb atoms.

21 FCC platinum has a density of 21.45 Mg m^{-3} and a lattice parameter of 3.9231 Å. On average, what fraction of the lattice points contain vacancies?

22 Suppose one out of every 500 atoms is missing in FCC copper, which has a lattice parameter of 3.6153 Å. Calculate the density of copper.

23 Suppose one out of every 750 atoms is missing in BCC chromium, which has a lattice parameter of 2.8844 Å. Calculate the density of chromium.

24 Nickel has a FCC structure with a lattice parameter of 3.5238 Å. Suppose one carbon atom fits into an interstitial site in one out of every ten unit cells. Assume that carbon atoms do not change a_0. Estimate the density of the Ni-C alloy and compare to the theoretical density of nickel.

25 Suppose that a hydrogen atom just fits into an interstitial position of BCC iron. If one hydrogen atom is present for every 200 iron atoms calculate (a) the average packing factor and (b) the density of the Fe-H alloy and compare to the theoretical density of iron. Assume $a_0 = 2.860$ Å, $r_{Fe} = 1.241$ Å, and $r_H = 0.36$ Å.

26 Suppose that a hydrogen atom fits into the interstitial site in BCC iron, which has a lattice parameter of 2.866 Å. If the measured density of an Fe-H alloy is 7.9010 Mg m^{-3}, calculate the atomic percent of hydrogen in the alloy.

27 The density of magnesium oxide, or magnesia, MgO, is 3.58 Mg m^{-3} and its lattice parameter is 4.20 Å. Determine the number of Schottky defects per unit cell in MgO.

28 Determine the constants σ_0 and K in the Hall-Petch equation for steel, using Figure 4-18.

29 The ASTM grain size number of a metal is O. How many grains would be observed per square meter in a photograph taken at magnification ×100.

30 Figure 4-25 shows the grain structure of brass at magnification ×250. Determine the ASTM grain size number.

31 Figure 4-19 shows the grain structure of nickel at magnification ×100. Determine the ASTM grain size number.

32 (a) Calculate the approximate average width of grains having ASTM numbers 1, 5, and 9. Assume the grains are square in shape. (b) Calculate the number of grains/cm^3 for ASTM numbers 1, 5, and 9,

FIG. 4-25   Microstructure for Problem 4-30.

assuming the grains are cubes. (c) Calculate the total grain boundary area in square centimeters per cubic centimeter of material for ASTM numbers 1, 5, and 9, assuming the grains are cubes.

33 Calculate the angle θ of a small angle grain boundary in FCC nickel when the dislocations in the boundary are 20,000 Å apart. (Each dislocation is produced by inserting an extra (110) plane of atoms.)

34 If a small angle grain boundary that is tilted 1° is produced in BCC iron by inserting extra (111) planes, calculate the average distance between the dislocations.

5
Atom Movement in Materials

5-1 Introduction

Diffusion is the movement of atoms within a material. Atoms move in an orderly fashion to eliminate concentration differences and produce a homogeneous uniform composition. Atoms can also be forced to move by applying voltages or external forces to the material. In fact, atoms even move about randomly in pure metals when no external forces are applied or no concentration differences exist.

Movement of atoms is required for many of the treatments that we perform on materials. Diffusion is required for the heat treatment of metals, the manufacture of ceramics, the solidification of materials, the manufacture of transistors and solar cells, and even the electrical conductivity of many ceramic materials.

In this chapter, we will concentrate on understanding the diffusion of atoms in solid materials. We should recognize, however, that diffusion occurs in gases and liquids as well. But convection, mixing, and other phenomena assist the diffusion of atoms in the gaseous and liquid states; furthermore, diffusion is much more rapid in gases and liquids because of less efficient packing of the atoms.

5-2 Self-diffusion

Even in absolutely pure solid materials, atoms move from one lattice position to another. This process, known as *self-diffusion*, can be detected by using radioactive tracers. Suppose we introduce a radioactive isotope of gold (Au^{198}) onto the surface of normal gold (Au^{197}). After a period of time the radioactive atoms move into the regular gold. If we wait long enough, the radioactive atoms are uniformly distributed throughout the entire regular gold. Although self-diffusion occurs continually in all materials, the effect on the material's behavior is not significant.

5-3 Diffusion in Alloys

In metal alloys and ceramics, diffusion of unlike atoms also occurs. If a nickel sheet is bonded to a copper sheet, nickel atoms gradually diffuse into the copper and copper atoms migrate to the nickel. Again if we wait long enough, the nickel and copper atoms are uniformly distributed throughout the entire metal and the metal is homogeneous, with the same concentration of copper and nickel atoms everywhere (Figure 5-1).

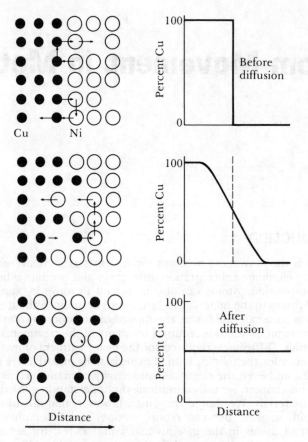

FIG. 5-1 Diffusion of copper atoms into nickel.
Eventually, the copper atoms are randomly distributed
throughout the nickel.

5-4 Diffusion Mechanisms

There are several mechanisms by which atoms diffuse (Figure 5-2).

Vacancy diffusion.   In self-diffusion and diffusion involving substitutional atoms, an atom leaves its lattice site to fill a nearby vacancy (thus creating a new vacancy at the original lattice site). As diffusion continues, we have a countercurrent flow of atoms and vacancies. This mechanism was assumed in Figure 5-1.

Interstitial diffusion.   When a small interstitial atom is present in the crystal structure, the atom moves from one interstitial site to another. No vacancies are required for this mechanism to work.

Other diffusion mechanisms.   Sometimes a substitutional atom leaves its normal lattice site and enters an interstitial position. This *interstitialcy* diffusion

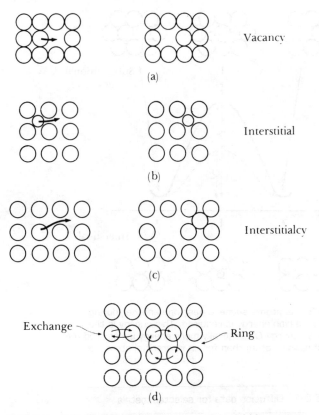

FIG. 5-2   Diffusion mechanisms in materials. (a) Vacancy or substitutional atom diffusion, (b) interstitial diffusion, (c) interstitialcy diffusion, and (d) exchange and ring diffusion.

mechanism is uncommon because the atom does not easily fit into the small interstitial site. Atoms also move by a simple *exchange* mechanism or by a *ring* mechanism. However, the vacancy and interstitial mechanisms appear to be responsible for diffusion in most cases.

5-5   Activation Energy for Diffusion

A diffusing atom must squeeze past the surrounding atoms to reach its new site. In order to do this, we must supply energy to force the atom to its new position. This is shown schematically for vacancy and interstitial diffusion in Figure 5-3. The atom is originally in a low energy, relatively stable location. In order to move to a new location, the atom must pass over an energy barrier. The energy barrier is the *activation energy Q*. Heat supplies the atom with the energy to exceed this barrier.

Normally less energy is required to squeeze an interstitial atom past the surrounding atoms; consequently, activation energies are lower for interstitial diffusion than for vacancy diffusion. Typical values for activation energies are shown in Table 5-1: a low activation energy indicates easy diffusion.

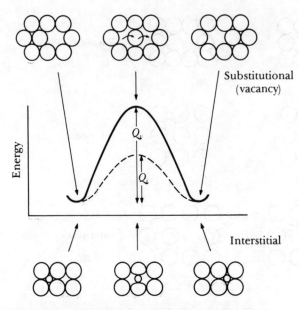

FIG. 5-3   As atoms squeeze past one another during diffusion, a high energy is required. This energy is the activation energy Q. Generally more energy is required for a substitutional atom than for an interstitial atom.

TABLE 5-1   Diffusion data for selected metals

Diffusion Couple	Q ($J\,mol^{-1}$)	D_0 ($m^2\,s^{-1}$)
Interstitial diffusion		
C in FCC iron	137,850	0.23×10^{-4}
C in BCC iron	87,570	0.011×10^{-4}
N in FCC iron	144,970	0.0034×10^{-4}
N in BCC iron	76,680	0.0047×10^{-4}
H in FCC iron	43,160	0.0063×10^{-4}
H in BCC iron	15,080	0.0012×10^{-4}
Self-diffusion		
Au in Au	183,520	0.13×10^{-4}
Al in Al	134,920	0.10×10^{-4}
Ag in Ag	188,550	0.80×10^{-4}
Cu in Cu	206,570	0.36×10^{-4}
Fe in FCC iron	279,470	0.65×10^{-4}
Pb in Pb	108,520	1.27×10^{-4}
Pt in Pt	283,240	0.27×10^{-4}
Mg in Mg	134,920	1.0×10^{-4}
Zn in Zn	91,340	0.1×10^{-4}
Ti in HCP Ti	95,950	0.4×10^{-4}
Fe in BCC iron	246,790	4.1×10^{-4}
Heterogeneous diffusion		
Ni in Cu	242,600	2.3×10^{-4}

TABLE 5-1   (Continued)

Diffusion Couple	Q ($J \, mol^{-1}$)	D_0 ($m^2 \, s^{-1}$)
Cu in Ni	257,690	0.65×10^{-4}
Zn in Cu	183,940	0.78×10^{-4}
Ni in FCC iron	268,160	4.1×10^{-4}
Au in Ag	190,650	0.26×10^{-4}
Ag in Au	168,440	0.072×10^{-4}
Al in Cu	165,510	0.045×10^{-4}

Adapted from Y. Adda and J. Philibert, *La Diffusion dans les Solides*, vol. 2, 1966.

5-6   Rate of Diffusion (Fick's First Law)

The rate at which atoms diffuse in a material can be measured by the *flux J*, which is defined as the number of atoms passing through a plane of unit area per unit time (Figure 5-4). *Fick's first law* explains the net flux of atoms,

$$J = -D \frac{\Delta c}{\Delta x} \tag{5-1}$$

where J is the flux [atoms $(m^2 \cdot s)^{-1}$], D is the diffusivity or *diffusion coefficient* ($m^2 \, s^{-1}$), and $\Delta c / \Delta x$ is the *concentration gradient* [atoms $(m^3 \cdot m)^{-1}$]. Several factors affect the flux of atoms during diffusion.

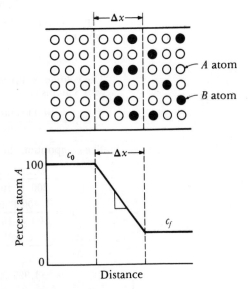

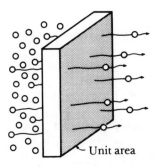

FIG. 5-4   The flux during diffusion is defined as the number of atoms passing through a plane of unit area per unit time.

FIG. 5-5   Illustration of the concentration gradient.

Concentration gradient.   The concentration gradient shows how the composition of the material varies with distance; Δc is the difference in concentration over the distance Δx (Figure 5-5). No net flux occurs for self-diffusion in pure gold, since the concentration of gold atoms is identical everywhere. However, if some of the gold atoms are radioactive, a net flux of atoms occurs until the radioactive atoms are uniformly distributed and no concentration gradient remains. Likewise a net flux of copper and nickel atoms develops when copper is joined to nickel until the metal is homogeneous. We should note that the flux is initially high when the concentration gradient is high and gradually decreases as the gradient is reduced.

EXAMPLE 5-1

One way to manufacture transistors, which amplify electrical signals, is to diffuse impurity atoms into a semiconductor material such as silicon. Suppose a silicon wafer 1 mm thick, which originally contains one phosphorous atom for every 10,000,000 Si atoms, is treated so that there are 400 P atoms for every 10,000,000 Si atoms at the surface. Calculate the concentration gradient (a) in atomic percent/m^{-1} and (b) in atoms $(m^3 \cdot m)^{-1}$. The lattice parameter of silicon is 5.4307 Å.

Answer:

First, let's calculate the initial and surface compositions in atomic percent.

$$c_i = \frac{1 \text{ P atom}}{10,000,000 \text{ atoms}} \times 100 = 0.00001 \text{ at\% P}$$

$$c_s = \frac{400 \text{ P atoms}}{10,000,000 \text{ atoms}} \times 100 = 0.004 \text{ at\% P}$$

$$\frac{\Delta c}{\Delta x} = \frac{0.00001 - 0.004 \text{ at \% P}}{0.001 \text{ m}} = -3.99 \text{ at \% P m}^{-1}$$

To find the gradient in terms of atoms $(m^3 \cdot m)^{-1}$, we must find the volume of the unit cell.

$$V_{\text{cell}} = (5.4307 \times 10^{-10})^3 = 160.2 \times 10^{-30} \text{ m}^3 \text{ cell}^{-1}$$

The volume occupied by 10,000,000 Si atoms, which are arranged in a DC structure with 8 atoms/cell, is

$$V = \frac{10,000,000 \text{ atoms}}{8 \text{ atoms/cell}} (160.2 \times 10^{-30} \text{ m}^3 \text{ cell}^{-1}) = 200 \times 10^{-24} \text{ m}^3$$

The compositions in atoms/cm^3 are

$$c_i = \frac{1 \text{ P atom}}{200 \times 10^{-24} \text{ m}^3} = 5 \times 10^{21} \text{ P atoms m}^{-3}$$

$$c_s = \frac{400 \text{ P atoms}}{200 \times 10^{-24} \text{ m}^3} = 2 \times 10^{24} \text{ P atoms m}^{-3}$$

$$\frac{\Delta c}{\Delta x} = \frac{5 \times 10^{21} - 2100 \times 10^{21} \text{ P atoms m}^{-3}}{0.001 \text{ m}}$$

$$= -1.995 \times 10^{27} \text{ P atoms } (m^3 \cdot m^{-1})$$

EXAMPLE 5-2

A thick-walled pipe 30 mm in diameter contains a gas including 0.5×10^{26} N atoms per m^3 on one side of a 0.01 mm thick iron membrane. The gas is continuously introduced to the pipe. The gas on the other side of the membrane contains 1×10^{24} N atoms per m^3. Calculate the total number of nitrogen atoms passing through the iron membrane at 700°C if the diffusion coefficient for nitrogen in iron is 4×10^{-11} m^2 s^{-1}.

Answer:

$$c_1 = 0.5 \times 10^{26} \text{ N atoms m}^{-3} = 50 \times 10^{24} \text{ N atoms m}^{-3}$$

$$c_2 = 1 \times 10^{24} \text{ N atoms m}^{-3}$$

$$\Delta c = (1 - 50) \times 10^{24} = -49 \times 10^{24} \text{ N atoms m}^{-3}$$

$$\Delta x = 0.01 \text{ mm} = 0.01 \times 10^{-3} \text{ m}$$

$$J = -D\frac{\Delta c}{\Delta x} = -(4 \times 10^{-11} \frac{m^2}{s})\left(\frac{-49 \times 10^{24} \text{ N atoms m}^{-3}}{0.01 \times 10^{-3} \text{ m}}\right)$$

$$J = 1.96 \times 10^{20} \text{ N atoms } (m^2 \cdot s)^{-1}$$

$$\text{Total atoms/s} = JA = J\left(\frac{\pi}{4}d^2\right) = (1.96 \times 10^{20})\left(\frac{\pi}{4}\right)(0.03)^2 = 1.39 \times 10^{17} \text{ atoms s}^{-1}$$

Obviously, if the gas on the high nitrogen side of the membrane were not continuously replenished, that side would soon be depleted in nitrogen atoms.

Temperature and the diffusion coefficient. The diffusion coefficient is given by the expression

$$D = D_0 \exp\left(\frac{-Q}{RT}\right) \tag{5-2}$$

where Q is the activation energy (J mol^{-1}), R is the gas constant (8.314 J mol^{-1} K^{-1}) and T is the absolute temperature (K). D_0 is a constant for a given diffusion system. Typical values for D_0 are given in Table 5-1, while the temperature dependence of D is shown in Figure 5-6 for several materials.

When the temperature of a material increases, the diffusion coefficient and the flux of atoms increase as well. At higher temperatures, the thermal energy supplied to the diffusing atoms permits the atoms to overcome the activation energy barrier and more easily move to new lattice sites. At low temperatures, often below about 0.4 times the absolute melting temperature of the material, diffusion is very slow and may not be significant. For this reason, the heat treatment of metals and the processing of ceramics are done at high temperatures, where atoms move rapidly to complete reactions or to reach equilibrium conditions.

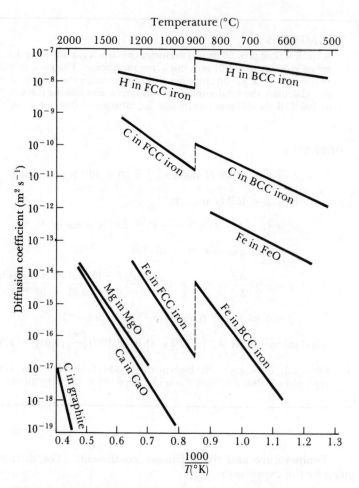

FIG. 5-6   The diffusion coefficient D as a function of reciprocal temperature for several metals and ceramics.

EXAMPLE 5-3

The diffusion coefficient for aluminum in copper is found to be 2.5×10^{-24} m^2 s^{-1} at 200°C and 3.1×10^{-17} m^2 s^{-1} at 500°C. Calculate the activation energy for the diffusion of aluminum in copper.

Answer:

Let's make a ratio between the diffusion coefficients at the two temperatures, 200°C = 473 K and 500°C = 773 K.

$$\frac{D_{773}}{D_{473}} = \frac{D_0 \exp\left(\dfrac{-Q}{(8.314)(773)}\right)}{D_0 \exp\left(\dfrac{-Q}{(8.314)(473)}\right)} = \frac{3.1 \times 10^{-17}}{2.5 \times 10^{-24}}$$

$$\exp\left[Q\left(\frac{-1}{773} - \frac{-1}{473}\right)\left(\frac{1}{8.314}\right)\right] = 1.24 \times 10^7$$

$$\exp(9.8544 \times 10^{-5}Q) = 1.24 \times 10^7$$

$$9.8544 \times 10^{-5}Q = \ln(1.24 \times 10^7) = 16.33$$

$$Q = 166{,}000 \text{ J mol}^{-1}$$

Activation energy and diffusion mechanism.   A small activation energy Q increases the diffusion coefficient and flux, since less thermal energy is required to overcome the smaller activation energy barrier. Interstitial diffusion, with a low activation energy, usually occurs an order of magnitude or more faster than vacancy, or substitutional, diffusion.

Activation energies are usually lower for atoms diffusing through open crystal structures compared to close-packed crystal structures. The activation energy for carbon diffusing in FCC iron is 137,850 J mol^{-1}, but only 87,570 J mol^{-1} for carbon diffusing in BCC iron.

Activation energies are also lower for diffusion of atoms in materials with a low melting temperature and are usually lower for small substitutional atoms compared to larger atoms.

EXAMPLE 5-4

Compare the diffusion coefficients for hydrogen and nickel in FCC iron at 1000°C and explain the difference.

Answer:

From Table 5-1, at $T = 1000°C = 1273$ K

$$D_H = 0.0063 \times 10^{-4} \exp\left(\frac{-43{,}160}{RT}\right) = 0.0063 \times 10^{-4} \exp\left[\frac{-43{,}160}{(8.314)(1273)}\right]$$

$$= 0.0063 \times 10^{-4} \exp(-4.07) = (0.0063 \times 10^{-4})(0.017) = 1.07 \times 10^{-8} \text{ m}^2 \text{ s}^{-1}$$

$$D_{Ni} = 4.1 \times 10^{-4} \exp\left(\frac{-268160}{RT}\right) = 4.1 \times 10^{-4} \exp\left(\frac{-268{,}160}{(8.314)(1273)}\right)$$

$$= 4.1 \times 10^{-4} \exp(-25.3) = (4.1 \times 10^{-4})(1.03 \times 10^{-11}) = 4.2 \times 10^{-15} \text{ m}^2 \text{ s}^{-1}$$

Because hydrogen is a tiny interstitial atom and nickel is a larger substitutional atom, the activation energy for diffusion of hydrogen in iron is small and the rate of diffusion is seven orders of magnitude greater than that of nickel.

EXAMPLE 5-5

Compare the diffusion coefficients of carbon in BCC and FCC iron at the allotropic transformation temperature of 910°C and explain the difference.

Answer:

From Table 5-1, at $T = 910°C = 1183$ K

$$D_C \text{ (BCC)} = 0.011 \times 10^{-4} \exp\left[\frac{-87,570}{(8.314)(1183)}\right] = 0.011 \times 10^{-4} \exp(-8.89)$$

$$= (0.011 \times 10^{-4})(1.38 \times 10^{-4}) = 1.52 \times 10^{-10} \text{ m}^2 \text{ s}^{-1}$$

$$D_C \text{ (FCC)} = 0.23 \times 10^{-4} \exp\left[\frac{-137,850}{(8.314)(1183)}\right] = 0.23 \times 10^{-4} \exp(-14.00)$$

$$= (0.23 \times 10^{-4})(8.31 \times 10^{-7}) = 1.91 \times 10^{-11} \text{ m}^2 \text{ s}^{-1}$$

Diffusion is faster in the BCC iron than in the FCC iron, even at the same temperature, because the packing factor of BCC structures is less than that of FCC structures.

EXAMPLE 5-6

From the data in Table 5-1, determine the relationship between activation energy and melting temperature for self-diffusion in FCC metals.

Answer:

Let's compare the activation energy and melting temperature for each metal.

	Melting Temperature (°C)	Activation Energy J mol^{-1}
Pb	327	108,500
Al	660	134,900
Ag	961	188,600
Au	1063	183,500
Cu	1083	206,600
Fe	1539	279,400

The activation energy increases almost linearly with the melting temperature of the metal (Figure 5-7).

Time. Diffusion requires time; the units for flux are atoms $(\text{m}^2 \cdot \text{s})^{-1}$! If a large number of atoms must diffuse to produce a uniform structure, long times may be required, even at high temperatures. Times for heat treatments may be reduced by using higher temperatures or by making the diffusion distances (related to Δx) as small as possible.

We find that some rather remarkable structures and properties are obtained if we prevent diffusion. Steels quenched rapidly from high temperatures to prevent diffusion form nonequilibrium structures which provide the basis for sophisticated heat treatments.

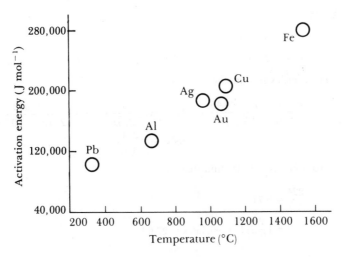

FIG. 5-7 The activation energy for self-diffusion increases as the melting point of the metal increases.

5-7 Composition Profile (Fick's Second Law)

Fick's second law, which describes the dynamic, or nonsteady state, diffusion of atoms, is the differential equation $\frac{dc}{dt} = D \frac{d^2c}{dx^2}$, whose solution depends on the boundary conditions for a particular situation. One solution is

$$\frac{c_s - c_x}{c_s - c_0} = \mathrm{erf}\left(\frac{x}{2\sqrt{Dt}}\right) \tag{5-3}$$

where c_s is a constant concentration of the diffusing atoms at the surface of the material, c_0 is the initial uniform concentration of the diffusing atoms in the material, and c_x is the concentration of the diffusing atom at location x below the surface after time t. These concentrations are illustrated in Figure 5-8. The function erf is the error function and can be evaluated from Figure 5-9.

The solution to Fick's second law permits us to calculate the concentration of one diffusing species near the surface of the material as a function of time and distance, provided that the diffusion coefficient D remains constant and the concentrations of the diffusing atom at the surface c_s and within the material c_0 remain unchanged.

EXAMPLE 5-7

The surface of a 0.1% C steel is to be strengthened by carburizing. In carburizing, the steel is placed in an atmosphere that provides a maximum of 1.2% C at the surface of the steel at a high temperature. Carbon then diffuses into the surface of the steel. For optimum properties, the steel must contain 0.45% C at a depth of 2 mm below the surface. How long will carburizing take if the diffusion coefficient is 2×10^{-11} m^2 s^{-1}

Answer:

From the problem,

$$c_s = 1.2\% \text{ C} \qquad c_0 = 0.1\% \text{ C} \qquad c_x = 0.45\% \text{ C} \qquad x = 2 \text{ mm} = 0.002 \text{ m}$$

From Fick's second law,

$$\frac{c_s - c_x}{c_s - c_0} = \frac{1.2 - 0.45}{1.2 - 0.1} = 0.68 = \text{erf}\left(\frac{x}{2\sqrt{Dt}}\right) = \text{erf}\left(\frac{0.002}{2\sqrt{(2 \times 10^{-11})t}}\right)$$

$$0.68 = \text{erf}\left(\frac{224}{\sqrt{t}}\right)$$

From Figure 5-9, we find that

$$\frac{224}{\sqrt{t}} = 0.71$$

$$t = \left(\frac{224}{0.71}\right)^2 = 99{,}536 \text{ s} = 27.6 \text{ h}$$

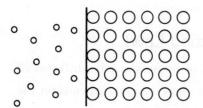

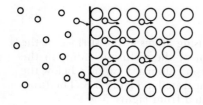

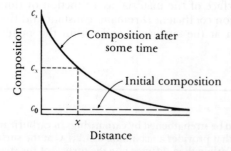

FIG. 5-8 Diffusion of atoms into the surface of a material, illustrating the use of Fick's second law.

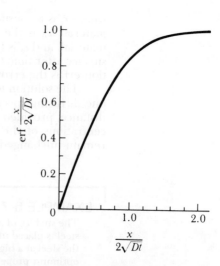

FIG. 5-9 The error function.

One of the consequences of Fick's second law is that the same concentration profile can be obtained for different conditions, so long as the term Dt is constant. This permits us to determine the effect of temperature on the time required for a particular heat treatment to be accomplished.

EXAMPLE 5-8

We find that 10 h are required to cause carbon to diffuse 1 mm into the surface of a steel gear at 800°C. How long is required to achieve the same carbon depth at 900°C? The activation energy for the diffusion of carbon atoms in FCC iron is 137 850 J mol^{-1}.

Answer:

The temperatures are $800 + 273 = 1073$ K and $900 + 273 = 1173$ K. Since we want the carbon profile to remain the same at both temperatures,

$$D_{1073}t_{1073} = D_{1173}t_{1173}$$

$$t_{1173} = \frac{D_{1073}t_{1073}}{D_{1173}} = t_{1073}\frac{\exp(-Q/1073R)}{\exp(-Q/1173R)}$$

$$t_{1173} = (10\text{ h})\frac{\exp(-137,850/(8.314)(1073))}{\exp(-137,850/(8.314)(1173))}$$

$$= 10\exp(-1.315) = (10)(0.268) = 2.68\text{ h}$$

5-8 Interdiffusion and the Kirkendall Effect

When we wish to determine the concentration profile of two atom species concurrently diffusing into one another, a different solution to Fick's second law is required,

$$\frac{c_x - c_m}{c_1 - c_m} = \text{erf}\left(\frac{x}{2\sqrt{Dt}}\right) \tag{5-4}$$

where c_1 is the concentration of atom A in material 1, c_m is the average concentration of atom A in materials 1 and 2, and c_x is the concentration of atom A in material 2 at a distance x from the original interface after time t (Figure 5-10). D is the diffusion coefficient of atom A in material 2. A similar equation is required to give the composition of atom B in material 1.

EXAMPLE 5-9

A diffusion couple is produced by joining silver and gold at 1000°C. Plot the concentration profiles for both silver and gold after 10 h.

Answer:

For gold diffusing into silver:

$$c_1 = 100\%\text{ Au} \qquad c_m = 50\%\text{ Au} \qquad D = 0.26 \times 10^{-4}\exp\left(\frac{-190,650}{1273R}\right) = 4 \times 10^{-13}\text{ m}^2\text{ s}^{-1}$$

$$\frac{c_{Au} - 50}{100 - 50} = \text{erf}\left(\frac{x}{2 - \sqrt{(4 \times 10^{-13})(10\text{ h})(3600\text{ s/h})}}\right) = \text{erf}(4166x)$$

For silver diffusing into gold:

$$c_1 = 100\% \text{ Ag} \qquad c_m = 50\% \text{ Ag} \qquad D = 0.072 \times 10^{-4} \exp\left(\frac{-168,440}{1273R}\right) = 9 \times 10^{-13} \text{ m}^2\text{ s}^{-1}$$

$$\frac{c_{Ag} - 50}{100 - 50} = \text{erf}\left(\frac{x}{2\sqrt{(9 \times 10^{-13})(10 \text{ h})(3600 \text{ s/h})}}\right) = \text{erf}(2778x)$$

The following table is obtained by evaluating each equation using different values of x. Note that for gold diffusing into silver, x is positive as we move from the interface into the gold; for silver diffusing into gold, x is positive as we move from the interface into the silver.

Gold Diffusing into Silver				Silver Diffusing into Gold			
x(m)	$4166x$	$\dfrac{c - 50}{100 - 50}$	c_{Au}	x(m)	$2778x$	$\dfrac{c - 50}{100 - 50}$	c_{Ag}
$+0.04 \times 10^{-2}$	1.68	0.98	99	$+0.04 \times 10^{-2}$	1.11	0.88	94
0.02×10^{-2}	0.84	0.76	88	0.02×10^{-2}	0.55	0.56	78
0.01×10^{-2}	0.42	0.44	72	0.01×10^{-2}	0.28	0.31	65.5
0.005×10^{-2}	0.21	0.23	61.5	0.005×10^{-2}	0.14	0.16	58
0	0	0	50	0	0	0	50
-0.005×10^{-2}	-0.21	-0.23	38.5	-0.005×10^{-2}	-0.14	-0.16	42
-0.01×10^{-2}	-0.42	-0.44	28	-0.01×10^{-2}	-0.28	-0.31	34.5
-0.02×10^{-2}	-0.84	-0.76	12	-0.02×10^{-2}	-0.55	-0.56	22
-0.04×10^{-2}	-1.68	-0.98	1	-0.04×10^{-2}	-1.11	-0.88	6

The profiles for the gold and silver diffusion couple are plotted in Figure 5-11.

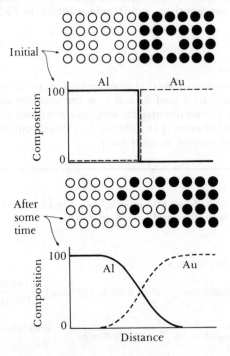

FIG. 5-10 Interdiffusion of aluminum and gold, illustrating Fick's second law.

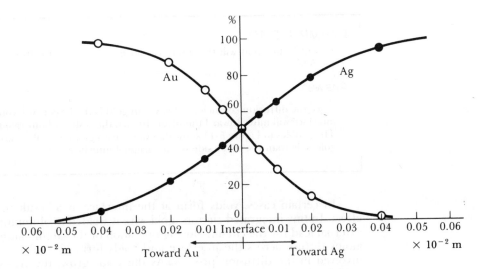

FIG. 5-11   The interdiffusion of silver and gold atoms, as calculated in Example 5-9.

The Kirkendall effect. When two atom species are diffusing concurrently in a diffusion couple with the same diffusion coefficients, symmetrical concentration profiles arise and the original interface between material 1 and 2 is fixed. But the diffusion coefficients are seldom the same; the two diffusing atoms do not simply exchange positions, but instead move by filling vacancies. If we examine the Al-Au diffusion couple, we find that aluminum atoms diffuse faster into gold than gold atoms diffuse into aluminum. Consequently more total atoms will eventually be on the original gold side of the interface than on the original aluminum side (Figure 5-12). This causes the physical location of the original interface to move towards the aluminum side of the couple. Any foreign particles originally trapped at the interface also move with the interface. This movement of the interface of the diffusion couple due to unequal diffusion rates is called the *Kirkendall effect.*

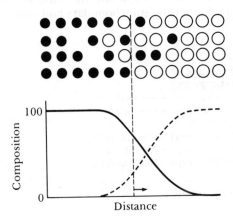

FIG. 5-12   The Kirkendall effect. Unequal rates of diffusion cause the original interface between two metals to move.

EXAMPLE 5-10

In which direction will the interface move in the Au-Ag diffusion couple in Example 5-9?

Answer:

Since the diffusion coefficient of silver in gold is the larger, we would expect that the interface will move toward the silver. In fact, the results of our calculations verify this. The profiles in Figure 5-11 show that there are a greater total number of atoms on the gold side than the silver side of the original interface.

In certain cases, voids form at the interface as a result of the Kirkendall effect. In tiny integrated circuits, gold wire is welded to aluminum to provide an external lead for the circuit. During operation of the circuit, aluminum and gold atoms diffuse across the interface and voids form by coalescence of vacancies involved in the diffusion process. As the voids grow, the Au-Al connection is weakened and eventually may fail. Because the area around the connection turns purple due to the alloying that occurs, this premature failure is called the *purple plague*.

One solution to the purple plague problem is to expose the welded connection to hydrogen. The hydrogen dissolves in the aluminum, fills the vacancies, and prevents self-diffusion of aluminum atoms. This keeps the aluminum atoms from diffusing into the gold, preventing the Kirkendall effect and the purple plague.

5-9 Types of Diffusion

In *volume diffusion*, the atoms move through the crystal from one lattice or interstitial site to another. Because of the surrounding atoms, the activation energy is large and the rate of diffusion is relatively slow.

However, atoms can also diffuse along boundaries, interfaces, and surfaces in the material. Atoms diffuse more easily by *grain boundary diffusion* because the atom packing is poor in the grain boundaries. Because atoms can more easily squeeze their way through the disordered grain boundary, the activation energy is low. *Surface diffusion* is easier still. Consequently, the activation energy is lower and the diffusion coefficient is higher for grain boundary and surface diffusion (Table 5-2 and Figure 5-13).

TABLE 5-2   The effect of the type of diffusion on the diffusivity of thorium in tungsten

Diffusion Mechanism	Diffusion Coefficient
Surface	$0.47 \times 10^{-4} \exp(-278,216/RT)$
Grain boundary	$0.74 \times 10^{-4} \exp(-377,100/RT)$
Volume	$1.00 \times 10^{-4} \exp(-502,800/RT)$

I. Langmuir, *J. Franklin Inst.*, vol. 217, 1934.

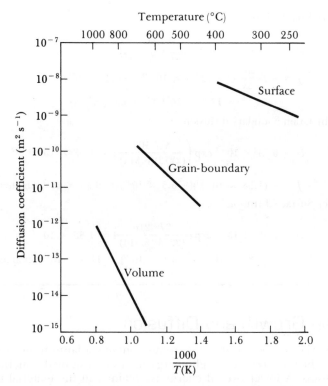

FIG. 5-13   A comparison of the diffusion coefficients for silver by volume diffusion, grain boundary diffusion, and surface diffusion.

EXAMPLE 5-11

Consider a diffusion couple set up between pure tungsten and a tungsten-1 at% thorium alloy. After several minutes of exposure at 2000°C, a transition zone of 0.1 mm thickness is established. What is the flux of thorium atoms at this time if diffusion is due to (a) volume diffusion, (b) grain boundary diffusion, or (c) surface diffusion?

Answer:

The lattice parameter of BCC tungsten in 3.165 A. Thus, the number of tungsten atoms m^{-3} is

$$\frac{W \text{ atoms}}{m^3} = \frac{2 \text{ atoms/cell}}{(3.165 \times 10^{-10})^3 \text{ m}^3/\text{cell}} = 6.3 \times 10^{28}$$

In the tungsten-1 at % thorium alloy, the number of thorium atoms is

$$c_{Th} = (0.01)(6.3 \times 10^{28}) = 6.3 \times 10^{26} \text{ Th atoms m}^{-3}$$

In the pure tungsten, the number of thorium atoms is zero. Thus, the concentration gradient is

$$\frac{\Delta c}{\Delta x} = \frac{0 - 6.3 \times 10^{26}}{0.0001 \text{ m}} = -6.3 \times 10^{30} \text{ Th atoms (m}^3 \cdot \text{m)}^{-1}$$

(a) Volume diffusion

$$D = 1.0 \times 10^{-4} \exp\left(\frac{502,800}{(2273)(8.314)}\right) = 2.75 \times 10^{-16}\,\text{m}^2\,\text{s}^{-1}$$

$$J = -D\frac{\Delta c}{\Delta x} = -(2.75 \times 10^{-16})(-6.3 \times 10^{30})$$

$$= 17.3 \times 10^{14}\,\text{Th atoms (m}^2 \cdot \text{s)}^{-1}$$

(b) Grain boundary diffusion

$$D = 0.74 \times 10^{-4} \exp\left(\frac{377,100}{(2273)(8.314)}\right) = 1.58 \times 10^{-13}\,\text{m}^2 \cdot \text{s}^{-1}$$

$$J = -(1.58 \times 10^{-13})(-6.3 \times 10^{30}) = 9.95 \times 10^{17}\,\text{Th atoms (m}^2 \cdot \text{s)}^{-1}$$

(c) Surface diffusion

$$D = 0.47 \times 10^{-4} \exp\left(\frac{278,216}{(2273)(8.314)}\right) = 1.88 \times 10^{-11}\,\text{m}^2\,\text{s}^{-1}$$

$$J = -(1.88 \times 10^{-11})(-6.3 \times 10^{30}) = 11.8 \times 10^{19}\,\text{Th atoms (m}^2 \cdot \text{s)}^{-1}$$

5-10 Grain Growth and Diffusion

A material composed of many grains contains a large number of grain boundaries, which represent a high-energy area because of the inefficient packing of the atoms. A lower overall energy is obtained in the material if the amount of grain boundary area is reduced by grain growth.

Grain growth involves the movement of grain boundaries, permitting some grains to grow at the expense of others. Diffusion of atoms across the grain boundary (Figure 5-14) is required for grain growth to occur. Consequently, the

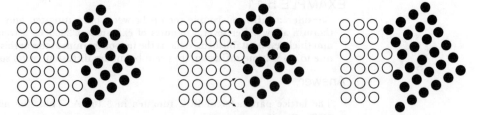

FIG. 5-14 Grain growth occurs as atoms diffuse across the grain boundary from one grain to another.

growth of the grain boundaries is related to the activation energy for an atom to jump across the boundary. The average diameter d of the grains can be given by the equation

$$d = (kt)^n \tag{5-5}$$

where t is the time permitted for grain growth, n is an exponent usually less than 0.5, and k is a constant which depends on several factors, including the diffusion coefficient. High temperatures or low activation energies increase k and thus increase the size of the grains. Many heat treatments of metals, which include holding the metal at high temperatures, must be carefully controlled to avoid excessive grain growth.

We can impede grain growth by introducing obstacles at the grain boundaries that provide a "drag" force which prevents the grains from enlarging (Figure 5-15). The grain size is related to the number and size of the obstacles by the equation

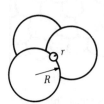

$$R = \frac{4r}{f} \tag{5-6}$$

where R is the radius of the grain, r is the radius of the obstacle, and f is the volume fraction of the obstacles. We will find that we can introduce these particles in a variety of ways to control grain growth during high temperature use and processing of many materials.

FIG. 5-15
Small particles can pin the grain boundaries and impede grain growth.

EXAMPLE 5-12

A spheroidized steel contains tiny round hard iron carbide spheres, each 1×10^{-7} m in diameter. The steel contains 0.08 volume fraction of these particles. Estimate the expected grain size of the steel.

Answer:

The radius of each particle is $r = \frac{1}{2}(1 \times 10^{-7}$ m$) = 0.5 \times 10^{-7}$ m. The radius of the grains is expected to be

$$R = \frac{4r}{f} = \frac{(4)(0.5 \times 10^{-7} \text{ m})}{0.08} = 25 \times 10^{-7} \text{ m}$$

5-11 Diffusion Bonding

Diffusion bonding, a method used to join materials, occurs in three steps (Figure 5-16). The first step forces the two surfaces together at a high pressure, flattening the surface, fragmenting impurities, and producing a high atom-to-atom contact area. Atomic bonding across the interface establishes a joint. Normally, the pressure is applied at high temperatures, where the metal is softer and more ductile.

As the surfaces remain pressed together at high temperatures, atoms diffuse along grain boundaries to the remaining voids; the atoms condense and reduce the size of any voids in the interface. Because grain boundary diffusion is rapid, this second step may occur rapidly. Eventually, however, grain growth isolates remaining voids from the grain boundaries. For the third step—final elimination of the voids—volume diffusion must occur, which is comparatively slow.

The diffusion bonding process is often used with some of the more exotic alloys, such as titanium, for joining dissimilar metals and materials, and even for joining ceramics.

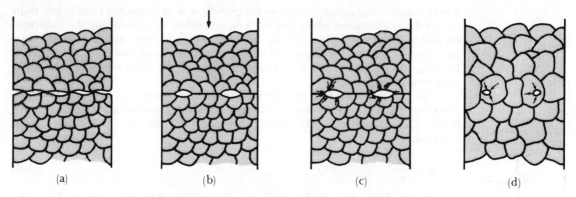

FIG. 5-16   The steps in diffusion bonding. (a) Initially the contact area is small, (b) application of pressure deforms the surface, increasing the bonded area, (c) grain boundary diffusion permits voids to shrink, and (d) final elimination of the voids requires volume diffusion.

5-12   Sintering and Powder Metallurgy

Sintering is a high-temperature treatment that causes particles of a material to join together and gradually reduce the volume of pore space between them. Sintering is a frequent step in the manufacture of ceramic components as well as for the production of metallic parts by *powder metallurgy*.

When a powdered material is compacted into a shape, the powder particles are in contact with one another at numerous sites, with a significant amount of pore space between the particles. In order to reduce the boundary energy, atoms diffuse to the boundaries, permitting the particles to be bonded together and eventually causing the pores to shrink. If sintering is carried out for a long time, the pores may be eliminated and the material becomes dense.

Surfaces that have a small radius of curvature grow rapidly. The points of contact have the smallest radius and thus grow first. Atoms diffuse to these points while vacancies diffuse away from the interface. The net movement of vacancies permits the particles to move closer together (Figure 5-17), causing the pore size to decrease and the density to increase. The rate of sintering depends on the temperature, the activation energy and diffusion coefficient for diffusion, and the original size of the particles.

In powder metallurgy processes, summarized in Figure 5-18, metal powders with a diameter of 0.001 to 0.3 mm are produced. Techniques for producing the powders include *atomization,* in which a jet of gas or liquid fragments a tiny metal stream into spherical droplets; *pulverization,* in which a brittle metal is crushed and ground to a fine size; and *chemical reduction,* in which metal compounds are chemically converted to metal particles. The powders are then compacted at high pressures in dies to produce intricate, precise green compacts. The compacts have sufficient strength to be handled and placed into a furnace, where they are sintered. Although the powder compact does shrink in size during sintering, small components requiring a minimum of machining can often be economically produced.

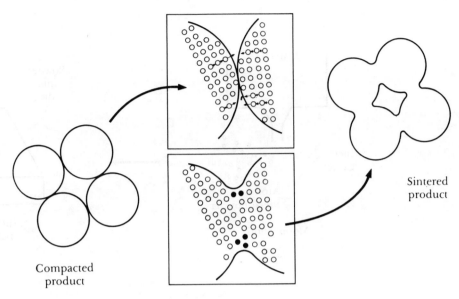

Compacted
product

Sintered
product

FIG. 5-17   Diffusion processes during sintering and powder metallurgy. Atoms diffuse to points of contact, creating bridges and eventually filling in the pores.

5-13   Diffusion in Ionic Compounds

In metals and alloys, atoms can move into any nearby vacancy or interstitial site. However, in ionic materials, a diffusing ion only enters sites having the same charge. In order to reach that site, the ion must physically squeeze past adjoining ions, pass by a region of opposite charge, and move a relatively long distance (Figure 5-19). Consequently, the activation energies are high and the rates of diffusion are low for ionic materials compared to metals (see Figure 5-6).

We also find that cations (with a positive charge) have higher diffusion coefficients than anions (with a negative charge). The cations, because they have given up their valence electrons, typically have the smaller size and thus can diffuse more easily than the larger anions. In sodium chloride, for instance, the activation energy for diffusion of chloride ions is about twice that for diffusion of sodium ions.

5-14   Diffusion in Polymers

Many polymers contain long chains of molecules, with the bonds within the chains being strong covalent bonds. Polymer chains may either be rigidly tied together in a framework or network structure by covalent bonds, or may be more weakly bonded by Van der Waals bonds. In the latter type of polymers, the long molecular chains can move, affecting the deformation and stability of the polymer.

In addition, the properties and processing of many polymers require the diffusion of small molecules between the long chains of the polymer. Molecules can diffuse more easily in amorphous polymers, which have no long-range order, than in polymers whose chains form a glassy or crystalline pattern. Diffusion

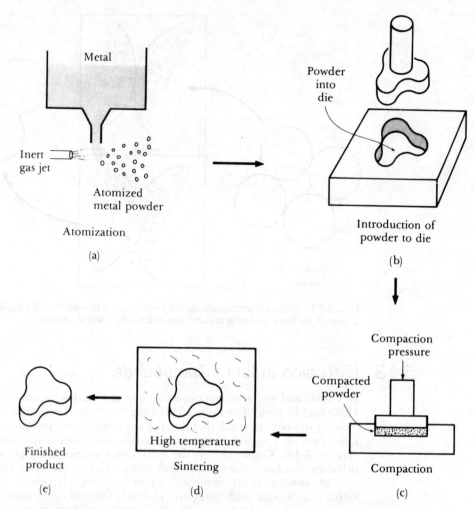

FIG. 5-18 Powder metallurgy processing flow diagram: (a) molten metal is atomized by a jet of inert gas, (b) the powder is poured into a die, (c) the powder is pressed into a compact, and (d) the compact is sintered to produce the final product (e).

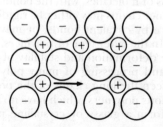

FIG. 5-19 Diffusion in ionic compounds. Anions can only enter other anion sites.

is required to enable dyes to uniformly enter many of the synthetic polymer fabrics. Selective diffusion of polymer membranes is used to cause desalinization of water; while water molecules pass through the polymer membrane, the ions in the salt are trapped.

However, diffusion in polymers is often undesirable. If air diffuses through the plastic wrapping used to store food, the food may spoil; if air diffuses through the rubber inner tube of an automobile tire, obvious problems result; some molecules can diffuse into and cause swelling of certain polymers.

SUMMARY

Atoms move through a solid material, particularly at high temperatures when a concentration gradient is present, by diffusion mechanisms. Atom diffusion is of paramount importance to us, because many of the materials processing techniques, such as sintering, powder metallurgy, and diffusion bonding, require diffusion. Furthermore, many of the heat treatments and strengthening mechanisms used to control structures and properties in materials are diffusion-controlled processes. The stability of the structure and the properties of materials during use at high temperatures depend on diffusion. Finally, many important materials are produced by deliberately preventing diffusion, thereby forming nonequilibrium structures. In the following chapters, we will encounter many instances where diffusion plays a significant role.

GLOSSARY

Activation energy   The energy required to cause a particular reaction to occur. In diffusion, the activation energy is related to the energy required to move an atom from one lattice site to another.

Concentration gradient   The rate of change of composition with distance in a nonuniform material, expressed as atoms $(m^3 \cdot m)^{-1}$ or at % m^{-1}.

Diffusion   The movement of atoms within a material.

Diffusion bonding   A joining technique in which two surfaces are pressed together at high pressures and temperatures. Diffusion of atoms to the interface fills in voids and produces a strong bond.

Diffusion coefficient   A temperature-dependent coefficient that is related to the rate at which atoms diffuse. The diffusion coefficient depends on the temperature and activation energy.

Diffusivity   Another term for diffusion coefficient.

Fick's first law   The equation relating the flux of atoms by diffusion to the diffusion coefficient and the concentration gradient.

Fick's second law   The partial differential equation that describes the rate at which atoms are redistributed in a material by diffusion.

Flux   The number of atoms passing through a plane of unit area per unit time. This is related to the rate at which mass is transported by diffusion in a solid.

Grain boundary diffusion   Diffusion of atoms along grain boundaries. This is faster than volume diffusion because the atoms are less closely packed in grain boundaries.

Grain growth   Movement of grain boundaries by diffusion in order to reduce the amount of grain boundary area. As a result, small grains shrink and disappear while other grains become larger.

Interstitial diffusion   Diffusion of small atoms from one interstitial position to another in the crystal structure.

Interstitialcy diffusion   A diffusion mechanism by which an atom leaves its regular lattice position to fill an interstitial position.

Kirkendall effect   Due to unequal rates of diffusion in a diffusion couple, the original interface moves.

Sintering   A high-temperature treatment used to join small particles. Diffusion of atoms to points of contact causes bridges to form between the particles. Further diffusion eventually fills in any remaining voids.

Surface diffusion   Diffusion of atoms along surfaces, such as cracks or particle surfaces.

Vacancy diffusion   Diffusion of atoms when an atom leaves a regular lattice position to fill a vacancy in the crystal. This process creates a new vacancy and the process continues.

Volume diffusion   Diffusion of atoms through the interior of grains.

PRACTICE PROBLEMS

1   The diffusion coefficient for zinc in copper is 1.4×10^{-21} m^2 s^{-1} at 300°C and 8.0×10^{-16} m^2 s^{-1} at 600°C. Determine (a) the activation energy Q and (b) the diffusion constant D_0 for zinc in copper.

2   The diffusion coefficient for Fe^{2+} ions in ferrous oxide (FeO) is 5×10^{-14} m^2 s^{-1} at 600°C and 1.5×10^{-12} m^2 s^{-1} at 900°C. Determine (a) the activation energy Q and (b) the diffusion constant D_0 for Fe^{2+} ions in FeO.

3   From Figure 5-6, calculate (a) the diffusion constant D_0 and (b) the activation energy Q for the diffusion of calcium ions in calcium oxide, or lime, CaO.

4   From Figure 5-6, calculate (a) the diffusion constant D_0 and (b) the activation energy Q for the diffusion of magnesium ions in magnesium oxide, or magnesia, MgO.

5   A germanium crystal 0.1 mm thick contains one arsenic atom for every 10^8 Ge atoms at one surface and 1000 As atoms for every 10^8 Ge atoms at the other surface. The lattice parameter of DC germanium is 5.66 Å. Calculate (a) the at% arsenic at each surface, (b) the composition gradient in terms of at%/cm, and (c) the composition gradient in terms of atoms/cm$^3 \cdot$ cm.

6   A silicon crystal 0.5 mm thick contains two gallium atoms per 10^7 Si atoms at one surface. The other surface is treated to produce a higher concentration of gallium. How many gallium atoms per 10^7 silicon atoms must be present at this surface to produce a concentration gradient of -2×10^{26} atoms (m$^3 \cdot$ m)$^{-1}$? The lattice parameter of silicon is 5.407 Å.

7   Copper and nickel are joined to produce a diffusion couple. After some period of time, the couple contains 78% Cu 1 mm to the left of the interface and 28% Cu 1 mm to the right of the interface. Assume an average lattice parameter of 3.58 Å. Determine (a) the concentration gradient between the two points in weight percent copper per cm, atomic percent copper per cm, and copper atoms/cm$^3 \cdot$ cm, (b) the flux of copper atoms across the interface at 100°C, and (c) the flux of copper atoms across the interface at 500°C.

8   A sheet of gold is diffusion-bonded to a sheet of silver. After 2 h, a concentration gradient of -165 at% silver per cm is produced. If the gradient is uniform, and assuming an average lattice parameter of 4.08 Å, (a) determine the concentration gradient in silver atoms (m$^3 \cdot$ m)$^{-1}$, and (b) calculate the flux of silver atoms across the interface at 500°C.

9   An iron foil 0.1 mm thick separates a reaction chamber into two sections. A gas containing 5×10^{26} H atoms m^{-3} is present on one side of the foil and a gas containing 3×10^{25} H atoms m^{-3} is present on the other side of the foil. (a) Determine the concentration gradient of hydrogen in the foil, (b) calculate the flux of hydrogen atoms if the system is held at 911°C, where iron is BCC, and (c) calculate the flux of hydrogen atoms if the system is held at 913°C, where iron is FCC.

10   An iron sheet 10 mm thick separates a reaction chamber into two sections. A gas containing 1×10^{28} N atoms m^{-3} is present on one side of the sheet and a gas containing 2×10^{26} N atoms m^{-3} is present on the other side of the sheet. (a) Determine the concentration gradient of nitrogen in the sheet, (b) calculate the flux of nitrogen atoms in BCC iron at 911°C, and (c) calculate the flux of nitrogen atoms in FCC iron at 913°C.

11 Determine the concentration gradient required to transport 100×10^4 N atoms $(m^2 \cdot s)^{-1}$ through an iron foil at 1000°C, where iron is FCC.

12 We want to obtain the flux of 1000×10^4 H atoms $(m^2 \cdot s)^{-1}$ through a BCC iron foil at 800°C. The composition of hydrogen on either side of the foil is 2×10^{24} H atoms m^{-3} and 5×10^{22} H atoms m^{-3}, respectively. Determine (a) the concentration gradient that is required and (b) the thickness of the iron foil that is required.

13 A concentration gradient of -10^{-24} H atoms $(m^3 \cdot m)^{-1}$ is produced across a BCC iron foil. Calculate the temperature required to produce a flux of 2×10^{16} H atoms $(m^2 \cdot s)^{-1}$.

14 From Table 5-1, what can you deduce concerning the relative atomic size of hydrogen, carbon, and nitrogen atoms? From the analysis, would you expect the activation energy for the diffusion of boron atoms in iron to be closer to that of carbon or hydrogen? (See Appendix B.)

15 Estimate the activation energy for self diffusion of Mg, Zn, and Mo from Figure 5-7.

16 An iron sheet separates a gas containing 10^{26} H atoms m^{-3} from a gas containing 10^{25} H atoms m^{-3} at 1200°C, where the iron is FCC. How thick should the iron sheet be if we want to permit a flux of 10^{20} H atoms $(m^2 \cdot s)^{-1}$?

17 During carburizing of a 0.2% C steel, 1.0% C is introduced at the surface. Calculate the amount of carbon at a distance of 1 mm below the surface after 1 h at 1000°C, when the steel has the FCC structure.

18 During carburizing of a 0.10% C steel, 1.0% C is introduced at the surface. The final carbon content 1 mm beneath the surface must be 0.6% C. Determine the diffusion coefficients and the required carburizing times if carburizing is done at (a) 900°C and (b) 1000°C. (In both cases, the iron is FCC.)

19 During carburizing of a 0.10% C steel, 1.2% C is introduced at the surface when the iron is FCC. The carbon content 0.16 mm below the surface is 0.90% after 1 h. Calculate the carburizing temperature.

20 Steels may be strengthened by nitriding, or diffusing nitrogen into the surface at 700°C, when the iron is BCC. If the nitriding atmosphere produces 0.1% N at the surface of a steel initially containing 0.001% N, determine the percent of nitrogen at a distance of 1 mm beneath the surface after 5 h.

21 Nitrogen is introduced to the surface of a steel which contains 0.002% N at 650°C, when the iron is BCC. We want to obtain 0.15% N at a distance of 0.126 mm below the surface in 1 h. Calculate the composition of nitrogen that must be introduced to the surface.

22 A nitriding atmosphere introduces 0.4% N at the surface of a BCC steel containing 0.003% N. We define the *case depth* as the distance below the surface that contains more than 0.03% N. If nitriding is accomplished at 700°C for 3 h, what is the case depth?

23 A satisfactory nitrogen case depth is produced in BCC iron after 1 h at 700°C. How much time is required to produce a satisfactory case depth if nitriding is carried out at 600°C?

24 A satisfactory carbon case depth is produced in FCC iron after 8 h at 800°C. What temperature is required if we want to obtain the same case depth in 1 h?

25 A steel containing 0.4% C is exposed to an oxidizing atmosphere containing 0% C but high oxygen. In this case, carbon diffuses from the BCC steel to the surface, reacts with the oxygen and leaves. This is called *decarburizing*. What will the percent of carbon be at a point 2 mm below the surface of a 0.4% C steel if the steel is held for 10 h at 900°C?

26 A steel containing 1.0% C is placed in an oxidizing atmosphere. How long can the FCC steel be held at 1000°C before a point 1 mm below the surface is decarburized to 0.2% C?

27 Suppose a steel containing 0.03% N is placed in a vacuum at 680°C. After 6 h, how much N is present 0.7 mm below the surface?

28 A diffusion couple is produced by joining copper to nickel. Plot the concentration of both copper and nickel across the interface after holding 1 h and 100 h, respectively, at 1000°C. Will the original interface move toward copper or nickel?

29 A diffusion couple is produced by joining pure copper to a Cu-50% Ni alloy. How long will it take to produce a composition of 60% Cu in the Cu-50% Ni alloy at a distance of 1 mm from the original interface if we diffusion bond the two materials at 850°C?

30 A 50% Au-50% Ag alloy forms a diffusion couple with pure silver. Calculate the concentration of each element across the interface after the couple has been heated for 12 h at 900°C.

31 Pure silver is bonded to pure gold. After heating for 2 days, enough gold diffuses into the silver so that there is 25% Au in the silver at a distance of 0.03 mm from the original interface. What temperature must have been used?

32 Calculate the activation energies and diffusion constant D_0 for volume, grain boundary, and surface diffusion of silver in silver (Figure 5-13).

PART **II**

CONTROLLING THE MICROSTRUCTURE AND MECHANICAL PROPERTIES OF MATERIALS

In Part II, we will examine the methods used to control the microstructure and macrostructure of materials. We will find that by controlling the structure we in turn can control the mechanical properties.

There are six important mechanisms used to control structure and properties—grain size strengthening, solid solution strengthening, strain hardening, dispersion strengthening, age hardening, and phase transformations. All introduce barriers to slip. In the first three methods, we rely on the three types of lattice imperfections. By controlling surface defects such as grain boundaries, we obtain *grain size strengthening*. Controlling point defects such as substitutional atoms gives *solid solution strengthening*. Increasing the number of line defects, or dislocations, provides *strain hardening*.

We obtain strengthening in the other three mechanisms by introducing multiple phases, where each phase has a different composition or crystal structure. The boundaries between the phases can provide strengthening by interfering with the deformation mechanisms. *Dispersion strengthening* is a general term indicating strengthening by multiple phases. *Age hardening* is a special technique that provides an optimum, fine dispersion of phases. *Phase transformations* include more sophisticated treatments, often relying on allotropic transformations.

We will discuss the strengthening mechanisms, at least partly, from the point of view of processing of the material. In particular, we will examine solidification, alloying, deformation, and heat treatment. *Solidification* helps determine grain size, grain shape, and the fineness and distribution of phases in many multiple-phase alloys. *Alloying* produces solid solution strengthening and provides the basis for dispersion strengthening. *Deformation* processing produces strain hardening and helps control grain size and shape. *Heat treatment* permits us to perform the dispersion strengthening, age hardening, and phase transformation strengthening techniques.

Before looking at strengthening mechanisms and the processes used to control these mechanisms, we will first briefly examine the mechanical testing of materials and understand the results of these tests, which are the mechanical properties of a material.

6

Mechanical Testing and Properties

6-1 Introduction

We select materials for many components and applications by matching the properties of the material to the service conditions required of the component. The first step in the selection process requires that we analyze the application to determine the most important characteristics that the material must possess. Should the material be strong, or stiff, or ductile? Will it be subjected to repeated application of a high force, a sudden intense force, a high stress at elevated temperature, or abrasive conditions? Once we have determined the required properties, we can select the appropriate material using data listed in handbooks. However, we must know how the properties listed in the handbook are obtained, know what the properties mean, and realize that the properties listed are obtained from idealized tests that may not apply exactly to real-life engineering applications.

In this chapter we will study several tests that are used to measure how a material withstands an applied force. The results of these tests are the mechanical properties of the material.

TENSILE TEST

6-2 The Stress-Strain Diagram

The *tensile test* measures the resistance of a material to a static or slowly applied force. A test setup is shown in Figure 6-1; a typical specimen has a diameter of 12.5 mm and a gage length of 50.0 mm. The specimen is placed in the testing machine and a force F, called the *load*, is applied. A strain gage or extensometer is used to measure the amount that the specimen stretches between the gage marks when the force is applied.

The results of a tensile test are shown in Table 6-1 and Figure 6-2 as load versus gage length. Displaying the results of the test in this manner describes only how a material having this particular diameter behaves. The force required to produce a given amount of stretching is much greater if a larger diameter specimen is used.

Engineering stress and strain. The results of a single test apply to all sizes and shapes of specimens for a given material if we convert the force to stress and

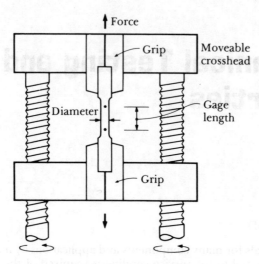

FIG. 6-1   A unidirectional force is
applied to a specimen in the tensile
test by means of the moveable
crosshead.

the distance between gage marks to strain. *Engineering stress* and *engineering strain*
are defined by the following equations.

$$\begin{array}{c}\text{Engineering}\\\text{stress}\end{array} = \sigma = \dfrac{F}{A_0} \tag{6-1}$$

$$\begin{array}{c}\text{Engineering}\\\text{strain}\end{array} = \varepsilon = \dfrac{l - l_0}{l_0} \tag{6-2}$$

where A_0 is the original cross-sectional area of the specimen before the test begins,
l_0 is the original distance between the gage marks, and l is the distance between the
gage marks after force F is applied. The stress-strain curve is usually used to
record the results of a tensile test.

TABLE 6-1   The load-gage length
data obtained from a tensile test
of a 12.5 mm diameter aluminum
alloy test bar

Load (kN)	Gage Length (mm)
0	50.00
4.5	50.02
13.4	50.07
22.3	50.13
31.2	50.18
33.4	50.75
35.2	52.00
35.7	53.00
35.7 (max)	54.00
33.8 (fracture)	55.13

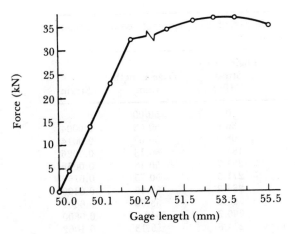

FIG. 6-2 Graph of the load-gage length data from Table 6-1 for an aluminum alloy test specimen 12.5 mm in diameter.

EXAMPLE 6-1

Convert the load-gage length data in Table 6-1 to engineering stress and strain and plot a stress-strain curve.

Answer:

For the 1000-lb load

$$\sigma = \frac{F}{A_0} = \frac{4.5 \times 10^3}{(\pi/4)(0.0125)^2} = \frac{4.5 \times 10^3}{1.23 \times 10^{-4}} = 36.5 \text{ MPa}$$

$$\varepsilon = \frac{l - l_0}{l_0} = \frac{50.02 - 50.00}{50.00} = 0.0004$$

The results of similar calculations for each of the remaining loads are given in Table 6-2 and are plotted in Figure 6-3.

EXAMPLE 6-2

Compare the force required to produce a stress of 170 MPa in a 25 mm diameter bar and in a 50 mm diameter bar.

Answer:

$$F = \sigma A_0 = (170 \times 10^6)\left(\frac{\pi}{4}\right)(0.025)^2 = 83.4 \text{ kN for a 25 mm bar}$$

$$F = \sigma A_0 = (170 \times 10^6)\left(\frac{\pi}{4}\right)(0.050)^2 = 333.8 \text{ kN for a 50 mm bar}$$

TABLE 6-2   Conversion of load-gage length data to stress and strain data (for Example 6-1)

Load (kN)	Engineering Stress (MPa)	Gage Length (mm)	Strain
0	0	50.00	0
4.5	36.6	50.02	0.0004
13.4	108.9	50.07	0.0014
22.3	181.3	50.13	0.0026
31.2	253.7	50.18	0.0036
33.4	271.5	50.75	0.0150
35.2	286.2	52.00	0.0400
35.7	290.2	53.00	0.0600
35.7 (max)	290.2	54.00	0.0800
33.8 (fracture)	274.8	55.13	0.1062

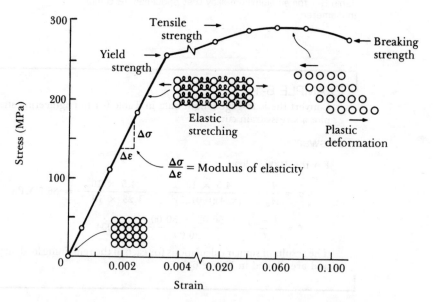

FIG. 6-3   The stress-strain curve for an aluminum alloy from Table 6-2.

The engineering strain tells us how much each inch of the metal will stretch for a given applied stress. If the metal part is 10 m long, we can multiply the strain by 10 to determine the total amount that the part will stretch, assuming that the part stretches uniformly.

EXAMPLE 6-3

Suppose a 20 kN force is applied to a 12.5 mm diameter bar that is 1.27 m long. The bar is made from the same aluminum alloy we have discussed previously. Determine the length of the bar when the force is applied.

Answer:

$$\sigma = \frac{F}{A_0} = \frac{2 \times 10^3}{(\pi/4)(0.0125)} = 163 \text{ MPa}$$

From Figure 6-3, $\epsilon = 0.0025$

$$\frac{l - l_0}{l_0} = 0.004$$

$$\frac{l - 1.27}{1.27} = 0.004$$

$$l = 1.27 + (0.004)(1.27) = 1.275 \text{ m}$$

Units. Some conversion factors for stress are summarized in Table 6-3. Because strain is really dimensionless, no conversion factors are required to change the system of units.

TABLE 6-3 Units and conversion factors for stress

1 MPa = megapascal
1 MN/m² = 1 MPa = meganewton per square meter
                      = newton per square millimeter
1 GPa = 1000 MPa = gigapascal
1 psi = pounds per square inch
1 ksi = 1000 psi
1 ksi = 6.893 MPa
1 psi = 0.006895 MPa
1 MPa = 0.145 ksi = 145 psi

EXAMPLE 6-4

Determine the stress in megapascals when a 5000-lb force is applied to a 0.505-in. diameter bar.

Answer:

$$\sigma = \frac{F}{A_0} = \frac{5000}{(\pi/4)(0.505)^2} = 25,000 \text{ psi}$$

$$\sigma = (25,000 \text{ psi})(0.006895 \text{ MPa/psi}) = 172.4 \text{ MPa}$$

6-3 Elastic versus Plastic Deformation

When a force is first applied to the specimen, the bonds between the atoms are stretched and the specimen elongates. When we remove the force, the bonds return to their original length and the specimen returns to its initial size. Stretching of the metal in this *elastic* portion of the stress-strain curve is recoverable.

At higher forces the material behaves in a *plastic* manner. As the stress increases, dislocations begin to move, slip occurs, and the material begins to plastically deform. Removal of the force permits the elastic deformation to be recovered, but the deformation caused by slip is permanent. The stress at which slip begins is the dividing point between elastic and plastic behavior.

6-4 Yield Strength

The yield strength is the stress at which slip becomes noticeable and significant. If we are designing a component that must support a force during use, we must be sure that the component does not plastically deform. Crankshafts do not operate very well in engines when they are deformed out of their intended shape! We must therefore select a material that has a high yield strength or we must make the component large so that the applied force produces a stress that is below the yield strength.

On the other hand, if we are manufacturing shapes or components by some deformation process, the applied stress must exceed the yield strength to produce a permanent change in the shape of the material.

EXAMPLE 6-5

You are to design a cable that must support an elevator cab that weighs 45 kN. The cable is made from the aluminum alloy in Example 6-1. Calculate the minimum diameter of the cable required to support the cab without permanent deformation.

Answer:

We must not exceed the yield strength of 240 MPa.

$$A_0 = \frac{F}{\sigma} = \frac{45 \times 10^3}{240 \times 10^6} = 1.875 \times 10^{-4} \text{ m}^3$$

$$d = \sqrt{\frac{4}{\pi} A_0} = \sqrt{\frac{4}{\pi} (1.875 \times 10^{-4})} = 15.5 \text{ mm}.$$

EXAMPLE 6-6

You wish to bend an aluminium bar which is 10 mm × 150 mm in cross section into a bracket by applying a tensile force. What is the minimum force that must be exerted by your forming equipment?

Answer:

We must exceed the yield strength of 240 MPa.

$$F = \sigma A_0 = (240 \times 10^6)(0.01)(0.150) = 366 \text{ kN}$$

Offset yield strength.   In some materials, the stress at which the material changes from elastic to plastic behavior is not easily detected. In this case we may determine an *offset yield* or proof stress. We decide that a small amount of permanent deformation, such as 0.2% or 0.002 strain, might be allowable without damaging the performance of our component. We can construct a line parallel to the initial portion of the stress-strain curve but offset by 0.002 strain from the origin. The 0.2% proof stress is the stress at which our constructed line intersects the stress-strain curve (Figure 6-4).

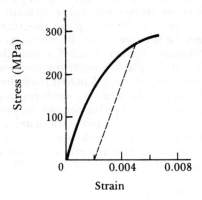

FIG. 6-4   Determination of the 0.2% proof stress for gray cast iron.

EXAMPLE 6-7

Determine the 0.2% proof stress for gray cast iron (Figure 6-4).

Answer:

By constructing a line starting at 0.002 strain which is parallel to the elastic portion of the stress-strain curve, we find that the 0.2% proof stress is 275 MPa.

Double yield point.   On the other hand, the stress-strain curve for certain low-carbon steels displays a *double yield point* (Figure 6-5). The material is expected

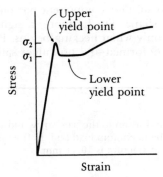

FIG. 6-5   The stress-strain behavior of a metal that displays the double yield point.

to plastically deform at stress σ_1. However, small interstitial atoms clustered around the dislocations interfere with slip and raise the yield point to σ_2. Only after we apply the higher stress σ_2 does the dislocation slip. After slip begins at σ_2, the dislocation moves away from the clusters of small atoms and continues to move very rapidly at the lower stress σ_1. We can easily determine the lower (σ_1) yield strengths for metals having this type of stress-strain behavior.

6-5   Tensile Strength

The *tensile strength* is the stress obtained at the highest applied force and thus is the maximum stress on the engineering stress-strain curve. In many ductile materials, deformation does not remain uniform. At some point, one region deforms more than other areas and a large local decrease in the cross-sectional area occurs (Figure 6-6). This locally deformed region is called a *neck*. Because the cross-sectional area becomes smaller at this point, a lower force is required to continue its deformation, and the engineering stress, calculated from the original area A_0, will decrease. The tensile strength is the stress at which necking begins.

Tensile strengths are often reported in handbooks because they are easy to measure; they are useful in comparing the behavior of materials and they permit us to estimate other properties which are more difficult to measure. However, the tensile strength is relatively unimportant for materials selection or materials fabrication—the yield strength determines whether the metal will or will not deform.

6-6   True Stress-True Strain

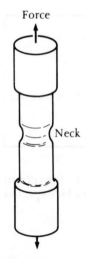

Force

Neck

FIG. 6-6
Localized
deformation of a
ductile metal
during a tensile
test produces a
necked region.

The decrease in engineering stress beyond the tensile point occurs because of our definition of engineering stress. We used the original area A_0 in our calculations, which is not precise because the area continually changes. We define *true stress* and *true strain* by the following equations.

$$\text{True stress} = \sigma_t = \frac{F}{A} \tag{6-3}$$

$$\text{True strain} = \varepsilon_t = \int \frac{dl}{l_0} = \ln\left(\frac{l}{l_0}\right) = \ln\left(\frac{A_0}{A}\right) \tag{6-4}$$

where A is the actual area at which the force F is applied. The expression $\ln(A_0/A)$ must be used after necking begins. The true stress-strain curve is compared to the engineering stress-strain curve in Figure 6-7. The true stress continues to increase after necking because, although the load required decreases, the area decreases even more.

We seldom require true stress and true strain. As soon as we exceed the yield strength, the metal begins to deform. Our component has failed because it no longer has the original intended shape. Furthermore, a significant difference develops between the two curves only when necking begins. But when necking begins, our component is grossly deformed and no longer satisfies its intended use.

EXAMPLE 6-8

Compare the engineering stress and strain to the true stress and strain for the aluminum alloy in Example 6-1 at (a) the maximum load and (b) fracture. The diameter at maximum load is 12.62 mm and at fracture is 10.11 mm.

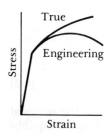

FIG. 6-7
The relationship
between the true
stress-true strain
diagram and the
engineering
stress-strain
diagram.

Answer:

(a) At the tensile or maximum load

$$\text{Engineering stress} = \frac{F}{A_0} = \frac{35.7 \times 10^3}{(\pi/4)(0.0125)^2} = 290 \text{ MPa}$$

$$\text{True stress} = \frac{F}{A} = \frac{35.7 \times 10^3}{(\pi/4)(0.0123)^2} = 313 \text{ MPa}$$

$$\text{Engineering strain} = \frac{l - l_0}{l_0} = \frac{53.00 - 50.00}{50.00} = 0.060$$

$$\text{True strain} = \ln\left(\frac{l}{l_0}\right) = \ln\left(\frac{53.00}{50.00}\right) = 0.058$$

(b) At fracture

$$\text{Engineering stress} = \frac{F}{A_0} = \frac{33.8 \times 10^3}{(\pi/4)(0.0125)^2} = 275 \text{ MPa}$$

$$\text{True stress} = \frac{F}{A} = \frac{33.8 \times 10^3}{(\pi/4)(0.0099)^2} = 443 \text{ MPa}$$

$$\text{Engineering strain} = \frac{l - l_0}{l_0} = \frac{53.13 - 50.00}{50.00} = 0.1025$$

$$\text{True strain} = \ln\left(\frac{A_0}{A}\right) = \ln\left[\frac{(\pi/4)(0.0125)^2}{(\pi/4)(0.0099)^2}\right] = \ln(1.594) = 0.466$$

The true stress becomes much greater than the engineering stress only after necking begins.

6-7 Brittle Behavior

Ductile materials display an engineering stress-strain curve that goes through a maximum at the tensile strength. In more brittle materials, the maximum load or tensile strength occurs at the point of failure. In extremely brittle materials, such as ceramics, the yield strength, tensile strength, and breaking strength are all the same (Figure 6-8).

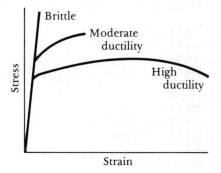

FIG. 6-8   The stress-strain behavior for brittle materials compared to more ductile materials.

6-8   Modulus of Elasticity

The *modulus of elasticity*, or *Young's modulus*, is the slope of the stress-strain curve in the elastic region. This relationship is *Hooke's law*.

$$E = \frac{\sigma}{\varepsilon} = \text{modulus of elasticity} \tag{6-5}$$

The modulus is closely related to the forces bonding the atoms in the material (Figure 6-9). A steep slope in the force-interatomic spacing graph at the equilib-

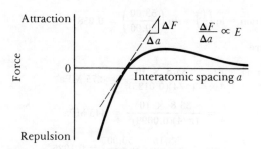

FIG. 6-9   The slope of the force-interatomic spacing curve is related to the modulus of elasticity.

rium spacing indicates that high forces are required to separate the atoms and cause the metal to stretch elastically. Thus, the metal has a high modulus of elasticity. Binding forces, and consequently the modulus of elasticity, are higher for high melting point metals (Table 6-4).

The modulus is a measure of the *stiffness* of the material. A stiff material, with a high modulus of elasticity, maintains its size and shape even under an elastic load. If we are designing a shaft and bearing, we may need very close tolerances. But if the shaft deforms elastically, those close tolerances may cause excessive rubbing, wear, or seizing. Figure 6-10 shows the elastic behavior of iron and aluminium. If a stress of 200 MPa is applied to the shaft, the steel deforms

TABLE 6-4   Relationship between the modulus of elasticity and the melting temperature of metals

Metal	Melting Temperature (°C)	Modulus of Elasticity (GPa)
Pb	327	13.79
Mg	650	44.82
Al	660	68.97
Ag	962	71.03
Au	1064	77.93
Cu	1085	124.83
Ni	1453	206.21
Fe	1538	206.90
Mo	2610	299.31
W	3410	403.45

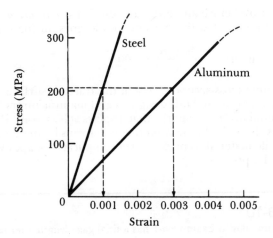

FIG. 6-10 Comparison of the elastic behavior of steel and aluminum.

elastically 0.001 (strain) while, at the same stress, aluminum deforms 0.003 (strain). Iron has a modulus of elasticity three times greater than that of aluminum.

EXAMPLE 6-9

From the data in Example 6-1, calculate the modulus of elasticity of the aluminum alloy. Use the modulus to determine the length of a 1.25 m bar to which a stress of 210 MPa is applied.

Answer:

When a stress of 238 MPa is applied, a strain of 0.0035 is produced. Thus

$$\text{Modulus of elasticity} = E = \frac{\sigma}{\varepsilon} = \frac{238 \times 106}{0.0035} = 689 \text{ Pa}$$

From Hooke's law

$$\varepsilon = \frac{\sigma}{E} = \frac{210 \times 10^6}{68 \times 10^9} = 0.003 = \frac{l - l_0}{l_0}$$

$$l = l_0 + \varepsilon l_0 = 1.25 + (0.003)(1.25) = 1.254 \text{ m}.$$

6-9 Ductility

Ductility measures the amount of deformation that a material can withstand without breaking. There are two ways to express ductility. First, we could measure the distance between the gage marks on our specimen before and after the test. The *% elongation* describes the amount that the specimen stretches before fracture.

$$\% \text{ Elongation} = \frac{l_f - l_0}{l_0} \times 100 \tag{6-6}$$

where l_f is the distance between gage marks after the specimen breaks.

A second approach is to measure the percent change in cross-sectional area at the point of fracture before and after the test. This *% reduction in area* describes the amount of thinning that the specimen undergoes during the test.

$$\% \text{ Reduction in area} = \frac{A_0 - A_f}{A_0} \times 100 \tag{6-7}$$

where A_f is the final cross-sectional area at the fracture surface.

Ductility is important to both designers and manufacturers. The designer of a component would prefer a material that displays at least some ductility so that, if the applied stress is too high, the component deforms before it breaks. A fabricator wants a ductile material so he can form complicated shapes without breaking the material in the process.

EXAMPLE 6-10

The aluminum alloy in Example 6-1 has a final gage length after failure of 54.88 mm and a final diameter of 9.87 mm at the fractured surface. Calculate the ductility of this alloy.

Answer:

$$\% \text{ Elongation} = \frac{l_f - l_0}{l_0} \times 100 = \frac{54.88 - 50.00}{50.00} \times 100 = 9.76\%$$

$$\% \text{ Reduction in area} = \frac{A_0 - A_f}{A_0} \times 100 = \frac{(\pi/4)(12.5)^2 - (\pi/4)(9.87)^2}{(\pi/4)(12.5)^2} \times 100$$

$$= 37.5\%$$

The final gage length is less than 55.13 mm (see Table 6-2) since, after fracture, the elastic strain is recovered.

6-10   Temperature Effects

The tensile properties are significantly affected by temperature (Figure 6-11). The yield strength, tensile strength, and modulus of elasticity decrease at higher temperatures, whereas the ductility, as measured by the amount of strain at failure, commonly increases. A materials fabricator may wish to deform a material at a high temperature (known as *hot working*) to take advantage of the higher ductility and lower required stress.

IMPACT TEST

6-11   Nature of the Impact Test

In order to select a material to withstand a sudden intense blow, we must measure a material's resistance to failure in an *impact test*. Many test procedures have been devised, including the *Charpy test* (Figure 6-12). The test specimen may either be

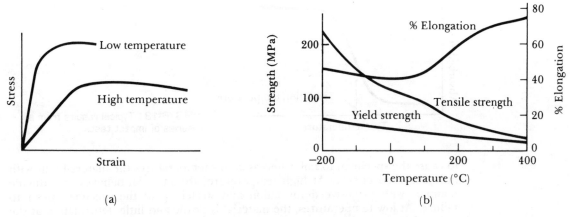

FIG. 6-11   The effect of temperature (a) on the stress-strain curve and (b) on the tensile properties of an aluminum alloy.

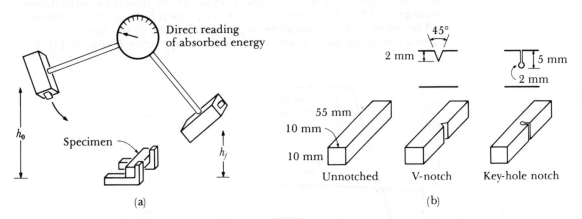

FIG. 6-12   (a) The Charpy impact test setup. (b) Typical specimens.

notched or unnotched; the V-notched specimens better measure the resistance of the material to crack propagation.

In the test, a heavy pendulum which starts at an elevation h_0 swings through its arc, strikes and breaks the specimen, and reaches a lower final elevation h_f. By knowing the initial and final elevations of the pendulum, the difference in potential energy can be calculated. This difference is the *impact energy* absorbed by the specimen during failure. The energy is usually expressed in foot·pounds (ft·lb) or joules (J), where 1 ft·lb = 1.356 J. The ability of a material to withstand an impact blow is often referred to as the *toughness* of the material.

6-12   Temperature Effects of the Impact Test

The results of a series of impact tests performed at various temperatures are shown in Figure 6-13. At high temperatures, a large absorbed energy is required

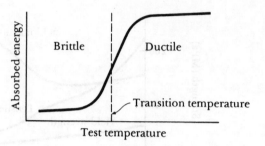

FIG. 6-13 Typical results from a series of impact tests.

to cause the specimen to fail, whereas at low temperatures the material fails with little absorbed energy. At high temperatures the material behaves in a ductile manner, with extensive deformation and stretching of the specimen prior to failure. At low temperatures, the material is brittle and little deformation at the point of fracture is observed. The *transition temperature* is the temperature at which the material changes from ductile to brittle failure.

A material that may be subjected to an impact blow during service must have a transition temperature *below* the temperature of the material's surroundings. For example, the transition temperature of a steel used for a carpenter's hammer should be below room temperature to prevent chipping of the steel.

Not all materials have a distinct transition temperature (Figure 6-14). BCC

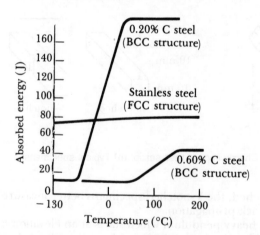

FIG. 6-14 The Charpy V-notch properties for two plain-carbon steels (BCC structure) and a FCC stainless steel.

metals have transition temperatures but most FCC metals do not. FCC metals have high absorbed energies, with the energy decreasing gradually and slowly as the temperature decreases.

6-13 Notch Sensitivity

Notches caused by poor machining, fabrication, or design cause stresses to be concentrated, reducing the toughness of the material. The *notch sensitivity* of a material can be evaluated by comparing the absorbed energies of notched versus unnotched specimens. The absorbed energies are much lower in notched specimens if the material is notch sensitive, as in ductile cast iron (Figure 6-15). However, some materials, such as gray cast iron, are not notch sensitive.

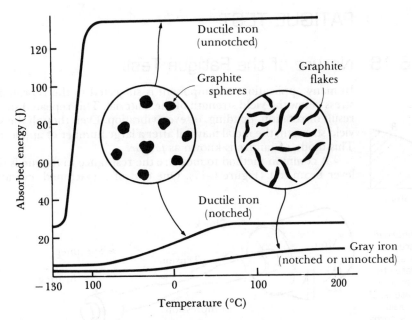

FIG. 6-15   The effect of internal and external notches on impact properties. Gray iron structures contain sharp graphite flakes that act as notches and produce low energies. Ductile iron structures contain spherical graphite nodules that do not act as notches. An external notch has a significant effect only on ductile iron.

6-14 Relation of Impact Energy to True Stress-True Strain

The impact energy corresponds to the area contained within the true stress-true strain diagram. Materials that have both high strength and high ductility have a good toughness (Figure 6-16). Ceramics, on the other hand, have poor toughness because they display virtually no ductility.

6-15 Use and Precautions of Impact Properties

The absorbed energy and the transition temperature are very sensitive to the loading conditions. For example, a higher rate at which energy is applied to the specimen will reduce the absorbed energy and increase the transition temperature. The size of the specimen also affects the results—smaller energies might be required to break thicker materials. Finally, the configuration of the notch may affect the behavior—a surface crack permits lower absorbed energies than a V-notch. Because we often cannot predict or control all of these conditions, the impact test is best used for comparison and selection of materials rather than as a design criterion.

FATIGUE TEST

6-16   Nature of the Fatigue Test

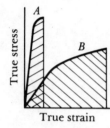

FIG. 6-16
The area contained within the true stress-true strain curve is related to the impact energy. Although material *B* has a lower yield strength, it absorbs a greater energy than material *A*.

In many applications a component is subjected to the repeated application of a stress below the yield strength of the material. This repeated stress may occur as a result of rotation, bending, or even vibration. Even though the stress is below the yield strength, the metal may fail after a large number of applications of the stress. This mode of failure is known as *fatigue*.

A common method to measure the resistance to fatigue is the rotating cantilever beam test (Figure 6-17). One end of a machined, cylindrical specimen is

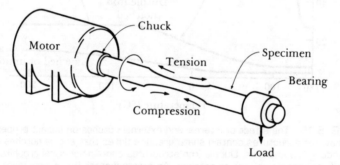

FIG. 6-17   The rotating cantilever beam fatigue tester.

mounted in a motor-driven chuck. A weight is suspended from the other end. The specimen initially has a tensile force acting on the top surface, while the bottom surface is compressed. After the specimen turns 90°, the locations that were originally in tension and compression have no stress acting on them. After a total revolution of 180°, the material that was originally in tension is now in compression. Thus, the stress at any one point goes through a complete cycle from zero stress to maximum tensile stress to zero stress to maximum compressive stress.

After a sufficient number of cycles, the specimen may fail. Generally, a series of specimens are tested at different applied stresses and the stress plotted versus the number of cycles to failure (Figure 6-18).

6-17   Results of the Fatigue Test

The two most important results from a series of fatigue tests are the fatigue life at a particular stress and the endurance limit for the material. The *fatigue life* tells us how long a component survives when a stress σ is repeatedly applied to the material. If we are designing a tool steel part that must undergo 100,000 cycles

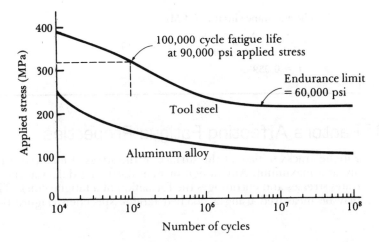

FIG. 6-18   The stress-number of cycles to failure curve for a tool steel and an aluminum alloy.

during its lifetime, then the part must be designed so that the applied stress is lower than 620 MPa (Figure 6-18). .

The *endurance limit* is the stress below which failure by fatigue never occurs. To prevent the tool steel part from failing, we must be sure that the applied stress is below 414 MPa (Figure 6-18).

Some materials, including many aluminum alloys, have no true endurance limit. For these materials, fatigue life is a more critical consideration; applied stresses must be low enough so that fatigue does not occur in the lifetime of the component. Often the *fatigue strength*, or stress below which failure does not occur within 500 million cycles, is specified.

EXAMPLE 6-11

The maximum stress acting at the surface of a cyclindrical bar when a bending force is applied at one end is

$$\sigma = \frac{95.01\ lW}{d^3}$$

where l is the length of the bar, W is the load, and d is the diameter. A 300 kg force is applied to a tool steel bar rotating at 50 cycles $\cdot$ s^{-1}. The bar is 25 mm in diameter and 300 mm long. (a) Estimate the time before the bar fails and (b) calculate the diameter of the shaft that would prevent fatigue failure.

Answer:

(a) $\sigma = \dfrac{95.01\ lW}{d^3} = \dfrac{(95.01)(0.3)(300)}{(0.025)^3} = 547.25$ MPa

From Figure 6-18, the number of cycles to failure is 150,000. The time to failure is

$$t = \frac{150,000}{50 \times 60} = 50 \text{ min}$$

(b) **The endurance limit is 414 MPa.**

$$d^3 = \frac{(95.01)(0.3)(300)}{414 \times 10^6} = 2.07 \times 10^{-5}$$

$$d = 0.039 \text{ m}$$

6-18 Factors Affecting Fatigue Properties

Fatigue cracks initiate at the surface of the stressed material, where the stresses are at a maximum. Any design or manufacturing defect at the surface concentrates stresses and encourages the formation of a fatigue crack. This susceptibility may be measured using a notched fatigue specimen (Figure 6-19). Sometimes

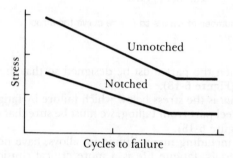

FIG. 6-19 The effect of a notch on the fatigue properties of a metal.

highly polished surfaces are prepared in order to minimize the likelihood of a fatigue failure.

The fatigue resistance is also related to the strength of the material at the surface. In many ferrous, or iron-base, alloys the endurance limit is approximately one-half the tensile strength of the material. This ratio of endurance limit to tensile strength is the *endurance ratio*.

$$\text{Endurance ratio} = \frac{\text{endurance limit}}{\text{tensile strength}} \approx 0.5 \qquad (6\text{-}8)$$

If the tensile strength at the surface of the material increases, the resistance to fatigue also increases.

Similarly, temperature influences the fatigue resistance. As the temperature of the material increases, the strength decreases and consequently both fatigue life and endurance limit decrease.

CREEP TEST

6-19 Nature of the Creep Test

If we apply a stress to a material at a high temperature, the material may stretch and eventually fail, even though the applied stress is less than the yield strength at

TABLE 6-5   Approximate temperatures at which creep
becomes pronounced for selected metals and alloys

Metal	Temperature (°C)
Aluminum alloys	200
Titanium alloys	325
Low-alloy steels	375
High-temperature steels	550
Nickel and cobalt superalloys	650
Refractory metals (tungsten, molybdenum)	1000–1550

that temperature. Plastic deformation at high temperatures is known as *creep*.
Table 6-5 gives the approximate temperatures at which several metals begin to
creep.

To determine the creep characteristics of a material, a constant stress is
applied to a cylindrical specimen placed in a furnace (Figure 6-20). As soon as the
stress is applied, the specimen stretches elastically a small amount ε_0 (Figure 6-21)
depending on the applied stress and the modulus of elasticity of the material at
the high temperature.

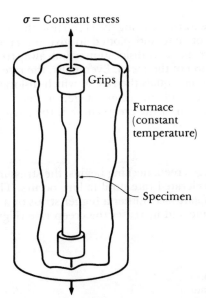

σ = Constant stress

Grips

Furnace
(constant
temperature)

Specimen

FIG. 6-20   A specimen is placed in a furnace at
an elevated temperature under a constant
applied stress in the creep test.

Dislocation climb.   The high temperature permits dislocations in the metal
to *climb*. In climb, atoms move either to or from the dislocation line by diffusion,
causing the dislocation to move in a direction that is perpendicular, not parallel, to
the slip plane (Figure 6-22). The dislocation can now escape from lattice imperfections that block the slip process. The dislocation, after climbing away from the

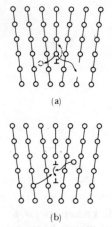

FIG. 6-22
Dislocations can climb away from obstacles when atoms leave the dislocation line to create interstitials or to fill vacancies (a) or when atoms are attached to the dislocation line by creating vacancies or eliminating interstitials (b).

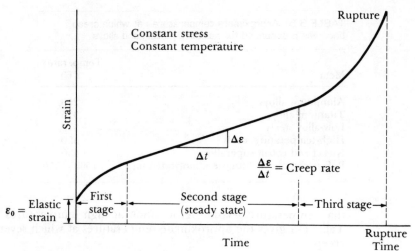

FIG. 6-21   A typical creep curve showing the strain produced as a function of time for a constant stress and temperature.

imperfection, continues to slip and causes additional deformation of the specimen even at low applied stresses.

Creep rate and rupture time.   During the creep test, the strain or elongation is measured as a function of time and plotted to give the creep curve (Figure 6-21). In the first stage of creep, many dislocations climb away from obstacles, slip, and contribute to deformation of the metal. Eventually, the rate at which dislocations climb away from obstacles equals the rate at which dislocations are blocked by other imperfections. This leads to second-stage, or steady-state, creep. The slope of the steady-state portion of the creep curve is the *creep rate*.

$$\text{Creep rate} = \frac{\Delta\text{strain}}{\Delta\text{time}} \qquad (6\text{-}9)$$

Eventually, during third-stage creep, necking begins, the stress increases, and the specimen deforms at an accelerated rate until failure occurs. The time required for failure to occur is the *rupture time*. Either a higher stress or a higher temperature reduces the rupture time and increases the creep rate (Figure 6-23).

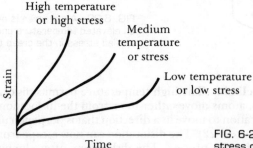

FIG. 6-23   The effect of temperature or applied stress on the creep curve.

EXAMPLE 6-12

The results of a creep test are given in Table 6-6. Calculate the creep rate in strain/hour and in s⁻¹.

TABLE 6-6 Data from a creep test for Example 6-12

Strain	Time (h)
0.003	0
0.006	250
0.009	1000
0.012	2250
0.015	3500
0.018	4750
0.021	6000
0.024	7100
0.027	7500
0.030 (fracture)	7750

Answer:

The data in Table 6-6 are plotted in Figure 6-24. From the slope of the steady-state portion of the curve

$$\text{Creep rate} = \frac{\Delta \varepsilon}{\Delta t} = \frac{0.021 - 0.009}{6000 - 1000} = \frac{0.012}{5000} = 2.4 \times 10^{-6} \, h^{-1}$$
$$= 6.7 \times 10^{-10} \, s^{-1}$$

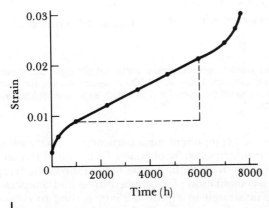

FIG. 6-24 Graph of the data in Table 6-6 to produce a creep curve. The slope of the steady-state portion of the graph is the answer to Example 6-12.

6-20 Use of Creep Data

Four ways to present the results from a series of creep tests are shown in Figure 6-25. The *stress-rupture curve* shown in Figure 6-25(a) permits us to estimate the

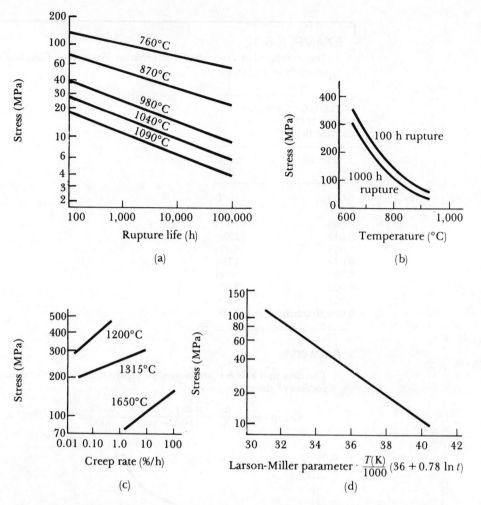

FIG. 6-25   Results from a series of creep tests. (a) Stress-rupture curves for an iron-chromium-nickel alloy, (b) 100-h and 1000-h rupture curves for a nickel heat-resistant alloy, (c) minimum creep rate curves for a tantalum alloy, and (d) Larson-Miller parameter for ductile cast iron.

expected lifetime of a component for a particular combination of stress and temperature. Similar information is obtained from the 100-h or 1000-h rupture curves displayed in Figure 6-25(b). Figure 6-25(c) shows the creep rate that would be obtained for a combination of an applied stress and temperature. The *Larson-Miller parameter*, illustrated in Figure 6-25(d), is used to consolidate the stress-temperature-rupture time relationship into a single curve.

EXAMPLE 6-13

Using the Larson-Miller parameter for ductile cast iron, as shown in Figure 6-25(d), determine the time required before the metal fails at an applied stress of 40 MPa and temperatures of 400°C and 600°C.

Answer:

The Larson-Miller parameter for 40 MPa is 34.3.

(a) At 400°C

$$34.3 = \frac{(400 + 273 \text{ K})}{1000}(36 + 0.78 \ln t)$$

$$0.78 \ln t = (34.3)\left(\frac{1000}{673}\right) - 36 = 14.97$$

$$t = 2.2 \times 10^8 \text{ h} = 25,100 \text{ years}$$

(b) At 600°C

$$34.3 = \frac{(600 + 273 \text{ K})}{1000}(36 + 0.78 \ln t)$$

$$0.78 \ln t = (34.3)\left(\frac{1000}{873}\right) - 36 = 3.29$$

$$t = 67.9 \text{ h} = 2.8 \text{ days}$$

HARDNESS TEST

6-21 Nature of the Hardness Test

The *hardness test* measures the resistance to penetration of the surface of a material by a hard object. A variety of hardness tests have been devised, but the most commonly used are the Rockwell test and the Brinell test (Figure 6-26).

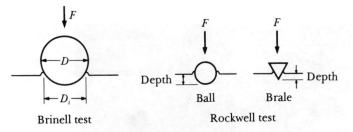

Brinell test

Rockwell test

FIG. 6-26   The Brinell and Rockwell hardness tests.

In the *Brinell hardness test* a hard steel sphere, usually 10 mm in diameter, is forced into the surface of the material. The diameter of the impression left on the surface is measured and the Brinell hardness number (*BHN*) is calculated from the following equation.

$$BHN = \frac{F}{(\pi/2)D(D - \sqrt{D^2 - D_i^2})} \tag{6-10}$$

where F is the applied load in kilograms, D is the diameter of the indentor in millimeters, and D_i is the diameter of the impression in millimeters.

The *Rockwell hardness test* uses either a small diameter steel ball for soft materials or a diamond cone, or Brale, for harder materials. The depth of penetration of the indentor is automatically measured by the testing machine and converted to a Rockwell hardness number. Several variations of the Rockwell test are used, as shown in Table 6-7.

TABLE 6-7   Comparison of typical hardness tests

Test	Indentor	Load	Application
Brinell	10-mm ball	3000 kg	Cast iron and steel
Brinell	10-mm ball	500 kg	Nonferrous alloys
Rockwell A	Brale	60 kg	Very hard materials
Rockwell B	$\frac{1}{16}$-in. ball	100 kg	Brass, low-strength steel
Rockwell C	Brale	150 kg	High-strength steel
Rockwell D	Brale	100 kg	High-strength steel
Rockwell E	$\frac{1}{8}$-in. ball	100 kg	Very soft materials
Rockwell F	$\frac{1}{16}$-in. ball	60 kg	Aluminum, soft materials
Vickers	Diamond pyramid	10 kg	Hard materials
Knoop	Diamond pyramid	500 g	All materials

The Vickers and Knoop tests are *microhardness* tests; they form such small indentations that a microscope is required to obtain the measurement (Figure 6-27).

The hardness numbers are used primarily as a basis for comparison of materials, specifications for manufacturing and heat treatment, quality control, and

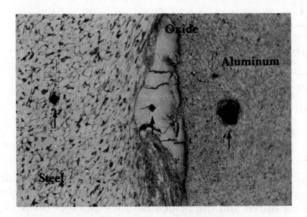

FIG. 6-27 Microhardness impressions in an explosive bond joining aluminum to steel. The small size of the impression indicates that the inclusion trapped at the interface is harder than either aluminum or steel.

correlation with other properties and behavior of materials. For example, Brinell hardness is closely related to the tensile strength of steel by the relationship

Tensile strength $= 3.45 \times 10^6\ BHN$ (6-11)

A Brinell hardness number can be obtained in just a few minutes with virtually no preparation of the specimen and without destroying the component, yet provides a close approximation for the tensile strength.

Hardness correlates well with wear resistance. A material used to crush or grind ore should be very hard to assure that the material is not eroded or abraded by the hard feed materials. Similarly, gear teeth in a transmission or drive system of a vehicle should be hard so that the teeth do not wear out.

EXAMPLE 6-14

A Brinell hardness test is performed on a steel using a 10-mm indentor with a load of 3000 kg. A 3.2-mm impression is measured on the surface of the steel. Calculate the BHN, the tensile strength, and the endurance limit of the steel.

Answer:

$$BHN = \frac{3000}{(\pi/2)(10)(10 - \sqrt{10^2 - 3.2^2})} = 363\ \text{kg/mm}^2$$

Tensile strength $= 500\ BHN = (3.45 \times 10^6)(363) = 1.25$ GPa

Endurance limit $= 0.5$ tensile strength $= (0.5)(1.25 \times 10^9)$

$$= 626\ \text{MPa}$$

SUMMARY

The mechanical behavior of materials is described by their mechanical properties, which are simply the results of idealized, simple tests. These tests are designed to represent different types of loading conditions. The tensile test describes the resistance of the material to a slowly applied stress; the results define the yield strength, ductility, and stiffness of the material. The fatigue test permits us to understand how a material performs when a cyclical stress is applied, the impact test indicates the shock resistance of the material, and the creep test provides information on the load-carrying ability of the material at high temperatures. Finally, the hardness test, besides providing a measure of the wear and abrasion resistance of the material, can be correlated to a number of other mechanical properties.

Although many other tests, including very specialized ones, are used to characterize mechanical behavior, the properties obtained from these five tests are most commonly found in handbooks. However, we should always recall that the properties listed in handbooks are average values from idealized tests and must be used with care.

GLOSSARY

Climb   Movement of a dislocation perpendicular to its slip plane by diffusion of atoms to or from the dislocation line.

Creep rate   The rate at which a material continues to stretch as a function of time when a stress is applied at a high temperature.

Creep test   Measures the resistance of a material to deformation and failure when subjected to a static load below the yield strength at an elevated temperature.

Double yield point   In certain materials, dislocations are initially pinned by lattice imperfections. A high stress is required to start plastic deformation, giving an upper yield point, while subsequent deformation occurs at a lower stress, giving a lower yield point.

Ductility   The ability of a material to be permanently deformed without breaking when a force is applied.

Elastic deformation   Deformation of the material that is recovered when the applied load is removed.

% Elongation   The total percent increase in the length of a specimen during a tensile test.

Endurance limit   The stress below which a material will not fail in a fatigue test.

Endurance ratio   The endurance limit divided by the tensile strength of the material. The ratio is about 0.5 for many ferrous metals.

Engineering strain   The amount that a material deforms per unit length in a tensile test.

Engineering stress   The applied load, or force, divided by the original cross-sectional area of the material.

Fatigue life   The number of cycles at a particular stress before a material fails by fatigue.

Fatigue strength   The stress required to cause failure by fatigue in 500 million cycles.

Fatigue test   Measures the resistance of a material to failure when a stress below the yield strength is repeatedly applied.

Hardness test   Measures the resistance of a material to penetration by a sharp object. Common tests include the Brinell test, Rockwell test, Knoop test, and Vickers test.

Hooke's law   The relationship between stress and strain in the elastic portion of the stress-strain curve.

Impact energy   The energy required to fracture a standard specimen when the load is suddenly applied.

Impact test   Measures the ability of a material to absorb a sudden application of a load without breaking. The Charpy test is a commonly used impact test.

Larson-Miller parameter   One of several parameters used to relate the stress, temperature, and rupture time in creep.

Modulus of elasticity   Young's modulus, or the slope of the stress-strain curve in the elastic region.

Necking   Local deformation of a tensile specimen. Necking begins at the tensile point.

Notch sensitivity   Measures the effect of a notch on the impact energy.

Offset yield strength   A yield strength obtained graphically that describes the stress that gives no more than a specified amount of plastic deformation.

Plastic deformation   Permanent deformation of the material when a load is applied, then removed.

% Reduction in area   The total percent decrease in the cross-sectional area of a specimen during the tensile test.

Rupture time   The time required for a specimen to fail by creep at a particular temperature and stress.

Stiffness   A qualitative measure of the elastic deformation produced in a material. A stiff material has a high modulus of elasticity.

Stress-rupture curve   A method of reporting the results of a series of creep tests by plotting the applied stress versus the rupture time.

Tensile strength   The stress that corresponds to the maximum load in a tensile test.

Tensile test   Measures the response of a material to a slowly applied uniaxial force. The yield strength, tensile strength, modulus of elasticity, and ductility are obtained.

Toughness   A qualitative measure of the impact properties of a material. A material that resists failure by impact is said to be tough.

Transition temperature The temperature below which a material behaves in a brittle manner in an impact test.

True strain The actual strain produced when a load is applied to a material.

True stress The load divided by the actual area at that load in a tensile test.

Yield strength The stress applied to a material that just causes permanent plastic deformation.

PRACTICE PROBLEMS

1 A 227.3 kg force is applied to a 2.54 mm diameter copper wire having a yield strength of 137.9 MPa. Will the wire plastically deform?

2 A force of 70,000 N is applied to a 10-mm diameter steel rod having a yield strength of 550 MPa. Will the steel rod plastically deform?

3 A 2.5 mm diameter copper wire with a modulus of elasticity of 117 GPa is 460 m long. Calculate the length when a 91 kg load acts on the wire.

4 A 3-mm diameter beryllium wire with a modulus of elasticity of 250 GPa is 2500 cm long. Calculate the length of the wire when a force of 20,000 N acts on the wire.

5 The yield strength of an aluminum alloy is 41.4 MPa. Its modulus of elasticity is 69 GPa. (a) Determine the maximum load that a 25 mm × 3 mm plate will withstand without permanently deforming. (b) How much does each inch of the specimen stretch when this load is applied?

6 The yield strength of a magnesium alloy is 180 MPa and the modulus of elasticity is 45 GPa. (a) Calculate the maximum load in newtons that a 10 mm × 2 mm strip will withstand without permanently deforming. (b) How much does each millimeter of the specimen stretch when this load is applied?

7 We wish to plastically deform a 12.7 mm × 76.2 mm bar of magnesium having a yield strength of 124 MPa. The forming press can exert a force of 270 kN. Is the press large enough to perform the deformation?

8 We wish to reduce a titanium plate to 6 mm in thickness. The modulus of elasticity of the titanium is 110 GPa and the yield strength is 621 MPa. In order to compensate for elastic deformation, to what thickness should we initially deform the plate?

9 A square 25.4 mm × 25.4 mm tensile bar with a 50.8 mm gage length is pulled to failure. The final distance between the gage marks is 69.8 mm and the final dimensions at the fracture are 20.8 mm × 20.8 mm. Calculate the % elongation and % reduction in area.

10 A chain link is manufactured from a 25 mm thick steel rod having a yield strength of 828 MPa. Assuming that each half of the link supports half the total load, calculate the maximum load that the chain will support without permanently deforming.

11 The following data are obtained from a standard 12.8 mm diameter tensile specimen of nodular cast iron.

Load (kN)	Gage Length (mm)
0	50.80
8.9	50.82
17.8	50.84
26.8	50.86
29.2	50.89
35.7	50.94
44.6	51.12
51.3	51.41
53.1 (max)	60.96
52.6 (fracture)	65.02

The diameter at maximum load is 12.2 mm, the final diameter at fracture is 11.3 mm, and the final length between the gage marks is 57.9 mm. Plot the engineering stress-strain curve and calculate (a) the modulus of elasticity, (b) the 0.2% offset yield strength, (c) the tensile strength, (d) the % elongation, and (e) the % reduction in area.

12 A 12.8 mm diameter tensile bar machined from a titanium alloy is inscribed with gage marks 50.8 mm apart. The results of the tensile test are shown below.

Load	Gage Length
0	50.8
28.5	50.9
57.1	51.0
85.6	51.1
114.2	51.2
121.3	51.3
125.7	51.8
129.3	52.8

Load (kN)	Gage Length (mm)
131.1	53.8
129.3	54.9
121.3	55.9
110.6 (fracture)	56.5

Plot the stress-strain curve and calculate (a) the modulus of elasticity, (b) the 0.1% offset yield strength, and (c) the tensile strength.

13　A 6.4 mm × 25.4 mm plate tensile specimen is machined from a temperature-resistant alloy steel. The specimen has a 50.8 mm gage length. The results of the tensile test are shown in the table.

Load	Gage Length
0	50.8
72.4	50.9
144.9	51.0
189.5	51.1
211.8	51.2
240.8	51.6
255.3	52.3
256.4	53.8
231.9	55.9
189.5	57.9
165.0	58.9

From the stress-strain curve, calculate (a) the modulus of elasticity, (b) the 0.2% offset yield strength, and (c) the tensile strength.

14　A 20-mm diameter magnesium alloy tensile specimen has a 50-mm gage length. The following results are obtained from a tensile test.

Load (N)	Gage Length (mm)
0	50.00
14,100	50.05
28,200	50.10
42,100	50.15
54,700	50.20
65,600	50.25
74,100	50.30
84,560	50.40
91,700	50.50
95,300 (fracture)	50.57

From the stress-strain curve, calculate (a) the tensile strength in megapascals, (b) the 0.1% offset yield strength in megapascals, and (c) the modulus of elasticity in gigapascals.

15　A 10 mm × 10 mm square tensile bar obtained from a nickel superalloy has a 40-mm gage length. The results of the tensile test are as follows.

Load (N)	Gage Length (mm)
0	40.00
43,100	40.10
86,200	40.20
102,000	40.40
104,800	40.80
109,600	41.60
113,800	42.40
121,300	44.00
126,900	46.00
127,600	48.00
113,800 (fracture)	50.20

From the stress-strain curve, calculate (a) the tensile strength in megapascals, (b) the 0.2% offset yield strength in megapascals, (c) the modulus of elasticity in gigapascals and (d) the approximate % elongation.

16　Using the data in Problem 11, plot the approximate true stress-strain curve and compare to the engineering stress-strain curve by calculating the true stress at the tensile and breaking loads.

17　A 25 mm diameter bar 0.6 mm long is subjected to a load of 90 kN. Its length must not increase more than 1.3 mm under load. Will an aluminum bar perform adequately? Will an iron bar perform adequately?

18　Figure 6-28 shows the initial portion of the stress-strain curves for a titanium alloy at three temperatures. (a) Calculate the modulus of elasticity at each temperature. (b) Calculate the 0.2% offset yield strength at each temperature. (c) Suppose a 254 mm long specimen with a 25.4 mm diameter is subjected to a load of 175 kN. Calculate the final length of the specimen at each temperature (assuming that dimensional changes are due only to elastic deformation).

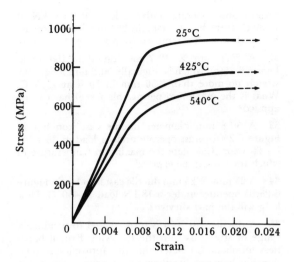

FIG. 6-28   Stress-strain curves for a titanium alloy at three temperatures. (See Problem 6-18.)

19   A tool steel bar (Figure 6-18) which is 12.7 mm in diameter and 152 mm long must survive 1 million cycles without failing in a rotating beam test. Calculate the maximum load that can be applied.

20   A tool steel bar (Figure 6-18) which is 457 mm long is repeatedly loaded with a 44.6 kN force in a rotating beam test. Calculate the minimum diameter of a bar that will never fail by fatigue.

21   A 6.35 mm diameter rod made from a high-strength aluminum alloy (Figure 6-18) is subjected to a repeated application of a 7.58 kN load applied along

the axis of the rod (not a rotating beam test). Estimate the fatigue life of the alloy.

22   A high-strength aluminum alloy (Figure 6-18) is subjected to a repeated application of a 134 kN load applied along the axis of the specimen (not a rotating beam test). Calculate the minimum diameter of a bar that will never fail by fatigue.

23   The relationship between the tensile strength and endurance limit of gray cast iron is shown in the table. Plot the data and determine the endurance ratio in the notched and unnotched conditions. Discuss the notch sensitivity of gray cast irons.

Tensile Strength (MPa)	Endurance Limit (MPa)	
	Notched	Unnotched
138	63	63
172	79	82
206	94	103
232	108	152
294	131	161

Data from *Gray and Ductile Iron Castings Handbook*, Ed. C. Walton, Gray and Ductile Iron Founders Society, 1971.

24   The following data are obtained from a series of impact tests for several low-carbon steels. (a) Estimate the transition temperature for each steel. (b) Which steel(s) should be selected for service under Arctic conditions? (c) What is the effect of manganese on the toughness of steel?

Test Temperature (°C)	Impact Energy (J)			
	0% Mn	0.5% Mn	1.0% Mn	2.0% Mn
−50	10	10	10	10
−25	15	15	20	290
0	18	20	40	290
25	20	35	100–275	290
50	25	60	275	290
75	30	230	275	290
100	45	230	275	290
125	180	230	275	290

25 The following data are obtained from a series of notched impact tests for three ductile cast irons containing different silicon contents.

Test Temperature (°C)	Impact Energy (J) 1% Si	2% Si	3% Si
−200	7	6	5
−150	8	6	5
−100	15	6	5
−50	43	10	5
0	47	37	8
50	47	38	27
100	47	38	35
150	47	38	35

(a) Estimate the transition temperature for each cast iron. (b) How does silicon affect the transition temperature? (c) Estimate the impact energy in both joules for each cast iron in the ductile range. (d) How does silicon affect the absorbed energy in ductile iron?

26 A 12.7 mm diameter bar of the iron-chromium-nickel alloy in Figure 6-25(a) is operated under a load of 22.3 kN. Estimate (a) the time required before the bar fails at 870°C and (b) the maximum temperature to which the alloy can be exposed if it is to survive five years with a load of 22.3 kN.

27 The iron-chromium-nickel alloy in Figure 6-25(a) is to survive for five years at 870°C under a 22.3 kN load. Estimate the minimum diameter of the rod.

28 A 2 cm diameter bar of the iron-chromium-nickel alloy in Figure 6-25(a) must withstand a force of 6000 N. Estimate (a) the time required for the bar to fail at 980°C and (b) the maximum temperature if the bar is to survive 10 years before failing.

29 The nickel heat-resistant alloy in Figure 6-25(b) must survive 1000 h at 700°C. Calculate the minimum cross-sectional area that the alloy part can have if a load of 3.57 kN is applied.

30 The nickel heat-resistant alloy in Figure 6-25(b) has a cross-sectional area of 516 mm². Calculate the maximum load that can be applied if the alloy is to survive 1000 h at 700°C.

31 A 25.4 mm diameter tantalum alloy in Figure

6-25(c) must operate under a load of 44.6 kN at 1650°C. If the bar is originally 254 mm long, estimate its length when it is removed from service after 100 h.

32 A 25.4 mm × 12.7 mm tantalum fixture in Figure 6-25(c), which is originally 50.8 mm long, may stretch no more than 2.50 mm in 10 h at 1315°C. What is the maximum permissible force that can be applied?

33 A 50.8 mm diameter ductile cast iron bar in Figure 6-25(d) must operate unde a load of 35.7 kN for 20 years. Calculate the maximum temperature to which the bar can be exposed.

34 A 25 mm × 25 mm ductile cast iron bar in Figure 6-25(d) operates under a 18 kN load at 450°C. How long will the part survive?

35 A ductile cast iron is heat treated to produce a range of strengths with the following Brinell hardness numbers. (a) Determine the approximate relationship between hardness and tensile strength for the alloy. (b) How does this relationship compare to that for steel?

Tensile Strength (MPa)	Brinell Hardness Number
676	200
828	250
966	300
1103	350
1310	400

36 A steel has a Brinell hardness number of 250. Estimate its tensile strength and endurance limit.

37 A steel has a Brinell hardness number of 400. Estimate the tensile strength in megapascals and pounds per square inch.

38 A steel has a tensile strength of 1931 MPa. Estimate the Brinell hardness number.

39 A Brinell hardness test using a 500-kg load with a 10-mm diameter indentor produces a 5.5-mm impression on aluminum. Calculate the Brinell hardness number.

40 A Brinell hardness test using a 3000-kg load with a 10-mm diameter indentor produces a 4.1-mm impression on steel. Calculate the Brinell hardness number and estimate the tensile strength and endurance limit of the steel.

CHAPTER 7

Solidification and Grain Size Strengthening

7-1 Introduction

In almost all metals and alloys, as well as in some ceramics and polymers, the material at one point in the processing is a liquid. The liquid then solidifies as it cools below the freezing temperature. The material may be used in the as-solidified condition or may be further processed by mechanical working or heat treatment. The structures produced during the solidification process affect the mechanical properties and influence the type of further processing needed to achieve the required properties. In particular, the grain size and shape may be controlled by solidification.

In this chapter we will introduce the fundamental principles of solidification, concentrating on the behavior of pure materials. In subsequent chapters we will see how solidification differs in alloys and multiple-phase materials.

7-2 Nucleation

During solidification, the atomic arrangement changes from at best a short-range order to a long-range order, or crystal structure. Solidification requires two steps—nucleation and growth. *Nucleation* occurs when a small piece of solid forms from the liquid. The solid must achieve a certain minimum critical size before it is stable. *Growth* of the solid occurs as atoms from the liquid are attached to the tiny solid until no liquid remains.

We expect a material to solidify when the liquid cools to just below the freezing temperature because the energy associated with the crystalline structure of the solid is then less than the energy of the liquid. As the temperature falls further below the freezing temperature, the energy difference becomes larger, making the solid even more stable (Figure 7-1). We might refer to this energy difference as the *volume free energy* ΔF_v.

However, in order for the solid to form, an interface must be created separating the solid from the liquid (Figure 7-2). A *surface free energy* σ is associated with this interface; the larger the surface, the greater the increase in surface energy. When the liquid cools to the freezing temperature, atoms in the liquid cluster together to produce a small region that resembles the solid material. This small solid particle is called an *embryo*. The total change in free energy produced when

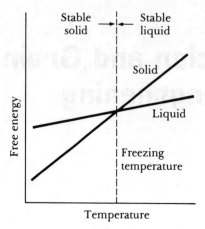

FIG. 7-1 The volume free energy versus temperature for a pure metal. Below the freezing temperature, the solid has a lower free energy and is stable.

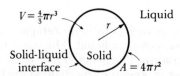

FIG. 7-2 An interface is created when a solid forms from the liquid.

the embryo forms is the sum of the decrease in volume free energy and the increase in surface free energy.

$$\Delta F = \frac{4}{3} \pi r^3 \Delta F_v + 4\pi r^2 \sigma$$

(7-1)

where $\frac{4}{3}\pi r^3$ is the volume of a spherical embryo of radius r, $4\pi r^2$ is the surface area of a spherical embryo, σ is the surface free energy, and ΔF_v is the volume free energy, which is a negative change.

The total change in the free energy depends on the size of the embryo (Figure 7-3). If the embryo is very small, further growth of the embryo would cause the free energy to increase. Instead of growing, the embryo remelts and causes the free energy to decrease. Thus, the metal remains liquid. Since the liquid is present below the equilibrium freezing temperature, the liquid is undercooled. The *undercooling* is the equilibrium freezing temperature minus the actual temperature of the liquid. Nucleation has not occurred and growth cannot begin, even though the temperature is below the equilibrium freezing temperature!

If the embryo is large, the total energy decreases when the size of the embryo increases. The solid that now forms is stable, nucleation has occurred, and growth of the solid particle, which is now called a *nucleus*, begins.

Nucleation only occurs when enough atoms spontaneously cluster together to produce a solid with a radius greater than the *critical radius r**, corresponding to the maximum on the total free energy curve.

Homogeneous nucleation. As the temperature of the liquid cools further below the equilibrium freezing temperature, there is a greater probability that atoms will cluster together to form an embryo larger than the critical radius. In

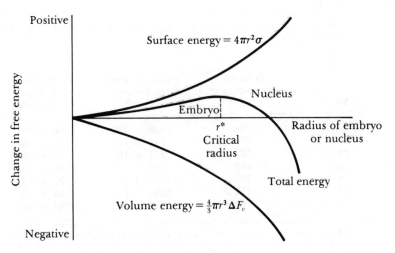

FIG. 7-3 The total free energy of the solid-liquid system changes with the size of the solid. The solid is an embryo if its radius is less than the critical radius and is a nucleus if its radius is greater than the critical radius.

addition, at larger undercoolings there is a larger volume free energy difference between the liquid and the solid; this reduces the critical size of the nucleus. *Homogeneous nucleation* occurs when the undercooling becomes large enough to permit the embryo to exceed the critical size.

We can estimate the size of the critical nucleus if we differentiate the total free energy equation. The differential with respect to r is zero when $r = r^*$, since the free energy curve is then at a maximum.

$$\frac{d}{dr}(\Delta F) = \frac{d}{dr}\left(\frac{4}{3}\pi r^3 \Delta F_v + 4\pi r^2 \sigma\right) = 0$$

$$4\pi r^{*2}\Delta F_v + 8\pi r^* \sigma = 0$$

$$r^* = \frac{-2\sigma}{\Delta F_v} \tag{7-2}$$

The volume free energy is given by the expression

$$\Delta F_v = \frac{-\Delta H_f \Delta T}{T_m} \tag{7-3}$$

where ΔH_f is the latent heat of fusion of the metal, T_m is the equilibrium freezing temperature in K, and $\Delta T = T_m - T$ is the undercooling when the liquid temperature is T. The *latent heat of fusion* represents the heat that is given up during the liquid-solid transformation. By combining Equations (7-2) and (7-3)

$$r^* = \frac{2\sigma T_m}{\Delta H_f \Delta T} \tag{7-4}$$

As the undercooling increases, the critical radius required for nucleation decreases. Table 7-1 presents values for σ and ΔH_f for selected metals. As an approximation, homogeneous nucleation occurs when

$$\Delta T = 0.2 T_m(K) \tag{7-5}$$

TABLE 7-1   Values for freezing temperature, latent heat of fusion, surface energy, and maximum undercooling for selected metals

Metal	Freezing Temperature (°C)	Latent Heat of Fusion ($J\ m^{-3}$)	Surface Energy ($J\ m^{-2}$)	Maximum Undercooling Observed (°C)
Ga	30	488×10^6	56×10^{-3}	76
Bi	271	543×10^6	54×10^{-3}	90
Pb	327	237×10^6	33×10^{-3}	80
Ag	962	965×10^6	126×10^{-3}	250
Cu	1085	1628×10^6	177×10^{-3}	236
Ni	1453	2756×10^6	255×10^{-3}	480
Fe	1538	1737×10^6	204×10^{-3}	420

Adapted from B. Chalmers, *Principles of Solidification*, John Wiley & Sons, 1964.

EXAMPLE 7-1

The freezing temperature of pure copper is 1085°C. Estimate the undercooling required for homogeneous nucleation.

Answer:

$$\Delta T = 0.2 T_m = (0.2)(1085 + 273) = 272°C$$

Undercoolings of this magnitude are never observed in the normal processing of molten copper.

EXAMPLE 7-2

Calculate the size of the critical radius and the number of atoms in the critical nucleus when solid copper forms by homogeneous nucleation.

Answer:

$$\Delta T = 0.2 T_m = 272°C, \quad T_m = 1358\ K$$
$$\Delta H_f = 1628 \times 10^6\ J\ m^{-3}$$
$$\sigma = 177 \times 10^{-3}\ J\ m^{-2}$$
$$r^* = \frac{2\sigma T_m}{\Delta H_f \Delta T} = \frac{(2)(177 \times 10^{-3})(1358)}{(1628 \times 10^6)(272)} = 10.85 \times 10^{-10}\ m$$

The lattice parameter for FCC copper is $a_0 = 3.615$ Å ($\times 10^{-10}$ m)

$$V_{unit\ cell} = (a_0)^3 = (3.615 \times 10^{-10})^3 = 47.24 \times 10^{-30}\ m^3$$
$$V_{r^*} = \frac{4}{3}\pi r^3 = \left(\frac{4}{3}\pi\right)(10.85 \times 10^{-10}\ m^3)^3 = 5350 \times 10^{-30}\ m^3$$

The number of unit cells in the critical nucleus is

$$\frac{5350 \times 10^{-30}}{47.24 \times 10^{-30}} = 113 \text{ unit cells}$$

Since there are four atoms in each unit cell of FCC metals, the number of atoms in the critical nucleus must be

$$(4 \text{ atoms/cell})(113 \text{ cells/nucleus}) = 452 \text{ atoms/nucleus}$$

We do not expect this many atoms to spontaneously cluster together to form a nucleus.

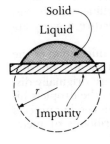

FIG. 7-4
A solid forming on an impurity can assume the critical radius with a smaller increase in the surface energy. Thus, heterogeneous nucleation can occur with relatively low undercoolings.

Heterogeneous nucleation. Except in unusual laboratory experiments, homogeneous nucleation never occurs in liquid metals. Instead impurities in contact with the liquid, either suspended in the liquid or on the walls of the container that holds the liquid, provide a surface on which the solid can form (Figure 7-4). Now, a radius of curvature greater than the critical radius is achieved with very little total surface between the solid and liquid. Only a few atoms must cluster together to produce a solid particle that has the required radius of curvature. Much less undercooling is required to achieve the critical size, so nucleation occurs more readily. Nucleation on impurity surfaces is known as *heterogeneous nucleation*. All engineering metals and alloys nucleate heterogeneously during solidification.

EXAMPLE 7-3

Calculate the number of copper atoms that must cluster together to produce the heterogeneous nucleus shown in Figure 7-5.

Answer:

From solid geometry, we find that the volume of the spherical cap is

$$V_{cap} = \frac{\pi h^2}{3}(3r - h)$$

In the figure, the height of the cap is specified as $0.2r$.

$$V_{cap} = \frac{\pi(0.2\ r)^2}{3}(3r - 0.2r)$$

$$= 0.117r^3$$

If $r = r^*$, from Example 7-2, then

$$V_{cap} = (0.117)(10.85 \times 10^{-10})^3 = 149 \times 10^{-30} \text{ m}^3$$

$$\frac{V_{cap}}{V_{unit\ cell}} = \frac{149 \times 10^{-30}}{47.24 \times 10^{-30}} = 3.15 \text{ unit cells}$$

Therefore, the number of atoms in the critical nucleus is

$$(4 \text{ atoms/cell})(3.15 \text{ cells/nucleus}) = 12 \text{ atoms/nucleus}$$

We expect that 12 atoms can spontaneously cluster together to produce heterogeneous nucleation even at small undercoolings.

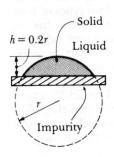

FIG. 7-5
A spherical cap forms on an impurity for heterogeneous nucleation. (See Example 7-3.)

Grain size strengthening by nucleation. Sometimes we may intentionally introduce impurity particles into the liquid. Such practices in metals are called *grain refinement* or *inoculation*. For example, a combination of 0.02% to 0.05% titanium and 0.01% to 0.03% boron is added to many liquid aluminum alloys. Solid titanium boride particles form and serve as effective sites for heterogeneous nucleation. The grain refining or inoculation procedure produces a large number of grains, each grain beginning to grow from one nucleus. The greater grain boundary surface area more effectively blocks slip, or movement of dislocations, and provides *grain size strengthening*.

Glasses. In extreme cases of very rapid cooling, nucleation of the crystalline solid never occurs. Instead an unstable amorphous, or noncrystalline, solid forms. A short-range order of the atoms in this solid gives the structure a *glassy* appearance.

Cooling rates of 10^6 °C/s or faster are required to suppress nucleation of the crystal structure in metals. This rapid cooling rate is obtained by directing the molten metal onto a chilled copper surface. These materials were originally discovered by firing small droplets of liquid at a cold surface; when the droplets hit the surface, they spread out as a thin film and cooled rapidly. This cooling process was termed "splat" cooling. More recently, continuous thin ribbons of metal glasses, about 0.04 mm in thickness, have been produced.

Metal glasses include complex iron-nickel-boron-phosphorus alloys. The metal glasses combine high strength and good ductility with some excellent physical properties, including ferromagnetic behavior, that warrant their continued development. '

In many ceramic and polymer materials, nucleation of the solid crystalline structure is prevented at normal or even slow cooling rates. The ability to produce ceramic and polymer glasses by relatively simple and economical manufacturing processes gives us the transparent materials we need for so many uses.

7-3 Growth

Once solid nuclei have formed, growth occurs as atoms are attached to the solid surface. In pure metals, the nature of the growth of the solid during solidification depends on how heat is removed from the solid-liquid system. Two types of heat must be removed—the specific heat of the liquid and the latent heat of fusion. The *specific heat* is the heat required to change the temperature of a unit weight of the material one degree. The specific heat must be removed first, either by radiation into the surrounding atmosphere or by conduction into the surrounding mold, until the liquid cools to the freezing temperature. The latent heat of fusion, which represents the energy that is evolved as the disordered liquid structure transforms to a more stable crystal structure, must be removed from the solid-liquid interface before solidification is completed. The manner in which we remove the latent heat of fusion determines the growth mechanism and final structure.

Planar growth. Let's suppose that a well-inoculated liquid cools slowly, under equilibrium conditions. The temperature of the liquid metal is greater than the freezing temperature and the temperature of the solid is at or below the freezing temperature. The latent heat of fusion must be removed by conduction from the solid-liquid interface through the solid to the surroundings for solidification to continue. Any small protuberance that begins to grow on the interface is

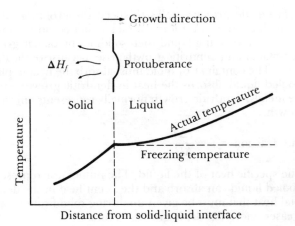

FIG. 7-6 When the temperature of the liquid is above the freezing temperature, a protuberance on the solid-liquid interface will remelt, leading to maintenance of a planar interface. Latent heat is removed from the interface through the solid.

surrounded by liquid metal above the freezing temperature (Figure 7-6). The growth of the protuberance then stops until the remainder of the interface catches up. This growth mechanism, known as *planar growth*, occurs by the movement of a smooth solid-liquid interface into the liquid.

Dendritic growth. When nucleation is poor, the liquid undercools to a temperature below the freezing temperature before the solid forms (Figure 7-7). Under these conditions, a small solid protuberance called a *dendrite*, which forms at the interface, is encouraged to grow. As the solid dendrite grows, the latent heat of fusion is conducted into the undercooled liquid, raising the temperature of the liquid towards the freezing temperature. Secondary and tertiary dendrite arms

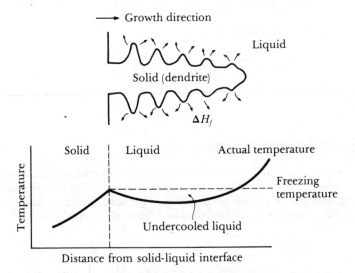

FIG. 7-7 If the liquid is undercooled, a protuberance on the solid-liquid interface can rapidly grow as a dendrite. The latent heat of fusion is removed by raising the temperature of the liquid back to the freezing temperature.

can also form on the primary stalks to speed the evolution of the latent heat. Dendritic growth continues until the undercooled liquid warms to the freezing temperature. Any remaining liquid then solidifies by planar growth. The difference between planar and dendritic growth arises because of the different sinks for the latent heat. The container or mold must absorb the heat in planar growth, but the undercooled liquid absorbs the heat in dendritic growth.

In pure metals, dendritic growth normally represents only a small fraction of the total growth.

$$\text{Dendrite fraction} = f = \frac{c\Delta T}{\Delta H_f} \tag{7-6}$$

where c is the specific heat of the liquid. The numerator represents the heat that the undercooled liquid can absorb and the latent heat in the denominator represents the total heat that must be given up during solidification. As the undercooling ΔT increases, more dendritic growth occurs.

EXAMPLE 7-4

Calculate the fraction of growth that occurs dendritically in copper which (a) nucleates homogeneously and (b) nucleates heterogeneously with 10°C undercooling.

Answer:

The latent heat of fusion for copper is $1628 \times 10^6 \, \text{J m}^{-3}$, the specific heat is $4.4 \times 10^6 \, \text{J} \, (\text{m}^3 \cdot \text{K})^{-1}$, and from Example 7-1, the undercooling for homogeneous nucleation is 272°C.

(a) For homogeneous nucleation

$$f = \frac{c\Delta T}{\Delta H_f} = \frac{(4.4 \times 10^6)(272)}{1628 \times 10^6} = 0.735$$

(b) For 10°C undercooling

$$f = \frac{(4.4 \times 10^6)(10)}{1628 \times 10^6} = 0.027$$

7-4 Solidification Time

The rate at which growth of the solid occurs during solidification depends on the cooling rate, or rate of heat extraction. A fast cooling rate produces rapid solidification or short solidification times. The time required for a simple casting to solidify completely can be calculated using *Chvorinov's rule*.

$$t_s = B\left(\frac{V}{A}\right)^2 \tag{7-7}$$

where t_s is the time required for the casting to solidify, V is the volume of the casting, A is the surface area of the casting in contact with the mold, and B is a *mold constant*. The mold constant depends on the properties and initial temperatures of both the metal and the mold. Almost always, a shorter solidification time produces a finer grain size and a stronger casting.

EXAMPLE 7-5

Two castings are produced under identical conditions. One casting, C1, has the dimensions 20 mm × 80 mm × 160 mm, and the dimensions of the second casting, C2, are 30 mm × 60 mm × 80 mm. Which casting will be stronger?

Answer:

The casting that freezes in the shortest time should have the higher strength. From Chvorinov's rule

$$t_{s(C1)} = B\left[\frac{(20)(80)(160)}{(2)(20)(80) + (2)(20)(160) + (2)(80)(160)}\right]^2$$

$$= B\left(\frac{256000}{352000}\right)^2 = 53B$$

$$t_{s(C2)} = B\left[\frac{(30)(60)(80)}{(2)(30)(60) + (2)(30)(80) + (2)(60)(80)}\right]^2$$

$$= B\left(\frac{144000}{18000}\right)^2 = 64B$$

Because $t_{(C1)} < t_{(C2)}$, then casting C1 freezes faster and is stronger.

The solidification time also affects the size of the dendrites that grow. Normally, the dendrite size is characterized by measuring the distance between the secondary dendrite arms (Figure 7-8). The *secondary dendrite arm spacing*, or *SDAS*,

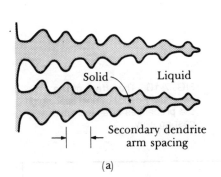

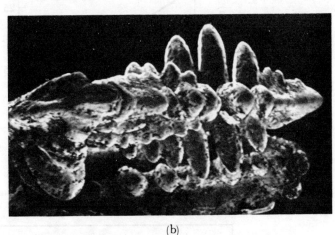

(a)   (b)

FIG. 7-8 (a) The secondary dendrite arm spacing SDAS. (b) Scanning electron micrograph of dendrites in steel, ×15.

is reduced when the casting freezes more rapidly. Because there is less time available to transfer heat, additional dendrite arms develop and grow to assist with the evolution of the latent heat. The finer, more extensive dendritic network serves as a more efficient conductor of the latent heat to the undercooled liquid. The *SDAS* is related to the solidification time by

$$SDAS = kt_s^n \tag{7-8}$$

where n and k are constants depending on the composition of the metal. This relationship is shown in Figure 7-9 for several alloys. Small secondary dendrite arm spacings are associated with higher strengths and improved ductility (Figure 7-10).

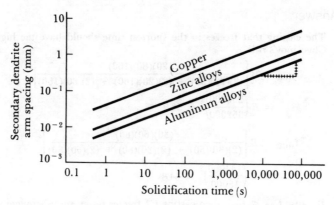

FIG. 7-9   The effect of solidification time on the secondary dendrite arm spacing of copper, zinc, and aluminum.

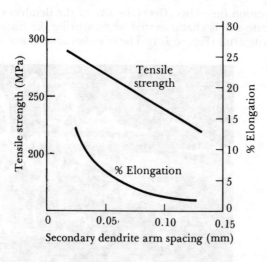

FIG. 7-10   The effect of the secondary dendrite arm spacing on the properties of an aluminum casting alloy.

EXAMPLE 7-6

Determine the constants in the equation that describes the relationship between secondary dendrite arm spacing and solidification time for aluminum alloys (Figure 7-9).

Answer:

We can obtain the slope n on a log-log plot by measuring the slope on the graph. In Figure 7-9, five equal units are marked on the vertical scale and 12 equal units on the horizontal scale. The slope is

$$n = \frac{5}{12} = 0.42$$

The constant k is the value of $SDAS$ when $t_s = 1$, since

$$\log SDAS = \log k + n \log t_s$$

If $t_s = 1$, $n \log t_s = 0$, and $SDAS = k$. From Figure 7-9

$$k = 8 \times 10^{-3} \text{ mm}$$

EXAMPLE 7-7

Calculate the SDAS and the tensile strength you expect to find at the centre of an aluminium alloy casting with dimensions 25 mm × 200 mm × 300 mm. The mould constant in Chvorinov's rule for aluminium alloys is 0.072 min mm^{-2}.

Answer:

$$V_{casting} = (25)(200)(300) = 15 \times 10^5 \text{ mm}^3$$
$$A_{casting} = (2)(25)(200) + (2)(25)(300) + (2)(200)(300) = 1.45 \times 10^5 \text{ mm}^2$$

From Chvorinov's rule

$$t_s = B\left(\frac{V}{A}\right)^2 = 0.072 \left[\frac{15 \times 10^5}{1.45 \times 10^5}\right]^2 = 7.7 \text{ min} = 462 \text{ s}$$

For Example 7-6 for aluminium alloys, $n = 0.42$ and $k = 8 \times 10^{-3}$ mm

$$SDAS = (8 \times 10^{-3}) t_s^{0.42} = (8 \times 10^{-3})(462)^{0.42}$$
$$= 105 \times 10^{-3} \text{ mm} = 0.105 \text{ mm}$$

From Figure 7-10

$$\text{Tensile strength} = 270 \text{ MPa}$$

7-5 Cooling Curves

We can summarize our discussion to this point by examining a cooling curve, or how the temperature of the metal changes with time (Figure 7-11). The liquid metal is poured into a mold at the *pouring temperature*. The difference between the pouring temperature and the freezing temperature is the *superheat*. The liquid metal cools as the specific heat of the liquid is extracted by the mold until the liquid reaches the freezing temperature. The slope of the cooling curve before solidification begins is the *cooling rate* $\Delta T/\Delta t$.

If effective heterogeneous nuclei are present in the liquid metal, solidification begins at the freezing temperature, as shown in Figure 7-11(a). A *thermal arrest*, or plateau, is produced because of the evolution of the latent heat of fusion. The latent heat keeps the remaining liquid at the freezing temperature until all of the liquid has solidified and no more heat can be evolved. Growth under these conditions is planar. The *total solidification time* of the casting is the time required to remove both the specific heat of the superheated liquid and the latent heat of fusion. This is measured from the time of pouring until solidification is complete and is given by Chvorinov's rule. The *local solidification time* is the time required to

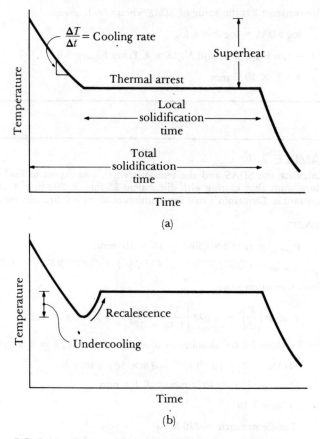

FIG. 7-11   Cooling curves for (a) liquids that nucleate with no undercooling and (b) liquids that require large undercoolings for nucleation.

remove only the latent heat of fusion at a particular location in the casting and is measured from when solidification begins until solidification is completed.

If undercooling develops due to poor nucleation, the cooling curve dips below the freezing temperature, as shown in Figure 7-11(b). After the solid finally nucleates, dendritic growth occurs. The latent heat, however, is absorbed by the undercooled liquid, raising the temperature of the liquid back to the freezing temperature. This phenomenon is known as *recalescence*. After the temperature of the remaining liquid is raised to the freezing temperature, a thermal arrest occurs until solidification is completed by planar growth.

7-6   Casting or Ingot Structure

Molten metals are poured into molds and permitted to solidify. Often the mold produces a finished shape, or casting. In other cases, the mold produces a simple shape, called an *ingot*, that requires extensive plastic deformation or machining before a finished product is created.

In either a casting or an ingot, a *macrostructure* is produced which can consist of as many as three parts (Figure 7-12).

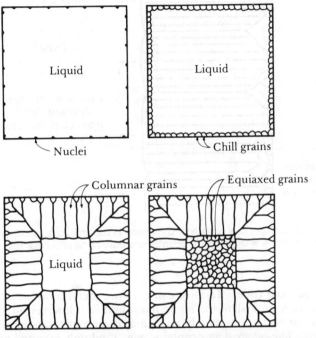

FIG. 7-12 Development of the macrostructure of a casting during solidification. (a) Nucleation begins. (b) The chill zone forms. (c) Preferred growth produces the columnar zone. (d) Additional nucleation creates the equiaxed zone.

Chill zone. The *chill zone* is a narrow band of randomly oriented grains at the surface of the casting. The metal at the mold wall is the first to cool to or below the freezing temperature. The mold wall also provides many surfaces at which heterogeneous nucleation may take place. Therefore, a large number of grains begins to nucleate and grow along the mold wall.

Columnar zone. The *columnar zone* contains elongated grains oriented in a particular crystallographic direction. As heat is removed from the casting by the mold material, the grains in the chill zone begin to grow in the direction opposite to the heat flow, or from the coldest towards the hottest areas of the casting. This usually means that the grains grow perpendicular to the mold wall.

Grains grow fastest in certain crystallographic directions. In metals with a cubic crystal structure, grains in the chill zone that have a $\langle 100 \rangle$ direction perpendicular to the mold wall grow faster than other less favorably oriented grains (Figure 7-13). Eventually, the grains in the columnar zone have $\langle 100 \rangle$ directions that are parallel to one another, giving the columnar zone anisotropic properties.

The formation of the columnar zone is influenced primarily by growth, rather than nucleation, phenomena. The grains may be composed of many dendrites if the liquid is originally undercooled. Or solidification may proceed by planar growth of the columnar grains if no undercooling has occurred.

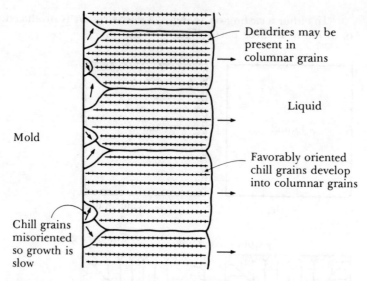

FIG. 7-13 Competitive growth of the grains in the chill zone results in only those grains with favorable orientations developing into columnar grains.

Equiaxed zone.   In most cases, a pure metal continues to grow in a columnar manner until all of the liquid has solidified. However, in alloys and in special circumstances in pure metals, an equiaxed zone forms in the center of the casting or ingot. The *equiaxed zone* contains new, randomly oriented grains, often caused by a low pouring temperature, alloying elements, or grain refining or inoculating agents. These grains grow as relatively round, or equiaxed, grains with a random orientation and stop the growth of the columnar grains. The formation of the equiaxed zone is a nucleation-controlled process and causes that portion of the casting to have isotropic behavior.

7-7 Solidification Defects

Although there are a large number of potential defects that can be produced during solidification, two deserve special mention.

Shrinkage.   Almost all materials are more dense in the solid than in the liquid state (Figure 7-14). During solidification, the material contracts, or shrinks, as much as 7% (Table 7-2).

If the shrinkage is unidirectional (Figure 7-15) only one dimension of the solid casting would be smaller than the dimensions of the mold. The mold could then be made oversized by the appropriate amount in order to compensate for the shrinkage.

However, in most situations, the bulk of the shrinkage occurs as *cavities*, if solidification begins at all surfaces of the casting, or as *pipes*, if one surface solidifies more slowly than the others. In either case, the casting is defective. A common technique for controlling cavity and pipe shrinkage is to place a *riser*, or an extra reservoir of metal, adjacent and connected to the casting. As the casting solidifies

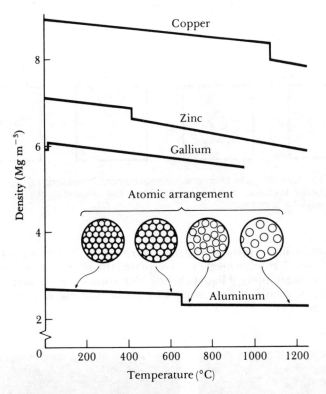

FIG. 7-14   The density of selected metals versus temperature. Most metals have a higher density as solids than as liquids and thus contract during solidification. Note that gallium has the opposite behavior.

TABLE 7-2   Shrinkage during solidification for selected materials

Material	Shrinkage (%)
Al	7.0
Cu	5.1
Mg	4.0
Zn	3.7
Fe	3.4
Pb	2.7
Ga	+3.2 (expansion)
H_2O	+8.3 (expansion)

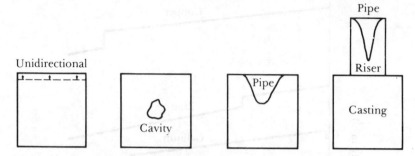

Unidirectional

Cavity

Pipe

Pipe

Riser

Casting

FIG. 7-15   Several types of macroshrinkage can occur including (a) unidirectional, (b) cavity, and (c) pipe. (d) Risers can be used to help compensate for shrinkage.

and shrinks, liquid metal flows from the riser into the casting to fill the shrinkage void (Figure 7-16). We need only assure that the riser freezes after the casting and that there is an internal liquid channel that connects the liquid in the riser to the last liquid to solidify in the casting. Chvorinov's rule can be used to help design the size of the riser.

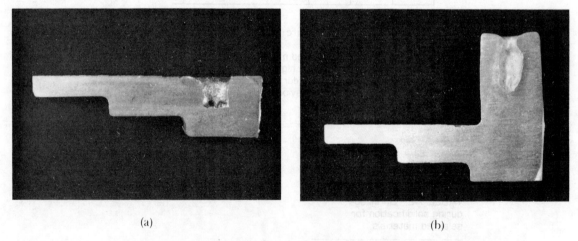

(a)   (b)

FIG. 7-16   Sections through an aluminum casting. (a) Because no riser is used, concentrated shrinkage is present in the thick part of the casting. (b) Shrinkage is contained in the riser, thus producing a sound casting.

EXAMPLE 7-8

A cylindrical riser with a height equal to twice its diameter is to compensate for shrinkage in a 20 mm × 80 mm × 160 mm casting. Estimate the minimum size of the riser.

Answer:

We know that the riser must freeze after the casting.

$$t_{riser} > t_{casting} \quad \text{so} \quad B\left(\frac{V}{A}\right)_r^2 > B\left(\frac{V}{A}\right)_c^2$$

$$\left(\frac{V}{A}\right)_r > \left(\frac{V}{A}\right)_c$$

$$V_c = (20)(80)(160) = 256 \times 10^3 \text{ mm}^3$$

$$A_c = (2)(20)(80) + (2)(20)(160) + (2)(80)(160) = 35.2 \times 10^3 \text{ mm}^2$$

$$V_r = \frac{\pi}{4} D^2 H = \frac{\pi}{4} D^2(2D) = \frac{\pi}{2} D^3$$

$$A_r = 2\left(\frac{\pi}{4} D^2\right) + \pi DH = 2\left(\frac{\pi}{4} D^2\right) + \pi D(2D) = \frac{5}{2}\pi D^2$$

$$\frac{(\pi/2)(D)^3}{(5\pi/2)(D)^2} > \frac{256}{35.2}$$

$$\frac{D}{5} > 7.27$$

$$D > 36.4 \text{ mm}$$

$$H > 72.8 \text{ mm}$$

$$V_r > 75\,800 \text{ mm}^3$$

Although the volume of the riser is much smaller than that of the casting, the riser freezes more slowly due to its compact shape.

EXAMPLE 7-9

A casting, shown in Figure 7-17, is to be poured without shrinkage. Will a riser placed at location A be adequate if it is large enough?

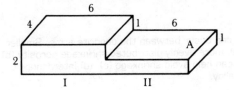

FIG. 7-17 Casting geometry for Example 7-9.

Answer:

We divide the casting into two sections, I and II, and determine the freezing time for each.

Section I

$$V_I = (2)(4)(6) = 48 \text{ units}^3$$

$$A_I = (1)(2)(4) + (2)(2)(6) + (2)(4)(6) + (1)(1)(4) = 84 \text{ units}^2$$

$$t_s = B\left(\frac{V}{A}\right)^2 = B\left(\frac{48}{84}\right)^2 = 0.33B$$

Section II

$$V_{II} = (1)(4)(6) = |\ 24 \text{ units}^3$$

$$A_{II} = (1)(1)(4) + (2)(1)(6) + (2)(4)(6) = |\ 64 \text{ units}^3$$

$$t_s = B\left(\frac{V}{A}\right)^2 = B\left(\frac{24}{64}\right)^2 = 0.14B$$

Section II freezes before Section I. Therefore liquid metal from the riser cannot flow to and compensate for shrinkage in Section I. The riser probably should be placed at the end of Section I rather than at Section II.

Interdendritic shrinkage is found when extensive dendritic growth occurs (Figure 7-18). Liquid metal may be unable to flow from a riser through the fine

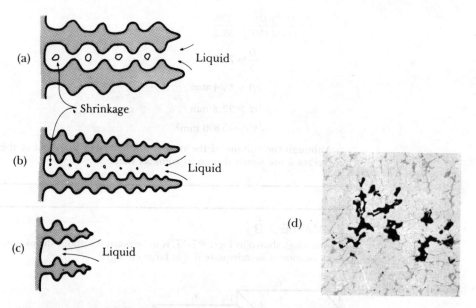

FIG. 7-18   (a) Shrinkage can occur between the dendrite arms. Smaller dendrites result in smaller, more evenly distributed shrinkage porosity (b), while short primary arms can help avoid shrinkage (c). (d) Interdendritic shrinkage in an aluminum alloy.

dendritic network to the solidifying metal. Consequently, small shrinkage pores are produced throughout the casting. This defect, also called *microshrinkage* or *shrinkage porosity*, is difficult to prevent by the use of risers. Fast cooling rates may reduce problems with interdendritic shrinkage; the dendrites may be shorter, permitting liquid to flow through the dendritic network to the solidifying solid interface. In addition, any shrinkage that remains may be finer and more uniformly distributed.

Gas porosity.   Many metals dissolve a large quantity of gas when they are liquid. Aluminum, for example, dissolves hydrogen. However, when the aluminum solidifies, the solid metal retains in its structure only a small fraction of the hydrogen (Figure 7-19). The excess hydrogen forms bubbles that may be trapped in the solid metal, producing *gas porosity*. The porosity may be spread uniformly throughout the casting or may be trapped between dendrite arms.

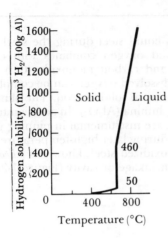

FIG. 7-19 The solubility of hydrogen gas in aluminum. Although solid aluminum contains very little hydrogen, the amount of hydrogen that dissolves in liquid aluminum is large and increases rapidly with temperature.

The amount of gas that can be dissolved in the molten metal is given by Sievert's law.

$$\text{Percent of gas} = K\sqrt{p_{gas}} \tag{7-9}$$

where p_{gas} is the partial pressure of the gas in contact with the metal and K is a constant which, for a particular metal-gas system, increases with increasing temperature. We can minimize gas porosity in castings by keeping the liquid temperature low, adding materials to the liquid to combine with the gas and form a solid, or by assuring that the partial pressure of the gas remains low. The latter may be achieved by placing the molten metal in a vacuum chamber or bubbling an inert gas through the metal. Because p_{gas} is low in the vacuum or inert gas, the gas leaves the metal, enters the vacuum or inert gas, and is carried away.

EXAMPLE 7-10

After melting at atmospheric pressure, molten copper contains 0.01% O. How much oxygen would remain if the molten copper were placed in a vacuum at 10^{-6} bar?

Answer:

The ratio of the partial pressure of oxygen before and during the vacuum treatment will be the same as the ratio of the total pressures. Thus

$$\frac{p_{initial}}{p_{vacuum}} = \frac{1}{10^{-6}} = 10^6$$

By forming a ratio, the constant K in Sievert's law cancels.

$$\frac{\% \, O_{initial}}{\% \, O_{vacuum}} = \frac{K\sqrt{p_{initial}}}{K\sqrt{p_{vacuum}}} = \sqrt{\frac{p_{initial}}{p_{vacuum}}} = \sqrt{10^6}$$

$$\% \, O_{vacuum} = \frac{\% \, O_{initial}}{\sqrt{10^6}} = (0.01)(10^{-3}) = 1 \times 10^{-5}\%$$

Oxygen is often dissolved in liquid steel during the steel-making process. During solidification, the dissolved oxygen combines with carbon, which is present as an alloying element, and carbon monoxide (CO) gas bubbles are trapped in the steel casting. The dissolved oxygen can be completely eliminated, however, if aluminum is added prior to solidification. The aluminum combines with the oxygen, producing solid alumina (Al_2O_3). In addition to eliminating gas porosity, the tiny Al_2O_3 inclusions are instrumental in pinning grain boundaries and thus preventing grain growth during later high-temperature heat treatments. Unfortunately, the completely deoxidized steel, known as *killed* or *fine-grained* steel, often displays a deep pipe shrinkage or cavity [Figure 7-20(a)].

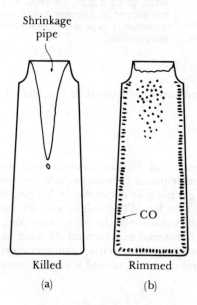

FIG. 7-20 The ingot structure of (a) a killed and (b) a rimmed steel after different degrees of deoxidation with aluminum. The killed steel, which is free of gas porosity, contains a deep pipe, and the rimmed steel, which contains distributed porosity, is relatively free of shrinkage.

Sometimes steels are only partly deoxidized. By adding a small amount of aluminum, a *rimmed* steel is produced in which enough CO is precipitated to just offset the solidification shrinkage [Figure 7-20(b)]. In addition to having less concentrated shrinkage, a rimmed steel helps produce smooth attractive surfaces on steel sheet after subsequent processing.

EXAMPLE 7-11

Castings for steel pump housings used in the chemical industry are normally killed, whereas steel used to make sheet metal for automobile fenders is rimmed. Explain why the different deoxidation practices are used.

Answer:

Steel castings cannot be treated to eliminate gas bubbles after solidification; therefore gas bubbles must be prevented by killing the steel to remove all oxygen. In addition, the inclusions produced during deoxidation may prevent grain growth when the castings are heat treated.

Steel sheet, on the other hand, is formed from a casting by a series of plastic deformation steps, in which the thickness of the steel is continually reduced. During deformation, the gas bubbles produced in the original casting are squeezed shut and eliminated by diffusion. The rimmed steel permits a smooth surface during the final deformation steps.

7-8 Control of Casting Structure

As a general rule, we control solidification so that we produce a casting macrostructure containing a large number of small equiaxed grains. This permits the casting to have isotropic properties and improved strength due to grain size strengthening. In addition, we wish to make any dendrites as small as possible, which again improves the strength of the casting and refines both microshrinkage and gas porosity.

In order to obtain this desired structure, we must first assure that widespread nucleation occurs by using appropriate grain refining or inoculating agents. Second, we may encourage rapid solidification to assure that the secondary dendrite arm spacing within the grains is very small. The rate of solidification for any given metal can be influenced by the size of the casting, the mold material, and the casting process. Thick castings solidify more slowly than thin castings. Mold materials having a high density, thermal conductivity, and heat capacity produce more rapid solidification.

Figure 7-21 summarizes four of the dozens of casting processes. The processes are divided into several groups—sand molds, ceramic molds, and metal molds. The processes using metal molds tend to give the highest strength castings due to rapid solidification. Ceramic molds, because they are good insulators, give the slowest cooling and lowest strength castings.

Directional solidification. There are some applications for which a small equiaxed grain structure in the casting is not desired. Castings used for blades and vanes in turbine engines are an example (Figure 7-22). These castings are often made of cobalt or nickel superalloys by investment casting.

In conventionally cast parts, an equiaxed grain structure is produced. However, blades and vanes for turbine and jet engines fail along transverse grain boundaries. Better creep and fracture resistance are obtained using the directionally solidified (DS) technique. In the DS process, the mold is heated from one end and cooled from the other, producing a columnar microstructure with all of the grain boundaries running in the longitudinal direction of the part. No grain boundaries are present in the transverse direction [Figure 7-22(b)].

Still better properties are obtained by using a single crystal (SC) technique. Solidification of columnar grains again begins at a cold surface; however, due to the helical connection, only one columnar grain is able to grow to the main body of the casting [Figure 7-22(c)]. The single crystal casting has no grain boundaries at all and has its crystallographic planes and directions in an optimum orientation.

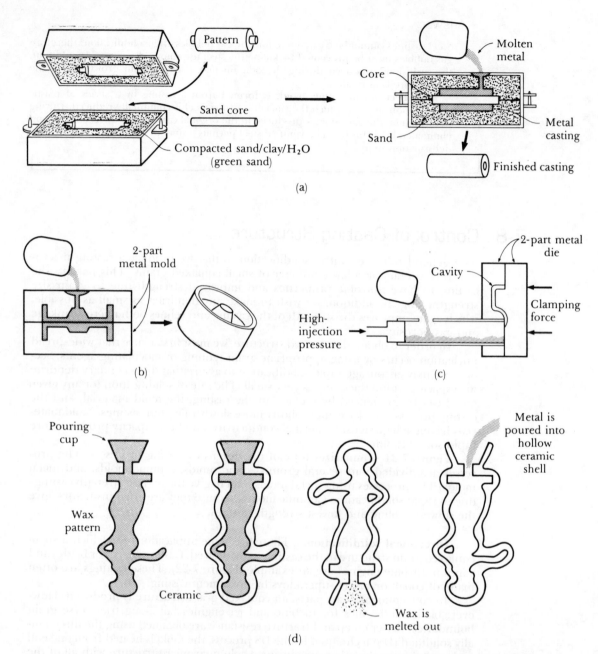

FIG. 7-21 Four of the typical casting processes. (a) Green sand molding, in which clay-bonded sand is packed around a pattern. Sand cores can produce internal cavities in the casting. (b) The permanent mold process, in which metal is poured into an iron or steel mold. (c) Die casting, in which metal is injected at high pressures into a steel die. (d) Investment casting, in which a wax pattern is surrounded by a ceramic; after the wax is melted and drained, metal is poured into the mold.

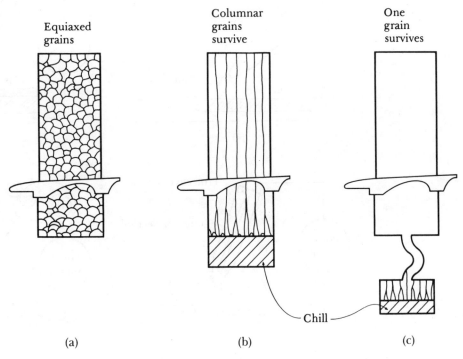

FIG. 7-22 Controlling grain structure in turbine blades. (a) Conventional equiaxed grains, (b) directionally solidified columnar grains, and (c) single crystal.

7-9 Solidification and Metals Joining

Solidification is also important in the joining of metals by fusion welding. Some typical fusion-welding processes are illustrated in Figure 7-23. In the fusion-welding processes, a portion of the metals to be joined are melted, and in many instances additional molten filler metal is added. The pool of liquid metal is called the *fusion zone*. When the fusion zone subsequently solidifies, the original pieces of metal are joined together.

During solidification of the fusion zone, nucleation is not required. Heat is extracted most rapidly from the fusion zone through the original pieces of metal, which act as heat sinks. The liquid in the fusion zone first cools to the freezing temperature at the edges of the weld (Figure 7-24). But there are already solid grains of the original material at those locations! Consequently, the solid simply begins to grow from these grains, frequently in a columnar manner. Growth of the solid grains in the fusion zone from the preexisting grains is called *epitaxial growth*.

The structure and properties in the fusion zone depend on many of the same variables as in a metal casting. Addition of inoculating agents to the fusion zone reduces the grain size. Fast cooling rates or short solidification times promote a finer microstructure and improved properties. Factors that increase the cooling rate include increasing the thickness of the metal, smaller fusion zones, low original metal temperatures, and the type of welding process. Oxyacetylene welding, for example, uses relatively low intensity flames; consequently, welding times are

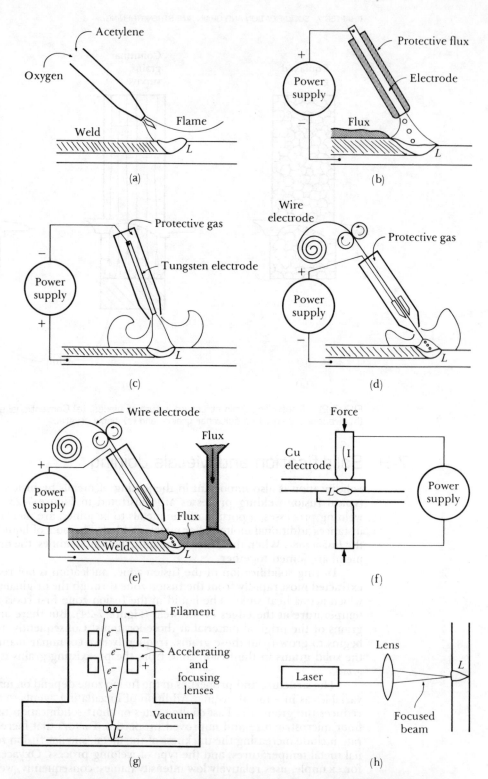

FIG. 7-23 Typical fusion-welding processes. (a) Oxyactylene welding, (b) shielded-metal arc welding, (c) gas-tungsten arc welding, (d) gas-metal arc welding, (e) submerged arc welding, (f) resistance welding, (g) electron-beam welding, and (h) laser welding.

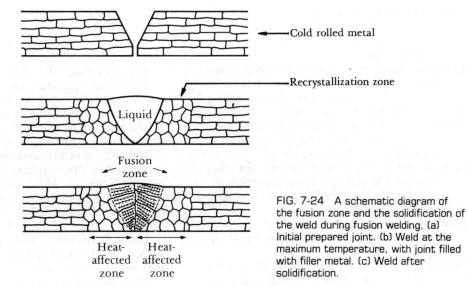

FIG. 7-24 A schematic diagram of the fusion zone and the solidification of the weld during fusion welding. (a) Initial prepared joint. (b) Weld at the maximum temperature, with joint filled with filler metal. (c) Weld after solidification.

long and the surrounding solid metal, which becomes very hot, is not an effective heat sink. Arc-welding processes provide a more intense heat source, thus minimizing heating of the surrounding metal and providing faster cooling. Resistance welding, laser welding, and electron-beam welding are exceptionally intense heat sources and produce very rapid cooling rates and potentially strong welds.

SUMMARY

One of the first opportunities that we have to control the mechanical properties of a material occurs during solidification of the liquid melt. During solidification we control the size and shape of the grains to improve overall properties, to obtain uniform properties, or, if we wish, to obtain anisotropic behavior. We exercise this control by assuring proper nucleation and growth through inoculation or grain refining, proper solidification time, and correct metal temperature—in other words, we must control the materials processing technique. We find that improved nucleation of grains and rapid cooling give smaller grains and thus provide grain size strengthening. Furthermore, by proper treatment of the molten material and correct casting procedures, we are able to prevent or control gas and shrinkage voids. In addition to providing an improved casting, these precautions also improve our ability to further process the material by deformation techniques and heat treatment.

GLOSSARY

Cavity shrinkage A large void within a casting caused by the volume contraction that occurs during solidification.

Chill zone A region of small randomly oriented grains that forms at the surface of a casting due to heterogeneous nucleation.

Chvorinov's rule The solidification time of a casting is directly proportional to the square of the volume to surface area ratio of the casting.

Columnar zone A region of elongated grains having a preferred orientation that forms as a result of competitive growth during the solidification of a casting.

Cooling rate The change in temperature for a given change in time. Rapid cooling rates normally give stronger castings.

Critical radius r^* The minimum size that must be formed by atoms clustering together in the liquid before the solid particle is stable and begins to grow.

Dendrite The treelike structure of the solid that grows when an undercooled liquid nucleates.

Dendritic growth Rapid growth of a solid dendrite when an undercooled liquid nucleates and grows.

Directional solidification Assuring that a casting grows from one direction only. This technique is used to cast high temperature-resistant turbine blades.

Embryo A tiny particle of solid that forms from the liquid as atoms cluster together. The embryo is too small to grow.

Epitaxial growth Growth of a liquid onto an existing solid material without the need for nucleation.

Equiaxed zone A region of randomly oriented grains in the center of a casting produced as a result of widespread nucleation.

Fusion zone The portion of a weld heated to produce all liquid during the welding process. Solidification of the fusion zone provides joining.

Gas porosity Bubbles of gas trapped within a casting during solidification due to the lower solubility of the gas in the solid compared to the liquid.

Grain refinement The addition of heterogeneous nuclei in a controlled manner to increase the number of grains in a casting.

Grain size strengthening By reducing the size of the grains, causing an increase in the amount of grain boundary area, a material may be strengthened.

Heterogeneous nucleation Formation of a critically sized solid from the liquid on an impurity surface.

Homogeneous nucleation Formation of a critically sized solid from the liquid by the clustering together of a large number of atoms at a high undercooling.

Ingot structure The macrostructure of a casting, including the chill zone, columnar zone, and equiaxed zone.

Inoculation The addition of heterogeneous nuclei in a controlled manner to increase the number of grains in a casting.

Interdendritic shrinkage Small, frequently isolated pores between the dendrite arms formed by the shrinkage that accompanies solidification. Also known as microshrinkage or shrinkage porosity.

Latent heat of fusion ΔH_f The heat evolved when a liquid solidifies. The latent heat of fusion is related to the energy difference between the solid and the liquid.

Local solidification time The time required for a particular location in a casting to solidify once nucleation has begun.

Nucleus A tiny particle of solid that forms from the liquid as atoms cluster together. When the nucleus is large enough to be stable, nucleation has occurred and growth of the solid can begin.

Pipe shrinkage A large conical-shaped void at the surface of a casting caused by the volume contraction that occurs during solidification.

Planar growth The growth of a smooth solid-liquid interface during solidification when no undercooling of the liquid is present.

Pouring temperature The temperature of the metal when it is poured into a mold during the casting process.

Recalescence The increase in the temperature of a solidifying liquid that is growing dendritically. The increase is caused by the transfer of the latent heat of fusion into the undercooled liquid.

Riser An extra reservoir of liquid metal connected to a casting. If the riser freezes after the casting, the riser can provide liquid metal to compensate for shrinkage.

Secondary dendrite arm spacing The distance between the centers of two adjacent secondary dendrite arms.

Sievert's law The amount of a gas that dissolves in a metal is proportional to the partial pressure of that gas in the surroundings.

Solidification   The transformation of a liquid to a solid material.

Specific heat   The heat required to change the temperature of a unit weight of the material one degree.

Superheat   The pouring temperature minus the freezing temperature.

Surface free energy σ The increase in energy associated with the surface between a growing solid and a liquid.

Thermal arrest   A plateau on the cooling curve during the solidification of a material. The thermal arrest is due to the evolu-

tion of the latent heat of fusion during solidification.

Total solidification time   The time required for the casting to completely solidify after the casting has been poured.

Undercooling   The temperature to which the liquid metal must cool below the equilibrium freezing temperature before nucleation occurs.

Volume free energy ΔF_v The change in free energy of a material when the material solidifies.

PRACTICE PROBLEMS

1   Calculate the temperature and the number of degrees of undercooling at which nickel nucleates homogeneously.

2   Calculate the temperature and the number of degrees of undercooling at which silver nucleates homogeneously.

3   Calculate the size of the critical radius when iron nucleates homogeneously.

4   Calculate the size of the critical radius when silver nucleates homogeneously.

5   Calculate the number of atoms in the critical radius when nickel nucleates homogeneously. The lattice parameter of FCC nickel is 3.5167 Å.

6   Calculate the number of atoms in the critical radius when iron nucleates homogeneously. The iron is BCC with a lattice parameter of 2.866 Å.

7   Calculate the number of atoms in the critical radius of nickel if it nucleates as a sphere at 10°C undercooling. (a_0 = 3.5167 Å)

8   Calculate the number of atoms in the nucleus if nickel nucleates heterogeneously with 10°C undercooling on an impurity that gives a height of a spherical cap of $\frac{1}{4}r$. (a_0 = 3.5167 Å)

9   Calculate the number of atoms in the nucleus if lead nucleates heterogeneously with 20°C undercooling on an impurity that gives a height of a spherical cap of $\frac{1}{3}r$. (a_0 = 4.9489 Å)

10   In Table 7-1, check to see how closely the approximation of $\Delta T = 0.2T_m$ correlates with the maximum observed undercooling for each metal.

11   From the data in Table 7-1, calculate and plot how the total free energy ΔF changes with the radius of a spherical nucleus of iron at undercoolings of 20°C and 400°C.

12   How would you expect an increase in the following variables to affect the solidification time for a casting?

mold temperature

thermal conductivity of mold

heat capacity of mold

latent heat of fusion

melting point of metal

pouring temperature of metal

superheat

thickness of metal mold

volume to surface area ratio of the casting

13   Measure the secondary dendrite arm spacing of the dendrites in Figure 7-25.

14   The following temperatures are measured in molten copper just ahead of the solid-liquid interface. (a) Do you expect the solid-liquid interface to grow in a planar or dendritic manner? Explain. (b) If dendritic growth occurs, estimate the length of the primary dendrite stalks.

Distance from Solid-Liquid Interface (mm)	Temperature (°C)
0	1085
0.01	1074
0.02	1069
0.03	1070
0.04	1075
0.05	1089
0.06	1110

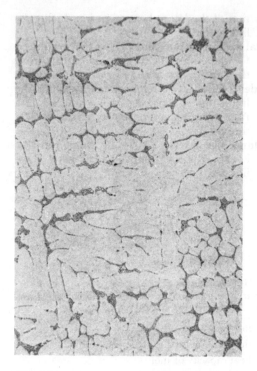

FIG. 7-25   The dendritic structure in an aluminum alloy for Problem 7-13, ×50.

15   The following temperatures are measured in molten lead just ahead of the solid-liquid interface. Will solidification occur by planar or dendritic growth? Explain.

Distance from Solid-Liquid Interface (mm)	Temperature (°C)
0	327
0.005	328
0.010	330
0.015	345
0.020	370

16   Estimate the percent dendritic growth that occurs when iron, which has a specific heat of 3.5×10^6 J $(m^3 \cdot K)^{-1}$, nucleates at 75°C undercooling.

17   Calculate the amount of planar growth and the amount of dendritic growth you expect in lead at undercoolings of 10°C, 40°C, and 80°C. (The specific heat of lead is 1.58×10^6 J $(m^3 \cdot K)^{-1}$

18   The amount of dendritic growth in a pure copper ingot is roughly 0.5 of the total growth. The specific heat of copper is 3.44×10^6 J $(m^3 \cdot K)^{-1}$. Estimate the undercooling of the copper prior to nucleation and growth.

19   Determine the constants k and n in Equation (7-8) that describe the effect of the solidification time on the secondary dendrite arm spacing of copper. See Figure 7-9.

20   Determine the constants k and n in Equation (7-8) that describe the effect of the solidification time on the secondary dendrite arm spacing of zinc. See Figure 7-9.

21   A 25 mm × 125 mm × 225 mm plate casting solidifies in 10 min. What time is required for a 100 mm diameter sphere to solidify?

22   A 50 mm diameter cylindrical casting 100 mm tall solidifies in 15 min. What is the solidification time for a 20 mm × 120 mm × 160 mm casting?

23   Calculate the minimum dimensions of a riser required to prevent shrinkage in a 50 mm × 200 mm × 250 mm casting. Assume that the riser is cylindrical, with a H/D ratio of 1.

24   Calculate the minimum dimensions of a riser required to prevent shrinkage in a 60 mm × 80 mm × 140 mm casting (a) if the riser is a cylinder with $H/D = 1$ and (b) if the riser is a cylinder with $H/D = 2$.

25   Calculate the V/A ratio for a riser that has a volume of 100 units² if the riser shape is (a) a cube, (b) a sphere, and (c) a cylinder with $H/D = 1$. What is the most efficient riser shape?

26   In the 118 mm casting in Figure 7-26, two cylindrical risers witn $H = D$ are needed to prevent shrinkage in the casting. Calculate the size of each riser.

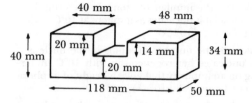

FIG. 7-26   Casting geometry for Problem 7-26.

27   In Figure 7-27, two identical cylindrical risers with $H = D$ located at points x and y are required to prevent shrinkage in the casting. Calculate the size of each riser.

28   Which casting will have the higher strength if poured under identical conditions—a 100 mm × 150

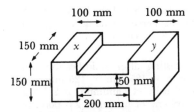

FIG. 7-27 Casting geometry for Problem 7-27.

mm × 300 mm casting or a 125 mm diameter × 250 mm tall cylindrical casting?

29 The temperature at various locations in an aluminum casting is obtained, with cooling curves shown in Figure 7-28. By measuring the local solidification time, estimate the secondary dendrite arm spacing and the tensile strength at each location in the casting.

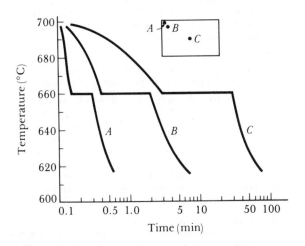

FIG. 7-28 Cooling curves for Problem 7-29.

30 A 75 mm copper cube and an 200 mm copper cube are poured under identical conditions. Calculate the solidification time and estimate the secondary dendrite arm spacing at the center of each casting. The mold constant B for copper is about 0.024 min mm^{-2}.

31 Estimate the secondary dendrite arm spacing in the center of a 250 mm zinc cube and the center of a 12.5 mm zinc cube. The mold constant B is about 0.072 min mm^{-2}.

32 The following data are obtained from a series of zinc castings. Determine the mold constant B and

the exponent n in Chvorinov's rule for zinc, using a log (V/A) versus log (t) plot.

Casting Dimensions (mm)	Solidification Time (min)
75 × 75 × 75	16.1
150 × 37.5 × 41	8.8
75 diameter × 94 height	18.0
144 × 56 × 375	10.5

33 The following data are obtained from a series of steel castings. Determine the mold constant B and the exponent n in Chvorinov's rule for steel.

Casting Diameter (mm)	Solidification Time (min)
75 mm sphere	3
112 mm sphere	7
150 mm sphere	12.5
225 mm sphere	28

From R. Ruddle, *The Solidification of Castings*, The Institute of Metals, 1950.

34 A cooling curve is shown in Figure 7-29. Determine (a) the pouring temperature, (b) the freezing temperature, (c) the probable identity of the metal, (d) the cooling rate prior to solidification, and (e) the local solidification time.

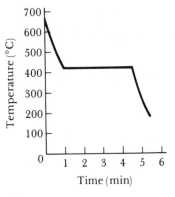

FIG. 7-29 Cooling curve for Problem 7-34.

35 A cooling curve is shown in Figure 7-30. Determine (a) the pouring temperature, (b) the freezing

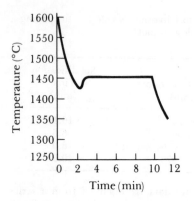

FIG. 7-30   Cooling curve for
Problem 7-35.

temperature, (c) the superheat, (d) the local solidification time, (e) the total solidification time, and (f) the probable identity of the metal.

36 Suppose the entire shrinkage of an aluminum casting is in the form of a concentrated shrinkage cavity at the center. At room temperature the dimensions of the casting are 75 mm × 75 mm × 75 mm. (a) Based on Table 7-2, how much does the casting weigh? (b) How much would it weigh if no shrinkage had occurred? (The density of aluminum is 2.7 Mg m^{-3})

37 Suppose a 50 mm diameter × 250 mm long aluminum cylinder experiences only unidirectional shrinkage in the longitudinal direction during freezing. Based on Table 7-2, what is the length of the cylinder immediately after solidification?

38 The density of solid aluminum just below the melting point is 2.56 Mg m^{-3} and the density of liquid aluminum just above the melting point is 2.38 Mg m^{-3}. Calculate the volume percent shrinkage that occurs during solidification. How does this compare to the value in Table 7-2?

39 The density of solid mercury at the melting point is 14.193 Mg m^{-3} and that of the liquid is 14.43 Mg m^{-3}. Calculate the percent volume change on freezing of mercury. Does mercury expand or contract during freezing?

40 An aluminum casting after solidification has the dimensions 125 mm × 125 mm × 125 mm and weighs 5.4 kg. The density of aluminum is 2.7 Mg m^{-3}. Calculate the volume of hydrogen gas porosity assuming that gas porosity is responsible for the lighter than expected weight.

41 Liquid aluminum containing 500 mm^3 hydrogen per 100 g aluminum at atmospheric pressure is placed in a vacuum chamber at 10^{-6} bar. Calculate the amount of hydrogen that remains in the aluminum. Will gas bubbles form during freezing?

42 Liquid aluminum contains 700 mm^3 hydrogen per 100 g aluminum at atmospheric pressure. In order to be sure that hydrogen bubbles do not form during solidification, the aluminum should be placed in a vacuum of how many atmospheres?

43 Sometimes castings contain a machining allowance; the casting is made oversized so some metal can be removed by machining. Explain why a very generous machining allowance may reduce the strength of the final machined casting.

44 An aluminum casting may be made either by sand casting or by permanent mold casting. Which process would give a stronger casting? Explain.

45 Some foundrymen place a block of copper against one surface of a sand casting, with the copper eventually being in contact with the liquid metal in the mold. (a) What effect does the copper have on the structure and properties of the metal at that surface? (b) Suppose that without the copper, a shrinkage cavity forms at the exact center of the casting. Will the shrinkage cavity move towards or away from the copper block? (The copper block is called a *chill*.)

46 Oxygen can be eliminated from copper by the addition of phosphorus. Speculate on the mechanism by which phosphorus causes deoxidation of copper.

CHAPTER 8

Solidification and Solid Solution Strengthening

8-1 Introduction

The mechanical properties of materials can be controlled by the addition of point defects, in particular substitutional and interstitial atoms. The point defects disturb the atomic arrangement in the lattice and interfere with the movement of dislocations, or slip. The point defects cause the material to be solid solution strengthened.

In addition, the introduction of point defects changes the composition of the material and influences the solidification behavior. We will examine this effect by introducing the equilibrium phase diagram. From the phase diagram we can predict how a material will solidify both under equilibrium and nonequilibrium conditions.

8-2 Phases, Solutions, and Solubility

Pure materials have many engineering applications but frequently, particularly when improved mechanical properties are required, alloys or mixtures of materials are used. There are two types of alloys—single-phase alloys and multiple-phase alloys. In this chapter we will examine the behavior of single-phase alloys. First, however, we must define a phase and a solid solution. The best way of doing this is to list the characteristics of a phase and show some examples.

Phase. A *phase* has the following characteristics: (a) a phase has the same structure or atomic arrangement throughout; (b) a phase has roughly the same composition and properties throughout; and (c) there is a definite interface between the phase and any surrounding or adjoining phases. For example, we could enclose a block of ice in a vacuum chamber [Figure 8-1(a)]. The ice would begin to melt and, in addition, some of the water might vaporize. Under these conditions we would have three phases coexisting—solid H_2O, liquid H_2O, and gaseous H_2O. Each of these forms of H_2O is a distinct phase; each has a unique atomic arrangement, unique properties, and a definite boundary between each form. In this case the phases have identical compositions, but that is not sufficient to permit us to call the entire system one phase.

Unlimited solubility. On the other hand, we could begin with a glass of water and a glass of alcohol. The water is one phase and the alcohol is a second

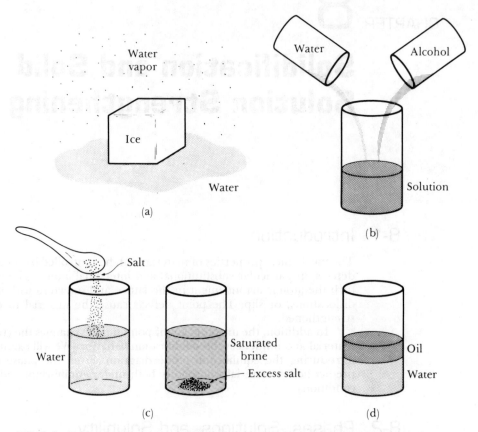

FIG. 8-1 Illustration of phases and solubility. (a) The three forms of water—gas, liquid, and solid—are each a phase. (b) Water and alcohol have unlimited solubility. (c) Salt and water have limited solubility. (d) Oil and water have virtually no solubility.

phase. If we pour the water into the alcohol and stir, only one phase is produced [Figure 8-1(b)]. The glass contains a solution of water and alcohol that has a unique structure, properties, and composition. Water and alcohol are soluble in each other. Furthermore, they display *unlimited solubility*—regardless of the ratio of water and alcohol, only one phase is produced by mixing them together.

Similarly, if we were to mix a container of liquid copper and a container of liquid nickel, only one liquid phase would be produced. The liquid alloy has the same composition, properties, and structure everywhere [Figure 8-2(a)]. Liquid nickel and copper also have unlimited solubility; regardless of the relative amounts of nickel and copper, only one liquid phase is produced.

If the liquid copper-nickel alloy solidifies and cools to room temperature, only one solid phase is produced. After solidification the copper and nickel atoms do not separate, but instead are randomly located at the lattice points of a FCC lattice. Within the solid phase, the structure, properties, and composition are uniform and no interface exists between the copper and nickel atoms.

Copper and nickel also have unlimited solid solubility. The solid phase may be called a *solid solution* [Figure 8-2(b)].

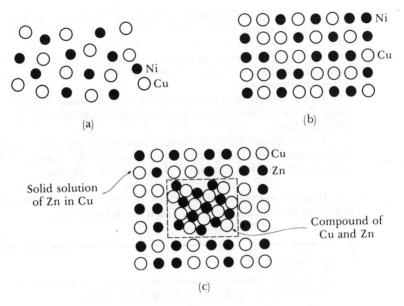

FIG. 8-2   (a) Liquid copper-nickel alloys are completely soluble in one another. (b) Solid copper-nickel alloys display complete solid solubility, with copper and nickel atoms occupying random lattice sites. (c) In copper-zinc alloys containing more than 40% Zn, a second phase forms due to limited solubility of zinc in copper.

Limited solubility.   When we add a small quantity of salt (one phase) to a glass of water (a second phase) and stir, the salt dissolves completely in the water [Figure 8-1(c)]. Only one phase—salty water or brine—is found. However, if we add too much salt to the water, the excess salt sinks to the bottom of the glass. Now we have two phases—water that is saturated with salt plus excess solid salt. We find that salt has a *limited solubility* in water.

If we add a small amount of liquid zinc to liquid copper, a single liquid solution is produced. When that copper-zinc solution cools and solidifies, a single solid solution having a FCC structure results, with copper and zinc atoms randomly located at the normal lattice points. However, if the liquid solution contains more than about 40% Zn, the excess zinc atoms combine with some of the copper atoms to form a CuZn compound [Figure 8-2(c)]. Two solid phases now coexist—a solid solution of copper saturated with about 40% Zn plus a CuZn compound. The solubility of zinc in copper is limited.

In the extreme case, there may be no solubility of one material in another. This is the case for oil and water [Figure 8-1(d)] or for copper-lead alloys.

8-3 Conditions for Unlimited Solid Solubility in Metals

In order for an alloy system, such as copper-nickel, to have unlimited solid solubility, certain conditions must be satisfied. These conditions are known as the Hume-Rothery rules and are as follows.

1. The atoms of the metals must be of similar size, with no more than a 15% difference in atomic radius. Otherwise the lattice strain produced by the differently sized atoms is too great to permit unlimited solubility.

2. The metals must have the same crystal structure; if not, there must be some point at which there is a transition from one phase to a second phase with a different structure.

3. The atoms of the metals must have the same valence; otherwise the valence electron difference may encourage the formation of compounds rather than solutions.

4. The atoms of the metals must have about the same electronegativity. If the electronegativities differ significantly, compounds again tend to form, as when sodium and chlorine combine to form sodium chloride.

Hume-Rothery's conditions must be met, but are not necessarily sufficient, in order for two metals to have unlimited solid solubility.

EXAMPLE 8-1

Determine which of the following alloy systems might be expected to display unlimited solid solubility: Ag-Au, Al-Si, Ca-Al, K-Ba, Ag-Cu.

Answer:

From Hume-Rothery's conditions

Ag-Au: Both have a valence of 1, both are in the same column of the periodic table and have about the same electronegativity, and both are FCC. $r_{Ag} = 1.445 \times 10^{-8}$ cm, $r_{Au} = 1.442 \times 10^{-8}$ cm, $\Delta r/r = 0.2\% < 15\%$. Silver and gold have complete solid solubility.

Al-Si: Aluminum is FCC, silicon is DC. Aluminum and silicon have limited solid solubility.

Ca-Al: $r_{Ca} = 1.97 \times 10^{-8}$ cm, $r_{Al} = 1.43 \times 10^{-8}$ cm, $\Delta r/r = 37.8\% > 15\%$. Calcium and aluminum have limited solid solubility.

K-Ba: Potassium has a valence of 1 and barium has a valence of 2. They have limited solid solubility.

Ag-Cu: Both have a valence of 1, both are in the same column of the periodic table and have similar electronegativities, and both are FCC. $r_{Ag} = 1.445 \times 10^{-8}$ cm, $r_{Cu} = 1.28 \times 10^{-8}$ cm, $\Delta r/r = 12.9\% < 15\%$. Silver and copper satisfy all of Hume-Rothery's conditions, yet silver and copper do not display unlimited solid solubility!

8-4 Solid Solution Strengthening

By producing solid solution alloys, we cause *solid solution strengthening*. In the copper-nickel system, we have intentionally introduced a solid substitutional atom (say nickel) into the original lattice (say copper). The copper-nickel alloy has a strength that is greater than that of pure copper. Similarly, by adding less than 40% Zn to copper, the zinc behaves as a substitutional atom which strengthens the copper-zinc alloy compared to pure copper.

Degree of solid solution strengthening. The degree of solid solution strengthening depends on two factors. First, a large difference in atomic size between the original (or solvent) atom and the added (or solute) atom will increase the strengthening effect. A larger size difference produces a greater disruption of the initial lattice, making slip of dislocations more difficult (Figure 8-3).

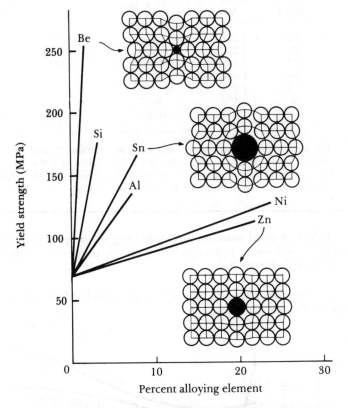

FIG. 8-3 The effect of several alloying elements on the yield strength of copper. Nickel and zinc atoms are about the same size as copper atoms, but beryllium and tin atoms are much different than copper atoms. Both increasing atomic size difference and amount of alloying element increase solid solution strengthening.

Second, the greater the amount of alloying element added, the greater the strengthening effect (Figure 8-3). A Cu-20% Ni alloy is stronger than a Cu-10% Ni alloy. Of course, if too much of a large or small atom is added, the solubility limit may be exceeded and a different strengthening mechanism—*dispersion strengthening*—may be produced. This mechanism will be discussed in Chapter 10.

EXAMPLE 8-2

From the atomic radii, show whether the size difference between copper atoms and alloying atoms accurately predicts the amount of strengthening found in Figure 8-3.

Answer:

The atomic radii and percent size difference are shown below.

Metal	Radius (A)	$\frac{r - r_{Cu}}{r_{Cu}} \times 100$
Cu	1.278	
Zn	1.332	+4.2%
Al	1.432	+12.0%
Sn	1.509	+18.1%
Ni	1.243	−2.7%
Si	1.176	−8.0%
Be	1.14	−10.8%

For atoms larger than copper, namely zinc, aluminum, and tin, increasing the size difference increases the strengthening effect. Likewise for smaller atoms, increasing the size difference increases the strengthening. Note that the larger atoms generally produce less strengthening than the smaller atoms.

Effect of solid solution strengthening on properties. The results of the effects of solid solution strengthening on the properties of the material include the following (Figure 8-4).

1. The yield strength, tensile strength, and hardness of the alloy are greater than for the pure metals.

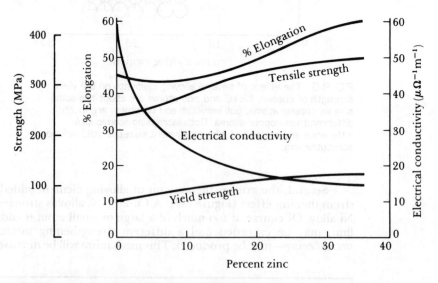

FIG. 8-4 The effect of additions of zinc to copper on the properties of the solid solution-strengthened alloy. The increase in % elongation with increasing zinc content is not typical of solid solution strengthening.

2. Almost always, the ductility of the alloy is less than that of the pure metal. Only rarely, as in copper-zinc alloys, does solid solution strengthening increase both strength and ductility.

3. Electrical conductivity of the alloy is much lower than that of the pure metal. Solid solution strengthening of copper or aluminum wires used for transmission of electrical power is not recommended due to this pronounced effect.

4. The resistance to creep, or loss of strength at elevated temperatures, is improved by solid solution strengthening. High temperatures do not cause a catastrophic change in the properties of solid solution-strengthened alloys.

8-5 Isomorphous Phase Diagrams

A *phase diagram* shows the phases and their compositions at any combination of temperature and alloy composition. When only two elements are present in the alloy, a *binary phase diagram* can be constructed. A simple *isomorphous*, meaning only one solid phase, binary phase diagram for the copper-nickel system is shown in Figure 8-5. There are several valuable pieces of information that we can obtain from the phase diagram.

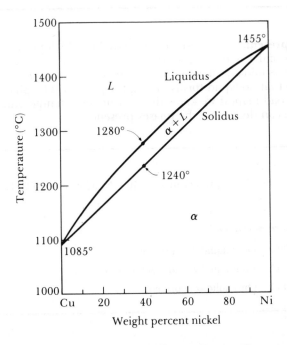

FIG. 8-5 The equilibrium phase diagram for the copper-nickel system. The liquidus and solidus temperatures are shown for a Cu-40% Ni alloy.

Liquidus and solidus temperatures.   The upper curve on the diagram represents the *liquidus* temperature for all of the copper-nickel alloys. We must heat a copper-nickel alloy above the liquidus to produce a completely liquid alloy that can then be cast into a useful shape. The liquid alloy begins to solidify when the temperature cools to the liquidus temperature.

The *solidus* temperature for the copper-nickel alloys is the lower curve. A copper-nickel alloy is not completely solid until the metal cools below the solidus temperature. If we use a copper-nickel alloy at high temperatures, we must be sure that the service temperature is below the solidus so that no melting occurs.

The copper-nickel alloys melt and freeze over a range of temperatures, between the liquidus and solidus. The temperature difference between the liquidus and solidus is the *freezing range* of the alloy. Within the freezing range, two phases coexist—a liquid and a solid. The solid is a solution of copper and nickel atoms; solid phases are typically designated by a lowercase Greek letter, such as α.

EXAMPLE 8-3

Determine the liquidus temperature, solidus temperature, and freezing range for the Cu-40% Ni alloy shown in Figure 8-5.

Answer:

$$T_{\text{liquidus}} = 1280°C$$
$$T_{\text{solidus}} = 1240°C$$
$$\text{Freezing range} = 1280 - 1240 = 40°C$$

Phases present.   Often we are interested in which phases are present in an alloy at a particular temperature. If we plan to make a casting, we must be sure that the metal is initially all liquid; if we plan to heat treat an alloy component, we must be sure that no liquid forms during the process. The phase diagram can be treated as a road map; if we know the coordinates—temperature and alloy composition—we can determine the phases present.

EXAMPLE 8-4

Determine the phases present in a Cu-40% Ni alloy at 1300°C, 1250°C, and 1200°C.

Answer:

From Figure 8-5 we find

    1300°C: Only liquid L is present

    1250°C: Both liquid L and solid α are present

    1200°C: Only solid α is present

Composition of each phase.   Each phase present in an alloy has a composition, expressed as the percentage of each element in the phase. Usually the composition is expressed in weight percent (wt%). When only one phase is present in the alloy, the composition of the phase equals the overall composition of the alloy. If the original composition of the alloy changes, then the composition of the phase must also change.

However, when two phases coexist, such as liquid and solid, the compositions of the two phases differ from one another and also differ from the original overall composition. In this case, if the original composition changes slightly, the composition of the two phases is unaffected, providing that the temperature remains constant.

This difference is explained by the *Gibbs phase rule*, which for a constant pressure is

$$F = C - P + 1 \qquad\qquad (8\text{-}1)$$

where F is the number of degrees of freedom, C is the number of components in the system, and P is the number of phases present. The components C are the smallest number of elements or compounds present in the system. The degrees of freedom F are the number of variables, such as temperature or phase composition, that must be fixed to completely describe the system.

EXAMPLE 8-5

Determine the degrees of freedom in a Cu-40% Ni alloy at 1300°C, 1250°C, and 1200°C.

Answer:

At 1300°C, $P = 1$, since only one phase, liquid, is present. $C = 2$ since both copper and nickel atoms are present. Thus

$$F = 2 - 1 + 1 = 2$$

We must fix both the temperature and the composition of the liquid phase to completely describe the state of the copper-nickel alloy in the liquid region.

At 1250°C, $P = 2$ since both liquid and solid are present. $C = 2$ since copper and nickel atoms are present. Now

$$F = 2 - 2 + 1 = 1$$

If we fix the temperature in the two-phase region, the compositions of the two phases are also fixed. Or, if the composition of one phase is fixed, the temperature and composition of the second phase are automatically fixed.

At 1200°C, $P = 1$, since only one phase, solid, is present. $C = 2$ since both copper and nickel atoms are present. Again

$$F = 2 - 1 + 1 = 2$$

and we must fix both temperature and composition to completely describe the solid.

Because there is only one degree of freedom in a two-phase region of a binary phase diagram, the compositions of the two phases are always fixed when we specify the temperature. This is true even if the overall composition of the alloy changes. This permits us to use a tie line to determine the composition of the two phases. A *tie line* is a horizontal line within a two-phase region drawn at the temperature of interest (Figure 8-6). Tie lines are not used in single-phase regions. In an isomorphous system, the tie line connects the liquidus and solidus points at the specified temperature. The ends of the tie line represent the compositions of the two phases in equilibrium.

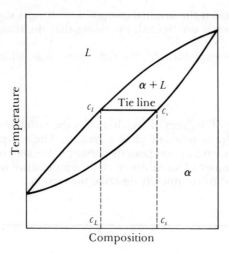

FIG. 8-6 When an alloy is present in a two-phase region, a tie line at the temperature of interest fixes the composition of the two phases. This is a consequence of Gibbs phase rule, which provides only one degree of freedom.

EXAMPLE 8-6

Determine the composition of each phase in a Cu-40% Ni alloy at 1300°C, 1270°C, 1250°C, and 1200°C. (See Figure 8-7.)

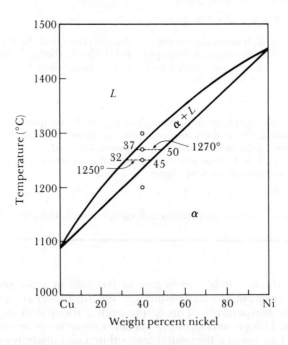

FIG. 8-7 Tie lines and phase compositions for a Cu-40% Ni alloy at several temperatures.

Answer:

A vertical line at 40% Ni represents the overall composition of the alloy.

1300°C: The only phase present is liquid. The liquid must contain 40% Ni, the overall composition of the alloy.

1270°C: Two phases are present. A horizontal line within the $\alpha + L$ field is drawn. The endpoint at the liquidus, which is in contact with the liquid region, is at 37% Ni. The endpoint at the solidus, which is in contact with the α region, is at 50% Ni. Therefore, the liquid contains 37% Ni and the solid contains 50% Ni.

1250°C: Again two phases are present. The tie line drawn at this temperature shows that the liquid contains 32% Ni and the solid contains 45% Ni.

1200°C: Only solid α is present, so the solid must contain 40% Ni.

In Example 8-6, we find that the solid α contains more nickel than the overall alloy and the liquid L contains more copper than the original alloy. Generally, the higher melting point element, in this case nickel, is concentrated in the first solid that forms.

Amount of each phase (the lever law).   Lastly, we are interested in the relative amounts of each phase present in the alloy. These amounts are normally expressed as weight percent (wt%).

In single-phase regions, the amount of the single phase is 100%. However, in two-phase regions we must calculate the amount of each phase. One technique is to perform a materials balance, as shown in Example 8-7.

EXAMPLE 8-7

Calculate the amount of α and L at 1250°C in the Cu-40% Ni alloy shown in Figure 8-8.

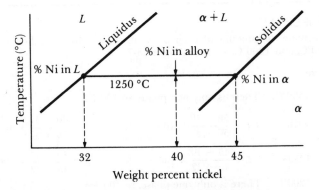

FIG. 8-8   A tie line at 1250°C in the copper-nickel system that is used in Example 8-7 to find the amount of each phase.

Answer:

Let's say that x = fraction of the alloy that is solid.

$$(\% \text{ Ni in } \alpha)(x) + (\% \text{ Ni in } L)(1 - x) = (\% \text{ Ni in alloy})$$

By multiplying and rearranging

$$x = \frac{(\% \text{ Ni in alloy}) - (\% \text{ Ni in } L)}{(\% \text{ Ni in } \alpha) - (\% \text{ Ni in } L)}$$

From the phase diagram at 1250°C

$$x = \frac{40 - 32}{45 - 32} = \frac{8}{13} = 0.62$$

If we convert from weight fraction to weight percent, the alloy at 1250°C contains 62% α and 38% L.

To calculate the amounts of liquid and solid, we construct a lever on our tie line, with the fulcrum of our lever being the original composition of the alloy. The leg of the lever opposite to the composition of the phase whose amount we are calculating is divided by the total length of the lever to give the amount of that phase. In Example 8-7, note that the denominator represents the total length of the tie line and the numerator is the portion of the lever that is opposite the composition of the solid which we are trying to calculate.

The *lever law* in general can be written as

$$\text{Phase percent} = \frac{\text{opposite arm of lever}}{\text{total length of tie line}} \times 100 \qquad (8\text{-}2)$$

We can work the lever law in any two-phase region of a binary phase diagram. The lever law calculation is not used in single-phase regions because the answer is trivial—there is 100% of that phase present.

EXAMPLE 8-8

Determine the amount of each phase in the Cu-40% Ni alloy at 1300°C, 1270°C, 1250°C, and 1200°C, shown in Figure 8-7.

Answer:

1300°C: There is only one phase, so 100% L

1270°C: $\% L = \dfrac{50 - 40}{50 - 37} \times 100 = 77\%$ $\% \alpha = \dfrac{40 - 37}{50 - 37} \times 100 = 23\%$

1250°C: $\% L = \dfrac{45 - 40}{45 - 32} \times 100 = 38\%$ $\% \alpha = \dfrac{40 - 32}{45 - 32} \times 100 = 62\%$

1200°C: There is only one phase, so 100% α

EXAMPLE 8-9

Sometimes we wish to express composition as atomic percent (at%) rather than weight percent (wt%). Express the composition of a phase containing 40 wt% Ni-60 wt% Cu in atomic percent.

Answer:

Select as a base 100 g of the alloy, so there are 40 g of nickel and 60 g of copper in the base. From this data, the atomic weights of nickel and copper, and the Avogadro number N_A, we can calculate the atomic percent.

$$\frac{40 \text{ g Ni}}{58.71 \dfrac{\text{g}}{\text{g} \cdot \text{mole}}} \times 6.02 \times 10^{23} \frac{\text{atoms}}{\text{g} \cdot \text{mole}} = 4.1 \times 10^{23} \text{ Ni atoms}$$

$$\frac{60 \text{ g Cu}}{63.54 \dfrac{\text{g}}{\text{g} \cdot \text{mole}}} \times 6.02 \times 10^{23} \frac{\text{atoms}}{\text{g} \cdot \text{mole}} = 5.7 \times 10^{23} \text{ Cu atoms}$$

$$\text{at\% Ni} = \frac{4.1 \times 10^{23}}{4.1 \times 10^{23} + 5.7 \times 10^{23}} = 42\%$$

$$\text{at\% Cu} = \frac{5.7 \times 10^{23}}{4.1 \times 10^{23} + 5.7 \times 10^{23}} = 58\%$$

8-6 Relationship between Strength and the Phase Diagram

We have previously mentioned that a copper-nickel alloy may be stronger than either pure copper or pure nickel due to solid solution strengthening. The change in mechanical properties of a series of copper-nickel alloys is shown in conjunction with the phase diagram in Figure 8-9.

The strength of the copper increases by solid solution strengthening until about 60% Ni is added. On the other hand, pure nickel is solid solution strengthened by the addition of copper until 40% Cu is added. The maximum strength is obtained for a Cu-60% Ni alloy, known as *Monel*. The maximum is closer to the pure nickel side of the phase diagram because pure nickel is stronger than pure copper.

8-7 Solidification of a Solid Solution Alloy

When an alloy such as Cu-40% Ni is melted and cooled, solidification requires that both nucleation and growth occur. Heterogeneous nucleation permits little or no undercooling, so solidification begins when the liquid reaches the liquidus temperature. The phase diagram (Figure 8-10), with a tie line drawn at the liquidus temperature, tells us that the first solid to form has a composition of Cu-52% Ni. Growth of the solid requires that the latent heat of fusion, which evolves as the liquid solidifies, be removed from the solid-liquid interface. In addition, diffusion must occur so that the compositions of the solid and liquid phases follow the solidus and liquidus curves during cooling. The latent heat of fusion is removed over a range of temperatures so that the cooling curve shows a change in slope, rather than a flat plateau (Figure 8-11).

At the start of freezing, the liquid contains Cu-40% Ni and the first solid contains Cu-52% Ni. Nickel atoms must have diffused to and concentrated at the first solid to form. But after cooling to 1250°C, solidification has advanced and the

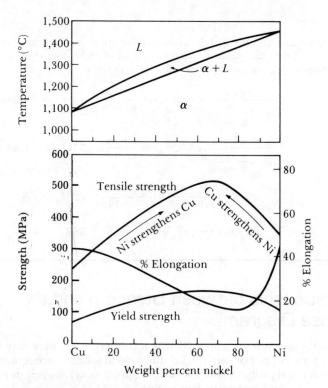

FIG. 8-9 The mechanical properties of copper-nickel alloys.
Copper is strengthened by up to 60% Ni and nickel is
strengthened by up to 40% Cu.

phase diagram tells us that now all of the liquid must contain 32% Ni and all of the
solid must contain 45% Ni. On cooling from the liquidus to 1250°C, some nickel
atoms must diffuse from the first solid to the new solid, reducing the nickel in the
first solid. Additional nickel atoms diffuse from the solidifying liquid to the new
solid. Meanwhile copper atoms have concentrated, by diffusion, into the remaining liquid. This process must continue until the last liquid, which contains Cu-28%
Ni, solidifies and forms a solid containing Cu-40% Ni. Just below the solidus, all of
the solid must contain a uniform concentration of 40% Ni throughout.

In order to achieve this equilibrium final structure, the cooling rate must be
extremely slow. Sufficient time must be permitted for the copper and nickel atoms
to diffuse and produce the composition given by the phase diagram. In most
practical casting situations, the cooling rate is too rapid to permit equilibrium.

8-8 Nonequilibrium Solidification of Solid Solution Alloys

When cooling is too rapid for atoms to diffuse and produce the equilibrium
conditions, then unusual structures are produced in the casting. Let's see what
happens to our Cu-40% Ni alloy on rapid cooling.

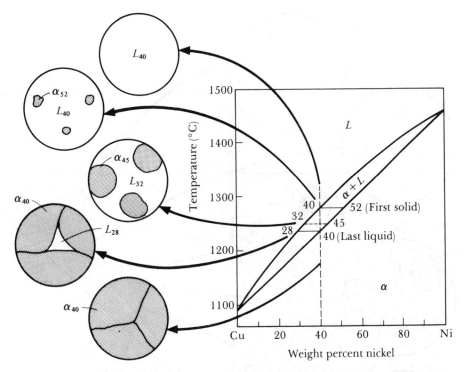

FIG. 8-10   The change in structure of a Cu-40% Ni alloy during equilibrium solidification. The nickel and copper atoms must diffuse during cooling in order to satisfy the phase diagram and produce a uniform, equilibrium structure.

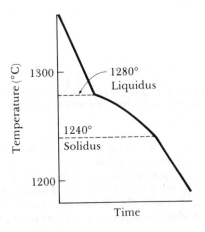

FIG. 8-11   The cooling curve for an isomorphous alloy during solidification. The changes in slope of the cooling curve indicate the liquidus and solidus temperatures, in this case for a Cu-40% Ni alloy.

Again the first solid, containing 52% Ni, forms on reaching the liquidus temperature (Figure 8-12). On cooling to 1260°C, the tie line tells us that the liquid contains 34% Ni and the solid which forms at that temperature contains 46% Ni. Since diffusion occurs rapidly in liquids, we expect the tie line to accu-

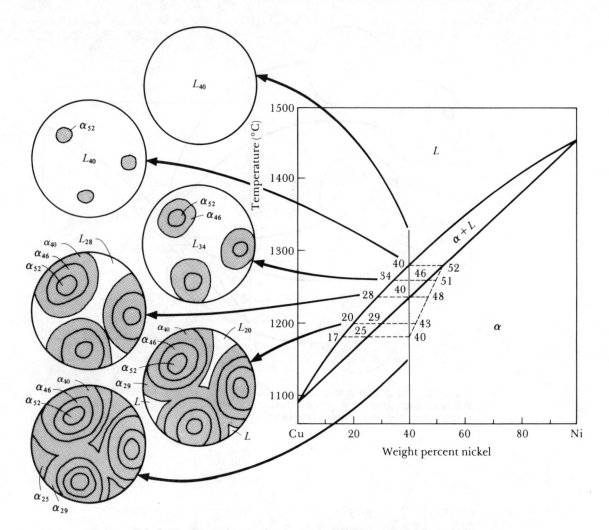

FIG. 8-12 The change in structure of a Cu-40% Ni alloy during nonequilibrium solidification. Insufficient time for diffusion in the solid produces a segregated structure.

rately predict the liquid composition. However, diffusion in solids is comparatively slow. The first solid that forms still has about 52% Ni but the new solid contains only 46% Ni. The average composition of the solid, although difficult to actually calculate, may be approximated at 51% Ni. This gives a different nonequilibrium solidus than that given by the phase diagram. As solidification continues, the nonequilibrium solidus line continues to separate from the equilibrium solidus.

When the temperature reaches 1240°C, the equilibrium solidus line, a significant amount of liquid remains. We could estimate the amount of liquid by performing a lever law calculation, where the ends of the lever are given by the liquidus point and the nonequilibrium solidus point. The liquid will not completely solidify until we cool to 1190°C, where the nonequilibrium solidus intersects the original composition of 40% Ni. At that temperature liquid containing 17% Ni solidifies, giving solid containing 25% Ni. The average composition of the solid is 40% Ni, but the composition is not uniform.

The actual location of the nonequilibrium solidus line and the final nonequilibrium solidus temperature depend on the cooling rate. Faster cooling rates cause greater departures from equilibrium.

EXAMPLE 8-10

Calculate the composition and amount of each phase in a Cu-40% Ni alloy that is present under the nonequilibrium conditions shown in Figure 8-12 at 1300°C, 1280°C, 1260°C, 1240°C, 1200°C, and 1150°C. Compare to the equilibrium compositions and amounts of each phase.

Answer:

Temperature		Equilibrium			Nonequilibrium	
1300	L:	40% Ni	100% L	L:	40% Ni	100% L
1280	L:	40% Ni	100% L	L:	40% Ni	100% L
	α:	52% Ni	~0% α	α:	52% Ni	~0% α
1260	L:	34% Ni	$\dfrac{46-40}{46-34}=50\%\,L$	L:	34% Ni	$\dfrac{51-40}{51-34}=65\%\,L$
	α:	46% Ni	$\dfrac{40-34}{46-34}=50\%\,\alpha$	α:	51% Ni	$\dfrac{40-34}{51-34}=35\%\,\alpha$
1240	L:	28% Ni	0% L	L:	28% Ni	$\dfrac{48-40}{48-28}=40\%\,L$
	α:	40% Ni	100% α	α:	48% Ni	$\dfrac{40-28}{48-28}=60\%\,\alpha$
1200	α:	40% Ni	100% α	L:	20% Ni	$\dfrac{43-40}{43-20}=13\%\,L$
				α:	43% Ni	$\dfrac{40-20}{43-20}=87\%\,\alpha$
1150	α:	40% Ni	100% α	α:	40% Ni	100% α

8-9 Segregation

The nonuniform composition produced by nonequilibrium solidification is known as *segregation*. Figure 8-13 shows the development of *interdendritic segregation* or *microsegregation*, sometimes known as *coring*, which occurs over short distances between small dendrite arms. Dendritic growth is typical during the solidification of solid solution alloys, even when no thermal undercooling occurs.

Because of nonequilibrium solidification, the centers of the dendrites, which represent the first solid to freeze, are rich in the higher melting point element in the alloy. The regions between the dendrites are rich in the lower melting point element, since these regions represent the last liquid to freeze. Although we still have just one solid phase α, with a FCC structure, the composition and properties of α differ from one region to the next. We would expect the casting to have poorer properties as a result.

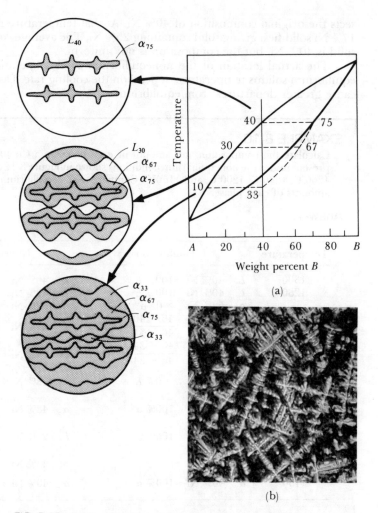

FIG. 8-13   (a) Development of interdendritic segregation during solidification. (b) Photomicrograph of segregation in a copper-nickel alloy.

Hot shortness.   Microsegregation can cause *hot shortness*, or melting of the lower melting point interdendritic material at temperatures below the equilibrium solidus. When we heat the Cu-40% Ni alloy to 1225°C, below the equilibrium solidus but above the nonequilibrium solidus, the low-nickel regions between the dendrites melt.

Homogenization.   We can reduce the interdendritic segregation and problems with hot shortness by using a *homogenization heat treatment*. If we heat the casting to a temperature below the nonequilibrium solidus and hold at that temperature for a long time, diffusion occurs (Figure 8-14). The nickel atoms in the center of the dendrites diffuse to the interdendritic regions; copper atoms diffuse in the opposite direction. Since the diffusion distances are relatively short, only a few hours are required to eliminate most of the composition differences.

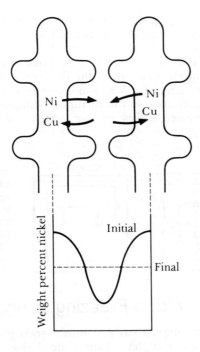

Weight percent nickel

Initial

Final

FIG. 8-14   Microsegregation between dendrites can be reduced by an homogenization heat treatment. Counterdiffusion of nickel and copper atoms may eventually eliminate the composition gradients and produce a homogeneous composition.

EXAMPLE 8-11

Suppose that segregation in a copper-nickel alloy has produced a dendritic network, with the centers of the dendrites containing 52% Ni and the interdendritic region containing only 20% Ni. The secondary dendrite arm spacing is 0.005 cm. We wish to homogenize the casting and raise the % Ni from 20% to 30% in the interdendritic regions. Estimate the time required if the homogenization treatment is carried out at 1000°C. The diffusion coefficient of nickel in copper at 1000°C is about 2.63×10^{-10} cm²/s.

Answer:

As an approximation, let's use the solution to Fick's second law given by Equation 5-4.

$$\frac{c_x - c_m}{c_1 - c_m} = \text{erf} \frac{x}{2\sqrt{Dt}}$$

We want $c_x = 30\%$ Ni at $x = 0.0025$ cm, which is directly between two dendrite arms. c_1 is 52% Ni, the original composition at the center of the dendrites. $c_m = (52 + 20)/2 = 36\%$ Ni.

$$\frac{30 - 36}{52 - 36} = \text{erf}\left(\frac{0.0025}{2\sqrt{2.63 \times 10^{-10})(t)}}\right)$$

$$-0.375 = \text{erf}\left(\frac{77}{\sqrt{t}}\right)$$

From Figure 5-9

$$0.34 = \frac{77}{\sqrt{t}} \quad \text{or } t = 51,290 \text{ s} = 14.2 \text{ h}$$

Macrosegregation.   *Macrosegregation* occurs over a larger distance, between the surface and the center of the casting (Figure 8-15). The surface, which freezes first, contains slightly more than the average amount of the higher melting point metal (nickel is an example). The center of the casting contains more of the lower melting point metal (say copper). Fortunately, the difference in composition is rather small and usually is not significant. We cannot eliminate macrosegregation by a homogenization treatment because the diffusion distances are too great. Macrosegregation can be reduced by *hot working*, which will be discussed in Chapter 9.

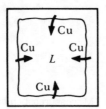 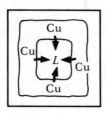 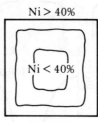

FIG. 8-15
Macrosegregation occurs over a large distance in a casting and cannot be eliminated by homogenization in a practical period of time.

8-10 Castability of Alloys with a Freezing Range

In alloys with a long freezing range, a pasty or mushy solid plus liquid region develops before solidification is complete. Liquid metal does not easily flow through this mushy region to compensate for solidification shrinkage; massive risers may not even be sufficient. Consequently, widespread interdendritic shrinkage is typically found. The pure metals and alloys with a short freezing range, on the other hand, tend to produce little or no mushy region; consequently, concentrated shrinkage voids which can be controlled by risers are typically found (Figure 8-16).

EXAMPLE 8-12

Determine the freezing range for each of the following alloys and comment on the shrinkage that occurs during solidification: Cu-30% Ni, Cu-15% Zn, Cu-5% Al, Cu-4% Si, Cu-10% Sn. The phase diagrams are shown in Figure 12-7.

Answer:

From the phase diagrams, Figures 12-7 and 8-5, the freezing range for each alloy is

$$(Cu\text{-}30\%\ Ni) = 1245 - 1200 = 45°C$$
$$(Cu\text{-}15\%\ Zn) = 1025 - 1010 = 15°C$$
$$(Cu\text{-}5\%\ Al) = 1065 - 1055 = 10°C$$
$$(Cu\text{-}4\%\ Si) = 1000 - 920 = 80°C$$
$$(Cu\text{-}10\%\ Sn) = 1015 - 830 = 185°C$$

The alloys containing nickel, zinc, and aluminum have short freezing ranges. Shrinkage will be concentrated as cavities, but can be eliminated easily by effective use of risers. The alloys containing silicon and tin have long freezing ranges and will have dispersed shrinkage. Even numerous large risers may not eliminate the dispersed shrinkage.

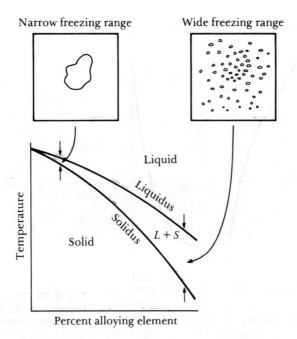

FIG. 8-16 The effect of freezing range on shrinkage.
Narrow freezing range alloys display concentrated
shrinkage, while wide freezing range alloys contain
dispersed shrinkage.

8-11 Ceramic and Polymer Systems

Although we have used the copper-nickel alloy system as an example for discussing solid solution strengthening and the isomorphous phase diagram, similar behavior is observed in many other metallic and ceramic alloy systems. Figure 8-17 shows the phase diagram for the NiO-MgO system; this phase diagram is treated in exactly the same way as the copper-nickel system.

Polymers behave in a different manner although, as shown in Figure 8-18, we can show the softening temperature for a complete series of solid solutions of

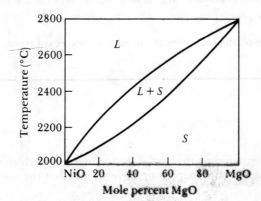

FIG. 8-17 The equilibrium phase
diagram for the NiO-MgO system.

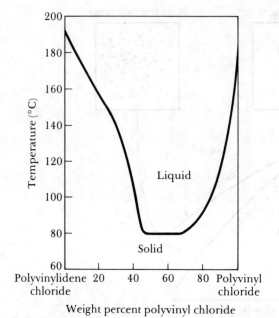

FIG. 8-18 The softening temperature for the vinyl chloride-vinylidene chloride system.

vinyl chloride and vinylidene chloride. These two organic molecules can combine in any percentage to form long polymer chains.

EXAMPLE 8-13

Determine the number of degrees of freedom in the NiO-50% MgO system at 2400°C. (See Figure 8-17.)

Answer:

The Gibbs phase rule, Equation (8-1), is

$$F = C - P + 1$$

Although there are three types of atoms—nickel, oxygen, and magnesium—present, there are only two components, NiO and MgO. Thus $C = 2$. At 2400°C, the NiO-50% MgO material is in a two-phase region since liquid and solid are present, so $P = 2$.

$$F = 2 - 2 + 1 = 1$$

Just as in the copper-nickel system, the number of degrees of freedom is one in a two-phase region.

SUMMARY

Solid solution strengthening is the first of the techniques by which we strengthen a material by alloying. In this chapter we have shown that the strength of single-phase alloys is improved by the controlled addition of point defects. The response of the material to solid solution strengthening is determined by the type of element, in particular the atomic size difference and, at least up to a point, the amount of that element. Solid solution strengthening may provide a significant increase in the strength of the material which may be preserved at elevated temperatures, providing effective barriers against creep.

The amount of solid solution strengthening may be limited by the solubility of the alloying element. However, even in this case the original phase is strengthened by whatever amount of alloying element is soluble. In more complex multiple-phase materials, each phase is often strengthened by solid solution strengthening, even though the bulk of strengthening is by another mechanism.

Finally, solid solution strengthening complicates the solidification process of the material. The material now solidifies over a range of temperatures, providing an opportunity for composition differences to be produced from point to point in the material. This change in the solidification pattern, which is predicted by the phase diagram, introduces additional problems in controlling the solidification process and may even necessitate a homogenization heat treatment or restrict the temperatures at which a material can be used.

GLOSSARY

Binary phase diagram A phase diagram in which there are only two components.

Freezing range The temperature difference between the liquidus and solidus temperatures.

Gibbs phase rule This describes the number of degrees of freedom, or the number of variables that must be fixed to specify the temperature and composition of a phase.

Homogenization heat treatment The heat treatment used to reduce the segregation caused during nonequilibrium solidification.

Hot shortness Melting of the lower melting point nonequilibrium material that forms due to segregation, even though the temperature is below the equilibrium solidus temperature.

Hume-Rothery rules The conditions that an alloy system must meet if the system is to display unlimited solid solubility. Hume-Rothery's rules are necessary but not sufficient.

Isomorphous phase diagram A phase diagram that displays unlimited solid solubility.

Lever law A technique for determining the amount of each phase in a two-phase system.

Limited solubility When only a maximum amount of a solute material can be dissolved in a solvent material.

Liquidus The temperature at which the first solid begins to form during solidification.

Macrosegregation Presence of composition differences in a material over large distances due to nonequilibrium solidification.

Microsegregation Presence of concentration differences in a material over short distances due to nonequilibrium solidification. Also known as interdendritic segregation or coring.

Phase A material having the same composition, structure, and properties everywhere under equilibrium conditions.

Phase diagram A diagram showing the phases and the phase compositions at each combination of temperature and overall composition.

Segregation Presence of nonequilibrium composition differences in a material, often caused by insufficient time for diffusion during solidification.

Solid solution A solid phase that contains a mixture of more than one element, with the elements combining to give a uniform composition everywhere.

Solid solution strengthening Increasing the

strength of a material by introducing point defects into the structure in a deliberate and controlled manner.

Solidus   The temperature below which all liquid has completely solidified.

Solubility   The amount of one material that will completely dissolve in a second material without creating a second phase.

Tie line   A horizontal line drawn in a two-phase region of a phase diagram to assist in determining the compositions of the two phases.

Unlimited solubility   When the amount of one material that will dissolve in a second material without creating a second phase is unlimited.

PRACTICE PROBLEMS

1   Air contains about 80% nitrogen and 20% oxygen. How many phases are present in air?

2   Which of the following alloy systems might be expected to display unlimited solid solubility?

  (a) Cd-Mg   (b) Cu-Sn   (c) W-V

3   Which of the following alloy systems might be expected to display unlimited solid solubility?

  (a) Mo-W   (b) Ge-Si   (c) Al-Au

4   Suppose 2 at% of the following elements are added to copper without exceeding the solubility limit. Which will give the higher strength alloy? Which alloys might have unlimited solid solubility in copper?

  (a) Au   (b) Ag   (c) Co   (d) Ge
  (e) In   (f) Li   (g) Mn   (h) Pd

5   Suppose 2 at% of the following elements are added to BCC iron without exceeding the solubility limit. Which element is expected to give the higher strength alloy? Which alloys might have unlimited solid solubility in iron?

  (a) Mo   (b) Ni   (c) W   (d) Sn
  (e) V   (f) Co   (g) Be   (h) Al

6   Determine the solubility of copper in solid nickel at 500°C, 1100°C, and 1400°C.

7   Determine the solubility of nickel oxide in solid magnesium oxide at 1800°C, 2200°C, and 2400°C.

8   Determine the liquidus temperature, solidus temperature, and freezing range of the following alloys. At what composition is the largest freezing range obtained?

  (a) 100% Ni   (b) Ni-20% Cu

  (c) Ni-50% Cu   (d) Ni-80% Cu

  (e) 100% Cu

9   Determine the liquidus temperature, solidus temperature, and freezing range for the following alloys (Figure 8-19). At what composition is the largest freezing range obtained?

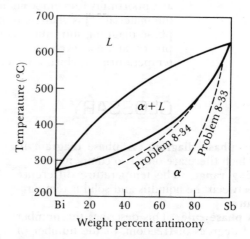

FIG. 8-19   The bismuth-antimony phase diagram. (The two dashed lines are the nonequilibrium solidus lines for Problems 8-33 and 8-34.)

  (a) 100% Bi   (b) Bi-20% Sb
  (c) Bi-40% Sb   (d) Bi-60% Sb
  (e) Bi-80% Sb   (f) 100% Sb

10   Determine the phases, compositions of each phase, and amounts of each phase in wt% for a Cu-80% Ni alloy at 1500°C, 1400°C, and 1300°C.

11   Determine the phases, compositions of each phase, and amounts of each phase for a Cu-20% Ni alloy at 1100°C, 1180°C, and 1300°C.

12   Determine the phases, compositions of each phase, and amounts of each phase in a Bi-40% Sb alloy at 300°C, 350°C, 400°C, 450°C, and 500°C. See Figure 8-19.

13   Determine the phases, compositions of each phase, and amounts of each phase for a Bi-60% Sb alloy at 300°C, 400°C, 500°C, and 600°C. See Figure 8-19.

14 Determine the phases, compositions of each phase, and amounts of each phase in a NiO-40% MgO ceramic at 2200°C, 2400°C, and 2600°C.

15 Determine the phases, compositions of each phase, and amounts of each phase for a NiO-70% MgO ceramic at 2400°C, 2600°C, and 2800°C.

16 A Cu-48% Ni alloy and a Cu-55% Ni alloy are heated to 1300°C. (a) What phases are present in each alloy? (b) What is the composition of each phase in both of the alloys? (c) How does this result compare with the statement of the phase rule?

17 A copper-nickel alloy heated to 1200°C contains two phases—60% L and 40% α. Determine (a) the composition of each phase and (b) the overall composition of the alloy.

18 A copper-nickel alloy heated to 1300°C contains 30% L and 70% α. Determine (a) the composition of each phase and (b) the overall composition of the alloy.

19 A bismuth-antimony alloy heated to 500°C contains 80% L and 20% α. Determine (a) the composition of each phase and (b) the overall composition of the alloy. See Figure 8-19.

20 What is the maximum temperature to which a homogeneous Cu-60% Ni alloy can be heat treated without melting? What would happen at this temperature if the alloy were not homogeneous?

21 A Bi-60% Sb alloy (Figure 8-19) is heated into the $\alpha + L$ region. The composition of the liquid is determined to be 30% Sb. Determine (a) the temperature to which the alloy is heated, (b) the composition of the solid, and (c) the amount of each phase.

22 A NiO-40% MgO ceramic is heated into the $\alpha + L$ region until the solid contains 50 mol% MgO. Determine (a) the temperature to which the ceramic is heated, (b) the composition of the liquid, and (c) the amount of each phase.

23 A Bi-60% Sb alloy (Figure 8-19) is heated into the $\alpha + L$ region until 30% L forms. By a trial and error method, estimate (a) the temperature to which the alloy is heated and (b) the composition of each phase.

24 A Cu-50% Ni alloy is heated to the $\alpha + L$ region until only 40% α remains. By a trial and error method, determine (a) the temperature to which the alloy is heated and (b) the composition of each phase.

25 For a Cu-80% Ni alloy at 1400°C, determine (a) the amounts and compositions in wt% of each phase and (b) the amounts and compositions in at% of each phase for the same conditions.

26 For a Bi-40% Sb alloy at 400°C (See Figure 8-19), determine (a) the amount of each phase in wt% and (b) the amount of each phase in at%.

27 For a NiO- mol% MgO ceramic at 2400°C, determine (a) the amount of each phase in mol% and (b) the amount of each phase in wt%.

28 A copper-nickel alloy is prepared by mixing together equal numbers of copper and nickel atoms. Determine (a) the wt% Ni in the alloy and (b) the amounts and compositions of each phase present at 1300°C.

29 A bismuth-antimony alloy (Figure 8-19) is prepared by mixing together twice as many bismuth atoms as antimony atoms. Determine (a) the wt% Sb in the alloy and (b) the amounts and compositions of each phase present at 400°C.

30 Develop an equation or expression to convert weight percent to volume percent.

31 A Bi-40% Sb alloy (Figure 8-19) is melted, then cooled. Determine (a) the composition of the first solid to freeze and (b) the composition of the last liquid to solidify.

32 A NiO-60% MgO ceramic is melted, then cooled. Determine (a) the composition of the first solid to form and (b) the composition of the last liquid to solidify.

33 A Bi-80% Sb alloy is solidified rapidly, producing the nonequilibrium solidus curve shown in Figure 8-19. (a) What is the composition of the first solid to form? (b) What are the amounts and compositions of each phase under equilibrium conditions at 500°C and at 400°C? (c) What are the amounts and compositions of each phase under nonequilibrium conditions at 500°C and 400°C? (d) Under equilibrium conditions, what are the compositions of the last liquid to freeze and the last solid to form? (e) Under nonequilibrium conditions, what are the compositions of the last liquid to freeze and the last solid to form? (f) Compare the equilibrium and nonequilibrium solidus temperatures. (g) What will happen if the nonequilibrium Bi-80% Sb alloy is heat treated at 400°C?

34 Figure 8-19 shows the nonequilibrium solidus line for a Bi-40% Sb alloy. (a) What is the composition of the first solid to form? (b) What are the amounts and compositions of each phase under equilibrium conditions at 450°C, 400°C, and 350°C? (c) What are the amounts and compositions of each phase under nonequilibrium conditions at 450°C, 400°C, and 350°C? (d) Under equilibrium conditions, what are the compositions of the last liquid to freeze and the last solid to form? (e) Under nonequilibrium conditions, what are the compositions of the last liquid to freeze and the last solid to form? (f) Compare the equilibrium and nonequilibrium solidus temperatures. (g) What will happen if the nonequilibrium alloy is heat treated at 325°C?

35   Describe how a change in the size of the secondary dendrite arm spacing of a copper-nickel alloy will affect the time required for homogenization.

36   From Figure 8-20, determine (a) the pouring temperature, (b) the liquidus temperature, (c) the superheat, (d) the solidus temperature, (e) the freezing range, (f) the solidification time, and (g) the local solidification time. If the alloy is a bismuth-antimony alloy, estimate its composition.

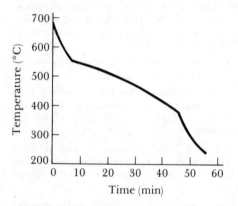

FIG. 8-20   The cooling curve for Problem 8-36.

37   The cooling curves shown in Figure 8-21 are obtained from a series of molybdenum-vanadium alloys. Reproduce the phase diagram for this alloy system.

38   A tungsten-molybdenum alloy produces microsegregation on cooling, with a secondary dendrite arm spacing of 0.20 mm. The diffusion coefficient of tungsten in molybdenum is $D = 0.54 \exp(-120,500/$

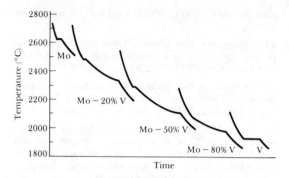

FIG. 8-21   The cooling curves for several molybdenum-vanadium alloys for Problem 8-37

RT). The center of the dendrites contains 55% W and the interdendritic region contains 20% W. Estimate the homogenization time required to increase the % W at the interdendritic regions to 30% W if homogenization is done at (a) 2500°C and (b) 1500°C.

39   A Cu-Zn alloy produces microsegregation on cooling with a secondary dendrite arm spacing of 0.015 cm. The diffusion coefficient for zinc in copper can be found in Table 5-1. The center of the dendrites contains 17% Zn and the interdendritic regions contain 25% Zn. Estimate the time required to increase the % Zn to 19% Zn in the center of the dendrites at (a) 800°C and (b) 500°C.

40   Draw two isomorphous phase diagrams, one with a large freezing range between the liquidus and solidus, the other with a small freezing range between the liquidus and solidus. In which of the two alloy systems do you expect a greater amount of segregation? Explain.

41   Suppose a 3-in. thick copper-nickel casting has macrosegregation. At the surface the % Ni is 55% and at the center the % Ni is 45%. How long must we homogenize at 800°C to raise the % Ni in the center to 48%?

CHAPTER 9

Deformation, Strain Hardening, and Annealing

9-1 Introduction

In this chapter we will discuss three main topics—*cold working,* by which an alloy is simultaneously deformed and strengthened, *hot working,* by which an alloy is deformed at high temperatures without strengthening, and *annealing,* during which the effects of strengthening caused by cold working are eliminated or modified by heat treatment. The strengthening we obtain during cold working, which is brought about by increasing the number of dislocations, is called *strain hardening* or *work hardening.* By controlling these processes of deformation and heat treatment, we are able to process the material into a usable shape yet still improve and control the properties.

COLD WORKING

9-2 Relationship to the Stress-Strain Curve

A stress-strain curve is shown in Figure 9-1(a). As long as the stress does not exceed the yield strength σ_y, no permanent plastic deformation occurs and the elastic deformation is recovered. This is the condition we want to maintain after the finished part is put into service. However, when we wish to manufacture a part by deformation processing, the applied stress must exceed the yield strength, causing the metal to be permanently deformed into a useful shape.

If we apply a stress σ_1 that is greater than the yield strength [Figure 9-1(a)] we cause a permanent deformation, or strain ε_1, when the stress is removed. If we remove a sample from the metal that had been stressed to σ_1 and we retest that metal, we would obtain a different stress-strain curve [Figure 9-1(b)]. Our new test specimen would have a yield strength at σ_1 and would also have a higher tensile strength and a lower ductility. If we continue to apply a stress until we reach σ_2, then release the stress and again retest the metal, the new yield strength would be σ_2. Each time we apply a higher stress to the metal, the yield strength and tensile strength increase while the ductility decreases. We may eventually strengthen the metal until the yield, tensile, and breaking strengths are equal and there is no ductility [Figure 9-1(c)]. At this point the metal can be plastically deformed no further.

By applying a stress that exceeds the original yield strength of the metal, we have *strain hardened* or *cold worked* the metal while simultaneously deforming the metal into a more useful shape.

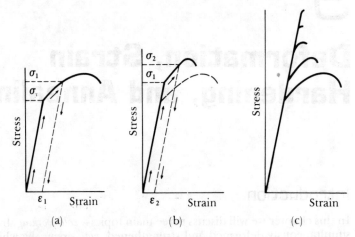

FIG. 9-1   Development of strain hardening from the stress-strain diagram. (a) A specimen is stressed beyond the yield strength before the stress is removed. (b) Now the specimen has a higher yield strength and tensile strength, but lower ductility. (c) By repeating the procedure, the strength continues to increase and the ductility continues to decrease until the alloy becomes very brittle.

Strain-hardening coefficient.   The response of the metal to cold working is given by the *strain-hardening coefficient n*, which is the slope of the plastic portion of the true stress-true strain curve in Figure 9-2 when a log-log scale is used.

$$\sigma_t = K\varepsilon_t^n \tag{9-1}$$

or

$$\log \sigma_t = \log K + n \log \varepsilon_t$$

The constant K is equal to the stress when $\varepsilon_t = 1$.

The strain-hardening coefficient is relatively low for HCP metals but is higher for BCC and particularly for FCC metals (Table 9-1). Metals with a low strain-hardening coefficient have a poor response to cold working.

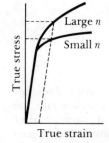

FIG. 9-2
The true stress-true strain curves for metals with large and small strain-hardening coefficients. Larger degrees of strengthening are obtained for a given strain for the metal with the larger *n*.

TABLE 9-1   Strain-hardening coefficients of typical metals and alloys

Metal	Crystal Structure	n	K (MPa)
Titanium	HCP	0.05	1200
Annealed alloy steel	BCC	0.15	640
Quench and Tempered medium-carbon steel	BCC	0.10	1570
Molybdenum	BCC	0.13	725
Copper	FCC	0.54	315
Cu-30% Zn	FCC	0.50	895
Austenitic stainless steel	FCC	0.52	1515

Adapted from G. Dieter, *Mechanical Metallurgy*, McGraw-Hill, 1961, and other sources.

9-3   Dislocation Multiplication

We obtain strengthening during deformation by increasing the number of dislocations in the metal. Before deformation a metal contains about 10^4 mm of dislocation line per cubic millimeter of metal. This is a relatively small number of dislocations.

When we apply a stress greater than the yield strength, dislocations begin to slip. Eventually, a dislocation moving on its slip plane encounters obstacles which pin the ends of the dislocation line. As we continue to apply the stress, the dislocation attempts to move by bowing in the center. The dislocation may move so far that a loop is produced (Figure 9-3). When the dislocation loop finally touches itself, a new dislocation is created. The original dislocation is still pinned and can create additional dislocation loops. This mechanism for generating dislocations is called a *Frank-Read* source.

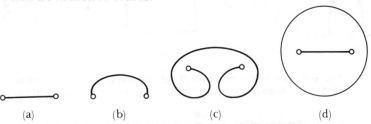

(a)   (b)   (c)   (d)

FIG. 9-3   The Frank-Read source can generate dislocations. (a) A dislocation is pinned at its ends by lattice defects. (b) As the dislocation continues to move, the dislocation bows, eventually bending back on itself, (c). Finally the dislocation loop forms (d) and a new dislocation is created.

The number of dislocations may increase to about 10^{10} mm of dislocation line per cubic millimeter of metal. We know that the more dislocations we have, the more likely they are to interfere with one another, and the stronger the metal becomes.

Ceramic materials may contain dislocations and can even be strain hardened to a small degree. However, ceramics are normally so brittle that significant deformation and strengthening are not possible.

Certain polymers have an excellent response to strengthening by cold working. But the mechanism for strain hardening is quite different in polymers and involves alignment and possibly crystallization of the long chainlike molecules. We will discuss this mechanism in more detail in Chapter 15.

9-4   Properties versus Percent Cold Work

Many techniques are used to simultaneously shape and strengthen a metal by cold working (Figure 9-4). By controlling the amount of deformation we achieve using these processes, we control the amount of strain hardening. We normally measure the amount of deformation by defining the percent cold work.

$$\text{Percent cold work} = \frac{A_0 - A_f}{A_0} \times 100 \qquad (9\text{-}2)$$

where A_0 is the original cross-sectional area of the metal and A_f is the final cross-sectional area after deformation.

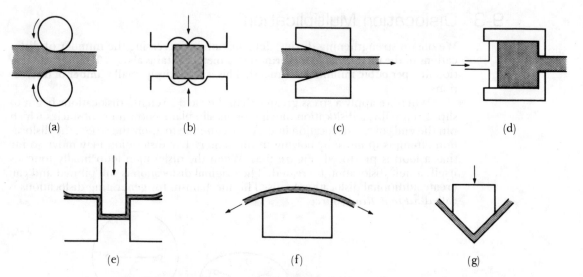

(a)   (b)   (c)   (d)

(e)   (f)   (g)

FIG. 9-4   Schematic drawings of deformation processing techniques. (a) Rolling, (b) forging, (c) drawing, (d) extrusion, (e) deep drawing, (f) stretch forming, and (g) bending.

EXAMPLE 9-1

Calculate the percent cold work when a 10 mm × 60 mm × 200 mm plate is deformed by rolling to a 5 mm plate (Figure 9-5).

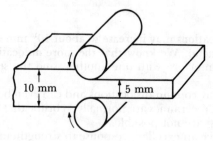

10 mm   5 mm

FIG. 9-5   Diagram showing the rolling of a 10 mm plate to a 5 mm plate.

Answer:

The final dimensions of the plate will be 5 mm × 60 mm × 400 mm since, due to friction between the plate and the rolls, the width does not change during deformation.

$$\text{Percent cold work} = \frac{A_0 - A_f}{A_0} \times 100 = \frac{(10)(60) - (5)(60)}{(10)(60)} \times 100 = 50\%$$

If we measure the properties of the rolled plate, we find that the strength has increased but the ductility has decreased.

The effect of cold work on the mechanical properties of commercially pure copper is shown in Figure 9-6. As the percent cold work increases, both the yield

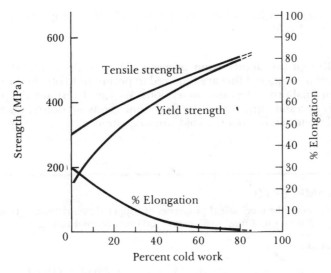

FIG. 9-6 The effect of cold work on the mechanical properties of copper.

and the tensile strength increase; however, the ductility decreases and approaches zero. The metal will break if more cold work is attempted. Therefore, there is a maximum amount of cold work or deformation that we can do to a metal.

EXAMPLE 9-2

Determine if a 10 mm × 60 mm × 200 mm copper plate can be cold rolled to a 1 mm × 60 mm × 2000 mm plate.

Answer:

The amount of cold work required is

$$\text{Percent cold work} = \frac{A_0 - A_f}{A_0} \times 100 = \frac{(10)(60) - (1)(60)}{(10)(60)} \times 100 = 90\%$$

The required cold work exceeds the maximum of 85% cold work for the alloy in Figure 9-6. Therefore, we cannot make the 1 mm thick material with a single rolling operation.

We can use the relationship between cold work and mechanical properties in several ways.

Determine properties after deformation. We can predict the properties of a metal if we know the amount of cold work that is achieved. We can then decide whether the component has adequate strength at critical locations.

Determine necessary cold work. When we wish to select a material for a component that has certain minimum mechanical properties, we can specify the amount of cold work that must be done.

Design the deformation process. We can determine the original dimensions of the material that are required to permit us to obtain the desired combination of final properties and dimensions. We can determine the percent cold work that is necessary and, using the final dimensions we desire, calculate the original metal dimensions from the cold work equation.

EXAMPLE 9-3

Estimate the mechanical properties of copper that is deformed from 10 mm $\times$ 60 mm $\times$ 200 mm to 4 mm $\times$ 60 mm $\times$ 500 mm by cold rolling.

Answer:

$$\text{Percent cold work} = \frac{A_0 - A_f}{A_0} \times 100 = \frac{(10)(60) - (4)(60)}{(10)(60)} \times 100 = 60\%$$

From Figure 9-6, we can determine the properties that are obtained for this amount of deformation.

Yield strength = 490 MPa

Tensile strength = 510 MPa

% Elongation = 2%

EXAMPLE 9-4

A copper rod 12 mm in diameter is to be reduced to 7 mm in diameter. Determine the expected mechanical properties.

Answer:

$$\text{Percent cold work} = \frac{A_0 - A_f}{A_0} \times 100 = \frac{(\pi/4)(d_0^2) - (\pi/4)(d_f^2)}{(\pi/4)(d_0^2)} \times 100$$

$$= \frac{(12)^2 - (7)^2}{(12)^2} \times 100$$

$$= 66\%$$

From Figure 9-6

Yield strength = 500 MPa

Tensile strength = 520 MPa

% Elongation = 1.5%

EXAMPLE 9-5

Suppose we want to produce a copper bar that has at least 415 MPa tensile strength, 380 MPa yield strength, and 5% elongation. How much cold work is required?

Answer:

From Figure 9-6, we need at least 25% cold work to produce a tensile strength of 415 MPa and 30% cold work to produce a yield strength of 380 MPa, but less than 42% cold work to meet the 5% elongation requirement. Any cold work between 30% and 42% will be satisfactory.

EXAMPLE 9-6

Determine the original thickness of a 60 mm wide copper strip that, after cold working to 1 mm, will have greater than 415 MPa yield strength and 5% elongation.

Answer:

From Figure 9-6, the percent cold work must be greater than 40% to meet the minimum yield strength but must be less than 45% to meet the % elongation specification. Any percent cold work within this range will be satisfactory. Thus

$$\text{Percent cold work} = 40 = \frac{(5_{min})(60) - (1)(60)}{(5_{min})(60)} \times 100 \qquad t_{min} = 1.67 \text{ mm}$$

$$\text{Percent cold work} = 45 = \frac{(5_{max})(60) - (1)(60)}{(5_{max})(60)} \times 100 \qquad t_{max} = 1.82 \text{ mm}$$

Any original thickness between 0.167 cm and 0.182 cm will be acceptable.

9-5 Microstructure of Cold-Worked Metals

During deformation, a fibrous microstructure is produced as the grains within the metal become elongated (Figure 9-7).

Anisotropic behavior. During deformation, the grains rotate as well as elongate, causing certain crystallographic directions and planes to become aligned. Consequently, *preferred orientations*, or *textures*, are developed. The effect, as in the columnar zone of a casting, is to produce anisotropic behavior.

In processes such as wire drawing, a *fiber texture* is produced. In BCC metals, ⟨110⟩ directions line up with the axis of the wire [Figure 9-8(a)]. In FCC metals, ⟨111⟩ or ⟨100⟩ directions are aligned. Fortunately, this gives the highest strength along the axis of the wire, which is desired.

In processes such as rolling, both a preferred direction and plane are produced [(Figure 9-8(b)], giving a *sheet texture*. In BCC metals the texture may be {100} ⟨110⟩ and in FCC metals the texture may be {112} ⟨111⟩ or {100} ⟨112⟩.

The properties of a rolled sheet or plate depend on the direction in which we apply the stress (Table 9-2). If the sheet is properly oriented to the applied stress

(a)

(b)

(c)

(d)

FIG. 9-7   The fibrous grain structure of copper after cold working. (a) 0% cold work; (b) 31% cold work; (c) 67% cold work; (d) 82% cold work.

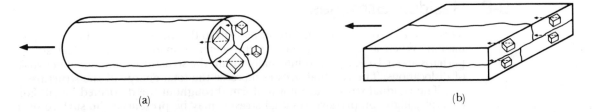

(a)   (b)

FIG. 9-8 Development of anisotropic behavior in (a) drawn and (b) rolled products. Note the orientation of the unit cells in the grains.

TABLE 9-2 Anisotropic properties produced by sheet textures during cold rolling

Metal	Tensile Strength (MPa)			% Elongation		
	0°	45°	90°	0°	45°	90°
Deoxidized Cu						
0% Cold work	214	214	214	50	50	50
50% Cold work	365	365	379	6.0	5.2	5.7
90% Cold work	441	434	462	5.7	5.0	4.0
90% Cu-10% Zn						
0% Cold work	255	255	255	45	45	45
37% Cold work	407	414	434	5	4	3
56% Cold work	510	510	531	4	3	3
95% Cold work	565	600	655	2.7	2.7	3.2
Zn	172		214	45		28

Note: 0° = parallel to rolling direction; 90° = perpendicular to rolling direction.

Adapted from *ASM Metals Handbook, Vol. 1, 8th Ed.*, 1961.

during use, high strengths may be achieved. However, premature failure may occur if we apply a stress from a different direction.

Textures, as one would expect, become more pronounced as the amount of deformation increases.

Alignment and deformation of second phases. Any inclusions (foreign particles such as oxides) or second-phase grains that are present in the original structure are also aligned during deformation. Soft inclusions normally deform and elongate; hard inclusions may not deform but are still aligned in the direction of deformation. Elongated inclusions, called *stringers*, act as tiny internal notches and reduce the mechanical properties of the cold-worked metal. When the part is further processed or put into service, we must be sure that high tensile stresses do not act on the sharp inclusions, causing nucleation of cracks.

9-6 Residual Stresses

Residual stresses, or stresses that are stored within the metal, develop during deformation. When a stress is applied to the metal, a small portion of that stress, perhaps about 10%, is stored internally within the structure as a tangled network of dislocations. The residual stresses increase the total energy of the structure.

The residual stresses are not uniform throughout the deformed metal. For example, high compressive residual stresses may be present at the surface of a rolled plate and high tensile stresses may be stored in the center (Figure 9-9). If

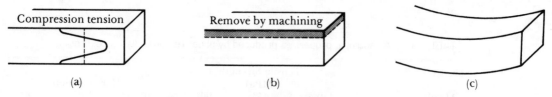

FIG. 9-9   (a) Compressive residual stresses at the surface are balanced by tensile stresses in the center of a cold-worked bar. (b) If one surface is machined, part of the compressive stresses in that surface are removed, upsetting the stress balance. (c) To restore the balance, the bar distorts.

we machine a small amount of metal from one surface of a cold-worked part, we remove metal that only contains compressive residual stresses. To restore the balance, the plate must distort.

Residual stresses also affect the ability of the part to carry a load (Figure 9-10). If a tensile stress is applied to a material that already contains tensile

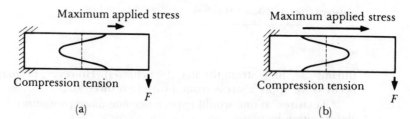

FIG. 9-10   The compressive residual stresses can be harmful or beneficial. In (a), a bending force applies a tensile stress on the top of the beam. Since there are already tensile residual stresses at the top, the load-carrying characteristics are poor. In (b), the top contains compressive residual stresses. Now the load-carrying characteristics are very good.

residual stresses, the total stress acting on the part is the sum of the applied and residual stresses. However, if compressive stresses are stored at the surface of a metal part, an applied tensile stress must first balance the compressive residual stresses. Now the part may be capable of withstanding a larger than normal load.

EXAMPLE 9-7

A cold-worked part has a yield strength of 140 MPa. Calculate the tensile stress that can be applied without permanent deformation if tensile residual stresses of 70 MPa are present at the surface of the part.

Answer:

If both the residual stress and the applied stress are tensile, then the stresses are additive. The applied stress must be less than $140 - 70 = 70$ MPa. The part can withstand only half of the expected load due to the presence of the residual stresses.

EXAMPLE 9-8

Calculate the tensile stress that a cold-worked part, having a yield strength of 140 MPa, can withstand if compressive residual stresses of 70 MPa are present.

Answer:

In this case, the total applied stress that can be supported is $140 + 70 = 210$ MPa. In this example, the compressive residual stresses are helpful.

Sometimes components that are subject to fatigue failure can be strengthened by *shot peening*. Bombarding the surface with steel shot propelled at a high velocity introduces compressive residual stresses at the surface; the compressive stresses significantly increase the resistance of the metal surface to fatigue failure.

9-7 Characteristics of Cold Working

There are a number of advantages and limitations of strengthening a metal by cold working or strain hardening.

1. We can simultaneously strengthen the metal while producing a desired final shape.

2. We can obtain excellent dimensional tolerances and surface finishes by the cold-working process.

3. The cold-working process is an inexpensive method of producing large numbers of small parts. However, for large parts, the amount of cold work is limited. If too much deformation is attempted, the metal may fail during processing. In addition, high forces, requiring large and expensive forming equipment, must be applied to exceed the yield strength in large parts.

4. Some metals, such as HCP magnesium, are rather brittle at room temperature. Only a small degree of cold working can be accomplished without causing the part to embrittle and fail.

5. Ductility, electrical conductivity, and corrosion resistance are impaired by the cold-working process. However, cold working reduces electrical conductivity less than many of the other strengthening processes, such as solid solution

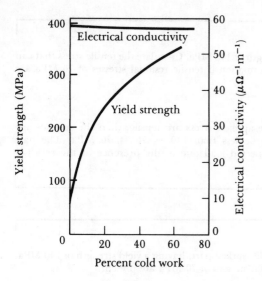
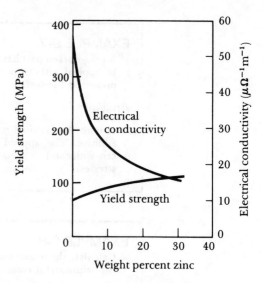

FIG. 9-11   A comparison of strengthening copper by (a) cold working and (b) solid solution strengthening with zinc. Note that cold working produces a greater strengthening effect yet has little effect on electrical conductivity.

strengthening (Figure 9-11). This makes cold working a more satisfactory way to strengthen conductor materials, such as copper wires used for transmission of electrical power.

EXAMPLE 9-9

We wish to increase the yield strength of a copper wire from 70 MPa to 100 MPa Compare the change in electrical conductivity if we achieve strengthening by cold working rather than by adding zinc.

Answer:

From Figure 9-11, we can obtain the required yield strength either with 2% cold work or by adding 19% Zn. For 2% cold work, the electrical conductivity is virtually unaffected, but 19% Zn reduces the electrical conductivity from $58 \times 10^6 \ \Omega^{-1}m^{-1}$ to $20 \times 10^6 \ \Omega^{-1}m^{-1}$.

---

6. Residual stresses and anisotropic behavior may be introduced during cold working. These characteristics may be either harmful or beneficial, depending on how they can be controlled.

7. Some deformation processing techniques can only be accomplished if cold working occurs. For example, wire drawing requires that a rod be pulled through a die to produce a smaller cross-sectional area (Figure 9-12). For a given draw force F_d, a different stress is produced in the original and final wire. The stress on the initial wire must exceed the yield strength of the metal to cause deformation. The stress on the final wire must be less than its yield strength to prevent failure. This is accomplished only if the wire strain hardens during drawing.

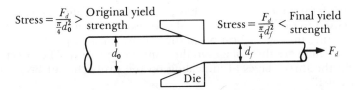

FIG. 9-12   The wire-drawing process. The force F_d acts on both the original and final diameters. Thus, the stress produced in the final wire is greater than that in the original. If the wire did not strain harden during drawing, the final wire would break before the original wire is drawn through the die.

EXAMPLE 9-10

A copper rod 6.4 mm in diameter is to be drawn through a 5.0 mm diameter die in a wire-drawing process. What is the force required to deform the metal? Is the force sufficient to break the wire after it has been formed?

Answer:

$$\text{Percent cold work} = \frac{A_0 - A_f}{A_0} \times 100 = \frac{(6.4)^2 - (5.0)^2}{(6.4)^2} \times 100 = 39\%$$

The initial yield strength, from Figure 9-6, with 0% cold work is 150 MPa. The final yield strength with 39% cold work is 390 MPa. The force required to deform the initial wire is

$$F = \sigma_y A_0 = (150)\left|\left(\frac{\pi}{4}\right)(6.4)^2\right| = 4825 \text{ N}$$

The stress acting on the wire after passing through the die is

$$\sigma = \frac{F}{A_f} = \frac{4825}{(\pi/4)(5.0)^2} = 246 \text{ MPa}$$

Since 246 MPa < 390 MPa, the wire will not break.

ANNEALING

9-8   Three Stages of Annealing

Annealing is a heat treatment designed to eliminate the effects of cold working and to restore the cold-worked metal to the original soft ductile condition. Several applications for annealing are employed. First, annealing may be used to completely eliminate the strain hardening achieved during cold working; the final part is soft and ductile but still has a good surface finish and dimensional accuracy. Second, after annealing, additional cold work could be done since the ductility is restored. By combining repeated cycles of cold working and annealing, large total deformations may be achieved. Finally, annealing at a low temperature may be used to eliminate the residual stresses produced during cold working without affecting the mechanical properties of the finished part.

There are three stages in the annealing process. The effects of cold working and the three stages of annealing on the properties of brass are shown in Figure 9-13.

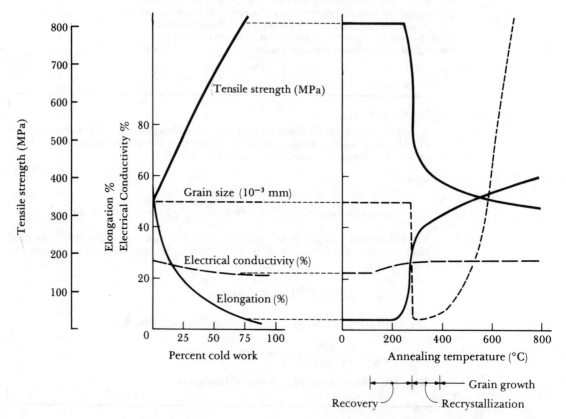

FIG. 9-13 The effect of cold work and annealing on the properties of a Cu-35% Zn alloy.

Recovery. *Recovery,* or a *stress-relief anneal,* is a low-temperature heat treatment designed to reduce residual stresses. The microstructure contains deformed grains which contain a very large number of dislocations in a tangled network. When we heat the metal to slightly elevated temperatures, the dislocations move and are rearranged, while the residual stresses are reduced and eventually eliminated.

However, the number of dislocations is not significantly reduced; they are instead rearranged in a polygonized structure (Figure 9-14). The *polygonized structure* appears to be a subgrain structure within the normal deformed grains, with dislocations forming the subgrain boundaries. Because the number of dislocations is not reduced, the mechanical properties of the metal are relatively unchanged. Recovery restores high electrical conductivity to cold-worked copper or alu-

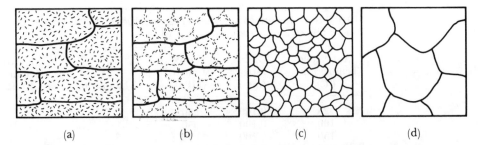

(a)   (b)   (c)   (d)

FIG. 9-14 The effect of annealing temperature on the microstructure of cold-worked metals. (a) Cold worked, (b) after recovery, (c) after recrystallization, and (d) after grain growth.

minum wire used for transmission of electrical power. The wire is strong due to the cold work and can be strung between poles placed far apart without failing; yet the wire has good electrical conductivity.

Recrystallization. *Recrystallization* occurs by the nucleation and growth of new grains containing few dislocations. When the metal is heated above the *recrystallization temperature,* approximately 0.4 times the absolute melting temperature of the metal, rapid recovery eliminates residual stresses and produces the polygonized dislocation structure. New grains then nucleate at the cell boundaries of the polygonized structure, eliminating most of the dislocations (Figure 9-14). Because the number of dislocations is greatly reduced, the recrystallized metal has a low strength but a high ductility.

Grain growth. At still higher annealing temperatures, both recovery and recrystallization rapidly occur, producing the fine recrystallized grain structure. However, the energy associated with the large amount of grain boundary area makes the fine structure unstable at high temperatures. To reduce this energy, the grains begin to grow, with favored grains consuming the smaller grains (Figure 9-14). This phenomenon is called *grain growth* and was described in Chapter 5. Grain growth is almost always undesirable.

EXAMPLE 9-11

From the data in Table 9-3, determine the temperature at which recovery, recrystallization, and grain growth begin in a Cu-12.5% Zn alloy.

Answer:

The start of recovery is indicated by an increase in the electrical conductivity, or at a temperature between 150°C and 200°C.

The start of recrystallization is indicated by a decrease in grain size and tensile strength and an increase in % elongation. These changes begin between 300°C and 350°C.

Grain growth is indicated by the change in grain size. Grain growth occurs at an accelerated rate near 600°C.

TABLE 9-3   The effect of annealing temperature on the properties of a Cu-12.5% Zn alloy. (See Example 9-11.)

Annealing Temperature (°C)	Grain Size (mm)	Tensile Strength (MPa)	% Elongation	Electrical Conductivity ($\times 10^6 \ \Omega^{-1}\mathrm{m}^{-1}$)
25	0.100	550	5	16
100	0.100	550	5	16
150	0.100	550	5	17
200	0.100	550	5	19
250	0.100	550	5	20
300	0.005	515	9	20
350	0.008	380	30	21
400	0.012	330	40	21
500	0.018	275	48	21
600	0.025	270	48	22
700	0.050	260	47	22

9-9   Control of Annealing

There are three important factors that must be considered when selecting an annealing heat treatment—the recrystallization temperature, the size of the recrystallized grains, and the grain growth temperature.

Recrystallization temperature.   The recrystallization temperature is affected by a variety of processing variables.

1. The recrystallization temperature decreases when the amount of cold work increases. Greater amounts of cold work make the metal less stable and encourage nucleation of recrystallized grains. There is a minimum amount of cold work, about 30% to 40%, below which recrystallization will not occur.
2. A smaller original cold-worked grain size also reduces the recrystallization temperature by providing more sites—the former grain boundaries—at which new grains can nucleate.
3. Pure metals recrystallize at lower temperatures than solid solution-strengthened alloys. Often this is helpful. For example, alloys that are to be brazed or soldered may resist annealing and softening during the joining operation more effectively than would a pure metal.
4. Increasing the annealing time reduces the recrystallization temperature (Figure 9-15). Thus, more time is available for nucleation and growth of the new recrystallized grains. However, temperature is far more important. Doubling the annealing time reduces the recrystallization temperature by only about 10°C.
5. The recrystallization temperature depends on the alloy. Generally, higher melting point alloys have a higher recrystallization temperature. Since recrystallization is a diffusion-controlled process, the recrystallization temperature is roughly proportional to $0.4T_\mathrm{m}$ (absolute). Typical recrystallization temperatures for selected metals are included in Table 9-4.

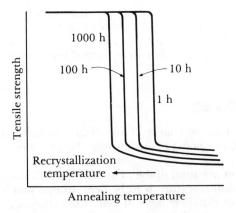

FIG. 9-15 The effect of annealing time on the recrystallization temperature.

TABLE 9-4 Typical recrystallization temperatures for selected metals

Metal	Melting Temperature (°C)	Recrystallization Temperature (°C)
Sn	232	< room temperature
Cd	321	< room temperature
Pb	327	< room temperature
Zn	420	< room temperature
Al	660	150
Mg	650	200
Ag	962	200
Au	1064	200
Cu	1085	200
Fe	1538	450
Pt	1769	450
Ni	1453	600
Mo	2610	900
Ta	2996	1000
W	3410	1200

Adapted from R. Brick, A. Pense, and R. Gordon, *Structure and Properties of Engineering Materials*, McGraw-Hill, 1977.

Recrystallized grain size. A number of factors also influence the size of the recrystallized grains. Reducing the annealing temperature, the time required to heat to the annealing temperature, or the annealing time will reduce the grain size by minimizing the opportunity for grain growth. Increasing the initial cold work also reduces the final grain size by providing a greater number of nucleation sites for new grains. Finally, the presence of a second phase in the microstructure helps prevent grain growth and keeps the recrystallized grain size small.

Grain growth temperature. Heat treatment times can be reduced, often at considerable economic savings, by performing the heat treatment at higher temperatures. However, care must be taken to prevent grain growth. This applies to other types of heat treatments besides annealing. Occasionally, inclusions are deliberately introduced into an alloy in order to pin the grain boundaries and prevent grain growth, even at higher than normal temperatures.

9-10 Annealing Textures

Although annealing eliminates most of the effects of cold working, a preferred orientation or texture may persist, causing the annealed metal to be anisotropic.

Generally, textures are undesirable. For example, if an annealed sheet containing the recrystallization texture is to be deep drawn into a cup or beverage can, certain segments of the sheet deform more than other segments. This anisotropic deformation causes local thinning of the metal, with the extra metal forming "ears" (Figure 9-16). In cubic metals, the ears appear at 90° intervals around the cup.

Sometimes, however, the annealing texture is useful, as in the silicon-iron sheet material used for electrical transformers. The alloy magnetizes most readily in the $\langle 001 \rangle$ directions. Fortunately, this orientation corresponds to the annealing texture produced in the silicon-iron alloy.

FIG. 9-16 Nonuniform deformation, or earing, may occur during deep-drawing operations of certain metals due to anisotropic behavior, or textures.

9-11 Control of Properties by Combining Cold Working and Annealing

By taking advantage of the annealing heat treatment, we can increase the amount of deformation we can accomplish. If we are required to reduce a 125 mm thick plate to a 1.25 mm thick sheet, we can do the maximum prmissible cold work, anneal to restore the metal to its soft ductile condition, then cold work again. We can repeat the cold work-anneal cycle repeatedly until we approach the proper thickness. The final cold-working step can be designed to produce both the final desired dimensions and properties.

EXAMPLE 9-12

Design the processing sequence required to reduce a 10 mm × 60 mm copper strip to a 1.0 mm × 60 mm strip having greater than 415 MPa yield strength and 5% elongation.

Answer:

We determined in Example 9-6 that, to obtain the required properties, a cold work of 40% to 45% is required. An intermediate thickness of 1.67 mm to 1.82 mm must be obtained before this final cold work. Let's check to see if we can cold work from 10 mm to the intermediate thickness.

$$\text{Percent cold work} = \frac{t_0 - t_f}{t_0} \times 100 = \frac{10 - 1.82}{10} \times 100 = 81.8\%$$

$$\text{Percent cold work} = \frac{t_0 - t_f}{t_0} \times 100 = \frac{10 - 1.67}{10} \times 100 = 83.3\%$$

A maximum of 85% cold work is permitted for the copper in Figure 9-6. Thus, our complete process is as follows.

(a) Cold work the 10 mm strip, which must originally be in the annealed condition, 81.8% to 83.3% to an intermediate thickness of 1.67 mm to 1.82 mm.

(b) Anneal above the recrystallization temperature to eliminate the strengthening caused by the large amount of cold work.

(c) Cold work 40% to 45% from 1.67 mm or 1.82 mm to the final dimension of 1.0 cm. This gives the correct final dimensions and properties.

If, in Example 9-12, we had started with an original plate thickness of 100 mm, a large number of cold work-anneal cycles would be required. We will find later that we might hot work rather than cold work the plate from 100 mm to about 1.70 mm, then finish the deformation process by cold working.

9-12 Implications of Annealing on High-Temperature Properties

Neither strain hardening nor grain size strengthening would be appropriate for an alloy that is to be used at elevated temperatures, as in creep-resistant applications. Suppose we were to produce a strong, fine-grained structure by cold working the metal. When the metal is placed into service at a high temperature, recrystallization immediately causes a catastrophic decrease in strength. In addition, if the temperature were high enough, the strength would continue to decrease due to grain growth of the newly recrystallized grains. Neither control of cold working nor grain size are suitable strengthening mechanisms for high-temperature applications!

A similar problem occurs when we weld a cold-worked metal. The metal adjacent to the weld heats above the recrystallization and grain growth temperatures. This region is called the *heat-affected zone*. The structure and properties in the heat-affected zone of the weld are shown in Figure 9-17. The properties are catastrophically reduced by the heat of the welding process. In order to maintain properties when welding cold-worked metals, the time of exposure to the high temperatures must be as short as possible. Electron-beam welding and laser welding, which provide high rates of heat input for brief times, may be used in extreme cases.

HOT WORKING

9-13 Characteristics of the Hot-Working Process

We can deform a metal into a useful shape by hot working rather than cold working the metal. *Hot working* is defined as plastically deforming the metal at a temperature above the recrystallization temperature. During hot working, the metal is continually recrystallized (Figure 9-18). Hot working provides several advantages and introduces several limitations when compared to cold working.

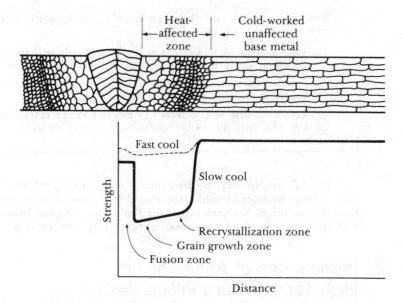

FIG. 9-17 The structure and properties surrounding a fusion weld in a cold-worked metal. Note the loss in strength due to recrystallization and grain growth in the heat-affected area.

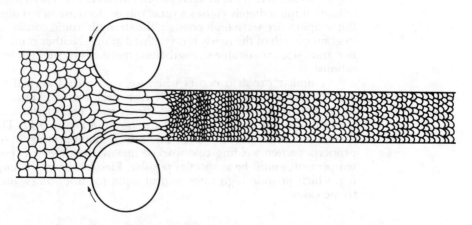

FIG. 9-18 During hot working, the elongated, anisotropic grains immediately recrystallize. If the hot-working temperature is properly controlled, the final hot-worked grain size can be very fine.

1. No strengthening occurs during deformation by hot working; consequently, the amount of plastic deformation is almost unlimited. A very large ingot can be reduced to a small size in a continuous series of operations. The first steps in the deformation process are carried out at temperatures well above the recrystallization temperature to take advantage of the lower strength of the metal. The

last step in the hot-working process is performed just above the recrystallization temperature, using a large percent deformation, in order to produce the finest possible grain size.

2. Hot working is well suited for forming large parts, since the metal has a low yield strength and high ductility at elevated temperatures.

3. Some original casting defects in the ingot can be eliminated or their effects minimized. Gas pores can be closed and welded shut during hot working—the internal lap formed when the pore is closed is eliminated by diffusion during the forming and cooling process. Macrosegregation can also be reduced, as the surface and center of an ingot are brought closer together during forming. The high temperatures during deformation can also help break down and homogenize the columnar structure produced during solidification, thus reducing microsegregation and producing a fine, equiaxed isotropic microstructure.

4. At high temperatures the HCP metals, such as magnesium, have more active slip systems; the higher ductility permits larger deformations than are possible by cold working.

5. A fibrous structure is produced because the deformation elongates or strings out inclusions and second-phase particles. In addition, textures similar to annealing textures may develop. As in cold-worked parts, the preferred orientation may cause anisotropic behavior which, if properly developed during forming, can be highly desirable.

6. The final properties in hot-worked parts may be less uniform than in cold-worked parts. The forming rolls or dies, which are normally at a lower temperature than the metal, cool the surface more rapidly than the center of the part. The surface then has a finer grain size than the center.

7. The surface finish is usually poorer than that obtained by cold working. Oxygen may react with the metal at the surface, forming oxides which are forced into the surface during forming. Some metals, like tungsten, react so severely with oxygen that they cannot be hot worked except in a protective atmosphere.

8. Dimensional accuracy is more difficult to obtain during hot working, since the metal contracts during cooling. The combination of elastic strain and thermal contraction requires that the part be made oversized during deformation, with the hope that the metal will then contract to the proper dimensions during cooling. Precise temperature control is required for accurate dimensions to be obtained.

EXAMPLE 9-13

An aluminum plate 25 mm thick is to be made by rolling. Calculate the size of the opening between the rolls if the plate is (a) made by cold rolling and (b) made by hot rolling at 400°C. The properties of the alloy are listed below.

Yield strength at 25°C = 450 MPa

Yield strength at 400°C = 69 MPa

Modulus of elasticity at 25°C = 70 GPa

Modulus of elasticity at 400°C = 45 GPa

Coefficient of thermal expansion = $24.7 \times 10^{-6} \, K^{-1}$

Answer:

(a) At room temperature, the elastic deformation which is recovered is

$$\text{Strain} = \varepsilon = \frac{\sigma_y}{E} = \frac{450}{70 \times 10^3} = 6.43 \times 10^{-3}$$

$$\text{Elastic recovery} = (6.43 \times 10^{-3})(25) = 0.161 \text{ mm}$$

No thermal contraction is expected since the metal remains at room temperature. Thus, the opening between the rolls must be $25 - 0.00643 = 24.994$ mm. For elastic recovery, the rolls must be undersized, since the stresses are compressive during rolling.

(b) At 400°C, the elastic recovery is

$$\text{Strain} = \varepsilon = \frac{\sigma_y}{E} = \frac{69}{45 \times 10^3} = 1.53 \times 10^{-3}$$

$$\text{Elastic recovery} = (1.53 \times 10^{-3})(25) = 0.038 \text{ mm}$$

The thermal contraction of the plate after cooling is

$$(400 - 25)(25)(24.7 \times 10^{-6}) = 0.232 \text{ mm}$$

The opening between the rolls is

$$25 - 0.038 + 0.232 = 25.194 \text{ mm}$$

This time, the opening between the rolls must be oversized because of the extra contribution of the thermal contraction.

9-14 Deformation Processing by Hot Working

Although we can produce large deformations by repeated cold work-anneal cycles, normally most of the deformation is done by hot working. After the hot-worked surface has been cleaned and oxides removed, the metal can be cold worked to give the final dimensions and properties.

EXAMPLE 9-14

Describe a process by which a 100 mm × 60 mm copper strip can be reduced to 1 mm × 60 mm, having a yield strength of 415 MPa and 5% elongation.

Answer:

From the results obtained in the previous examples, the quickest way to produce the finished product is (a) hot work the strip from 10 cm to between 1.67 mm and 1.82 mm, then (b) cold work the strip from between 1.67 mm and 1.82 mm to the final 1.0 mm, using between 40% and 45% cold work.

9-15 Deformation Bonding Processes

Both cold working and hot working are used to join or weld metals. *Deformation bonding* requires that (a) impurities on the joining surfaces be removed or broken up into discrete particles by the deformation process and (b) the pressure applied be sufficient to bring the atoms on each surface into intimate contact. By satisfying these two requirements, a large contact area is produced at the mating surfaces across which atom-to-atom attraction causes bonding to occur.

In *cold identation welding* [Figure 9-19(a)], two thin sheets of metal are joined

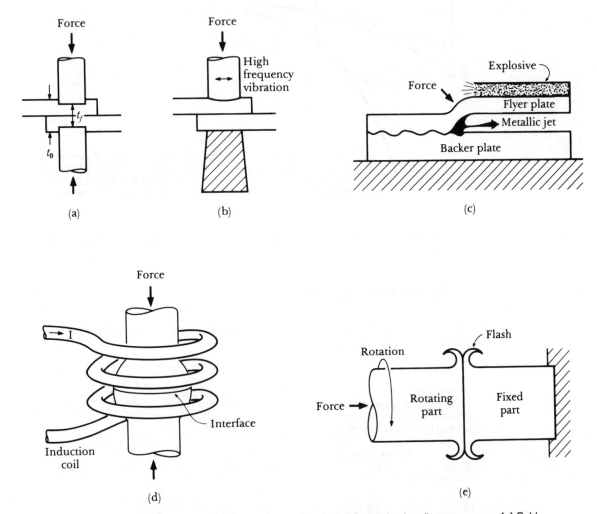

FIG. 9-19 Schematic diagrams of typical deformation bonding processes. (a) Cold indentation welding, (b) ultrasonic welding, (c) explosive bonding, (d) induction welding, and (e) friction or inertia welding.

as the sheets are deformed between punches. The strength of the bond may be greater than the surrounding base material, even though the joint is reduced in thickness, because of cold working of the joint during welding (Figure 9-20). However, when too much deformation is attempted, the joint becomes too thin to support a large applied load.

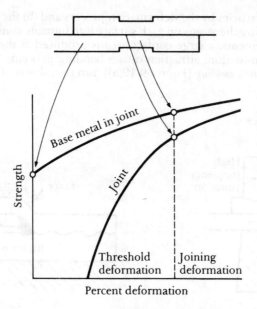

FIG. 9-20   The effect of percent deformation on the strength of the base metal and the welded joint in a cold indentation welding process.

The optimum thickness of the joint can, for a particular metal, be estimated from the figure of merit,

$$\text{Figure of merit} = \frac{t_f}{t_0} \times 100 \tag{9-3}$$

where t_f is the final thickness of the joint and t_0 is the total original thickness of the two sheets. Typical values for the figure of merit are shown in Table 9-5. The

TABLE 9-5   Figure of merit values for cold indentation welding of several materials

Metal	Figure of Merit (%)
Al	33
Pb	16
Cu	14
Fe	8
Ag	6
Al to Cu	16
Fe to Ni	6

From H. Udin, E. Funk, and J. Wulff, *Welding for Engineers*, John Wiley & Sons, 1954.

amount of plastic deformation required to make the weld can then be calculated from

$$\text{Percent deformation} = \frac{t_0 - t_f}{t_0} \times 100 \qquad (9\text{-}4)$$

If the deformation is too large, the calculated figure of merit is less than the optimum; the joint has a high yield strength because of extensive strain hardening but the joint is too thin to support a large load. If the deformation is too small, the calculated figure of merit is greater than the optimum and the bond is inadequate.

EXAMPLE 9-15

Two aluminum sheets, each 2.5 mm thick, are joined by the cold indentation process. Calculate the final thickness of the optimum joint and the percent deformation required to produce that joint.

Answer:

The figure of merit for aluminum is 33%. Thus

$$\text{Figure of merit} = 33 = \frac{t_f}{t_0} \times 100 = \frac{t_f}{(2)(2.5)} \times 100$$

$$t_f = \frac{(2)(2.5)(33)}{100} = 1.65 \text{ mm}$$

$$\text{Percent deformation} = \frac{t_0 - t_f}{t_0} \times 100 = \frac{(2)(2.5) - 1.65}{(2)(2.5)} \times 100 = 67\%$$

Most deformation-bonding processes use high temperatures to assist in the bonding process. *Ultrasonic welding* of thin sheets of metal resembles the cold indentation process [Figure 9.19(b)]. However, the vibrating motion caused by an ultrasonic probe breaks up oxides with much less deformation and thinning of the joint than the cold indentation process. The vibrating action, often at a frequency of 10,000 Hz or more, may also raise the temperature of the joint to the recrystallization temperature. The metals joined in *explosive bonding* also begin at room temperature; the intense energy produced during explosive bonding strips away surface impurities and forces the surfaces together at high pressures [Figure 9-19(c)]. But the energy also increases the temperature at the bonded interface, often up to the melting temperature of the metal.

Hot pressure bonding and *friction* or *inertia* welding rely on localized heating and deformation for the joining process [Figure 9-19(d) and (e)]. An induction coil might be used to supply the heat for hot pressure bonding. Heat develops in friction or inertia welding due to the frictional forces between the rotating surfaces.

9-16   Superplastic Forming

Some alloys, when specially heat treated and processed, can be uniformly deformed an exceptional amount, perhaps as much as 1000%; this behavior is called *superplasticity*. These parts can be formed into very complicated shapes in one or only a few forming dies. Common superplastic alloys include Ti-6% Al-4% V and Zn-23% Al. Several conditions are required for an alloy to display superplastic behavior.

1. The metal must have a very fine grain structure, with grain diameters being less than about 0.005 mm.

2. The alloy must be deformed at a high temperature, often near 0.5 to 0.65 times the absolute melting point of the alloy.

3. A very slow rate of forming, or *strain rate*, must be employed. In addition, the stress required to deform the alloy must be very sensitive to the strain rate (Figure 9-21). If necking begins to occur, the necked region strains at a higher

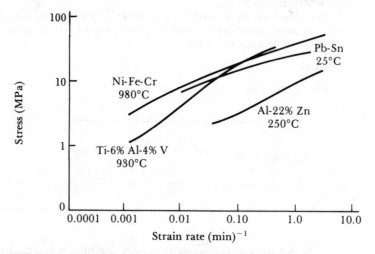

FIG. 9-21   The effect of strain rate on the flow stress at which an alloy deforms. The flow stress increases when the strain rate increases; when this effect is large enough, superplastic behavior might be observed.

rate; the higher strain rate in turn strengthens the necked region, stops the necking, and uniform deformation continues.

4. The grain boundaries in the alloy should allow grains to easily slide over one another and rotate when a stress is applied (Figure 9-22). The proper temperature and fine grain size are necessary for this to occur.

SUMMARY

The properties of materials, and in particular metals, can be controlled by combinations of deformation and simple heat treatments. When a metal is deformed by cold working, strain hardening occurs because additional dislocations are introduced into the structure. Very large increases in strength can be obtained in this

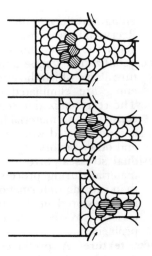

FIG. 9-22 In superplastic forming, the alloy deforms by grain boundary sliding rather than by elongation and distortion of the grains.

manner. This process combines a materials processing technique with a materials strengthening technique.

However, because ductility is simultaneously reduced, the amount of strain hardening is limited. Furthermore, residual stresses, which often are detrimental, may be introduced. By annealing, we can eliminate all or a portion of the effects of strain hardening. A low-temperature recovery treatment causes stress relief without reducing strength; a higher temperature recrystallization treatment eliminates all of the effects of strain hardening. Excessive temperatures, however, cause grain growth.

Deformation and annealing can be combined in one step—hot working—to improve our materials processing ability. Large changes in shape are possible at high temperatures since the material is not strain hardened. Combining hot working and cold working may permit us to achieve both processing of the material to a useful shape along with controlled improvement of properties.

GLOSSARY

Annealing A heat treatment used to eliminate part or all of the effects of cold working.

Cold indentation welding One method of joining metals by a cold-working process. The two surfaces are squeezed together between two punches, bringing the two surfaces into atom-to-atom contact by extensive deformation.

Cold working Deformation of a metal below the recrystallization temperature. During cold working, the number of dislocations increases, causing the metal to be strengthened as its shape is changed.

Deformation bonding A group of materials joining techniques by which the two surfaces are forced together at high pressures and often high temperatures. The pressure breaks up impurities at the surfaces and brings the materials into atom-to-atom contact, permitting bonding to occur.

Explosive bonding A deformation bonding technique by which the high pressures and temperatures are produced by the detonation of a layer of explosive spread on one of the surfaces.

Fiber texture A preferred orientation obtained in drawing processes by which grains are preferentially elongated in the drawing direction. Certain crystallographic directions in each grain also line up with the drawing direction, causing anisotropic behavior.

Figure of merit　The expression that describes the optimum amount of deformation needed to obtain the best properties in a cold indentation welding process.

Frank-Read source　A pinned dislocation which, under an applied stress, produces additional dislocations. This mechanism is at least partly responsible for strain hardening.

Friction welding　A group of deformation bonding processes, which includes inertia welding, by which two surfaces are heated by friction caused as the parts rotate against one another and are finally joined when the heated surfaces are forced together under high pressures.

Heat-affected zone　The area adjacent to a weld that is heated during the welding process above some critical temperature at which a change in the structure, such as grain growth or recrystallization, occurs.

Hot pressure bonding　Producing a weld by forcing two heated metal surfaces into intimate contact at a high pressure.

Hot working　Deformation of a metal above the recrystallization temperature. During hot working only the shape of the metal changes; the strength remains relatively unchanged because no strain hardening occurs.

Polygonized structure　A subgrain structure produced in the early stages of annealing. The subgrain boundaries are a network of dislocations rearranged during heating.

Preferred orientation　An alignment of grains, inclusions, or other microstructural features in a particular direction or plane in a material as a result of its processing.

Recovery　A low-temperature annealing heat treatment designed to eliminate residual stresses introduced during deformation without reducing the strength of the cold-worked material.

Recrystallization　A medium-temperature annealing heat treatment designed to eliminate all of the effects of the strain hard-

ening produced during cold working. Recrystallization must be accomplished above the recrystallization temperature.

Recrystallization temperature　The temperature above which the effects of strain hardening are eliminated during annealing. The recrystallization temperature is not a constant for a material but depends on the amount of cold work, the annealing time, and other factors.

Residual stresses　Stresses introduced in the material during processing which, rather than causing deformation of the material, remain stored in the structure. Later release of stresses as deformation can be a problem.

Sheet texture　A preferred orientation obtained in rolling processes. The grains are aligned so that a preferred crystallographic direction rotates parallel to the rolling direction and a preferred crystallographic plane rotates parallel to the sheet surface.

Shot peening　Introducing compressive residual stresses at the surface of a part by bombarding that surface with steel shot. The residual stresses may improve the overall performance of the material.

Strain hardening　Strengthening of a material by increasing the number of dislocations by deformation, or cold working. Also known as work hardening.

Strain-hardening coefficient　The effect that strain has on the resulting strength of the material. A material with a high strain-hardening coefficient obtains high strength with only small amounts of deformation or strain.

Strain rate　The rate at which the material is deformed. A material may behave much differently if it is slowly pressed into a shape rather than smashed rapidly into a shape by an impact blow.

Stress relief anneal　The recovery stage of the annealing heat treatment, during which residual stresses are relieved without reducing the mechanical properties of the material.

Stringer　Inclusions that are deformed or aligned with the direction of deformation during hot or cold working.

Superplasticity　The ability of a material to deform uniformly by an exceptionally large amount. Careful control over temperature, grain size, and strain rate are required for a material to behave in a superplastic manner.

Ultrasonic welding A special deformation welding technique in which the load that forces the two surfaces together is partly introduced by a very high frequency vibration. Bonding is achieved with very little total deformation.

PRACTICE PROBLEMS

1 The true stress and true strain obtained for two materials during a tensile test are given below. Using a log-log plot, determine the strain-hardening coefficient n and the constant K in Equation (9-1). Which material do you consider to be more responsive to strain hardening?

Material A		Material B	
σ_t (MPa)	ε_t	σ_t (MPa)	ε_t
0	0	0	0
14	0.0002	35	0.00025
28	0.0004	70	0.0005
42	0.0006	112	0.0008
56	0.0013	140	0.0011
70	0.0024	210	0.0016
105	0.0070	350	0.0026
140	0.0165	490	0.0037
210	0.0500	700	0.0053

2 Based on the relationship between σ_t and ε_t in Equation (9-1), would you expect that a similar relationship might be found for yield strength versus percent cold work? Using a log-log plot, see if there is such a relationship for the commercially pure copper in Figure 9-6 and the aluminum alloy in Figure 9-23. Determine the constants in the equations that describe the relationship.

3 A strip of an aluminum alloy (Figure 9-23) is cold rolled from 67 mm to 23 mm in thickness. (a) Calculate the percent cold work and (b) estimate the mechanical properties of the final strip.

4 A Cu-30% Zn brass plate (Figure 9-24) is cold rolled from 190 mm to 50 mm (a) Calculate the percent cold work and (b) estimate the mechanical properties of the final plate.

5 A Cu-30% Zn alloy wire (Figure 9-24) is cold worked 10% to a diameter of 5 mm. Calculate (a) the total percent cold work and (b) the properties of the alloy if the wire is further deformed through a 4.6 mm diameter die?

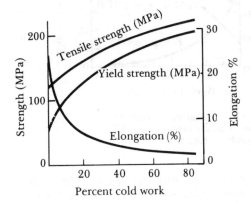

FIG. 9-23 The effect of percent cold work on the properties of a 3105 aluminum alloy.

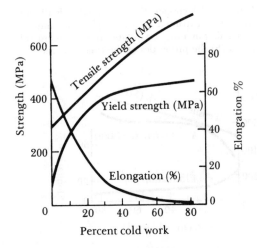

FIG. 9-24 The effect of percent cold work on the properties of a Cu-30% Zn brass.

6 An oxygen-free (OF) copper plate (Figure 9-25) is cold rolled 25% to a thickness of 10 mm. If the plate is then reduced to 6 mm, (a) determine the total percent cold work and (b) estimate the final mechanical properties.

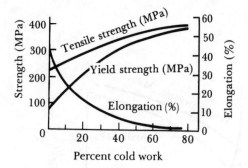

FIG. 9-25 The effect of percent cold work on the properties of OF (oxygen-free) copper.

7 A Cu-30% Zn alloy wire (Figure 9-24) is to be made into a spring with a yield strength greater than 350 MPa. A 5 mm wire, originally annealed, is pulled through a 4.6 mm diameter die. Will the wire have adequate properties to perform its function as a spring?

8 A low carbon steel plate (Figure 9-26) must have an elongation of at least 10% after it is cold worked from 50 mm to 9.5 mm. Is the minimum % elongation obtained?

9 A billet of OF copper (Figure 9-25) that is 125 mm in diameter is extruded cold through a 50 mm diameter opening in a die. Calculate (a) the percent cold work and (b) the properties of the final extrusion.

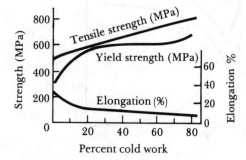

FIG. 9-26 The effect of percent cold work on the properties of a Carbon steel.

10 A steel sheet 2.7 mm thick is to have a yield strength of at least 550 MPa (Figure 9-26). Determine (a) the minimum percent cold work that can be done and (b) the minimum thickness of the steel before it is rolled to the final thickness of 2.7 mm.

11 A 20 mm diameter OF copper rod (Figure 9-25) must have a maximum tensile strength of 350 MPa Determine (a) the maximum percent cold work that can be done to produce the rod and (b) the maximum original diameter of the rod before cold working.

12 A Cu-30% Zn brass (Figure 9-24) is to be rolled into a thin sheet 1.3 mm thick. If more than 40% cold work is done, the sheet will corrode rapidly. What is the maximum thickness of the starting material?

13 An aluminum alloy (Figure 9-23) must have a final diameter of 38 mm with at least 175 MPa tensile strength. Determine (a) the minimum percent cold work that must be used to make the rod and (b) the minimum diameter of the original rod.

14 A 50 mm × 50 mm square billet of OF copper (Figure 9-25) is to be extruded into a 25 mm diameter bar. Calculate the percent cold work involved in the process.

15 A 30 mm diameter aluminum alloy bar (Figure 9-23) is to be extruded into a 10 mm × 10 mm square bar. Calculate the percent cold work involved in the process.

16 An aluminum alloy (Figure 9-23) is drawn into a 2.5 mm diameter wire with greater than 6% elongation and 124 MPa yield strength. Calculate (a) the percent cold work required and (b) the original diameter of the wire before drawing.

17 A cylindrical billet of steel (Figure 9-26) is to be extruded into a 50 mm diameter bar having a tensile strength greater than 690 MPa and an elongation of at least 10%. Calculate the required diameter of the cylindrical billet used in this cold extrusion process.

18 A Cu-30% Zn bar (Figure 9-24) originally 100 mm in diameter is to be cold worked into a smaller diameter bar with a minimum of 550 MPa tensile strength and 5% elongation. What range of final diameters would be satisfactory?

19 An OF copper billet (Figure 9-25) originally 60 mm × 60 mm is to be deformed into a square billet with a minimum of 207 MPa yield strength and 10% elongation. What range of dimensions would be satisfactory?

20 A 50 mm diameter Cu-30% Zn brass rod (Figure 9-24), originally in the annealed condition, is to be deformed to a 2 mm bar. The final properties must exceed 550 MPa tensile strength, 414 MPa yield strength, and 5% elongation. Describe the steps, including percent cold work and intermediate thicknesses, required. A maximum of 85% cold work is permitted.

21 An aluminum alloy (Figure 9-23) is to be rolled from a 150 mm thick plate to a 3.8 mm thick plate, producing properties of at least 178 MPa tensile strength, 159 MPa yield strength, and 4% elongation. Describe the steps, including percent cold work and intermediate thicknesses, required for the process. A maximum of 85% cold work is permitted.

22 A rolling mill is set to give an opening of 25 mm. A 70% Cu-30% Zn brass alloy (Figure 9-24), which has a yield strength of 70 MPa at room temperature, 45 MPa at 600°C, and a coefficient of thermal expansion of $20 \times 10^{-6} \text{ K}^{-1}$, is rolled, starting at 50 mm. The modulus of elasticity is 117 GPa at room temperature and 69 GPa at 600°C. (a) Estimate the yield strength of the plate after cold rolling and after hot rolling at 600°C. (b) Estimate the thickness of the plate after it emerges from the roll for both cold and hot rolling.

23 An aluminum alloy (Figure 9-23) is to be extruded through a circular die to produce a 50 mm diameter cylindrical bar. The alloy has a yield strength of 12 MPa at 400°C, 55 MPa at room temperature, and the coefficient of thermal expansion is $25.6 \times 10^{-6} \text{ K}^{-1}$. The modulus of elasticity is 69 GPa at 25°C and 41.4 GPa at 400°C. Calculate the size of the die needed to compensate for contraction and elastic recovery.

24 A steel rod (Figure 9-26) is drawn through a die. The rod is originally 5 mm in diameter and is intended to finally be 4 + 0.02 mm. The modulus of elasticity is 207 GPa. If the die opening is exactly 4 mm, will the final wire meet the specifications?

25 An OF copper rod (Figure 9-25) 2.5 mm in diameter is drawn through a 2.0 mm die. The original yield strength is 98 MPa. (a) Calculate the draw stress and draw force, assuming no friction. (b) Will the draw force cause the drawn wire to break? (Prove by calculating the maximum force that the drawn wire can withstand.)

26 A steel wire 2.5 mm in diameter (Figure 9-26) is to be made having a tensile strength of 690 MPa. Determine (a) the original diameter of the wire required, (b) the required draw stress and draw force, and (c) whether the as-drawn wire will survive the drawing process.

27 Determine the optimum thickness of the joint and the percent deformation required for a cold indentation weld joining two 1.5 mm sheets of (a) iron, (b) lead, and (c) aluminum.

28 A cold indentation process is used to join two 2 mm sheets of OF copper (Figure 9-25). The total final thickness of the joint is 1 mm. What is the maximum force that the individual sheets and the weld should be able to support for each inch of width?

29 (a) From the data below, estimate the recovery, recrystallization, and grain growth temperatures. (b) How will the recrystallization temperature vary when the percent cold work increases, when the time of annealing increases, and when the purity of the metal increases? (c) Estimate the melting temperature of the material.

Annealing Temperature (°C)	Residual Stresses (MPa)	Yield Strength (MPa)	Grain Size (mm)
50	13.8	69	0.070
75	13.8	69	0.070
100	13.8	68.9	0.070
125	3.45	68.9	0.070
150	0	68.7	0.070
175	0	38	0.015
200	0	37.8	0.020
225	0	37.6	0.090
250	0	37.3	0.200

30 (a) From the data below, estimate the recovery, recrystallization, and grain growth temperatures of the cold-worked metal. (b) At what temperature would you stress relieve the metal? (c) At what temperature would you hot work the metal?

Annealing Temperature (°C)	Electrical Conductivity ($\times 10^7 \, \Omega^{-1} \text{ m}^{-1}$)	Tensile Strength (MPa)	Grain Size (mm)
125	4.5	586	0.150
175	4.5	586	0.150
225	6.0	586	0.150
275	6.1	586	0.150
325	6.1	393	0.005
375	6.2	372	0.007
425	6.2	359	0.010
475	6.3	338	0.030
525	6.3	324	0.050
575	6.4	317	0.080
625	6.4	310	0.125
700	6.4	303	0.200

31 If you deform lead or tin at room temperature, have you hot or cold worked the metal? Explain.

32 From Table 9-4, plot the measured recrystallization temperature versus the melting point of the alloy and compare to a line drawn for $0.4T_m$. Comment on the suitability of the $0.4T_m$ approximation for the recrystallization temperature.

33 Phosphorus-deoxidized copper ingots contain phosphorus oxide inclusions. Describe how you could determine the rolling direction in a small annealed plate of this material.

34 A cold-worked copper plate is properly annealed at 250°C in 2 h. In order to speed up the annealing process, you suggest raising the temperature to 300°C. What is the minimum time you must anneal to obtain the proper structure?

35 A cold-worked copper plate is welded. The hardness of the original cold-worked plate is R_B62.

When the plate is welded by the oxyacetylene process, the hardness in the heat-affected zone is R_B10, by arc welding is R_B35, and by electron-beam welding is R_B61. Explain the difference in hardness in the heat-affected zone.

36 Two cold-worked aluminum bars are joined by friction welding. After being placed into service, the joined bars fail 0.01 mm from the friction-welded joint. Explain why failure occurred at this location.

37 A Cu-10% Zn alloy is cold rolled 95% to a 5 mm thick sheet. Compare the maximum load that will be observed in tensile specimens cut parallel to and perpendicular to the rolling direction. See Table 9-2.

38 Draw a sketch of the structure of the fusion zone and heat-affected zone in a weld of an annealed copper plate.

CHAPTER 10

Solidification and Dispersion Strengthening

10-1 Introduction

When the solubility of a metal is exceeded by adding too much of an alloying element, a second phase forms and a two-phase alloy is produced. The boundary between the two phases is a surface at which the atomic arrangement is not perfect. As a result, this boundary interferes with the slip of dislocations and strengthens the metal. The general term for strengthening by the introduction of a second phase is *dispersion strengthening*.

In this chapter we will first discuss the fundamentals of dispersion strengthening to determine the structure we should aim to produce. Next we will examine the types of reactions that produce multiple-phase alloys. Finally, we will look in some detail at how we control dispersion strengthening by controlling the solidification process.

10-2 Principles of Dispersion Strengthening

More than one phase must be present in any dispersion-strengthened alloy. We call the continuous phase, which usually is present in larger amounts, the *matrix*. The second phase, usually present in smaller amounts, is called the *precipitate*. In some cases, two phases form simultaneously. We will define these structures differently, often calling the intimate mixture of phases a *microconstituent*.

There are some general considerations for determining how the characteristics of the matrix and precipitate affect the overall properties of the alloy (Figure 10-1).

1. The matrix should be soft and ductile, while the precipitate should be hard and brittle. The precipitate acts as a very strong obstacle to slip of dislocations in the matrix. However, the matrix provides at least some ductility to the overall alloy.

2. The hard brittle precipitate should be discontinuous, while the soft ductile matrix should be continuous. If the precipitate were continuous, cracks could propagate through the entire structure. However, cracks in the discontinuous brittle precipitate are arrested by the precipitate-matrix interface.

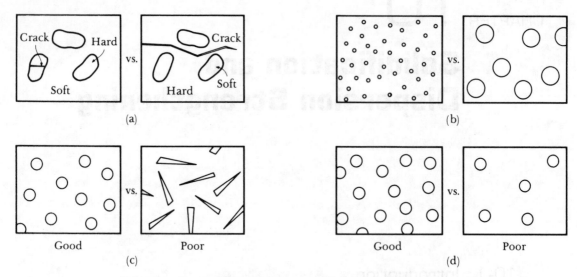

FIG. 10-1   Considerations for effective dispersion strengthening. (a) The precipitate should be hard and discontinuous, (b) the precipitate particles should be small and numerous, (c) the precipitate particles should be round rather than needlelike, and (d) larger amounts of precipitate increase strengthening.

3. The precipitate particles should be small and numerous, increasing the likelihood that they interfere with the slip process.

4. The precipitate particles should be round, rather than needlelike or sharp-edged. The rounded shape is less likely to initiate a crack or to act as a notch.

5. Larger amounts of the precipitate increase the strength of the alloy.

10-3   Intermetallic Compounds

Often dispersion-strengthened alloys contain an intermetallic compound. An *intermetallic compound* is made up of two or more elements, producing a new phase with its own composition, crystal structure, and properties. Intermetallic compounds are almost always very hard and brittle but can provide excellent dispersion strengthening of the softer matrix.

Stoichiometric intermetallic compounds have a fixed composition. Steels are strengthened by a stoichiometric intermetallic compound, Fe_3C, that has a fixed ratio of three iron atoms to one carbon atom. Stoichiometric intermetallic compounds are usually easy to detect in phase diagrams because they are represented by a vertical line [Figure 10-2(a)].

Nonstoichiometric intermetallic compounds can have a range of compositions. In the molybdenum-rhodium system, the γ phase is an intermetallic compound [Figure 10-2(b)]. Because the molybdenum-rhodium atom ratio is not rigidly fixed, the γ phase can contain from 45 wt% to 83 wt% Rh at 1600°C.

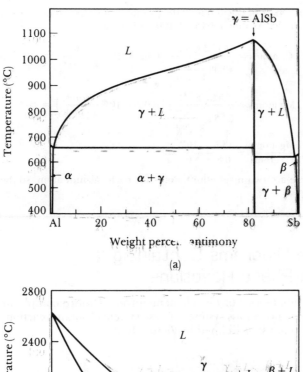

FIG. 10-2   (a) The aluminum-antimony phase diagram
includes a stoichiometric intermetallic compound γ. (b) The
molybdenum-rhodium phase diagram includes a
nonstoichiometric intermetallic compound γ.

EXAMPLE 10-1

A stoichiometric compound forms in the calcium-magnesium system at 45.3% Ca.
What is the formula of the intermetallic compound?

Answer:

We must convert weight percent to atomic percent to obtain the atom ratio. The atomic mass of calcium is 40.08 g/g · mole and the atomic mass of magnesium is 24.31 g/g · mole. Assume a base of 100 g of the alloy.

$$\text{Ca atoms} = \frac{45.3 \text{ g Ca}}{40.08 \dfrac{\text{g}}{\text{g} \cdot \text{mole}}}\left(6.02 \times 10^{23} \frac{\text{atoms}}{\text{g} \cdot \text{mole}}\right) = 6.8 \times 10^{23}$$

$$\text{Mg atoms} = \frac{54.7 \text{ g Mg}}{24.31 \dfrac{\text{g}}{\text{g} \cdot \text{mole}}}\left(6.02 \times 10^{23} \frac{\text{atoms}}{\text{g} \cdot \text{mole}}\right) = 13.5 \times 10^{23}$$

$$\frac{\text{Mg atoms}}{\text{Ca atoms}} = \frac{13.5 \times 10^{23}}{6.8 \times 10^{23}} = 2$$

There are two magnesium atoms for each calcium atom, so the compound must be Mg_2Ca.

10-4 Phase Diagrams Containing Three-Phase Reactions

Many combinations of two elements produce more complicated phase diagrams than the isomorphous systems. These systems contain reactions that involve three separate phases, as defined in Figure 10-3.

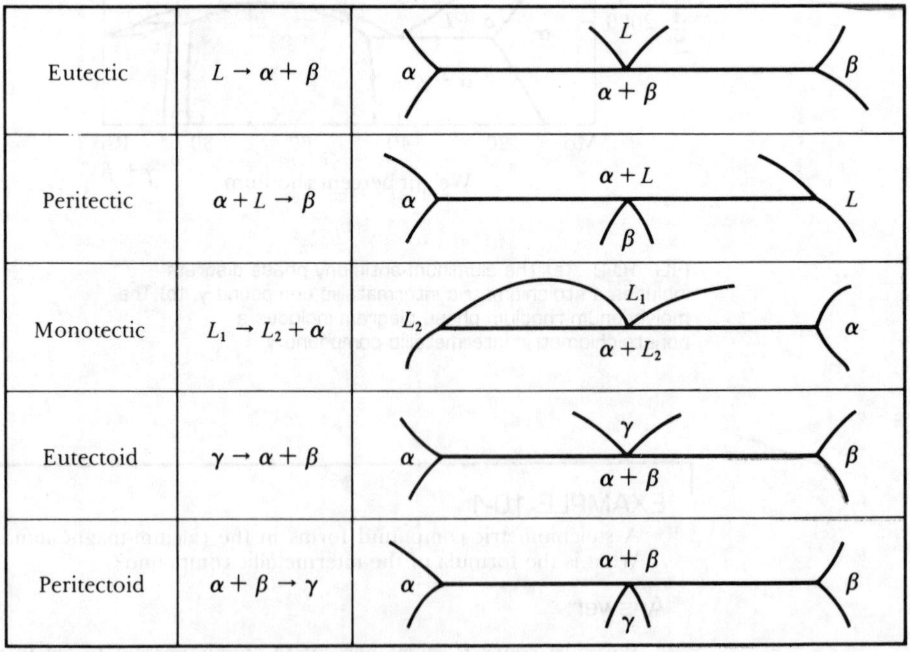
Eutectic
L → α + β
L
α
β
α + β
Peritectic
α + L → β
α + L
α
L
β
Monotectic
L₁ → L₂ + α
L₁
L₂
α
α + L₂
Eutectoid
γ → α + β
γ
α
β
α + β
Peritectoid
α + β → γ
α + β
α
β
γ

FIG. 10-3 The five most important three-phase reactions in binary phase diagrams.

Each of the reactions can be identified in a complex phase diagram by the following procedure.

1. Locate a horizontal line on the phase diagram. The horizontal line, which indicates the presence of a three-phase reaction, represents the temperature at which the reaction occurs under equilibrium conditions.

2. Locate three distinct points on the horizontal line; the two endpoints plus a third point, often near the center of the horizontal line. The center point represents the composition at which the three-phase reaction occurs.

3. Look immediately above the center point and identify the phase or phases present; look immediately below the center point and identify the phase or phases present. Then write in reaction form the phase(s) above the center point transforming to the phase(s) below the point. Compare this reaction to those in Figure 10-3 to identify the reaction.

EXAMPLE 10-2

Consider the phase diagram in Figure 10-4. Identify the three-phase reactions that occur.

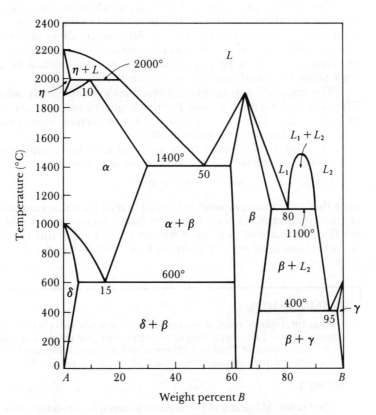

FIG. 10-4 A hypothetical phase diagram for Example 10-2.

Answer:

We find horizontal lines at 2000°C, 1400°C, 1100°C, 600°C, and 400°C.

2000°C: The center point is at 10% B; $\eta + L$ are present above the point, α is present below. The reaction is

$$\eta + L \rightarrow \alpha \quad \text{or a } peritectic$$

1400°C: This reaction occurs at 50% B.

$$L \rightarrow \alpha + \beta \quad \text{or an } eutectic$$

1100°C: This reaction occurs at 80% B.

$$L_1 \rightarrow \beta + L_2 \quad \text{or a } monotectic$$

600°C: This reaction occurs at 15% B.

$$\alpha \rightarrow \delta + \beta \quad \text{or an } eutectoid$$

400°C: This reaction occurs at 95% B.

$$L_2 \rightarrow \beta + \gamma \quad \text{or an } eutectic$$

The eutectic, peritectic, and monotectic reactions are a part of the solidification process. Alloys used for casting or soldering often take advantage of the low melting point of the eutectic reaction. Monotectic alloys improve the machining characteristics of the finished product, as in many brass and bronze alloys. Peritectic reactions are avoided if possible because nonequilibrium solidification and segregation frequently accompany the reaction.

The eutectoid and peritectoid reactions are completely solid-state reactions. The eutectoid forms the basis for the heat treatment of several alloy systems, including steel. The peritectoid reaction is extremely slow, producing undesirable, nonequilibrium structures in alloys.

All of these three-phase reactions occur at a fixed temperature and composition. The Gibbs phase rule for a three-phase reaction is

$$F = C - P + 1 = 2 - 3 + 1 = 0$$

since there are two components C in a binary phase diagram and three phases P are involved in the reaction. When the three phases are in equilibrium during the reaction, there are no degrees of freedom. The temperature and the composition of each phase involved in the three-phase reaction are fixed.

EXAMPLE 10-3

What three-phase reaction occurs at the aluminum-rich end of the Al-Cu phase diagram, Figure 11-1? According to the phase rule, the temperature and composition of each phase must be fixed in the three-phase reaction. What are the temperature of the reaction and the compositions of each phase in the reaction?

Answer:

The horizontal line at 548°C represents a three-phase reaction. Above the center point at 33.2% Cu, the alloy is all liquid. Below the center point, $\alpha + \theta$ form. Thus the reaction $L \rightarrow \alpha + \theta$ is an eutectic.

The temperature of the reaction is 548°C. The compositions of each phase are given by the three points on the line.

α: 5.65% Cu θ: 52.5% Cu L: 33.2% Cu

10-5 The Eutectic Phase Diagram

The lead-tin system contains only a simple eutectic reaction (Figure 10-5). This alloy system is the basis for the most common alloys used for soldering. Let's examine four classes of alloys in this system.

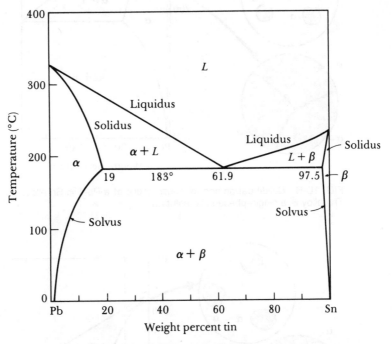

FIG. 10-5 The lead-tin equilibrium phase diagram.

Solid solution alloys. Alloys that contain 0% to 2% Sn behave exactly like the copper-nickel alloys; a single-phase solid solution α forms during solidification and remains after the alloy has cooled to room temperature (Figure 10-6). These alloys are strengthened by solid solution strengthening, by strain hardening, and by controlling the solidification process to refine the grain structure.

Alloys that exceed the solubility limit. Alloys containing between 2% and 19% Sn also solidify to produce a single solid solution α. However, as the alloy continues to cool, a solid-state reaction occurs, permitting a second solid phase, β, to precipitate from the original α phase (Figure 10-7).

The α is a solid solution of tin in lead. Because Hume-Rothery's conditions are not satisfied, however, the solubility of tin in the α solid solution is limited. At 0°C, only 2% Sn can dissolve in α. As the temperature increases, more tin dissolves in the lead until, at 183°C, the solubility of tin in lead has increased to 19% Sn.

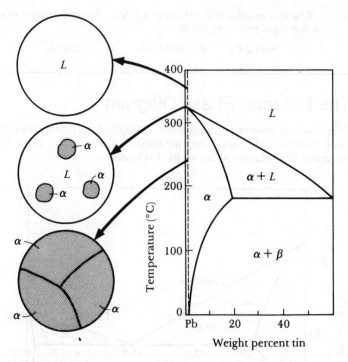

FIG. 10-6   Solidification and microstructure of a Pb-1% Sn alloy.
The alloy is a single-phase solid solution.

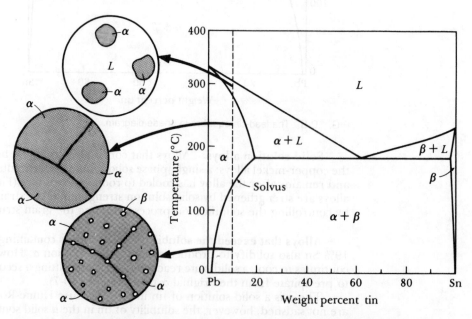

FIG. 10-7   Solidification, precipitation, and microstructure of a Pb-10% Sn alloy. Some
dispersion strengthening occurs as the β solid precipitates.

This is the *maximum solubility* of tin in lead. The solubility of tin in solid lead at any temperature is given by the *solvus* line. Any alloy containing between 2% and 19% Sn cools past the solvus, the solubility limit is exceeded, and a small amount of β forms to satisfy the phase diagram.

We control the properties of this type of alloy by several techniques, including solid solution strengthening of the α portion of the structure, controlling the microstructure produced during solidification, and controlling the amount and characteristics of the β phase. This latter mechanism is one type of dispersion strengthening.

EXAMPLE 10-4

Determine (a) the solubility of tin in solid lead at 100°C, (b) the maximum solubility of lead in solid tin, and (c) the amount of β that forms if a Pb-10% Sn alloy is cooled to 0°C.

Answer:

(a) The 100°C temperature intersects the solvus line at 5% Sn. The solubility of tin in lead at 100°C is therefore 5% Sn.

(b) The maximum solubility of lead in tin, which is found from the tin-rich side of the phase diagram, occurs at the eutectic temperature of 183°C and is 2.5% Pb.

(c) At 0°C, the 10% Sn alloy is in an α + β region of the phase diagram. By drawing a tie line at 0°C and working the lever law, we find that

α: 2% Sn   β: 100% Sn

$$\% \; \beta = \frac{10 - 2}{100 - 2} \times 100 = 8.2\%$$

Eutectic alloys.   The alloy containing 61.9% Sn has the eutectic composition (Figure 10-8). Above 183°C the alloy is all liquid and therefore must contain 61.9% Sn. After the liquid cools to 183°C, the eutectic reaction begins.

$$L_{61.9\% \; Sn} \rightarrow \alpha_{19\% \; Sn} + \beta_{97.5\% \; Sn}$$

Two solid solutions—α and β—are formed during the eutectic reaction. The compositions of the two solid solutions are given by the ends of the eutectic line.

Diffusion must occur during the eutectic reaction, as liquid containing 61.9% Sn transforms to the lead-rich solid α and the tin-rich solid β. During solidification, growth of the eutectic requires both removal of the latent heat of fusion and redistribution of the two different atom species by diffusion. Since solidification occurs completely at 183°C, the cooling curve (Figure 10-9) is similar to that of a pure metal; that is, a thermal arrest or plateau occurs at the eutectic temperature.

In order for atoms to be redistributed during eutectic solidification, a characteristic microstructure must develop. In the lead-tin system, the solid α and β phases grow from the liquid in a *lamellar*, or platelike, arrangement (Figure 10-10). The lamellar structure permits the lead and tin atoms to move through the liquid, in which diffusion is easy, without having to move an appreciable distance. This lamellar structure is characteristic of numerous other eutectic systems.

The product of the eutectic reaction is a unique and characteristic arrangement of the two solid phases called the *eutectic microconstituent.* In the Pb-61.9% Sn alloy, 100% of the eutectic microconstituent is formed, since all of the liquid goes through the reaction.

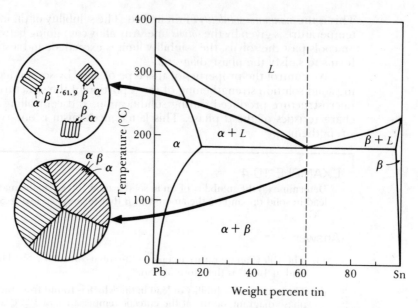

FIG. 10-8 Solidification and microstructure of the eutectic alloy Pb-61.9% Sn.

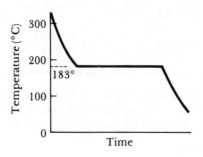

FIG. 10-9 The cooling curve for an eutectic alloy is a simple thermal arrest, since eutectics freeze or melt at a single temperature.

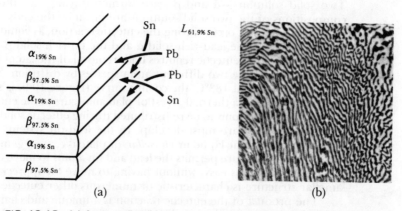

FIG. 10-10 (a) Atom redistribution during lamellar growth of a lead-tin eutectic. Tin atoms from the liquid preferentially diffuse to the β plates and lead atoms diffuse to the α plates. (b) Photomicrograph of the lead-tin eutectic microconstituent.

EXAMPLE 10-5

Determine the amount and composition of each phase that is present in the eutectic microconstituent in a lead-tin alloy.

Answer:

The eutectic microconstituent contains 61.9% Sn. We work the lever law at a temperature just below the eutectic, say at 182°C, since that is the temperature at which the eutectic reaction is just completed. The fulcrum of our lever is 61.9% Sn. The ends of the tie line coincide approximately with the ends of the eutectic line.

$$\alpha: \text{ Pb-19\% Sn} \qquad \% \; \alpha = \frac{97.5 - 61.9}{97.5 - 19} \times 100 = 45\%$$

$$\beta: \text{ Pb-97.5\% Sn} \qquad \% \; \beta = \frac{61.9 - 19}{97.5 - 19} \times 100 = 55\%$$

Hypoeutectic and hypereutectic alloys. As an alloy containing between 19% and 61.9% Sn cools, the liquid begins to solidify at the liquidus temperature. However, solidification is completed by going through the eutectic reaction (Figure 10-11). This solidification sequence occurs any time the vertical line corresponding to the original composition of the alloy crosses both the liquidus and the eutectic.

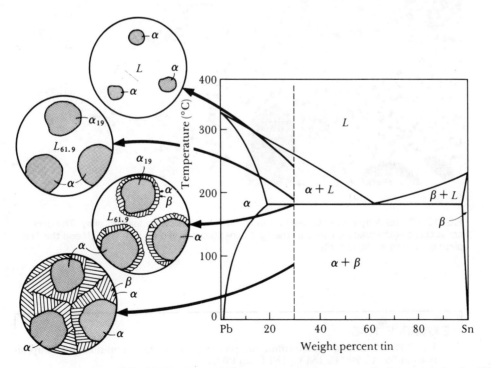

FIG. 10-11 The solidification and microstructure of a hypoeutectic alloy (Pb-30% Sn).

Alloys with compositions of 19% to 61.9% Sn are called *hypoeutectic* alloys, or alloys containing less than the eutectic amount of tin. An alloy to the right of the eutectic composition, between 61.9% and 97.5% Sn, is *hypereutectic*.

Let's consider a hypoeutectic alloy containing Pb-30% Sn and follow the changes in structure during solidification (Figure 10-11). On reaching the liquidus temperature of 260°C, solid α containing about 12% Sn nucleates. The solid α grows until the alloy cools to just above the eutectic temperature. At 184°C, we draw a tie line and find that the solid α contains 19% Sn and the remaining liquid contains 61.9% Sn. We note that, at 184°C, the liquid contains the eutectic composition! When the alloy is cooled below 183°C, all of the remaining liquid goes through the eutectic reaction and transforms to a lamellar mixture of α and β. The microstructure shown in Figure 10-12(a) results. Notice that the eutectic microconstituent surrounds the solid α that formed between the liquidus and eutectic temperatures. The eutectic microconstituent is continuous.

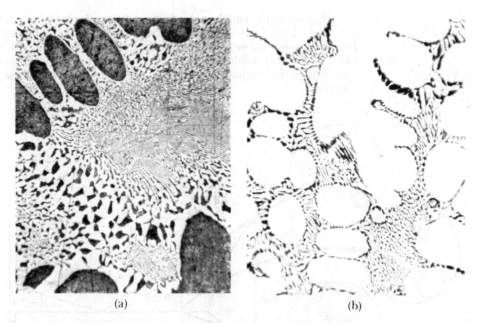

(a)   (b)

FIG. 10-12   (a) A hypoeutectic lead-tin alloy. (b) A hypereutectic lead-tin alloy. The dark constituent is the lead-rich solid α, the light constituent is the tin-rich solid β, and the fine plate structure is the eutectic.

EXAMPLE 10-6

For a Pb-30% Sn alloy, determine the phases present, their amounts and compositions at 300°C, 200°C, 184°C, 182°C, and 0°C.

Answer:

Temperature (°C)	Composition		Amount
300	L	L: 30% Sn	$L = 100\%$
200	$\alpha + L$	L: 55% Sn	$L = \dfrac{30 - 18}{55 - 18} \times 100 = 32\%$
		α: 18% Sn	$\alpha = \dfrac{55 - 30}{55 - 18} \times 100 = 68\%$
184	$\alpha + L$	L: 61.9% Sn	$L = \dfrac{30 - 19}{61.9 - 19} \times 100 = 26\%$
		α: 19% Sn	$\alpha = \dfrac{61.9 - 30}{61.9 - 19} \times 100 = 74\%$
182	$\alpha + \beta$	α: 19% Sn	$\alpha = \dfrac{97.5 - 30}{97.5 - 19} \times 100 = 86\%$
		β: 97.5% Sn	$\beta = \dfrac{30 - 19}{97.5 - 19} \times 100 = 14\%$
0	$\alpha + \beta$	α: 2% Sn	$\alpha = \dfrac{100 - 30}{100 - 2} \times 100 = 71\%$
		β: 100% Sn	$\beta = \dfrac{30 - 2}{100 - 2} \times 100 = 29\%$

We call the solid α phase that formed when we cooled from the liquidus to the eutectic the *primary* or *proeutectic microconstituent*. This solid α did not take part in the eutectic reaction. Often we find that the amounts and compositions of the microconstituents are of more use to us than the amounts and compositions of the phases.

EXAMPLE 10-7

Determine the amounts and compositions of each microconstituent in a Pb-30% Sn alloy immediately after the eutectic reaction has been completed.

Answer:

The microconstituents are primary α and eutectic. We can determine their amounts and compositions if we look at how they form. The primary α is all of the solid α that forms before the alloy cools to the eutectic temperature; the eutectic microconstituent is all of the liquid that goes through the eutectic reaction. At a temperature just above the eutectic, say 184°C, the amounts and compositions of the two phases are

$$\alpha: \quad 19\% \text{ Sn} \qquad \% \, \alpha = \frac{61.9 - 30}{61.9 - 19} \times 100 = 74\% = \% \text{ primary } \alpha$$

$$L: \quad 61.9\% \text{ Sn} \qquad \% \, L = \frac{30 - 19}{61.9 - 19} \times 100 = 26\% = \% \text{ eutectic}$$

When the alloy cools below the eutectic to 182°C, nothing changes except that the liquid at 184°C transforms to eutectic. The solid α present at 184°C remains unchanged and is the primary microconstituent.

The cooling curve for the type of alloy in Example 10-7 is a composite of those for solid solution alloys and "straight" eutectic alloys (Figure 10-13). A

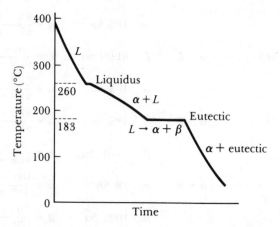

FIG. 10-13   The cooling curve for a hypoeutectic Pb-30% Sn alloy.

change in slope occurs at the liquidus as primary α begins to form. Evolution of the latent heat of fusion slows the cooling rate as the solid α grows. When the alloy cools to the eutectic temperature, a thermal arrest is produced as the eutectic reaction proceeds at 183°C. The solidification sequence is similar in a hypereutectic alloy, giving the microstructure shown in Figure 10-12(b).

10-6 Strength of Eutectic Alloys

Each phase in the eutectic alloy is, to some degree, solid solution strengthened. In the lead-tin system, α, which is a solid solution of tin in lead, is stronger than pure lead. Some eutectic alloys can be strengthened by cold working. We also control grain size by adding appropriate inoculants or grain refiners. However, we can best influence the properties by controlling the amount, size, shape, and distribution of the two solid phases in the eutectic.

Eutectic grain size. *Eutectic grains* each nucleate and grow independently. Within each grain, the orientation of the lamellae in the eutectic microconstituent is identical. The orientation changes on crossing a eutectic grain boundary [Figure 10-17(a)]. We are able to refine the eutectic grain size, and consequently improve the strength of the eutectic alloy, by inoculation.

Interlamellar spacing. The *interlamellar spacing* of an eutectic is the distance from the center of one α lamella to the center of the next α lamella (Figure 10-14). A small interlamellar spacing indicates that the individual lamellae are thin and consequently the amount of α-β interface area is large. A small interlamellar spacing therefore suggests that the strength of the eutectic is high.

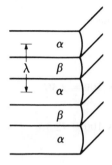

FIG. 10-14
The interlamellar
spacing in an
eutectic
microstructure.

The interlamellar spacing is determined primarily by the growth rate of the eutectic.

$$\lambda = cR^{-1/2} \tag{10-1}$$

where R is the growth rate (cm/s) and c is a constant. The interlamellar spacing for the lead-tin eutectic is shown in Figure 10-15. We can increase the growth rate R,

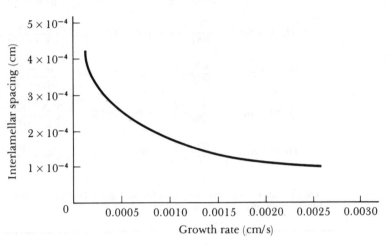

FIG. 10-15   The effect of growth rate on the interlamellar spacing in the lead-tin eutectic.

and consequently reduce the interlamellar spacing, by increasing the cooling rate or reducing the solidification time.

Amount of eutectic.   By changing the composition of the alloy away from the eutectic composition, a primary microconstituent will form. We control the properties of the alloy by controlling the relative amounts of the primary microconstituent and the eutectic. In the lead-tin system, we find that as the tin content increases from 19% to 61.9%, the amount of the eutectic microconstituent changes from 0% to 100% and the amount of the primary α is reduced accordingly. With increasing amounts of the stronger eutectic microconstituent, the strength of the alloy increases (Figure 10-16). Similarly, when we increase the lead added to tin from 2.5% to 38.1% Pb, the amount of primary β in the hypereutectic alloy decreases, the amount of the strong eutectic increases, and the strength increases. Whenever both individual phases have about the same strength, the eutectic alloy is expected to have the highest strength due to effective dispersion strengthening.

EXAMPLE 10-8

Calculate the total % β and the % eutectic for the following lead-tin alloys: 10% Sn, 20% Sn, 50% Sn, 60% Sn, 80% Sn, 95% Sn.

Answer:

From the phase diagram (Figure 10-5) we want to calculate the total amount of β at room temperature, about 25°C, while we calculate the eutectic just above the eutectic temperature.

Alloy	% β	% Eutectic
Pb-10% Sn	$\dfrac{10 - 2}{100 - 2} \times 100 = 8.2\%$	0%
Pb-20% Sn	$\dfrac{20 - 2}{100 - 2} \times 100 = 18.4\%$	$\dfrac{20 - 19}{61.9 - 19} \times 100 = 2.3\%$
Pb-50% Sn	$\dfrac{50 - 2}{100 - 2} \times 100 = 49\%$	$\dfrac{50 - 19}{61.9 - 19} \times 100 = 72.3\%$
Pb-60% Sn	$\dfrac{60 - 2}{100 - 2} \times 100 = 59.2\%$	$\dfrac{60 - 19}{61.9 - 19} \times 100 = 95.6\%$
Pb-80% Sn	$\dfrac{80 - 2}{100 - 2} \times 100 = 79.6\%$	$\dfrac{97.5 - 80}{97.5 - 61.9} \times 100 = 49.2\%$
Pb-95% Sn	$\dfrac{95 - 2}{100 - 2} \times 100 = 94.9\%$	$\dfrac{97.5 - 95}{97.5 - 61.9} \times 100 = 7\%$

The % β and % eutectic are included in Figure 10-16 to show their relationship to the tensile strength of lead-tin alloys.

Shape of the eutectic.   The shape of the two phases in the eutectic micro-constituent is influenced by the cooling rate, the presence of impurity elements, and the nature of the alloy (Figure 10-17).

In some alloys, the eutectic microconstituent is composed of thin discontinuous plates of a brittle phase. The aluminum-silicon eutectic phase diagram (Figure 10-18) forms the basis for a number of important commercial alloys. Unfortunately, the silicon portion of the eutectic grows as thin flat plates that appear needlelike in a photomicrograph [Figure 10-17(b)]. The silicon platelets concentrate stresses and reduce ductility and toughness.

The nature of the eutectic structure in aluminum-silicon alloys is altered by modification. *Modification* causes the silicon phase to grow as thin interconnected rods between aluminum dendrites [Figure 10-17(c)], improving both tensile strength and % elongation (Figure 10-19). In two dimensions, the modified silicon appears to be composed of tiny round particles. Rapidly cooled alloys, such as those in die casting, are naturally modified during solidification. At slower cooling rates, however, 0.02% Na or 0.01% Sr must be added to cause modification.

The shape of the primary phase is also important. Often the primary phase grows in a dendritic manner; decreasing the secondary dendrite arm spacing of the primary phase may improve the properties of the alloy. Sometimes, however, the primary phase may take a different shape. In hypereutectic aluminum-silicon alloys, β is the primary phase. Because β is hard, the hypereutectic alloys are wear resistant; consequently, the hypereutectic aluminum-silicon alloys are considered for engine blocks. However, the primary silicon is normally very coarse [Figure 10-20(a)], causing poor castability, poor machinability, and gravity segregation (where the primary β floats to the surface of the casting during freezing). Addition of 0.05% P encourages nucleation of primary silicon, refines its size, and minimizes its deleterious qualities [Figure 10-20(b)].

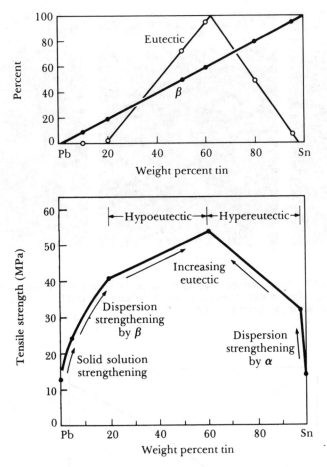

FIG. 10-16 The effect of composition and strengthening mechanism on the tensile strength of lead-tin alloys.

EXAMPLE 10-9

Aluminum engine blocks for automobiles are manufactured using a hypereutectic Al-17% Si alloy. Estimate the amount of primary silicon and total amount of silicon phase present in the microstructure.

Answer:

From a tie line at 578°C in Figure 10-18

$$\% \text{ Primary } \beta = \frac{17 - 12.6}{99.83 - 12.6} \times 100 = 5.0\%$$

From a tie line at 576°C

$$\% \beta = \frac{17 - 1.65}{99.83 - 1.65} \times 100 = 15.6\%$$

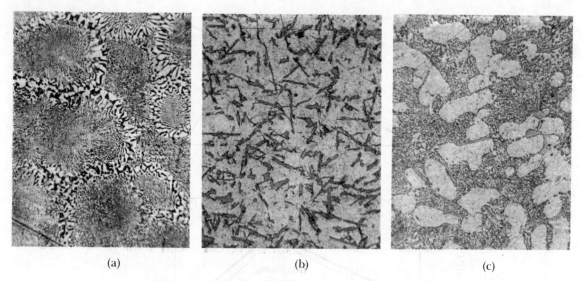

(a)   (b)   (c)

FIG. 10-17   Typical eutectic microstructures. (a) Grains in the lead-tin eutectic, (b) needlelike silicon plates in the aluminum-silicon eutectic, and (c) rounded silicon rods in the modified aluminum-silicon eutectic.

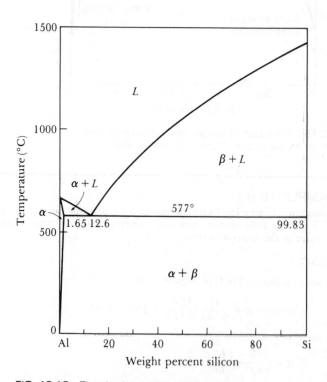

FIG. 10-18   The aluminum-silicon phase diagram.

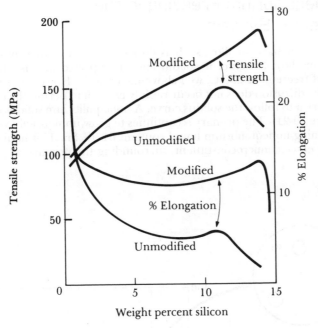

FIG. 10-19   The effect of silicon content and modification on the tensile strength and % elongation of aluminum-silicon alloys.

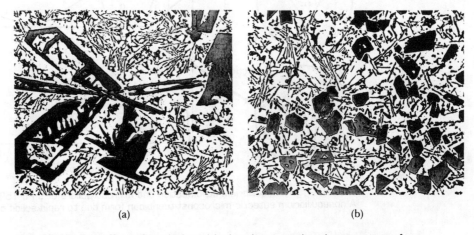

(a)   (b)

FIG. 10-20   The effect of hardening with phosphorus on the microstructure of hypereutectic aluminum-silicon alloys. (a) Coarse primary silicon; (b) fine primary silicon as refined by phosphorus addition. From *Metals Handbook*, Vol. 7, 8th Ed., American Society for Metals, 1972.

10-7 Nonequilibrium Freezing in the Eutectic System

Suppose we have an alloy, such as Pb-15% Sn, that ordinarily solidifies as a solid solution alloy. No eutectic solidification is expected in this alloy; the last liquid should freeze near 230°C, well above the eutectic. However, if the alloy cools too quickly, diffusion does not occur rapidly enough to permit the composition of the primary α to follow the solidus curve. A nonequilibrium solidus curve is produced (Figure 10-21). The primary α continues to grow until just above 183°C, when the remaining nonequilibrium liquid contains 61.9% Sn. This liquid then transforms to the eutectic microconstituent, surrounding the primary α.

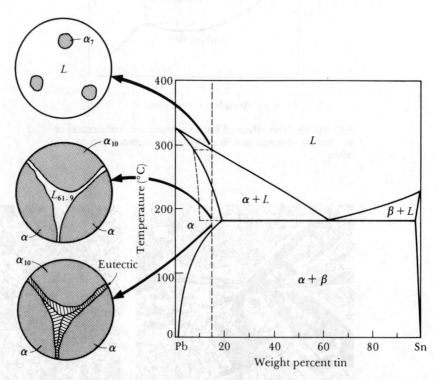

FIG. 10-21 Nonequilibrium solidification and microstructure of a Pb-15% Sn alloy. A nonequilibrium eutectic microconstituent can form due to rapid solidification.

EXAMPLE 10-10

Calculate the amount of nonequilibrium eutectic that forms in a Pb-15% Sn alloy for the nonequilibrium solidus in Figure 10-21.

Answer:

If we draw a tie line just above the eutectic line that runs from the nonequilibrium solidus to the liquidus curve, we can perform the lever law calculations.

$$\% \text{ Primary } \alpha = \frac{61.9 - 15}{61.9 - 10} \times 100 = 90.4\%$$

$$\% \text{ Eutectic} = \frac{15 - 10}{61.9 - 10} \times 100 = 9.6\%$$

The presence of the nonequilibrium eutectic can make the alloy *hot short*, or melt at temperatures below the equilibrium melting point. When heat treating an alloy such as Pb-15% Sn, we must keep the maximum temperature below the eutectic temperature of 183°C to prevent hot shortness.

10-8   The Peritectic Reaction

Peritectic reactions are found in a variety of alloys, including low-carbon steels (Figure 10-22). In the peritectic reaction, a liquid reacts with solid δ to form a second solid phase γ. After the reaction begins, the γ phase separates the two reacting phases. For the peritectic reaction to continue, atoms must diffuse

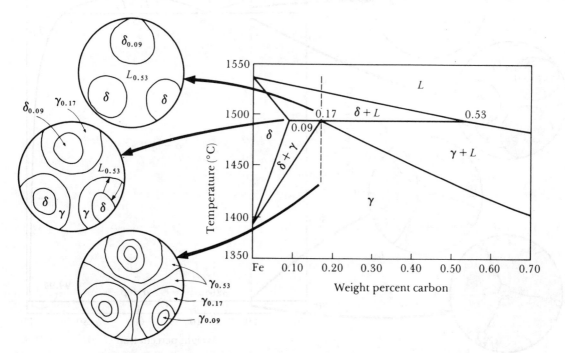

FIG. 10-22   Solidification and microstructure that develop as a result of the peritectic reaction.

through solid γ. But diffusion through a solid is much slower than diffusion through a liquid. As the peritectic reaction continues, the γ layer gets thicker and the reaction slows down even more. Unless the cooling rate is very slow, a segregated microstructure is produced.

In steels, segregation during the peritectic reaction is not a severe problem. Most steels eventually undergo significant plastic deformation and heat treatment, which reduce the segregation effects.

10-9 The Monotectic Reaction

We are aware that oil and water do not mix. They have very little solubility in one another and can be regarded as immiscible liquids. Certain alloy systems behave in the same manner (Figure 10-23). Liquid copper and liquid lead are completely soluble at high temperatures. However, alloys containing between 36% and 87% Pb separate into two liquids during cooling. The two liquids coexist in the *miscibility gap*, or dome, that is characteristic of all alloys that undergo the monotectic reaction.

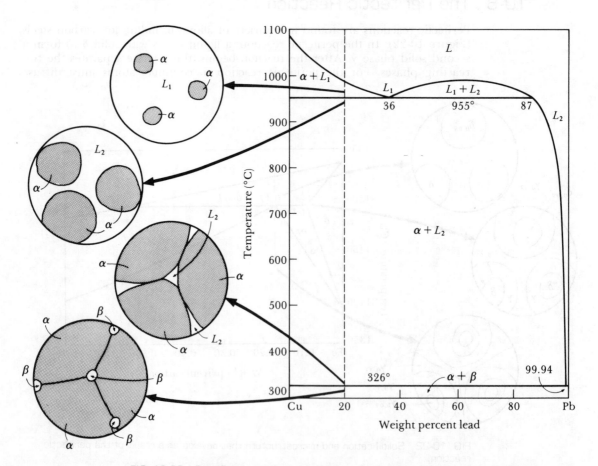

FIG. 10-23 Solidification and microstructure of a hypomonotectic copper-lead alloy.

During the solidification of a hypomonotectic copper-lead alloy, we find that solid α, almost pure copper, forms first. The liquid composition shifts to the monotectic composition, 36% Pb. Next the liquid transforms to more solid α plus a second liquid containing 87% Pb. From the lever law, we find that only a relatively small amount of the second liquid is present. On further cooling, the second liquid eventually goes through an eutectic reaction, producing α and β, where β is nearly pure lead.

The microstructure of the copper-lead alloy contains spherical β particles randomly distributed throughout a copper matrix. Little dispersion strengthening occurs because the lead-rich β phase is very soft.

EXAMPLE 10-11

Calculate the amount of lead-rich liquid that forms immediately after the monotectic reaction in a Cu-5% Pb alloy. How much lead-rich solid β is present at room temperature?

Answer:

From a tie line at 954°C

$$L_2 = \frac{5 - 0}{87 - 0} \times 100 = 5.7\%$$

From a tie line at 25°C

$$\beta = \frac{5 - 0}{100 - 0} \times 100 = 5.0\%$$

10-10 Ternary Phase Diagrams

There are many alloy systems that are based on three or even more elements. When three elements are present, we have a *ternary* alloy. To describe the changes in structure with temperature, we must draw a three-dimensional phase diagram (Figure 10-24).

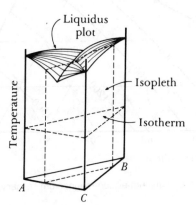

FIG. 10-24 A schematic drawing of a ternary phase diagram, showing how the three types of two-dimensional plots are obtained.

However, we can present the information from ternary diagrams in two dimensions by three different techniques—the liquidus plot, the isothermal plot, and the isopleth plot.

Liquidus plot.   The liquidus plot for the aluminum-silicon-magnesium alloy system (Figure 10-25) is the basis for several casting alloys. The liquidus plot is composed of isothermal contours showing the liquidus temperature at each combination of elements. This presentation is helpful in predicting the freezing temperature of the alloy, just as in binary phase diagrams. The plot shows the sequence in which phases precipitate during solidification. The liquidus plot also shows how the composition of the liquid changes during solidification.

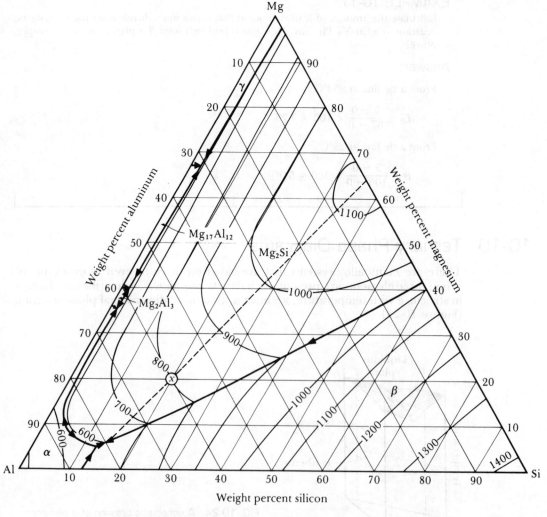

FIG. 10-25   The liquidus plot for the aluminum-silicon-magnesium ternary phase diagram.
(Point *x* refers to Example 10-12.)

EXAMPLE 10-12

Locate the Al-20% Si-20% Mg alloy on the liquidus plot, determine the liquidus temperature, find the primary solid phase, and predict how the composition of the liquid will change during solidification.

Answer:

The coordinates of the ternary permit us to locate the alloy at point x on the diagram. The phase diagram shows that the liquidus temperature is 800°C. The first solid phase is Mg_2Si. The liquidus temperatures become lower as we move towards a higher aluminum content; thus, the liquid is enriched in aluminum during solidification. As Mg_2Si continues to form, the liquid follows the dashed line drawn from pure Mg_2Si through point x until the composition intersects the solid line separating Mg_2Si and β (nearly pure silicon).

Solidus plots, showing the temperature at which each alloy begins to melt on heating, and solvus plots are also available.

Isothermal plot. A second common presentation is the isothermal plot. The isothermal plot shows the phases present in each alloy at a particular temperature and is useful in predicting the phases and their amounts and compositions at that temperature. Figure 10-26 shows the isothermal plot at 1250°C for the nickel-chromium-molybdenum system.

EXAMPLE 10-13

Determine the phases present at 1250°C for each of the following nickel-chromium-molybdenum ternary alloys: Ni-40% Cr-30% Mo, Ni-40% Cr-20% Mo, Ni-60% Cr-10% Mo.

Answer:

The compositions are shown as points x, y, and z on the phase diagram (Figure 10-26).

x: Ni-40% Cr-30% Mo—σ

y: Ni-40% Cr-20% Mo—$\alpha + \sigma$

z: Ni-60% Cr-10% Mo—$\alpha + \beta + \sigma$

Amounts and compositions of phases can be calculated much as in the binary phase diagrams. In single-phase regions, the results are trivial—the phase has the original composition and its amount is 100%. In two-phase regions, a tie line can be used to determine the composition of each phase.

A three-phase region is characterized by a triangle. The three corners of the triangle give the compositions of the three phases in equilibrium. Use of the lever law to determine the amount of each phase is illustrated in Example 10-14.

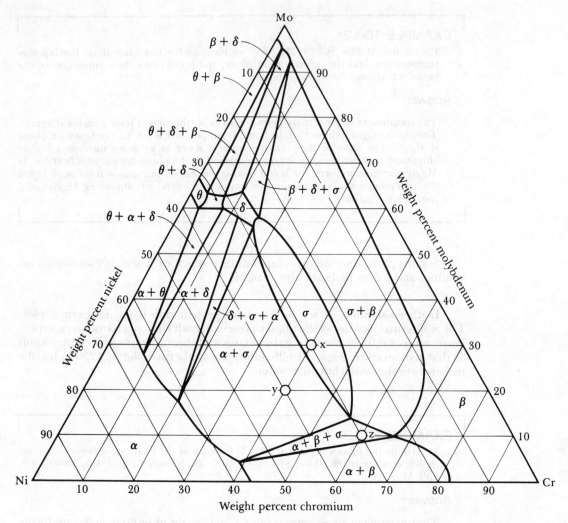

FIG. 10-26   An isothermal plot at 1250°C for the nickel-chromium-molybdenum ternary phase diagram.

EXAMPLE 10-14

Determine the amounts and compositions of each phase in the following nickel-chromium-molybdenum alloys: Ni-40% Cr-30% Mo, Ni-40% Cr-20% Mo, Ni-60% Cr-10% Mo.

Answer:

We found the phases present in each alloy in Example 10-13. Figure 10-27 shows these regions of the phase diagram in greater detail.

x: Ni-40% Cr-30% Mo—since only one phase, σ, is present, the amount of σ is 100% and the σ contains 30% Ni-40% Cr-30% Mo.

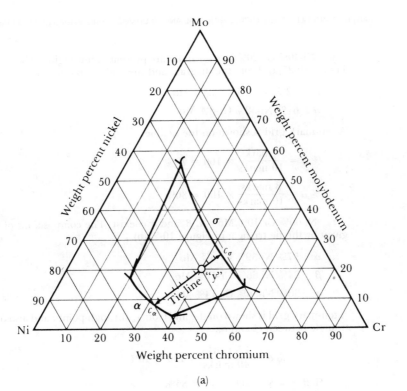

(a)

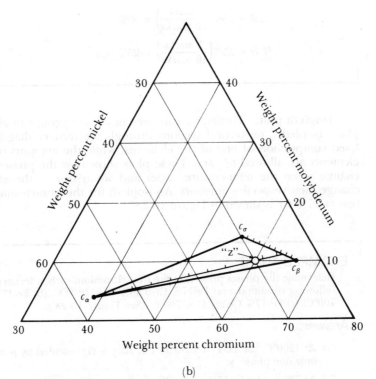

(b)

FIG. 10-27 Sections of the nickel-chromium-molybdenum phase
diagram for Example 10-14. (a) A tie line in a two-phase region. (b)
The tie triangle in a three-phase region.

y: Ni-40% Cr-20% Mo—α and σ are present in this region. The tie line is shown in Figure 10-27(a). From the diagram and the ends of the tie line

   σ: 32% Ni-43% Cr-25% Mo

   α: 61% Ni-32% Cr-8% Mo

A ten-mark grid is shown on the tie line.

$$\% \ \sigma = \frac{7 \ \text{marks}}{10 \ \text{marks}} \times 100 = 70\%$$

$$\% \ \alpha = \frac{3 \ \text{marks}}{10 \ \text{marks}} \times 100 = 30\%$$

z: Ni-60% Cr-10% Mo—α, β, and σ coexist. The compositions of each phase are given by the tie triangle in Figure 10-27(b).

   α: 57% Ni-39% Cr-4% Mo

   β: 24% Ni-66% Cr-10% Mo

   σ: 30% Ni-56% Cr-14% Mo

A ten-mark grid is shown on the line from c_{α} through point z and also on the leg of the triangle between c_{σ} and c_{β}.

$$\% \ \alpha = \frac{1.5 \ \text{marks}}{10 \ \text{marks}} \times 100 = 15\%$$

$$\% \ \beta + \% \ \gamma = 100 - 15 = 85\%$$

$$\% \ \beta = 85\% \left(\frac{7 \ \text{marks}}{10 \ \text{marks}} \right) = 60\%$$

$$\% \ \sigma = 85\% \left(\frac{3 \ \text{marks}}{10 \ \text{marks}} \right) = 25\%$$

Isopleth plot.   Finally, we can present certain groups of alloys by isopleth plots. Isopleths, or vertical sections through the ternary diagram, represent a fixed composition of one of the elements, while the amounts of the other two elements are allowed to vary. These plots show how the phases and structures change when the temperature varies and when two of the elements present change their respective amounts. An isopleth for the iron-chromium-carbon system at 17% Cr is shown in Figure 10-28.

EXAMPLE 10-15

Determine the phases present in the iron-chromium-carbon ternary diagram for the following conditions: (a) Fe-17% Cr-0.50% C at 1200°C, (b) Fe-17% Cr-0.50% C at 700°C, (c) Fe-17% Cr-2% C at 700°C. (See Figure 10-28.)

Answer:

(a) At 1200°C, the Fe-17% Cr-0.50% C alloy is represented by point x and contains only one phase, γ.

(b) At 700°C, the Fe-17% Cr-0.50% C alloy, represented by point y, has cooled and

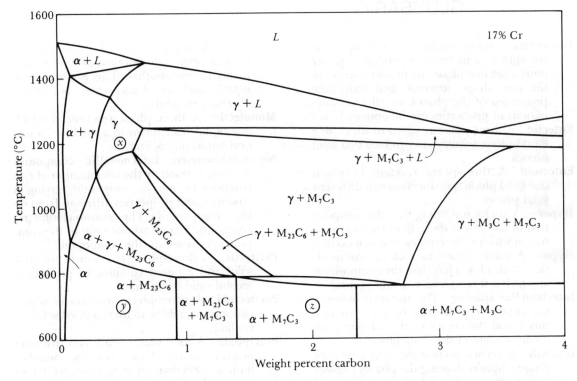

FIG. 10-28 An isopleth section through a portion of the iron-chromium-carbon ternary phase diagram at a constant 17% Cr.

transformed to α and $M_{23}C_6$. The M in this compound indicates some mixture of iron and chromium in the intermetallic compound.

(c) At 700°C, the Fe-17% Cr-2% C alloy, point z, contains α plus M_7C_3. The higher carbon content has caused a different intermetallic compound to form.

SUMMARY

By taking advantage of three-phase reactions and solubility limits, we control the properties of a material by dispersion strengthening. The eutectic reaction permits us to control dispersion strengthening by solidification techniques. As a result many common casting, brazing, and soldering alloys are based on eutectics. We are able to control the structure and properties by standard solidification techniques—inoculation and cooling rate or solidification time. Now we have the additional flexibility of controlling the amount, size, shape, and distribution of the two phases in the eutectic structure.

GLOSSARY

Dispersion strengthening Increasing the strength of a material by mixing together more than one phase. By proper control of the size, shape, amount, and individual properties of the phases, excellent combinations of properties can be obtained.

Eutectic A three-phase reaction in which one liquid phase solidifies to produce two solid phases.

Eutectoid A three-phase reaction in which one solid phase transforms to two different solid phases.

Hyper- A prefix indicating that the composition of an alloy is more than the composition at which a three-phase reaction occurs.

Hypo- A prefix indicating that the composition of an alloy is less than the composition at which a three-phase reaction occurs.

Interlamellar spacing The distance between the center of a lamella or plate of one phase and the center of the adjoining lamella or plate of the same phase.

Isopleth A vertical section through a ternary phase diagram showing the phases present at any temperature when the amount of one of the components is fixed.

Isothermal plot A horizontal section through a ternary phase diagram showing the phases present at a particular temperature.

Lamella A thin plate of a phase that forms during certain three-phase reactions, such as the eutectic or eutectoid.

Liquidus plot A two-dimensional plot showing the temperatures at which a three-component alloy system begins to solidify on cooling.

Matrix Typically the first solid material to form during cooling of an alloy. Usually, the matrix is continuous and a second phase precipitates from it. However, in some complex alloys, the matrix is more difficult to define.

Microconstituent A phase or mixture of phases in an alloy that has a distinct appearance. Frequently, we describe a microstructure in terms of the microconstituents rather than the actual phases.

Miscibility gap A region in a phase diagram in which two phases, with essentially the same structure, do not mix, or have no solubility in one another. This is common in liquids, such as oil and water, but also is observed in solids.

Monotectic A three-phase reaction in which one liquid transforms to a solid and a second liquid on cooling.

Nonstoichiometric intermetallic compound A phase formed by the combination of two components into a compound having a structure and properties different from either component. The nonstoichiometric compound has a variable ratio of the components present in the compound.

Peritectic A three-phase reaction in which a solid and a liquid combine to produce a second solid on cooling.

Peritectoid A three-phase reaction in which two solids combine to form a third solid on cooling.

Precipitate A solid phase that forms from the original matrix phase when the solubility limit is exceeded. In most cases, we try to control the formation of the precipitate to produce the optimum dispersion strengthening.

Primary A prefix used to indicate that a microconstituent has formed before the start of a three-phase reaction. Generally, a primary microconstituent will be a single phase.

Solvus A solubility line that separates a single solid phase region from a two solid phase region in the phase diagram.

Stoichiometric intermetallic compound A phase formed by the combination of two components into a compound having a structure and properties different from either component. The stoichiometric intermetallic compound has a fixed ratio of the components present in the compound.

Ternary phase diagram A phase diagram between three components showing the phases present and their compositions at various temperatures. This requires a three-dimensional plot.

PRACTICE PROBLEMS

1 A hypothetical phase diagram is shown in Figure 10-29. (a) What intermetallic compound is present? Is it stoichiometric or nonstoichiometric? (b) Identify the four solid solutions present in the system. (c)

3 The copper-tin phase diagram is shown in Figure 12-7. (a) Identify the intermetallic compounds in the phase diagram. Are the compounds stoichiometric or nonstoichiometric? (b) Identify the three-phase reac-

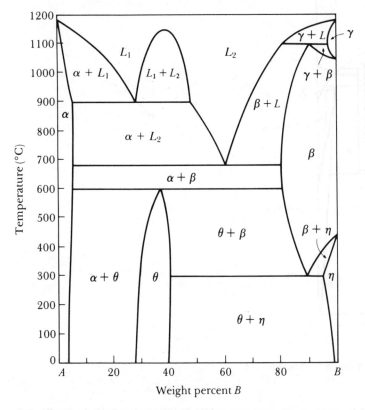

FIG. 10-29 Hypothetical phase diagram for Problem 10-1.

Identify the three-phase reactions by writing down the temperature, the reaction in equation form, the composition of each phase in the reaction, and the name of the reaction.

2 The copper-zinc phase diagram is shown in Figure 12-7. (a) Identify the intermetallic compounds in the phase diagram. Are the compounds stoichiometric or nonstoichiometric? (b) Identify the three-phase reactions by writing down the temperatures, the reaction in equation form, the composition of each phase in the reaction, and the name of the reaction.

tions by writing down the temperatures, the reaction in equation form, the composition of each phase in the reaction, and the name of the reaction.

4 The magnesium-tin phase diagram is shown in Figure 12-5. (a) Identify the intermetallic compound in the phase diagram. Is the compound stoichiometric or nonstoichiometric? (b) Determine the formula for the intermetallic compound. (c) Identify the three-phase reactions by writing down the temperatures, the reaction in equation form, the composition of each phase in the reaction, and the name of the reaction.

5 The titanium-manganese phase diagram is shown in Figure 10-30. (a) Identify the intermetallic compounds in the phase diagram. Are the compounds stoichiometric or nonstoichiometric? (b) Identify the

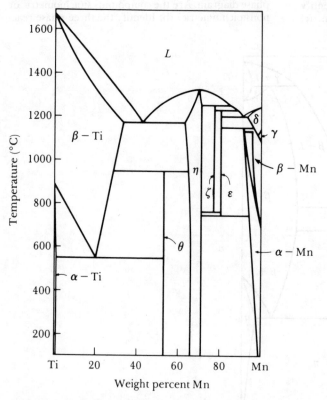

FIG. 10-30   The titanium-manganese phase diagram, Problem 10-5.

three-phase reactions by writing down the temperatures, the reaction in equation form, the composition of each phase in the reaction, and the name of the reaction.

6 A hypothetical phase diagram is shown in Figure 10-31. (a) Identify the stoichiometric and nonstoichiometric intermetallic compounds. (b) One of the stoichiometric compounds has the formula A_5B_2. Which compound is it? (Assume that the atomic mass of A is 33 g/g · mole and the atomic mass of B is 72 g/g · mole.)

7 Calculate the amount of β phase at 0°C for a Pb-5% Sn alloy and for a Pb-12% Sn alloy (Figure 10-5). Which alloy is strengthened more by dispersion strengthening?

8 Determine the amount of each phase at 0°C in an

Al-5% Si alloy and an Al-10% Si alloy (Figure 10-18). Which alloy is strengthened more by dispersion strengthening?

9 From Figure 12-7, determine the solubility of zinc in copper at 0°C, 500°C, and 1000°C. Determine the solubility of copper in zinc at 0°C, 200°C, and 400°C.

10 Consider a Pb-10% Sn alloy (Figure 10-5). During solidification, determine (a) the composition of the first solid to form, (b) the amounts and compositions of each phase at 290°C, (c) the liquidus, solidus, and solvus temperatures, (d) the amounts and compositions of each phase at 200°C, and (e) the amounts and compositions of each phase at 0°C. (f) Suppose, due to rapid cooling, that the composition of the last primary α was 8% Sn. Calculate the amount of nonequilibrium eutectic microconstituent.

11 Consider a Pb-15% Sn alloy (Figure 10-5). During solidification, determine (a) the composition of the first solid to form, (b) the amounts and compositions of each phase at 250°C, (c) the liquidus, solidus,

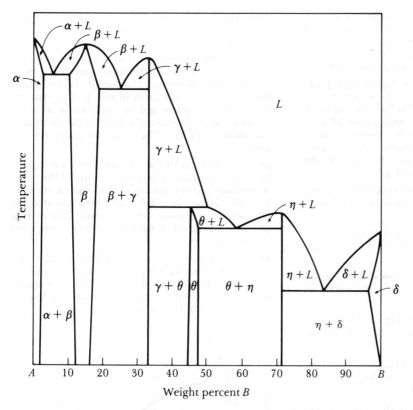

FIG. 10-31 Hypothetical phase diagram for Problem 10-6.

and solvus temperatures, (d) the amounts and compositions of each phase at 200°C, and (e) the amounts and compositions of each phase at 0°C.

12 Locate the eutectic composition in the aluminum-silicon phase diagram (Figure 10-18). Calculate the amount of each phase in the eutectic microconstituent. Based on the amount of each phase, which should be continuous? Would you expect the eutectic microconstituent to be ductile or brittle?

13 Locate the eutectic composition at the aluminum-rich end of the aluminum-copper phase diagram (Figure 11-1). Calculate the amount of each phase in the eutectic microconstituent. Based on the amount of each phase, which should be continuous? Would you expect the eutectic microconstituent to be ductile or brittle?

14 Two eutectic reactions occur in the aluminum-magnesium phase diagram (Figures 12-1 and 12-5). Calculate the amount of each phase in the eutectic microconstituents in both eutectic reactions. Based on the amount of each phase, which should be continu-

ous in each reaction? Would you expect the eutectic microconstituents to be ductile or brittle?

15 Consider a Pb-80% Sn alloy (Figure 10-5). Determine (a) if the alloy is hypo- or hypereutectic, (b) the composition of the first solid to form during solidification, (c) the amounts and compositions of each phase at 184°C, (d) the amounts and compositions of each phase at 182°C, (e) the amounts and compositions of each microconstituent at 182°C, and (f) the amounts and compositions of each phase at 0°C.

16 Consider a Pb-30% Sn alloy (Figure 10-5). Determine (a) if the alloy is hypo- or hypereutectic, (b) the composition of the first solid to form during solidification, (c) the amounts and compositions of each phase at 184°C, (d) the amounts and compositions of each phase at 182°C, (e) the amounts and compositions of each microconstituent at 182°C, and (f) the amounts and compositions of each phase at 0°C.

17 Consider an Al-5% Si alloy (Figure 10-18). Determine (a) if the alloy is hypo- or hypereutectic, (b) the composition of the first solid to form during solidification, (c) the amounts and compositions of each phase at 578°C, (d) the amounts and compositions of each phase at 576°C, (e) the amounts and compositions of each microconstituent at 576°C, and (f) the amounts and compositions of each phase at 0°C.

18 Consider an Al-18% Si alloy (Figure 10-18). Determine (a) if the alloy is hypo- or hypereutectic, (b) the composition of the first solid to form during solidification, (c) the amounts and compositions of each phase at 578°C, (d) the amounts and compositions of each phase at 576°C, (e) the amounts and compositions of each microconstituent at 576°C, and (f) the amounts and compositions of each phase at 0°C.

19 A lead-tin alloy (Figure 10-5) contains 75% eutectic and 25% primary β. Determine the overall composition of the alloy.

20 A lead-tin alloy (Figure 10-5) contains 30% eutectic and 70% primary α. Determine the overall composition of the alloy.

21 A lead-tin alloy (Figure 10-5) contains 25% α and 75% β at 0°C. Determine the overall composition of the alloy.

22 An aluminum-silicon alloy (Figure 10-18) contains 30% primary β and 70% eutectic. Determine the overall composition of the alloy.

23 An aluminum-silicon alloy (Figure 10-18) contains 45% primary α and 55% eutectic. Determine the overall composition of the alloy.

24 Prepare a graph showing how the percent eutectic changes with silicon content in the aluminum-silicon system (Figure 10-18).

25 Prepare a graph showing how the percent eutectic changes with copper content in the aluminum-copper system (Figure 11-1).

26 The densities of the two phases in the aluminum-silicon system (Figure 10-18) are 2.65 Mg m^{-3} for α and 2.35 g/cm^3 for β. (a) Estimate the density of the eutectic microconstituent. (b) If an aluminum-silicon alloy contains 60 vol% eutectic and 40 vol% primary α, calculate the composition of the alloy in weight percent.

27 The densities of the two phases in the lead-tin system (Figure 10-5) are 10.2 Mg m^{-3} for α and 7.30 Mg m^{-3} for β. (b) Estimate the density of the eutectic microconstituent. (b) If a lead-tin alloy contains 30 vol% primary β and 70 vol% eutectic, calculate the composition of the alloy in weight percent. (c) What is the liquidus temperature of the alloy?

28 Plot the freezing range of lead-tin alloys (Figure 10-5) (a) as a function of percent tin and (b) as a function of percent eutectic. (The solder alloys with short freezing ranges are used for delicate work, but

those with long freezing ranges are used as "wiping" solders for body-work on automobiles.)

29 In Figure 11-13 for the Fe-Fe$_3$C phase diagram, locate the composition and temperature of the eutectic reaction. (a) What are the phases and their compositions during the eutectic reaction? (b) Do you expect the eutectic to be ductile or brittle? (c) Calculate the amount of each phase and each microconstituent immediately after the eutectic reaction occurs in an alloy containing 3.6% C. (These eutectic alloys are cast irons.)

30 In Figure 12-13 for the zinc-aluminum system, locate the composition and temperature of the eutectic reaction. (Alloys containing this eutectic are common die casting alloys.) A typical structure immediately after the reaction occurs contains 75% eutectic and 25% primary α. What is the overall composition of the alloy?

31 A cooling curve is shown in Figure 10-32. Determine (a) the pouring temperature, (b) the superheat, (c) the liquidus temperature, (d) the eutectic temperature, (e) the freezing range, (f) the total solidification time, and (g) the local solidification time.

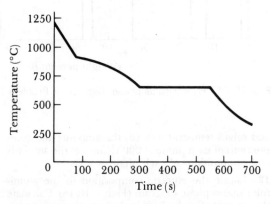

FIG. 10-32   Cooling curve for Problem 10-31.

32 Cooling curves are obtained for a series of silver-copper alloys (Figure 10-33). Use the cooling curves to produce the silver-copper phase diagram. The maximum solubility of copper in silver is 7.9%, the maximum solubility of silver in copper is 8.8%, and the solubilities at room temperature are about zero.

33 The densities of α and β in the lead-tin system are given in Problem 10-27. Estimate the relative thickness of the α and β lamellae in the eutectic microconstituent.

34 Determine the constant c in Equation (10-1) for the effect of growth rate on the interlamellar spacing

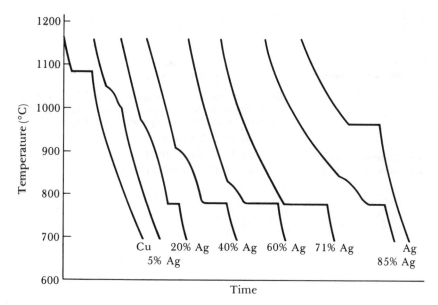

FIG. 10-33   Cooling curves for a series of silver-copper alloys. (See Problem 10-32.)

of the lead-tin eutectic, using the data in Figure 10-15.

35 A cylinder of liquid lead-tin eutectic is placed in a crucible inside a furnace, then is withdrawn at a slow rate from the furnace into a quenching medium. This arrangement permits the solid-liquid interface to move at a rate equal to the withdrawal rate. If it takes 30 min to cool a 5-cm long specimen, what is the interlamellar spacing that we expect to find?

36 Draw a sketch showing how the hardness of aluminum-silicon alloys should change as the silicon content increases from 0% to 20%. (The hardness of silicon is much higher than that of pure aluminum.)

37 Determine the amount of each phase in a Fe-0.35% C alloy (Figure 10-22) at 1520°C, 1500°C, 1480°C, and 1400°C.

38 Calculate the amount of δ and liquid that must combine to produce 100% γ in the peritectic reaction in the iron-carbon system (Figure 10-22).

39 In a Fe-0.30% C alloy (Figure 10-22), calculate the fraction of the total alloy that solidifies by going through the peritectic reaction.

40 In a Fe-0.12% C alloy (Figure 10-22), calculate the fraction of the total alloy that solidifies by going through the peritectic reaction.

41 Determine the composition of the alloy that produces a structure of 50% δ and 50% γ just below the peritectic reaction temperature in the iron-carbon system (Figure 10-22). If this composition solidified under nonequilibrium conditions, would you expect to have more or less γ forming by the peritectic reaction?

42 In the copper-lead system (Figure 10-23), determine the composition of the two liquids in equilibrium at 960°C.

43 How much L_2 is present in a Cu-5% Pb alloy (Figure 10-23) at 960°C, 900°C, and 400°C?

44 How much L_2 is present in a Cu-60% Pb alloy (Figure 10-23) at 975°C, 956°C, 954°C, and 600°C?

45 For each of the following aluminum-silicon-magnesium alloys, determine the liquidus temperature and the primary phase that forms during solidification (Figure 10-25).

(a) Al-20% Si-60% Mg

(b) Al-50% Si-20% Mg

(c) Al-5% Si-5% Mg

46 An aluminum-silicon-magnesium alloy (Figure 10-25) contains equal amounts of silicon and aluminum. What must be the composition of the alloy in terms of % Al, % Si, and % Mg so that primary Mg_2Si first forms at a liquidus of 1000°C?

47 Suppose an aluminum-silicon-magnesium alloy (Figure 10-25) contains equal amounts of aluminum and magnesium. What must be the composition of the alloy if primary β forms at a liquidus of 1120°C?

48 For each of the following nickel-chromium-molybdenum alloys at 1250°C, determine the phases present and estimate the amounts and compositions of each phase (Figure 10-26).

  (a) Ni-10% Cr-20% Mo

  (b) Ni-10% Cr-40% Mo

  (c) Ni-5% Cr-55% Mo

49 For each of the following nickel-chromium-molybdenum alloys at 1250°C, determine the phases present and estimate the amounts and compositions of each phase (Figure 10-26).

  (a) Ni-60% Cr-5% Mo

  (b) Ni-10% Cr-60% Mo

  (c) Ni-55% Cr-10% Mo

50 Determine the temperatures at which each of the following iron-chromium-carbon alloys (a) first start to freeze and (b) are completely solid (Figure 10-28).

  (a) Fe-17% Cr-0.5% C

  (b) Fe-17% Cr-1.0% C

  (c) Fe-17% Cr-2.0% C

11

Dispersion Strengthening by Phase Transformation and Heat Treatment

11-1 Introduction

In this chapter we will discuss dispersion strengthening further as we introduce a variety of solid-state transformation processes, including age hardening and the eutectoid reaction. We will also examine how nonequilibrium phase transformations, in particular the martensitic reaction, can provide strengthening. These dispersion-strengthening techniques require a heat treatment.

As we discuss these strengthening mechanisms, we must keep in mind the characteristics that produce the most desirable dispersion strengthening, as discussed in Chapter 10. The matrix should be relatively soft and ductile and the precipitate or second phase should be hard and brittle; the precipitate should be round and discontinuous; the precipitate particles should be small and numerous; and in general the more precipitate we have, the stronger will be the alloy.

11-2 Nucleation and Growth in Solid-State Reactions

In order for a precipitate to form from a solid matrix, both nucleation and growth must occur. The total change in free energy required for nucleation of a spherical solid precipitate from the matrix is

$$\Delta F = \frac{4}{3}\pi r^3 \Delta F_v + 4\pi r^2 \sigma + \frac{4}{3}\pi r^3 \varepsilon \qquad (11\text{-}1)$$

The first two terms include the volume free energy change and the surface energy change, just as in solidification [Equation (7-1)]. However, the third term takes into account the *strain energy* ε introduced when the precipitate forms in a solid rigid matrix. The precipitate does not occupy the same volume that is displaced, so additional energy is required to permit the precipitate to be accommodated in the matrix.

As in solidification, nucleation occurs most easily on surfaces already present in the structure, thereby minimizing the surface energy term. Thus, the precipitate nucleates and grows most easily at the grain boundaries of the matrix or at other lattice defects. Increasing the number of lattice defects permits us to exercise some control over the number of nuclei produced.

Growth of the precipitate normally occurs by long-range diffusion and redistribution of the atoms to satisfy the phase diagram. These reactions proceed relatively slowly, since the atoms must diffuse in the solid, but occur more readily at high temperatures, where diffusion is more rapid. Thus, the growth rates are usually controlled primarily by controlling temperature. The relationship between growth and nucleation plays an important role in the phase transformation in solid-state reactions, just as in the solidification of materials.

11-3 Alloys Strengthened by Exceeding the Solubility Limit

In Chapter 10 we pointed out that lead-tin alloys containing between about 2% and 19% Sn can be dispersion strengthened because the solubility of tin in lead is exceeded.

A similar situation occurs in aluminum-copper alloys. For example, the Al-4% Cu alloy, shown in Figure 11-1, is completely α, or an aluminum solid solution,

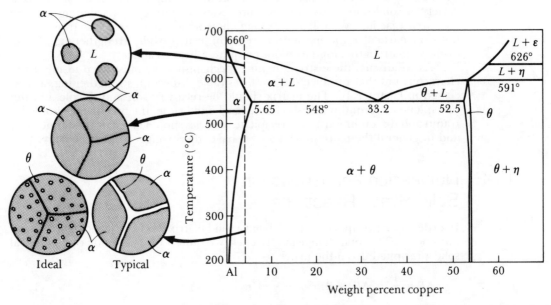

FIG. 11-1 The aluminum-copper phase diagram and the microstructures that may develop during cooling.

above 500°C. On cooling below the solvus temperature, a second phase, θ, precipitates. The θ phase, which is a hard brittle intermetallic compound $CuAl_2$, provides dispersion strengthening.

EXAMPLE 11-1

Calculate the amount of θ that forms at room temperature when an Al-4% Cu alloy slowly cools.

Answer:

From a tie line at 25°C

α: 0.02% Cu

θ: 53.5% Cu

$$\% \; \theta = \frac{4 - 0.02}{53.5 - 0.02} \times 100 = 7.4\%$$

Even this amount of θ is capable of providing effective dispersion strengthening if properly controlled.

EXAMPLE 11-2

Calculate the amount of θ in the aluminum-copper eutectic microconstituent. Explain why most aluminum-copper alloys are designed to avoid the eutectic reaction.

Answer:

We obtain all eutectic when the alloy contains Al-33.2% Cu. Thus

$$\% \; \theta = \frac{33.2 - 5.65}{52.5 - 5.65} \times 100 = 58.8\%$$

Most of the eutectic is composed of the hard brittle compound θ. The eutectic microconstituent will be brittle and, since the eutectic is continuous, the overall alloy will be brittle.

Unfortunately, we often are unable to control the precipitation of the second phase so that the requirements of good dispersion strengthening are satisfied. The second phase, such as θ in the aluminum-copper system, may not have a desirable size, shape, or distribution. Several factors influence the shape of the precipitate.

Widmanstatten structure. The second phase may grow so that certain planes and directions in the precipitate are parallel to preferred planes and directions in the matrix. This growth mechanism minimizes strain and surface energies and permits faster growth rates. Widmanstatten growth produces a characteristic appearance for the precipitate, such as plates, needles, rods, or even cubes (Figure 11-2). Particularly when the needlelike shape is produced, the Widmanstatten precipitate may embrittle the alloy.

Interfacial energy relationships. We expect the precipitate to have a spherical shape in order to minimize surface energy. However, the shape of the precipitate is also influenced by the *interfacial energy* associated with both the boundary between the matrix grains (γ_m) and the boundary between the matrix and the precipitate (γ_p). The interfacial surface energies fix a *dihedral angle* θ between the matrix-precipitate interface that in turn determines the shape of the precipitate [Figure 11-3(a)]. The relationship is

$$\gamma_m = 2\gamma_p \cos \frac{\theta}{2} \qquad (11\text{-}2)$$

FIG. 11-2 The Widmanstatten structure in a medium-carbon steel. The platelike precipitate is the BCC form of iron that forms during rapid cooling.

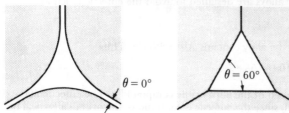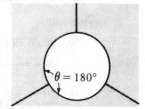

$\theta = 0°$

$\theta = 60°$

$\theta = 180°$

FIG. 11-3(a) The effect of surface energy and the dihedral angle on the shape of a precipitate.

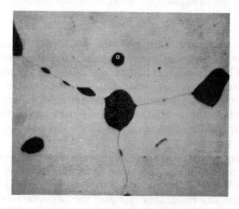

FIG. 11-3(b) A small precipitate of lead in copper.

If the dihedral angle is small, the precipitate may be continuous. If the precipitate is also hard and brittle, the thin film that surrounds the matrix grains causes the alloy to be very brittle. On the other hand, discontinuous and even spherical precipitates form when the dihedral angle is large [Figure 11-3(b)].

EXAMPLE 11-3

Calculate the ratio between γ_m and γ_p required to produce a continuous precipitate at grain boundaries.

Answer:

We obtain a continuous precipitate when $\theta = 0°$.

$$\frac{\gamma_m}{2\gamma_p} = \cos 0 = 1$$

$$\frac{\gamma_p}{\gamma_m} = \frac{1}{2}$$

EXAMPLE 11-4

From the photomicrograph of the copper-lead alloy in Figure 11-3(b), determine the dihedral angle and calculate the interfacial energy between the copper matrix and the lead precipitate. The energy of copper grain boundaries is $646 \times 10^{-3}\,\mathrm{J\ m^{-2}}$.

Answer:

The lead precipitate is round, so $\theta = 180°$. From Equation (11-2)

$$\gamma_m = 2\gamma_p \cos\frac{\theta}{2} = 2\gamma_p \cos\frac{180}{2} = 2\gamma_p \cos 90 = 0$$

$$\gamma_p = \frac{646 \times 10^{-3}}{0} = \infty$$

The energy really isn't infinity. This curious result means that the energy of the copper-lead interface is so large that the lead will produce the smallest possible surface area, or will be spherical.

Cooling rate.   The rate at which the alloy cools past the solvus line determines the time available for diffusion and consequently affects the shape of the precipitate. Fast cooling rates help offset the effect of very low dihedral angles and permit a discontinuous rather than a continuous grain boundary precipitate to form. Figure 11-4 compares the microstructure of the Al-4% Cu alloy for two different cooling rates.

In our example of the Al-4% Cu alloy, slow cooling permits the hard brittle θ phase to form as a thin almost continuous film at the α grain boundaries. The slow-cooled Al-4% Cu alloy does not have a desirable microstructure. Some improvement is obtained by increasing the rate of cooling as the alloy crosses the solvus line; however, optimum properties are still not obtained.

Coherent precipitate.   Even if we produce a uniform distribution of discontinuous θ precipitate, the precipitate may not significantly disrupt the surrounding matrix structure. Consequently, the precipitate blocks slip only if it lies directly in the path of the dislocation [Figure 11-5(a)].

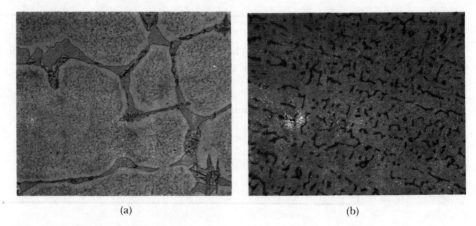

(a)   (b)

FIG. 11-4 Photomicrographs of Al-4% Cu alloy. Slow cooling (a) produces a more continuous, embrittling θ than fast cooling (b).

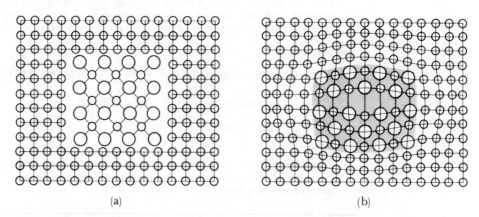

(a)   (b)

FIG. 11-5 (a) A noncoherent precipitate has no relationship with the crystal structure of the surrounding matrix. (b) A coherent precipitate forms so that there is a definite relationship between the precipitate's and the matrix's crystal structure.

But when a *coherent precipitate* forms, the planes of atoms in the lattice of the precipitate are related to, or even continuous with, the planes in the lattice of the matrix [Figure 11-5(b)]. Now a widespread disruption of the matrix lattice is created and the movement of a dislocation is impeded even if the dislocation merely passes near the coherent precipitate. A special heat treatment, such as age hardening, may be required to produce the coherent precipitate.

11-4 Age Hardening or Precipitation Hardening

Age hardening, or *precipitation hardening*, is designed to produce a uniform dispersion of a fine hard coherent precipitate in a softer, more ductile matrix. The Al-4% Cu alloy is a classical example of an age hardenable alloy. There are three steps in the age-hardening heat treatment (Figure 11-6).

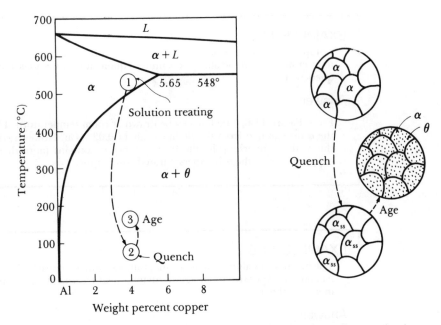

FIG. 11-6   The aluminum-rich end of the aluminum-copper phase diagram showing the three steps in the age-hardening heat treatment and the microstructures that are produced.

Step 1: solution treatment.   The alloy is first heated to a temperature above the solvus temperature and held until a homogeneous solid solution α is produced. This step dissolves the θ precipitate and reduces any segregation present in the original alloy.

We could heat the alloy to just below the solidus temperature and increase the rate of homogenization. However, the presence of a nonequilibrium eutectic microconstituent may cause hot shortness. Thus, the aluminum-copper alloy is solution treated between the solvus and eutectic temperatures, assuring that any eutectic microconstituent in the alloy does not melt. In the Al-4% Cu alloy, this treatment would be done between 500°C and 548°C.

Step 2: quench.   After solution treatment, the alloy, which contains only α in its structure, is rapidly cooled, or quenched. The atoms do not have time to diffuse to potential nucleation sites and permit the θ phase to form. After the quench the structure still contains only α. The α is a *supersaturated solid solution*, containing excess copper, and is not an equilibrium structure.

Step 3: age.   Finally, the supersaturated α is heated to a temperature below the solvus temperature. At this aging temperature, atoms are able to diffuse short distances. Because the supersaturated α is not stable, the extra copper atoms diffuse to numerous nucleation sites and a precipitate forms and grows. Eventually, if we hold the alloy for a sufficient time at the aging temperature, the equilibrium α and θ structure is produced.

EXAMPLE 11-5

Compare the composition of the α solid solution in the Al-4% Cu alloy at room temperature when the alloy cools under equilibrium conditions and when the alloy is quenched.

Answer:

From Figure 11-6, a tie line can be drawn at room temperature. The composition of the α determined from the tie line is about 0.02% Cu. However, the composition of the α after quenching is still 4% Cu. Since α contains more than the equilibrium copper content, the α is supersaturated with copper.

EXAMPLE 11-6

The magnesium-aluminum phase diagram is shown in Figure 12-1(a). Suppose a Mg-8% Al alloy is responsive to an age-hardening heat treatment. Recommend a heat treatment for the alloy.

Answer:

Step 1: Solution treat at a temperature between the solvus and the eutectic to avoid hot shortness. Thus, heat between 340°C and 437°C.

Step 2: Quench to room temperature fast enough to prevent the precipitate from forming.

Step 3: Age at a temperature below the solvus, or at some temperature below 340°C.

Nonequilibrium precipitates during aging. During aging of aluminum-copper alloys, a series of precipitates form before the equilibrium θ is produced. At the start of aging, the copper atoms concentrate on {100} planes in the α matrix and produce very thin clusters of copper atoms called *Guinier-Preston,* or GP-I, zones. As aging continues, more copper atoms diffuse to the precipitate and the GP-I zones grow into thin disks, or GP-II zones. Later, the GP-II zones dissolve and θ', which is similar to the stable θ, forms. Finally, θ' dissolves and the stable θ phase precipitates.

The nonequilibrium precipitates—GP-I, GP-II, and θ'—are coherent precipitates. The strength of the alloy increases with aging time as these coherent phases grow in size during the initial stages of the heat treatment. When these coherent precipitates are present, the alloy is in the *aged* condition. Figure 11-7(a) shows the structure of the aged alloy.

When the stable noncoherent θ phase precipitates, the strength of the alloy decreases. Now the alloy is in the *overaged* condition. The θ still provides some dispersion strengthening, but with increasing time, the θ grains grow larger and less numerous [Figure 11-7(b)] and even the simple dispersion strengthening effect diminishes.

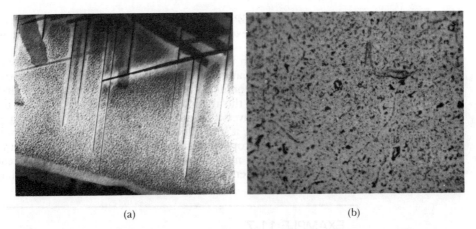

(a)   (b)

FIG. 11-7   (a) An electron micrograph of aged Al-15% Ag showing coherent γ′ plates and round GP zones, ×40,000. (Courtesy J. B. Clark.) (b) Overaged Al-4% Cu alloy containing coarse θ precipitates, ×500.

11-5 Effects of Aging Temperature and Time

The properties of an age hardenable álloy depend both on the temperature and the time for aging (Figure 11-8). At an aging temperature of 260°C, diffusion in the Al-4% Cu alloy is rapid and precipitates quickly form. As aging continues, we

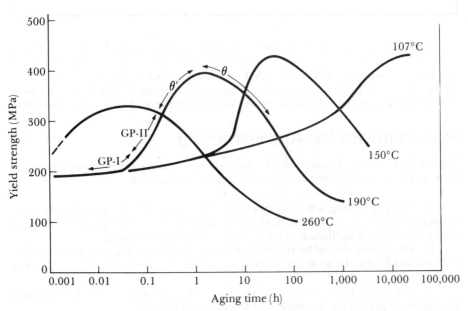

FIG. 11-8   The effect of aging temperature and time on the yield strength of an Al-4% Cu alloy.

progress from tiny GP-I and GP-II zones to θ'. The strength reaches a maximum after less than 0.1 h exposure. Overaging occurs if the alloy is held for longer than 0.1 h.

At 190°C, which is a typical aging temperature for many of the aluminum alloys, longer times are required to produce the optimum strength. However, there are several benefits to using the lower temperature. First, the maximum strength increases as the aging temperature decreases. Second, the strength maintains its maximum over a longer period of time. This broader peak permits the heat treater to make a small miscalculation in temperature or time but still produce the required properties. Third, the properties are more uniform. If the alloy is aged for only 10 min at 260°C, the surface of the part reaches the proper temperature and strengthens, but the center remains cool and ages only slightly.

EXAMPLE 11-7

The operator of a furnace left for his hour lunch break without removing the Al-4% Cu alloy from the aging furnace. Compare the effect on the yield strength of the extra hour of aging for aging temperatures of 190°C and 260°C.

Answer:

At 190°C, the peak strength of 400 MPa occurs at 6 h (Figure 11-8). After 7 h, the strength is essentially the same.

At 260°C, the peak strength of 340 MPa occurs at 0.06 h. However, after 1.06 h, the strength decreases to 250 MPa.

Thus, the higher aging temperature gives a lower peak strength and makes the strength more sensitive to aging time.

Aging at either 190°C or 260°C is called *artificial aging* because the alloy is heated to produce precipitation. Some solution-treated and quenched alloys age at room temperature and this is called *natural aging*. Natural aging requires long times, often several days, to reach the maximum strength. However, the peak strength is higher than that obtained in artificial aging and no overaging occurs.

11-6   Requirements for Age Hardening

Not all alloys are age hardenable. Four conditions must be satisfied for an alloy to have a true age-hardening response during heat treatment.

1. The phase diagram must display decreasing solid solubility with decreasing temperature. In other words, the alloy must form a single phase on heating above the solvus line, then enter a two-phase region on cooling.

2. The matrix should be relatively soft and ductile and the precipitate should be hard and brittle. In most age hardenable alloys, the precipitate is a hard brittle intermetallic compound.

3. The alloy must be quenchable. We cannot quench some alloys rapidly enough to suppress the formation of the second phase.

4. The precipitate that forms must be coherent with the matrix structure in order to develop the maximum strength and hardness. Furthermore, its size, shape, and distribution must be controlled.

EXAMPLE 11-8

Discuss the likelihood that each alloy shown in Figure 11-9 will display an age-hardening response.

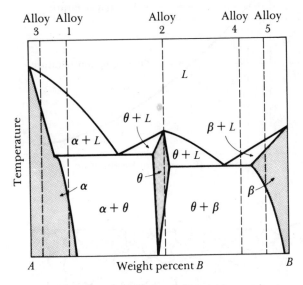

FIG. 11-9 A hypothetical phase diagram for use in Example 11-8.

Answer:

Alloy 1:   The solvus displays increasing solid solubility as the temperature decreases; thus, an age-hardening response is impossible.

Alloy 2:   The matrix is a hard brittle intermetallic compound but the precipitate is more likely to be softer and more ductile. Thus, an age-hardening response is unlikely.

Alloy 3:   This alloy is single phase up to the solidus temperature. No age hardening can occur.

Alloy 4:   This is a two-phase alloy up to the melting or eutectic temperature. A slight aging response might be possible in this type of alloy but the effect will be only small.

Alloy 5:   We now have decreasing solid solubility with decreasing temperature and the precipitate is a hard brittle intermetallic compound. This alloy might be a potential candidate for age hardening.

11-7 Use of Age Hardenable Alloys at High Temperatures

Based on our previous discussion, we would not select an age-hardened Al-4% Cu alloy for use at high temperatures. At service temperatures ranging from above

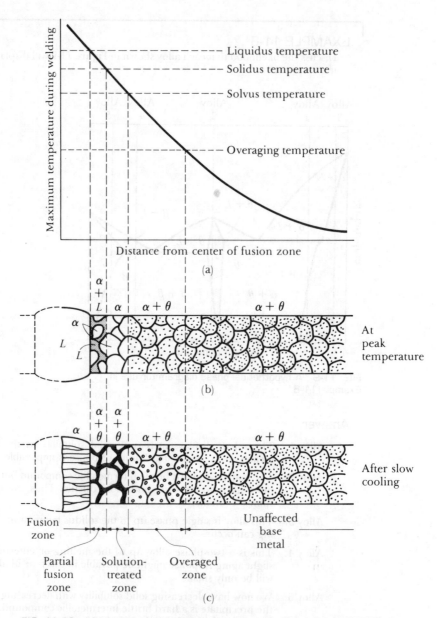

FIG. 11-10 Microstructural changes that occur in age-hardened alloys during fusion welding. (a) Peak temperature that occurs during welding, (b) microstructure in the weld at the peak temperature, and (c) microstructure in the weld after slowly cooling to room temperature.

room temperature to 500°C, the alloy overages and loses its strength rapidly. Above 500°C, the second phase redissolves in the matrix and we do not even obtain dispersion strengthening. In general, the aluminum age hardenable alloys are suited only for service near room temperature. However, some magnesium alloys may maintain their strength to about 250°C and certain nickel superalloys resist overaging at 1000°C.

We also have problems when welding age hardenable alloys (Figure 11-10). During welding the metal adjacent to the weld is heated. The heat-affected area contains two principle zones. The lower temperature zone near the unaffected base metal is exposed to temperatures just below the solvus and may overage. The higher temperature zone is solution treated, eliminating the effects of age hardening. If the solution-treated zone cools slowly, stable θ may form at the grain boundaries, embrittling the weld area. Very fast welding processes such as electron-beam welding, complete reheat treatment of the area after welding, or welding the alloy in the solution-treated condition improve the quality of the weld.

EXAMPLE 11-9

Describe the microstructure obtained in the heat-affected zone of an Al-4% Cu alloy welded in the solution-treated condition when (a) the weld cools slowly and (b) the weld cools rapidly. (See Figure 11-11.)

Answer:

Figure 11-11(a) shows the structure at the maximum temperature. In zone II, some aging may begin. However, most of the heat-affected area is all α.

(a) On slow cooling, grain boundary θ precipitates from zone I and embrittles the alloy. A complete reheat treatment, including solution treating, quenching, and aging, is needed to restore ductility.

(b) On fast cooling, zone I cools rapidly and produces supersaturated α. Now either natural or artificial aging will strengthen both the heat-affected zone and the original base metal.

11-8 Residual Stresses During Quenching

When an age hardenable alloy is quenched, the center of the part cools more slowly than the surface. The rapidly cooled surface contracts due to the coefficient of thermal expansion and contraction. The contracting surface applies a compressive stress to the center which, because the center is still hot, soft, and ductile, deforms. Later, the center cools and attempts to contract. But the contraction of the center is restrained by the cold, strong surface. The center is placed in tension, while the surface is compressed. Consequently, a residual stress pattern is produced in the quenched part (Figure 11-12). The residual stresses cause distortion, warpage, or even cracking of the part.

To help minimize problems due to residual stresses, age hardenable alloys are not quenched any more rapidly than necessary. Aluminum-base alloys are normally quenched in hot water, at about 80°C, rather than cold water.

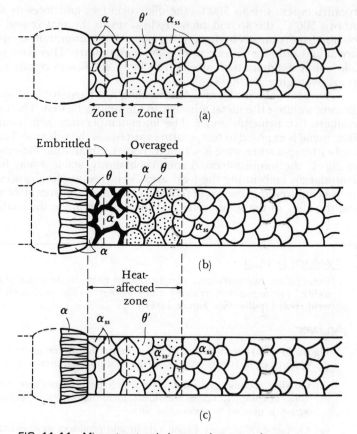

FIG. 11-11 Microstructural changes that occur in age hardenable alloys during fusion welding if the alloy is originally solution treated. (a) At the peak temperature, (b) after slow cooling, and (c) after rapid cooling.

11-9 The Eutectoid Reaction

In Chapter 10, we defined the eutectoid as a solid-state reaction in which one solid phase transforms to two other solid phases.

$$S_1 \rightarrow S_2 + S_3 \tag{11-3}$$

The formation of the two solid phases permits us to obtain dispersion strengthening. As an example of how we can use the eutectoid reaction to control the microstructure and properties of an alloy, let's examine the iron-carbon system, which is the basis for steels and cast irons.

The iron-cementite phase diagram.   Figure 11-13 shows the Fe-Fe$_3$C phase diagram. The following features should be noted.

Solid solutions.   Iron goes through two allotropic transformations during

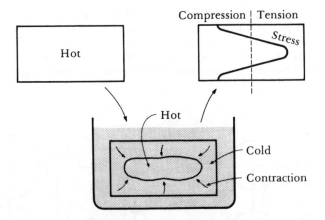

FIG. 11-12   The residual stress pattern produced in a solution-treated and quenched alloy during the age-hardening process. Note that the surface is in compression and the center in tension.

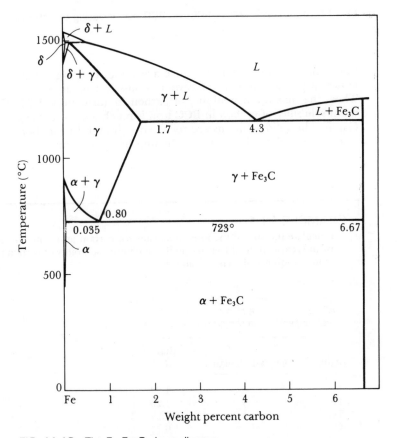

FIG. 11-13   The Fe-Fe$_3$C phase diagram.

heating or cooling. Immediately after solidification, iron forms a BCC structure called δ-*ferrite*. On further cooling, the iron transforms to a FCC structure called γ, or *austenite*. Finally, iron transforms back to the BCC structure at lower temperatures; this structure is called α, or *ferrite*. Both of the ferrites and the austenite are solid solutions of interstitial carbon atoms in iron (Figure 11-14). Because

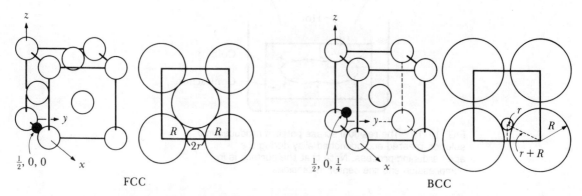

FIG. 11-14   The unit cells of FCC and BCC iron, including the interstitial sites for carbon.

interstitial holes in the FCC lattice are somewhat larger than holes in the BCC lattice, a greater number of carbon atoms can be accommodated in FCC iron. Thus, the maximum solubility of carbon in austenite is 1.7% C while the maximum solubility of carbon in BCC iron is much lower—0.035% C in α and 0.08% C in δ. The solid solutions are relatively soft and ductile, but stronger than pure iron due to solid solution strengthening by the carbon.

EXAMPLE 11-10

Calculate the size of the interstitial sites for carbon atoms in δ, α, and γ. From these results, explain the difference in the maximum solubility of carbon in each phase. The atomic radii are shown in Table 11-1.

TABLE 11-1   Size of the atoms in steel, depending on the crystal structure

Atom	Crystal Structure	Radius (Å)
Fe	α	1.24
Fe	γ	1.29
Fe	δ	1.27
C		0.71

Answer:

The largest interstitial site in BCC iron has the coordinates $\frac{1}{2}$, 0, $\frac{1}{4}$. The ratio between the radius of the iron atom and the radius of the interstitial site can be calculated with the aid of Figure 11-14.

$$(r_{Fe} + r_{interstitial})^2 = \left(\frac{a_0}{4}\right)^2 + \left(\frac{a_0}{2}\right)^2 = \left(\frac{5}{16}\right) a_0^2 = \left(\frac{5}{16}\right)\left(\frac{4r_{Fe}}{\sqrt{3}}\right)^2 = \frac{5\,r_{Fe}^2}{3}$$

$$r_{Fe} + r_{interstitial} = \frac{\sqrt{5}\,r_{Fe}}{\sqrt{3}}$$

$$\frac{r_{interstitial}}{r_{Fe}} = 0.291$$

The largest interstitial site in FCC iron has the coordinates $\frac{1}{2}$,0,0. From Figure 11-14

$$2r_{Fe} + 2r_{interstitial} = a_0 = \frac{4r_{Fe}}{\sqrt{2}}$$

$$r_{Fe} + r_{interstitial} = \sqrt{2}\,r_{Fe}$$

$$\frac{r_{interstitial}}{r_{Fe}} = 0.414$$

Therefore

$$\alpha \text{ site} = (0.291)(1.24) = 0.36 \text{ Å}$$

$$\gamma \text{ site} = (0.414)(1.29) = 0.53 \text{ Å}$$

$$\delta \text{ site} = (0.291)(1.27) = 0.37 \text{ Å}$$

The interstitial sites are all smaller than the carbon atom, $r_C = 0.71$ Å, causing low solubility and good solid solution strengthening. The solubility is about 100 times greater in austenite than in ferrite because of the larger hole in FCC unit cells.

Intermetallic compounds. A stoichiometric intermetallic compound Fe_3C, or *cementite*, forms when the solubility of carbon in solid iron is exceeded. The Fe_3C contains 6.67% C, is extremely hard and brittle, and is present in all of the commercial steels. By properly controlling the amount, size, and shape of Fe_3C, we control the degree of dispersion strengthening and the properties of the steel.

Eutectoid reaction. If we heat an alloy containing the eutectoid composition of 0.80% C above 723°C, we produce a structure containing only austenite grains. When austenite cools to 723°C, the eutectoid reaction begins.

$$\gamma_{0.80\% \text{ C}} \rightarrow \alpha_{0.035\% \text{ C}} + Fe_3C_{6.67\% \text{ C}} \tag{11-4}$$

As in the eutectic reaction, the two phases that form have different compositions, so atoms must diffuse during the reaction (Figure 11-15). Most of the carbon in the austenite diffuses to the Fe_3C, but a greater percentage of iron atoms diffuse to α. This redistribution of atoms is easiest if the diffusion distances are short, which is the case when the α and Fe_3C grow as thin lamellae, or plates.

Pearlite. The lamellar structure of α and Fe_3C that develops in the iron-carbon system is called *pearlite*. Pearlite is a microconstituent in steel. The lamellae

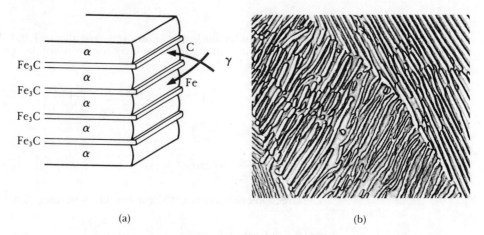

(a)

(b)

FIG. 11-15   Growth and structure of pearlite. (a) Redistribution of carbon and iron. (b) Photomicrograph of the pearlite lamellae (×2000). From *Metals Handbook*, Vol. 7, 8th Ed., American Society for Metals, 1972.

in pearlite are much finer than the lamellae in the lead-tin eutectic because the iron and carbon atoms must diffuse through solid austenite rather than through liquid.

EXAMPLE 11-11

Calculate the amounts of ferrite and cementite present in **pearlite at 722°C**

Answer:

Since pearlite must contain 0.80%C, then using the lever rule

$$\% \; \alpha = \frac{(6.67 - 0.80)}{(6.67 - 0.035)} \times 100 = 88\%$$

$$\% \; Fe_3C = \frac{(0.80 - 0.035)}{(6.67 - 0.035)} \times 100 = 12\%$$

From Example 11-11, we find that most of the pearlite is composed of ferrite. In fact, if we examine the pearlite closely, we find that the Fe_3C lamellae are surrounded by α. The pearlitic structure therefore produces effective dispersion strengthening—the continuous ferrite phase is relatively soft and ductile and the hard brittle cementite is dispersed.

Primary microconstituents. Hypoeutectoid steels contain less than 0.80% C and hypereutectoid steels contain more than 0.80% C. Ferrite is the primary or proeutectoid microconstituent in hypoeutectoid alloys and cementite is the primary or proeutectoid microconstituent in hypereutectoid alloys. If we heat a hypoeutectoid alloy containing 0.60% C above 750°C, only austenite remains in the microstructure. Figure 11-16 shows what happens when the austenite cools. Just below 750°C, ferrite precipitates and grows, usually at the austenite grain

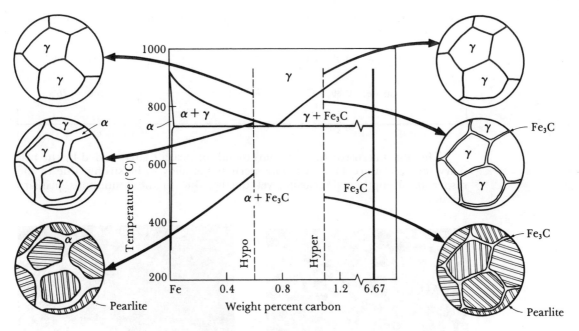

FIG. 11-16   The evolution of the microstructure of hypoeutectoid and hypereutectoid steels during cooling in relationship to the phase diagram.

boundaries. Primary ferrite continues to grow until the temperature falls to 723°C. The remaining austenite at that temperature is now surrounded by ferrite and has changed in composition from 0.60% C to 0.80%C. Subsequent cooling to below 723°C causes all of the remaining austenite to transform to pearlite by the eutectoid reaction. The final structure contains two phases—ferrite and cementite—arranged as two microconstituents—primary ferrite and pearlite.

EXAMPLE 11-12

Calculate the amounts and compositions of phases and microconstituents in a Fe-0.60% C alloy at 722°C.

Answer:

The phases are ferrite and cementite. Using a tie line and working the lever law at 722°C, we find

$$\alpha: \quad 0.035\% \text{ C} \qquad \% \, \alpha = \frac{6.67 - 0.60}{6.67 - 0.035} \times 100 = 91.5\%$$

$$\text{Fe}_3\text{C}: \quad 6.67\% \text{ C} \qquad \% \, \text{Fe}_3\text{C} = \frac{0.60 - 0.035}{6.67 - 0.035} \times 100 = 8.5\%$$

The microconstituents are primary ferrite and pearlite. If we construct a tie line just above 723°C, we can calculate the amounts and compositions of ferrite and austenite just before the eutectoid reaction starts. All of the austenite at that temperature will transform to pearlite; all of the ferrite will remain as primary ferrite.

$$\text{Primary } \alpha: \quad 0.035\% \text{ C} \qquad \% \text{ Primary } \alpha = \frac{0.80 - 0.60}{0.80 - 0.035} \times 100 = 26\%$$

$$\text{Pearlite: } \quad 0.80\% \text{ C} \qquad \% \text{ Pearlite} = \frac{0.60 - 0.035}{0.80 - 0.035} \times 100 = 76\%$$

The final microstructure contains islands of pearlite surrounded by the primary ferrite [Figure 11-17(a)]. This permits the alloy to be strong, due to the dispersion-strengthened pearlite, yet ductile, due to the continuous primary ferrite.

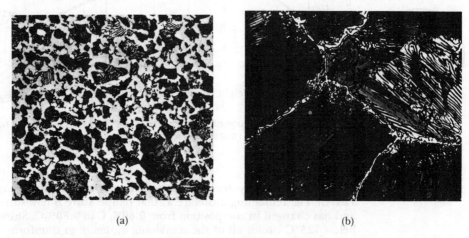

(a)   (b)

FIG. 11-17   (a) A hypoeutectoid steel showing primary α (white) and pearlite. (b) A hypereutectoid steel showing primary Fe_3C surrounding pearlite. From *Metals Handbook*, Vol. 7, 8th Ed., American Society for Metals, 1972.

In hypereutectoid alloys, however, the primary phase is Fe_3C, which again forms at the austenite grain boundaries. After the austenite cools through the eutectoid reaction, the steel contains hard brittle cementite surrounding islands of pearlite [Figure 11-17(b)]. Now, because the hard brittle microconstituent is continuous, the steel is also brittle. Fortunately, we can improve the microstructure and properties of the hypereutectoid steels by heat treatment.

11-10 Controlling the Eutectoid Reaction

We can control dispersion strengthening in the eutectoid alloys in much the same way that we did in eutectic alloys.

Controlling the amount of the eutectoid.   By changing the composition of the alloy, we change the amount of the hard second phase. As the carbon content of a steel increases towards the eutectoid composition of 0.80% C, the amounts of

Fe_3C and pearlite increase, thus increasing the strength. However, this strengthening effect eventually peaks and the properties decrease when the carbon content is too high (Table 11-2).

TABLE 11-2   The effect of carbon on the strength of steels

Carbon (%)	Slow Cooling			Fast Cooling		
	Yield Strength (MPa)	Tensile Strength (MPa)	% Elon- gation	Yield Strength (MPa)	Tensile Strength (MPa)	% Elon- gation
0.15	285	386	37	324	424	37
0.20	295	395	36.5	347	441	36
0.30	341	464	31	345	521	32
0.40	353	519	30	374	590	28
0.50	366	636	24	428	748	20
0.60	372	626	23	421	776	18
0.80	376	616	25	524	1010	11
0.95	379	657	13	500	1014	9.5

After *Metals Progress Materials and Processing Databook, 1981.*

EXAMPLE 11-13

Calculate the amount of Fe_3C and pearlite in steels at room temperature containing 0.2% C, 0.4% C, 0.8% C, and 1.2% C. Then plot the strength, % Fe_3C, and % pearlite versus the carbon content.

Answer:

Using the tie line and lever law in the iron-carbon diagram we obtain the following.

Carbon (%)	Fe_3C (%)	Pearlite (%)
0.20	$\dfrac{0.20 - 0.035}{6.67 - 0.035} \times 100 = 2.5\%$	$\dfrac{0.2 - 0.035}{0.80 - 0.035} \times 100 = 21.6\%$
0.40	$\dfrac{0.40 - 0.035}{6.67 - 0.035} \times 100 = 5.5\%$	$\dfrac{0.4 - 0.035}{0.80 - 0.035} \times 100 = 47.7\%$
0.80	$\dfrac{0.80 - 0.035}{6.67 - 0.035} \times 100 = 11.5\%$	$\dfrac{6.67 - 0.8}{6.67 - 0.80} \times 100 = 100\%$
1.20	$\dfrac{1.2 - 0.035}{6.67 - 0.035} \times 100 = 17.6\%$	$\dfrac{6.67 - 1.2}{6.67 - 0.80} \times 100 = 93.2\%$

These amounts are plotted in Figure 11-18.

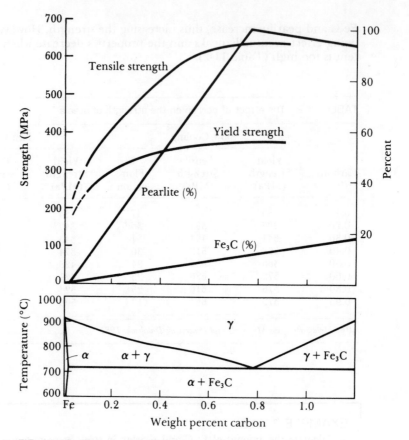

FIG. 11-18 The strength, % Fe₃C, and % pearlite versus the carbon content in slowly cooled steels.

Controlling the austenite grain size. Under normal conditions, pearlite grows as grains or *colonies*. Within each colony the orientation of the lamellae is identical. The colonies nucleate most easily at the grain boundaries of the original austenite grains. We can increase the number of pearlite colonies by reducing the prior austenite grain size, usually by using low temperatures to produce the austenite or by deoxidizing the steel with aluminum. Typically, we can increase the strength of the alloy by reducing the size or increasing the number of colonies.

Controlling cooling rate. By increasing the cooling rate during the eutectoid reaction, we reduce the distance that the atoms are able to diffuse. Consequently, the lamellae produced during the reaction are finer or more closely spaced. By producing a finer pearlite, we increase the strength of the alloy. The strength of the alloy is closely related to the interlamellar spacing, or distance between the lamellae (Figure 11-19).

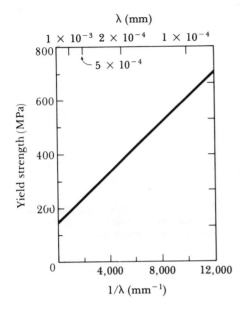

FIG. 11-19   The effect of interlamellar spacing of pearlite on the mechanical properties of pearlite.

EXAMPLE 11-14

Estimate the interlamellar spacing and strength of the pearlite shown in Figure 11-15.

Answer:

If we count the number of lamellar spacings in the upper right of Figure 11-15, remembering that interlamellar spacing is from one α plate to the next α plate, we find 14 spacings over a 20 mm distance. Due to magnification $\times 2000$, the distance is 0.01 mm. The interlamellar spacing λ is

$$\lambda = \frac{0.01}{14} = 0.714 \times 10^{-3} \text{ mm}$$

From Figure 11-19, we can estimate the yield strength of the pearlite to be 200 MPa.

Controlling the transformation temperature.   The solid-state eutectoid reaction is rather slow and the steel may cool below the equilibrium eutectoid temperature before the transformation begins. The transformation temperature affects the fineness of the structure (Figure 11-20), the time required for transformation, and even the arrangement of the two phases. This information is contained in the isothermal-transformation (IT) diagram (Figure 11-21). This diagram, also called the C-curve, permits us to predict the structure, properties, and heat treatment required in steels.

1. Nucleation and growth of pearlite: If we quench to just below the eutectoid temperature, the austenite is only slightly undercooled. Long times are re-

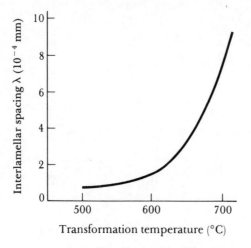

FIG. 11-20   The effect of the austenite transformation temperature on the interlamellar spacing in pearlite.

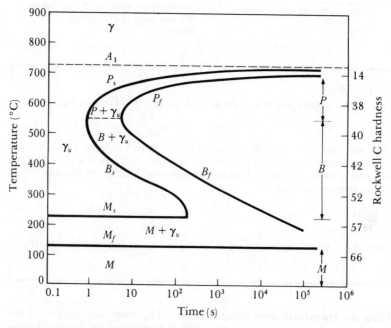

FIG. 11-21   The isothermal-transformation (*IT*) diagram for an eutectoid steel (0.80% carbon).

quired before stable nuclei for ferrite and cementite form; nucleation does not begin until near the pearlite start (P_s) time. After pearlite begins to grow, atoms diffuse rapidly and coarse pearlite is produced; the transformation is complete at the pearlite finish (P_f) time.

Austenite quenched to a lower temperature is more highly undercooled. Consequently, nucleation occurs more rapidly and the P_s is shorter. However, diffusion is also slower, so atoms diffuse only short distances and a finer pearlite is produced. Even though growth rates are slower, the overall time required for the transformation is reduced because of faster nucleation. Finer pearlite forms in shorter times as we reduce the isothermal transformation temperature to about 550°C, which is the *nose* or *knee* of the *IT* curve (Figure 11-21).

EXAMPLE 11-15

Describe the complete heat treatment and the microstructure after each step required to isothermally produce a hardness of R_c32 in an eutectoid steel.

Answer:

Note that Rockwell C hardnesses are shown as a function of transformation temperature in the *IT* diagram (Figure 11-21). A hardness of R_c32 is obtained by transforming at 650°C, where the P_s time is 5 s and the P_f time is 50 s. The heat treatment and microstructures are as follows.

1. Austenitize above 723°C and hold for about 1 h. The steel contains 100% austenite.

2. Quench to 650°C and hold for at least 50 s. After 5 s, pearlite nucleates from the unstable austenite. Pearlite grows until, after 50 s, the microstructure contains 100% pearlite. The pearlite has a medium fineness. (Note that we returned to time zero when we quenched!)

3. Cool in air to room temperature. The microstructure remains all pearlite.

---

2. Nucleation and growth of bainite: At a temperature just below the nose of the *IT* diagram, nucleation occurs rapidly but diffusion is slow. No transformation is detected until somewhat longer times, and total transformation times increase due to very slow growth.

In addition, we find a different microstructure. At low transformation temperatures, the lamellae in pearlite would have to be extremely thin and consequently the boundary area between the ferrite and Fe_3C lamellae would be very large. Because of the energy associated with the ferrite-cementite interface, the total energy of the steel would have to be very high. The steel can reduce its internal energy by permitting the cementite to precipitate as discrete rounded particles in a ferrite matrix. This new microconstituent, or arrangement of ferrite and cementite, is called *bainite*. Transformation begins at a bainite start (B_s) time and ends at a bainite finish (B_f) time.

EXAMPLE 11-16

Excellent combinations of hardness, strength, and toughness are obtained from bainite. One heat treater austenitized an eutectoid steel at 750°C, quenched and held the steel at 250°C for 15 min, and finally permitted the steel to cool to room temperature. Did he produce the required bainitic structure?

Answer:

Let's examine the heat treatment using Figure 11-21. After heating at 750°C, the microstructure is 100% γ. After quenching to 250°C, unstable austenite remains for slightly more than 100 s, when fine bainite begins to grow. After 15 min, or 900 s, about 50% fine bainite has formed and the remainder of the steel still contains unstable austenite. As we will see later, the unstable austenite transforms to martensite when the steel is cooled to room temperature and the final structure is a mixture of bainite and hard brittle martensite. The heat treatment was not successful. The heat treater should have held the steel at 250°C for at least 1 h.

The times required for austenite to begin and finish its transformation to bainite increase and the bainite becomes finer as the transformation temperature continues to decrease. The bainite that forms just below the nose of the curve is called *coarse bainite, upper bainite,* or *feathery bainite.* The bainite that forms at lower temperatures is called *fine bainite, lower bainite,* or *acicular bainite.* Figure 11-22 shows typical microstructures of bainite.

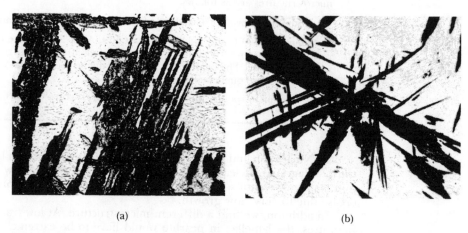

(a)   (b)

FIG. 11-22   (a) Upper bainite (gray feathery plates). (b) Lower bainite (dark needles). From *Metals Handbook,* Vol. 8, 8th Ed., American Society for Metals, 1973.

Figure 11-23 shows the effect of transformation temperature on the properties of an eutectoid steel. As the temperature decreases, there is a general trend towards higher strength and lower ductility due to the finer microstructure that is produced.

11-11   The Martensitic Reaction

Martensite is a phase that forms as the result of a diffusionless solid-state transformation. Cobalt, for example, transforms from a FCC to a HCP crystal structure by a slight shift in the atom locations which alters the stacking sequence of close-packed planes. Because the reaction does not depend on diffusion, the martensite reaction is *athermal,* or the reaction depends only on the temperature, not the time. The martensite reaction often proceeds rapidly, at speeds approaching the velocity of sound in the material.

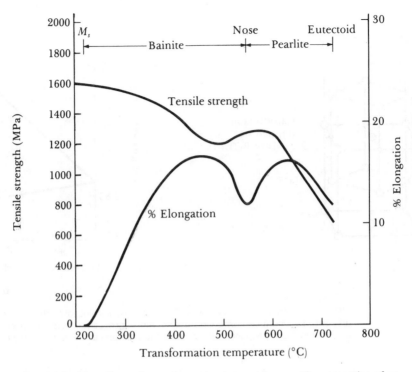

FIG. 11-23   The effect of transformation temperature on the properties of an eutectoid steel.

Martensite in steels.   In steels with less than about 0.2% C, the FCC austenite transforms to a supersaturated BCC martensite structure. In higher carbon steels, the martensite reaction occurs as FCC austenite transforms to BCT (body centered tetragonal) martensite. The relationship between the FCC austenite and the BCT martensite [Figure 11-24(a)] shows that carbon atoms in the $0,0,\frac{1}{2}$ type of interstitial sites in the FCC cell can be trapped during the transformation to the body centered structure, causing the tetragonal structure to be produced. As the carbon content of the steel increases, a greater number of carbon atoms are trapped in these sites, thereby increasing the difference between the a- and c-axes of the martensite [Figure 11-24(b)].

The steel must be quenched, or rapidly cooled, from the stable austenite region to prevent the formation of pearlite, bainite, or primary microconstituents. The martensite reaction begins in an eutectoid steel when austenite cools below 220°C, the martensite start (M_s) temperature (Figure 11-21). The amount of martensite increases as the temperature decreases. When the temperature passes below the martensite finish temperature (M_f), the steel should contain 100% martensite. At any intermediate temperature, the amount of martensite does not change as the time at that temperature increases.

The composition of martensite must be the same as that of the phase from which it forms. There is no long-range diffusion during the transformation that can change the composition. Thus in iron-carbon alloys, the initial austenite composition and the final martensite composition are the same.

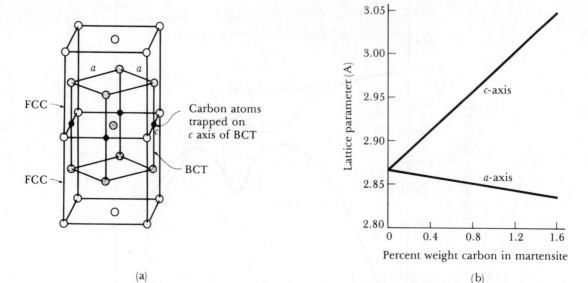

FIG. 11-24 (a) The unit cell of BCT martensite is related to the FCC austenite unit cell. (b) As the percent of carbon increases, more interstitial sites are filled by the carbon atoms and the tetragonal structure of the martensite becomes more pronounced.

EXAMPLE 11-17

A steel containing 0.40% C is heated to 740°C and then quenched. Determine the amount and composition of the martensite that forms.

Answer:

When the steel is heated to 740°C, a mixture of ferrite and austenite forms (Figure 11-13). We can use a tie line and the lever law to determine the amount and composition of austenite in the two-phase region, then equate the austenite to the martensite. Figure 13-1 shows the eutectoid region in greater detail.

$$\frac{\text{Austenite}}{\text{Composition}} = \frac{\text{Martensite}}{\text{Composition}} : \text{Fe-0.71\% C}$$

$$\% \text{ Martensite} = \% \text{ Austenite} = \frac{0.40 - 0.030}{0.71 - 0.030} \times 100 = 54\%$$

Properties of steel martensite. Martensite in steels is very hard and brittle. The BCT crystal structure has no close-packed slip planes in which dislocations can easily move. The martensite is highly supersaturated with carbon, since iron normally contains 0.003% C at room temperature, and martensite contains the amount of carbon present in the steel. Finally, martensite has a fine grain size and an even finer substructure within the grains. Consequently, martensite has little or no ductility and may be so hard that it can be cut only with special tools. Because of this behavior, steel martensite by itself is not normally used. In the next section we will discuss how we can "temper" the martensite to produce more desirable properties.

The structure and properties of the steel martensites depend significantly on the carbon content of the alloy (Figure 11-25). When the carbon content is low,

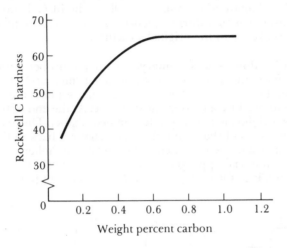

FIG. 11-25 The effect of carbon content on the hardness of martensite in steels.

the martensite grows in a "lath" shape, composed of bundles of flat narrow plates that grow side by side [Figure 11-26(a)]. This martensite is not very hard. At a higher carbon content, plate martensite grows, in which flat narrow plates grow individually rather than as bundles [Figure 11-26(b)]. The hardness is much greater in the higher carbon, plate martensite structure, partly due to the greater distortion, or large c/a ratio, of the crystal structure.

Martensite in other systems.   The characteristics of the martensite reaction are different in other alloy systems. For example, martensite can form in iron-

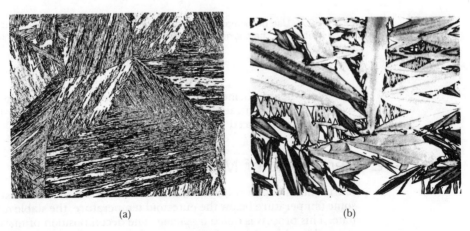

FIG. 11-26 (a) Lath martensite in low-carbon steel. (b) Plate martensite in high-carbon steel. From *Metals Handbook*, Vol. 8, 8th Ed., American Society for Metals, 1973.

base alloys that contain little or no carbon by a transformation of the FCC crystal structure to a BCC crystal structure. In certain high-manganese steels and stainless steels, the FCC structure changes to a HCP crystal structure during the martensite transformation.

The properties of martensite in other alloys are also different from the properties of steel martensite. In titanium alloys, the BCC titanium transforms to a HCP martensite structure during quenching. However, the titanium martensite is softer and weaker than the original structure.

Martensitic alloys with a memory.   A unique property possessed by some alloys that undergo the martensitic reaction is the "memory" effect. A Ni-50% Ti alloy and several copper-base alloys can be given a sophisticated thermomechanical treatment to produce a martensitic structure. At the end of the treatment, the metal has been deformed to a predetermined shape. The metal can then be deformed into a second shape; but the metal changes back to the original shape when the temperature is increased. The metal remembers its predetermined shape! One commercial application for the memory effect in these martensitic alloys is for couplings for tubing (Figure 11-27). The coupling is set into a small

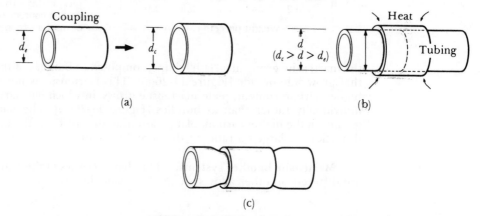

FIG. 11-27   Use of memory alloys for coupling tubing. A memory alloy coupling is expanded (a) so it fits over the tubing (b). When the coupling is heated, it shrinks back to its original diameter (c).

diameter, then deformed into a larger diameter. The coupling, which is slipped over the tubing, contracts back to its predetermined shape after heating. A strong bond is thus produced between the tubes.

11-12   Tempering of Martensite

Martensite is not an equilibrium structure. When martensite in a steel is heated to some temperature below the eutectoid temperature, the stable α and Fe_3C precipitate. This process is called *tempering*. The decomposition of martensite causes the strength and hardness of the martensite to decrease, while the ductility and impact properties are improved (Figure 11-28).

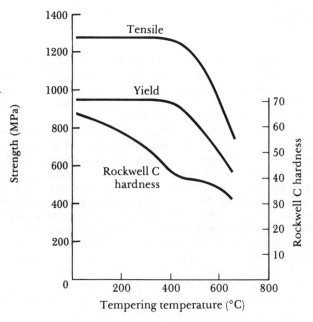

FIG. 11-28   Effect of tempering temperature on the properties of an eutectoid steel.

At low tempering temperatures, the martensite may form two transition phases—a lower carbon martensite and a very fine nonequilibrium ε-carbide, or $Fe_{2.4}C$. The steel is still strong, brittle, and perhaps even harder than before tempering. At higher temperatures, the stable α and Fe_3C form and the steel becomes softer and more ductile. If the steel is tempered just below the eutectoid temperature, the Fe_3C becomes very coarse and the dispersion-strengthening effect is greatly reduced. By selecting the appropriate tempering temperature, a wide range of properties can be obtained. The product of the tempering process is a microconstituent called *tempered martensite* (Figure 11-29).

FIG. 11-29   The structure of tempered martensite.

SUMMARY

Many types of solid-state transformations occur and can be controlled by proper heat treatments in materials. These heat treatments are designed to provide an optimum distribution of two or more phases in the microstructure. The resulting dispersion strengthening caused by the phases permits us to obtain a wide variety of structures and properties in materials.

In the most common of these transformations—exceeding the solubility limit, age hardening, control of the eutectoid, and the martensitic reaction—we strive to produce a final microstructure containing a uniform distribution of many tiny hard precipitate particles in a softer more ductile matrix. By doing this, we are able to provide effective obstacles to the movement of dislocations, thus providing strength but still maintaining at least usable ductility and toughness.

Careful control of the heat treatment temperatures and times is essential in obtaining the proper microstructure. Phase diagrams assist in selecting the appropriate temperatures, but experimental data is needed to finally obtain the optimum combination of times, temperatures, and compositions.

Finally, since optimum properties are obtained through heat treatment, we must bear in mind that the structure and properties may change when the material is used at elevated temperatures. Overaging, overtempering, and loss of coherency may occur as a natural extension of the phenomena governing these transformations when the material is placed into service.

GLOSSARY

Age hardening A special dispersion-strengthening heat treatment. By solution treatment, quenching, and aging, a coherent precipitate forms that provides a substantial strengthening effect. Also known as precipitation hardening.

Artificial aging Reheating a solution-treated and quenched alloy to a temperature below the solvus in order to provide the thermal energy required for a precipitate to form.

Athermal transformation When the amount of the transformation depends only on the temperature, not on the time.

Austenite The name given to the FCC crystal structure of iron.

Bainite A two-phase microconstituent, containing ferrite and cementite, that forms in steels that are isothermally transformed at relatively low temperatures.

Cementite The hard brittle intermetallic compound Fe_3C that when properly dispersed provides the strengthening in steels.

Coherent precipitate A precipitate whose crystal structure and atomic arrangement have a continuous relationship with the matrix from which the precipitate formed. The coherent precipitate provides excellent disruption of the atomic arrangement in the matrix and provides excellent strengthening.

Dihedral angle The angle that defines the shape of a precipitate particle in the matrix. The dihedral angle is determined by the relative surface energies.

Ferrite The name given to the BCC crystal structure of iron.

Guinier-Preston zones Tiny clusters of atoms that precipitate from the matrix in the early stages of the age-hardening process. Although the GP zones are coherent with the matrix, they are too small to provide optimum strengthening.

Interfacial energy The energy associated with the boundary between two phases.

Isothermal transformation When the amount of a transformation at a particular temperature depends on the time permitted for the transformation.

Martensite   A metastable phase formed in steel and other materials by a diffusionless, athermal transformation.

Natural aging   When a coherent precipitate forms from a solution-treated and quenched age hardenable alloy at room temperature, providing optimum strengthening.

Pearlite   A two-phase lamellar microconstituent, containing ferrite and cementite, that forms in steels that are cooled in a normal fashion or are isothermally transformed at relatively high temperatures.

Solution treatment   The first step in the age-hardening heat treatment. The alloy is heated above the solvus temperature to dissolve any second phase and to produce a homogeneous single-phase structure.

Strain energy   The energy required to permit a precipitate to fit into the surrounding matrix during nucleation and growth of the precipitate.

Supersaturated solid solution   The solid solution formed when a material is rapidly cooled from a high-temperature single-phase region to a low-temperature two-phase region without the second phase precipitating. Because the quenched phase contains more alloying element than the solubility limit, it is supersaturated in that element.

Tempering   A low-temperature heat treatment used to reduce the hardness of martensite by permitting the martensite to begin to decompose to the equilibrium phases.

Widmanstatten structure   The precipitation of a second phase from the matrix when there is a fixed crystallographic relationship between the precipitate and matrix crystal structures. Often needlelike or platelike structures form in the Widmanstatten structure.

PRACTICE PROBLEMS

1   Calculate the amount of each phase at room temperature that forms on equilibrium cooling of an Al-3% Cu alloy (Figure 11-1).

2   Suppose such θ particle that forms on equilibrium cooling of an Al-3% Cu alloy (Figure 11-1) is 1×10^{-4} mm in diameter. Estimate the number of these particles per cubic centimeter of the alloy. The density of θ is about 4.26 Mg m^{-3}.

3   How much copper must be added to an aluminum-copper alloy to produce 6 wt% θ? What is the vol% θ? (See Problem 11-2 for the density.) See Figure 11-1.

4   A dihedral angle of 45° is observed between the two phases in a microstructure of an alloy. Determine the ratio of the surface energies γ_p/γ_m.

5   Describe a heat treatment procedure for a Mg-5% Al alloy that might give age hardening. Include approximate temperatures and the amounts and compositions of each phase after each step of the treatment. The phase diagram is shown in Figure 12-5.

6   Describe a heat treatment procedure for a Cu-1% Be alloy that might give age hardening. Include approximate temperatures and the amounts and compositions of each phase after each step of the treatment. The phase diagram is shown in Figure 12-7.

7   Based only on the phase diagram and the properties of the phases, which of the following alloys are potential candidates for age hardening? Explain.

(a)   W-4% Mo   (Figure 12-21)

(b)   Al-20% Zn   (Figure 12-13)

(c)   Al-43% Cu   (Figure 11-1)

(d)   Cu-4.5% Si   (Figure 12-7)

(e)   Mg-1% Mn   (Figure 12-5)

8   A weld profile and the maximum temperatures reached during welding of an age-hardened Al-4% Cu alloy are shown in Figure 11-30. Compare the peak temperatures to the phase diagram for aluminum-copper. (a) Determine the width of each zone in the heat-affected area. Assume overaging begins when the temperature exceeds 450°C. (b) What phases do you expect to find in each area at the peak temperature? (c) What phases do you expect to find in each area after slow cooling of the weld? (d) What phases do you expect to find on fast cooling of the weld?

9   An age-hardened aluminum bracket that supports a heavy piece of equipment near a heat-treating furnace collapses after several weeks of use, but a similar bracket in another part of the room continues

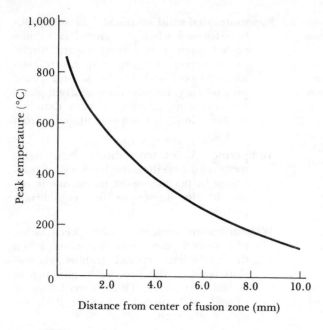

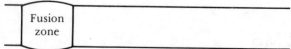

FIG. 11-30   The weld profile and maximum temperatures for an Al-4% Cu weldment. (See Problem 11-8.)

to support the same load. Give an explanation for the failure.

10 An age hardenable aluminum casting is found to be warped following the solution treatment, cold-water quench, and aging procedure. Suggest a cause and cure for the distortion.

11 Determine the amounts and compositions of each phase and microconstituent in a Fe-0.3% C alloy at 700°C. See Figure 11-13.

12 Determine the amounts and compositions of each phase and microconstituent in a Fe-1.4% C alloy at 700°C. See Figure 11-13.

13 The density of α-Fe is 7.87 Mg m^{-3} and that of Fe$_3$C is 7.66 Mg m^{-3}. Calculate the ratio of the thickness of a Fe$_3$C lamella to the α lamella in pearlite. See Figure 11-13.

14 A steel contains 20% Fe$_3$C and 80% α. Determine the % C in the steel. Is the steel hypoeutectoid or hypereutectoid? Use Figure 11-13.

15 A steel contains 5% Fe$_3$C and 95% α. Determine

the % C in the steel. Is the steel hypoeutectoid or hypereutectoid? Use Figure 11-13.

16 A steel contains 15% pearlite and 85% primary ferrite. Determine the % C in the steel. Is the steel hypoeutectoid or hypereutectoid? Use Figure 11-13.

17 A steel contains 12% primary cementite and 88% pearlite. Determine the % C in the steel. Is the steel hypoeutectoid or hypereutectoid? Use Figure 11-13.

18 Prove that Fe$_3$C should contain 6.67% C.

19 Calculate the density of pearlite. (The densities for ferrite and cementite are given in Problem 11-13.)

20 When we examine the microstructure of a steel, we find that the structure contains about 35 vol% primary α and 65 vol% pearlite. Estimate the % C in the steel. (You need the answer to Problem 11-19!)

21 When we examine the microstructure of a steel, we find that the structure contains about 10 vol% primary Fe$_3$C and 90 vol% pearlite. Estimate the % C in the steel.

22 Determine the amounts and compositions of

each phase in the Cu-11.8% Al eutectoid product (Figure 12-8) at 560°C. Do you expect the eutectoid product to be ductile or brittle? Explain.

23 Determine the amounts and compositions of each phase in the Cu-6% Be eutectoid product (Figure 12-7) at 600°C. Do you expect the eutectoid alloy to be ductile or brittle? Explain.

24 Calculate the amounts of each phase and microconstituent in a Cu-14% Al alloy at 560°C (Figure 12-8).

25 Calculate the amounts of each phase and microconstituent in a Cu-4% Be alloy at 600°C (Figure 12-7).

26 Describe the heat treatment you would use to control the reactions in a Cu-10% Al alloy (Figure 12-8). (a) What temperature would you have to reach to obtain a single-phase alloy? (b) Calculate the amount of each phase and microconstituent that forms immediately after the eutectoid reaction. (c) Which microconstituent will be continuous?

27 The Cu-9% Si alloy has an eutectoid reaction (Figure 12-7). What is the temperature at which the reaction occurs? Explain why the alloy would be of no use for structural purposes.

28 Calculate the size of the interstitial location at the $0,\frac{1}{2},0$ location in BCC α iron and compare to the normal $\frac{1}{2},0,\frac{1}{4}$ site (Example 11-10).

29 Calculate the size of the interstitial location at the $\frac{1}{4},\frac{1}{4},\frac{1}{4}$ location in FCC iron and compare to the normal $\frac{1}{2},0,0$ site (Example 11-10).

30 The effect of the transformation temperature on the interlamellar spacing λ in Figure 11-20 can be expressed by the equation $1/\lambda = AT^n$, where T is the absolute temperature and λ is in units of millimeters. Determine the constants A and n in this equation using a log-log plot.

31 A steel containing 0.80% C is transformed at 625°C to pearlite. Determine the interlamellar spacing and the yield strength that you would expect.

32 Determine from Figure 11-19 the equation that relates the interlamellar spacing and the yield strength of pearlite in MPa.

33 Combine the data in Figures 11-19 and 11-20 to obtain a relationship between yield strength and transformation temperature at which pearlite forms.

34 An eutectoid steel is heated to 800°C, then quenched to the following temperatures. For each temperature, determine the microconstituent that is produced and the minimum time required to completely transform the austenite.

(a) 700°C   (b) 600°C   (c) 500°C   (d) 300°C
(e) 100°C

35 Determine the microstructure present after the following treatments in an eutectoid steel.

(a) Heat at 780°C for 2 hours

(b) Heat at 780°C for 1 h, quench to 600°C and hold for 5 s

(c) Heat at 780°C for 1 h, quench to 600°C and hold for 10 s

(d) Heat at 780°C for 1 h, quench to 400°C and hold for 10 s

(e) Heat at 780°C for 1 h, quench to 400°C and hold for 1 h

(f) Heat to 780°C for 1 h, quench to 100°C and hold for 10 s

(g) Heat to 780°C for 1 h, quench to 100°C and hold for 10^5 s

36 Figure 11-24 shows the effect of carbon content on the lattice parameters of BCT martensite. Calculate the volume change that occurs during the martensite reaction for a Fe-0.2% C alloy, a Fe-0.6% C alloy, and a Fe-1.0% C alloy. Assume the lattice parameter of FCC iron is 3.60 Å in each case.

37 Calculate the amount and composition of martensite that forms in a Fe-0.3% C alloy quenched from the following temperatures. (a) 850°C (b) 780°C (c) 740°C (d) 700°C. See Figure 13-1 for your calculations.

38 Calculate the amount and composition of martensite in a Fe-1.2% C alloy quenched from the following temperatures. (a) 950°C (b) 850°C (c) 800°C (d) 730°C. See Figure 13-1.

39 Steel held above 723°C for some time is quenched to produce 64% martensite and 36% ferrite. The martensite contains 0.58% C. Estimate the % C in the alloy and the temperature at which the steel is held prior to quenching. See Figure 13-1.

40 Steel held above 723°C for 2 h is quenched to produce 97% martensite and 3% cementite. The martensite contains 0.95% C. Estimate the % C in the alloy and the temperature at which the steel is held prior to quenching. See Figure 13-1.

41 FCC cobalt has a lattice parameter of 3.5441 Å and HCP cobalt has lattice parameters of $a_0 = 2.5071$ Å and $c_0 = 4.0686$ Å. Determine the change in volume when cobalt changes from FCC to HCP during the martensite transformation.

42 In eutectic alloys, the eutectic microconstituent is generally the continuous one. But in the eutectoid structures, the primary microconstituent is normally continuous. Explain the difference, since the eutectoid and eutectic reactions do seem to be closely allied.

43 Pearlite appears to have a different interlamellar

spacing in different locations in the same steel. Can you explain why the spacing appears different, even though it actually is the same?

44 An eutectoid steel is quenched to produce martensite, then tempered at 400°C. (a) How does tempering affect the tensile strength, yield strength, and Rockwell C hardness number? (b) Suppose the steel is now put into service at a temperature of 600°C. What will happen to the properties of the steel?

45 We would like to produce an eutectoid steel that has a minimum yield strength of 690 MPa with a maximum hardness of R_c40. What tempering temperature would you recommend?

PART III

ENGINEERING MATERIALS

Although there is a tremendous variety of engineering materials, the mechanical properties of each can be predicted and controlled by understanding the atomic bonding, atomic arrangement, and strengthening mechanisms discussed in the previous sections. This is particularly evident in Chapters 12 and 13 in which we examine the characteristics of specific metals and alloys. We will extensively utilize the ideas of solid solution strengthening, strain hardening, and dispersion strengthening for the ferrous and nonferrous alloys.

In ceramics and polymers, Chapters 14 and 15, the importance of atomic bonding and atomic arrangement will also be very evident. Although the strengthening mechanisms that apply to metals are less applicable to these materials, we will be able to draw many parallels between them. For example, deformation of certain polymers produces a fibrous microstructure that provides strengthening, just as in metals, but for different reasons. By producing copolymers, we in a sense provide solid solution strengthening. Phase diagrams play an important part in understanding the behavior of ceramics, although we often utilize the phase diagrams differently than we do in metals. We explain and control the mechanical properties of ceramics and polymers by mechanisms that do not involve dislocation movement.

Composite materials, Chapter 16, are even more difficult to categorize because of the many types and intended uses of the materials. The behavior of some of the composites can be explained in terms of dispersion strengthening. However, many composites are designed to provide special characteristics that go beyond the conventional methods for controlling the structure-property relationship.

CHAPTER **12**

Nonferrous Alloys

12-1 Introduction

Metals and alloys are often divided into two categories—ferrous and nonferrous. Ferrous alloys are based on iron as the principal metal and include steels, stainless steels, and cast irons. Nonferrous alloys are based on metals other than iron. In this chapter, we will look at a cross section of the more important nonferrous engineering alloys, pointing out how the strengthening mechanisms introduced in Part II of the text are applied to specific types of alloys.

12-2 Aluminum Alloys

Aluminum is a lightweight metal, with a density of 2.70 Mg m^{-3} or one third the density of steel. Although aluminum alloys have relatively low tensile properties compared to steel, their strength-to-weight ratio, as defined below, is excellent.

$$\text{Strength-to-weight ratio} = \frac{\text{tensile strength}}{\text{density}} \tag{12-1}$$

Aluminum is often used when weight is an important factor, as in aircraft and automotive applications.

Aluminum also responds readily to strengthening mechanisms. Table 12-1 compares the strength of pure annealed aluminum to alloys strengthened by various techniques. The alloys may be 30 times stronger than pure aluminum.

On the other hand, aluminum often does not display an endurance limit in fatigue, so failure eventually occurs even at rather low stresses. Because of its low melting temperature, aluminum does not perform well at elevated temperatures. Finally, aluminum alloys have a low hardness, leading to poor wear resistance.

EXAMPLE 12-1

A steel cable 10 mm in diameter has a yield strength of 310 MPa. The density of steel is about 7.87 Mg m^{-3}. Based on the data in Table 12-4, determine (a) the maximum load that the steel cable can support, (b) the diameter of a cold-worked aluminum-manganese alloy (3003-H18) required to support the same load as the steel, and (c) the weight per meter of the steel cable versus the aluminum alloy cable.

Answer:

(a) Load $= F = \sigma_y A = 310 \times 10^6 \left(\dfrac{\pi}{4}\right)(0.010)^2 = 24.3$ kN

(b) The yield strength of the alloy is 186 MPa. Thus

$$A = \frac{\pi}{4} d^2 = \frac{F}{\sigma_y} = \frac{24.3 \times 10^3}{186 \times 10^6} = 130.6 \text{ mm}^2$$

$$d = 12.9 \text{ mm}$$

(c) Density of steel $= \rho = 7.87$ Mg m$^{-3} = 7870$ kg m^{-3}

Density of aluminum $= \rho = 2.70$ Mg m$^{-3} = 2700$ kg m^{-3}

Weight of steel $= A l \rho = \dfrac{\pi}{4}(0.01)^2(1)(7870) = 0.618$ kg m^{-1}

Weight of aluminum $= A l \rho = \dfrac{\pi}{4}(0.0129)^2(1)(2700) = 0.353$ kg m^{-1}

Although the yield strength of the aluminum is lower than that of the steel and the cable must be larger in diameter, the cable only weighs about half as much as the steel.

TABLE 12-1 The effect of strengthening mechanisms in aluminum and aluminum alloys

Material	Tensile Strength (MPa)	Yield Strength (MPa)	% Elon-gation	Yield Strength (alloy) / Yield Strength (pure)
Pure annealed Al (99.999% Al)	44.83	17.24	60	
Commercially pure Al (annealed, 99% Al)	89.67	34.48	45	2.0
Solid solution strengthened (1.2% Mn)	110.34	41.38	35	2.4
75% cold worked pure Al	165.52	151.72	15	8.8
Dispersion strengthened (5% Mg)	289.66	151.72	35	8.8
Age hardened (5.6% Zn-2.5% Mg)	572.41	503.54	11	29.2

Adapted from data in *Metals Handbook*, Vol. 2, 9th Ed., American Society for Metals, 1979.

Designation. Aluminum alloys can be subdivided into two major groups, wrought and casting alloys, based on their method of fabrication. *Wrought* alloys, which are shaped by plastic deformation, have compositions and microstructures significantly different from casting alloys, reflecting the different requirements of the manufacturing process. Within each major group we can divide the alloys into two subgroups—heat treatable and nonheat treatable alloys. Heat treatable alloys are age hardened, whereas nonheat treatable alloys are strengthened by solid solution strengthening, strain hardening, or dispersion strengthening.

Aluminum alloys are designated by the numbering system in Table 12-2. The first number specifies the principle alloying elements and the remaining numbers refer to the specific composition of the alloy. Representative phase diagrams are shown in Figures 10-18, 11-1, 12-1, and 12-2.

The degree of strengthening is given by the *temper designation* T or H, depending on whether the alloy is heat treated or strain hardened (Table 12-3). Other designations indicate if the alloy is annealed (O), solution treated (W), or used in the as-fabricated condition (F). The numbers following the T or H indicate the amount of strain hardening, the exact type of heat treatment, or other special aspects of the processing of the alloy.

Wrought alloys.   The 1xxx, 3xxx, 5xxx, and most of the 4xxx alloys are not age hardenable. Compositions and properties of typical alloys are shown in Table 12-4.

The 1xxx and 3xxx alloys are single-phase alloys except for the presence of small amounts of inclusions or intermetallic compounds (Figure 12-3). The properties of these alloys are controlled by strain hardening, solid solution strengthening, and grain size control. However, because the solubilities of the alloying elements in aluminum are small, the degree of solid solution strengthening is limited. For example, annealed 3003 alloy has a yield strength of only 6000 psi compared to 5000 psi for commercially pure aluminum.

The 5xxx alloys contain two phases at room temperature—α, a solid solution of magnesium in aluminum, and Mg_2Al_3, a hard brittle intermetallic compound. The aluminum-magnesium alloys are strengthened by a fine dispersion of Mg_2Al_3 as well as by strain hardening, solid solution strengthening, and grain size control. However, because the Mg_2Al_3 precipitate is not coherent, age-hardening treatments are not possible.

The 4xxx series alloys also contain two phases, α and nearly pure silicon, β.

TABLE 12-2   Designation system for aluminum alloys

Wrought alloys

1xxx	Commercially pure Al (>99% Al)	Not aged
2xxx	Al-Cu	Age hardenable
3xxx	Al-Mn	Not aged
4xxx	Al-Si and Al-Mg-Si	Age hardenable if magnesium is present
5xxx	Al-Mg	Not aged
6xxx	Al-Mg-Si	Age hardenable
7xxx	Al-Mg-Zn	Age hardenable

Casting alloys

1xx.x	Commercially pure Al	Not aged
2xx.x	Al-Cu	Age hardenable
3xx.x	Al-Si-Cu or Al-Mg-Si	Some are age hardenable
4xx.x	Al-Si	Not aged
5xx.x	Al-Mg	Not aged
7xx.x	Al-Mg-Zn	Age hardenable
8xx.x	Al-Sn	Age hardenable

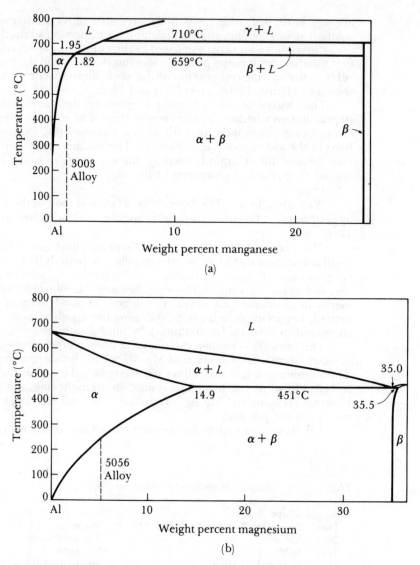

FIG. 12-1   Portions of the phase diagrams for (a) aluminum-manganese and (b) aluminum-magnesium.

Alloys which contain both silicon and magnesium can be age hardened by permitting Mg_2Si to precipitate.

The 2xxx, 6xxx, and 7xxx alloys are age hardenable ternary alloys. In each alloy, several coherent precipitates form before the final equilibrium phase is produced.

2xxx: $\alpha_{ss} \rightarrow \alpha + GP\text{-}I \rightarrow \alpha + GP\text{-}II \rightarrow \alpha + \theta' \rightarrow \alpha + \theta$

6xxx: $\alpha_{ss} \rightarrow \alpha + GP\ zones \rightarrow \alpha + \beta'(Mg_2Si) \rightarrow \alpha + \beta(Mg_2Si)$

7xxx: $\alpha_{ss} \rightarrow \alpha + GP\ zones \rightarrow \alpha + \eta'(MgZn_2) \rightarrow \alpha + \eta(MgZn_2)$

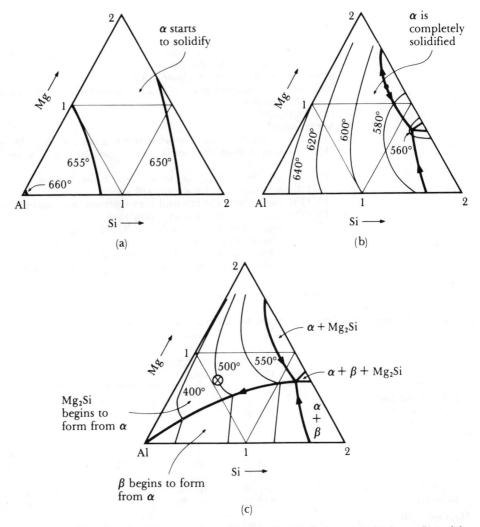

FIG. 12-2 Portions of the ternary phase diagrams for the 6xxx series aluminum alloys. (a) Liquidus plot, (b) solidus plot, and (c) solvus plot.

TABLE 12-3 Temper designations for aluminum alloys.
Alternative British designations are shown in parentheses.

F(M) As-fabricated (hot worked, forged, cast, etc.)
  O Annealed (in the softest possible condition)
  H Cold worked
    H1x — cold worked only. (x refers to the amount of cold work and strengthening.)
      H12 — cold work that gives a tensile strength midway between the O and H14 tempers.
      H14 — cold work that gives a tensile strength midway between the O and H18 tempers.
      H16 — cold work that gives a tensile strength midway between the H14 and H18 tempers.
      H18 — cold work that gives about 75% reduction.

(continued)

TABLE 12-3   (*Continued*)

        H19—cold work that gives a tensile strength greater than 2000 psi of that obtained by H18 temper.

        H2x—cold worked and partly annealed.

        H3x—cold worked and stabilized at a low temperature to prevent age hardening of the structure.

W   Solution treated

T   Age hardened

     T1   —cooled from the fabrication temperature and naturally aged.

     T2   —cooled from the fabrication temperature, cold worked, and naturally aged.

     T3   —solution treated, cold worked, and naturally aged.

     T4(TB)—solution treated and naturally aged.

     T5(TE)—cooled from the fabrication temperature and artifically aged.

     T6(TF)—solution treated and artifically aged.

     T7   —solution treated and stabilized by overaging.

     T8   —solution treated, cold worked, and artifically aged.

     T9   —solution treated, artifically aged, and cold worked.

     T10   —cooled from the fabrication temperature, cold worked, and artifically aged.

TABLE 12-4   Properties of typical wrought aluminum alloys

Alloy		Tensile Strength (MPa)	Yield Strength (MPa)	% Elongation	Applications
Nonheat treatable					
1100-O	>99% Al	89.66	34.48	40	Electrical components, foil,
1100-H18		165.52	165.52	10	corrosion resistance
3003-O	1.2% Mn	110.34	41.38	35	Beverage cans, architec-
3003-H18		200.00	186.21	7	tural uses
4043-O	5.2% Si	144.82	68.97	22	Filler metal for welding
4043-H18		282.76	268.97	1	
5056-O	5% Mg	289.66	165.50	35	Containers, marine
5056-H18		413.79	344.83	15	components
Heat treatable					
2024-O	4.4% Cu	186.21	75.86	20	
2024-T4		468.97	324.14	20	
4032-T6	12% Si-1% Mg	379.31	317.24	9	Transportation, aircraft,
6061-O	1% Mg-0.6% Si	124.14	55.17	27	aerospace, and other high
6061-T6		310.34	275.86	15	strength applications
7075-O	5.6% Zn-2.5% Mg	267.06	103.45	17	
7075-T6		572.41	503.45	11	

From data in *Metals Handbook*, Vol. 2, 9th Ed., American Society for Metals, 1979.

EXAMPLE 12-2

Determine the maximum amount of the precipitate that forms in aluminum alloy 5056 under equilibrium conditions at room temperature.

Answer:

From Table 12-4, the 5056 alloy contains 5% Mg. The phase diagram in Figure 12-1(b) permits us to calculate the amount of Mg_2Al_3 (β) precipitate.

Composition of α: 0% Mg

Composition of β: 35% Mg

$$\% \beta = \frac{5 - 0}{35 - 0} \times 100 = 14.3\%$$

The dispersed β precipitate provides a large increase in yield strength compared to the pure aluminum — 152 MPa versus 35 MPa.

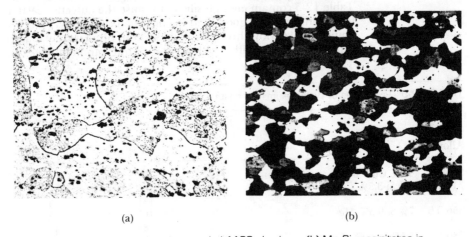

(a)   (b)

FIG. 12-3   (a) $FeAl_3$ inclusions in annealed 1100 aluminum; (b) Mg_2Si precipitates in annealed 5457 aluminum alloy. From *Metals Handbook*, Vol. 7, 8th Ed., American Society for Metals, 1972.

EXAMPLE 12-3

Compare the strength-to-weight ratio of commercially pure aluminum to an age-hardened 7xxx series aluminum alloy.

Answer:

The density for both is about 2.70 Mg m^{-3}. The tensile strength for commercially pure aluminum is 90 MPa and the tensile strength for age hardened 7075 Al-Mg-Zn alloy is 572 MPa.

$$\text{Strength-to-weight ratio (Al)} = \frac{90 \times 10^6}{2700} = 3.3 \times 10^4 \text{ m}$$

$$\text{Strength-to-weight ratio (7075)} = \frac{572 \times 10^6}{2700} = 2.1 \times 10^5 \text{ m}$$

EXAMPLE 12-4

The solvus curve for the aluminum-rich end of the aluminum-silicon-magnesium ternary diagram is shown in Figure 12-2(c). The recommended temperature for solution treatment for an Al alloy containing 0.4% Si and 0.7% Mg is 520°C. Is this reasonable?

Answer:

Yes! The solvus temperature for this alloy, shown as point x in the phase diagram, is 500°C.

Casting alloys.   Many of the common aluminum casting alloys shown in Table 12-5 contain enough silicon to cause the eutectic reaction, giving the alloys low melting points, good fluidity, and good castability. *Fluidity* is the ability of the liquid metal to flow through a mold without prematurely solidifying, and *castability* refers to the ease with which a good casting can be made from the alloy.

TABLE 12-5   Properties of typical aluminum casting alloys

Alloy		Tensile Strength (MPa)	Yield Strength (MPa)	% Elongation	Casting Process
295-T6	4.5% Cu-0.8% Si	248	166	5	Sand
319-F	6% Si-3.5% Cu	186	124	2	Sand
		234	131	2.5	Permanent mold
356-T6	7% Si-0.3% Mg	228	166	3.5	Sand
		262	186	5	Permanent mold
380-F	8.5% Si-3.5% Cu	317	159	3.5	Permanent mold
384-F	11.2% Si-4.5% Cu-0.6% Mg	331	166	2.5	Permanent mold
390-F	17% Si-4.5% Cu-0.6% Mg	282	241	1	Die casting
443-F	5.2% Si	131	55	8	Sand
		159	62	10	Permanent mold
		228	110	9	Die casting
413-F	12% Si	297	145	2.5	Die casting
518-F	8% Mg	310	193	7	Sand
713-T5	7.5% Zn-0.7% Cu-0.35% Mg	207	152	4	Sand
850-T5	6.2% Sn-1% Ni-1% Cu	159	76	10	Sand

After data in *Metals Handbook*, Vol. 2, 9th Ed., American Society for Metals, 1979.

The properties of the aluminum-silicon alloys are controlled by solid solution strengthening of the α aluminum matrix, dispersion strengthening by the β phase, and solidification, which controls the primary grain size and shape as well as the nature of the eutectic microconstituent. Fast cooling obtained in die casting or permanent mold casting normally increases strength by refining grain size and the eutectic microconstituent (Figure 12-4). Grain refinement using boron and titanium additions, modification using sodium or strontium to change the eutectic structure, or hardening with phosphorus to refine the primary silicon are all done in certain alloys to improve the microstructure and thus the degree of dispersion strengthening.

Many alloys also contain copper or magnesium, permitting an age-hardening reaction by precipitation of $CuAl_2$ or Mg_2Si. These alloys are dispersion strengthened by the β silicon phase and age hardened by the precipitates that form in the α.

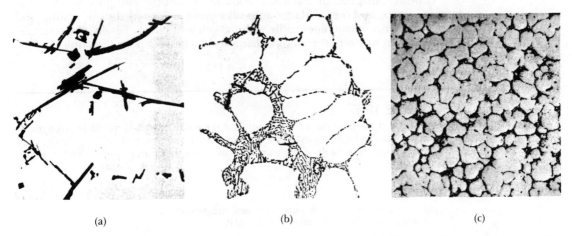

(a)   (b)   (c)

FIG. 12-4   (a) Sand-cast 443 aluminum alloy containing coarse silicon and inclusions; (b) permanent-mold 443 alloy containing fine dendrite cells and fine silicon due to faster cooling; (c) Die-cast 443 alloy with still a finer microstructure. From *Metals Handbook*, Vol. 7, 8th Ed., American Society for Metals, 1972.

EXAMPLE 12-5

Aluminum casting alloys such as 295-T6 and 518-F have poor fluidity and shrinkage characteristics compared to the 413-F alloy. Explain why this might be the case from the phase diagram.

Answer:

The 295 alloy contains 4.5% Cu, the 518 alloy contains 8% Mg, and the 413 alloy contains 12% Si. From the phase diagrams in Figures 10-18, 11-1, and 12-1, neither the aluminum-copper nor the aluminum-magnesium alloy passes through the eutectic reaction, but the 12% Si alloy is nearly a straight eutectic. We expect much better fluidity in alloys that contain the eutectic reaction.

In addition, shrinkage problems are accentuated for long freezing range alloys. From the phase diagrams

Freezing range of alloy 295 = 650 − 575 = 75°C

Freezing range of alloy 413 = 0°C

Freezing range of alloy 518 = 620 − 530 = 90°C

The short freezing range of the 413 alloy gives concentrated shrinkage that can be prevented by risers. The long freezing range of the other alloys gives microshrinkage that is difficult to control.

12-3 Magnesium Alloys

Magnesium is lighter than aluminum, with a density of $1.74 \ Mg \ m^{-3}$, and melts at a lower temperature. Although magnesium alloys are not as strong as aluminum alloys, their strength-to-weight ratios are comparable. Consequently, magnesium alloys are used in aerospace applications, high-speed machinery, and transportation and materials handling equipment.

However, magnesium has a low modulus of elasticity and poor resistance to fatigue, creep, and wear. Magnesium also poses a hazard during casting and machining, since it combines easily with oxygen and burns. Finally, the response of magnesium to strengthening mechanisms is relatively poor (Example 12-6).

EXAMPLE 12-6

From the data in Table 12-7, (a) estimate the ratio by which the yield strength of magnesium can be increased by alloying and heat treatment and (b) compare the yield strength-to-weight ratio of high-strength magnesium alloys to high-strength aluminum alloys.

Answer:

(a) The yield strength of pure annealed magnesium is 89.7 MPa and that of age-hardened AZ80A-T5 alloy is 275.9 MPa.

$$\text{Ratio} = \frac{275.9 \ MPa}{89.7 \ MPa} = 3$$

When compared to a ratio of 30 in aluminum, the response of magnesium to strengthening is rather poor.

(b) The strength-to-weight ratio of AZ80A-T5 compared to 7075-T6 aluminum alloy, Table 12-4, is

$$\text{AZ80A-T5:} \quad \frac{275.9 \ MPa}{1.74 \ Mg \ m^{-3}} = 1.6 \times 10^5 \ m^2 \ s^{-2}$$

$$\text{7075-T6:} \quad \frac{503.4 \ MPa}{2.70 \ Mg \ m^{-3}} = 1.9 \times 10^5 \ m^2 \ s^{-2}$$

Designation. Magnesium alloys can be either wrought or cast, and both heat treatable and nonheat treatable grades are available. Strengthening is achieved by solid solution strengthening, strain hardening, grain size control, dispersion strengthening, and age hardening. As in aluminum alloys, the solubility of the alloying elements in magnesium is rather low, so the degree of solid solution strengthening is limited. Some binary phase diagrams involving magnesium alloys are shown in Figure 12-5.

The alloys are normally designated by a combination of letters and numbers (Table 12-6). The letters refer to the two major alloying elements present in the alloy, and the numbers refer to the approximate amounts of the major alloying elements. The temper designations, such as T and H, are the same as for the aluminum alloys.

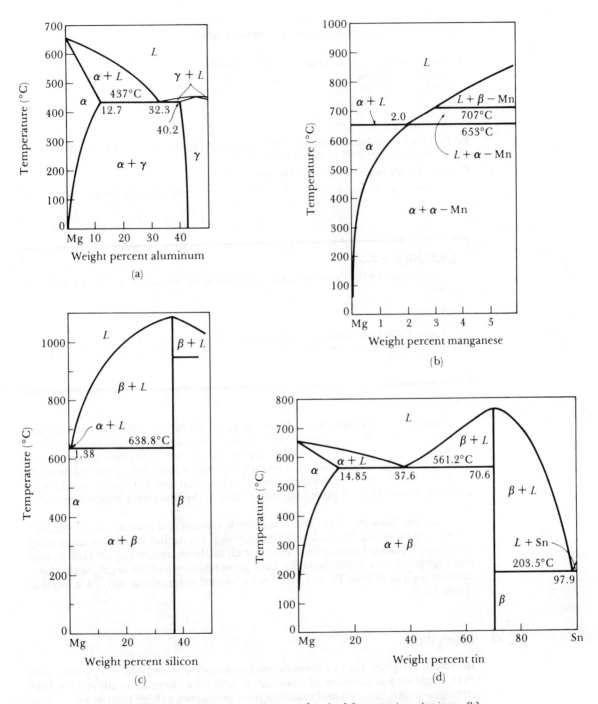

FIG. 12-5 Binary phase diagrams for the (a) magnesium-aluminum, (b) magnesium-manganese, (c) magnesium-silicon, and (d) magnesium-tin systems.

TABLE 12-6   Designation system for magnesium alloys

A. Two letters indicate the major alloying elements.

A—Al	Q—Ag
E—Ce	S—Si
H—Th	T—Sn
K—Zr	Z—Zn
M—Mn	

B. Two (and sometimes three) numbers indicate the approximate amounts of alloying elements, rounded off to the nearest percent.

C. A final letter gives variations to the normal alloy.

EXAMPLE 12-7

What would be the designation for a magnesium alloy containing 2.8% Th and 1.2% Mn?

Answer:

From Table 12-6, thorium is designated by H and manganese by M. To the nearest whole number, the alloy contains about 3% Th and 1% Mn. Thus, the alloy is designated HM31.

Structure and properties.   Magnesium, which has the HCP structure, is less ductile than aluminum. However, the alloys do have some ductility because alloying increases the number of active slip planes. Some deformation and strain hardening can be accomplished at room temperature, and the alloys can be readily deformed at elevated temperatures. However, strain hardening produces a relatively small effect in pure magnesium due to the low strain-hardening coefficient.

Casting alloys are often based on eutectic systems and most are heat treatable. Alloys may be strengthened by either dispersion strengthening or age hardening. Some age-hardened magnesium alloys, such as those containing zirconium, thorium, silver, or rare earth elements, have good resistance to overaging at temperatures as high as 300°C. Properties of some typical magnesium alloys are listed in Table 12-7.

12-4 Beryllium

Beryllium is lighter than aluminum, with a density of 1.848 Mg m^{-3}, yet is stiffer than steel, with a modulus of elasticity of 290 GPa. Beryllium alloys have high strength-to-weight ratios and maintain their properties to high temperatures (Figure 12-6). A number of aerospace and nuclear applications indicate that beryllium is an excellent engineering material.

Unfortunately, beryllium is expensive, brittle, reactive, and toxic. It has the HCP structure and has limited ductility at room temperature. When exposed to

TABLE 12-7 Properties of typical magnesium alloys

Alloy		Tensile Strength (MPa)	Yield Strength (MPa)	% Elongation
Casting alloys				
AM100-T6	10% Al-0.1% Mn	276	152	1
AZ81A-T4	7.6% Al-0.7% Zn	276	83	15
ZK61A-T6	6% Zn-0.7% Zr	310	193	10
AM60A-F	6% Al-0.13% Mn	220	131	6
ZE41A-T5	4.2% Zn-1.2% Ce	207	138	3.5
Wrought alloys				
AZ10A-F	1.2% Al-0.4% Zn	241	145	10
AZ80A-T5	8.5% Al-0.5% Zn	279	276	7
ZK40A-T5	4% Zn-0.45% Zr	276	255	4
HK31A-H24	3% Th-0.6% Zr	262	207	8
Pure Mg				
Annealed		159	90	3–15
Cold worked		179	117	2–10

After data in *ASM Metals Handbook*, Vol. 2, 9th Ed., 1979.

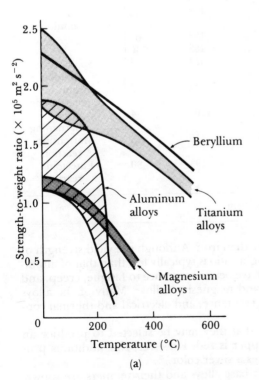

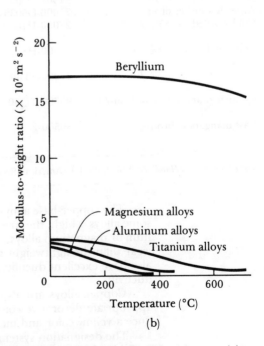

FIG. 12-6 A comparison of the strength-to-weight (a) and modulus of elasticity-to-weight (b) ratios of beryllium and other nonferrous alloys.

the atmosphere at elevated temperatures, beryllium rapidly oxidizes to BeO, which is toxic. Consequently, sophisticated manufacturing techniques, such as vacuum casting, vacuum forging, and powder metallurgy, must be used.

12-5   Copper Alloys

There are a myriad of copper-base alloys that take advantage of all of the strengthening mechanisms that we have discussed. Typical properties of alloys displaying these strengthening mechanisms are shown in Table 12-8.

TABLE 12-8   Properties of typical copper alloys obtained by different strengthening mechanisms

Material	Alloy and Temper Designation	Tensile Strength (MPa)	Yield Strength (MPa)	% Elongation	Strengthening Mechanism
Pure Cu, annealed		209	33	60	
Commercially pure Cu, annealed to coarse grain size	10100-O5050	220	69	55	
Commercially pure Cu, annealed to fine grain size	10100-O5025	234	76	55	Grain size
Commercially pure Cu, cold worked	10100-H10	393	366	4	Strain hardening
Annealed Cu-35% Zn	27000-OS050	324	103	62	
As-fabricated Cu-30% Ni	71500-M20	379	138	45	Solid solution
Annealed Cu-10% Sn	52400-O5035	455	193	68	
Annealed nickel silver	77000-O5035	413	186	40	
Cold-worked Cu-35% Zn	27000-H10	673	434	3	Solid solution +
Cold-worked Cu-30% Ni	71500-H80	578	544	3	Strain hardening
Age-hardened Cu-2% Be	17200-TF00	1310	1207	4	Age hardening
Quenched and tempered Cu-Al	95500-TQ50	759	413	5	Martensitic reaction
Cast manganese bronze	86500-F	490	193	30	Eutectoid reaction

Data from *Metals Handbook*, Vol. 2, 9th Ed., American Society for Metals, 1979.

The copper-base alloys are heavier than iron. Although the yield strength of some alloys is high, the strength-to-weight ratio is typically less than that of aluminum or magnesium alloys. The alloys have better resistance to fatigue, creep, and wear than the lightweight aluminum and magnesium alloys. Many of the alloys also have excellent ductility, corrosion resistance, and electrical and thermal conductivity.

Copper alloys are also unique in that they may be selected to produce an appropriate decorative color. Pure copper is red. However, zinc additions produce a yellow color and nickel produces a silver color.

The designation system for copper-base alloys and their tempers are shown in Tables 12-9 and 12-10. The temper designation system differs from that of aluminum and magnesium alloys.

TABLE 12-9 Designation system for copper-base alloys

Wrought alloys

100xx–159xx	Commercially pure Cu
160xx–199xx	Nearly pure Cu but age hardenable due to Cd, Be, or Cr
2xxxx	Cu-Zn (brass)
3xxxx	Cu-Zn-Pb (leaded brass)
4xxxx	Cu-Zn-Sn (tin bronze)
5xxxx	Cu-Sn and Cu-Sn-Pb (phosphor bronze)
6xxxx	Cu-Al (aluminum bronze), Cu-Si (silicon bronze), Cu-Zn-Mn (manganese bronze)
7xxxx	Cu-Ni (cupronickel), Cu-Ni-Zn (nickel silver)

Cast alloys

800xx–811xx	Commercially pure Cu
813xx–828xx	95–99% Cu
833xx–899xx	Cu-Zn alloys containing Sn, Pb, Mn, or Si
9xxxx	Other Cu alloys, including tin bronze, aluminum bronze, cupronickel, and nickel silver

---

TABLE 12-10 Temper designation for copper alloys

Hxx—cold worked. (xx indicates the degree of cold work.)

		Percent Reduction in Thickness or Diameter
H01	$\frac{1}{4}$ hard	10.9
H02	$\frac{1}{2}$ hard	20.7
H03	$\frac{3}{4}$ hard	29.4
H04	hard	37.1
H06	extra hard	50.1
H08	spring hard	60.5
H10	extra spring	68.6
H12	special spring	75.1
H14	super spring	80.3

HRxx —cold worked and stress relieved. (xx refers to initial percent cold work.)
HTxx —cold worked and heat treated to produce ordering. (xx refers to initial percent cold work.)
Mxx   —as-manufactured. (xx refers to the type of manufacturing process.)
Oxx   —annealed. (xx refers to annealing method.)
OSxxx—annealed to produce a particular grain size. (xxx refers to the grain diameter in 10^{-3} mm. Thus, OS025 gives a grain diameter of 0.025 mm.)
TB00 —solution treated.
TDxx —solution treated and cold worked. (xx refers to percent cold work.)
TF00 —age hardened.
THxx —cold worked and aged. (xx refers to degree of cold work.)
TLxx —aged and cold worked. (xx refers to degree of cold work.)
TQxx —quenched and tempered. (xx gives details of heat treatment.)

Commercially pure copper. Coppers containing less than 1% impurities are used for electrical applications. Small amounts of cadmium or silver improve high-temperature hardness, and tellurium or sulfur improve machinability. Some coppers are dispersion strengthened by small amounts of Al_2O_3, which improve the hardness of the copper without significantly impairing conductivity. Any of these coppers can be strengthened by strain hardening, producing large increases in strength with relatively small decreases in conductivity.

EXAMPLE 12-8

You wish to select a conductive material for the contacts of a switch or relay. Which of the coppers would you choose to maximize the life of the device?

Answer:

We wish a material that has a high hardness yet good conductivity. The hard ceramic particles of Al_2O_3 provide wear resistance yet do not enter into the lattice of the copper and so do not interfere with the electrical conductivity.

Strain hardening.   The single-phase copper alloys are strengthened significantly by cold working. Examples of this effect are shown in Table 12-8. The FCC copper has excellent ductility and a high strain-hardening coefficient.

Solid solution-strengthened alloys.   A number of copper-base alloys contain large quantities of alloying elements yet remain single phase. Important binary phase diagrams are shown in Figure 12-7.

The copper-zinc, or brass, alloys with less than 40% Zn form single-phase solid solutions of zinc in copper. The mechanical properties, even elongation, increase as the zinc content increases. These alloys can be cold formed into rather complicated yet corrosion-resistant components. *Manganese bronze* is a particularly high-strength alloy containing manganese as well as zinc for solid solution strengthening. A copper-nickel-zinc ternary alloy called *nickel silver* really has no silver present although it does have a silver color. The formability of these alloys approaches that of the copper-zinc alloys.

Tin bronzes, often called *phosphor bronzes,* may contain up to 10% Sn and remain single phase. The phase diagram predicts that the alloy will contain the Cu_3Sn (ε) compound. However, the kinetics of the reaction are so slow that the precipitate may not form, particularly in low-tin alloys.

Alloys containing less than about 9% Al or less than 3% Si are also single phase. These *aluminum bronzes* and *silicon bronzes* have good forming characteristics and are often selected for their good strength and excellent toughness.

EXAMPLE 12-9

We say that copper can contain up to 40% Zn or 9% Al and still be single phase. How do we explain this statement in view of the phase diagrams in Figure 12-7?

Answer:

According to the phase diagrams, the solubility of zinc in copper at room temperature is about 30% and the solubility of aluminum in copper is about 8%. The equilibrium precipitates do not form, however, because of the slow rates of diffusion at these temperatures, just as in the copper-tin alloys.

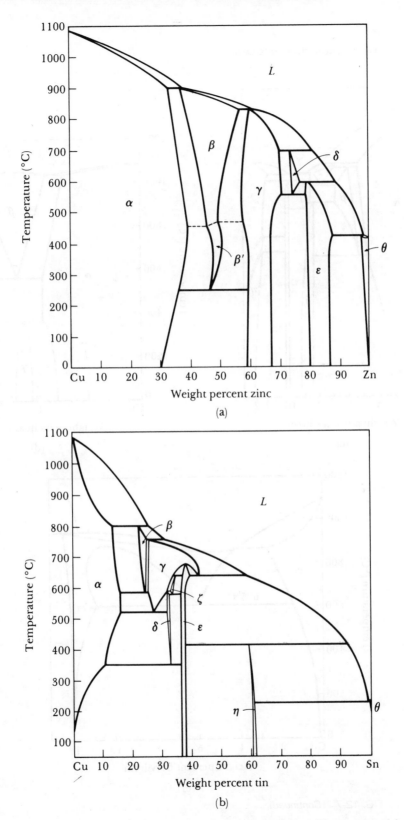

FIG. 12-7 Binary phase diagrams for the (a) copper-zinc, (b) copper-tin, (c) copper-silicon, (d) copper-aluminum, and (e) copper-beryllium systems.

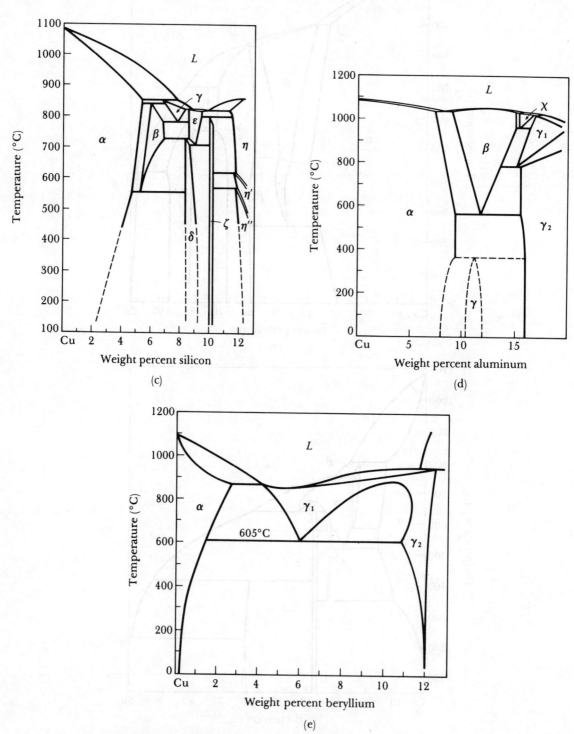

FIG. 12-7 (Continued)

EXAMPLE 12-10

Compare the percentage increase in the yield strength of commercially pure annealed aluminum, magnesium, and copper by strain hardening.

Answer:

From Tables 12-4, 12-7, and 12-8, we find that

Al increases from 35 MPa to 152 MPa

Mg increases from 90 MPa to 117 MPa

Cu increases from 69 MPa to 53 MPa

The percentage increases are

Al: $\dfrac{152 - 35}{35} \times 100 = 334\%$

Mg: $\dfrac{117 - 90}{90} \times 100 = 30\%$

Cu: $\dfrac{366 - 69}{69} \times 100 = 430\%$

Strain hardening is an effective way to strengthen copper and aluminum but produces little effect in HCP magnesium.

EXAMPLE 12-11

Suppose the effect of increasing the percent alloying element in copper produces a linear increase in yield strength. Determine the equations for the yield strength of copper as a function of zinc, nickel, and tin. Are the results consistent with the atomic radii of these elements?

Answer:

From Table 12-8, we find the yield strengths of pure copper and some typical alloys.

Pure copper: 69 MPa

Cu-35% Zn: 103 MPa

Yield strength $= (69 \times 10^6) + \dfrac{(103 \times 10^6) - (69 \times 10^6)}{35} \, (\% \text{ Zn})$

$= (69 \times 10^6) + (0.97 \times 10^6) \, (\% \text{ Zn})$

Cu-30% Ni: 138 MPa

Yield strength $= (69 \times 10^6) + \dfrac{(138 \times 10^6) - (69 \times 10^6)}{30} \, (\% \text{ Ni})$

$= (69 \times 10^6) + (2.3 \times 106) \, (\% \text{ Ni})$

Cu-10% Sn: 193 MPa

Yield strength $= (69 \times 10^6) + \dfrac{(193 \times 10^6) - (69 \times 10^6)}{10} \, (\% \text{ Sn})$

$= (69 \times 10^6) + (12.4 \times 10^6) \, (\% \text{ Sn})$

The atomic radii of the elements are $r_{Cu} = 1.278$ Å, $r_{Zn} = 1.332$ Å, $r_{Ni} = 1.243$ Å, and $r_{Sn} = 1.405$ Å. Copper, nickel, and zinc have about the same atomic radii, but tin atoms are much larger. Thus, addition of tin produces large lattice distortions and gives a larger solid solution-strengthening effect.

Age hardenable alloys. A number of copper-base alloys display an age-hardening response, including zirconium copper, chromium copper, and beryllium copper. The copper-beryllium alloys are used for their high strength, high stiffness (making them useful as springs), and their nonsparking qualities (making them useful for tools to be used near flammable gases and fluids).

Phase transformations. Aluminum bronzes that contain over 9% Al can form at least some β phase on heating above 565°C, the eutectoid temperature (Figure 12-8). On subsequent cooling, the eutectoid reaction produces a lamellar

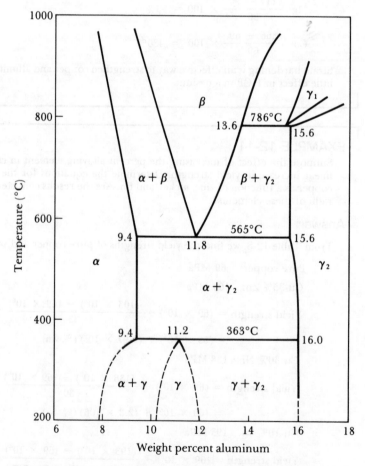

FIG. 12-8 The eutectoid portion of the copper-aluminum phase diagram.

structure, or pearlite, that contains a brittle γ_2 compound. The low temperature peritectoid reaction, $\alpha + \gamma_2 \rightarrow \gamma$, normally does not occur. The eutectoid product is relatively weak and brittle. However, the alloy can be heated to about 900°C, then rapidly quenched to produce martensite, or β', which has a high strength and low ductility. When β' is tempered at 400°C to 650°C, a combination of high strength, good ductility, and excellent toughness is obtained as α precipitates from the β' with a fine platelike shape (Figure 12-9).

FIG. 12-9 The microstructure of a quenched and tempered aluminum bronze containing alpha plates in a beta matrix. From *Metals Handbook*, Vol. 7, 8th Ed., American Society for Metals, 1972.

EXAMPLE 12-12

Consider an aluminum bronze alloy containing 10% Al (Figure 12-8). (a) Calculate the amount of brittle γ_2 that will form on equilibrium cooling past the eutectoid temperature and (b) determine the phases you would expect to have during each step of the quench and temper heat treatment.

Answer:

(a) The amount of γ_2 that forms after the eutectoid reaction is

$$\gamma_2 = \frac{10 - 9.4}{15.6 - 9.4} \times 100 = 9.7\%$$

(b) The steps in the quench and temper heat treatment are to hold at 900°C, quench, and temper between 400°C and 650°C.

The alloy is 100% β above 850°C. At 900°C, the alloy contains all β with a composition of 10% Al.

After quenching, the β transforms to β' martensite which also contains 10% Al. The β' is supersaturated in copper.

During tempering, the copper supersaturation is removed by precipitation of α in a matrix of β. The phase diagram shows that α and β are stable above 565°C and therefore the tempering may lead to the equilibrium phases. However, if we temper at 400°C, the phase diagram shows that γ_2 and α should be present. Normally, the equilibrium γ_2 is not produced in the usual tempering times.

Leaded copper alloys. Virtually any of the wrought alloys may contain up to 4.5% Pb. The lead forms a monotectic reaction with the copper and produces tiny lead spheres as the last liquid to solidify. The lead improves the machining characteristics by improving chip-forming tendencies.

Even larger amounts of lead are used for copper casting alloys. The lead helps provide lubrication and embeddability, by which hard particles or grit are embedded in the soft lead spheres, and therefore helps minimize wear. The large amounts of lead cannot be introduced to wrought alloys because the lead melts and embrittles the alloy during hot working.

12-6   Nickel and Cobalt

Nickel and cobalt alloys are used for corrosion protection and for high-temperature resistance, taking advantage of their high melting points and high strengths. Nickel is FCC and has good formability; cobalt is an allotropic metal, with the FCC structure above 417°C and the HCP structure at lower temperatures. Special cobalt alloys are used for exceptional wear resistance and, because of resistance to human body fluids, for prosthetic devices. Typical alloys and their applications are listed in Table 12-11.

Nickel and Monel.   Commercially pure nickel has excellent corrosion resistance and forming characteristics.

When copper is added to nickel, the maximum strength is obtained near 60% Ni. A number of alloys, called *Monels,* with approximately this composition are used for their strength and corrosion resistance in salt water and at elevated temperatures. Some of the Monels contain small amounts of aluminum and titanium. These alloys show an age-hardening response by the precipitation of γ', a coherent Ni_3Al or Ni_3Ti precipitate which nearly doubles the tensile properties. The precipitates resist overaging at temperatures up to 200°C to 425°C (Figure 12-10).

Superalloys.   Superalloys contain large amounts of alloying elements intended to produce a combination of high strength at elevated temperatures, resistance to creep at temperatures up to 1000°C, and resistance to corrosion. Yet these excellent high-temperature properties are obtained even though the melting temperature of the alloys is about the same as that for steels.

The three categories of superalloys—nickel base, iron-nickel base, and cobalt base—often have exotic names, befitting their use. Typical applications include vanes and blades for turbine and jet engines, heat exchangers, chemical reaction vessel components, and heat-treating equipment. Figure 12-11 shows the effect of temperature and intended use of the alloys on the stress-rupture properties.

To obtain high strengths and creep resistance, the alloying elements must produce a strong stable microstructure at high temperatures.

Solid solution strengthening.   Large additions of chromium, molybdenum, and tungsten and smaller additions of tantalum, zirconium, niobium and boron provide solid solution strengthening. Because no catastrophic metallurgical softening process occurs on heating, the effects of solid solution strengthening are stable and consequently make the alloy resistant to creep.

Carbide dispersion.   All of the alloys contain a small amount of carbon which, by combining with other alloying elements, produces a network of fine stable carbide particles. The carbide network interferes with dislocation move-

TABLE 12-11 Compositions, properties, and applications for selected nickel and cobalt alloys

Material	Tensile Strength (MPa)	Yield Strength (MPa)	% Elongation	Strengthening Mechanism	Applications
Pure Ni (99.9% Ni)	344	110	45	Annealed	Corrosion resistance
	655	62	4	Cold worked	
Ni-Cu alloys					
Monel 400 (Ni-31.5% Cu)	538	269	37	Annealed	Valves, pumps, heat exchangers
Monel K-500 (Ni-29.5% Cu-2.7% Al-0.6% Ti)	1034	759	30	Aged	Shafts, springs, impellors
Ni superalloys					
Inconel 600 (Ni-15.5% Cr-8% Fe)	621	200	49	Carbides	Heat treatment equipment
Hastelloy B-2 (Ni-28% Mo)	897	414	61	Carbides	Corrosion resistance
Hastelloy G (Ni-20% Cr-20% Fe-7% Mo + Nb, Ta)	690	324	50	Aged	Chemical processing
MAR-M246 (Ni-10% Co-9% Cr-10% W + Ti, Al, Ta)	966	862	5	Aged	Jet engines
DS-Ni (Ni-2% ThO$_2$)	490	331	14	Dispersion	Gas turbines
Fe-Ni superalloys					
Incoloy 800 (Ni-46% Fe-21% Cr)	614	282	37	Carbides	Heat exchangers
Co superalloys					
Haynes 25 (50% Co-20% Cr-15% W-10% Ni)	931	448	60	Carbides	Jet engines
Stellite 6B (60% Co-30% Cr-4.5% W)	1221	710	4	Carbides	Abrasive wear resistance

Data from *Metals Handbook*, Vol. 3, 9th Ed., American Society for Metals, 1980.

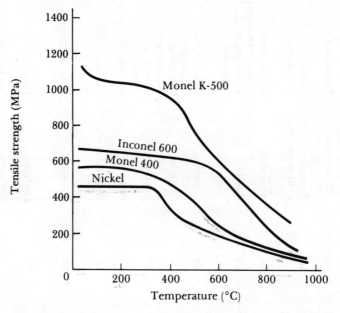

FIG. 12-10   The effect of temperature on the tensile strength of several nickel-base alloys.

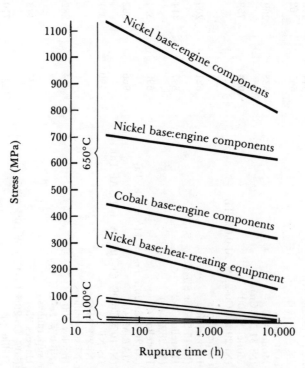

FIG. 12-11   The stress-rupture behavior of selected superalloys at 650°C and 1100°C.

ment and prevents grain boundary sliding. The carbides include TiC, BC, ZrC, TaC, Cr_7C_3, $Cr_{23}C_6$, Mo_6C, and W_6C, although often the carbides are more complex and contain several alloying elements. Stellite 6B, a cobalt-base superalloy, has unusually good wear resistance at high temperatures due to these carbides.

Precipitation hardening. Some of the nickel and nickel-iron superalloys which contain aluminum and titanium form the coherent precipitate γ' (Ni_3Al or Ni_3Ti) during aging. The γ' particles (Figure 12-12) increase strength and resistance to creep.

FIG. 12-12 The microstructure of a typical superalloy, showing a grain boundary carbide network and the dispersed γ' precipitate, 15,000×. From *Metals Handbook*, Vol. 7, 8th Ed., American Society for Metals, 1972.

12-7 Zinc Alloys

Pure zinc is nearly as heavy as steel, melts at only 420°C, has the HCP crystal structure, and has a strength less than that of many aluminum alloys. With these properties, we might expect that zinc would seldom be used. Yet many applications are found for both wrought and cast zinc. Table 12-12 summarizes the properties of several alloys.

TABLE 12-12 Properties of selected zinc alloys

Alloy	Tensile Strength (MPa)	Yield Strength (MPa)	% Elongation	Processing
Casting alloys				
Zn-4% Al	283		10	Die casting
Zn-12% Al	310	207	2	Sand casting
Zn-27% Al	421	356	5	Sand casting
Wrought alloys				
Zn-0.08% Pb	138		60	Hot work
	159		45	Cold work
Zn-1% Cu	200		40	Hot work
	248		30	Cold work
Zn-22% Al	400	352	11	Superplastic

Data from *Metals Handbook*, Vol. 2, 9th Ed., American Society for Metals, 1979.

Because zinc recrystallizes and creeps near room temperature, it has excellent ductility. However, strain hardening is negligible. Wrought zinc is used for batteries, photoengraving plates, and roofing components, such as gutters. A special wrought alloy, Zn-22% Al, displays superplastic behavior. The unusually large deformations permit complex panels and cabinets to be formed.

Zinc die castings are produced using an alloy that contains about 4.5% Al. This eutectic alloy (Figure 12-13) has a low melting temperature and excellent

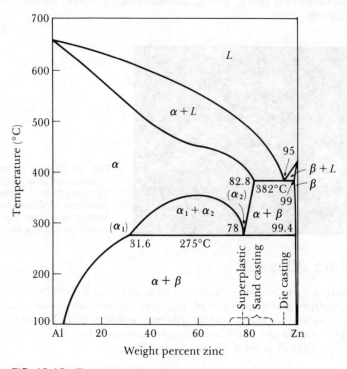

FIG. 12-13   The zinc-aluminum phase diagram.

fluidity, permitting the metal to fill thin complicated molds. The castings are easily cleaned, then coated or plated to produce a decorative effect. Zinc alloys containing 12% to 27% Al have also been developed which have sufficient strength to compete with many copper alloys and cast irons.

12-8   Titanium Alloys

Titanium is a relatively lightweight metal that provides excellent corrosion resistance, high strength-to-weight ratio, and good high-temperature properties. Strengths up to 1380 MPa coupled with a density of 4.505 Mg m^{-3} provide the excellent mechanical properties, while an adherent, protective TiO_2 film provides excellent resistance to corrosion and contamination below 535°C. Above 535°C, the oxide film breaks down and small atoms such as carbon, oxygen, nitrogen, and hydrogen embrittle the titanium.

Titanium is allotropic, with the HCP crystal structure (α) at low temperatures and a BCC structure (β) above 882°C. Alloying elements provide solid solution strengthening and change the allotropic transformation temperature. The alloying elements can be divided into four groups, as summarized in Figure 12-14. Additions such as tin provide solid solution strengthening without affecting the transformation temperature. Aluminum, oxygen, hydrogen, and other alpha stabilizing elements increase the temperature at which α transforms to β. Beta stabi-

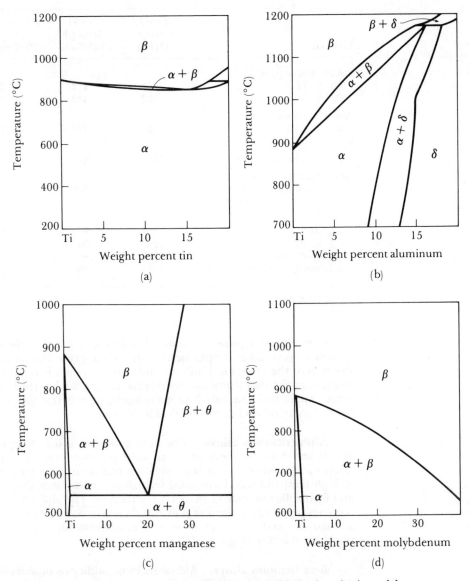

FIG. 12-14 The phase diagrams for (a) titanium-tin, (b) titanium-aluminum, (c) titanium-manganese, and (d) titanium-molybdenum systems.

lizers such as vanadium, tantalum, molybdenum, and niobium lower the transformation temperature, even causing β to be stable at room temperature. Finally manganese, chromium, and iron produce an eutectoid reaction, reducing the temperature at which the α-β transformation occurs and producing a two-phase structure at room temperature. There are several categories for titanium and its alloys, which are summarized in Table 12-13.

TABLE 12-13   Properties of selected titanium alloys

Material	Tensile Strength (MPa)	Yield Strength (MPa)	% Elongation
Commercially pure Ti			
99.5% Ti	241	172	24
99.0% Ti	552	483	15
Alpha Ti alloys			
5% Al-2.5% Sn	862	779	15
Beta Ti alloys			
13% V-11% Cr-3% Al	1290	1214	5
Near alpha Ti alloys			
8% Al-1% Mo-1% V	966	828	14
6% Al-4% Zr-2% Sn-2% Mo	1007	993	3
Alpha-beta Ti alloys			
8% Mn	966	862	15
6% Al-4% V	1034	966	8
7% Al-4% Mo	1172	1034	10
6% Al-6% V-2% Sn	1103	1034	12

Data from *Metals Handbook,* Vol. 3, 9th Ed., American Society for Metals, 1980.

Commercially pure titanium.   Unalloyed titanium is used for its superior corrosion resistance. Impurities, such as oxygen, dramatically increase the strength of the titanium. Commercially pure titanium is relatively weak, loses its strength at elevated temperatures (Figure 12-15), but has the best corrosion resistance. Applications include heat exchangers, piping, reactors, pumps, and valves for the chemical and petrochemical industries.

Alpha titanium alloys.   The most common of the all-alpha alloys contains 5% Al and 2.5% Sn, both of which solid solution strengthen the alpha. These alloys have good corrosion and oxidation resistance, maintain their strength well at high temperatures, have good weldability, and normally have good ductility and formability in spite of their HCP structure. The alpha alloys are annealed at high temperatures in the β region and then are cooled. Rapid cooling gives a fine acicular or needlelike α grain structure, whereas furnace cooling gives a more platelike structure (Figure 12-16).

Beta titanium alloys.   Although large additions of vanadium or molybdenum produce an entirely β structure at room temperature, none of the so-called beta alloys are actually alloyed to that extent. Instead, they are rich in β stabilizers, so that rapid cooling produces a metastable structure composed of all β. In the

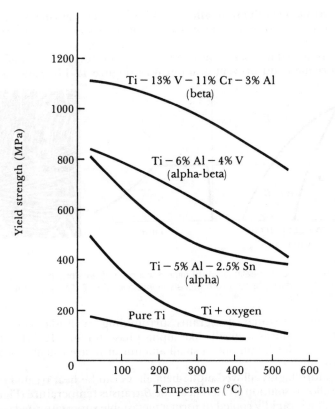

FIG. 12-15   The effect of temperature on the yield strength of selected titanium alloys.

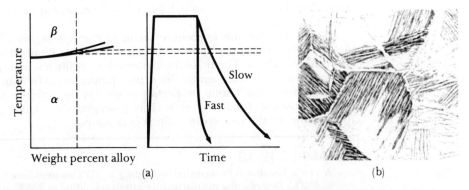

FIG. 12-16   (a) Annealing and (b) microstructure of rapidly cooled alpha titanium. Both the grain boundary precipitate and the Widmanstatten plates are alpha. From *Metals Handbook*, Vol. 7, 8th Ed., American Society for Metals, 1972.

annealed condition, where only β is present in the microstructure, strength is derived from solid solution strengthening. The alloys can also be aged to produce higher strengths. Applications include high strength fasteners, beams, and other fittings for aerospace applications.

Alpha-beta titanium alloys.   With proper balancing of the α and β stabilizers, a mixture of α and β is produced at room temperature. Annealing provides a combination of high ductility, uniform properties, and good strength. The alloy is heated just below the β-transus temperature, permitting a small amount of α to remain and prevent grain growth (Figure 12-17). Slow cooling permits the

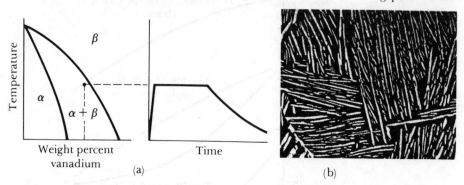

FIG. 12-17   (a) Annealing and (b) microstructure of an alpha-beta titanium alloy where alpha is the white phase. From *Metals Handbook*, Vol. 7, 8th Ed., American Society for Metals, 1972.

formation of equiaxed α grains surrounding small islands of the β phase. Fast cooling produces a needle-like alpha phase (Figure 12-17b). The "near-alpha" alloys, which contain only a small amount of β, are usually used in the annealed condition.

More highly alloyed alpha-beta alloys can be heat treated to high strengths. The alloy is solution treated near the β-transus temperature (Figure 12-18). Next, the alloy is rapidly cooled to form a metastable supersaturated solid solution β' or titanium martensite α'. The alloy is then aged or tempered near 500°C. During aging, finely dispersed α and β precipitate from β' or α', increasing the strength of the alloy (Figure 12-19). Note that α' is a soft phase, just the opposite to the case of steel martensite.

Normally, titanium martensite forms in the less highly alloyed alpha-beta alloys, whereas the supersaturated beta is more easily retained in alloys closer to the entirely β alloys. The titanium martensite typically has an acicular or needle-like appearance. During aging, the α precipitates in a Widmanstatten structure which improves the tensile properties as well as the fracture toughness of the alloy. Components for airframes, rockets, jet engines, and landing gear are typical applications for the heat-treated alpha-beta alloys.

EXAMPLE 12-13

A Ti-8% Mn alloy is heat treated by holding at 705°C, water quenching, and reheating to 480°C. Describe the microstructure after (a) holding at 705°C and (b) reheating to 480°C.

Answer:

(a) Let's perform a lever law calculation at 705°C, which is just below the β-transus temperature at 730°C, using Figure 12-14(c).

$$\beta = \frac{8 - 0.5}{12 - 0.5} \times 100 = 65.2\%$$

$$\alpha = \frac{12 - 8}{12 - 0.5} \times 100 = 34.8\%$$

The alloy is solution treated below 730°C in order to retain some α in the microstructure which in turn prevents grain growth of β. After quenching, the microstructure will remain the same.

(b) After aging at 480°C, the alloy is below the eutectoid and the stable phases are α and θ. If the aging step is long enough, the θ will precipitate from β and increase the strength.

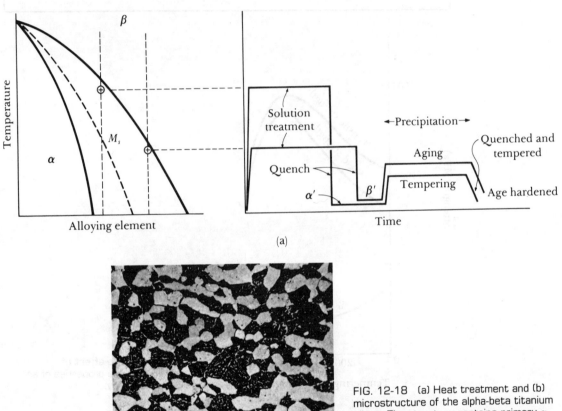

(a)

(b)

FIG. 12-18 (a) Heat treatment and (b) microstructure of the alpha-beta titanium alloys. The structure contains primary α (large white grains) and a dark β matrix with needles of α formed during aging. From *Metals Handbook*, Vol. 7, 8th Ed., American Society for Metals, 1972.

EXAMPLE 12-14

Compare the approximate yield strength-to-weight ratio of the 13% V-11% Cr-3% Al beta titanium alloy to the 7075-T6 aluminum alloy.

Answer:

From Tables 12-4 and 12-13, we find that the yield strengths are 1214 MPa for the titanium alloy and 503 MPa for the aluminum alloy. The ratios are

$$\text{Ti:} \quad \frac{1214 \text{ MPa}}{4.505 \text{ Mg m}^{-3}} = 2.69 \times 10^5 \text{ m}^2 \text{ s}^{-2}$$

$$\text{Al:} \quad \frac{503 \text{ MPa}}{2.70 \text{ Mg m}^{-3}} = 1.86 \times 10^5 \text{ m}^2 \text{ s}^{-2}$$

The high strength and moderate density of the titanium give it an excellent strength-to-weight ratio.

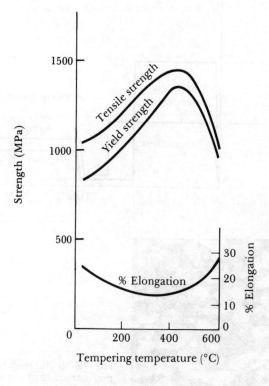

FIG. 12-19   The effect of tempering on the properties of an α-β titanium alloy.

Processing of titanium alloys.   Titanium alloys are processed into useful shapes by casting, forming, and joining techniques. However, care must be taken to prevent contamination when the temperature of the alloy exceeds 535°C.

The alloys for titanium castings are melted in a vacuum furnace and poured into molds constructed from a ceramic material or graphite. Special precautions are also required to minimize contamination during welding. For example, gas-tungsten arc welding produces welds with the same resistance to corrosion as the base material. However, the welded area must be protected with an inert gas until the weld cools to below 535°C.

Some alloys, including the Ti-6% Al-4% V alpha-beta alloy, are superplastic and can be deformed as much as 1000%. During the slow process of deformation, the alloy is protected by an argon atmosphere. The superplastic forming can also be coupled with simultaneous diffusion bonding to produce complicated aircraft parts.

12-9 Zirconium

Zirconium and its alloys resemble titanium, with a similar melting temperature, allotropic behavior, and corrosion resistance. Zirconium's chief applications are for nuclear reactor cores, due to its low neutron absorption cross section, and corrosion resistance. Zirconium is used for reactor towers, heat exchangers, pumps, piping, valves, and pressure vessels in the chemical industry.

Zirconium is naturally alloyed with up to 4.5% Hf. In specifying zirconium alloys, up to 4.5% Hf is considered commercially pure zirconium. Very few zirconium alloys are used; the most common is Zircaloy, which contains a small amount of tin.

12-10 Refractory Metals

The refractory metals—tungsten, molybdenum, tantalum, and niobium (or columbium)—have exceptionally high melting temperatures and consequently have the potential for high-temperature service. These metals find a variety of applications in aerospace and electronic components. Some important mechanical and physical properties of the refractory metals are given in Table 12-14.

TABLE 12-14 Properties of refractory metals

Metal	Melting Temperature (°C)	Density (Mg m^{-3})	Room Temperature			$T = 1000°C$	
			Tensile Strength (MPa)	Yield Strength (MPa)	% Elongation	Tensile Strength (MPa)	Yield Strength (MPa)
Nb	2470	8.66	310	138	25	117	55
Mo	2610	10.22	828	552	10	345	207
Ta	2996	16.6	345	241	35	186	165
W	3410	19.25	2069	1517	3	455	103

Oxidation. The refractory metals begin to oxidize between 200°C and 425°C and are rapidly contaminated or embrittled. Consequently, special precautions are required during casting, hot working, welding, or powder metallurgy. The metals must also be protected during service at elevated temperatures. For example, the tungsten filament in a light bulb is protected by a vacuum.

For some applications, the metals may be coated with a silicide or aluminide coating. The coating must (a) have a high melting temperature, (b) be compatible with the refractory metal, (c) provide a diffusion barrier to prevent contaminants from reaching the underlying metal, and (d) have a coefficient of thermal expansion similar to the refractory metal. Coatings are available that protect the metal to about 1650°C.

Forming characteristics.   The refractory metals, which have the BCC crystal structure, display a ductile-to-brittle transition temperature. Because the transition temperatures for niobium and tantalum are below room temperature, these two metals can readily be formed. However, annealed molybdenum and tungsten normally have a transition temperature above room temperature, causing them to be brittle. Fortunately, if these metals are hot worked to produce a fibrous microstructure, the transition temperature is lowered and the forming characteristics are significantly improved (Figure 12-20).

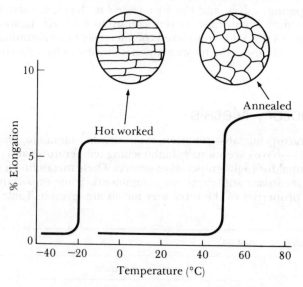

FIG. 12-20   The effect of temperature on the ductility of tungsten. Deformation reduces the transition temperature to below room temperature.

Alloys.   Large increases in both room temperature and high-temperature mechanical properties are obtained by alloying. However, a limited number of alloys is available for each metal and the alloying elements are quite exotic. These alloys typically are solid solution strengthened; in fact, tungsten and molybdenum form a complete series of solid solutions, much like copper and nickel (Figure 12-21). Some alloys, such as W-2% ThO_2, are dispersion strengthened by oxide particles during their manufacture by powder metallurgy processes.

SUMMARY

The nonferrous alloys take advantage of all of the strengthening mechanisms in materials. Depending on the characteristics of the base material, a tremendous

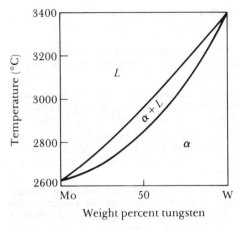

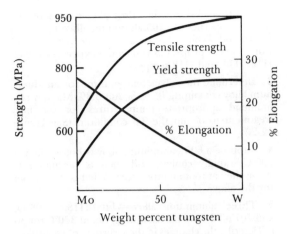

FIG. 12-21   The phase diagram and mechanical properties of tungsten-molybdenum alloys.

range of properties can be obtained, including strong, lightweight materials such as aluminum, beryllium, and magnesium and high-temperature materials such as titanium, nickel, cobalt, and refractory metals.

GLOSSARY

Castability   The ease with which a metal can be poured into a mold to make a casting without producing defects or requiring unusual or expensive techniques to prevent casting problems.

Fluidity   The ability of liquid metal to fill up a mold cavity without prematurely freezing.

Nonferrous alloy   An alloy based on some metal other than iron.

Refractory metals   Metals having a melting temperature above that of titanium.

Superalloys   A group of nickel, iron-nickel, and cobalt alloys that have exceptional heat resistance, creep resistance, and corrosion resistance.

Tempers   A shorthand notation using letters and numbers to describe the processing of an alloy. H tempers refer to cold-worked alloys; T tempers refer to age-hardening treatments.

Wrought alloys   Alloys that are shaped by a deformation process.

PRACTICE PROBLEMS

1   We wish to make a cable that can support a load of 4545.5 kg without permanent deformation. Determine the approximate diameter of the cable and the weight per metre of the cable if the cable is made from the following materials. Use the density of the pure metals to estimate the weight.

a. 3003-O aluminum
b. 3003-H18 aluminum
c. 7075-T6 aluminum
d. AZ80A-T5 magnesium
e. Annealed Cu-35% Zn
f. Age-hardened Cu-2% Be
g. Zn-22% Al
h. 99.0% titanium
i. Ti-13% V-11% Cr-3% Al

2 Figure 9-23 shows the effect of cold work on the tensile properties of 3105 aluminum alloy. From this figure and Table 12-3, determine the percent cold work and the properties you would expect for the O, H12, H14, and H18 tempers.

3 Determine the solvus temperature for an aluminum alloy containing 0.3% Si and 0.8% Mg and recommend appropriate temperatures for the heat treatment to obtain the T6 condition. (See Figure 12-2.)

4 Suppose a 6061 aluminum alloy is hot worked at 600°C, rapidly cooled, and then is permitted to strengthen at room temperature. What is the temper for the alloy?

5 Three aluminum alloys—1100-H18, 3003-O, and 7075-T6—are placed into service at 300°C for 36 h. Describe the changes in the structure of each alloy. Which one of the three alloys will retain the greatest percentage of its original strength? Explain.

6 The peak temperature reached during a slow arc-welding process is shown in Figure 12-22. If the alloy

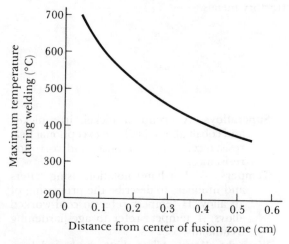

FIG. 12-22   The peak temperature reached during slow welding of an aluminum-magnesium-silicon alloy. (See Problem 12-6.)

contains Al-1% Mg-0.6% Si, estimate the width of the fusion zone, the partial fusion zone, and the solution-treated zone in the heat-affected area of the weld. (This requires that you determine the liquidus, solidus, and solvus temperatures for the alloy.)

7 Determine the formula Al_xMn_y of the intermetallic compound β that forms in aluminum-manganese alloys. (Convert wt% to at%.) How much β is normally present in a 3003-O aluminum alloy under equilibrium conditions?

8 Aluminum alloys containing primarily magnesium (the 5xxx series alloys) never contain more than 15% Mg. From the phase diagram in Figure 12-1, explain why this limit is necessary. (Review the requirements for good dispersion strengthening in Chapter 8.)

9 Aluminum casting alloys typically have much lower ductilities than the wrought alloys. Explain in terms of microstructure why this might be expected.

10 Compare the freezing ranges for the 413 and 443 aluminum casting alloys using Figure 10-18 and Table 12-5. Which alloy would you expect to have the better fluidity?

11 Figure 12-23 shows the approximate tie triangle in the aluminum-rich corner of the aluminum-silicon-copper phase diagram. Neglecting the effect of magnesium, calculate and compare the amounts of α (nearly pure aluminum), β (nearly pure silicon), and θ ($CuAl_2$) in (a) 319 alloy and (b) 390 alloy. Compare the as-fabricated properties from Table 12-5. What effect does an increased amount of β appear to have on properties?

12 Explain why aluminum die castings and permanent mold castings are stronger than aluminum sand castings. (See Table 12-5.)

13 Would you expect shrinkage to be concentrated rather than dispersed interdendritically in a 413 or a 443 aluminum alloy? Explain. (See Figure 10-18.)

14 From the phase diagrams in Figure 12-5, determine the maximum solubility of aluminum, manganese, silicon, and tin in magnesium. Is there a relationship between maximum solubility and the difference in atomic size between magnesium and the alloying element? See Appendix B.

15 How would you designate a magnesium alloy that contains 3.2% Ce and 1.6% Zn?

16 A magnesium alloy has the designation AK111. Estimate the approximate composition of the alloy.

17 Determine the formula Mg_xSi_y for the β compound in the magnesium-silicon system (Figure 12-5).

18 Compare the strength-to-weight ratio of annealed Cu-35% Zn brass (27000-OS050), aged Cu-2% Be (17200-TF00), and 7075-T6 aluminum. (Use units of inch.)

19 Two Cu-35% Zn brasses with designations 27000-OS025 and 27000-OS100 are obtained. Which do you expect to have the higher strength? Explain.

20 A commercially pure copper, C19400, is cold drawn from a diameter of 6.35 mm to 2.51 mm. Determine the temper of the alloy and, from Figure 9-6, estimate the mechanical properties of the final wire.

21 Calculate the wt% Be in the CuBe intermetallic compound. How does this compare to the phase dia-

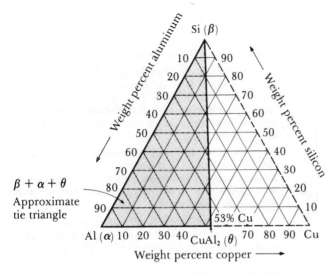

FIG. 12-23   The tie triangle in the aluminum-rich corner of the aluminum-silicon-copper phase diagram. (See Problem 12-11.)

gram? Calculate the amount of CuBe present in a Cu-1% Be alloy in the TB00 condition and in the O70 (equilibrium) condition.

22 Determine the solubility of tin in copper at room temperature and calculate the amount of Cu_3Sn (ε) that should be present in a Cu-10% Sn alloy under equilibrium conditions. If this amount of ε formed, what effect would it have on the % elongation expected for the alloy?

23 A copper plate 2 cm thick with the H06 temper is to be obtained by rolling. What should be the original thickness of the plate?

24 The compositions of three superalloys are listed below. (a) Which of the three major strengthening mechanisms utilized in superalloys are at work in each alloy? (b) Which elements are responsible for each strengthening mechanism in the alloys?

Alloy A	Alloy B	Alloy C
0.02% C	0.05% C	0.43% C
29.00% Cr	12.00% Cr	1.25% Mn
61.00% Ni	4.50% Mo	0.40% Si
9.00% Fe	2.00% Nb	20.50% Cr
	0.60% Ti	20.00% Ni
	5.90% Al	20.00% Co
	75.00% Ni	4.00% Mo
		4.00% W
		4.00% Nb
		25.40% Fe

25 Compare the percentage increase in the tensile strength of Zn-1% Cu produced by cold working to the increase in the tensile strength obtained in commercially pure aluminum, magnesium, nickel, and copper. What does this result suggest concerning the strain-hardening coefficient of HCP metals versus FCC metals?

26 Compare the strength-to-weight ratio of the cold-worked Zn-1% Cu alloy to that of 3003-H18 aluminum. Which material would you select for decorative trim on an automobile?

27 To produce superplasticity, an alloy should be heated to about 0.5 to 0.6 times its absolute melting temperature but remain in a two-phase region so that grain growth is prevented. What forming temperature would you recommend for the Zn-22% Al alloy?

28 Based on the binary phase diagrams, by what strengthening mechanism do aluminum and tin improve the mechanical properties in the Ti-5% Al-2.5% Sn alloy?

29 The titanium-vanadium phase diagram is shown in Figure 12-24. (a) For a Ti-8% V alloy, recommend a solution-treating temperature that will control grain growth by retaining 10% α in the microstructure. (b) What phase(s) should be present if the alloy is quenched to room temperature from the solution-treating temperature? (c) Calculate the amount of each phase after reheating to 400°C. Is this an aging or tempering treatment? Which phase would you expect to be continuous?

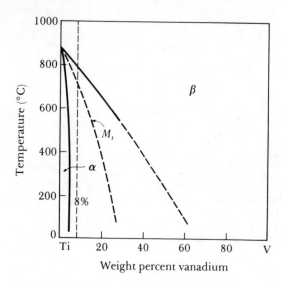

FIG. 12-24   The titanium-vanadium phase diagram.
(See Problem 12-29.)

30 Which of the titanium alloys in Table 12-13 would you select to produce a 25.4 mm diameter lever arm capable of withstanding a 356 kN force in a corrosive environment?

31 Compare the % elongation of the aluminum and copper alloys to that in the magnesium and titanium alloys. Which group generally tends to have the higher ductility? Why is this expected?

32 Determine the ratio between the yield strength of the strongest aluminum, magnesium, copper, titanium, and nickel-copper alloys to the yield strength of the pure metal. Compare the alloy systems and rank them in the order of their response to strengthening mechanisms.

33 Determine the strength-to-weight ratios (using the yield strength) of the strongest aluminum, magnesium, copper, titanium, tungsten, and Monel alloys. Rank the alloy systems according to their strength-to-weight ratio.

34 Explain why aluminum and magnesium alloys are commonly used in aerospace and transportation applications whereas copper and beryllium alloys are not.

35 Based on the phase diagrams, estimate the solubilities of nickel, zinc, silicon, aluminum, tin, and beryllium in copper at room temperature. Are these solubilities expected in view of Hume-Rothery's rules for solid solubility? Explain.

36 What happens when the protective coating on a tungsten part expands more than the tungsten? What happens when the protective coating on a tungsten parts expands less than the tungsten?

13

Ferrous Alloys

13-1 Introduction

Ferrous alloys, which are based primarily on iron-carbon alloys, include plain-carbon steels, alloy and tool steels, stainless steels, and cast irons. These groups of ferrous alloys have a wide variety of characteristics and applications. All of the strengthening mechanisms we have discussed apply to at least some of these materials.

In this chapter, we will discuss in some detail how we use the eutectoid reaction to control the heat treatment and properties of steels. The importance of alloying elements for these heat treatments will be pointed out. Finally, we will examine two special classes of ferrous alloys—stainless steels and cast irons.

13-2 Review of the Fe-Fe$_3$C Phase Diagram

The Fe-Fe$_3$C phase diagram provides the basis for understanding the ferrous alloys. We examined the phase diagram in Chapter 11, but let's quickly review it before going on.

Solid solutions.   There are three solid solutions of importance—δ-ferrite, austenite (γ), and ferrite (α)—and one intermetallic compound—cementite (Fe$_3$C). In addition, a metastable phase—martensite—can form on rapid cooling.

Three-phase reactions.   The three-phase reactions are

Peritectic: $L_{0.53\% \text{ C}} + \delta_{0.09\% \text{ C}} \rightarrow \gamma_{0.17\% \text{ C}}$

Eutectic: $L_{4.3\% \text{ C}} \rightarrow \gamma_{1.70\% \text{ C}} + \text{Fe}_3\text{C}_{6.67\% \text{ C}}$

Eutectoid: $\gamma_{0.8\% \text{ C}} \rightarrow \alpha_{0.035\% \text{ C}} + \text{Fe}_3\text{C}_{6.67\% \text{ C}}$

The dividing point between steels and cast irons is 1.7% C, or where the eutectic reaction becomes possible.

Microconstituents.   Several microconstituents may form depending on how we control the eutectoid reaction. Pearlite is a lamellar mixture of ferrite and cementite. Bainite is a nonlamellar mixture of ferrite and cementite obtained by transformation of austenite at a large undercooling. Either primary ferrite or primary cementite may form, depending on the original composition of the alloy.

Tempered martensite, a mixture of very fine cementite in ferrite, forms when martensite is reheated following its formation.

All of the heat treatments of a steel are directed towards producing the mixture of ferrite and cementite which gives the proper combination of properties. The next few sections describe some of the techniques by which we can exercise control in unalloyed steels.

We will concentrate on the eutectoid portion of the diagram (Figure 13-1).

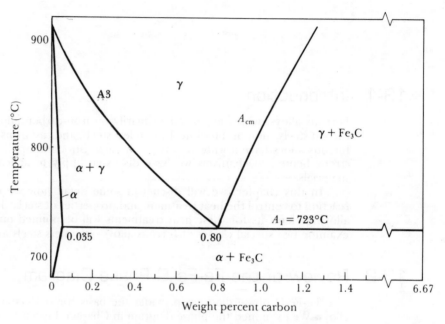

FIG. 13-1   The eutectoid portion of the Fe-Fe₃C phase diagram.

The solubility lines and the eutectoid isotherm are identified by A_3, A_{cm}, and A_1. The A_3 shows the temperature at which ferrite starts to form on cooling; the A_{cm} shows the temperature at which cementite starts to form; and the A_1 is the eutectoid temperature.

13-3 Designation and Typical Structures of Steels

Table 13-1 shows certain selected steels under the AISI and SAE designation together with typical BS 970 grades. For a full description of the classification systems the reader should refer to the appropriate standards. An AISI 1040 steel is a plain-carbon steel with 0.40% C. A SAE 10120 steel is a plain-carbon steel containing 1.20% C. An AISI 4340 steel is an alloy steel containing 0.40% C.

We described the microstructures of typical equilibrium-cooled steels in Chapter 11. These structures are shown in Figure 13-2. A hypoeutectoid steel, such as a 1050 steel, contains 100% γ after heating above the A_3 temperature. When the steel cools to between the A_3 and the A_1, primary α forms, normally

TABLE 13-1 Compositions of selected AISI-SAE steels

AISI-SAE Number	Typical BS Grade	% C	% Mn	% Si	% Ni	% Cr	Others
1020	040A20	0.18–0.23	0.30–0.60				
1040	080A40	0.37–0.44	0.60–0.90				
1060	080A62	0.55–0.65	0.60–0.90				
1080	070A78	0.75–0.88	0.60–0.90				
1095	060A96	0.90–1.03	0.30–0.50				
1140	212M44	0.37–0.44	0.70–1.00				0.10% S
1340	150M36	0.38–0.43	1.60–1.90	0.15–0.30			
1541		0.36–0.44	1.35–1.65				
4140	708M40	0.38–0.43	0.75–1.00	0.15–0.30		0.80–1.10	0.20% Mo
4340	817M40	0.38–0.43	0.60–0.80	0.15–0.30	1.65–2.00	0.70–0.90	0.25% Mo
4620	665H20	0.17–0.22	0.45–0.65	0.15–0.30	1.65–2.00		0.25% Mo
4820		0.18–0.23	0.50–0.70	0.15–0.30	3.25–3.75		0.25% Mo
5120	527A19	0.17–0.22	0.70–0.90	0.15–0.30		0.70–0.90	
52100	534A99	0.98–1.10	0.25–0.45	0.15–0.30		1.30–1.60	
6150	735A50	0.48–0.53	0.70–0.90	0.15–0.30		0.80–1.10	0.15% min V
8620	805H20	0.18–0.23	0.70–0.90	0.15–0.30	0.40–0.70	0.40–0.60	0.20% V
9260	250A58	0.56–0.64	0.75–1.00	1.80–2.20			

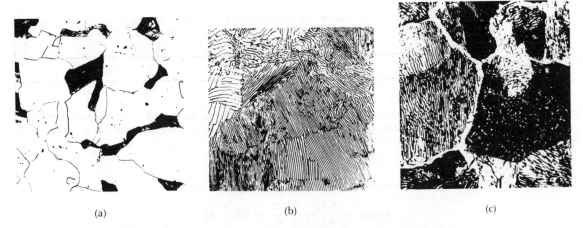

(a) (b) (c)

FIG. 13-2 (a) Hypoeutectoid steel with primary ferrite (white) and pearlite. (b) Eutectoid steel containing only pearlite colonies. (c) Hypereutectoid steel with primary Fe_3C and pearlite. From *Metals Handbook*, Vols. 7 and 8, 8th Ed., American Society for Metals, 1972, 1973.

outlining the austenite grain boundaries. When the steel cools to just above the A_1 temperature, a tie line tells us that the remaining austenite contains 0.80% C. This remaining austenite transforms through the eutectoid reaction to produce pearlite.

A 10120 steel is hypereutectoid and contains 100% γ above the A_{cm}. After cooling to between the A_{cm} and the A_1, primary Fe_3C forms at the austenite grain boundaries. The austenite present just above the A_1 temperature transforms to pearlite.

A 1077 steel only goes through the eutectoid reaction. Above the A_1 temperature, the structure is all austenite; below the A_1 the structure is all pearlite.

EXAMPLE 13-1

Most automobile axles are forged from a 1050 steel. Preliminary examination of the steel prior to forging and heat treatment reveals that the microstructure contains about 60% pearlite and 40% primary α. Calculate the amount of each phase and microconstituent that you expect to find in a 1050 steel and determine if the steel is indeed a 1050 steel.

Answer:

The amounts of each phase and microconstituent in a 1050 steel at 722°C can be calculated using the lever law. You may wish to review Example 11-12. These percentages will change only extremely slightly on cooling to room temperature.

$$\alpha = \frac{6.67 - 0.5}{6.67 - 0.035} \times 100 = 93.0\%$$

$$Fe_3C = \frac{0.5 - 0.035}{6.67 - 0.035} \times 100 = 7.0\%$$

$$Pearlite = \frac{0.5 - 0.035}{0.80 - 0.035} \times 100 = 60.8\%$$

$$Primary\ \alpha = \frac{0.80 - 0.5}{0.80 - 0.035} \times 100 = 39.2\%$$

The calculated amount of pearlite—60.8%—compares closely to the amount estimated from the microstructure—60%. The steel is probably a 1050 steel.

EXAMPLE 13-2

An as-received bar stock is observed to contain about 95% pearlite and 5% primary Fe_3C. Calculate the carbon content of the steel and determine the grade, or AISI-SAE number, of the steel.

Answer:

Let's calculate the percent carbon using the lever law.

$$Pearlite = \frac{6.67 - \%\ C}{6.67 - 0.80} \times 100 = 95$$

$$6.67 - \%\ C = 0.95\ (6.67 - 0.80)$$

$$\%\ C = 1.094$$

We notice from Table 13-1 that a range of carbon is permitted in AISI-SAE steels. Thus, the grade of our steel is closest to a 10110 steel, with approximately 1.10% C.

EXAMPLE 13-3

A steel to be used as a spring is estimated to contain 10% Fe_3C and 90% ferrite. Can we consider this steel to be an eutectoid steel?

Answer:

Let's calculate the percent carbon using the lever law.

$$\alpha = \frac{6.67 - \% \text{ C}}{6.67 - 0.035} \times 100 = 90$$

$$6.67 - \% \text{ C} = 0.9 (6.67 - 0.035)$$

$$\% \text{ C} = 0.70$$

A 1080 steel, which is approximately an eutectoid steel, should, from Table 13-1, contain 0.75% to 0.88% C. Since our steel contains only 0.69% C, it should not be considered an eutectoid steel.

13-4 Simple Heat Treatments

Four simple heat treatments are commonly used for steels. These heat treatments (Figure 13-3) are used to accomplish one of three purposes.

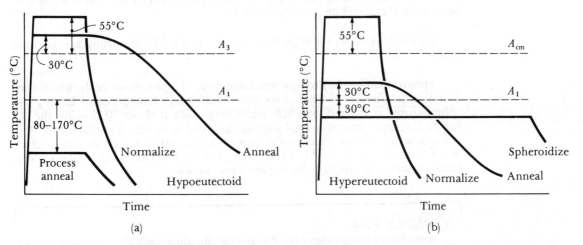

FIG. 13-3 Schematic summary of the simple heat treatments for (a) hypoeutectoid steels and (b) hypereutectoid steels.

Process anneal—eliminating cold work.   The ferrite in steels with less than 0.25% C is strengthened by cold working. The recrystallization heat treatment used to eliminate the effect of cold working is called a *process anneal*. The process anneal is done 80°C to 170°C below the A_1 temperature.

Annealing and normalizing—controlling dispersion strengthening.   Plain-carbon steels are dispersion strengthened by controlling the amount, size, shape, and distribution of Fe_3C. As the carbon increases, more Fe_3C is present and, up to a point, the strength of the steel increases.

We can refine the Fe_3C by controlling the cooling rate as the austenite transforms to pearlite. If we permit very slow cooling, the pearlite is coarse—this heat treatment is called *annealing* or a full anneal. Faster cooling produces fine pearlite—this heat treatment is *normalizing*.

1. Hypoeutectoid steels are annealed by heating the steel about 30°C above the A_3 temperature to produce homogeneous austenite. This is the *austenitizing treatment*. Then the steel is furnace cooled. By permitting both the furnace and the steel to cool together, slow cooling rates are produced. Because lots of time is available for diffusion, primary ferrite and pearlite are coarse and the steel has a low strength and good ductility.

2. Hypereutectoid steels are annealed by first heating to 30°C above the A_1. The steel is not heated above the A_{cm} to produce all austenite because, on slow cooling, Fe_3C would form as a continuous film on the austenite grains and cause embrittlement. Austenitizing just above the A_1 permits the Fe_3C to become rounded. After austenitizing, the steel is furnace cooled to produce discontinuous Fe_3C and coarse pearlite.

3. Steels are normalized by heating to 55°C above either the A_3 or A_{cm}, depending on the composition of the steel. After austenitizing, the steel is removed from the furnace and air cooled. Air cooling gives faster cooling rates and finer pearlite. The hypereutectoid steel can be normalized above the A_{cm} because, due to the faster cooling rate, the Fe_3C has less opportunity to form as a continuous film at the austenite grain boundaries.

Figure 13-4 shows the typical properties obtained by annealing and normalizing.

Spheroidizing—improving machinability. High-carbon steels, which contain a large amount of Fe_3C, have poor machining characteristics. During the spheroidizing treatment, which requires long times at about 30°C below the A_1, the Fe_3C changes into spherical particles in order to reduce boundary area. The microstructure, known as *spheroidite*, now has a continuous matrix of soft, machinable ferrite (Figure 13-5). After machining, the steel is given a more sophisticated heat treatment to produce the required properties.

EXAMPLE 13-4

Recommend temperatures for the process annealing, annealing, normalizing, and spheroidizing of 1020, 1080, and 10120 steels.

Answer:

If we consult Figure 13-1, we can determine the critical A_1, A_3, or A_{cm} temperatures for each steel. We can then specify the heat treatment based on these temperatures.

	1020	1080	10120
Critical temperatures	A_1 = 723°C A_3 = 830°C	A_1 = 723°C	A_1 = 723°C A_{cm} = 895°C
Process annealing	723 − (80 to 170) = 553°C to 643°C	Not done	Not done
Annealing	830 + 30 = 860°C	723 + 30 = 753°C	723 + 30= 753°C

Normalizing	830 + 55 = 885°C	723 + 55 = 778°C	895 + 55 = 950°C
Spheroidizing	Not done	723 − 30 = 693°C	723 − 30 = 693°C

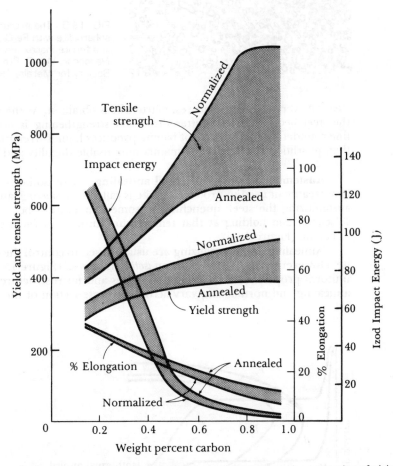

FIG. 13-4   The effect of carbon and heat treatment on the properties of plain-carbon steels.

13-5 Isothermal Heat Treatments and Dispersion Strengthening

The effect of transformation temperature on the properties of an entectoid steel was discussed in Chapter 11. As the isothermal transformation temperature decreases, pearlite becomes progressively finer before bainite begins to form in-

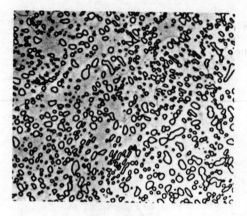

FIG. 13-5   The microstructure of spheroidite with Fe_3C particles dispersed in a ferrite matrix. From *Metals Handbook*, Vol. 7, 8th Ed., American Society for Metals, 1972.

stead. At very low temperatures, martensite is obtained. As the microstructure of the steel becomes finer, better dispersion strengthening is obtained. Also, the fine rounded microstructure of bainite produces higher strengths and hardnesses than pearlitic structures while maintaining usable ductility and toughness.

Austempering and isothermal annealing.   The isothermal transformation heat treatment used to produce bainite is called *austempering*, and simply involves austenitizing the steel, quenching to some temperature below the nose of the IT curve, and holding at that temperature until all of the austenite transforms to bainite (Figure 13-6).

Annealing and normalizing are usually used to control the fineness of pearlite. However, pearlite formed by an *isothermal anneal* (Figure 13-6) may give more uniform properties, since the cooling rates and microstructure obtained during annealing and normalizing vary across the cross section of the steel.

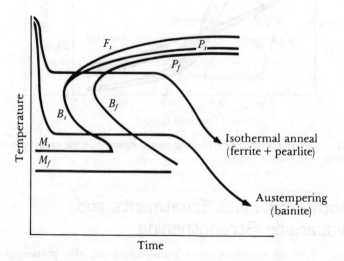

FIG. 13-6   The austempering and isothermal anneal heat treatments.

Isothermal transformations in hypoeutectoid and hypereutectoid steels.
In either a hypoeutectoid or hypereutectoid steel, the IT diagram must reflect the possible formation of a primary phase. The isothermal transformation diagrams for a 1050 (080M50) and a 10110 steel are shown in Figure 13-7. The most remark-

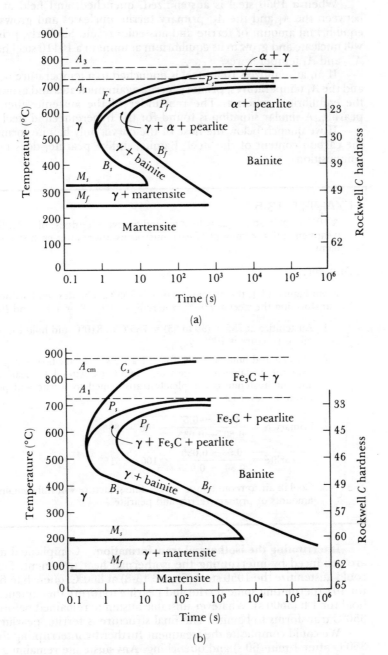

(a)

(b)

FIG. 13-7 The IT diagrams for (a) a 1050 and (b) a 10110 steel.

able change is the presence of a "wing" which begins at the nose of the curve and becomes asymptotic to the A_3 or A_{cm} temperature. The wing represents the ferrite start (F_s) time in hypoeutectoid steels or the cementite start (C_s) time in hypereutectoid steels.

When a 1050 steel is austenitized, quenched, and held at a temperature between the A_1 and the A_3, primary ferrite nucleates and grows; eventually an equilibrium amount of ferrite and austenite result. Similarly, primary cementite will nucleate and grow to its equilibrium amount in a 10110 steel held between the A_{cm} and A_1 temperatures.

If an austenitized 1050 steel is quenched to a temperature between the nose and the A_1 temperatures, primary ferrite again nucleates and grows until reaching the equilibrium amount. The remainder of the austenite then transforms to pearlite. A similar situation is found for the hypereutectoid steel.

If we quench below the nose of the curve, only bainite forms, regardless of the carbon content of the steel. Bainite, unlike pearlite, does not have a fixed composition.

EXAMPLE 13-5

A 1050 steel is isothermally heat treated to give a hardness of R_c23. Describe the heat treatment and the amount of each microconstituent after each step of the heat treatment.

Answer:

From Figure 13-1, the A_3 temperature is 755°C. The desired hardness is obtained by transforming the steel at 590°C, where $F_s = 0.9$ s, $P_s = 1.1$ s, and $P_f = 5$ s.

1. Austenitize at $755 + (30 \text{ to } 55) = 785°C$ to $810°C$ and hold for perhaps 1 h. The microstructure is 100% γ.

2. Quench and hold at 590°C for at least 5 s. Primary ferrite begins to precipitate from the unstable austenite after about 0.9 s. After 1.1 s pearlite begins to grow and the austenite is completely transformed to ferrite and pearlite after 5 s. From the lever law

$$\text{Primary } \alpha = \frac{0.80 - 0.5}{0.80 - 0.035} \times 100 = 39\%$$

$$\text{Pearlite} = \frac{0.5 - 0.035}{0.80 - 0.035} \times 100 = 61\%$$

3. Cool in air to room temperature. The structure will still contain the equilibrium amounts of primary ferrite and pearlite.

Interrupting the isothermal transformation. Complicated microstructures are produced by interrupting the isothermal heat treatment. For example, we could austenitize the 1050 steel (Figure 13-8) at 800°C, quench to 650°C and hold for 10 s (permitting some ferrite and pearlite to form), then quench to 350°C and hold for 1 h (3600 s). Whatever unstable austenite remained before quenching to 350°C transforms to bainite. The final structure is ferrite, pearlite, and bainite.

We could complicate the treatment further by interrupting the treatment at 350°C after 1 min (60 s) and quenching. Any austenite remaining after 1 min at 350°C forms martensite. The final structure now contains ferrite, pearlite, bainite,

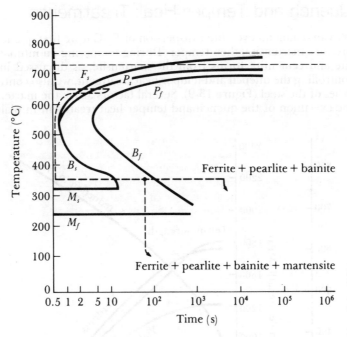

FIG. 13-8 Producing complicated structures by interrupting the isothermal heat treatment of a 1050 steel (080M50).

and martensite. Note that each time we change the temperature we start at zero time!

Because such complicated mixtures of microconstituents produce unpredictable properties, these structures are seldom produced intentionally.

EXAMPLE 13-6

A 1050 steel is held at 800°C for 1 h, quenched to 700°C and held for 50 s, quenched to 400°C and held for 20 s, and finally quenched to room temperature. What is the final microstructure of the steel? (Use the IT diagram (Figure 13-7a) for this steel.)

Answer:

1. After 1 h at 800°C, 100% austenite forms.

2. Ferrite begins to form after 20 s at 700°C but, after 50 s, the steel contains only ferrite and unstable austenite.

3. Immediately after quenching to 400°C, the steel is still only ferrite and austenite. Bainite begins to form after 3 s and, after 20 s, the steel contains ferrite, bainite, and still some unstable austenite.

4. After quenching to room temperature, the remaining austenite crosses the M_s and M_f temperatures and transforms to martensite. The final structure is ferrite, bainite, and martensite.

13-6   Quench and Temper Heat Treatments

We can obtain an even finer dispersion of Fe_3C if we first quench the austenite to produce martensite, then temper. During tempering an intimate mixture of ferrite and cementite forms from the martensite, as discussed in Chapter 11. By controlling the quench and temper heat treatment, we also control the final properties of the steel (Figure 13-9). Several factors affect the martensite reaction and the execution of the quench and temper heat treatment in plain-carbon steels.

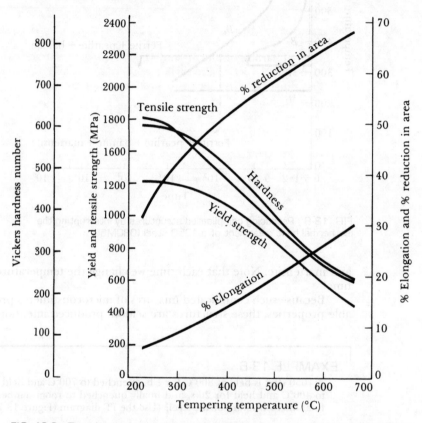

FIG. 13-9   The effect of tempering temperature on the mechanical properties of a 1050 steel.

EXAMPLE 13-7

Select a quench and temper heat treatment that will produce a yield strength of 1000 MPa and an elongation greater than 15% in a 1050 steel.

Answer:

From Figure 13-9, we find that the yield strength exceeds 1000 MPa if the steel is tempered below 460°C, while the elongation exceeds 15% if tempering is done above 425°C. One possible heat treatment is as follows.

1. Austenitize above the A_3 temperature of 755°C for 1 h. An appropriate temperature may be $755 + 55 = 810$°C.

2. Quench rapidly to room temperature. Since the M_f is about 250°C, martensite will form.

3. Temper by heating the steel to 440°C. Normally, 1 h will be sufficient if the steel is not too thick.

4. Cool to room temperature.

Effect of carbon on the M_s and M_f.   The martensite start and finish temperatures are reduced when the carbon content increases (Figure 13-10). High-carbon steels must be refrigerated to produce all martensite.

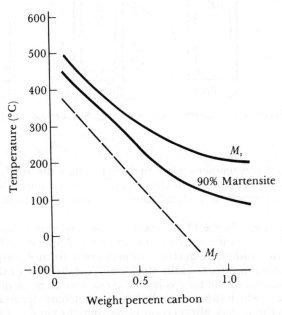

FIG. 13-10   The effect of carbon on the M_s and M_f temperatures in plain-carbon steels.

Retained austenite.   As the martensite plates form during quenching, they surround and isolate small pools of austenite (Figure 13-11). For the remaining austenite to transform, the surrounding martensite must deform, but the strong martensite resists the transformation. Either the existing martensite cracks or the austenite remains in the structure as *retained austenite*.

Retained austenite can be a serious problem. Martensite softens and becomes more ductile during tempering. After tempering, the retained austenite cools below the M_s and M_f temperatures and transforms to martensite, since the surrounding tempered martensite can deform. But now the steel contains more hard

Austenite

Martensite

FIG. 13-11
Formation of
retained austenite
in steel during
quenching. The
pools of austenite
cannot transform
due to the
restraint of the
strong surrounding
martensite.

brittle martensite! A second tempering step may be needed to eliminate the martensite formed from the retained austenite.

Residual stresses and cracking.　Residual stresses are also produced because of the volume change. The hard surface is placed in tension, while the center is compressed. If the residual stresses are high enough, quench cracks form at the surface (Figure 13-12). However, if we first cool to just above the M_s

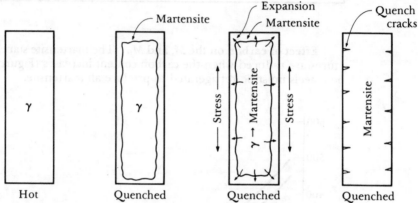

FIG. 13-12　Formation of quench cracks due to residual stresses produced during quenching.

and hold until the temperature equalizes through the steel, then subsequent quenching permits all of the steel to transform to martensite at about the same time. This heat treatment is called *marquenching* or *martempering* (Figure 13-13).

Quench rate.　In the IT diagram, we assumed that we could cool from the austenitizing temperature to the transformation temperature instantly. Because this is not true, undesired microconstituents may form during quenching. For example, pearlite may form as the steel cools past the nose of the curve, particularly since the time of the nose is less than one second in plain-carbon steels.

　　The rate at which the steel cools during quenching depends on two primary factors. First, the surface always cools faster than the center of the steel. In addition, as the size of the part increases, the cooling rate at any location is slower. Second, the cooling rate depends on the temperature and heat transfer characteristics of the quenching medium (Table 13-2).

Continuous cooling transformation diagrams.　We can develop a continuous cooling transformation (CCT) diagram by determining the microstructures produced in a steel at various rates of cooling. The CCT curve for an eutectoid steel (1080) is shown in Figure 13-14. The CCT diagram differs from the IT diagram (Figure 11-21) in that transformations begin at slightly longer times and no bainite region is observed.

　　If we cool an eutectoid steel at 5°C/s, the CCT diagram tells us that we obtain coarse pearlite; we have annealed the steel. Cooling at 35°C/s gives fine pearlite, or

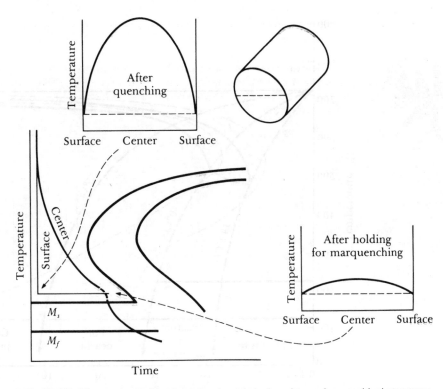

FIG. 13-13 The marquenching heat treatment designed to reduce residual stresses and quench cracking.

TABLE 13-2 The H coefficient, or severity of the quench, for several quenching media

Medium	H Coefficient
Oil (no agitation)	0.25–0.30
Oil (violent agitation)	0.80–1.10
H_2O (no agitation)	0.90–1.00
H_2O (violent agitation)	4.0
Brine (no agitation)	2.0
Brine (violent agitation)	5.0

A large H coefficient indicates a rapid cooling effect.

From *The Making, Shaping, and Treating of Steel, 9th Ed.*, United States Steel, 1971.

a normalizing heat treatment. Cooling at 100°C/s permits pearlite to start forming, but the reaction is incomplete and the remaining austenite changes to martensite. We obtain 100% martensite, and thus are able to perform a quench and temper heat treatment, only if we cool faster than 140°C/s.

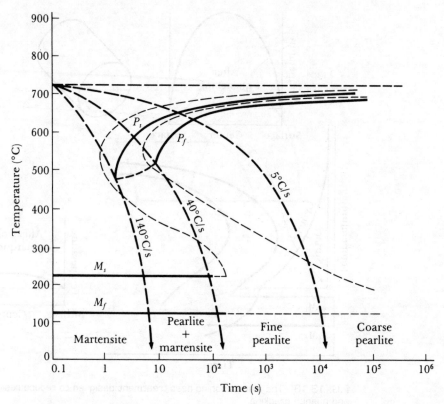

FIG. 13-14 The CCT diagram (solid lines) for an eutectoid steel (1080) compared to the IT diagram (dashed lines).

Other steels have more complicated CCT diagrams. Figure 13-15 shows the CCT diagram for a slightly alloyed 1020 steel. Cooling rates between 10°C/s and 20°C/s give a combination of ferrite, pearlite, bainite, and martensite.

EXAMPLE 13-8

A 1020 steel cools at 8°C/s when quenched in oil and 50°C/s when quenched in water. What microstructure is produced by each of the heat treatments?

Answer:

From Figure 13-15, an 8°C/s cooling curve crosses the F_s, P_s, B_s, and B_f lines. The structure is ferrite, pearlite and bainite.

At 50°C/s, the cooling curve crosses the F_s, B_s, and M_s lines. The structure is ferrite, bainite, and martensite. There may also be a small amount of retained austenite.

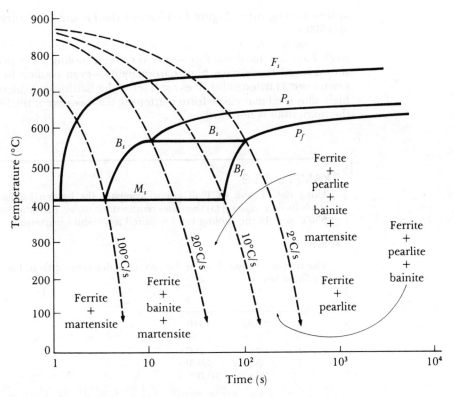

FIG. 13-15   The CCT diagram for a 1020 steel.

13-7   Purpose of Alloying Elements in Steels

Alloying elements are added to steels to (a) provide solid solution strengthening of ferrite, (b) cause the precipitation of alloy carbides rather than Fe_3C, (c) improve corrosion resistance and other special characteristics of the steel, and (d) improve the hardenability. Improving hardenability is most important in alloy and tool steels. An important purpose of alloying elements in stainless steels is to produce improved corrosion resistance.

13-8   Effect of Alloying Elements on the IT and CCT Diagrams

In plain-carbon steels, the nose of the IT and CCT curves occurs at very short times, hence very fast cooling rates are required to produce all martensite. In thin sections of steel, the rapid quench produces distortion and cracking. In thick steels, we are unable to produce martensite.

Hardenability.   All alloying elements in steel shift the IT and CCT diagrams to longer times, permitting us to obtain all martensite even in thick sections at slow cooling rates. Figure 13-16 shows the IT and CCT curves for a 4340 steel (817M40).

Hardenability refers to the ease with which martensite forms. Plain-carbon steels have low hardenability—only very high cooling rates produce all martensite. Alloyed steels have high hardenability—even cooling in air may produce martensite. Hardenability does *not* refer to the hardness of the steel. A low-carbon high-alloy steel may easily form martensite but, because of the low carbon content, the martensite is not hard.

EXAMPLE 13-9

Using the IT and CCT diagrams, compare the hardenabilities of 4340 and 1050 steels by determining (a) the times required for isothermal transformation to occur at 650°C and (b) the cooling rates required to produce martensite.

Answer:

The transformation times of the two steels determined from Figures 13-7 and 13-16 are listed below.

	1050	4340
F_s	2 s	200 s
P_s	5 s	3,000 s
P_f	20 s	10,000 s

From the CCT diagram for the 4340 steel, a cooling rate greater than 8°C/s is required to produce all martensite. A CCT curve for the 1050 steel is not available; however, if we examine the CCT diagrams for the 1020 and 1080 steels, we find that the 8°C/s cooling rate gives all pearlite in the 1080 steel and a mixture of ferrite, pearlite, and bainite in the 1020 steel. The alloying elements in the 4340 steel substantially increase the hardenability.

Effect on the phase diagram. When alloying elements are added to steel, the binary Fe-Fe$_3$C phase diagram is altered (Figure 13-17). First, alloying elements reduce the carbon content at which the eutectoid reaction occurs. Thus, a plain-carbon steel containing 0.6% C is hypoeutectoid whereas an alloy steel containing 0.6% C and 2% Mo is hypereutectoid. The alloy steel may contain primary cementite.

Some alloying elements, such as nickel, reduce the A_1, A_3, and A_{cm} temperatures. A tempering or service temperature that is satisfactory for plain-carbon steels may exceed the A_1 of the nickel alloy steel, permitting austenite to form. However, most alloying elements increase the transformation temperatures. When these steels are austenitized the usual 30°C to 55°C above the A_3 or A_{cm} lines for plain-carbon steels, the temperature may still be below the actual transformation temperatures of the alloy steel. Thus, 100% austenite does not form and the heat treatment is not successful. Figure 13-18 shows the effect of manganese.

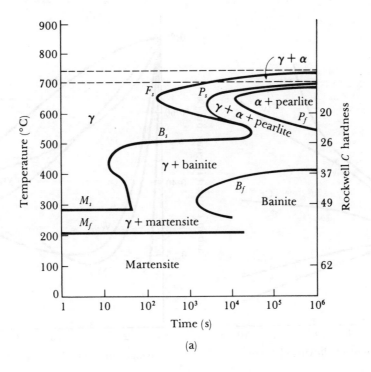

(a)

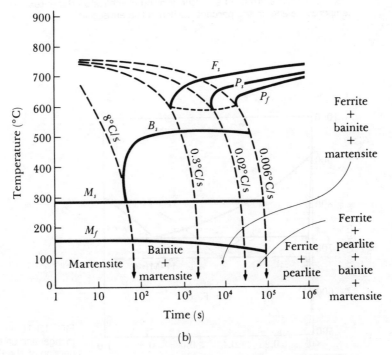

(b)

FIG. 13-16   (a) TTT and (b) CCT curves for a 4340 steel (817M40).

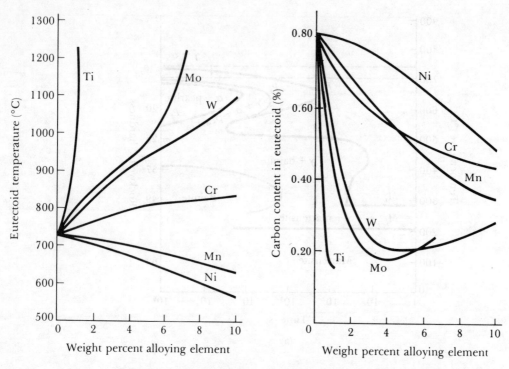

FIG. 13-17   The effect of selected alloying elements on the eutectoid temperature and on the percent carbon in the eutectoid.

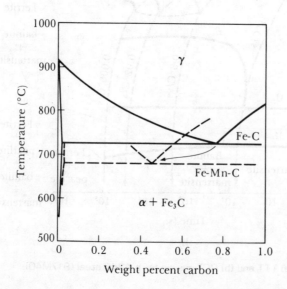

FIG. 13-18   The effect of 6% manganese on the eutectoid portion of the Fe-Fe₃C phase diagram.

EXAMPLE 13-10

A 1060 steel (080A62) can be given a subcritical anneal at 700°C; as a consequence of the heat treatment, the steel is softened and any effects of prior cold work are eliminated. Suppose a steel containing 0.6% C and 8% Ni is given such a heat treatment. Describe the structure and discuss the appropriateness of the heat treatment.

Answer:

In the nickel steel, the amount of carbon at the eutectoid composition is 0.54% C. The A_1 temperature for the nickel steel is 600°C. Consequently, when the nickel steel is held at 700°C, austenite will form. The steel is now hypereutectoid rather than hypoeutectoid. After cooling from the heat-treating temperature, primary cementite can form at the grain boundaries and embrittle the steel.

Martensite start and finish temperatures.   The alloying elements reduce the M_s and M_f temperatures. Thus, alloy steels may have to be quenched to lower than normal temperatures to produce martensite. One expression for the M_s temperature is

$$M_s = 561 - 474\% \text{ C} - 33\% \text{ Mn} - 17\% \text{ Cr} - 17\% \text{ Ni} - 21\% \text{ Mo} \qquad (13\text{-}1)$$

Shape of the IT diagram.   Alloying elements may introduce a "bay" region into the IT diagram, as in the case of the 4340 steel (Figure 13-16). The bay region is used as the basis for a thermomechanical heat treatment known as *ausforming*. A steel can be austenitized, quenched to the bay region, plastically deformed, and finally coded to room temperature (Figure 13-19).

Tempering.   Alloying elements reduce the rate of tempering compared to a plain-carbon steel (Figure 13-20). This effect may permit the alloy steels to operate successfully at higher temperatures than plain-carbon steels.

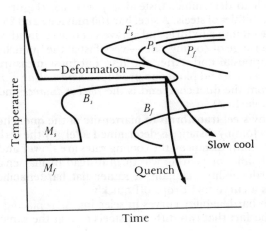

FIG. 13-19 When alloying elements introduce a bay region into the IT diagram, the steel can be ausformed.

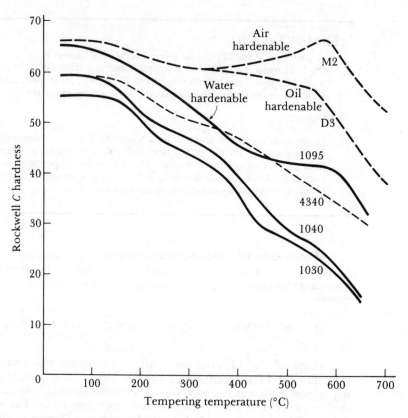

FIG. 13-20   The effect of alloying elements on the tempering curves of
steels. M2 and D3 refer to tool steels.

13-9 Hardenability Curves

CCT diagrams are unavailable for many steels; furthermore, accurate cooling
rates are difficult to determine. Instead, a *Jominy test* (Figure 13-21) is used to
compare hardenabilities of steels. A steel bar 100 mm long and 25 mm in diameter is
austenitized, placed into a fixture, and sprayed at one end with water. This procedure produces a range of cooling rates—very fast at the quenched end, almost air
cooling at the opposite end. After the test, hardness measurements are made
along the test specimen and plotted to produce a *hardenability curve* (Figure 13-22).
The distance from the quenched end is the *Jominy distance* and is related to the
cooling rate (Table 13-3).

Virtually any steel transforms to martensite at the quenched end. Thus, the
hardness at zero Jominy distance is determined solely by the carbon content of the
steel. At larger Jominy distances, the cooling rates are slower and there is a greater
likelihood that bainite or pearlite will form instead of martensite. An alloy steel
with a high hardenability maintains a rather flat hardenability curve; a plain-
carbon steel has a curve that drops off quickly.

We can use hardenability curves in selecting or replacing steels in practical
applications. The fact that two different steels cool at the same rate if quenched

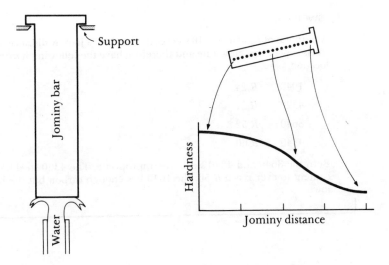

FIG. 13-21   The Jominy test for determining the hardenability of a steel.

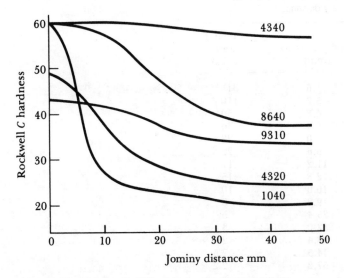

FIG. 13-22   The hardenability curves for several steels.

under identical conditions helps in this selection process. A simple example is described below.

EXAMPLE 13-11

A gear made from 9310 steel is found to wear at an excessive rate. The hardness at a critical location is R_c40. We decide that a hardness of at least R_c50 is needed at that location to resist wear. Which steel(s) in Figure 13-22 would be most appropriate?

Answer:

A hardness of R_c40 in a 9310 steel corresponds to a Jominy distance of 16 mm. The other steels cool at this same rate and therefore have the following hardnesses at the critical location.

1040	R_c24
4320	R_c31
8640	R_c52
4340	R_c60

Both the 8640 and 4340 steels are appropriate. The 4320 steel has too low a carbon content to ever reach R_c50; the 1040 has enough carbon but the hardenability is too low.

TABLE 13-3   The relationship between cooling rate and Jominy distance

Jominy Distance (mm)	Cooling Rate (°C/s)
1.6	315
3.2	110
4.8	50
6.4	36
8.0	28
9.6	22
11.2	17
12.8	15
16.0	10
19.2	8
25.4	5
32.0	3
38.4	2.8
44.5	2.5
57.2	2.2

In another simple technique, we utilize the severity of the quench and the Grossman chart (Figure 13-23) to determine the hardness at the center of a round bar. The bar diameter and H coefficient, or severity of the quench in Table 13-2, give the Jominy distance at the center of the bar. We can then determine the hardness from the hardenability curve of the steel. (See Example 13-12.)

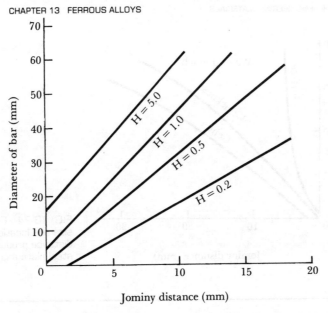

Jominy distance (mm)

FIG. 13-23 The Grossman chart used to determine the hardenability of a steel from the bar diameter and severity of the quench.

EXAMPLE 13-12

An 8640 steel bar 25 mm in diameter must have a minimum hardness of R_c58 throughout the cross section. Which of the following quenching media will be satisfactory: still oil, still water, and agitated brine?

Answer:

First, find the H coefficient from Table 13-2. Then, using the 25 mm diameter, determine the Jominy distance that corresponds to each quench medium from Figure 13-23. Finally, determine the hardness for each Jominy distance from the hardenability curve in Figure 13-22.

Medium	H Coefficient	Jominy Distance (mm)	R_c Hardness
Still oil	0.30	11	56
Still water	1.00	5	59
Agitated brine	5.00	2	60

The minimum R_c58 hardness is obtained by quenching in either still water or agitated brine.

Finally, a simplified technique can be demonstrated using the data in Figure 13-24 which shows the Jominy distance obtained at selected locations in a quenched steel plate.

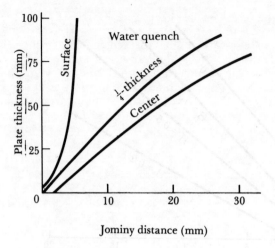

FIG. 13-24   Effect of section size and location on the Jominy distance produced when quenching steel plate and bars.

EXAMPLE 13-13

Determine the hardness profile across the cross section of a 4320 quenched plate 50 mm thick.

Answer:

From Figures 13-22 and 13-24 we can determine the hardness across the plate, which can then be plotted (Figure 13-25).

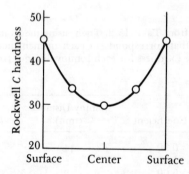

FIG. 13-25   Hardness profile across the cross section of a 4320 quenched plate 50 mm thick.

Location	Jominy Distance (mm)	R_c Hardness
Surface	5	45
¼-thickness	13	34
Center	18	30

13-10 Tool Steels

Most *tool steels* are high-carbon steels that obtain high hardnesses by a quench and temper heat treatment. Their applications include cutting tools in machining operations, dies for die casting, forming dies, and others in which a combination of high strength, hardness, toughness, or temperature resistance is needed.

A large variety of tool steels is available, ranging from plain-carbon tool steels such as 10100 to very high carbon, highly alloyed steels (Table 13-4). The tool steels differ in their hardenability and their resistance to tempering. British standards are prefixed by a B:- (AISI M1 : British standard BM1).

TABLE 13-4   Compositions of selected tool steels

Steel	% C	% Mn	% W	% Mo	% Cr	% V
W1 (water hardening)	0.6–1.4					
O1 (oil hardening)	0.9	1.0	0.5		0.5	
A2 (air hardening)	1.0			1.0	5.0	
S1 (shock resisting)	0.5		2.5	0.5	1.5	
D1 (chromium cold work)	1.0			1.0	12.0	
H11 (chromium hot work)	0.35			1.5	5.0	0.4
H20 (tungsten hot work)	0.35		9.0		2.0	
H41 (molybdenum hot work)	0.65		1.5	8.0	4.0	1.0
T1 (tungsten high speed)	0.75		18.0		4.0	1.0
M1 (molybdenum high speed)	0.85		1.5	8.5	4.0	1.0

Alloying elements also improve the high temperature stability of the tool steels (Figure 13-20). The water hardenable steels soften rapidly even at relatively low temperatures; oil hardenable steels temper more slowly but still soften at high temperatures. The air hardenable and special tool steels may not soften until near the A_1 temperature. In fact, the highly alloyed tool steels may pass through a *secondary hardening peak* near 500°C as the normal cementite dissolves and hard alloy carbides precipitate. The alloy carbides are particularly stable, resist growth or spheroidization, and are important in establishing the high temperature resistance of these steels.

13-11 Special Steels

There are many special categories of steels, including high strength-low alloy steels, microalloyed steels, and maraging steels.

High strength-low alloy (HSLA) steels are low-carbon steels specified on the basis of yield strength, with grades corresponding to 340 MPa to 550 MPa. The steels normally contain the least amount of alloying element that still provides the proper yield strength without heat treatment.

Microalloyed steels contain even less alloying elements than the HSLA steels and rely partly on a carefully controlled hot-rolling process to meet the minimum strength requirements.

Maraging steels are very low carbon, highly alloyed steels that derive their strength from a combination of solid solution strengthening and the precipitation hardening of a low-carbon martensite. The steels are austenitized and quenched to produce a soft martensite that contains less than 0.30% C. When the martensite is aged at about 500°C, a variety of intermetallic compounds, including Ni_3Ti, Fe_2Mo, and Ni_3Mo, may precipitate.

13-12   Surface Treatments

Low-carbon steels have low strength and hardness but good ductility and toughness, whereas high-carbon steels have the opposite behavior. We can, by proper heat treatment, produce a structure that is hard and strong at the surface, so that excellent wear and fatigue resistance are obtained, but at the same time gives a soft, ductile, tough core which provides good resistance to impact failure.

Selectively heating the surface.   We could begin by rapidly heating the surface of a medium-carbon steel above the A_3 temperature—the center remains below the A_1. After the steel is quenched, the center is still a soft mixture of ferrite and pearlite while the surface is martensite (Figure 13-26). The depth of the

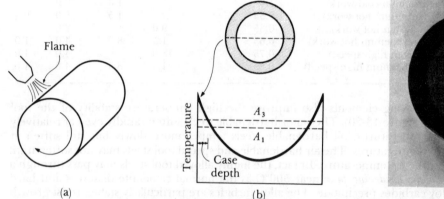 

FIG. 13-26   (a) Surface hardening by localized heating. (b) Only the surface heats above the A_1 temperature and is quenched to martensite. (c) The photograph shows the case depth in a surface-hardened 1050 steel automobile axle.

martensite layer is the *case depth*. Tempering produces the desired hardness at the surface. We can provide local heating of the surface by using a gas flame, an induction coil, a laser beam, or an electron beam. We can, if we wish, harden only selected areas of the surface that are most subject to failure by fatigue or wear.

Carburizing and nitriding.   For even better toughness, we start with a low-carbon steel, which usually contains alloying elements for improved hardenability. Carbon is diffused into the surface of the steel at a temperature above the A_3 (Figure 13-27). A high carbon content is produced at the surface due to rapid diffusion and the high solubility of carbon in austenite. When the steel is then quenched and tempered, the surface becomes a high-carbon tempered martensite, while the center remains soft and ductile. The thickness of the hardened surface, again called the *case depth*, depends on the amount of carbon present at the surface of the steel, the temperature, and the time (Figure 13-28). The case depth is much smaller in carburized steels than in flame or induction hardened steels.

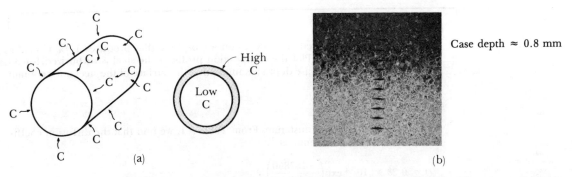

FIG. 13-27 (a) Carburizing of a low-carbon steel to produce a high-carbon wear-resistant surface. (b) The carburized case in a 1010 steel. The microhardness indentations show that the high-carbon case is harder than the low-carbon interior.

Case depth $\approx$ 0.8 mm

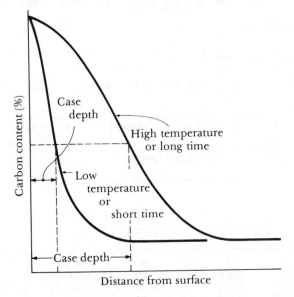

FIG. 13-28 The case depth depends on the temperature and time of carburizing.

Pack carburizing requires that the steel be surrounded by charcoal, which burns to produce an atmosphere of carbon monoxide from which carbon atoms diffuse into the steel. In *gas carburizing*, the steel is placed in a sealed furnace containing carbon monoxide. *Liquid carburizing* requires that the steel be placed in a bath of molten cyanide, which contains carbon.

Nitrogen provides a hardening effect similar to that of carbon. In *cyaniding*, the steel is immersed in a liquid cyanide bath which permits both carbon and nitrogen to diffuse into the steel. In *carbonitriding*, a gas containing carbon monoxide and ammonia is generated and both carbon and nitrogen diffuse into the steel. Finally, only nitrogen diffuses into the surface from a gas in *nitriding*. Nitriding is carried out below the A_1 temperature.

EXAMPLE 13-14

A 1010 steel is carburized at 900°C in an atmosphere that produces 1.0% C at the surface of the steel. (a) Plot the composition profile if the steel is carburized for 30 min. (b) Determine the case depth assuming that the surface of the steel must contain at least 0.40% C.

Answer:

At 900°C, the steel is all austenite. From Table 5-1, we find that the diffusion coefficient for carbon in austenite is

$$D = 0.23 \times 10^{-4} \exp\left(\frac{-137850}{RT}\right)$$

where $R = 8.314\,\text{J (mol. K)}^{-1}$ and $T = 900 + 273 = 1173$ K.

$$D = 0.23 \times 10^{-4} \exp\left[\frac{-137850}{(8.314)(1173)}\right] = 0.23 \times 10^{-4} \exp(-14.1) = 1.7 \times 10^{-11}\,\text{m}^2\,\text{s}^{-1}$$

From Equation (5-3)

$$\frac{c_s - c_x}{c_s - c_0} = \text{erf}\left(\frac{x}{2\sqrt{Dt}}\right)$$

where $c_s = 1.0\%$ C, $c_0 = 0.10\%$ C, and $t = 30$ min $= 1800$ s.

$$\frac{1 - c_x}{1 - 0.1} = \text{erf}\left[\frac{x}{2\sqrt{(1.7 \times 10^{-11})(1800)}}\right] = \text{erf}\left(\frac{x}{3.5 \times 10^{-4}}\right)$$

$$c_x = 1 - 0.9\,\text{erf}\left(\frac{x}{3.5 \times 10^{-4}}\right)$$

From Figure 5-9, the error function can be evaluated.

x	$\dfrac{x}{3.5 \times 10^{-4}}$	$\text{erf}\left(\dfrac{x}{3.5 \times 10^{-4}}\right)$	c_x
0	0	0	0
0.1×10^{-3}	0.285	0.29	0.74
0.2×10^{-3}	0.571	0.53	0.52
0.3×10^{-3}	0.857	0.70	0.37
0.4×10^{-3}	1.143	0.84	0.24
0.5×10^{-3}	1.429	0.93	0.16
1.0×10^{-3}	2.857	0.99	0.11
∞	∞	1	0.10

The results are plotted in Figure 13-29. The cash depth is 0.28 mm.

13-13 Weldability of Steel

During welding, the metal nearest the weld heats above the A_1 temperature and austenite forms (Figure 13-30). During cooling, the austenite in this heat-affected zone transforms to a new structure, dependent on the cooling rate and the CCT

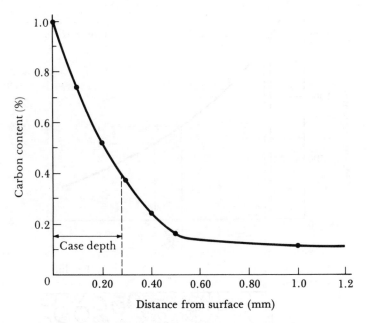

FIG. 13-29   The percent carbon versus distance below the surface of a carburized steel. (See Example 13-14.)

diagram for the steel. Plain low-carbon steels have such a low hardenability that normal cooling rates seldom produce martensite. However, an alloy steel may have to be preheated to slow up the cooling rate or postheated to temper any martensite that forms.

A steel that is originally quenched and tempered has two problems during welding. First, the portion of the heat-affected zone that heats above the A_1 may form martensite after cooling. Second, a portion of the heat-affected zone below the A_1 may overtemper. Normally, we should not weld a steel in the quenched and tempered condition.

EXAMPLE 13-15

Compare the structure in the heat-affected zones of welds in 1080 and 4340 steels, if the cooling rate in the heat-affected zone is 5°C/s.

Answer:

From the CCT diagrams, Figures 13-14 and 13-16, the cooling rate in the weld produces the following structures.

    1080:   100% pearlite

    4340:   Bainite and martensite

The high hardenability of the alloy steel reduces the weldability, permitting martensite to form and embrittle the weld.

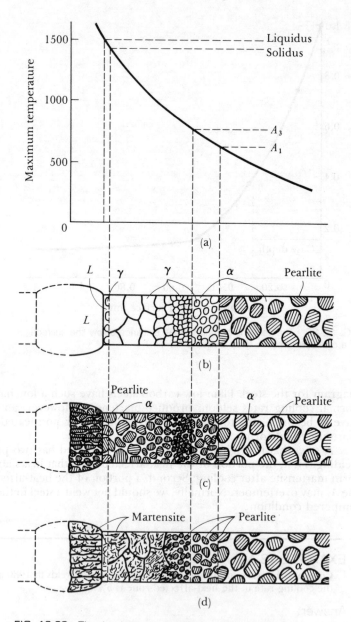

FIG. 13-30 The development of the heat-affected zone in a weld. (a) The maximum temperature at any point, (b) the structure at the maximum temperature, (c) the structure after cooling in a steel of low hardenability, and (d) the structure after cooling in a steel of high hardenability.

13-14 Stainless Steels

Stainless steels are selected for their excellent resistance to corrosion. All true stainless steels contain a minimum of about 12% Cr, which permits a thin protective surface layer of chromium oxide to form when the steel is exposed to oxygen.

There are four categories of stainless steels based on crystal structure and strengthening mechanism. Examples and characteristics of each type are included in Table 13-5.

The iron-chromium-carbon phase diagram. Figure 13-31 shows the iron-chromium phase diagram. The chromium produces a *gamma loop*. As the amount of chromium increases, the temperature range in which austenite is stable decreases until the austenite completely disappears. Thus with high chromium contents, the ferritic, or α, structure is present at all temperatures. Chromium is a *ferrite stabilizing element*.

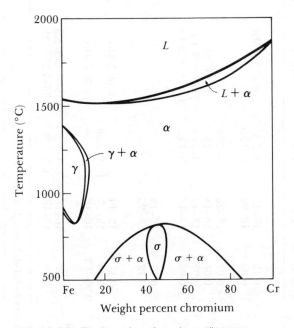

FIG. 13-31 The iron-chromium phase diagram.

Figure 13-32 illustrates the effect of chromium on the iron-carbon phase diagram. Small additions of chromium cause the austenite region to shrink while the ferrite region increases in size. For high-chromium, low-carbon compositions, ferrite is present as a single phase up to the solidus temperature.

TABLE 13-5 Compositions and properties of selected stainless steels

Steel	Typical BS grade	% C	% Cr	% Ni	Others	Tensile Strength (MPa)	Yield Strength (MPa)	% Elongation	Condition
Austenitic									
201		0.15	16–18	3.5–5.5	5.5–7.5% Mn	655	310	40	Annealed
304	304S15	0.08	18–20	8.0–10.5		517	207	30	Annealed
						1276	966	9	Cold worked
304L	304S12	0.03	18–20	8–12		517	207	30	Annealed
316	316S16	0.08	16–18	10–14	2–3% Mo	517	207	30	Annealed
321	321S12	0.08	17–19	9–12	Ti (5 × % C)	586	241	55	Annealed
347	347S17	0.08	17–19	9–13	Nb (10 × % C)	621	241	50	Annealed
Ferritic									
430	430S15	0.12	16–18			448	207	22	Annealed
442	442S19	0.12	18–23			517	276	20	Annealed
Martensitic									
416	416S21	0.15	12–14		0.60% Mo	1241	966	18	Quenched and tempered
431	431S29	0.20	15–17	1.25–2.50		1379	1035	16	
440C	—	0.95–1.2	16–18		0.75% Mo	1966	1897	2	
Precipitation hardening									
17-4		0.07	16–18	3–5	0.15–0.45% Nb	1310	1172	10	Age hardened
17-7		0.09	16–18	6.5–7.8	0.75–1.25% Al	1655	1586	6	

Adapted from *ASM Metals Handbook, Vol. 3, 9th Ed.,* 1980.

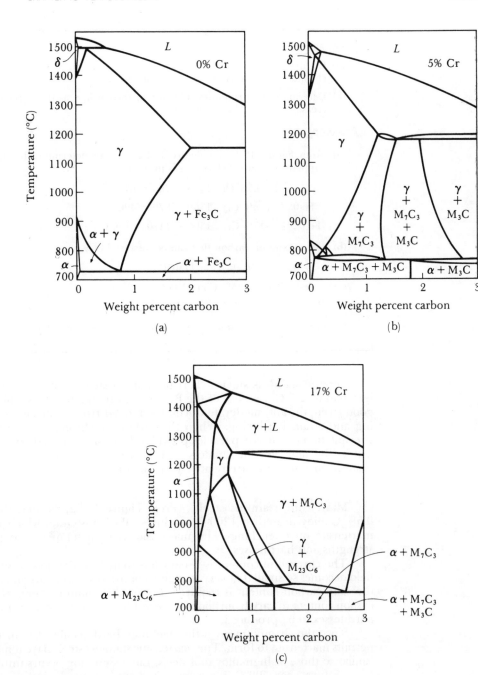

FIG. 13-32 The effect of chromium on the iron-carbon phase diagram, shown as isopleths at (a) 0% Cr, (b) 5% Cr, and (c) 17% Cr.

EXAMPLE 13-16

(a) Compare the temperature range over which only austenite is stable as the percent chromium increases in a 0.5% C stainless steel.

(b) Determine the maximum solubility of carbon in γ as the percent chromium increases.

Answer:

(a) From Figure 13-1 and Figure 13-32, we can determine the upper and lower temperatures at which only γ is present.

$$\text{Fe-0.5\% C:} \quad 1435 - 760 = 675°C$$
$$\text{Fe-0.5\% C-5\% Cr:} \quad 1360 - 870 = 490°C$$
$$\text{Fe-0.5\% C-17\% Cr:} \quad 1300 - 1170 = 130°C$$

(b) The maximum carbon that can be dissolved in γ is

$$\text{Fe-0\% Cr:} \quad 1.7\% \text{ C at } 1130°C$$
$$\text{Fe-5\% Cr:} \quad 1.25\% \text{ C at } 1195°C$$
$$\text{Fe-17\% Cr:} \quad 0.65\% \text{ C at } 1250°C$$

Both of these results indicate that chromium is a ferrite stabilizing element.

Ferritic stainless steels. Ferritic stainless steels contain up to 30% Cr and less than 0.12% C. Because of the BCC structure, the ferritic stainless steels have good strengths and moderate ductilities derived from solid solution strengthening and strain hardening. When the carbon or chromium contents are high, precipitation of carbide particles provides dispersion strengthening but also embrittles the alloy. Ferritic stainless steels have excellent corrosion resistance, moderate formability, and are relatively inexpensive.

Martensitic stainless steels. From Figure 13-32, we find that a 17% Cr-0.5% C alloy heated to 1200°C produces 100% austenite, which transforms to martensite on quenching. The martensite is then tempered to produce high strengths and hardnesses.

The chromium content is usually less than 17% Cr; otherwise the austenite field becomes so small that very stringent control over both austenitizing temperature and carbon content is required. Lower chromium contents also permit the carbon content to vary from about 0.1% to 1.0%, allowing martensites of different hardnesses to be produced.

Since the chromium gives the steel high hardenability, an air or oil quench permits martensite to form. The martensitic stainless steels have tempering curves similar to those of high-alloy tool steels. Little softening occurs until a tempering temperature near 500°C is reached. A secondary hardening peak may be observed if alloy carbides form (Figure 13-33). Figure 13-34 includes the structure of a typical martensitic stainless steel.

The low chromium content and the presence of two phases cause tempered martensitic stainless steels to have less corrosion resistance than the other stainless steels. However, the combination of hardness, strength, and corrosion resistance

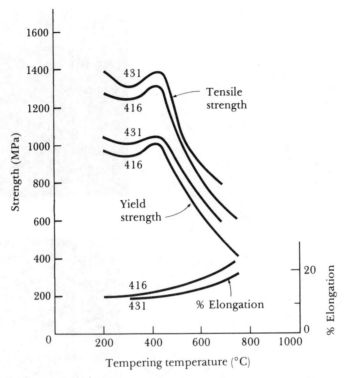

FIG. 13-33 Tempering curves for two martensitic stainless steels.

makes the alloys attractive for many applications, including high-quality knives, ball bearings, and valves.

Austenitic stainless steels. Nickel, which is an *austenite stabilizing element,* increases the size of the austenite field while nearly eliminating ferrite from the iron-chromium-carbon alloys (Figure 13-35). If the carbon content is below about 0.03%, the carbides do not form and the steel is virtually all austenite at room temperature (Figure 13-34).

The FCC austenitic stainless steels have excellent ductility, formability, and corrosion resistance. Strength is obtained by extensive solid solution strengthening, and the austenitic stainless steels may be cold worked to higher strengths than the ferritic stainless steels. The steels have excellent low-temperature impact properties, since they have no transition temperature. Furthermore, the austenitic stainless steels are not ferromagnetic. Unfortunately, the high nickel and chromium contents make the alloys expensive.

Austenitic stainless steels that contain more than 0.03% C may become *sensitized* to intergranular corrosion. When the steel cools slowly between about 870°C and 425°C, chromium carbide precipitates at the grain boundaries. Corrosion then occurs in these grain boundary regions. A *quench anneal* heat treatment can be used to prevent corrosion. The sensitized steel is heated above 870°C until the chromium carbides dissolve, then quenched rapidly to prevent the carbides from

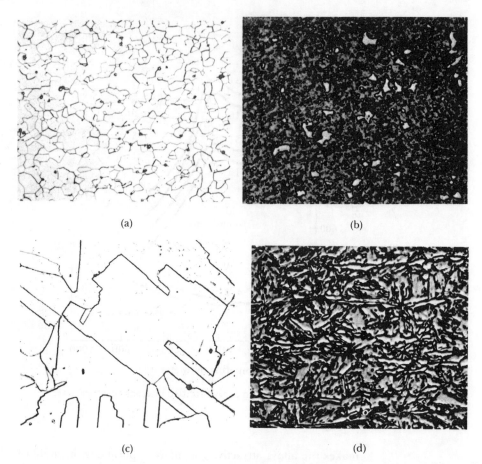

FIG. 13-34   (a) Annealed ferritic stainless steel. (b) Martensitic stainless steel containing large primary carbides and small carbides formed during tempering. (c) Austenitic stainless steel. (d) Precipitation-hardened stainless steel with ferrite plates in a martensite matrix. From *Metals Handbook*, Vols. 7 and 8, 8th ed., American Society for Metals, 1972, 1973.

reforming. An alternative solution is to add titanium or niobium to the steel. Carbon combines with the titanium or niobium, producing TiC or NbC in preference to the chromium. This is called *stabilization*.

EXAMPLE 13-17

Describe a simple test to separate high-nickel stainless steel from low-nickel stainless steel.

Answer:

The high-nickel stainless steels are austenitic, whereas the low-nickel stainless steels are probably ferritic or martensitic. An ordinary magnet will be attracted to the low-nickel ferritic and martensitic steels but will not be attracted to the high-nickel austenitic steel.

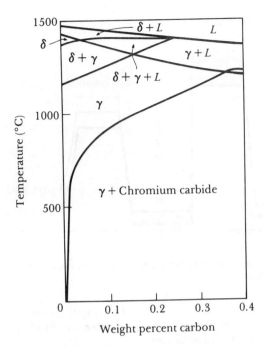

FIG. 13-35 A section of the iron-chromium-nickel-carbon phase diagram at a constant 18% Cr-8% Ni.

EXAMPLE 13-18

Which of the stainless steels would you select for a pump used to transport liquid helium at 4 K?

ANSWER:

Because of the extremely low temperature, a material with good low-temperature properties is necessary. The austenitic stainless steels might serve best, since they do not have a ductile-brittle transition temperature and thus have better low-temperature toughness.

Precipitation hardening (PH) stainless steels. The composition of the precipitation hardening, or PH, stainless steels is similar to the austenitic stainless steels except for the presence of aluminum, niobium, or tantalum. The PH stainless steels derive their properties from solid solution strengthening, strain hardening, age hardening, and the martensitic reaction. High mechanical properties are obtained even with low carbon contents.

A heat treatment for a typical 17-7 PH stainless steel is shown in Figure 13-36. After fabrication of the steel in the annealed condition, three steps are required.

1. Conditioning. This step, done at 760°C to 955°C, prepares the austenite for subsequent transformation to martensite.

2. Quenching and transformation. The steel is cooled to 15°C or below to permit austenite to transform to martensite.

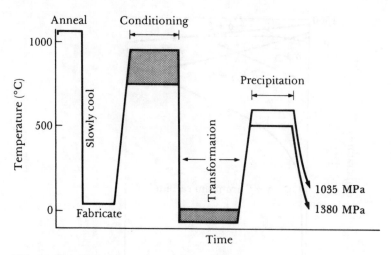

FIG. 13-36   Schematic diagram showing the steps in the heat treatment of a PH stainless steel.

3. Precipitation. The steel is reheated to 500°C to 600°C, permitting Ni_3Al and other precipitates to form from the martensite. Higher strengths are obtained with lower aging temperatures.

CAST IRONS

13-15 Solidification of Cast Irons

Cast irons are iron-carbon-silicon alloys, typically containing 2% to 4% C and 0.5% to 3% Si, that pass through the eutectic reaction during solidification.

The microstructures of the five important types of cast irons are shown in Figure 13-37. *Gray cast iron* contains small, interconnected graphite flakes that cause low strength and ductility. *White cast iron* is a hard, brittle, unmachinable alloy containing massive amounts of Fe_3C. *Malleable cast iron* is produced by the heat treatment of white iron, causing irregular but rounded clumps of graphite to precipitate. This graphite form permits good strength, ductility and toughness in the iron. *Ductile* or *nodular cast iron* contains spheroidal graphite particles obtained during solidification by the addition of small amounts of magnesium to the molten iron. Properties are similar to those of malleable iron. *Compacted graphite cast iron* contains rounded but interconnected (vermicular) graphite also produced during solidification by the addition of magnesium. The structure and properties are intermediate between gray and ductile irons.

To understand the origin of these cast irons, we must examine the phase diagram, solidification, and phase transformations of the alloys.

The iron-carbon-silicon phase diagram. Figure 13-38 shows the equilibrium between iron and graphite as solid lines. The dashed lines refer to the metastable relationship between iron and cementite. Although the iron-graphite lines represent the stable reactions, nucleation of graphite in iron-carbon alloys is so difficult that we almost always obtain the metastable reactions instead.

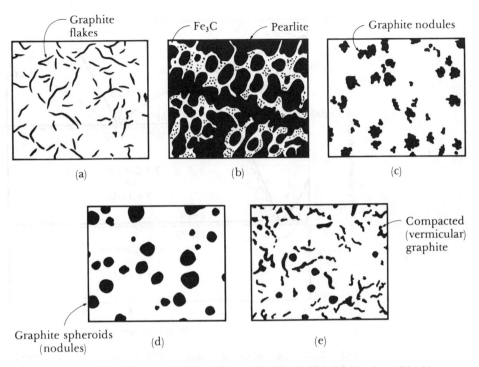

FIG. 13-37   Schematic drawings of the five types of cast irons: (a) gray iron, (b) white iron, (c) malleable iron, (d) ductile iron, and (e) compacted graphite iron.

However, when we add silicon to the iron, the temperature difference between the stable and metastable eutectic reactions increases (Figure 13-39). When the liquid iron-carbon-silicon alloy cools, there is a large temperature interval during which the stable iron-graphite eutectic reaction can nucleate and grow; thus silicon is a *graphite stabilizing element.*

Silicon also reduces the amount of carbon in the eutectic. A portion of the silicon can be treated as carbon, permitting us to define a *carbon equivalent* (CE).

$$CE = \% \ C + \tfrac{1}{3}\% \ Si \qquad\qquad\qquad\qquad (13\text{-}2)$$

The eutectic composition is always near 4.3% CE. Any cast iron with a carbon equivalent of less than 4.3% is hypoeutectic; any cast iron with a carbon equivalent greater than 4.3% is hypereutectic.

EXAMPLE 13-19

Determine whether a cast iron containing 3.6% C and 2.4% Si is hypoeutectic or hypereutectic.

Answer:

$$CE = \% \ C + \frac{\% \ Si}{3} = 3.6 + \frac{2.4}{3} = 4.4$$

Since the CE is greater than 4.3, the iron is hypereutectic.

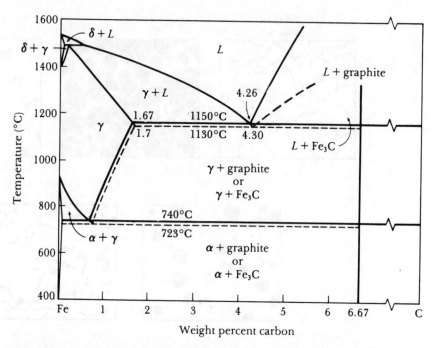

FIG. 13-38   The iron-carbon phase diagram showing the relationship between the stable iron-graphite equilibria (solid lines) and the metastable iron-cementite reactions (dashed lines).

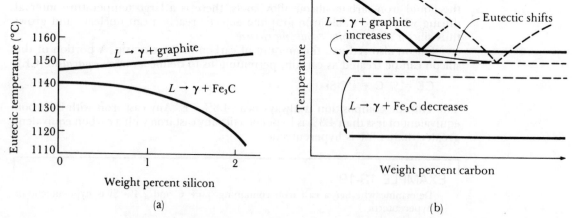

FIG. 13-39   The effect of silicon on (a) the eutectic temperature and (b) the phase diagram. Increasing silicon increases the difference between the two eutectic temperatures and reduces the percent carbon in the eutectic.

The eutectic reaction. We can relate the temperature at which the eutectic reaction occurs to the type of iron that forms: if solidification occurs between the two eutectic temperatures, a graphite-containing cast iron, such as gray, ductile, or compacted graphite iron, is produced. If solidification occurs below the lower eutectic temperature, a white cast iron forms (Figure 13-40). Some of the conditions that favor each of the two eutectic reactions are shown in Table 13-6.

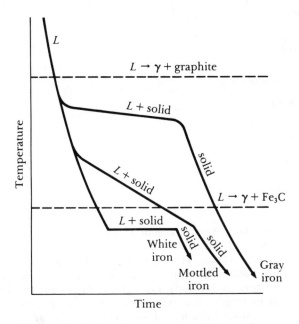

FIG. 13-40 Cooling curves superimposed on the eutectic temperatures of cast iron. If solidification occurs above the lower eutectic, graphite precipitates; if solidification occurs at or below the lower eutectic, Fe_3C forms. Both graphite and Fe_3C form in mottled iron.

EXAMPLE 13-20

Estimate the amount of graphite and the amount of Fe_3C in the stable and metastable eutectic microconstituents.

Answer:

If we assume a pure iron-carbon alloy, the lever laws are

$$\text{Graphite} = \frac{4.26 - 1.67}{100 - 1.67} \times 100 = 2.6\%$$

$$Fe_3C = \frac{4.30 - 1.70}{6.67 - 1.70} \times 100 = 52.3\%$$

When Fe_3C forms, it becomes the continuous phase in the eutectic microconstituent, making the white iron hard and brittle.

TABLE 13-6   Conditions favoring each of the eutectic reactions in cast irons

---

A. Formation of gray iron, ductile iron, compacted graphite iron

$$L \rightarrow \gamma + Gr$$

1. High CE
2. High silicon
3. Cu, Ni
4. Slow cooling
5. Thick castings
6. Inoculation

B. Formation of white iron

$$L \rightarrow \gamma + Fe_3C$$

1. Low CE
2. Low silicon
3. Cr, Mo, V, Bi, Te, others
4. Fast cooling
5. Thin castings
6. No inoculation

In some cases, both eutectic reactions may occur. *Mottled iron* begins to solidify above the lower eutectic temperature but completes solidification below that temperature, Figure 13-40. The mixture of white and gray cast iron, Figure 13-41(a), is undesirable under any conditions. *Chilled iron* is produced when the surface of the casting cools rapidly enough to cause white cast iron to form while the center of the casting, which cools more slowly, forms gray cast iron. This iron, Figure 13-41(b), is sometimes useful for inexpensive wear-resistant components.

The depth to which white cast iron forms can serve as a simple method for checking the carbon equivalent of a cast iron. By pouring specially designed castings which are subsequently broken, the *chill depth*, or depth of white iron

Chill depth ≈ 6 mm

(a)   (b)

FIG. 13-41   (a) Mottled iron containing graphite flakes, Fe_3C (white), and pearlite (gray). (b) Fracture surface of chilled iron, showing white iron at the rapidly cooled end of the casting.

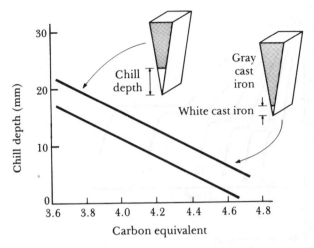

FIG. 13-42 The relationship between chill depth and carbon equivalent.

below the surface, can be measured and related to the carbon equivalent (Figure 13-42).

EXAMPLE 13-21

Liquid cast iron is poured into a mold that is cooled rapidly at one surface. Cooling curves from selected locations are shown in Figure 13-43. Predict the chill depth. The eutectic temperatures for this alloy are 1161°C for $L \rightarrow \gamma$ + graphite and 1131°C for $L \rightarrow \gamma$ + Fe$_3$C.

Answer:

The cooling curves at 3 mm and 6 mm from the surface show a thermal arrest at 1131°C, so white iron forms. The remaining locations have an arrest between 1131°C and 1161°C, indicating that gray iron forms. By plotting the arrest temperature versus distance, we can estimate the chill depth to be about 8 mm.

EXAMPLE 13-22

Inexpensive scissors, with a hard cutting edge, are sometimes made from gray iron. Describe how the appropriate hardness is obtained.

Answer:

The cross section of the blade of the scissors is a wedge. If the carbon equivalent is correctly adjusted, the tip of the wedge cools rapidly enough to produce white iron, while the remainder of the blade forms gray iron. The white iron, at the cutting surface, is hard and can be sharpened.

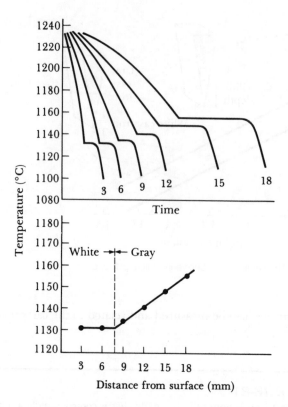

FIG. 13-43   Cooling curves and thermal arrest temperatures used to determine the chill depth for Example 13-21.

13-16   The Matrix Structure in Cast Irons

The matrix structure and properties of each type of cast iron are determined by how the austenite transforms during the eutectoid reaction. Under equilibrium conditions, austenite transforms to graphite and ferrite

$$\gamma \rightarrow \alpha + \text{graphite}$$

Carbon atoms diffuse from the austenite to existing graphite particles produced during solidification, leaving behind the low-carbon ferrite. The stable transformation to ferrite and graphite is encouraged if the cooling rate is slow (permitting time for diffusion) and if the graphite particles are fine and closely spaced (giving shorter diffusion distances). The reaction will not occur unless graphite is produced during solidification.

The transformation diagram in Figure 13-44 describes how the austenite might transform during heat treatment. *Annealing*, which requires furnace cooling after austenitizing, produces a slow cooling rate and a ferritic matrix. *Normalizing*, or air cooling, gives faster cooling and a pearlitic matrix. The cast irons can also be quenched and tempered, isothermally transformed to produce bainite (austempering), or surface hardened.

The matrix structure can also be controlled by the composition of the iron. Ferrite is encouraged by higher silicon contents. However, most alloying elements, even 0.05% Sn or 0.50% Cu, favor the formation of pearlite.

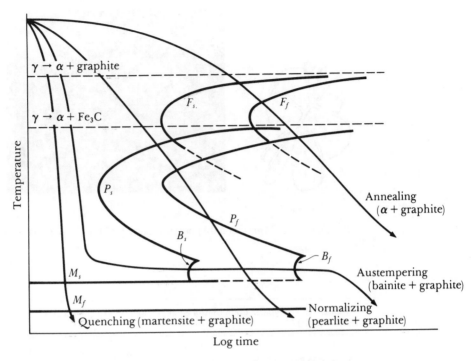

$\gamma \rightarrow \alpha + $ graphite

$\gamma \rightarrow \alpha + Fe_3C$

F_s.

F_f

P_s

P_f

B_s

B_f

Annealing
(α + graphite)

Austempering
(bainite + graphite)

M_s

M_f

Normalizing
(pearlite + graphite)

Quenching (martensite + graphite)

Log time

Temperature

FIG. 13-44   The transformation diagram for austenite in a cast iron.

13-17 Characteristics and Production of the Cast Irons

In order to produce each of the five important types of cast iron, we must carefully control the eutectic solidification, often by adding modifiers to encourage proper eutectic growth, as well as the eutectoid transformation or heat treatment of the iron.

Gray cast iron.   Gray iron is the most common of the cast irons. Solidification produces interconnected graphite flakes, which resemble a number of potato chips glued together at a single location (Figure 13-45). The point at which the flakes are connected is the original graphite nucleus. The gray cast iron contains many such clusters, or *eutectic cells*, of graphite flakes, with each cell representing one nucleation event. Inoculation or rapid cooling rates help produce finer graphite flakes with a smaller eutectic cell size, thus improving strength.

The graphite flakes, which resemble tiny cracks within the cast iron structure, concentrate stresses so that the gray iron possesses a low tensile strength and behaves in a brittle manner, with elongations of only 1% or less.

The gray irons are normally specified by a grade number that gives the minimum tensile strength of the iron (Table 13-7) in a 30 mm diameter test bar. A grade 180 iron would therefore have a minimum tensile strength of 180 MPa. Figure 13-46 illustrates that for any grade of iron the tensile strength can vary with

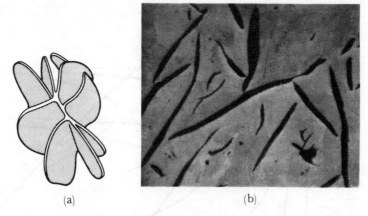

(a)   (b)

FIG. 13-45   (a) Sketch and (b) photomicrograph of the flake graphite in gray cast iron.

TABLE 13-7   Typical compositions and properties of gray irons
BS1452:1977

Grade/MPa	Approx. C.E.	0.2% Proof stress MPa	Elongation %	Un-notched impact J
150	4.5	100	0.6	8–13
180	4.3	120	0.5	8–13
220	4.1	140	0.4	8–16
260	3.85	175	0.5	13–23
300	3.65	205	0.5	16–31
350	3.5	235	0.5	24–47
400	3.4	270	0.5	24–47

cooling rate and/or casting size. If cooling is too fast, carbides may form.

At lower carbon equivalents, the nominal tensile strength of the cast iron is higher, since a smaller amount of graphite forms during solidification. Even higher strengths can be obtained by alloying or heat treatment.

In spite of its low tensile strength and ductility, gray iron has a number of attractive properties. The flakes do not act as stress raisers under compressive loading, so with proper design gray iron can carry large loads. The machinability of gray iron is excellent, since the graphite flakes act as chip breakers. Resistance to sliding wear is good; the porous graphite flakes absorb and hold lubricant and, because graphite is soft and slippery, may even provide self-lubrication. Vibration damping characteristics are exceptional, particularly when the flakes are coarse. Because of this characteristic, gray iron is used for engine blocks and machine tool bases. Gray iron does not make a very good bell, however.

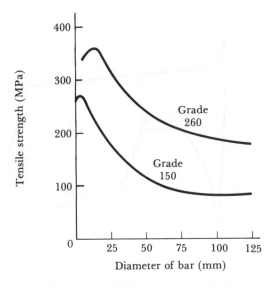

FIG. 13-46   The effect of cooling rate or casting size on the tensile properties of two gray cast irons.

White cast iron.   Low carbon equivalent white irons containing about 2.5% C and 1.5% Si are an intermediate product in the manufacture of malleable iron, as will be discussed in the next section. A group of highly alloyed white irons are used for their hardness and resistance to abrasive wear. Elements such as chromium, nickel, and molybdenum are added so that, in addition to the alloy carbides formed during solidification, martensite can be produced during subsequent heat treatment. White cast irons are shown in Figure 13-47.

Malleable cast iron.   Malleable iron is produced by heat treating unalloyed white iron. During the malleablizing heat treatment, the cementite formed during solidification is decomposed and graphite clumps, or nodules, are produced. The nodules, or temper carbon, often resemble popcorn. The rounded graphite shape permits the malleable iron to have a good combination of strength and ductility.

The production of malleable iron requires several steps (Figure 13-48).

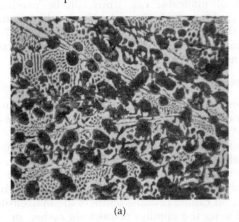

(a)

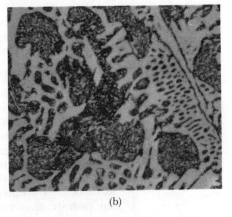

(b)

FIG. 13-47   (a) Normal white iron containing Fe_3C (white) and pearlite. (b) Martensitic white iron with Fe_3C (white) and a matrix of martensite needles and retained austenite.

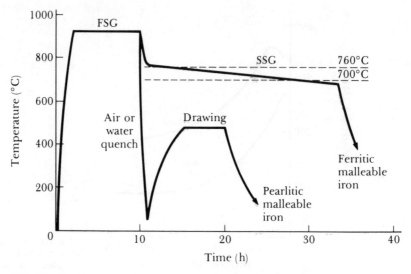

FIG. 13-48   The heat treatment for ferritic and pearlitic malleable iron.

1. A white cast iron must be produced that decomposes readily during heat treatment. A carbon equivalent of about 3% keeps graphite flakes from forming during solidification yet permits Fe_3C to decompose in short times.

2. Nucleation of the graphite nodules occurs as the white cast iron is slowly heated to the malleablizing temperature.

3. *First stage graphitization* (FSG) at about 925°C decomposes the cementite to the stable austenite and graphite phases.

$$Fe_3C \rightarrow \gamma + graphite \tag{13-3}$$

The carbon in Fe_3C diffuses to the graphite nuclei produced during heating, leaving behind an austenite matrix.

4. The austenite decomposes during subsequent cooling from the FSG temperature. Two types of malleable cast irons can be produced—ferritic and pearlitic—Figure 13-49. To make *ferritic malleable iron,* the casting is cooled at 5°C/h to 15°C/h through the eutectoid temperature range, about 760°C to 700°C, to cause *second stage graphitization* (SSG). The austenite transforms to ferrite, with the excess carbon diffusing to the existing graphite nodules.

$$\gamma \rightarrow \alpha + graphite \tag{13-4}$$

Ferritic malleable iron has exceptional toughness compared to other irons because its low silicon content reduces the transition temperature below room temperature.

Pearlitic malleable iron is obtained when the austenite is cooled in air or oil. The iron is called pearlitic malleable whether the matrix contains pearlite from air cooling or martensite from the oil quench. In either case, the matrix is hard and brittle. To improve the ductility, the pearlitic malleable iron is drawn at a temperature below the eutectoid range. *Drawing* tempers the martensite or spheroidizes the pearlite, thus reducing the amount of *combined carbon,* or cementite. As the amount of combined carbon decreases, the strength of the pearlitic malleable iron decreases and ductility and toughness increase. Some malleable iron grades are shown in Table 13-8.

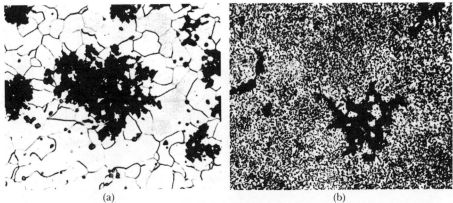

(a)   (b)

FIG. 13-49 (a) Ferritic malleable iron with graphite nodules and small MnS inclusions in a ferrite matrix. (b) Pearlitic malleable iron drawn to produce a tempered martensite matrix. From *Metals Handbook*, Vols. 7 and 8, 8th Ed., American Society for Metals, 1972, 1973.

EXAMPLE 13-23

Much longer heat treatment times are required to make malleable iron when the percent chromium exceeds 0.1%. Explain.

Answer:

The chromium is an excellent carbide stabilizer (Table 13-6). Consequently, the decomposition of the cementite is slow because the cementite is relatively stable.

TABLE 13-8 Grades of malleable cast iron

Grade	Minimum Tensile Strength (MPa)	Minimum 0.2% P.S. (MPa)	Minimum Elongation (%)	Matrix structure	
B 340/12	340	221	12	Graphite aggregate in ferrite	} Blackheart
B 310/10	310	202	10		
W 340/3	340	187	3	Ferrite case with core of pearlite and graphite	} Whiteheart
W 410/4	410	226	4		
P 440/7	440	274/277	7	Graphite aggregates in pearlite or tempered martensite	} Pearlitic
P 540/5	540	310/357	5		

BS310:1972 Blackheart malleable iron castings
BS309:1972 Whiteheart malleable iron castings
BS3333:1972 Pearlitic malleable iron castings

Ductile or nodular cast iron. Ductile iron is produced by treating a relatively high carbon equivalent liquid iron with magnesium or cerium, causing spheroidal graphite to grow during solidification. Several steps are required to produce this iron (Figure 13-50).

FIG. 13-50 Schematic diagram of the treatment of ductile iron.

1. *Desulfurization*. Sulfur causes graphite to grow in the flake rather than the spheroidal form. Low-sulfur irons are obtained by melting low-sulfur, high-quality charge materials; melting in furnaces that remove sulfur from iron during melting; or by mixing the iron with a desulfurizing agent such as CaO.

2. *Nodulizing*. Magnesium, added in the nodulizing step, removes any sulfur and oxygen still in the liquid metal and provides a residual of 0.03% Mg which causes growth of spheroidal graphite. The magnesium is added near 1500°C. Unfortunately, magnesium vaporizes near 1150°C. Many nodulizing alloys contain magnesium diluted with ferrosilicon to reduce the violence of the reaction and permit higher magnesium recoveries. Special techniques also are required to keep the lightweight nodulizing alloy below the surface of the metal.

EXAMPLE 13-24

A 400 kg ladle of low-sulfur molten iron is treated with 7 kg of a ferrosilicon nodulizing alloy containing 9% Mg. The final analysis shows that 0.035% Mg is dissolved in the iron. What fraction of the magnesium added was vaporized or oxidized?

Answer:

The amount of magnesium added is

$$kg\ Mg = (0.09)(7) = 0.63$$

The percent magnesium added to the iron is

$$\frac{0.63}{400} \times 100 = 0.16\%$$

The percent loss of magnesium is

$$\frac{0.16 - 0.035}{0.16} \times 100 = 78\%$$

This loss is typical during nodulizing. The recovery of the magnesium is only 100 − 78 = 22%.

Fading, or gradual, nonviolent vaporization or oxidation of the magnesium, must also be controlled. If the iron is not poured within a few minutes after nodulizing, the iron reverts to gray iron.

3. *Inoculation.* Magnesium is an effective carbide stabilizer; consequently nodulizing causes white iron to form during solidification. Although any graphite that nucleates is conditioned by the magnesium to grow as spheres, practically no graphite forms! Consequently, we must inoculate the iron after nodulizing. Inoculants are usually ferrosilicon alloys containing 50% to 85% Si and small amounts of calcium, aluminum, strontium, or barium which produce nucleation sites for graphite. The inoculant effect also fades with time.

Compared to gray iron, ductile cast iron has excellent strength, ductility, and toughness. Ductility and strength are also higher than in malleable irons, but due to the higher silicon content in ductile iron, the toughness may be slightly lower. Typical structures of ductile iron are shown in Figure 13-51 and properties are described in Table 13-9.

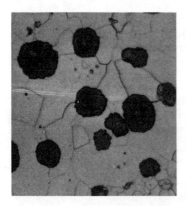
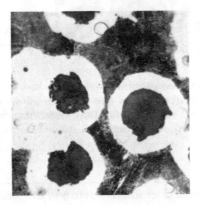

(a)   (b)   (c)

FIG. 13-51   (a) Annealed ductile iron with a ferrite matrix. (b) As-cast ductile iron with a matrix of ferrite (white) and pearlite. (c) Normalized ductile iron with a pearlite matrix.

TABLE 13-9   Grades of ductile cast iron

Grade	Tensile Strength (min) (MPa)	0.2% Proof stress (min) (MPa)	Elongation %	Matrix	Heat treatment
370/17	370	230	17	Ferrite	Annealed
420/12	420	250	12	Mainly ferrite	Annealed
500/7	500	310	7	Ferrite and pearlite	Annealed
800/2	800	460	2	Tempered martensite	Quenched and tempered

BS2789:1973 Iron castings with spheroidal or nodular graphite.

Compacted graphite cast iron.   Compacted graphite cast iron is obtained by undernodulizing the molten metal, producing a residual of about 0.015% Mg. The graphite shape is intermediate between flakes and spheres, with numerous rounded rods of graphite that are interconnected to the nucleus of the eutectic cell (Figure 13-52). This graphite, sometimes called *vermicular* graphite, also forms when ductile iron fades.

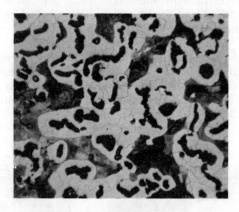

FIG. 13-52   The structure of compacted graphite cast iron.

The compacted graphite permits the cast iron to have strengths and ductilities that exceed that of gray cast iron but allows the iron to retain good thermal conductivity and vibration damping characteristics (Table 13-10). Furthermore, the ability to produce castings with smaller risers, which is typical of gray iron, is retained.

TABLE 13-10   Comparison of compacted graphite cast iron to gray and ductile cast iron

Cast Iron	Tensile Strength (MPa)	Yield Strength (MPa)	% Elongation	Relative Damping Capacity
Gray iron	140–490		<1.5	1
Compacted graphite iron				0.6
Grade A (<10% pearlite)	280	190	5	
Grade B (20–80% pearlite)	350	280	1	
Grade C (>80% pearlite)	450	380	1	
Ductile iron	420–830	280–620	2–18	0.3

The treatment for the compacted graphite iron is similar to that of ductile iron. Normally, low-sulfur base metals are required, the magnesium is introduced in dilute form in special alloys containing titanium, the metal must be poured shortly after treatment to avoid fading, and the iron must be inoculated to assure that graphite rather than cementite forms.

SUMMARY

At first glance, steels appear to be an exceptionally varied and complicated group of alloys. However, the properties of steels are determined primarily by the amount, size, shape, and distribution of the cementite, which in turn are controlled by heat treatment. Spheroidizing, which produces large spheroidal cementite, gives the softest, most ductile steel. Annealing gives a coarse pearlitic structure containing lamellar cementite. Normalizing produces a finer pearlite and improved strength. Austempering (giving bainite) or quench and temper heat treatments (giving tempered martensite) produce exceptionally fine dispersions of cementite. The most important function of alloying elements is to improve our ability to perform these heat treatments or to help make the structures that are produced more stable at high temperatures.

Stainless steels, which must contain a minimum of 12% Cr, are selected for their excellent corrosion resistance. By proper control of alloying elements and heat treatment, a variety of structures and properties can be produced.

Cast irons undergo the eutectic reaction during solidification. Depending on the reaction, either cementite and austenite or graphite and austenite form. The shape of the graphite is controlled by the addition of modifiers, such as magnesium and inoculants, producing spheroidal, compacted, or flake graphite cast irons. Cementite is decomposed by heat treatment to produce malleable iron. Furthermore, the decomposition of austenite to ferrite, pearlite, bainite, or martensite can be achieved by controlling alloying elements, cooling rate, or heat treatment.

GLOSSARY

Annealing (cast iron) A heat treatment used to produce a ferrite matrix in a cast iron by austenitizing, then furnace cooling.

Annealing (steel) A heat treatment used to produce a soft, coarse pearlite in a steel by austenitizing, then furnace cooling.

Ausforming A thermomechanical heat treatment in which austenite is plastically deformed below the A_1 temperature, then permitted to transform to bainite or martensite.

Austempering The isothermal heat treatment by which austenite transforms to bainite.

Austenitizing Heating a steel or cast iron to a temperature where homogeneous austenite can form. Austenitizing is the first step in most of the heat treatments for steel and cast irons.

Carbon equivalent Carbon plus one-third of the silicon in a cast iron.

Carburizing A group of surface-hardening techniques by which carbon diffuses into a steel. Some related processes by which nitrogen is introduced include carbonitriding, cyaniding, and nitriding.

Case depth The depth below the surface of a steel to which hardening occurs by surface hardening and carburizing processes.

Cast iron Ferrous alloys containing sufficient carbon so that the eutectic reaction occurs during solidification.

Chill depth The distance from the surface in which the cast iron freezes according to the cementite reaction rather than the graphite reaction.

Chilled iron Cast iron designed so that the surface is hard, wear-resistant white iron while the center is softer, tougher gray or ductile iron.

Compacted graphite cast iron A cast iron treated with small amounts of magnesium and titanium to cause graphite to grow during solidification as an interconnected, coral-shaped precipitate, giving properties midway between gray and ductile iron.

Drawing   Reheating a malleable iron in order to reduce the amount of carbon combined as cementite by spheroidizing pearlite, tempering martensite, or graphitizing both.

Ductile cast iron   Cast iron treated with magnesium or cerium to cause graphite to precipitate during solidification as spheres, permitting excellent strength and ductility. Also known as nodular cast iron.

Eutectic cell   A cluster of graphite flakes produced during solidification which are all interconnected back to a common nucleus.

Fading   The loss of the nodulizing or inoculating effect in cast irons as a function of time, permitting undesirable changes in microstructure and properties.

Ferrous alloys   Metal alloys based primarily on iron, including steel, stainless steel, and cast iron.

First stage graphitization   The first step in the heat treatment of a malleable iron, during which the massive carbides formed during solidification are decomposed to graphite and austenite.

Gamma loop   A number of alloying elements in iron, including chromium, reduce the temperature range over which austenite is stable, thus producing a loop in the phase diagram.

Gray cast iron   Cast iron which, during solidification, permits graphite flakes to grow, causing low strength and poor ductility.

Hardenability   The ease with which a steel can be quenched to form martensite. Steels with high hardenability form martensite even on slow cooling.

Hardenability curves   Graphs showing the effect of cooling rate on the hardness of a steel.

Inoculation   The addition of an agent to the molten cast iron that provides nucleation sites at which graphite precipitates during solidification.

Isothermal annealing   Heat treatment of a steel by austenitizing, cooling rapidly to a temperature between the A_1 and the nose of the IT curve, and holding until the austenite transforms to pearlite.

Jominy distance   The distance from the quenched end of a Jominy bar. The Jominy distance is related to the cooling rate.

Jominy test   The test used to evaluate hardenability. An austenitized steel bar is quenched at one end only, thus producing a range of cooling rates along the bar.

Malleable cast iron   Cast iron obtained by a lengthy heat treatment during which cementite decomposes to produce rounded clumps of graphite. Good strength and ductility are obtained as a result of this structure.

Marquenching   Quenching austenite to a temperature just above the M_s and holding until the temperature is equalized throughout the steel before further cooling to produce martensite. This process reduces residual stresses and quench cracking. Also known as *martempering*.

Mottled iron   Cast iron produced when both cementite and graphite precipitate during solidification. Mottled iron is not desirable under any conditions.

Nodulizing   The addition of magnesium to molten cast iron to cause the graphite to precipitate as spheres rather than as flakes during solidification.

Normalizing   A simple heat treatment obtained by austenitizing and air cooling to produce a fine pearlitic structure.

Process anneal   A low-temperature heat-treatment used to eliminate all or part of the effect of cold working.

Quench anneal   The heat treatment used to control carbide precipitation and intergranular corrosion in austenitic stainless steels. The iron is heated to dissolve carbides, then rapidly cooled to prevent their reformation.

Retained austenite   Austenite that is unable to transform into martensite during quenching because of the volume expansion associated with the reaction.

Second stage graphitization   The second step in the heat treatment of malleable irons that are to have a ferritic matrix. The iron is cooled slowly from the first stage graphitization temperature so that austenite transforms to ferrite and graphite rather than pearlite.

Sensitization Precipitation of chromium carbides at the grain boundaries in stainless steels, thus making the steel sensitive to intergranular corrosion.

Spheroidite A microconstituent containing coarse spheroidal cementite particles in a matrix of ferrite, permitting excellent machining characteristics in high carbon steels.

Stabilization The addition of titanium or niobium to stainless steels. The titanium and niobium combine preferentially with carbon, thus preventing the precipitation of chromium carbides and intergranular corrosion.

Stainless steels A group of ferrous alloys that contain at least 12% Cr, providing extraordinary corrosion resistance.

Tempered martensite The mixture of ferrite and cementite formed when martensite is tempered.

Tool steels A group of high-carbon steels which provide combinations of high hardness, toughness, or resistance to elevated temperatures.

Vermicular graphite The rounded, interconnected graphite that forms during the solidification of cast iron. This is the intended shape in compacted graphite iron but is a defective shape in ductile iron.

White cast iron Cast iron which produces massive amounts of cementite rather than graphite during solidification. The white irons are very hard but brittle.

PRACTICE PROBLEMS

1 Chemical analysis of a steel reveals a composition of 0.41% C, 0.58% Mn, 0.27% Si, 3.32% Ni, and 0.29% Mo. What is the probable AISI-SAE number for this steel?

2 In each BCC unit cell of iron are 12 unique interstitial sites at which carbon atoms can be located. In each FCC unit cell of iron are 4 unique interstitial sites at which carbon atoms can be located. (a) Calculate the atomic percent and weight percent of carbon in both BCC and FCC iron if all the interstitial sites are occupied by carbon atoms. (b) Based on the observed maximum solubility of carbon in iron, calculate the fraction of the interstitial sites of ferrite, austenite, and delta-ferrite that are actually occupied by carbon atoms.

3 Calculate the amount of ferrite, cementite, primary ferrite, and pearlite in a 0.6% carbon steel.

4 Calculate the amount of ferrite, cementite, primary cementite, and pearlite in a 1.1% carbon steel.

5 A plain-carbon steel contains 27% pearlite and 73% primary ferrite. Estimate the AISI-SAE number for the steel.

6 A spheroidized steel contains 10% cementite in a matrix of ferrite. Estimate the AISI-SAE number for this plain-carbon steel.

7 Complete the table below.

	1040 steel	10100 steel
A_1 temperature		
A_3 or A_{cm} temperature		
Full annealing temperature		
Normalizing temperature		
Process annealing temperature		
Spheroidizing temperature		

8 Suppose a 75 mm thick plate is heat treated. Will the structure and properties at the surface and center of the plate be closer if we anneal or normalize? Explain.

9 Describe the final microstructure obtained in a 1050 steel after each of the following heat treatments. (See Figure 13-7.)

(a) Heat at 820°C for 1 h, quench to 650°C and hold for 600 s, slowly cool to room temperature.

(b) Heat at 820°C for 1 h, quench to 700°C and hold 100 s, quench to 400°C and hold 1000 s, cool to room temperature.

(c) Heat at 820°C for 1 h, quench to 700°C and hold 100 s, quench to 400°C and hold 100 s, quench to room temperature.

(d) Heat at 820°C for 1 h, quench to 700°C and hold for 100 s, quench to room temperature.

(e) Heat at 820°C for 1 h, quench to 400°C and hold 1000 s, cool to room temperature.

(f) Heat at 820°C for 1 h, quench to 400°C and hold 10 s, cool to room temperature.

10 Describe the final microstructure obtained in a 10110 steel after each of the following heat treatments. (See Figure 13-7.)

(a) Heat at 920°C for 1 h, quench to 750°C and hold 10 s, quench to room temperature.

(b) Heat to 920°C for 1 h, quench to 500°C and hold 10 s, quench to room temperature.

(c) Heat to 920°C for 1 h, quench to 600°C and hold 10 s, quench to room temperature.

(d) Heat to 920°C for 1 h, quench to 300°C and hold 1000 s, quench to room temperature.

(e) Heat to 920°C for 1 h, quench to 750°C and hold 10 s, quench to 400°C and hold 10 s, quench to room temperature.

(f) Heat to 920°C for 1 h, quench to room temperature.

11 (a) A 1050 steel is austenitized, quenched to 600°C and held for 100 s before cooling to room temperature. What is this heat treatment called and what is the structure? (b) The same steel is austenitized, quenched to 400°C and held for 100 s before cooling to room temperature. What is this heat treatment called and what microstructure is obtained?

12 (a) A 10110 steel is austenitized, quenched to 600°C and held for 100 s before cooling to room temperature. What is this heat treatment called and what is the structure? (b) The same steel is austenitized, quenched to 400°C and held for 1000 s before cooling to room temperature. What is this heat treatment called and what microstructure is obtained?

13 Compare the minimum times required to austemper the following steels at 350°C: 1050, 1080, 10110, 4340.

14 For an AISI 4340 steel (Figure 13-16), describe the heat treatments required to produce the following microstructures: (a) Ferrite and fine pearlite; (b) Ferrite, fine pearlite, and martensite; (c) Bainite; (d) Ferrite and martensite; (e) Ferrite and lower bainite.

15 For an AISI 4340 steel (Figure 13-16), describe the microstructure obtained for the following heat treatments.

(a) Austenitize at 800°C for 1 h, quench to 600°C and hold 10^5 s, quench to room temperature.

(b) Austenitize at 800°C for 1 h, quench to 400°C and hold 10^6 s, cool to room temperature.

(c) Austenitize at 800°C for 1 h, quench to 600°C and hold 10^4 s, quench to room temperature.

(d) Austenitize at 800°C for 1 h, quench to 400°C and hold 10^3 s, quench to room temperature.

(e) Austenitize at 800°C for 1 h, quench to 300°C and hold 10 s, quench to room temperature.

(f) Austenitize at 800°C for 1 h, quench to 100°C, heat to 350°C and hold 10^4 s, quench to room temperature.

16 How much martensite will form if a 1040 steel is quenched from 750°C? How much martensite will form if a 10150 steel is quenched from 800°C? What is the percent carbon in the martensite in each case?

17 Suppose we austenitize and quench the following alloys: 1020, 1040, 4340, and 5160. Determine the percent carbon and the hardness of the martensite that forms in each case. (See Figure 11-25.)

18 Estimate the M_s temperature from Equation (13-1) for the following steels. Use average values for the alloying elements: 1080, 1541, 5120, 4820, 4340.

19 To reduce the chances for cracking in a 1060 steel, we may quench the steel in heated oil. What is the maximum temperature of the oil if we want to be sure of obtaining 90% martensite?

20 Suppose that after cooling from 800°C to 400°C in 15 s, a steel is quenched into water. What microstructure is expected if the steel is (a) 1020, (b) 1080, and (c) 4340?

21 Suppose that after cooling from 800°C to 400°C in 3 min., a steel is quenched into water. What microstructure is expected if the steel is (a) 1020, (b) 1080, and (c) 4340?

22 Are the following alloy steels hypoeutectoid or hypereutectoid?

(a) Fe-0.45% C-10% W (b) Fe-0.55% C-7% Cr
(c) Fe-0.60% C-5% Mo (d) Fe-0.50% C-6% Ni

23 Determine the approximate A_1 temperature and percent carbon in the eutectoid for steels containing (a) 2% Mo, (b) 5% Mn, and (c) 8% Cr.

24 From Figure 13-20, describe the effect of nickel and chromium on the rate of softening of a 0.4% C steel during tempering.

25 A 1040 steel quenched to produce a hardness at the surface of R_c50 and at the center of R_c25 is found to be too soft. What will be the corresponding hardnesses at the same locations if we use (a) 9310 steel or (b) 8640 steel?

26 A 25 mm diameter bar of 1040 steel is to be quenched. Determine the hardness of the bar if it is quenched in (a) agitated brine, (b) still water, and (c) still oil.

27 A 50 mm diameter steel bar must have a hardness of R_c50. Determine which of the steels in Figure 13-22 will be adequate if the steel is quenched in a medium with a H coefficient of 0.5.

28 A 37.5 mm diameter bar of 4320 steel is quenched to obtain a hardness of R_c35. What H coefficient is obtained during the quench?

29 A 62.5 mm diameter 8640 steel bar must have a hardness of R_c50 or higher. (a) What is the lowest possible H coefficient that can be used? (b) Approximately what is the cooling rate?

30 The cooling rate at the center of a certain gear made of 1040 steel is 10°C/s whereas the cooling rate is 110°C/s at the surface. What is the hardness at these locations if an 8640 steel is quenched under the same conditions?

31 Compare the hardness profile across a 75 mm steel plate quenched in water if (a) the steel is 4340 or (b) the steel is 1040.

32 Compare the hardness profile across a 50 mm diameter 1040 bar if the steel is quenched in (a) water and (b) oil. (See Figure 13-53 for the appropriate graphs.)

Location	Hardness
A	R_c55
B	R_c40
C	R_c30
D	R_c25
E	R_c20

34 A 75 mm diameter 1040 steel should have a minimum hardness at the surface of R_c50 and a maximum hardness at the center of R_c25. Can this be accomplished by (a) an oil quench and (b) a water quench? (See Figure 13-53.)

35 Compare the microstructures of a 1020, 1080, and 4340 steel cooled at 500°C/min.

36 A 1020 steel has a hardness of R_c15 after quenching. Predict the microstructure of the steel. (Figure 13-54 will be of help.)

37 A 1080 steel has a hardness of R_c50 at the surface and R_c35 in the center after quenching. Predict the microstructure at each location. (See Figure 13-54.)

38 Next to an arc weld, the heat-affected zone cools at a rate of 10°C/s. Determine the hardness and mi-

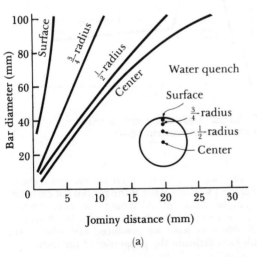

(a)

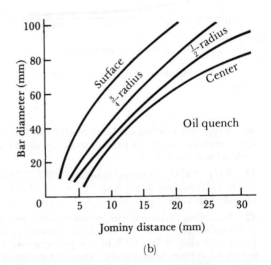

(b)

FIG. 13-53 The relationship between bar diameter, Jominy distance, and quenching medium. (a) Water quench. (b) Oil quench.

crostructure if the welded steels are (a) 1020, (b) 1080, and (c) 4340.

39 The heat-affected zone of an arc weld in 1080 steel has the following hardness profile.

33 A quenched 1040 steel bar has the following hardnesses at several locations. Determine the hardness at the same locations in a 9310 steel.

Distance from Edge of Fusion Zone (mm)	Hardness
2.5	R_c60
5.0	R_c55
7.5	R_c40
10.0	R_c35

Using Figure 13-54, determine the cooling rate and microstructure at each location in the 1080 steel. What would be the hardness profile if (a) a 1020 steel and (b) a 4340 steel were welded?

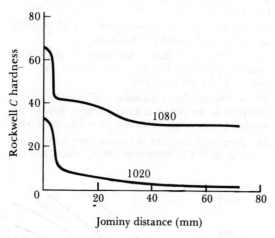

FIG. 13-54 Hardenability curves for selected steels.

40 A type 321 stainless steel contains 0.05% C. How much titanium should be present to tie up all of the carbon? See Table 13-5.

41 A type 347 stainless steel contains 0.06% C. How much niobium should be present to tie up all of the carbon? See Table 13-5.

42 According to the iron-chromium phase diagram, how much sigma phase (σ) should be present in an Fe-20% Cr alloy at 500°C? Why is σ phase normally not found at room temperature? Above what temperature would we have to anneal to eliminate the sigma phase?

43 An austenitic stainless steel is to be forged at 1000°C. Should the forging be permitted to cool slowly or rapidly? Explain. If it cools slowly, what precautions should you take with regard to the composition of the steel? If it cools rapidly, recommend a stress-relief heat-treatment temperature.

44 Why are the yield and tensile strengths higher for quenched and tempered 431 than for quenched and tempered 416 martensitic stainless steel (Figure 13-33)? Where would you expect the tempering curve for 440C alloy to fit into Figure 13-33?

45 A gray cast iron contains 3.5% C and 1.8% Si. Determine the carbon equivalent, the primary microconstituent that forms during solidification, and whether the iron is hypoeutectic or hypereutectic. If the iron solidifies as a white cast iron, what is the primary microconstituent?

46 A gray cast iron containing 3.3% C and 2.3% Si is produced. (a) Estimate the class number for the gray iron. What is the nominal tensile strength that is expected of this iron? (b) Estimate the tensile strength of the iron if the casting has a 100 mm diameter.

47 The heat treatment for a malleable cast iron is shown in Figure 13-55. (a) What is the purpose of each step (1-3) in the treatment? (b) What is the microstructure before step 1, after step 2, and after step 3? (c) What type of malleable iron have we produced and what is the grade? (d) Estimate the properties of the iron.

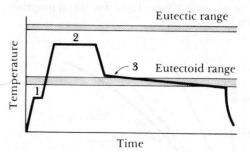

FIG. 13-55 Heat treatment for the malleable iron in Problem 13-47.

48 The heat treatment for a malleable iron is shown in Figure 13-56. (a) What is the purpose of each step (1-3) in the treatment? (b) What is the microstructure after each step of the treatment? (c) What type of malleable iron have we produced and what is the grade? (d) Estimate the properties of the iron.

49 Sulfur is less of a problem in gray iron than in ductile iron because manganese ties up with the sulfur to produce MnS inclusions. If a gray iron contains 0.12% S, how much manganese should be present to just neutralize the sulfur as MnS?

50 The number of graphite spheres that form in ductile iron during solidification is the nodule count. If we doubled the nodule count from 50 nodules/

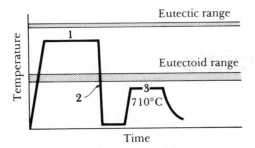

FIG. 13-56 Heat treatment for the malleable iron in Problem 13-48.

mm² to 100 nodules/mm², would the final matrix contain more or less ferrite? Explain.

51 Estimate the chill depth for a gray iron containing 3.4% C and 2.1% Si.

52 Suppose we allow an Fe-4.3% C iron to solidify. Calculate (a) the amount of cementite in weight percent and volume percent that forms if white iron is made, (b) the amount of graphite in weight percent and volume percent that forms if gray iron is made, and (c) the change in volume during solidification for gray and white cast irons. (The densities of the phases are 7.0 Mg m⁻³ for liquid, 7.69 Mg m⁻³ for austenite, 7.66 Mg m⁻³ for cementite, and 1.5 Mg m⁻³ for graphite.)

53 Calculate the percent carbon in an iron-carbon alloy that will give a 0.5% expansion when the liquid transforms to austenite plus graphite during solidification. (Use data in Problem 52.)

54 1500 kg of liquid metal for making ductile cast iron contains 0.10% S. If we add 0.15% Mg to the iron to provide nodulizing, how much magnesium is lost by reacting with sulfur to form MgS inclusions? How much magnesium can then be lost by vaporization and still provide the necessary residual magnesium of 0.03%?

55 Suppose we examine the microstructure of an iron casting that was intended to be ductile iron and find many carbides in the structure. Did we forget to nodulize or inoculate? Explain your answer.

14

Ceramic Materials

14-1 Introduction

Ceramic materials, which are joined by ionic or covalent bonds, are complex compounds and solutions containing both metallic and nonmetallic elements. Ceramics typically are hard, brittle, high melting point materials with low electrical and thermal conductivity, good chemical and thermal stability, and high compressive strengths.

Ceramic materials have a wide range of applications, ranging from pottery, brick, tile, cooking ware, and soil pipe to glass, refractories, magnets, electrical devices, and abrasives. The tiles that protect the space shuttle are silica, a ceramic material. In this chapter we will examine the structure of both crystalline and glassy ceramic materials, then summarize their processing, mechanical properties, and applications. In later chapters, the electrical, magnetic, thermal, and optical properties of ceramics will be discussed and contrasted to other materials.

14-2 Short-Range Order in Crystalline Ceramic Materials

Virtually all of the ceramic materials, even glasses, have at least a short-range order between the atoms in the structure. Several factors affect the type of short-range order and the coordination number for each atom.

Atomic bonding. The number of nearest neighbors in covalently bonded ceramics must assure that the proper number of covalent bonds are produced. In crystalline silica (SiO_2), for example, covalent bonding demands that the silicon atoms have four nearest neighbors—four oxygen atoms—thus creating a tetrahedral structure (Figure 14-1). The silicon-oxygen tetrahedra are the basic building blocks for more complicated silicate structures.

When ceramic materials contain an ionic bond, the distribution of electrical charge around the ions must be balanced and uniform. If the charges on the negative anion and positive cation are identical, the ceramic compound has the formula AX and the coordination number for each ion is identical to assure a proper balance of charge. However, if the valence of the cation is +1 and that of the anion is −2, then the structure of the AX_2 compound must assure that the coordination number of the anion is twice the coordination number of the cation.

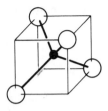

○ O^{2-} ● Si^{4+}

FIG. 14-1
The SiO_4 tetrahedron, which is the basic building block for silicate structures and many glasses.

EXAMPLE 14-1

Determine the coordination numbers for the silicon cation and the oxygen anion in silica which has the β-cristobalite crystal structure shown in Figure 14-2.

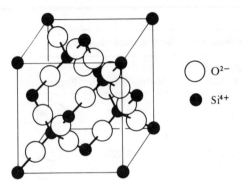

○ O^{2-}

● Si^{4+}

FIG. 14-2   The unit cell for β-cristobalite, SiO_2.

Answer:

From the figure, we see that each silicon atom is surrounded by four oxygen atoms, or the coordination number is four. However, each oxygen atom is placed between two silicon atoms. The coordination number of oxygen is therefore two. This is consistent with the valences of +4 for silicon and −2 for oxygen.

Ionic radii.   The ratio of the sizes of the ionic radii of the anion and cation also affects the coordination number (Table 14-1). The central ion may have a radius slightly larger than the radius of the hole into which it fits, thus pushing the surrounding ions slightly apart. However, the radius of the central ion cannot be smaller than the hole, or the ion will "rattle" around.

EXAMPLE 14-2

Calculate the minimum radius ratio for ions that fit into (a) the cubic site and (b) the octahedral site.

Answer:

(a) Figure 14-3(a) shows the arrangement of the ions when the smaller ion just fits into the center of the cube.

$$2R + 2r = 2R\sqrt{3}$$

$$r = \sqrt{3}R - R = (\sqrt{3} - 1)R$$

$$\frac{r}{R} = 0.732$$

(b) Figure 14-3(b) shows the arrangement of the ions when the smaller ion just fits into the center of an octahedron.

$$2R + 2r = 2R\sqrt{2}$$

$$r = \sqrt{2}R - R = (\sqrt{2} - 1)R$$

$$\frac{r}{R} = 0.414$$

TABLE 14-1   The coordination number and the radius ratio

Coordination Number	Location of Interstitial	Radius Ratio	Representation
2	Linear	0–0.155	
3	Corners of triangle	0.155–0.225	
4	Corners of tetrahedron	0.225–0.414	
6	Corners of octahedron	0.414–0.732	
8	Corners of cube	0.732–1.000	

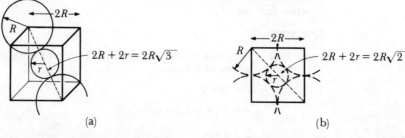

FIG. 14-3   Illustration for Example 14-2 to determine the radius ratio for (a) cubic holes and (b) octahedral holes.

EXAMPLE 14-3

Determine the expected coordination number for each ion in NiO.

Answer:

From Appendix B, the ionic radii are

$$r_{Ni} = 0.69 \text{ Å} \qquad r_0 = 1.32 \text{ Å}$$

$$\frac{r_{Ni}}{r_0} = \frac{0.69}{1.32} = 0.523$$

From Table 14-1, we find that the coordination number should be six, since $0.414 < 0.523 < 0.732$.

Interstitial sites. The crystal structure or Bravais lattice must permit the correct coordination to be achieved. If we place the cations at the regular lattice points of the unit cell, the anions occupy the interstitial positions that satisfy the proper coordination. Only the number of interstitial positions that are required to satisfy covalent bonding or ion charge balance are occupied.

Figure 14-4 shows the interstitial locations for the SC, FCC, BCC, and HCP structures. The "cubic" interstitial, with a coordination number of eight, occurs in the SC structure. Octahedral sites, with a coordination number of six, and tetrahedral sites, with a coordination number of four, are present in other unit cells. We

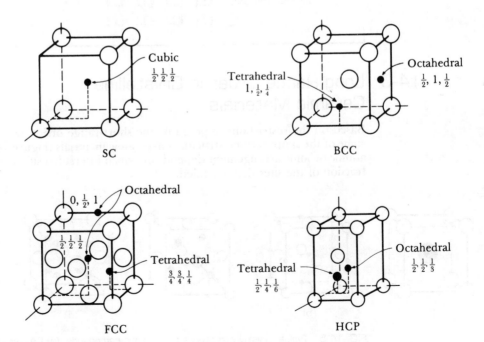

FIG. 14-4 The location of the interstitial sites in common unit cells. Only representative sites are shown.

expect the silicon ions in SiO_2 to fit into tetrahedral interstitial sites between oxygen ions to assure the proper coordination number of four.

EXAMPLE 14-4

The zincblende and cristobalite unit cells are shown in Figures 14-2 and 14-5. Which cell might represent BeO? Which cell might represent SiO_2?

Answer:

In zincblende, we find four anion interstitial sites within the unit cell and four cation sites representing the normal FCC lattice positions. Therefore, BeO, which has equal numbers of cations and anions, might have this structure.

The cristobalite structure has 16 anion sites all within the unit cell. The cation sites include four sites within the cell plus the four normal FCC sites, or a total of eight. Thus, SiO_2, which has twice as many oxygen anions as silicon cations, could have the cristobalite structure.

EXAMPLE 14-5

Write the coordinates for all of the tetrahedral interstitial sites in the FCC crystal structure.

Answer:

Tetrahedral sites: $\frac{1}{4},\frac{1}{4},\frac{1}{4}$ $\frac{1}{4},\frac{1}{4},\frac{3}{4}$ $\frac{3}{4},\frac{1}{4},\frac{1}{4}$ $\frac{3}{4},\frac{1}{4},\frac{3}{4}$

$\frac{1}{4},\frac{3}{4},\frac{1}{4}$ $\frac{1}{4},\frac{3}{4},\frac{3}{4}$ $\frac{3}{4},\frac{3}{4},\frac{1}{4}$ $\frac{3}{4},\frac{3}{4},\frac{3}{4}$

14-3 Long-Range Order in Crystalline Ceramic Materials

Based on the restrictions imposed by the short-range order, we can now justify some of the simple crystal structures in ceramic materials (Figure 14-5). The exact atomic or ionic arrangement depends on which interstitial sites are filled and the fraction of the sites that are filled.

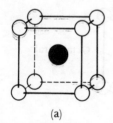 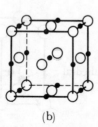 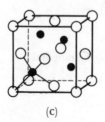 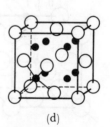 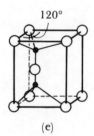

(a)   (b)   (c)   (d)   (e)

FIG. 14-5 Typical crystal structures for ceramic compounds. (a) Cesium chloride, (b) sodium chloride, (c) zincblende, (d) fluorite, and (e) wurtzite.

Cesium chloride structure. Cesium chloride is SC, with the "cubic" interstitial site filled by the anion. The radius ratio dictates that cesium chloride have a coordination number of eight.

EXAMPLE 14-6

Determine (a) whether CsCl may have the cesium chloride structure and (b) the packing factor for CsCl.

Answer:

(a) From Appendix B, $r_{Cs} = 1.67$ Å and $r_{Cl} = 1.81$ Å, so

$$\frac{r_{Cs}}{r_{Cl}} = \frac{1.67}{1.81} = 0.92$$

Since $0.732 < 0.92 < 1.000$, the coordination number is eight and the cesium chloride is a possible structure.

(b) The ions touch along the body diagonal of the unit cell, so

$$\sqrt{3}a_0 = 2r_{Cs} + 2r_{Cl} = 2(1.67) + 2(1.81) = 6.96 \text{ Å}$$

$$a_0 = \frac{6.96}{\sqrt{3}} = 4.0185 \text{ Å}$$

$$\text{Packing factor} = \frac{\frac{4}{3}\pi r_{Cl}^3 (1 \text{ Cl ion}) + \frac{4}{3}\pi r_{Cs}^3 (1 \text{ Cs ion})}{a_0^3}$$

$$= \frac{\frac{4}{3}\pi (1.81)^3 + \frac{4}{3}\pi (1.67)^3}{(4.0185)^3}$$

$$= 0.68$$

Sodium chloride structure. The sodium chloride structure is FCC; cations are located at normal FCC lattice points whereas anions occupy all of the octahedral positions. Thus, there are four anions and four cations in each unit cell. This particular arrangement of the ions assures that the coordination number of both sodium and chloride ions is six and that the electrical charge is balanced. Many ceramics, including MgO, CaO, and FeO, have this structure.

EXAMPLE 14-7

Show that MgO can have the sodium chloride crystal structure and calculate the density of MgO.

Answer:

From Appendix B, $r_{Mg} = 0.66$ Å and $r_O = 1.32$ Å, so

$$\frac{r_{Mg}}{r_O} = \frac{0.66}{1.32} = 0.50$$

Since $0.414 < 0.50 < 0.732$, the coordination number is six and the sodium chloride structure is possible.

The atomic weights are 24.3 and 16 g/g · mole for magnesium and oxygen, respectively. The ions touch along the edge of the cube, so

$$a_0 = 2r_{Mg} + 2r_O = 2(0.66) + 2(1.32) = 3.96 \text{ Å}$$

$$\rho = \frac{(4 \text{ Mg ions})(24.3) + (4 \text{ O ions})(16)}{(3.96 \times 10^{-8} \text{ cm})^3(6.02 \times 10^{23})} = 4.31 \text{ Mg m}^{-3}$$

Zincblende structure.   The zincblende structure is FCC, with four cations at normal FCC positions and four anions at half of the eight tetrahedral sites. Ions whose radius ratio predicts a coordination number of four can have this structure.

Fluorite structure.   The fluorite structure is FCC, with anions located at all of the tetrahedral positions. Thus, there are four cations and eight anions per cell and the ceramic compound must have the formula AX_2, as in calcium fluorite, or CaF_2. The coordination number of the calcium ions is eight but that of the fluoride ions is four.

EXAMPLE 14-8

Show that CaF_2 can have the fluorite structure.

Answer:

From Appendix B, $r_{Ca} = 0.99$ Å and $r_F = 1.36$ Å, so

$$\frac{r_{Ca}}{r_F} = \frac{0.99}{1.36} = 0.728$$

This predicts a coordination number of six, since $0.414 < 0.728 < 0.732$. However, this coordination number is not possible if the charge balance is to be maintained. There must be twice as many fluoride ions around the calcium ions as there are calcium ions around fluoride ions. The requirement for satisfying the charge balance offsets the ionic radius effect.

Wurtzite and hexagonal structures.   A series of crystal structures are formed when ions are placed in the interstitial locations of the HCP structure. The wurtzite structure has half of the tetrahedral sites filled with anions. Other structures based on HCP unit cells are produced when other interstitial sites are occupied.

Complicated oxide structures.   Some complicated ionic ceramics, including the perovskite structure, are also based on a cubic system (Figure 14-6). However, more than two types of ions are located in the unit cell. In these structures, both octahedral and tetrahedral positions are partially or completely filled with ions.

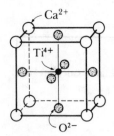

FIG. 14-6
The perovskite structure, with three different ions.

EXAMPLE 14-9

The perovskite structure of $TiCaO_3$ is shown in Figure 14-6. Does the Ti^{4+} ion in the center of the cell "rattle" around in its interstitial site? Calculate the size of the unit cell of $TiCaO_3$.

Answer:

Let's assume that the titanium ion is not present and determine the size of the interstitial hole. The ions will touch along a face diagonal in this case. The lattice parameter is

$$a_0 = \frac{2r_{Ca} + 2r_O}{\sqrt{2}} = \frac{2(0.99) + 2(1.32)}{\sqrt{2}} = 3.27 \text{ Å}$$

The size of the interstitial hole is

$$2r_{hole} = a_0 - 2r_O = 3.27 - 2(1.32) = 0.63$$

$$r_{hole} = 0.315 \text{ Å}$$

But the size of the hole, 0.315 Å, is less than that of the titanium ion, 0.68 Å. The titanium ion must push the surrounding ions apart.

Consequently, when the titanium ion is in place, the ions touch between the oxygen and titanium ions, and the actual lattice parameter is

$$a_0 = 2r_{Ti} + 2r_O = 2(0.68) + 2(1.32) = 4.00 \text{ Å}$$

Graphite. Graphite, or carbon, is sometimes considered to be a ceramic material, even though graphite is an element rather than a combination of a metal and a nonmetal. Graphite has a hexagonal, sheet-type structure (Figure 14-7).

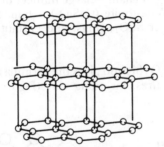

FIG. 14-7 The hexagonal (but not HCP) structure of crystalline graphite.

14-4 Silicate Structures

The silicate structures are based on the silica tetrahedron. Silica tetrahedra, SiO_4^{4-}, behave as ionic groups; the oxygen ions at the corners of the tetrahedra are attached to other ions or ionic groups to satisfy the charge balance. Figure 14-8 summarizes these structures.

Silicate compounds. When two Mg^{2+} ions are available to combine with one tetrahedron, a compound Mg_2SiO_4, or forsterite, is produced. The two Mg^{2+} ions satisfy the charge requirements and balance the SiO_4^{4-} ions. The Mg_2SiO_4 groups in turn produce a three-dimensional crystalline structure. Similarly, Fe^{2+} ions can combine with silica tetrahedra to produce Fe_2SiO_4. Mg_2SiO_4 and Fe_2SiO_4 form a series of solid solutions known as *olivines* or *orthosilicates*.

Two silicate tetrahedra can combine by sharing one corner to produce a double tetrahedron, or a $Si_2O_7^{6-}$ ion. This ionic group can then combine with other ions to produce *pyrosilicate*, or double tetrahedron, compounds.

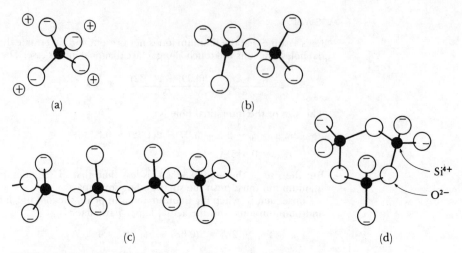

FIG. 14-8   Arrangement of silica tetrahedra. (a) Orthosilicate island, (b) pyrosilicate island, (c) chain, and (d) ring. Positive ions are attracted to the silicate groups.

Ring and chain structures.   When two corners of the tetrahedron are shared, rings and chains with the formula $(SiO_3)_n^{2n-}$ form, where n gives the number of SiO_3^{2-} groups in the ring or chain. A large number of ceramic materials have this *metasilicate* structure. Wollastonite ($CaSiO_3$) is built from Si_3O_9 rings; beryl ($Be_3Al_2Si_6O_{18}$) contains larger Si_6O_{18} rings; and enstatite ($MgSiO_3$) has a chain structure.

Sheet structures.   When the O : Si ratio gives the formula Si_2O_5, the tetrahedra combine to give sheet structures, including clay (Figure 14-9) and mica.

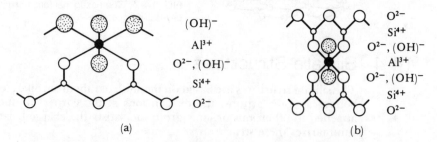

FIG. 14-9   The silicate sheet structure that forms the basis for clays. (a) Kaolinite clay and (b) montmorillonite clay.

Kaolinite, a common clay, is composed of a silicate sheet ionically bonded to a sheet composed of $AlO(OH)_2$, producing thin platelets of clay with the formula $Al_2Si_2O_5(OH)_4$. Montmorillonite, or $Al_2(Si_2O_5)_2(OH)_2$, contains two silicate sheets sandwiched around a central $AlO(OH)_2$ layer. The platelets are bonded to one another by weak Van der Waal's bonds.

Silica. Finally, when all four corners of the tetrahedra are shared, silica, or SiO_2, is produced (Figure 14-2). Silica can exist in several allotropic forms. As the temperature increases, silica changes from α-quartz to β-quartz to β-tridymite to β-cristobalite to liquid. The pressure-temperature equilibrium diagram in Figure 14-10 shows the stable forms of silica.

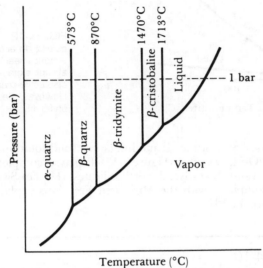

FIG. 14-10 The pressure-temperature phase diagram for SiO_2.

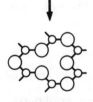

High-temperature form

Low-temperature form

FIG. 14-11 The displacive transformation in quartz.

The transformation from α-quartz to β-quartz is a *displacive transformation*. This transformation is identical to the martensite reaction; the quartz rapidly changes crystal structure by a slight distortion of the lattice involving second or further nearest neighbors (Figure 14-11). Similar transformations occur between different forms of tridymite and cristobalite. The higher temperature structure in the displacive type of transformation normally has a more open structure, a lower density, a higher heat capacity, and a more symmetrical crystal structure.

The transformation between β-quartz and β-tridymite and between β-tridymite and β-cristobalite are reconstructive rather than displacive. A *reconstructive transformation* requires that bonds between the atoms be broken and reestablished to produce a major change in crystal structure. Reconstructive transformations require a greater amount of energy, involve nucleation and growth, and occur more slowly than the displacive reaction.

When displacive transformations occur, there is an abrupt change in the dimensions of the ceramic crystal. These changes are shown in Figure 14-12 for quartz. High stresses and even cracking accompany these large volume changes in silica.

14-5 Imperfections in Crystalline Ceramic Structures

As in metals, the structure of ceramic materials contains a variety of imperfections.

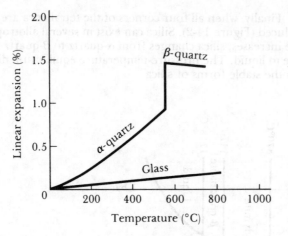

FIG. 14-12 The expansion of quartz. In addition to the regular almost linear expansion, a large abrupt expansion accompanies a displacive transformation. However, glasses expand uniformly.

Point defects. Substitutional and interstitial solid solutions form in ceramic materials. The NiO-MgO system (Figure 8-17) displays a complete series of substitutional solid solutions. Likewise, the orthosilicates, $(Mg,Fe)_2SiO_4$, display a complete range of solubility, with the Mg^{2+} and Fe^{2+} ions replacing one another completely (Figure 14-13).

EXAMPLE 14-10

Show that Mg_2SiO_4 and Fe_2SiO_4 should have complete solubility in one another.

Answer:

Both magnesium and iron have a valence of two, the crystal structures of the compound are identical, so we need to check the ionic radii. From Appendix B, $r_{Mg} = 0.66$ Å and $r_{Fe} = 0.74$ Å.

$$\frac{\Delta r}{r_{Fe}} = \frac{0.74 - 0.66}{0.74} \times 100 = 10.8\%$$

Thus, the ionic radii of magnesium and iron are within 15% of one another and complete solid solubility might be expected.

Note that metallic Fe and Mg are not soluble—their atomic radii are very different, as are their crystal structures and electronegativities.

Usually, the solid solubility of one phase in another is limited or even zero. Interstitial solid solutions are less likely in ceramics than in metals because the normal interstitial sites are already filled. For example, MgO (with the sodium chloride structure) has all of the octahedral sites filled and CaF_2 (the fluorite structure) has all of the tetrahedral sites filled.

Maintaining a balanced charge distribution is difficult when solid solution ions are introduced. However, charge deficiencies or excesses can be accommodated in ceramic materials in several ways. For example, if an Al^{3+} ion in the center of a montmorillonite clay platelet is replaced by a Mg^{2+} ion, the clay platelet has an extra negative charge. To equalize the charge, a positively charged ion,

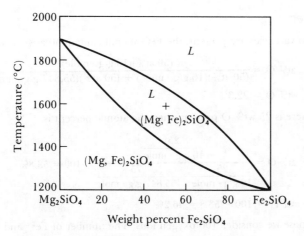

FIG. 14-13   The Mg_2SiO_4-Fe_2SiO_4 phase diagram
shows complete solid solubility.

such as sodium or calcium, is adsorbed onto the surface of the clay platelet (Figure
14-14).

A second way of accommodating the unbalanced charge is to create vacancies. Normally, FeO contains equal numbers of Fe^{2+} and O^{2-} ions in the sodium
chloride structure. However, two Fe^{3+} ions can substitute for three Fe^{2+} ions,
creating a vacancy where an iron ion would normally be (Figure 14-15).

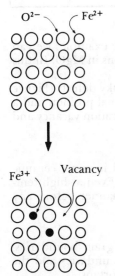

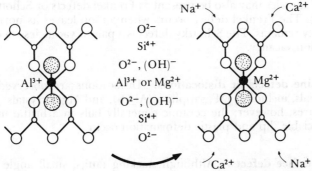

FIG. 14-14   Replacement of an Al^{3+} ion by a Mg^{2+} ion in a
montmorillonite clay platelet produces a charge imbalance
which permits cations, such as sodium or calcium, to be
attracted to the clay.

FIG. 14-15
Formation of
vacancies in FeO
when ions with a
different valence
are substituted
into the structure.
To maintain equal
charge, vacancies
must be created.

EXAMPLE 14-11

What wt% oxygen must be present in FeO to prevent any vacancies? What fraction of
the iron sites are vacancies if FeO contains 25 wt% O?

Answer:

If no vacancies are present, the FeO should contain 50 at% Fe and 50 at% O.

$$\text{at\% O} = \frac{(50 \text{ at\%})(16 \text{ g/g} \cdot \text{mole})}{(50 \text{ at\%})(16 \text{ g/g} \cdot \text{mole}) + (50 \text{ at\%})(55.847 \text{ g/g} \cdot \text{mole})} \times 100$$

$$\text{wt\% O} = 22.3\%$$

If there is 25 wt% O present, then the atomic percent is

$$\text{at\% O} = \frac{\dfrac{25}{16 \text{ g/g} \cdot \text{mole}}}{\dfrac{25}{16 \text{ g/g} \cdot \text{mole}} + \dfrac{75}{55.847 \text{ g/g} \cdot \text{mole}}} \times 100 = 53.8\%$$

$$\text{at\% Fe} = 100 - 53.8 = 46.2\%$$

Suppose we consider 100 oxygen ions. The number of Fe^{2+} and Fe^{3+} ions is

$$\frac{x}{46.2} = \frac{100}{53.8}$$

$$x = 86$$

The number of vacancies is $100 - 86 = 14$, so that the fraction of the iron sites which are vacant is $\frac{14}{100} = 0.14$

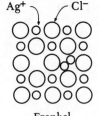

Frenkel
defect

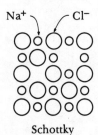

Schottky
defect

FIG. 14-16
The Frenkel and
Schottky defects
in ceramic crystal
structures.

We could also substitute more than one type of ion. For example, we can introduce a Li^+ ion and a Fe^{3+} ion to substitute for two Mg^{2+} ions in MgO. In this mechanism, vacancies do not need to be created.

Vacancies may also be present as Frenkel defects or Schottky defects (Figure 14-16). The Frenkel defect occurs when an ion leaves its normal position and a vacancy remains. The Schottky defect is a pair of vacancies—a cation vacancy and an anion vacancy.

Line defects or dislocations. Dislocations are observed in some ceramic materials, including LiF, sapphire (Al_2O_3), and MgO crystals. Even at high temperatures, however, the ceramic generally fails in a brittle manner before any appreciable slip and plastic deformation occur.

Surface defects. Although stacking faults, small angle grain boundaries, and twin boundaries are observed in some ceramics, grain boundaries and the surface of the ceramic particles are normally much more important.

Typically, ceramics with a small grain size have improved mechanical properties compared to coarse-grained ceramics. Finer grain sizes help reduce stresses at grain boundaries due to anisotropic expansion and contraction. Normally, a fine grain size is produced by beginning with finer ceramic raw materials.

Particle surfaces, which represent planes of broken, unsatisfied covalent or ionic bonds, are reactive. Gaseous molecules, for example, may be adsorbed onto the surface to reduce the surface energy. In clay deposits, foreign ions may be attracted to the platelet surface (Figure 14-17), altering the composition of the clay.

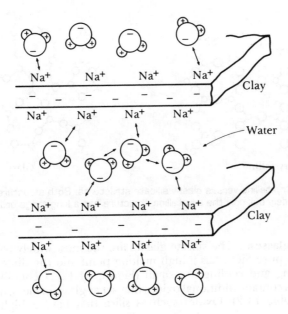

FIG. 14-17   The particle surface is important in the behavior and use of clays, adsorbing other ions and molecules and permitting the moist clay to bind coarser materials into ceramic bodies.

14-6   Noncrystalline Ceramic Materials

The most important of the noncrystalline ceramic materials are glasses. A glass is a solid material that has hardened and become rigid without crystallizing. A glass in some ways resembles an undercooled liquid. However, below the *fictive temperature* (Figure 14-18), the rate of volume contraction on cooling is reduced and the

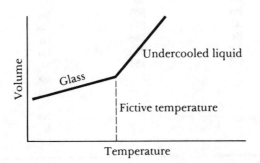

FIG. 14-18   The volume of silica versus temperature. The change in slope occurs at the fictive temperature and shows the transition from an undercooled liquid to a glass.

material can be considered a glass rather than an undercooled liquid. The glassy structures are produced by joining together silica tetrahedra or other ionic groups to produce a solid but noncrystalline framework structure (Figure 14-19).

We can also find noncrystalline structures in exceptionally fine powders, as in gels or colloids. In these materials, particle sizes may be 100 Å or less. These amorphous materials, which include some cements and adhesives, are produced by condensation of vapors, electrodeposition, or chemical reactions.

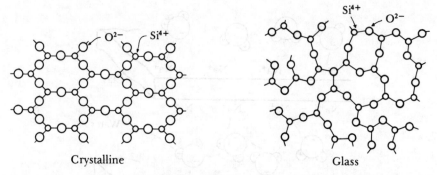

Crystalline

Glass

FIG. 14-19   Crystalline versus glassy silicate structures. Both structures have a short-range order, but only the crystalline structure has a long-range order.

Silicate glasses.   The silicate glasses are the most widely used. Fused silica, formed from pure SiO_2, has a high melting point and the dimensional changes during heating and cooling are small (Figure 14-12). Generally, however, the silicate glasses contain additional oxides that act as glass formers, intermediates, or modifiers (Table 14-2). Oxides such as silica that have a high bond strength

TABLE 14-2   Division of the oxides into glass formers, intermediates, and modifiers, and a comparison of the bond strengths

Metal in Oxide	Bond Strength (kJ mol^{-1})	Metal in Oxide	Bond Strength (kJ mol^{-1})	Metal in Oxide	Bond Strength (kJ mol^{-1})
Glass formers		**Intermediates**		**Modifiers**	
B	497	Ti	305	La	242
Si	443	Zn	301	Y	209
Ge	451	Pb	305	Sn	192
P	414	Al	251	Ga	188
V	414	Be	263	In	180
As	305	Zr	255	Pb	163
Sb	318	Cd	251	Mg	155
Zr	339			Ca	134
				Ba	138
				Sr	134
				Na	84
				K	54
				Cs	42

behave as *glass formers.* An *intermediate* oxide, such as lead or aluminum oxide, has a lower bond strength; the intermediate oxide does not form a glass by itself but is incorporated into the network structure of the glass formers. A final group of oxides, the *modifiers,* have a low bond energy. Modifiers eventually cause the glass to devitrify, or crystallize.

Modified silicate glasses.   Modifiers break up the silica network if the oxygen to silicon ratio increases significantly. When adding Na_2O, for example, the sodium ions enter holes within the network rather than becoming part of the network. However, the oxygen ion that enters with the Na_2O does become part of the network (Figure 14-20). When this happens, there aren't enough silicon ions to combine with the extra oxygen ions and keep the network intact. Eventually, a high O : Si ratio causes the remaining silica tetrahedra to form chains, rings, or compounds and the silica no longer transforms to a glass. When the O : Si ratio is above about 2.5, silica glasses are difficult to form and above a ratio of three, a glass forms only when special precautions are taken.

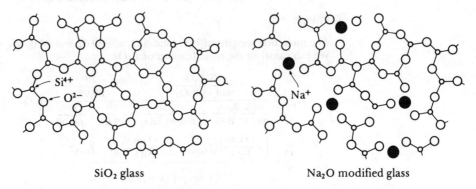

SiO₂ glass   Na₂O modified glass

FIG. 14-20 The effect of Na_2O on the silica glass network. Soda is a modifier, disrupting the glassy network and reducing the ability to form a glass.

EXAMPLE 14-12

How much Na_2O can be added to SiO_2 before the O : Si ratio exceeds 2.5 and the glass-forming tendencies are impaired?

Answer:

Let f_{Na_2O} be the mole fraction of Na_2O added to the glass. Then

$$\frac{\left(1 \frac{O \text{ ion}}{Na_2O}\right)(f_{Na_2O}) + \left(2 \frac{O \text{ ions}}{SiO_2}\right)(1 - f_{Na_2O})}{\left(1 \frac{Si \text{ ion}}{SiO_2}\right)(1 - f_{Na_2O})} = 2.5$$

$$\frac{f_{Na_2O} + 2(1 - f_{Na_2O})}{1 - f_{Na_2O}} = 2.5$$

$$f_{Na_2O} = 0.33 \qquad f_{SiO_2} = 0.67$$

$$wt\% \ Na_2O = \frac{(0.33)(\text{mol wt } Na_2O)}{(0.33)(\text{mol wt } Na_2O) + (0.67)(\text{mol wt } SiO_2)} \times 100$$

$$= \frac{(0.33)(61.98)}{(0.33)(61.98) + (0.67)(60.08)} \times 100$$

$$= 34\%$$

$$wt\% \ SiO_2 = 66\%$$

Nonsilicate glasses. Glasses produced from BeF_2, GeO_2, aluminum phosphate, and boron phosphate also have a tetrahedral short-range structure. However, borate, B_2O_3, glass is formed by combining triangular units rather than tetrahedra. Some glasses are formed by combining silica and borate. Pyrex, for example, contains appreciable quantities of B_2O_3 in silica.

EXAMPLE 14-13

Determine the O : Si ratio when 20 wt% B_2O_3 is added to SiO_2.

Answer:

The molecular weights of B_2O_3 and SiO_2 are 69.62 and 60.09 g/g · mole respectively. We can calculate the mole fraction of B_2O_3.

$$f_{B_2O_3} = \frac{\dfrac{wt\% \ B_2O_3}{mol \ wt \ B_2O_3}}{\dfrac{wt\% \ B_2O_3}{mol \ wt \ B_2O_3} + \dfrac{wt\% \ SiO_2}{mol \ wt \ SiO_2}} = \frac{\dfrac{20}{69.62}}{\dfrac{20}{69.62} + \dfrac{80}{60.09}} = 0.177$$

$$\frac{O}{Si} = \frac{\left(3 \dfrac{O \ ions}{B_2O_3}\right)(f_{B_2O_3}) + \left(2 \dfrac{O \ ions}{SiO_2}\right)(1 - f_{B_2O_3})}{\left(1 \dfrac{Si \ ion}{SiO_2}\right)(1 - f_{B_2O_3})}$$

$$\frac{O}{Si} = \frac{3(0.177) + 2(1 - 0.177)}{1(1 - 0.177)} = 2.65$$

A lead oxide glass forms if at least some silica or borate is present in PbO. Although lead oxide is not a glass former itself, it enters the glassy network of SiO_2 so effectively as an intermediate that some glasses may contain as much as 90% PbO.

14-7 Deformation and Failure

Crystalline ceramics generally fail in a brittle manner, even at high temperatures. At low temperatures, failure occurs through the grains but at high temperatures, failure may occur along grain boundaries. Movement of dislocations seldom plays an important role during failure.

In a glass, deformation occurs by isotropic *viscous flow*. Groups of atoms, such as silicate islands, rings, or chains, move past one another in response to the stress, permitting deformation. The resistance to an applied stress is offered by the attraction between these groups of atoms. This resistance is related to the viscosity η of the glass, which in turn is dependent on temperature.

$$\eta = \eta_0 \exp\left(\frac{E_\eta}{RT}\right) \tag{14-1}$$

As the temperature increases, the viscosity decreases, viscous flow is easier, and the glass is more easily deformed. The activation energy E_η is related to the ease with which the atom groups move past one another. The addition of modifiers, such as Na_2O, breaks up the network structure, permits the atom groups to move more easily, reduces E_η, and reduces the viscosity and strength of the glass (Figure 14-21).

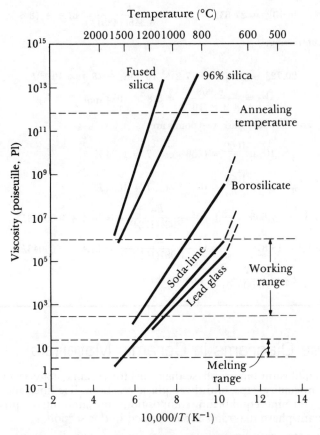

FIG. 14-21   The effect of temperature and composition on the viscosity of glass.

EXAMPLE 14-14

Estimate the activation energy for viscous flow in silica and soda-lime glass.

Answer:

From Figure 14-21, we can determine the slope of the curves for the two glasses. For SiO_2, two points are

$$\eta = 10^9 \text{ at } \frac{1000}{T} = 0.6$$

$$\eta = 10^{12} \text{ at } \frac{1000}{T} = 0.7 \quad \text{or} \quad T = 1429 \text{ K}$$

Since $\ln \eta = \ln A + \dfrac{E_\eta}{RT}$

$$\ln 10^9 = 20.723 = \ln A + \frac{E_\eta}{8.315(1667)} = \ln A + (7.2 \times 10^{-5})$$

$$\ln 10^{12} = 27.631 = \ln A + \frac{E_\eta}{8.315(1429)} = \ln A + (8.4 \times 10^{-5})$$

Subtracting,

$$20.723 - 27.631 = (7.2 \times 10^{-5})E_\eta - (8.4 \times 10^{-5})E_\eta$$

$$E_\eta = \frac{6.908}{1.2 \times 10^{-5}} = 5.8 \times 10^5 \, \text{J mol}^{-1}$$

For soda-lime glass, two points are

$$\eta = 10^2 \text{ at } \frac{1000}{T} = 0.58 \quad \text{or} \quad T = 1724 \text{ K}$$

$$\eta = 10^6 \text{ at } \frac{1000}{T} = 0.95 \quad \text{or} \quad T = 1053 \text{ K}$$

$$\ln 10^2 = \ 4.605 = \ln A + \frac{E_\eta}{1.987(1724)} = \ln A + 0.00029E_\eta$$

$$\ln 10^6 = 13.816 = \ln A + \frac{E_\eta}{1.987(1053)} = \ln A + 0.00048E_\eta$$

$$E_\eta = \frac{9.211}{(0.00019)} = 48,500 \text{ cal/mol}$$

14-8 Phase Diagrams in Ceramic Materials

The equilibrium binary phase diagrams for ceramic materials include solid solutions, miscibility gaps, and three-phase reactions. Tie lines and lever law calculations determine equilibrium compositions and amounts of phases. Some of the important phase diagrams are described in this section.

SiO_2-Na_2O.   Additions of other oxides to silica act as fluxes and reduce the liquidus temperature of the ceramic. The most powerful is soda, or Na_2O, which reduces the melting temperature from 1720°C to as low as 800°C at the eutectic composition [Figure 14-22(a)]. The soda is often added to silica in glass-making processes.

EXAMPLE 14-15

Glasses may be produced by melting silica and soda, then permitting the mixture to cool. What are the minimum glass-making temperatures for fused silica and for a window glass containing 15% Na_2O?

Answer:

From the silica-soda phase diagram, the melting point for fused silica, or pure SiO_2, is 1720°C, and the liquidus temperature for SiO_2-15% Na_2O is 1240°C. The silica-soda glass can be manufactured and shaped at a temperature nearly 500°C below that for fused silica.

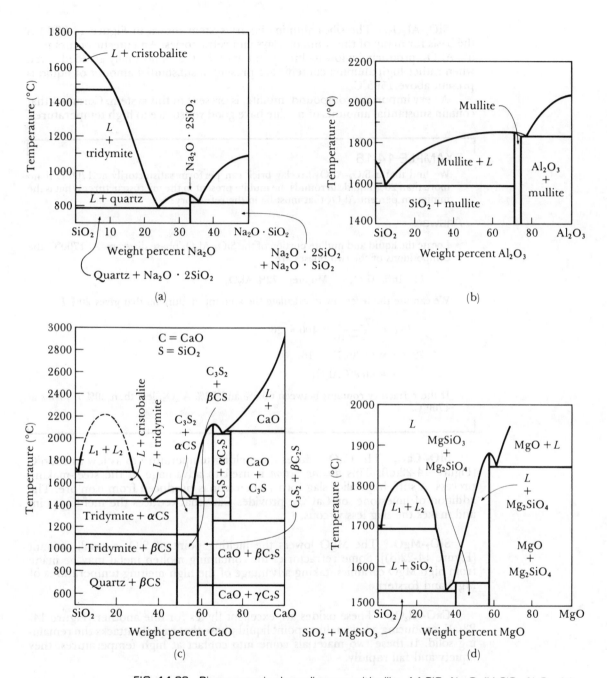

FIG. 14-22 Binary ceramic phase diagrams with silica. (a) SiO₂-Na₂O, (b) SiO₂-Al₂O₃, (c) SiO₂-CaO, and (d) SiO₂-MgO.

SiO$_2$-Al$_2$O$_3$.   The silica-alumina binary system shown in Figure 14-22(b) is the basis for many of the common clays and refractories. An eutectic occurs near 5% Al$_2$O$_3$, producing a low melting point material that is normally avoided. Even when rather high alumina contents are present, a substantial amount of liquid is present above 1595°C.

A very important compound, mullite, is present in this system. Ceramics that contain substantial amounts of mullite have good resistance to high temperatures.

EXAMPLE 14-16

We find that a SiO$_2$-Al$_2$O$_3$ fireclay brick can perform satisfactorily at 1700°C if no more than 20% liquid surrounds the mullite present in the microstructure. What is the minimum percent Al$_2$O$_3$ that must be in the refractory?

Answer:

Locate the liquid and mullite portion of the SiO$_2$-Al$_2$O$_3$ phase diagram. At 1700°C, the compositions of the two phases are

L: 16% Al$_2$O$_3$ Mullite: 72% Al$_2$O$_3$

We can use the lever law to calculate the amount of alumina that gives 20% L.

$$\% L = \frac{72 - x}{72 - 16} \times 100 = 20$$

$$72 - x = (0.20)(72 - 16)$$

$$x = 60.8\% \text{ Al}_2\text{O}_3$$

If the refractory contains between 60.8% and 72% Al$_2$O$_3$, less than 20% L forms at 1700°C.

SiO$_2$-CaO.   The CaO reduces the melting temperature to as low as 1436°C [Figure 14-22(c)]. This *fluxing action* is used to advantage in the steel-making process. A stiff, high silica slag forms when iron is produced from iron ore. The addition of limestone, or CaCO$_3$, provides CaO that reduces the melting point and makes the slag less viscous.

SiO$_2$-MgO.   The MgO lowers the melting point of SiO$_2$ a small amount [Figure 14-22(d)]. Some refractories for containing molten metal may be made from high MgO ceramics, taking advantage of the high melting temperatures of MgO and forsterite.

CaO-Al$_2$O$_3$.   These oxides are excellent fluxes for one another [Figure 14-23(a)], producing a low melting point liquid phase that rapidly attacks the remaining solid. If these two materials come into contact at high temperatures, they liquefy and fail rapidly.

ZrO$_2$-CaO.   The high melting temperature of ZrO$_2$ is used for special refractories. However, pure ZrO$_2$ transforms from a tetragonal to a monoclinic structure on cooling, often causing cracks to form. Addition of CaO produces a cubic form of zirconia, which is stable at all temperatures, and improves the crack resistance [Figure 14-23(b)].

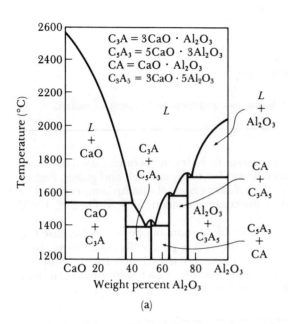

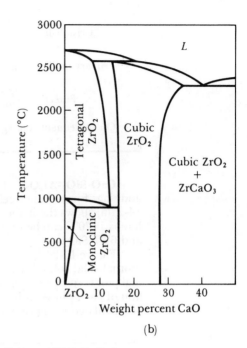

FIG. 14-23   Phase diagrams for (a) CaO-Al$_2$O$_3$ and (b) ZrO$_2$-CaO.

EXAMPLE 14-17

Describe in detail the equilibrium solidification and cooling of a SiO$_2$-40% Al$_2$O$_3$ ceramic.

Answer:

The important temperatures for the 40% Al$_2$O$_3$ ceramic are the liquidus temperature (1810°C) and the eutectic temperature (1595°C).

$T > 1810°C$:   The ceramic contains 100% L of composition 40% Al$_2$O$_3$.

$T = 1810°C$:   Solid mullite begins to crystallize from the melt. The mullite contains 72% Al$_2$O$_3$.

1595°C $< T <$ 1810°C: Mullite continues to grow and the composition of the liquid shifts toward 5% Al$_2$O$_3$. Just above the eutectic temperature, the amounts of each phase are

$$\text{Mullite} = \frac{40 - 5}{72 - 5} \times 100 = 52\%$$

$$\text{Liquid} = \frac{72 - 40}{72 - 5} \times 100 = 48\%$$

$T < 1595°C$:   When the eutectic solidifies, cristobalite (containing 0% Al$_2$O$_3$) and more mullite form. The amounts of each phase and microconstituent are

$$\text{Mullite} = \frac{40 - 0}{72 - 0} \times 100 = 55\%$$

$$\text{Cristobalite} = \frac{72 - 40}{72 - 0} \times 100 = 45\%$$

$$\text{Primary mullite} = \frac{40 - 5}{72 - 5} \times 100 = 52\%$$

$$\text{Eutectic} = \frac{72 - 40}{72 - 5} \times 100 = 48\%$$

On subsequent cooling, the cristobalite may transform to tridymite or quartz.

CaO-SiO$_2$-Al$_2$O$_3$.   This ternary system includes the ceramics used as fluxes and slags for making steel, for cements, and for many enamels and glazes. Figure 14-24(a) shows the liquidus plot for the ternary system. Liquidus temperatures as low as 1170°C can be produced when the ceramic contains 15% Al$_2$O$_3$, 23% CaO, and 62% SiO$_2$.

Figure 14-24(b) shows the isothermal plot at room temperature. Because the mutual solid solubility of the ceramic compounds is practically zero, the isothermal plot is comprised of tie triangles. The corners of the triangle fix the composition of each phase within that triangle. The amount of each solid phase can then be calculated using tie lines and lever laws, as described in Chapter 10.

EXAMPLE 14-18

Determine the sequence of solidification and the amounts of the final phases for a 57% SiO$_2$-38% CaO-5% Al$_2$O$_3$ composition. Figure 14-25 helps illustrate the necessary technique.

Answer:

The composition of the ceramic is shown in the figure at point x.

a. CaO $\cdot$ SiO$_2$ precipitates from the liquid first at a temperature of about 1450°C.

b. As the temperature falls, CaO $\cdot$ SiO$_2$ continues to grow and the composition of the liquid shifts along line $A-B$ towards the solubility line.

c. When the liquid composition reaches point B, SiO$_2$ (tridymite) begins to form and both CaO $\cdot$ SiO$_2$ and SiO$_2$ grow. The liquid composition now shifts toward point C.

d. At point C, a ternary eutectic reaction occurs, since all of the arrows converge at point C. The remaining liquid transforms to CaO $\cdot$ SiO$_2$, CaO $\cdot$ Al$_2$O$_3$ $\cdot$ 2 SiO$_2$, and SiO$_2$.

e. The final amounts of each phase can be calculated using the tie triangle and lever laws.

$$\text{CaO} \cdot \text{SiO}_2 = \frac{7.5 \text{ marks}}{10 \text{ marks}} \times 100 = 75\%$$

$$\text{SiO}_2 = (25)\left(\frac{5 \text{ marks}}{10 \text{ marks}}\right) = 12.5\%$$

$$\text{CaO} \cdot \text{Al}_2\text{O}_3 \cdot 2\text{SiO}_2 = (25)\left(\frac{5 \text{ marks}}{10 \text{ marks}}\right) = 12.5\%$$

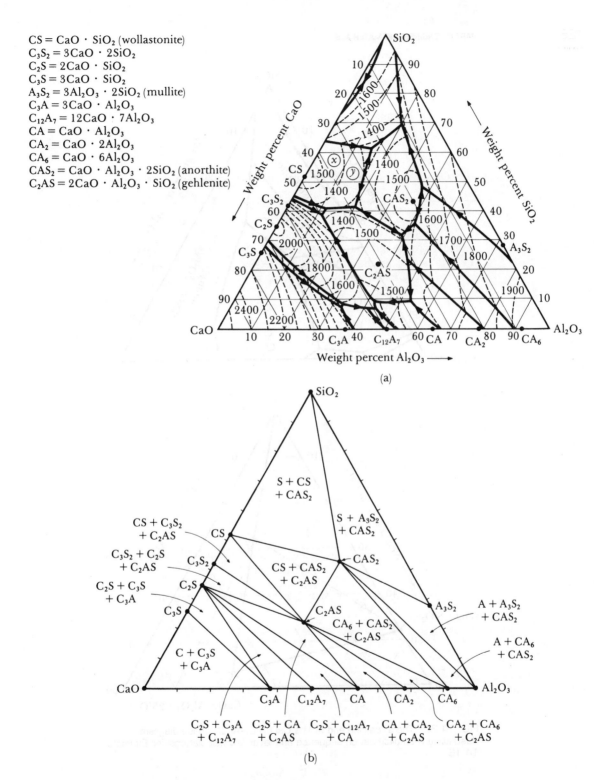

CS = CaO · SiO₂ (wollastonite)
C₃S₂ = 3CaO · 2SiO₂
C₂S = 2CaO · SiO₂
C₃S = 3CaO · SiO₂
A₃S₂ = 3Al₂O₃ · 2SiO₂ (mullite)
C₃A = 3CaO · Al₂O₃
C₁₂A₇ = 12CaO · 7Al₂O₃
CA = CaO · Al₂O₃
CA₂ = CaO · 2Al₂O₃
CA₆ = CaO · 6Al₂O₃
CAS₂ = CaO · Al₂O₃ · 2SiO₂ (anorthite)
C₂AS = 2CaO · Al₂O₃ · SiO₂ (gehlenite)

FIG. 14-24   The CaO-SiO₂-Al₂O₃ ternary phase diagram. (a) The liquidus plot, and (b) the tie triangles at room temperature.

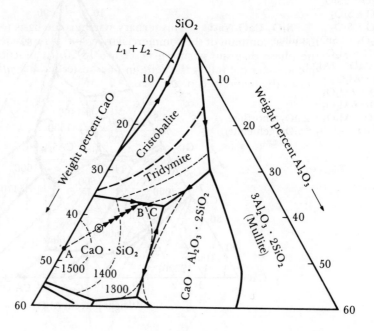

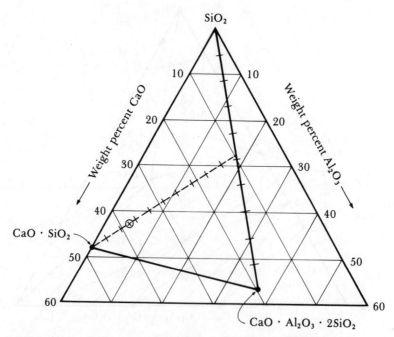

FIG. 14-25   The silica-rich end of the CaO-SiO₂-Al₂O₃ phase diagram, illustrating the solidification sequence and lever law calculations for Example 14-18.

SiO₂-CaO-Na₂O.   This ternary system is the basis for the soda-lime glasses, the most common of the commercial sheet and plate glasses. The silica-rich end of the phase diagram is shown in Figure 14-26. In practice, the solidification sequence does not follow the liquidus plot; instead, the modifying effect of the soda causes the liquid to cool as a glass.

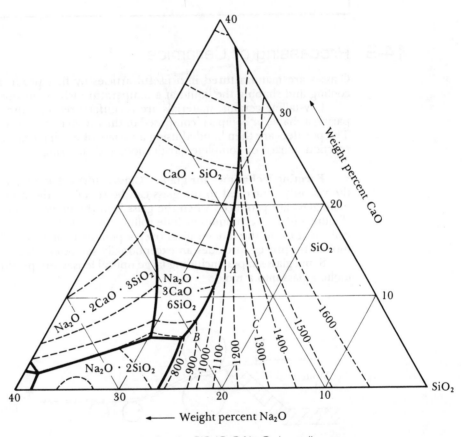

FIG. 14-26 The liquidus plot for the SiO₂-CaO-Na₂O phase diagram.

EXAMPLE 14-19

Compare the liquidus temperatures of soda-lime glasses of the following compositions:

Glass *A*:   74% SiO₂-13% CaO-13% Na₂O

Glass *B*:   74% SiO₂-7% CaO-19% Na₂O

Glass *C*:   80% SiO₂-7% CaO-13% Na₂O

Answer:

The locations of the three glasses are marked in Figure 14-26. The liquidus temperatures are

Glass A: Liquidus = 1200°C

Glass B: Liquidus = 900°C

Glass C: Liquidus = 1300°C

14-9   Processing of Ceramics

Glasses are manufactured into useful articles by first producing a liquid, then cooling and shaping the liquid at a temperature where viscous flow is possible.

Crystalline ceramic materials are manufactured into useful articles by preparing a shape, or compact, composed of the raw materials in a fine powder form. The powders are then bonded using a variety of mechanisms, including chemical reaction, partial or complete vitrification, and sintering.

Forming techniques for glass.   Glass is formed at a high temperature, with the viscosity controlled so the glass can be shaped without breaking. Sheet and plate glass are produced when the glass is in the molten state, or melting range (Figure 14-21). Techniques include rolling the molten glass through water-cooled rolls or floating the molten glass over a pool of molten tin (Figure 14-27). The molten tin process provides an exceptionally smooth surface on the glass.

Some glass shapes, including large optical lenses, are produced by casting the molten glass into a mold.

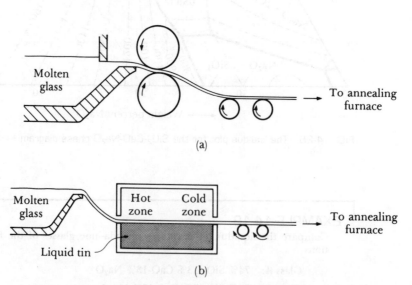

FIG. 14-27   Techniques for manufacturing sheet and plate glass. (a) Rolling, and (b) floating the glass on molten tin.

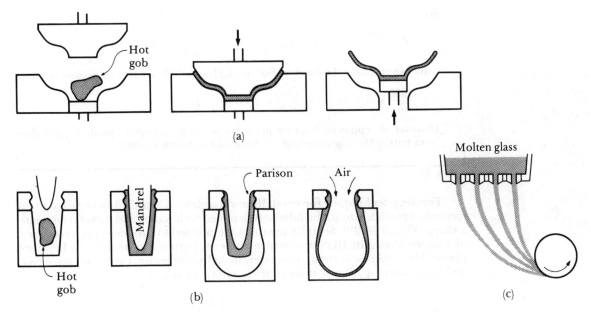

FIG. 14-28 Techniques for forming glass products. (a) Pressing, (b) press and blow process, and (c) drawing of fibers.

Shapes such as containers or light bulbs can be formed by pressing, drawing, or blowing the glass into molds (Figure 14-28). The glass is heated to the working range (Figure 14-21) so that the glass is formable but not "runny."

EXAMPLE 14-20

Compare typical glass-forming temperatures for plate glass and for glass bottles made from pure silica versus soda-lime glasses.

Answer:

From Figure 14-21, we find that soda-lime glasses can be worked into bottles in a temperature range corresponding to

$$0.72 < \frac{1000}{T} < 1.0$$

or

$$T = \frac{1000}{0.72} = 1389 \text{ K} = 1116°C \quad \text{to} \quad T = \frac{1000}{1.0} = 1000 \text{ K} = 727°C$$

For making plate glass from soda-lime glass, the melting temperature range corresponds to

$$0.55 < \frac{1000}{T} < 0.62$$

so

$$T = \frac{1000}{0.55} = 1818 \text{ K} = 1545°C \quad \text{to} \quad T = \frac{1000}{0.62} = 1613 \text{ K} = 1340°C$$

By extrapolating the curve in Figure 14-21, fused silica might be worked at about

$$\frac{1000}{T} = 0.5 \quad \text{or} \quad T = \frac{1000}{0.5} = 2000 \text{ K} = 1727°C$$

However, exceptionally high temperatures would be needed to produce plate glass from pure SiO_2. The advantage of the soda and lime is evident.

Forming techniques for crystalline ceramics.   Ceramic powders having a controlled particle size are blended, often mixed with water, and then formed into a shape (Figure 14-29). Semidry mixtures are pressed to produce green compacts of suitable strength. Higher moisture contents permit the mixture to be more plastic. Hydroplastic-forming processes, including pressing, extrusion, jiggering, and hand working, can be done to these plastic mixes.

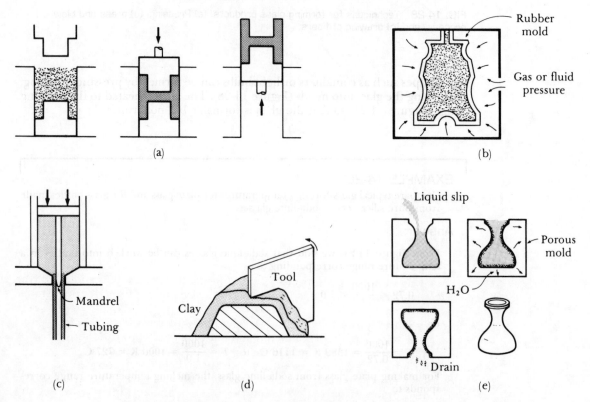

FIG. 14-29 Processes for shaping crystalline ceramics. (a) Pressing, (b) isostatic pressing, (c) extrusion, (d) jiggering, and (e) slip casting.

Even higher moisture contents permit the formation of a *slip*, or pourable slurry, containing the ceramic powder. The slip is poured into a porous mold. The water in the slip is drawn into the mold, leaving behind a soft solid which contains a low moisture content. When enough water has been drawn from the slip to produce a desired thickness of solid, the remaining liquid slip is poured from the mold. This leaves behind a hollow shell.

Drying of powder compacts. After the ceramic bodies are formed, the excess moisture is removed. During drying, large dimensional changes occur (Figure 14-30). Most of the shrinkage occurs during the initial stages of drying as the interparticle water evaporates. The temperature and humidity are carefully controlled during drying to minimize stresses, distortion, or cracking.

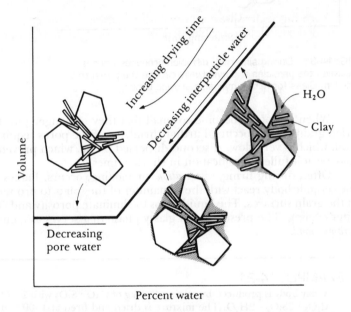

FIG. 14-30 The change in the volume of a ceramic body as the moisture is removed during drying. Dimensional changes cease after the interparticle water is gone.

Firing of powder compacts. During firing, the rigidity and strength of the ceramic increase. Firing, or sintering, causes additional shrinkage of the ceramic body as the pore size between the particles is reduced.

We must control four features of the final sintered microstructure—grain size, pore size, pore shape, and the amount of glass. We do this primarily by controlling the sintering temperature, the initial particle size, and fluxes.

During sintering, ions first diffuse along grain boundaries and surfaces to the points of contact between particles, providing bridging and connection of the individual grains (Figure 14-31). Further grain boundary diffusion closes the pores and increases the density, while the pores become more rounded. Finer initial particle sizes and higher temperatures accelerate the rate of pore shrinkage.

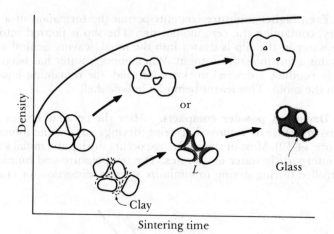

FIG. 14-31   During sintering, diffusion produces bridges between the particles and eventually causes the pores to be filled in. Glass formation may also provide bonding.

When the pores become so small that they no longer pin the grain boundaries, grain growth occurs. Further shrinkage of the pores requires volume diffusion, which is very slow. A second dispersed phase, which prevents grain growth, may permit fuller densification in shorter times.

Often during firing, *vitrification,* or melting, occurs. Fluxes or impurities in the ceramic body react with the remainder of the solids to produce a liquid phase at the grain surfaces. The liquid helps to eliminate porosity and changes to a glass after cooling. The presence of a glassy phase that serves as a binder is called the *ceramic bond.*

EXAMPLE 14-21

A clay body is produced by mixing 45.5 kg of CaO · SiO$_2$ with 22.7 kg of kaolinite clay, Al$_2$O$_3$ · 2SiO$_2$ · 2H$_2$O. The mixture is dried and fired at 1300°C until equilibrium is achieved. Determine the liquidus temperature, the phases present, and the amounts of each phase at equilibrium at 1300°C.

Answer:

First we must determine the final composition of the ceramic after firing. We will need the molecular weights of the oxides to do this.

$$\text{Mol wt CaO} = 40.08 + 16 = 56.08 \text{ g/g} \cdot \text{mole}$$

$$\text{Mol wt SiO}_2 = 28.09 + 2(16) = 60.09 \text{ g/g} \cdot \text{mole}$$

$$\text{Mol wt Al}_2\text{O}_3 = 2(28.09) + 3(16) = 101.96 \text{ g/g} \cdot \text{mole}$$

$$\text{Mol wt H}_2\text{O} = 2(1) + 16 = 18.00 \text{ g/g} \cdot \text{mole}$$

In 45.5 kg of CaO · SiO$_2$, the weight of each oxide is

$$\text{CaO} = \frac{56.08}{56.08 + 60.09} \times 45.5 = 22.0 \text{ kg}$$

$$\text{SiO}_2 = 45.5 - 22.0 = 23.5 \text{ kg}$$

In 22.7 kg kaolinite, the weight of each oxide is

$$SiO_2 = \frac{2(60.09)}{2(60.09) + 101.96 + 2(18)} \times 22.7 = 10.6 \text{ kg}$$

$$Al_2O_3 = \frac{101.96}{2(60.09) + 101.96 + 2(18)} \times 22.7 = 8.9 \text{ kg}$$

$$H_2O = 22.7 - 10.6 - 8.9 = 3.2 \text{ kg}$$

After drying and firing, the water is driven off, so the total final weight of the ceramic is

$$\text{Total weight} = 45.5 + 22.7 - 3.2 = 65 \text{ kg}$$

The composition of the fired ceramic is

$$CaO = \frac{22.0}{65} \times 100 = 33.8\%$$

$$SiO_2 = \frac{23.5 + 10.6}{65} \times 100 = 52.5\%$$

$$Al_2O_3 = \frac{8.9}{65} \times 100 = 13.7\%$$

This composition is shown in Figures 14-24 and 14-32 as point y.

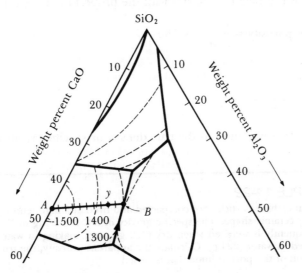

FIG. 14-32   A portion of the CaO-SiO$_2$-Al$_2$O$_3$ phase diagram showing the lever law calculations for Example 14-21.

    a. The liquidus temperature for this composition is about 1340°C.

    b. The phases at 1300°C are CaO · SiO$_2$ and L.

    c. We can calculate the amount of each phase using the lever law along the tie line A–B in Figure 14-32.

$$CaO \cdot SiO_2 = \frac{2 \text{ marks}}{10 \text{ marks}} \times 100 = 20\%$$

$$L = \frac{8 \text{ marks}}{10 \text{ marks}} \times 100 = 80\%$$

Most of the $CaO \cdot SiO_2$ has been converted to liquid by the fluxing action of the clay. After cooling to room temperature, we might expect to find the same structure, with $CaO \cdot SiO_2$ grains surrounded by a silica-lime-alumina glass.

Porosity.   The pores in the fired ceramic may be either interconnected or closed. The *apparent porosity* measures the interconnected porosity and determines the permeability, or the ease with which gases and fluids seep through the ceramic component. The apparent porosity is determined by weighing the dry ceramic (W_d), then reweighing the ceramic both when it is suspended in water (W_s) and after it is removed from the water (W_w). Then

$$\text{Apparent porosity} = \frac{W_w - W_d}{W_w - W_s} \times 100 \qquad (14\text{-}2)$$

The *true porosity* includes both interconnected and closed pores. The true porosity, which better correlates with the properties of the ceramic, is

$$\text{True porosity} = \frac{\rho - B}{\rho} \times 100 \qquad (14\text{-}3)$$

where

$$B = \frac{W_d}{W_w - W_s} \qquad (14\text{-}4)$$

B is the *bulk density* and ρ is the true density or specific gravity of the ceramic.

EXAMPLE 14-22

Silicon carbide particles are compacted and fired at a high temperature to produce a strong ceramic shape. The specific gravity of SiC is 3.2 Mg m^{-3}. The ceramic shape subsequently is weighed when dry (360 g), after soaking in water (385 g), and suspended in water (224 g). Calculate the apparent porosity, the true porosity, and the fraction of the pore volume that is closed.

Answer:

$$\text{Apparent porosity} = \frac{W_w - W_d}{W_w - W_s} \times 100 = \frac{385 - 360}{385 - 224} \times 100 = 15.5\%$$

$$\text{Bulk density} = B = \frac{W_d}{W_w - W_s} = \frac{360}{385 - 224} = 2.24$$

$$\text{True porosity} = \frac{\rho - B}{\rho} \times 100 = \frac{3.2 - 2.24}{3.2} \times 100 = 30\%$$

The closed pore percentage is the true porosity minus the apparent porosity, or $30 - 15.5 = 14.5\%$. Thus

$$\text{Fraction closed pores} = \frac{14.5}{30} = 0.483$$

Cementation. By *cementation,* the ceramic raw materials are joined using a binder that does not require firing or sintering. A liquid resin, such as sodium silicate, aluminum phosphate, or Portland cement, coats the ceramic particles and provides bridges (Figure 14-33). A chemical reaction produces a solid that joins the particles together. Some typical cementation reactions are shown in Table 14-3. The locations of some typical cements in the CaO-SiO_2-Al_2O_3 phase diagram are shown in Figure 14-34.

FIG. 14-33 A photograph of silica sand grains bonded with sodium silicate through the cementation mechanism.

Because cemented ceramic materials often have a high porosity and permeability, they may be used as ceramic filters. The binder systems are also used to make molds for metal castings. The binder produces strong, rigid bonds between sand (silica) grains, yet permits mold gases to escape through the permeable mold rather than be trapped as gas defects in the casting.

Heat treatment. Some ceramics are heat treated. Annealing, or heating the ceramic to high temperatures (Figure 14-21), reduces stresses. Large glass castings, for example, are often annealed and slowly cooled to prevent cracking. Glasses are also heat treated to cause *devitrification,* or precipitation of a crystalline phase from the glass.

Tempered glass is produced by quenching the surface of plate glass with air, causing the surface layers to cool and contract. When the center cools, its contraction is restrained by the already rigid surface, which is placed in compression. Prestressed glass is capable of withstanding much higher tensile stresses and impact blows than untempered glass.

TABLE 14-3   Typical cementation reactions in ceramic systems

Plaster of paris

$$CaSO_4 \cdot \tfrac{1}{2}H_2O + \tfrac{3}{2}H_2O \rightarrow CaSO_4 \cdot 2H_2O$$

    Calcined   Solid
    plaster of   crystals
    paris

Calcium aluminate cement

$$3CaO \cdot Al_2O_3 + 6H_2O \rightarrow Ca_3Al_2(OH)_{12}$$

    Tricalcium   Solid gel
    aluminate

Aluminum phosphate cement

$$Al_2O_3 + 2H_3PO_4 \rightarrow 2AlPO_4 + 3H_2O$$

Alumina   Phosphoric   Aluminum
         acid   phosphate

Sodium silicate cement

$$xNa_2O \cdot ySiO_2 \cdot zH_2O + CO_2 \rightarrow glass$$

    Liquid sodium
    silicate

Portland cement

$$3CaO \cdot Al_2O_3 + 6H_2O \rightarrow Ca_3Al_2(OH)_{12} + heat$$

$$2CaO \cdot SiO_2 + xH_2O \rightarrow Ca_2SiO_4 \cdot xH_2O + heat$$

$$3CaO \cdot SiO_2 + (x + 1)H_2O \rightarrow Ca_2SiO_4 \cdot xH_2O + Ca(OH)_2 + heat$$

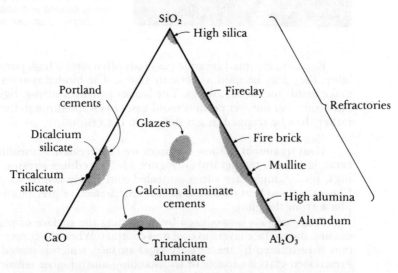

FIG. 14-34   The approximate locations of typical cements, glazes, and refractories in the $CaO\text{-}SiO_2\text{-}Al_2O_3$ phase diagram.

14-10 Applications and Properties of Ceramics

There are a large variety of ceramic materials and applications for these materials.

Clay products. Many ceramics are based primarily on clay to which a coarser material, such as quartz, and a flux material, such as feldspar, are added (Figure 14-35). Feldspars are a group of minerals including $(K,Na)_2O \cdot$

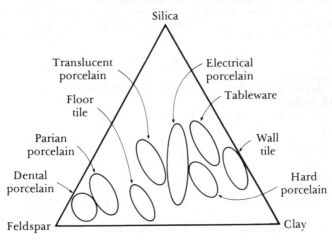

FIG. 14-35 The location of typical clay products in the silica-clay-feldspar phase diagram.

$Al_2O_3 \cdot 6SiO_2$. The materials are mixed with water and the product is formed, dried, and fired. High clay contents improve the forming characteristics, permitting more complicated ceramic bodies to be produced. High feldspar contents reduce the liquidus temperature and consequently the firing temperature. To a certain degree, the silica is a "filler" material.

Brick and tile are pressed or extruded into shape, dried, and fired to produce a ceramic bond. Higher firing temperatures or finer original particle sizes produce more vitrification, less porosity, and higher density. The higher density improves mechanical properties but reduces the insulating qualities of the brick or tile.

Earthenware are porous clay bodies fired at relatively low temperatures; little vitrification occurs, the porosity is very high and interconnected, and the earthenware ceramics leak. These products must be covered with an impermeable glaze.

Higher firing temperatures, giving greater vitrification and less porosity, produce stoneware. The stoneware, which is used for drainage and sewer pipe, contains only 2% to 4% closed pores.

China and porcelain require even higher firing temperatures to cause complete vitrification and virtually no porosity.

Refractories. Refractory materials must withstand high stresses at high temperatures. Most of the high melting point pure ceramic materials qualify as refractory materials; however, the pure oxide refractories are expensive and difficult to form into useful shapes. Instead, typical refractories are composed of coarse oxide particles, or *grog*, bonded by a finer refractory material. The finer material melts during firing, providing bonding. Typical refractory bricks contain about 20% to 25% apparent porosity, improving the thermal insulation.

The oxide refractory materials can be classed into three types—acidic, basic, and neutral (Table 14-4).

TABLE 14-4 Compositions of typical refractories

Refractory	SiO_2	Al_2O_3	MgO	Fe_2O_3	Cr_2O_3
Acidic					
Silica	95–97				
Superduty fire brick	51–53	43–44			
High-alumina fire brick	10–45	50–80			
Basic					
Magnesite			83–93	2–7	
Olivine	43		57		
Neutral					
Chromite	3–13	12–30	10–20	12–25	30–50
Chromite-magnesite	2–8	20–24	30–39	9–12	30–50

From *Ceramic Data Book*, Cahners Publishing Co., 1982.

Common acidic refractories include fireclays, or silica-alumina ceramics. Pure silica is a good refractory material and is used to contain molten metal. In some applications, the silica may be bonded with small amounts of boron oxide, which melts and produces the ceramic bond.

When 3% to 8% alumina is added to silica, the ceramic material has a very low melting temperature and is not suited for refractory applications. Increasing the alumina content, perhaps by using more kaolinite clay, improves the refractoriness of the fireclay.

The fireclays are typically vitreous. In the high-alumina fireclays, however, substantial amounts of mullite form, permitting the refractories to have a good combination of high temperature resistance along with high hardness and mechanical properties. Figure 14-34 shows the location of some typical refractories in the phase diagram.

Basic refractories include periclase (pure MgO), magnesite (MgO rich), dolomite (MgO plus CaO), and olivine (Mg_2SiO_4). The basic refractories are more expensive than the acidic refractories. However, in steelmaking and certain other high-temperature applications, the basic refractories must be used to provide compatability with the metal.

Neutral refractories include chromite and chromite-magnesite. These refractories might be used to separate acidic and basic refractories, since the acidic and basic refractories attack one another.

Other refractory materials include zirconia (ZrO_2), zircon ($ZrO_2 \cdot SiO_2$), and a variety of nitrides, carbides, borides, and graphite. Most of the carbides, such as TiC and ZrC, do not resist oxidation well and their high-temperature applications are best suited to reducing conditions. However, silicon carbide is an exception—

when SiC is oxidized at high temperatures, a thin layer of SiO_2 forms at the surface, protecting the SiC from further oxidation to about 1500°C. Nitrides and borides also have high melting temperatures and are less susceptible to oxidation. Some of the oxides and nitrides are possible materials for use in jet engines. Graphite is unique in that its strength increases as the temperature increases.

EXAMPLE 14-23

How much kaolinite clay must be added to 100 g of quartz to produce a SiO_2-30% Al_2O_3 fireclay brick after firing?

Answer:

Kaolinite is $2SiO_2 \cdot Al_2O_3 \cdot 2H_2O$. Let's first assume that the H_2O has been driven off and find the amount of $2SiO_2 \cdot Al_2O_3$ required. The molecular weights of the oxides were obtained in Example 14-21. Suppose we let x = the grams of $2SiO_2 \cdot Al_2O_3$. The weights of the oxides in the final refractory are

$$Al_2O_3 = (x)\frac{101.96}{101.96 + 2(60.09)} = 0.459x$$

$$SiO_2 = 100 + (x)\frac{2(60.09)}{101.96 + 2(60.09)} = 100 + 0.541x$$

Thus

$$30\% \; Al_2O_3 = \frac{0.459x}{0.459x + 100 + 0.541x} \times 100$$

$$(0.3)(x + 100) = 0.459x$$

$$x = \frac{30}{0.159} = 188.7 \text{ g } 2SiO_2 \cdot Al_2O_3$$

The amount of clay, which includes water, is

$$\text{Kaolinite} = (188.7)\frac{222.14 + 2(18)}{222.14} = 219.3 \text{ g}$$

Electrical and magnetic ceramics. Ceramics display a variety of useful electrical and magnetic properties. Some ceramics, including SiC, can serve as resistors and heating elements for furnaces. Other ceramics have semiconducting behavior and are used for thermistors and rectifiers.

Another group of ceramics, including barium titanate, display excellent dielectric, piezoelectric, and ferroelectric behavior. In particular, the piezoelectric properties of barium titanate make it an attractive material for capacitors and transducers.

Many of the clay bodies have excellent electrical insulation characteristics. Generally, electrical insulators should have a very low amount of porosity; thus, clay bodies that have completely vitrified, such as porcelain or glass, are used for electrical insulators for high voltages. However, when high temperature resistance at high frequencies is needed, as in automotive spark plugs, crystalline alumina performs better.

Glasses. Most of the commercial glasses are based on silica, with modifiers such as soda added to break down the network structure and reduce the melting point. Calcium oxide is added to counteract the higher solubility of the glass in water, also caused by the soda. The most common commercial glass is soda-lime glass, containing approximately 75% SiO_2, 15% Na_2O, and 10% CaO.

Improved optical qualities are obtained when the glass contains about 30% PbO. Borosilicate glasses, containing about 15% B_2O_3, have excellent stability; their use includes laboratory glassware or Pyrex. Aluminosilicate glass, with 20% Al_2O_3 and 12% MgO, and high-silica glasses with 3% B_2O_3 are excellent for high temperature resistance and for protection against heat or thermal shock. Typical glass compositions are shown in Table 14-5.

TABLE 14-5 Compositions of typical glasses

Glass	SiO_2	Al_2O_3	CaO	Na_2O	B_2O_3	MgO	PbO	Others
Fused silica	99							
Vycor	96				4			
Pyrex	81	2		4	12			
Glass jars	74	1	5	15		4		
Window glass	72	1	10	14		2		
Plate glass	73	1	13	13				
Light bulbs	74	1	5	16		4		
Fibers	54	14	16		10	4		
Thermometer	73	6		10	10			
Lead glass	67			6			17	10% K_2O
Optical flint	50			1			19	13% BaO, 8% K_2O, ZnO
Optical crown	70			8	10			2% BaO, 8% K_2O

After data collected by L. H. Van Vlack, *Physical Ceramics for Engineers,* Addison-Wesley, 1964.

Pyroceram. Some unusual characteristics are obtained when devitrification of a glass can be controlled to cause many small crystallites to nucleate and grow. The first stage of the treatment is to heat the glass to a low temperature so that many nuclei form within the glass. Then, after nucleation has occurred, the temperature is increased to promote growth of the nuclei into crystals.

Enamels and glazes. Enamels and glazes are added to the surface to make the ceramics impermeable, to provide protection and decoration, or for special purposes. Enamels and glazes may also be applied to metals. The enamels and glazes are clay products that vitrify easily during firing. A common composition is $CaO \cdot Al_2O_3 \cdot 2SiO_2$ (Figure 14-34).

Special colors can be produced in glazes and enamels by the addition of other minerals. Zirconium silicate gives a white glaze, cobalt oxide makes the glaze blue, chromium oxide produces green, lead oxide gives a yellow color, and a red glaze may be produced by adding a mixture of selenium and cadmium sulfides.

One of the problems encountered with a glaze or enamel is surface cracking, or crazing, which occurs when the glaze has a coefficient of thermal expansion different from that of the underlying ceramic material.

Slags and fluxes. When metals are melted or refined, oxides are produced that float to the surface of the molten metal as a *slag*. The slags help purify or refine the metal provided that the composition of the slag is properly adjusted. Consequently, ceramic fluxes are added to the slag for several reasons.

1. To reduce the melting temperature so the slag is more fluid.

2. To assure that a good refining action is produced; for example, CaO and CaF remove sulfur from molten steel and cast iron.

3. To assure that the refractories containing the molten metal are not attacked. If acidic refractories are used to contain metal, acid oxides such as SiO_2 or Al_2O_3 may be added to assure that the liquid slag is also acid and compatible with the refractory.

EXAMPLE 14-24

Suppose a high-alumina refractory containing SiO_2-72% Al_2O_3 (mullite) is used to contain liquid cast iron at 1500°C. Large amounts of CaO are added to the iron in an attempt to lower the sulfur content of the iron. What effect will the CaO have on the refractory?

Answer:

From Figure 14-24, we see that CaO additions reduce the liquidus temperature of the refractory, which normally is 1850°C. A straight line drawn on the phase diagram from mullite to CaO shows that the liquidus temperature falls to about 1790°C when 10% CaO dissolves in the refractory, to 1650°C for 20% CaO, to 1550°C for 30% CaO, and to 1480°C for 50% CaO. The fluxing action of the basic CaO can eventually cause the acid SiO_2-Al_2O_3 refractory to fail.

SUMMARY

The structure and properties of crystalline ceramic materials can be interpreted in terms of their complex crystal structures and phase diagrams. Because of their brittle behavior, they are normally manufactured into useful components by pressing moist aggregates or powders into shapes, followed by drying and firing. This permits the particles to sinter and become solid. The crystalline ceramics typically have high melting temperatures, high hardness, and are suitable for many high-temperature or corrosion-resistant applications.

The glassy ceramics have a nonequilibrium structure that, because of the effect on viscosity, permits us to shape the material into useful components by forming processes.

Modification of the structure of the ceramic materials cannot be interpreted in terms of slip, as we found in metals and alloys. Instead, additions to ceramics change the melting temperatures, influence the amount of glass that forms during sintering, or affect the physical properties of the ceramic.

GLOSSARY

Apparent porosity   The percent of a ceramic body that is composed of interconnected porosity.

Bulk density   The mass of a ceramic body per unit volume, including closed and interconnected porosity.

Cementation   Bonding ceramic raw materials into a useful product using binders that form a glass or gel without firing at high temperatures.

Ceramic bond   Bonding ceramic materials by permitting a glassy product to form at high firing temperatures.

Devitrification   The precipitation of a crystalline product from the glassy product, usually at high temperatures.

Displacive transformation   A change in crystal structure with changes in temperature or pressure without the necessity for significant amounts of diffusion.

Fictive temperature   The temperature at which an undercooled liquid becomes a glass.

Firing   Heating a ceramic body at a high temperature to cause a ceramic bond to form.

Flux   Additions to ceramic raw materials that reduce the melting temperature.

Glass formers   Oxides with a high bond strength which easily produce a glass during processing.

Grog   Coarse oxide particles that are bonded by finer minerals to produce refractory products.

Intermediates   Oxides that, when added to a glass, help to extend the glassy network,

although the oxides normally do not form a glass themselves.

Metasilicates   A group of silicate structures having a ring or chain structure.

Modifiers   Oxides that, when added to a glass, disrupt the glassy network, eventually causing crystallization.

Orthosilicates   A group of silicate structures based on a single silicate tetrahedral unit. Also known as olivines.

Pyrosilicates   A group of silicate structures based on a pair of silicate tetrahedral units.

Reconstructive transformation   Transformation from one crystal structure to another by a process requiring diffusion and rearrangement of atoms.

Refractories   A group of ceramic materials capable of withstanding high temperatures for prolonged periods of time.

Slag   An oxide product formed from a molten metal during melting or refining.

Slip   A liquid slurry that is poured into a mold. When the slurry begins to harden at the mold surface, the remaining liquid slurry is decanted, leaving behind a hollow ceramic casting.

Tempered glass   Glass that is prestressed during cooling to improve its strength.

True porosity   The percent of a ceramic body that is composed of both closed and interconnected porosity.

Viscous flow   Deformation of a glassy material at high temperatures.

Vitrification   Melting or formation of a glass.

PRACTICE PROBLEMS

1   Calculate the size of the cube containing the SiO_4^{4-} tetrahedron.

2   Write the coordinates for all of the octahedral sites in the BCC unit cell.

3   Write the coordinates for all of the octahedral sites in the FCC unit cell.

4   In FeO, should the ions be Fe^{2+} or Fe^{3+}? Determine whether FeO is more likely to have the cesium

chloride or sodium chloride structure by calculating the expected coordination number.

5   Determine whether MgS is more likely to have the cesium chloride, the sodium chloride, or the zincblende structure by calculating the expected coordination number.

6   What is the predicted coordination number for

sulfur ions in cerium sulfide? Based on valence, what is the formula for cerium sulfide?

7 Based on the calculated coordination number, should MgF_2 have the fluorite structure? Is the fluorite structure appropriate based on valence?

8 Calculate the lattice parameter, density, and packing factor for NiO, which has the sodium chloride structure.

9 ThO_2 has the fluorite structure. Is this reasonable based on ion charge and coordination number? Calculate the lattice parameter, packing factor, and density of ThO_2.

10 $SrZrO_3$ has the perovskite crystal structure. Determine the direction in which the ions are touching, then calculate the lattice parameter and density you expect.

11 Determine the size of the unit cell and the packing factor for ZnS, which has the zincblende structure.

12 What is the expected coordination number for P_2O_5?

13 CaF_2 has the fluorite structure, in which fluoride ions fit into tetrahedral sites between calcium ions. What is the largest ion that can fit into an octahedral site in CaF_2?

14 The density of orthorhombic forsterite, Mg_2SiO_4, is $3.21\ Mg\,m^{-3}$ and the lattice parameters are $a_0 = 4.76$ Å, $b_0 = 10.20$ Å, and $c_0 = 5.99$ Å. Calculate the number of Mg^{2+} ions and the number of SiO_4^{4-} ionic groups in each unit cell.

15 Do you expect the ceramic compound $FeO \cdot SiO_2$ to have an orthosilicate, pyrosilicate, or metasilicate structure? Explain.

16 Do you expect the ceramic compound $MgO \cdot CaO \cdot SiO_2$ to have an orthosilicate, pyrosilicate, or metasilicate structure? Explain.

17 Do you expect the ceramic compound $CaO \cdot Al_2O_3 \cdot 2SiO_2$ to have an orthosilicate, pyrosilicate, or metasilicate structure? Explain.

18 Suppose the ratio of Fe^{3+} to Fe^{2+} ions in FeO is $1:24$. Calculate the number of vacancies and the at% O in FeO.

19 In FeO, which has the sodium chloride structure, a Fe^{2+} ion is replaced by a Fe^{3+} ion in one of ten unit cells. Calculate the number of vacancies per cubic centimeter and the at% and wt% O in FeO.

20 Suppose $\frac{1}{16}$ of the Al^{3+} ions in montmorillonite are replaced by Mg^{2+} ions. How many grams of Na^+ ions will be attracted to the clay per 100 grams of clay?

21 Suppose 10 at% Na_2O is added to SiO_2. Calcu-

late the O : Si ratio. Is this combination capable of giving good glass-forming tendencies?

22 Suppose 50 at% Na_2O is added to SiO_2. Calculate the O : Si ratio. Is this combination capable of giving a good glass?

23 How much CaO can be added to 100 g of SiO_2 before the O : Si ratio exceeds 2.5?

24 Wollastonite, $CaSiO_3$, is built up of Si_3O_9 rings. Estimate the number of rings in each unit cell. The density of wollastonite is $2.905\ Mg\,m^{-3}$ and the lattice parameters are $a_0 = 7.88$ Å, $b_0 = 7.27$ Å, $c_0 = 7.03$ Å, $\alpha = 90°$, $\beta = 95.3°$, and $\gamma = 103.4°$. The volume of a triclinic unit cell is

$$V_{triclinic} = a_0 b_0 c_0 \sqrt{1 - \cos^2\alpha - \cos^2\beta - \cos^2\gamma + 2\cos\alpha\cos\beta\cos\gamma}$$

25 The viscosity of Pyrex glass is 10^9 Pl at 1400°C and 10^{13} Pl at 840°C. (a) What is the activation energy for viscous flow? (b) What temperature is required before the glass reaches a viscosity of 10^5 Pl, which permits the material to be easily formed?

26 (a) Identify the three three-phase reactions in the ZrO_2-$ZrCaO_3$ phase diagram. (b) Is cubic zirconia a stoichiometric or nonstoichiometric compound? (c) Are tetragonal zirconia and monoclinic zirconia solid solutions? (d) Compare the zirconias to the silica-containing phase diagrams in Figure 14-22. Are solid solubility and nonstoichiometric compounds uncommon in ceramic systems?

27 (a) Under equilibrium conditions, calculate the amount of solid in a SiO_2-10% Na_2O mixture at 1600°C, 1400°C, 1200°C, 1000°C, and 800°C. (b) Would we expect more or less solids under non-equilibrium conditions? What significance does this have for glass-forming technology?

28 A refractory material made from a SiO_2-45% Al_2O_3 material is used to contain molten steel at 1600°C. (a) What percentage of the refractory will melt under these conditions? (b) Does this appear to be a good choice of a refractory material?

29 A high-alumina refractory sintered to produce an equilibrium structure contains 35% mullite and 65% Al_2O_3. What is the composition of the refractory?

30 What type of three-phase reaction occurs in SiO_2-1% CaO ceramics at 1700°C?

31 A forsterite refractory, Mg_2SiO_4, is selected to contain molten steel at 1650°C. In preparing the refractory, 227 kg of SiO_2 are combined with 249 kg of MgO. If equilibrium is obtained, determine the phases and their amounts during service.

32 A refractory is produced by combining 200 kg of $3CaO \cdot Al_2O_3$ with 500 kg of $3CaO \cdot 5Al_2O_3$. Determine (a) the composition of the refractory in wt% Al_2O_3, (b) the liquidus temperature, and (c) the phases present at 1500°C.

33 Suppose iron ore contains 50% SiO_2, with the remainder being iron oxide. How many grams of CaO per kilogram of iron ore should be added to a blast furnace operating at 1600°C to produce a completely liquid CaO-SiO_2 slag. Assume all of the iron oxide is reduced to produce liquid steel.

34 Consider a 40% CaO-35% SiO_2-25% Al_2O_3 ceramic. Determine (a) the liquidus temperature, (b) the composition of the first solid to form, (c) the sequence of solidification, and (d) the amount of each phase at room temperature under equilibrium conditions.

35 Consider a 50% SiO_2-5% CaO-45% Al_2O_3 system. Determine (a) the liquidus, (b) the composition of the first solid to form, (c) the composition of the remaining liquid as the melt cools to 1700°C, 1600°C, and 1500°C, and (d) the composition of the remaining liquid at the temperature where a second solid phase should precipitate. (e) How much of the primary phase and the liquid phase are present just before the second solid begins to form? (f) How much of each solid phase is present at room temperature?

36 Suppose 50 kg of montmorillonite, $Al_2O_3 \cdot 4SiO_2 \cdot H_2O$, are combined with 50 kg of anorthite, $CaO \cdot Al_2O_3 \cdot 2SiO_2$, to produce a clay body. The clay body is dried, then fired at 1500°C. Determine the composition, the liquidus temperature, the phases present at 1500°C, and the amount of each phase in equilibrium at 1500°C.

37 Suppose 9.1 kg of kaolinite clay, $Al_2O_3 \cdot 2SiO_2 \cdot 2H_2O$, are combined with 45.5 kg of $2CaO \cdot Al_2O_3 \cdot SiO_2$ to produce a clay body. The body is dried and fired at 1450°C. Determine the composition, liquidus temperature, phases present at 1450°C, and amount of each phase at 1450°C.

38 The specific gravity of Al_2O_3 is 3.96 Mg m^{-3}. An alumina sintered product is weighed when dry (75 g), after soaking in water (85 g), and while immersed in water (55 g). Calculate the apparent porosity, the true porosity, and the fraction of the body that consists of closed pores.

39 4.55 kg of tricalcium aluminate are combined with water to produce a calcium aluminate cement. (a) How many pounds of cement are produced? (b) How much water is required?

CHAPTER 15

Polymers

15-1 Introduction

Polymers are giant organic molecules having molecular weights of 10,000 to 1,000,000 $g(g \cdot mole)^{-1}$. *Polymerization* is the process by which small molecules are joined to create these giant molecules. As the polymer grows in size, the melting or softening point increases and the polymer becomes stronger and more rigid.

Polymers are lightweight, corrosion resistant, electrical insulators but have relatively low tensile strength and are not suitable for use at high temperatures. Polymers are used in an amazing number of applications, including toys, home appliances, structural and decorative items, coatings, paint, adhesives, automobile tires, packaging, and many others.

15-2 Fitting Polymers into Categories

There are several ways that polymers can be classified (Table 15-1).

Polymerization mechanism. *Addition polymers* are produced by covalently joining the individual molecules, producing chains that may be thousands of molecules long. *Condensation polymers* are produced when two or more types of molecules are joined by a chemical reaction that releases a by-product, such as water.

Polymer structure. *Linear polymers* form long chains containing thousands of molecules. These chains may be formed by either the addition or condensation reaction. *Network polymers* are three-dimensional framework structures produced by a cross-linking process that involves either an addition or condensation reaction.

Polymer behavior. The most commonly used method to describe polymers is by their behavior when heated. *Thermoplastic polymers,* as the name suggests, behave in a plastic manner at high temperatures. Furthermore, the nature of their bonding is not dramatically changed when the temperature is increased. For example, thermoplastic polymers can be formed at elevated temperatures, cooled, and then reheated or reformed without affecting the behavior of the polymer. Thermoplastic polymers are linear.

Thermosetting polymers are network polymers formed by a condensation reaction. The thermosetting polymers cannot be reprocessed after they have been

TABLE 15-1   Summary of the classification methods for polymers

Behavior	General Structure	Joining Mechanism
Thermoplastic	Flexible linear chains	Addition or condensation
Thermosetting	Rigid three-dimensional network	Usually condensation but sometimes addition may produce the framework
Elastomers	Linear cross-linked chains	Usually addition produces both the chains and cross-links

formed because part of the molecules—the by-product of the condensation reaction—has left the material.

Elastomers, or rubbers, have an intermediate behavior but, more importantly, have the ability to elastically deform enormous amounts without being permanently changed in shape.

15-3   Representing the Structure of Polymers

Figure 15-1 shows four ways we could represent a segment of polyethylene, a simple linear addition thermoplastic polymer. The two-dimensional model in Figure 15-1(d) includes the essential elements of the polymer structure and, because of its simplicity, is the representation we will use.

We will also encounter ring structures, such as the benzene ring found in styrene and phenol molecules. Rather than showing all of the atoms in the benzene ring, we may use a hexagon (Figure 15-2). When we encounter more complex rings, we will include all of the atoms.

15-4   Chain Formation by the Addition Mechanism

As an example of simple addition polymerization, let's look at the formation of polyethylene from ethylene molecules. Ethylene, a gas, has the formula C_2H_4. The carbon atoms are joined by double covalent bonds. Each carbon atom shares two of its electrons with the second carbon atom, while two hydrogen atoms are bonded to each of the carbon atoms (Figure 15-3). The ethylene molecule is called a *monomer.*

In the presence of heat, pressure, or a catalyst, the double bond between the carbon atoms is broken, causing each carbon atom to have an unsatisfied bond. Now the ethylene molecule is called a *mer*. These bonds are satisfied when the mer combines with additional ethylene mers, producing a chain composed of a backbone of carbon atoms. As the process continues, more ethylene mers are attached to the chain. The long molecular chains, or *polymers,* are called polyethylene.

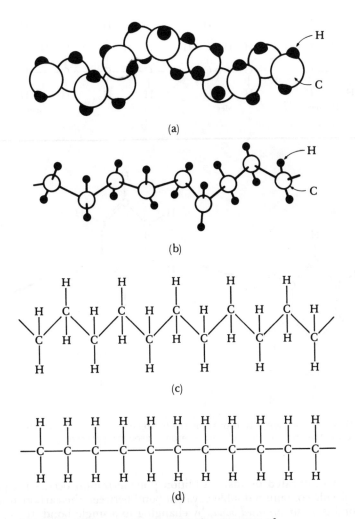

FIG. 15-1 Four ways to represent the structure of polyethylene. (a) Solid three-dimensional model, (b) three-dimensional "space" model, (c) two-dimensional model showing the kinked nature of the polymer chain, and (d) simple two-dimensional model.

EXAMPLE 15-1

The monomer for propylene is shown in Table 15-6. Show how the monomers join to produce polypropylene.

Answer:

Polypropylene forms when the double covalent bond in the monomer is replaced by a single bond between the carbon atoms. The structure is then

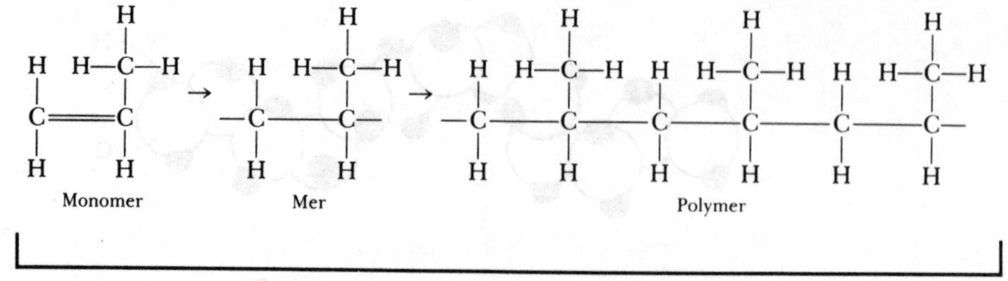

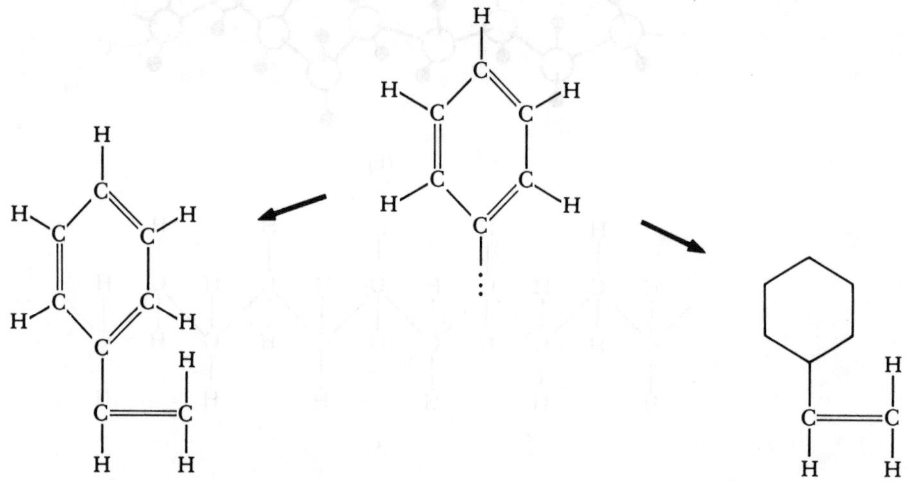

FIG. 15-2 Two ways to represent the benzene ring. In this case, the benzene ring is shown attached to a pair of carbon atoms, producing styrene.

Unsaturated bonds. Addition polymerization occurs because the original molecule contains a double covalent bond between the carbon atoms. The double bond is an *unsaturated bond;* by changing to a single bond, the carbon atoms are still joined but additional molecules can be added.

Tetrahedral structure of carbon. The structure of addition polymer chains is based on the nature of the covalent bonding in carbon. Carbon, like silicon, has a valence of four. The carbon atom shares its valence electrons with four surrounding atoms, producing a tetrahedral structure in which the four covalent bonds have a fixed angular relationship of 109° (Figure 15-4). In diamond, all of the atoms in the tetrahedron are carbon and the special FCC structure, diamond cubic, is produced.

However, in organic molecules, some of the positions in the tetrahedron are occupied by hydrogen, chlorine, fluorine, or even groups of atoms. Since the hydrogen atom has only one electron to share, the tetrahedron cannot be further extended. The structure in Figure 15-4(b) shows an organic molecule—methane—that could not undergo a simple addition polymerization process, since all four bonds are satisfied by hydrogen atoms.

The initial carbon atom could be joined with one covalent bond to a second

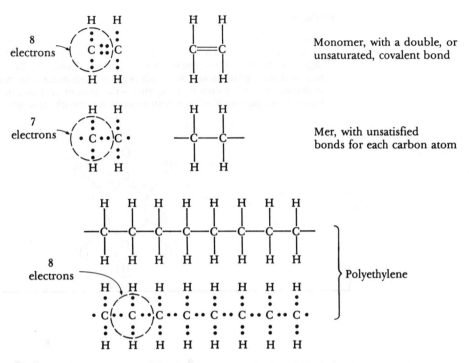

FIG. 15-3 The addition reaction for producing polyethylene from ethylene molecules. The unsaturated double bond in the monomer is broken to produce a mer, which can then attract additional mers to either end to produce a chain.

carbon atom, with all of the other bonds involving hydrogen, as in ethane [Figure 15-4(c)]. But again, polymerization cannot occur.

However, in ethylene, the carbon atoms are joined by an unsaturated double bond, while the other sites are occupied by hydrogen atoms [Figure 15-4(d)]. During polymerization the double bond is broken and each carbon atom in the molecule attracts a new mer.

Functionality. The *functionality* is the number of sites at which new molecules can be attached to the mer. In ethylene, there are two locations—each carbon atom—at which molecules can be attached. Thus, ethylene is bifunctional and only chains will form.

If there are three sites at which molecules can be attached, the mer is trifunctional and a three-dimensional network can form. Normally, trifunctional mers produce stronger polymers than bifunctional mers.

EXAMPLE 15-2

Phenol molecules have the structure shown below. The phenol molecules can be joined to one another when a hydrogen atom is removed from the ring and participates in a condensation reaction. What is the maximum functionality of phenol? Do you expect that a chain or network structure will be produced?

Answer:

The phenol molecule will give up a hydrogen atom from any of the five corners containing only hydrogen atoms. The sixth hydrogen in the OH group is bonded too tightly to the ring. Thus, the functionality is five. However, due to geometrical limitations—if all five functional sites underwent a condensation reaction, the participating molecules would be crowded together—the maximum functionality is actually only three. Since phenol is at least trifunctional, a network structure is produced.

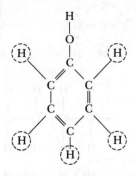

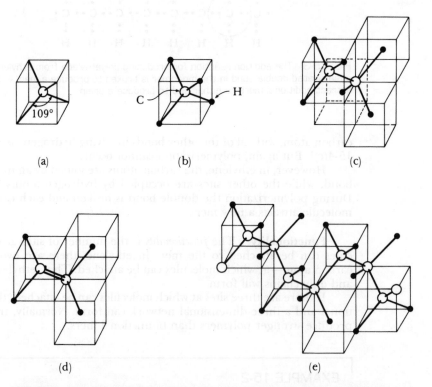

(a)   (b)   (c)

(d)   (e)

FIG. 15-4 The tetrahedral structure of carbon can be combined in a variety of ways to produce solid crystals, nonpolymerizable gas molecules, and polymers. (a) Carbon tetrahedron, (b) methane, with no unsaturated bonds, (c) ethane, with no unsaturated bonds, (d) ethylene, with an unsaturated bond, and (e) polyethylene.

Initiation of the addition mechanism. To begin the addition polymerization process, an initiator, such as hydrogen peroxide, H_2O_2, is added to ethylene (Figure 15-5). The covalent bonds between the oxygen atoms in the peroxide and between the carbon atoms in the ethylene are broken and an OH group is attached to one end of the ethylene mer. By terminating one end of the mer, the OH group acts as the nucleus for a chain.

Growth of the addition chain. Once the chain is initiated, the reaction proceeds spontaneously. Bonding energies between atoms are shown in Table 15-2. We must introduce 721 kJ $(g \cdot mole)^{-1}$ to break the double carbon bond, but $2 \times 369 = 738$ kJ $(g \cdot mole)^{-1}$ are released when the two single covalent bonds combine to extend the chain. Since 17 kJ $(g \cdot mole)^{-1}$ are evolved, this energy difference favors the continuation of the polymerization process.

TABLE 15-2 The strength of bonds in polymers

Type of Bond	Bond Strength [kJ $(g \cdot mole)^{-1}$]
C—C	369
C=C	721
C≡C	964
C—H	436
C—Cl	360
C—N	306
C—O	360
C=O	532
N—H	461
O—H	499
H—H	436
N—O	306
O—O	214
C—F	432
O—S	553

Data from *Handbook of Chemistry and Physics, 56th Ed.,* CRC Press, 1975.

The kinetics of growth are depicted in Figure 15-6. Growth is initially slow but speeds up dramatically after initiation. Since energy is released during polymerization, the temperature may rise, increasing the growth rate still further.

When polymerization is nearly complete, the remaining unattached mers must diffuse a long distance before reaching an active end of a chain. Consequently, the growth rate again decreases.

Termination of the addition chain. The chains may be terminated by two mechanisms (Figure 15-7). First, the ends of two growing chains may join to produce a single large chain. Second, the active end of the chain may attract an initiator group, OH, which then terminates the chain. We can control the length of the chain by controlling the amount of initiator—if small amounts of initiator are added, less is available to terminate the chains and the chains are longer.

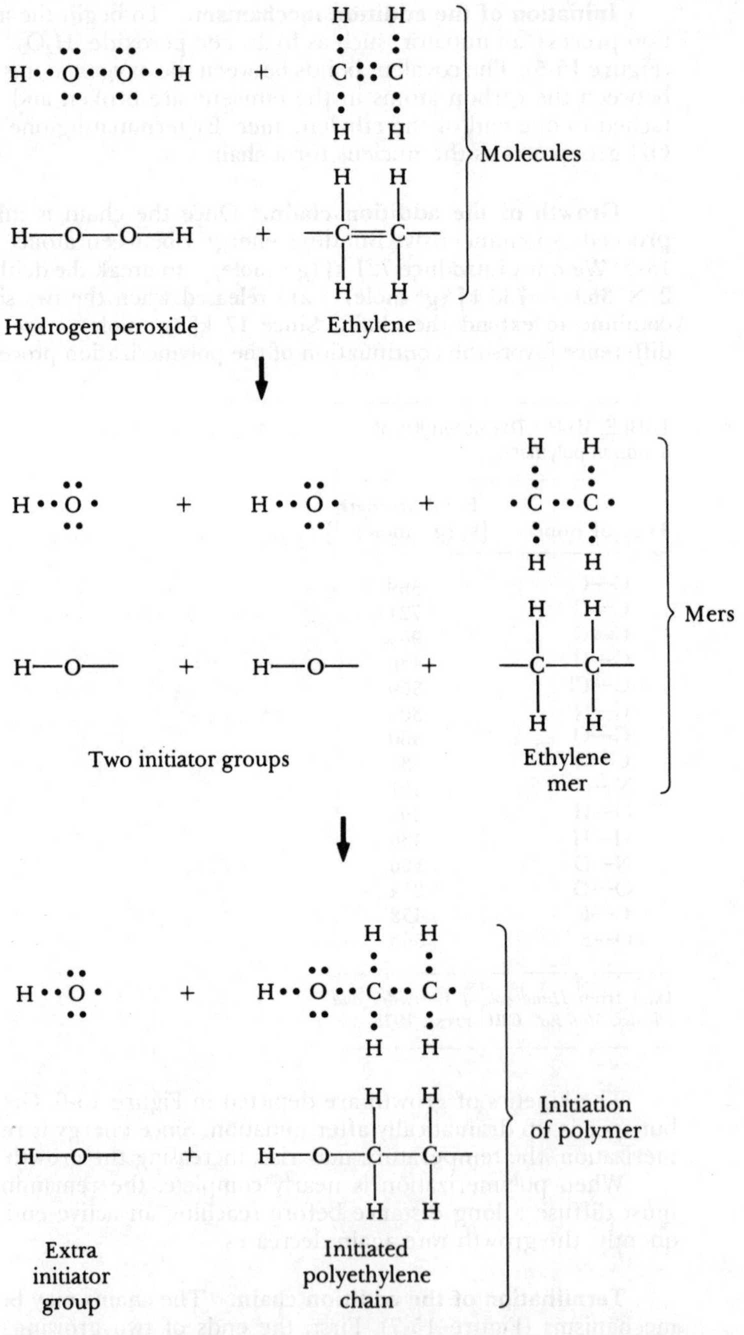

Hydrogen peroxide   Ethylene

Molecules

Two initiator groups   Ethylene mer

Mers

Extra initiator group   Initiated polyethylene chain

Initiation of polymer

FIG. 15-5 A polyethylene chain is initiated when hydrogen peroxide splits into two OH groups and a mer is formed. One of the OH groups is attached to the mer, initiating the chain.

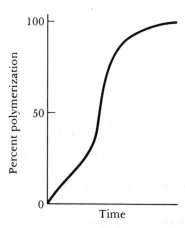

FIG. 15-6 The rate of growth of the chains and the overall rate of polymerization is initially slow, but then continues at a high rate. When polymerization is nearly complete, the rate slows down again.

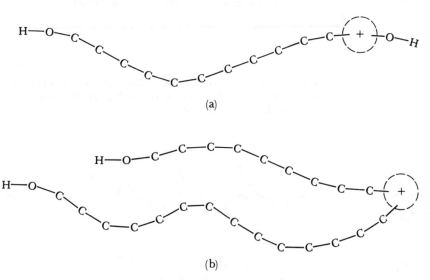

(a)

(b)

FIG. 15-7 Termination of the addition chains can occur when (a) OH groups are attached to the end of the chains or (b) two chains combine.

EXAMPLE 15-3

One gram of hydrogen peroxide is added to 10,000 g of ethylene to serve as the initiator and terminator. Calculate the average molecular weight of the polymer if all of the hydrogen peroxide is consumed.

Answer:

Each hydrogen peroxide molecule will initiate and terminate one polymer chain, since the hydrogen peroxide changes into two OH groups. The molecular weight of hydrogen peroxide is

$$\text{Mol wt } H_2O_2 = 2(M_H) + 2(M_O)$$

$$= 2(1) + 2(16) = 34 \text{ g (g} \cdot \text{mole)}^{-1}$$

$$\text{Number of } H_2O_2 \text{ molecules} = \frac{1}{34}(6.02 \times 10^{23})$$

$$= 1.77 \times 10^{22}$$

For ethylene

$$\text{Mol wt } C_2H_4 = 2(12) + 4(1) = 28 \text{ g (g} \cdot \text{mole)}^{-1}$$

$$\text{Number of } C_2H_4 \text{ molecules} = \frac{10,000}{28}(6.02 \times 10^{23})$$

$$= 2.15 \times 10^{26} \text{ mers of } C_2H_4$$

$$\text{Number of mers/chain} = \frac{2.15 \times 10^{26}}{1.77 \times 10^{22}} = 12,147$$

Therefore, the molecular weight of the polymer is

$$\text{Mol wt polymer} = 12,147(28) = 340,116 \text{ g (g} \cdot \text{mole)}^{-1}$$

Chain shape.   During growth, the polymer chains twist and turn due to the tetrahedral nature of the covalent bond. Figure 15-8 illustrates two possible geometries in which a chain might grow.

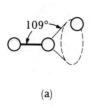 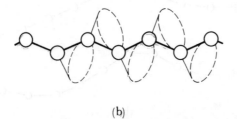 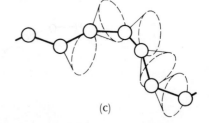

(a)   (b)   (c)

FIG. 15-8   The angular relationship between the bonds in the carbon chain can be satisfied when the third carbon atom is placed anywhere on the circle in (a). Depending on how the atoms are placed, the chain may be either straight (b) or highly twisted (c).

The third atom in Figure 15-8(a) can be located at any position on the circle and still preserve the directionality of the covalent bond. A straight chain, as in Figure 15-8(b), could be produced but it is more likely that the chain will be highly twisted, as in Figure 15-8(c).

The chains twist and turn in response to external factors, such as temperature or the availability and location of the next mer to be attached to the chain. Eventually, the chains become intertwined with other chains growing simultaneously. The appearance of the polymer chains may resemble that of a bucket of earthworms or a plate of spaghetti.

15-5 Chain Formation by the Condensation Mechanism

Linear polymers are also formed by condensation reactions, giving structures and properties that resemble those of linear addition polymers. Normally, heat, pressure, or a catalyst cause the reaction to proceed. The polymerization of dimethyl terephthalate and ethylene glycol to produce dacron, a polyester fiber, is an important example (Figure 15-9).

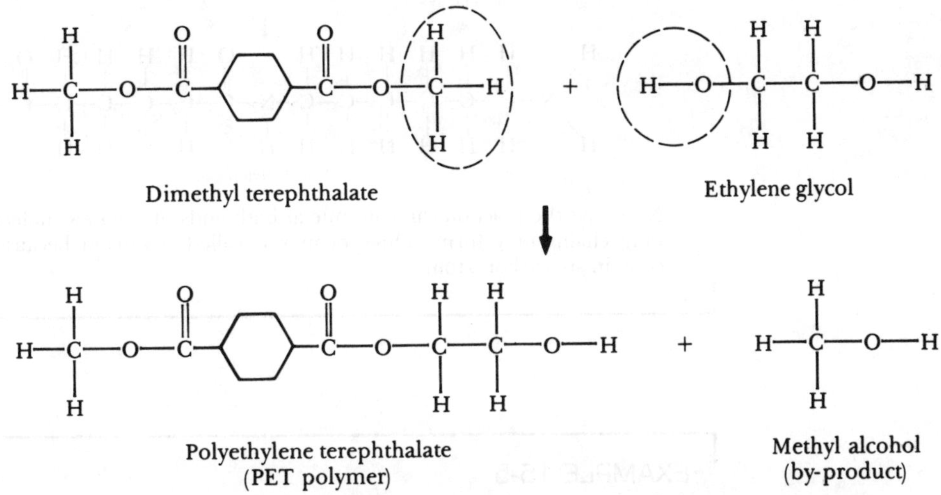

FIG. 15-9 The condensation reaction for polyethylene terephthalate (PET), more commonly known as dacron. A CH_3 and an OH group are removed from the monomers, producing methyl alcohol as a by-product.

During polymerization, the OH group of ethylene glycol combines with the CH_3 group of dimethyl terephthalate. A by-product, methyl alcohol, is driven off and the two monomers combine to produce a larger molecule. The monomers in this example are bifunctional. Consequently, polymerization and growth occur at both ends of our new molecule by the same reaction. Eventually, a long polymer chain—dacron—is produced.

The length of the chain depends on the ease with which mers can diffuse to the ends and undergo the condensation reaction. Termination occurs when no more monomers reach the end of the chain to continue the reaction.

EXAMPLE 15-4

The linear polymer 6,6-nylon is formed when adipic acid and hexamethylene diamine combine by a condensation reaction. Show how this reaction occurs and determine the by-product that forms.

Answer:

The molecular structures of the monomers are shown below. The linear nylon chain is produced when a hydrogen atom from the hexamethylene diamine combines with an OH group from adipic acid to form a water molecule.

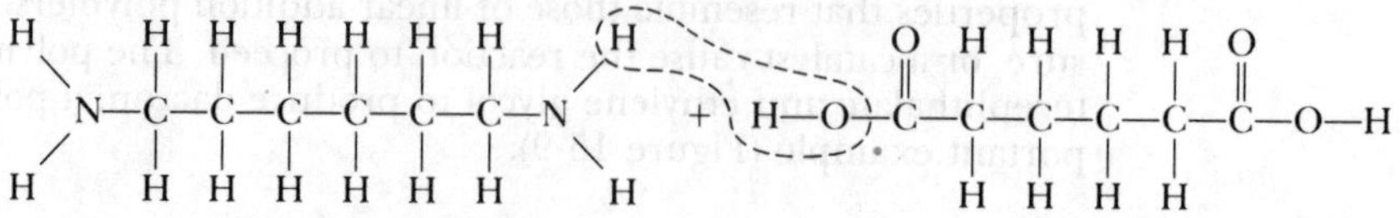

Hexamethylene diamine   Adipic acid

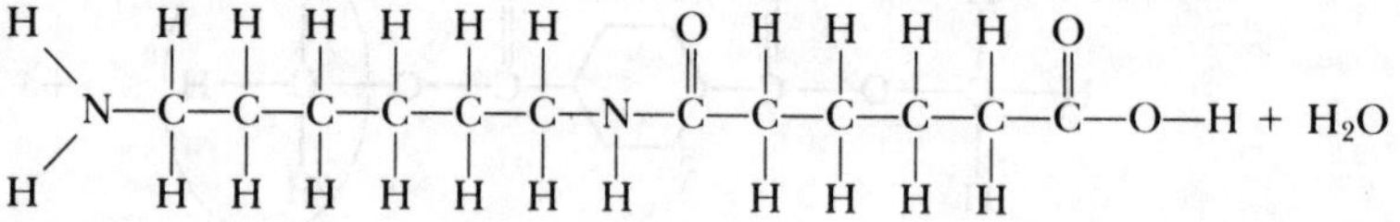

6,6-Nylon   Water

Note that the reaction can continue at both ends of the new molecule; consequently, long chains may form. This polymer is called 6,6-nylon because both monomers contain six carbon atoms.

EXAMPLE 15-5

How much energy is consumed or evolved when 6,6-nylon is formed? Will the reaction be spontaneous?

Answer:

6,6-Nylon forms when a N—H bond and an O—C bond are broken on the hexamethylene diamine and adipic acid monomers. (See Example 15-4.) A C—N bond forms to join the two monomers into a chain and an O—H bond is formed to produce the by-product water. The change in energy is

$$\Delta \text{ Energy} = \text{energy consumed} - \text{energy evolved}$$
$$= (E_{N-H} + E_{O-C}) - (E_{C-N} + E_{O-H})$$
$$= (461 + 360) - (306 + 499) = +16 \text{ kJ (g} \cdot \text{mole)}^{-1}$$

Since energy is consumed, the reaction is not spontaneous.

EXAMPLE 15-6

Suppose 1000 g of dimethyl terephthalate are combined with ethylene glycol to produce dacron. Determine how many grams of ethylene glycol are required and the total weight of the polymer after polymerization is completed.

Answer:

We need equal numbers of moles of the two monomers.

Mol wt dimethyl terephthalate = (10 C atoms)(12) + (4 O atoms)(16)

$$+ \text{ (10 H atoms)(1)}$$

$$= 194 \text{ g (g} \cdot \text{mole)}^{-1}$$

$$\begin{array}{c}\text{Number of dimethyl}\\\text{terephthalate molecules}\end{array} = \frac{1000 \text{ g}}{194}(6.02 \times 10^{23})$$

$$= 31.03 \times 10^{23}$$

We need the same number of molecules of ethylene glycol.

Mol wt ethylene glycol = (2 C atoms)(12) + (2 O atoms)(16)

$$+ \text{ (6 H atoms)(1)}$$

$$= 62 \text{ g (g} \cdot \text{mole)}^{-1}$$

$$\text{Grams of ethylene glycol} = \frac{31.03 \times 10^{23}}{6.02 \times 10^{23}}(62) = 319.6 \text{ g}$$

For each of the monomers, one methyl alcohol will be evolved.

Mol wt methyl alcohol = (1 C atom)(12) + (1 O atom)(16)

$$+ \text{ (4 H atoms)(1)}$$

$$= 32 \text{ g (g} \cdot \text{mole)}^{-1}$$

$$\text{Grams of methyl alcohol} = \frac{(31.03 \times 10^{23})(2)}{(6.02 \times 10^{23})}(32)$$

$$= 329.9 \text{ g}$$

The (2) is in the above equation because there are 31.03×10^{23} mol of dimethyl terephthalate and an equal number of molecules of ethylene glycol.

Total weight of polymer = 1000 + 319.6 − 329.9 = 989.7 g

15-6   Degree of Polymerization

The degree of polymerization describes the average length to which a chain grows. If the polymer contains only one type of monomer, the degree of polymerization is the average number of molecules or mers that are present in the chain. We can also define the degree of polymerization as:

$$\frac{\text{Degree of}}{\text{polymerization}} = \frac{\text{molecular weight of polymer}}{\text{molecular weight of mer}} \qquad (15\text{-}1)$$

When the chain is composed of more than one type of mer, we can define the average molecular weight of the mer as

$$\bar{M} = \Sigma f_i M_i \qquad (15\text{-}2)$$

where f_i is the molecular fraction of mers having molecular weight M_i.

If the polymer chain is formed by condensation, the molecular weight of the by-product must be subtracted from that of the mer

$$\bar{M} = \Sigma(f_i M_i - M_{\text{by-product}}) \tag{15-3}$$

EXAMPLE 15-7

Calculate the degree of polymerization if polyethylene has a molecular weight of 100,000 g (g · mole)$^{-1}$.

Answer:

The molecular weight of the ethylene mer is

$$\text{Mol wt mer} = 2(12) + 4(1) = 28 \text{ g (g · mole)}^{-1}$$

$$\text{Degree of polymerization} = \frac{100,000}{28} = 3571$$

EXAMPLE 15-8

Calculate the degree of polymerization if 6,6-nylon has a molecular weight of 120,000 g (g · mole)$^{-1}$.

Answer:

The nylon forms when hexamethylene diamine and adipic acid combine and release a molecule of water. When a long chain forms, there is on average one water molecule released for each reacting molecule. The molecular weights of the molecules are

$$\text{Mol wt diamine} = (6 \text{ C atoms})(12) + (2 \text{ N atoms})(14) + (16 \text{ H atoms})(1)$$

$$= 116 \text{ g (g · mole)}^{-1}$$

$$\text{Mol wt adipic acid} = (6 \text{ C atoms})(12) + (4 \text{ O atoms})(16) + (10 \text{ H atoms})(1)$$

$$= 146 \text{ g (g · mole)}^{-1}$$

$$\text{Mol wt water} = (2 \text{ H atoms})(1) + (1 \text{ O atom})(16) = 18 \text{ g (g · mole)}^{-1}$$

$$\text{Av. mol wt mer} = 0.5(116) + 0.5(146) - 18 = 113 \text{ g (g · mole)}^{-1}$$

$$\text{Degree of polymerization} = \frac{120,000}{113} = 1062$$

The degree of polymerization refers to the total number of monomers in the chain. Half of the 1062 mers are hexamethylene diamine and the other half are adipic acid.

Actually, the chains in the linear polymers are not all of the same length and thus each chain may have a different molecular weight. Two methods are used to calculate an average molecular weight for linear polymers.

The *weight average molecular weight* is obtained by dividing the chains into size ranges and determining the fraction of chains having molecular weights within that range (Table 15-3). The weight average molecular weight $\bar{M}_w$ is

$$\bar{M}_w = \frac{\Sigma f_i M_i}{\Sigma f_i} \tag{15-4}$$

where M_i is the mean molecular weight of each range and f_i is the weight fraction of the polymer having chains within that range.

The *number average molecular weight* $\bar{M}_n$ is based on the number fraction, rather than weight fraction, of the chains within each size range (Table 15-4).

$$\bar{M}_n = \frac{\Sigma x_i M_i}{\Sigma x_i} \tag{15-5}$$

TABLE 15-3 Data for the weight average molecular weight for the polymer in Example 15-9

Molecular Weight Range [g (g · mole)$^{-1}$]	Mean M_i	f_i	$f_i M_i$
0–5,000	2,500	0.01	25
5,000–10,000	7,500	0.05	375
10,000–15,000	12,500	0.07	875
15,000–20,000	17,500	0.23	4025
20,000–25,000	22,500	0.28	6300
25,000–30,000	27,500	0.22	6050
30,000–35,000	32,500	0.10	3250
35,000–40,000	37,500	0.03	1125
40,000–45,000	42,500	0.01	425
		$\Sigma = 1.00$	$\Sigma = 22,450$

TABLE 15-4 Data for the number average molecular weight for the polymer in Example 15-9

Molecular Weight Range [g (g · mole)$^{-1}$]	Mean M_i	x_i	$x_i M_i$
0–5,000	2,500	0.02	50
5,000–10,000	7,500	0.08	600
10,000–15,000	12,500	0.11	1375
15,000–20,000	17,500	0.19	3325
20,000–25,000	22,500	0.23	5175
25,000–30,000	27,500	0.25	6875
30,000–35,000	32,500	0.08	2600
35,000–40,000	37,500	0.03	1125
40,000–45,000	42,500	0.01	425
		$\Sigma = 1.00$	$\Sigma = 21,550$

where M_i is again the mean molecular weight of each size range but x_i is the fraction of the total number of chains within each range. Either $\bar{M}_w$ or $\bar{M}_n$ can be used to calculate the degree of polymerization.

EXAMPLE 15-9

The weight fraction f and the number fraction x showing the distribution of chains of differing molecular weight of a polymer are shown in Tables 15-3 and 15-4. Calculate the weight and number average molecular weights from the data.

Answer:

$$\bar{M}_w = \frac{\Sigma f_i M_i}{\Sigma f_i} = \frac{22,450}{1} = 22,450 \text{ g (g} \cdot \text{mole)}^{-1}$$

$$\bar{M}_n = \frac{\Sigma x_i M_i}{\Sigma x_i} = \frac{21,550}{1} = 21,500 \text{ g (g} \cdot \text{mole)}^{-1}$$

15-7   Deformation of Thermoplastic Polymers

The mechanical behavior of metals is often described in terms of dislocation slip. At stresses below the yield strength, elastic deformation occurs as the bond length increases. Above the yield strength, the materials deform by slip or dislocation movement.

Polymers also exhibit both elastic and plastic behavior when a stress is applied, with the total strain given by

$$\varepsilon_t = \varepsilon_e + \varepsilon_v \tag{15-6}$$

The elastic deformation ε_e is due to two mechanisms—stretching and distortion of the bonds within the chain and recoverable movement of entire segments of the chains.

Plastic deformation ε_v of polymers occurs by viscous flow rather than slip. Viscous flow occurs when the chains in the polymer slide past one another. When the stress is removed, the chains remain in their new positions and the polymer is permanently deformed (Figure 15-10).

The polymer displays *viscoelastic* behavior, or deformation occurs by a combination of elastic deformation and viscous flow. The relative contributions of the elastic and viscous components of deformation determine the ultimate behavior of the polymer.

The ease with which permanent deformation occurs is related to the viscosity of the polymer. The viscosity η, as pictured in Figure 15-11, is

$$\eta = \frac{\tau}{\Delta v/\Delta x} \tag{15-7}$$

where τ is the shear stress causing the chains to slide and $\Delta v/\Delta x$ is the velocity gradient, which is related to how rapidly the chains are displaced relative to one another. If the viscosity is high, a large stress is required to cause a given displacement. Therefore, polymers with a high viscosity have less viscous deformation.

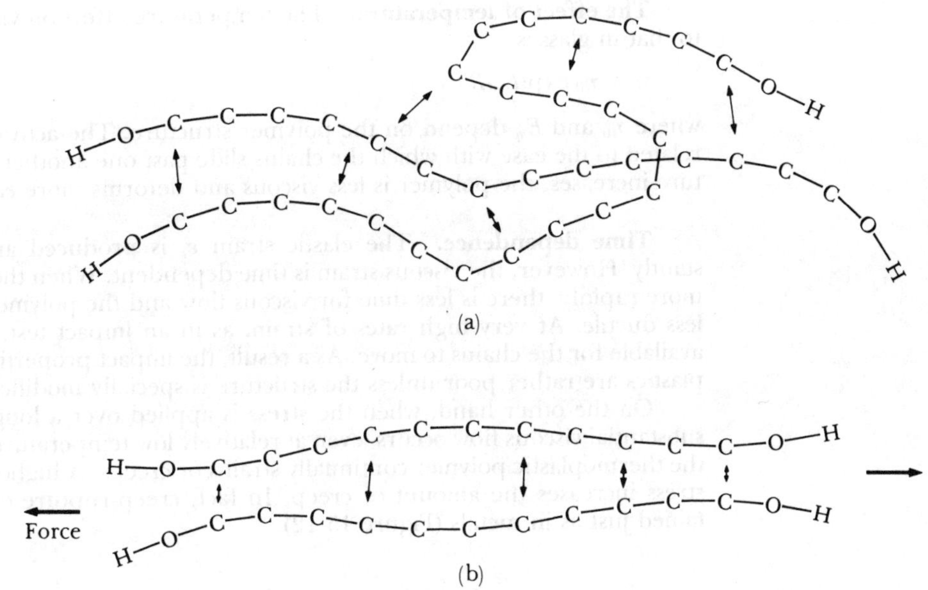

FIG. 15-10 (a) The individual chains are held together loosely by van der Waal's bonds and mechanical interlocking. (b) When the polymer is stretched, the chains are straightened out and eventually begin to slide past one another.

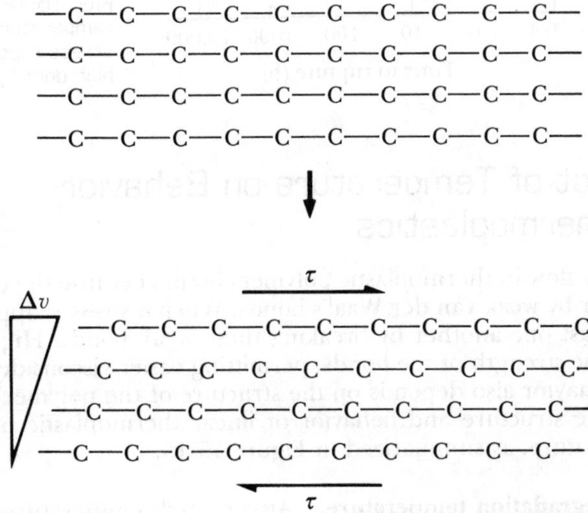

FIG. 15-11 A shear stress τ causes the polymer chains to slide over one another by viscous flow. The velocity gradient, $\Delta v/\Delta x$, produces a displacement of the chains that depends on the viscosity η of the polymer.

The effect of temperature.   The temperature effect on viscosity is identical to that in glasses,

$$\eta = \eta_0 \exp(E_\eta/RT) \tag{15-8}$$

where η_0 and E_η depend on the polymer structure. The activation energy E_η is related to the ease with which the chains slide past one another. As the temperature increases, the polymer is less viscous and deforms more easily.

Time dependence.   The elastic strain ε_e is produced and recovered instantly. However, the viscous strain is time dependent. When the strain is applied more rapidly, there is less time for viscous flow and the polymer is stronger and less ductile. At very high rates of strain, as in an impact test, there is no time available for the chains to move. As a result, the impact properties of the thermoplastics are rather poor unless the structure is specially modified.

On the other hand, when the stress is applied over a long period of time, substantial viscous flow occurs, even at relatively low temperatures. Consequently, the thermoplastic polymer continually strains or creeps. A higher temperature or stress increases the amount of creep. In fact, creep-rupture curves can be obtained just as in metals (Figure 15-12).

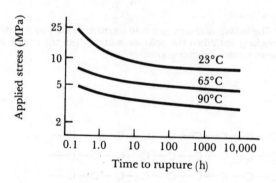

FIG. 15-12   The effect of temperature on the stress-rupture behavior of high-density polyethylene.

15-8 Effect of Temperature on Behavior of Thermoplastics

Viscous flow in thermoplastic polymers occurs because the chains are held to one another by weak van der Waal's bonds. When a stress is applied, the chains easily slide past one another by breaking these weak bonds. Higher temperatures reduce the strength of the bonds, permitting easier viscous deformation. However, this behavior also depends on the structure of the polymer chains; let's examine how the structure and behavior of linear thermoplastic polymers change with temperature, as summarized in Figure 15-13.

Degradation temperature.   At very high temperatures, the covalent bonds between the atoms in the linear chain may be destroyed—the polymer burns or chars. This temperature T_d, the *degradation temperature*, limits the usefulness of the polymer and represents the upper temperature at which the polymer can be formed into a useful shape.

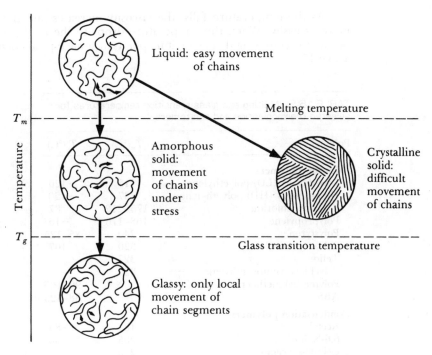

FIG. 15-13 The effect of temperature on the structure and behavior of thermoplastic polymers.

Liquid polymers (the melting temperature). When the temperature of the linear thermoplastic polymers is high, the viscosity is very low. Chains may even move without an external force and, if a force is applied, the polymer flows with virtually no elastic strain occurring. The strength and modulus of elasticity are nearly zero (Figure 15-14). The polymer is suitable for casting and many other forming processes.

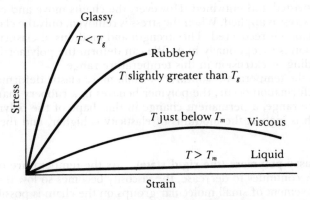

FIG. 15-14 The effect of temperature on the stress-strain behavior of the thermoplastic polymer.

As the temperature falls, the viscosity increases and the chains no longer move as easily. When the temperature drops to the *melting temperature* T_m, the polymer becomes rigid. The melting points of some typical polymers are shown in Table 15-5.

TABLE 15-5   Melting and glass transition temperatures for selected thermoplastics and elastomers

Polymer	$T_m(°C)$	$T_g(°C)$
Addition polymers		
Low-density (LD) polyethylene	115	−120
High-density (HD) polyethylene	137	−120
Polyvinyl chloride	175−212	87
Polypropylene	168−176	−16
Polystyrene	240	85−125
Polyacrylonitrile	320	107
Teflon	327	
Polychlorotrifluoroethylene	220	
Polymethyl methacrylate (acrylic)		90−105
ABS		88−125
Condensation polymers		
Acetal	181	−85
6,6-Nylon	265	50
Cellulose acetate	230	
Polycarbonate	230	145
Polyester	255	75
Elastomers		
Silicone		−123
Polybutadiene	120	−90
Polychloroprene	80	−50
Polyisoprene	30	−73

Amorphous polymers (the plastic state).   Just below the melting temperature, the polymer is rigid and maintains its shape, although the chains are still highly twisted and entwined. However, the chains move and cause deformation when a stress is applied. When the stress is removed, only the elastic portion of the deformation is recovered. The strength and modulus of elasticity are low but the elongation is exceptionally high. We can deform the polymer into useful shapes by molding or extrusion in this temperature range.

As the temperature continues to fall, more elastic deformation and less viscous deformation occur; the polymer behaves in a rubbery manner. In this temperature range, a permanent change in the shape of the polymer is limited. Its strength is higher, the modulus of elasticity is higher, and the elongation is reduced.

Glassy polymers (the rigid state).   As the temperature of the amorphous polymer continues to decrease, the viscosity becomes so low that only very localized movement of small molecular groups on the chain is possible; the chains no longer slide over one another. Below the *glass transition temperature* T_g, the linear polymer becomes hard and brittle and behaves much like a ceramic glass. Normally, the properties, such as the modulus of elasticity or the density, change at a different rate when the temperature falls below T_g (Figure 15-15). The glassy polymer is not easily formed into useful shapes and may even possess undesirable properties. Often polymers are chosen so that the glass transition temperature is below normal service temperatures.

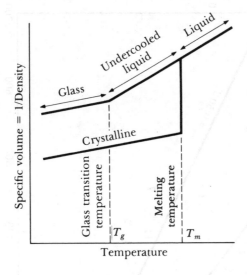

FIG. 15-15 The relationship between the specific volume and the temperature of the polymer shows the melting and glass transition temperatures.

The glass transition temperature is typically about 0.5 to 0.75 times the absolute melting temperature T_m. Simple addition chains composed of small symmetrical mers have a glass transition temperature about one-half of the melting temperature; chains produced from large or nonsymmetrical mers have higher transition temperatures (Figure 15-16).

Crystalline polymers. Some polymers crystallize when cooled below the melting temperature. Crystallization is easiest for simple addition polymers that have no large molecules or atom groups to upset the symmetry of the chain or to prevent close packing of the chains. Two models describing the shape of the chains in crystalline polymers are shown in Figure 15-17. These packing schemes produce higher densities in the crystalline polymers compared to the amorphous or glassy polymers.

The crystalline polymers have a unit cell that describes the packing of the chains. The crystal structure for polyethylene is shown in Figure 3-25. Some polymers may be allotropic; one of the nylons has three different crystal structures.

Several factors influence crystallization. Crystallization is more difficult when the polymer is composed of more than one type of monomer or nonsymmetrical molecules. Fast cooling prevents crystallization and encourages a glassy structure. Finally, deformation of the polymer between the melting and glass transition temperatures may promote crystallization by straightening the chains and bringing them into a parallel structure. Slow strain rates are more effective in causing crystallization than faster strain rates.

In crystalline polymers, the elastic deformation is low, since the chains are

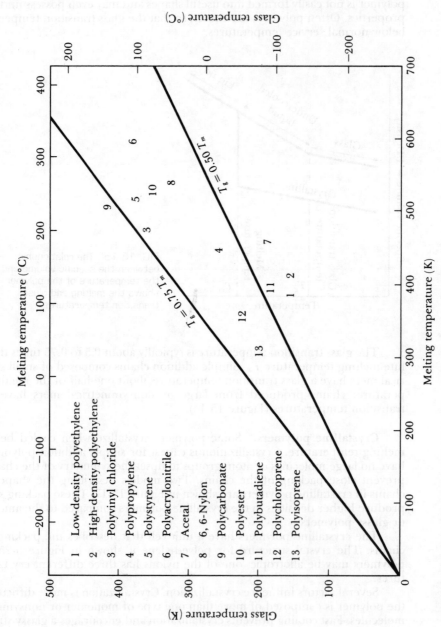

FIG. 15-16 The relationship between T_m and T_g for a number of polymers.

1 Low-density polyethylene
2 High-density polyethylene
3 Polyvinyl chloride
4 Polypropylene
5 Polystyrene
6 Polyacrylonitrile
7 Acetal
8 6,6-Nylon
9 Polycarbonate
10 Polyester
11 Polybutadiene
12 Polychloroprene
13 Polyisoprene

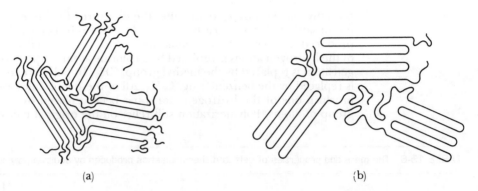

FIG. 15-17   The fringed micelle (a) and folded chain (b) models for the structure of crystalline polymers.

already straight and parallel to one another. Higher temperatures permit greater stretching of bonds but the modulus of elasticity remains high (Figure 15-18). In addition, the crystalline structure resists viscous flow until the temperature approaches the melting point.

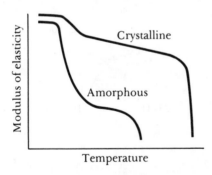

FIG. 15-18   The effect of crystallinity on the modulus of elasticity of thermoplastic polymers.

15-9   Controlling the Structure and Properties of Thermoplastics

There are many ways to modify and control the properties of linear thermoplastic polymers.

Degree of polymerization.   As the linear chains increase in length, the polymer has a higher melting temperature and improved mechanical properties. We can use the ethylene monomer to illustrate this point. Ethylene is a gas. As the degree of polymerization and the molecular weight increase, the polymerized ethylene becomes a liquid. Next, the liquid thickens into a wax. When the degree of polymerization is 200 to 500, the material is solid, has a high melting temperature, and is considered a polymer.

Type of monomer.   The ethylene molecule is the basic building block for polyethylene. However, we can alter the structure of the molecule while retaining the unsaturated pair of carbon atoms (Table 15-6). Polyvinyl chloride (PVC) is produced from vinyl chloride molecules, which resemble ethylene except that one of the hydrogen atoms is replaced by a chlorine atom. In polypropylene, a hydrogen atom is replaced by the methyl group, CH_3. In polystyrene, a hydrogen atom is replaced by the benzene ring, C_6H_5. All of these molecules are *vinyl compounds*, in which one of the hydrogen atoms has been replaced by a different atom or group of atoms. Polymerization still relies on the original pair of carbon atoms.

TABLE 15-6 The mers and properties of selected thermoplastics produced by addition polymerization

Polymer	Structure	Tensile Strength (MPa)	% Elongation	Modulus of Elasticity (GPa)	Density (Mg m⁻³)
Polyethylene low density (LD) high density (HD)		4–21 21–38	50–800 15–130	0.1–0.28 0.4–1.2	0.92 0.96
Polyvinyl chloride		34–62	2–100	2.1–4.1	1.40
Polypropylene		28–41	10–700	1.1–1.5	0.90
Polystyrene		22–55	1–60	2.6–3.1	1.06
Polymethyl methacrylate (acrylic-Plexiglass)		41–83	2–5	2.4–3.1	1.22

TABLE 15-6 (Continued)

Polymer	Structure	Tensile Strength (MPa)	% Elongation	Modulus of Elasticity (GPa)	Density (Mg m^{-3})
Polyvinylidene chloride	····—C—C—···· (H Cl / H Cl)	24–34	160–240	0.35–0.55	1.15
Polychloro-trifluoroethylene	····—C—C—···· (F Cl / F F)	31–41	80–250	1.0–2.1	2.15
Polytetra-fluoroethylene (teflon)	····—C—C—···· (F F / F F)	14–48	100–400	0.41–0.55	2.17

Vinylidene compounds are produced by substituting atoms or groups of atoms for two of the hydrogen atoms in ethylene. All four of the hydrogen atoms are replaced by fluorine in tetrafluoroethylene, or teflon.

The properties of these polymers are shown in Table 15-6. The larger atoms, such as chlorine, or groups of atoms, such as methyl or benzene groups, interfere with the sliding of the chains when a stress is applied, therefore increasing the strength of the polymer. However, if the groups are too large, they may impair properties by causing the main backbone of adjoining chains to be far apart. The poor packing prevents strong van der Waal's bonding and reduces the strength.

Replacing the hydrogen atoms also affects the strength of the bonds in the polymer. For example, the C—F bond permits teflon to have a high melting point. Teflon also has the added benefit of producing a low-friction polymer that makes teflon useful for bearings and for nonstick cooking ware.

Still more complicated thermoplastic polymers are formed from molecules having more than two carbon atoms or having atoms other than carbon in the backbone of the chain. Several important examples are shown in Table 15-7. These thermoplastics have a higher strength and stiffness than polymers with only a carbon backbone.

Branching. Branching occurs when an atom attached to the main linear chain is removed and replaced by another linear chain (Figure 15-19). Normally, branching reduces the tendency for crystallization and prevents dense packing of the chains. Low-density (LD) polyethylene, which has long branches, is weaker than high-density (HD) polyethylene, which has no branching (Table 15-6).

TABLE 15-7 The repeating units and properties for typical thermoplastics having complicated chain structures

Polymer	Structure	Tensile Strength (MPa)	% Elongation	Modulus of Elasticity (GPa)	Density (Mg m^{-3})
Polyether (acetal)	H H H ····—C—O—C—O—C—O—···· H H H	65–83	25–75	3.6	1.42
Polyamide (nylon)	H H H H H H H O H H H H H O H ····—C—C—C—C—C—C—N—C—C—C—C—C—C—N—···· H H H H H H H H H H	76–83	60–300	2.8–3.4	1.14
Polyester (dacron)	H O O H H ····—C—O—C—⬡—C—O—C—C—O—···· H H H	55–72	50–300	2.8–4.1	1.36
Polycarbonate	H—C—H ... C ... H—C—H with —O—C(=O)—O—···	62–76	110–130	2.1–2.8	1.2
Cellulose	ring structure	14–55	5–50	1.4–1.7	1.30

TABLE 15-7   (Continued)

Polymer	Structure	Tensile Strength (MPa)	% Elongation	Modulus of Elasticity (GPa)	Density (Mg m^{-3})
Polyimide	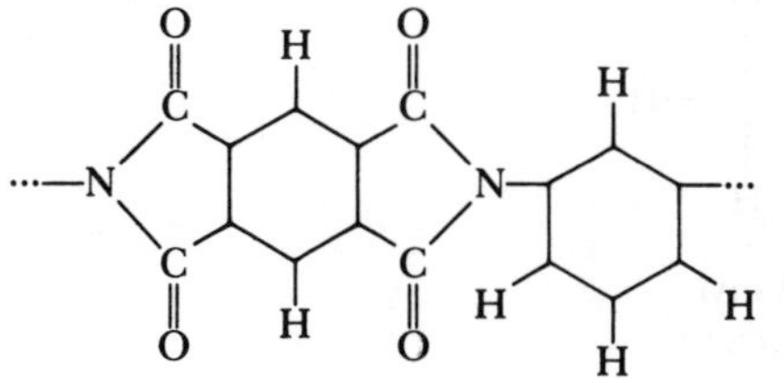	76–117	8–10	2.1	1.39

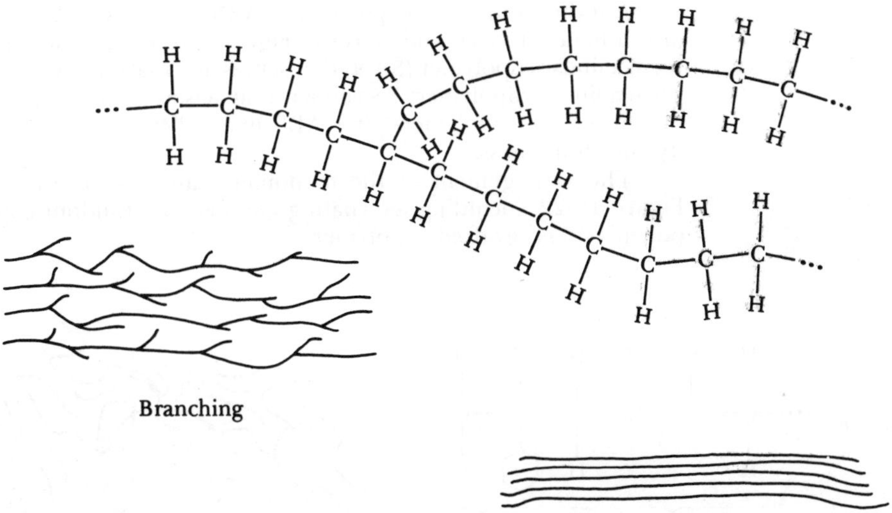

FIG. 15-19   Branching can occur in linear polymers. Branching makes crystallization more difficult.

Copolymers.   Linear addition chains composed of two or more types of molecules are called *copolymers*. For example, vinyl chloride and vinyl acetate can copolymerize (Figure 15-20). Both have unsaturated carbon bonds, permitting the mers to join without the formation of a by-product. Any ratio of the two molecules can be combined to give a range of structures, compositions, and properties.

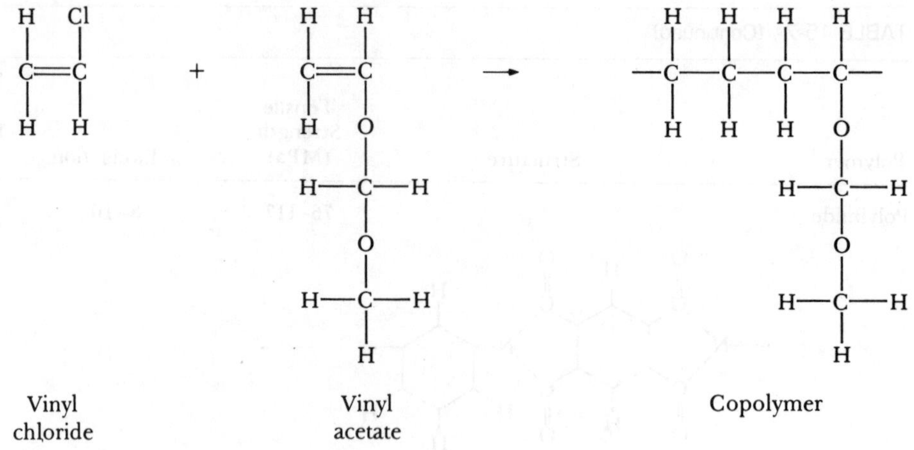

Vinyl
chloride

Vinyl
acetate

Copolymer

FIG. 15-20   Copolymerization of vinyl chloride and vinyl acetate.

Another important copolymer is ABS, composed of acrylonitrile, butadiene (a synthetic rubber), and styrene (Figure 15-21). The styrene and acrylonitrile form a linear copolymer (SAN) that serves as a matrix. Styrene and butadiene also form a linear copolymer, BS rubber, that acts as the filler material. The combination of the two copolymers gives ABS an excellent combination of strength, rigidity, and toughness.

The arrangement of the monomers can take several forms, as shown in Figure 15-22, including alternating copolymers, random copolymers, block copolymers, and grafted copolymers.

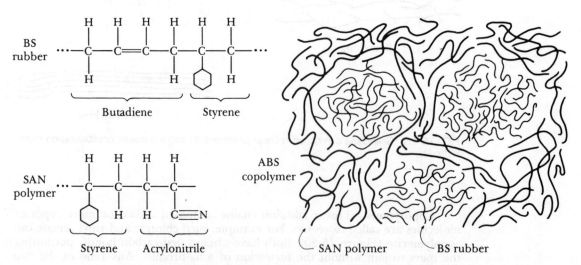

FIG. 15-21   Copolymerization produces the polymer ABS, which is really made up of two copolymers, SAN and BS.

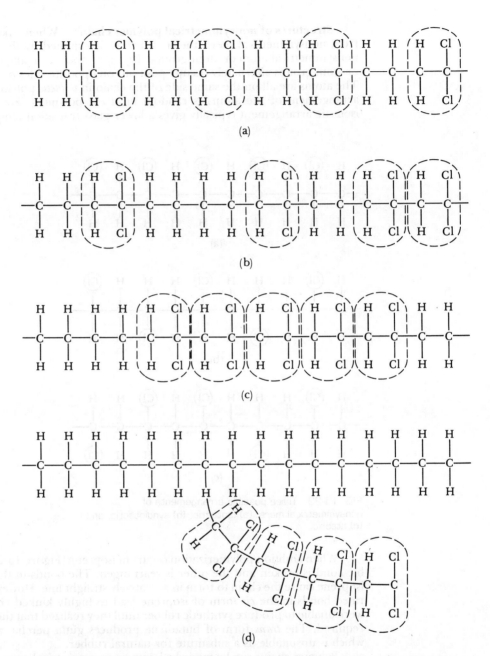

FIG. 15-22 Four types of copolymers. (a) Alternating monomers, (b) random monomers, (c) block copolymers, and (d) grafted copolymers.

Structures of nonsymmetrical polymer chains.   When a polymer is formed from nonsymmetrical mers, the structure is determined by the locations of the nonsymmetrical atom or atom groups. In the *syndiotactic* arrangement, the atoms or atom groups alternately occupy positions on opposite sides of the linear chain. The atoms are all on the same side of the chain in *isotactic* polymers, whereas the arrangement of the atoms is random in *atactic* polymers (Figure 15-23). The isotactic arrangement typically gives a lower glass transition temperature.

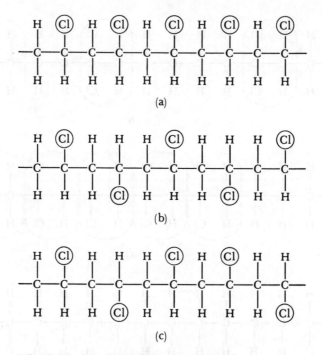

FIG. 15-23   Three possible arrangements of nonsymmetrical mers. (a) Isotactic, (b) syndiotactic, and (c) atactic.

When addition polymerization occurs in isoprene [Figure 15-24(a)], one double bond is broken and the other is rearranged. The bonds in the *trans* form of isoprene cause the chain to form in a relatively straight line. However, the unsaturated bonds in the *cis* form of isoprene lead to highly kinked chains. Chemists were unable to produce synthetic rubber until they realized that the *cis* structure is required. The *trans* form of butadiene produces gutta percha, a thermoplastic which is unsuitable as a substitute for natural rubber.

Polymer chains can be made by joining nonsymmetrical mers either head-to-head or tail-to-head, as in polyvinyl chloride [Figure 15-24(b)].

Crystallization.   Normally, crystalline polymers have higher densities and improved mechanical properties compared to amorphous or glassy polymers. Simple chains and slow cooling rates encourage crystallization. In addition, deformation straightens and aligns the chains, producing a preferred orientation and

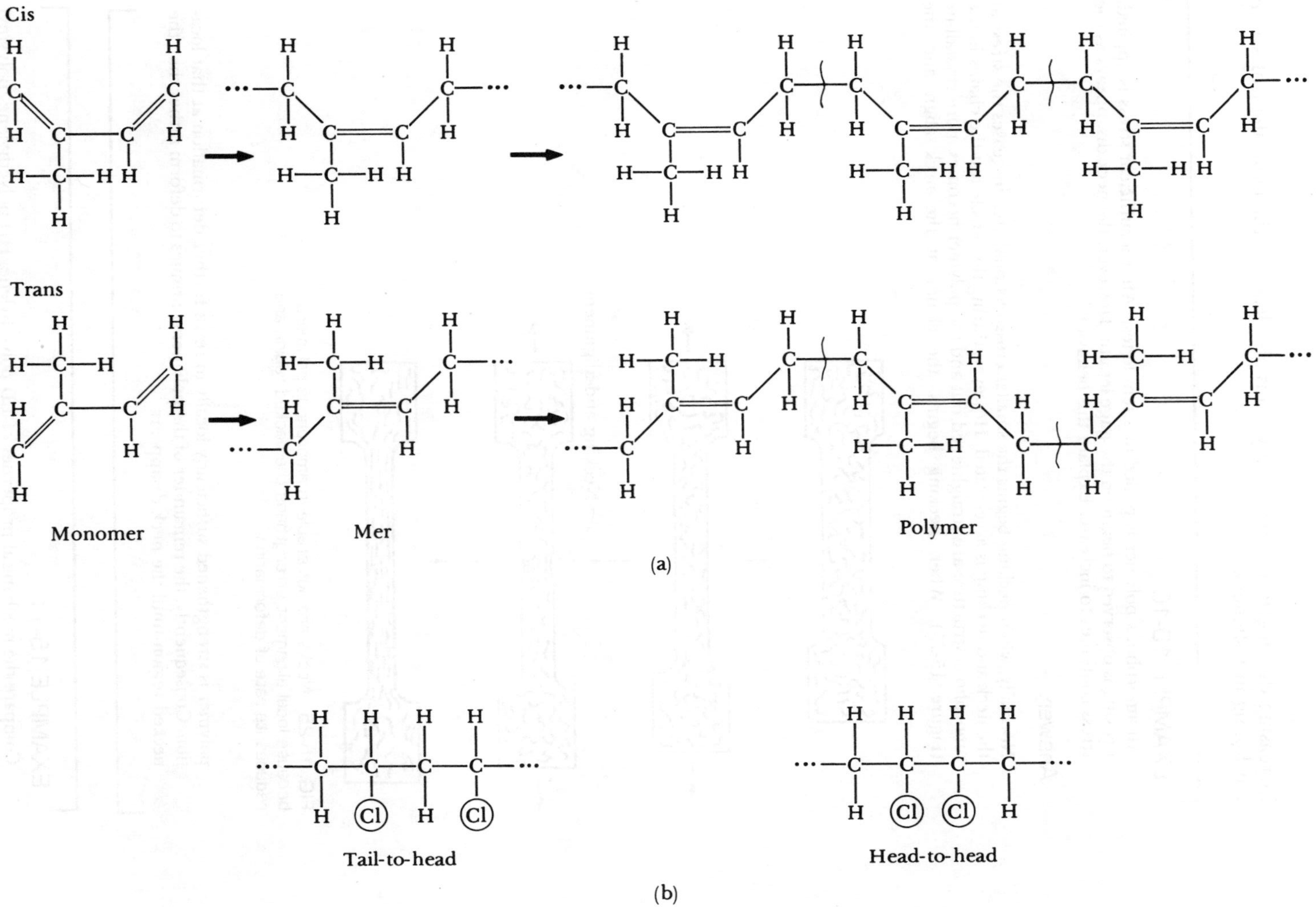

FIG. 15-24 (a) The *cis* and *trans* structures of isoprene. (b) The head-to-head versus tail-to-head arrangement of monomers in polyvinyl chloride.

increasing the degree of crystallinity. This also leads to the equivalent of strain hardening in polymers.

EXAMPLE 15-10

An amorphous polymer is pulled in a tensile test. After a sufficient stress is applied, necking is observed to begin on the gage length. However, the neck disappears as the stress continues to increase. Explain this behavior.

Answer:

Normally, when necking begins the smaller cross-sectional area increases the stress at the neck and necking is accelerated. However, during the tensile test the chains in the amorphous structure are straightened out and the polymer becomes more crystalline (Figure 15-25). When necking begins, the chains at the neck align and the

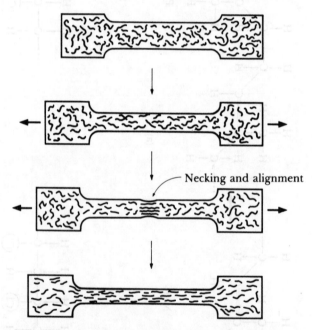

FIG. 15-25   Necks are not stable in amorphous polymers because local alignment strengthens the necked region and reduces its rate of deformation.

polymer is strengthened sufficiently locally to resist further deformation at that location. Consequently, the remainder of the polymer continues to deform rather than the necked region until the neck disappears.

EXAMPLE 15-11

Compare the mechanical properties of LD polyethylene, HD polyethylene, polyvinyl chloride, polypropylene, and polystyrene and explain their differences in terms of their structures.

Answer:

Let's look at the maximum tensile strength and modulus of elasticity for each polymer.

Polymer	Tensile Strength (MPa)	Modulus of Elasticity (GPa)	Structure
LD polyethylene	21	0.28	Highly branched amorphous structure with symmetrical mers
HD polyethylene	38	1.24	Amorphous structure with symmetrical mers but little branching
Polypropylene	41	1.52	Amorphous structure with small methyl side groups
Polystyrene	55	3.1	Amorphous structure with benzene side groups
Polyvinyl chloride	62	4.1	Amorphous structure with large chlorine atoms as side groups

We can conclude that

(a) Branching, which reduces the density and close packing of chains, reduces the mechanical properties of polyethylene.

(b) Adding atoms or atom groups other than hydrogen to the chain increases strength and stiffness. The methyl group in polypropylene provides some improvement, the benzene ring of styrene provides higher properties, and the chlorine atom in polyvinyl chloride provides a large increase in properties.

15-10   Elastomers (Rubbers)

A number of natural and synthetic linear polymers, called *elastomers,* display a large amount of elastic deformation when a force is applied. Yet deformation may be completely recovered when the stress is removed. A typical example is an elastic band. The structures of some typical elastomers are shown in Table 15-8.

In elastomers, the long polymer chain is coiled due to the *cis* arrangement of the bonds. Ideally, when a force is applied, the polymer stretches by uncoiling of the linear chains. When the stress is released, the chains recoil and the polymer returns to its original size and shape (Figure 15-26).

However, we find that the polymer does not behave in quite the way we have described. Not only do the chains uncoil, but they also slide over one another. We obtain a combination of viscous deformation (due to sliding) and elastic deformation (which, due to the coiling effect, is recovered).

Cross-linking.   We prevent viscous deformation while retaining large elastic deformation by cross-linking. *Vulcanization* joins the coiled chains with sulfur atoms using heat and pressure. The chains of natural and synthetic rubbers contain unsaturated carbon bonds. The unsaturated bonds can be broken in two adjacent chains; the bonds are satisfied by sulfur atoms, causing an addition type of cross-linking. Vulcanization of butadiene is shown in Figure 15-27.

TABLE 15-8   The repeating units and properties of selected elastomers

Polymer	Structure	Tensile Strength (MPa)	% Elongation	Density (Mg m^{-3})
Polyisoprene	(chemical structure)	21	800	0.93
Polybutadiene	(chemical structure)	24		0.94
Polybutylene	(chemical structure)	28	350	0.92
Polychloroprene (Neoprene)	(chemical structure)	24	800	1.24
Butadiene-styrene (BS or SBR rubber)	(chemical structure)	4–21	600–2000	1.0
Butadiene-acrylonitrile	(chemical structure)	5	400	1.0
Silicone	(chemical structure)	2.4–6.9	100–700	1.5

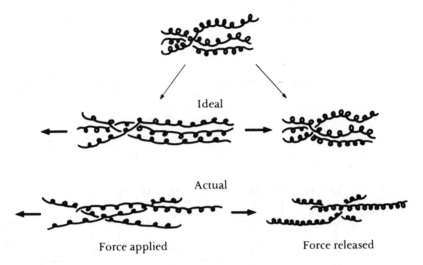

FIG. 15-26 Schematic diagram showing the recoil of an ideal elastomer versus the actual behavior.

The elasticity or rigidity of the rubber is determined by the number of cross-links, or the amount of sulfur. Low sulfur additions leave the rubber soft and flexible. Increasing the sulfur content restricts uncoiling of the chains and the rubber becomes harder, more rigid, and brittle. Up to 30% to 40% sulfur can be added to provide cross-linking in elastomers.

Polyisoprene is natural rubber and has been duplicated synthetically. A common synthetic is BS, or butadiene-styrene rubber (Figure 15-21). The BS rubber begins as a linear copolymer; since butadiene has an extra unsaturated carbon bond, cross-linking can be accomplished.

EXAMPLE 15-12

If 20% of the possible cross-linking sites in isoprene are used in vulcanization, calculate the wt% S that is present in the rubber.

Answer:

On inspection of the cross-linking process shown in Figure 15-27, we find that, on average, a maximum of one sulfur atom per monomer would be required to produce complete cross-linking.

Mol wt isoprene = (5 C atoms)(12) + (8 H atoms)(1) = 68 g (g · mole)$^{-1}$

Mol wt sulfur = 32 g (g · mole)$^{-1}$

$$\% \; S = \frac{(0.2)(1 \; S \; atom)(M_s)}{(0.2)(1 \; S \; atom)(M_s) + (1 \; mer)(M_{isoprene})} \times 100$$

$$= \frac{(0.2)(32)}{(0.2)(32) + 68} \times 100 = 8.6\%$$

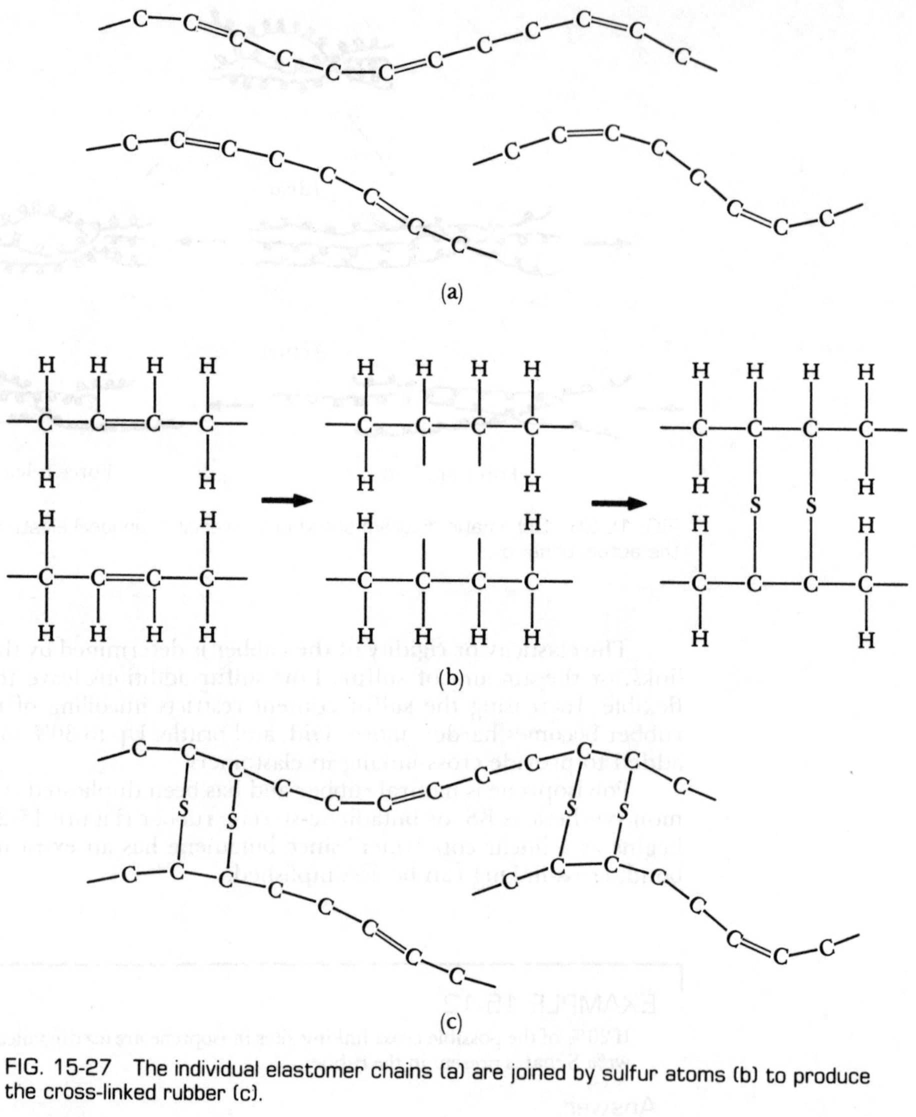

(a)

(b)

(c)

FIG. 15-27   The individual elastomer chains (a) are joined by sulfur atoms (b) to produce the cross-linked rubber (c).

Silicone rubbers are based on a linear chain composed of silicon and oxygen atoms. This type of chain provides high temperature resistance, permitting use of the rubber up to 315°C. Because the chain contains no unsaturated bonds, cross-linking is obtained by introducing a peroxide, which removes hydrogen atoms from the CH_3 groups and permits an addition-type cross-link (Figure 15-28).

Stress relaxation.   When an elastomer is subjected to a constant strain rather than a constant stress, viscous flow causes the stress acting on the polymer to decrease over time. When a stress is applied to an elastomer, its chains uncoil to

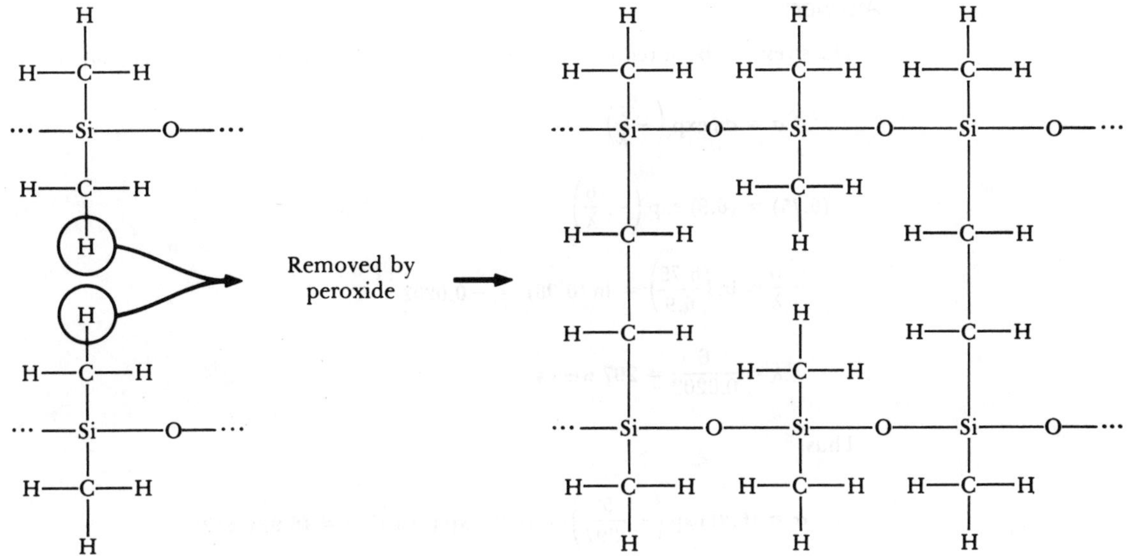

FIG. 15-28 Cross-linking in silicone rubbers.

produce a corresponding strain. Over time, some viscous flow occurs even in vulcanized rubber as the chains are rearranged. This permits the stretched chains to recoil. Since the dimensions are fixed due to the constant strain, the force or stress acting on the polymer is reduced, or relaxed.

The rate at which stress relaxation occurs depends on the *relaxation time* λ, which is a property of the polymer. The stress after time t is given by

$$\sigma = \sigma_0 \exp\left(-\frac{t}{\lambda}\right) \tag{15-9}$$

where σ_0 is the original stress. The relaxation time depends on the viscosity and therefore is affected by the temperature.

$$\lambda = \lambda_0 \exp\left(\frac{E_\eta}{RT}\right) \tag{15-10}$$

where E_η is the activation energy for viscous flow and λ_0 is a constant.

EXAMPLE 15-13

A stress of 6.9 MPa is required to stretch polyisoprene around a set of books. After six weeks, the stress acting on the rubber is only 6.75 MPa. How much stress will remain after one year?

Answer:

The stress has been reduced by stress relaxation. The strain is still unchanged. Thus,

$$\sigma = \sigma_0 \exp\left(-\frac{t}{\lambda}\right)$$

$$(6.75) = (6.9) \exp\left(-\frac{6}{\lambda}\right)$$

$$-\frac{6}{\lambda} = \ln\left(\frac{6.75}{6.9}\right) = \ln\,(0.98) = -0.0202$$

$$\lambda = \frac{6}{0.0202} = 297 \text{ weeks}$$

Thus

$$\sigma = (6.9) \exp\left(-\frac{52}{297}\right) = (6.9) \exp\,(-0.175) = (6.9)(0.839)$$

$$= 5.8 \text{ MPa}$$

15-11   Thermosetting Polymers

Thermosetting polymers typically are formed by first producing linear chains, then cross-linking the chains to produce a three-dimensional framework structure. A condensation reaction could be involved in forming the original chains or in cross-linking the chains. Often the thermosetting polymer materials are obtained in a two-part liquid resin. When the two parts are mixed, the cross-linking reaction begins. In other cases, heat and pressure are used to initiate the cross-linking action.

The functional groups for a number of thermosetting polymers are summarized in Table 15-9.

Phenolics.   The condensation reaction joining phenol and formaldehyde molecules provides the basis for the phenolic resins (Figure 15-29). The oxygen atom in the formaldehyde molecule reacts with a hydrogen atom on each of two phenol molecules and water is evolved as the by-product. The two phenol molecules are then joined by the remaining carbon atom in the formaldehyde. Since phenol is trifunctional, the same reaction can occur at other locations on each of the rings, binding each phenol molecule to several others.

Initially, a linear chain is produced. Two important resins are produced from this reaction. The *resole* resins, which contain an excess of formaldehyde, are linear chains with little or no cross-linking. When the resole resins are subsequently heated, the excess formaldehyde provides the cross-linking required to produce the three-dimensional structure.

TABLE 15-9 The functional groups for several common thermosetting polymers

Polymer	Structure	Tensile Strength (MPa)	% Elongation	Modulus of Elasticity (GPa)	Density (Mg m^{-3})
Phenolics		34–62	0–2	2.8–9.0	1.27
Amines	Melamine Urea	34–69	0–1	6.9–11.0	1.50
Polyesters		41–90	0–3	2.1–4.5	1.28
Epoxies		28–103	0–6	2.8–3.4	1.25
Urethanes		34–69	3–6		1.30

TABLE 15-9   (Continued)

Polymer	Structure	Tensile Strength (MPa)	% Elongation	Modulus of Elasticity (GPa)	Density (Mg · m⁻³)
Furan	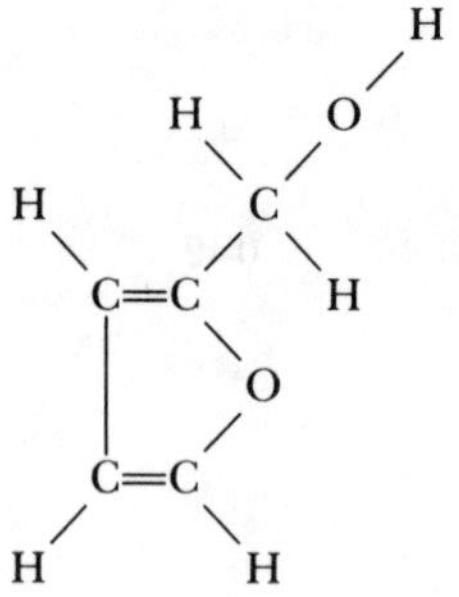	21–31		10.9	1.75
Silicone	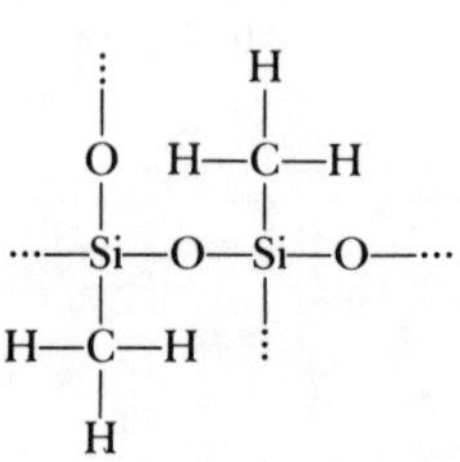	21–28	0	8.3	1.55

When phenol and formaldehyde are combined with excess phenol, a *novalac* chain is produced. The novalac resin can then be blended with hexamethylene-tetramine or other agents which, when heated, provide cross-linking.

The phenolic resins are commonly used in adhesives, coatings, laminates, and even as binders for foundry sands and cores.

Amines.   The amino resins, produced by combining urea-formaldehyde or melamine-formaldehyde, are similar to the phenolics. The urea or melamine molecules are joined by a formaldehyde link to produce linear chains. Excess formaldehyde can provide the cross-linking needed to give strong rigid polymers suitable for adhesives, laminates, and molding materials.

EXAMPLE 15-14

Describe the formation of an amino polymer using urea and formaldehyde.

Answer:

The molecular structure for the monomers of urea and formaldehyde are shown below. A linear chain can be formed when hydrogen atoms from the urea combine with the oxygen atom from the formaldehyde.

Urea   Formaldehyde   Urea

Urea-formaldehyde chain   Water

This chain can be extended indefinitely as the condensation reaction continues. If additional formaldehyde is present, other hydrogen atoms may also react with the formaldehyde to cross-link the chains into a framework structure.

Amino polymer

The functionality of the urea monomer may be as great as four.

Urethanes. Urethanes are produced when isocyanate molecules combine with other organic molecules to produce a linear polymer by an addition mechanism. The chains are then cross-linked. Depending on the exact composition and processing, the urethanes may behave as thermosetting polymers, thermoplastic polymers, or elastomers.

Polyesters. Polyesters form chains from acid and alcohol molecules by a condensation reaction, giving water as a by-product. The acid component of the chain contains an unsaturated carbon bond. A vinyl molecule, such as styrene, cross-links the chains by an addition reaction. Polyesters can be either thermosetting or thermoplastic.

EXAMPLE 15-15

A thermosetting polyester can be produced by combining adipic acid, ethylene glycol, and maleic acid to produce a linear condensation polymer chain containing an unsaturated carbon bond. Show how the chain develops and how styrene can provide cross-linking of the chains into a framework structure.

Answer:

The monomers are shown below. First, a chain forms,

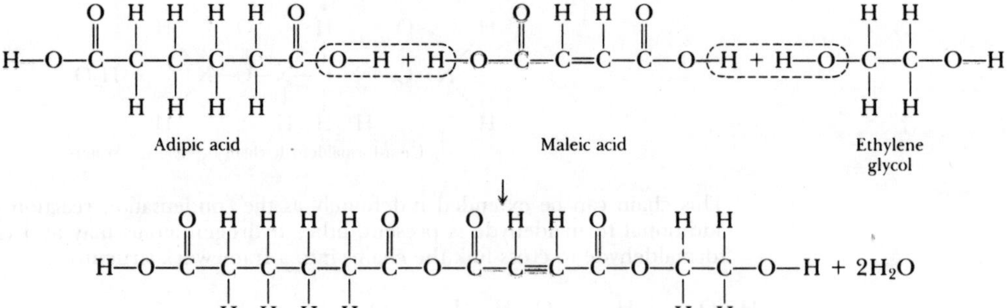

$$
\begin{array}{c}
\text{H—O—C—C—C—C—C—C(O—H + H)O—C—C=C—C—O(H + H—O)C—C—O—H} \\
\end{array}
$$

Adipic acid   Maleic acid   Ethylene glycol

$$\downarrow$$

$$
\text{H—O—C—C—C—C—C—C—O—C—C=C—C—O—C—C—O—H + 2H}_2\text{O}
$$

The linear chain contains an unsaturated carbon bond which, when styrene is added, is broken. The chains can then be joined by an addition mechanism.

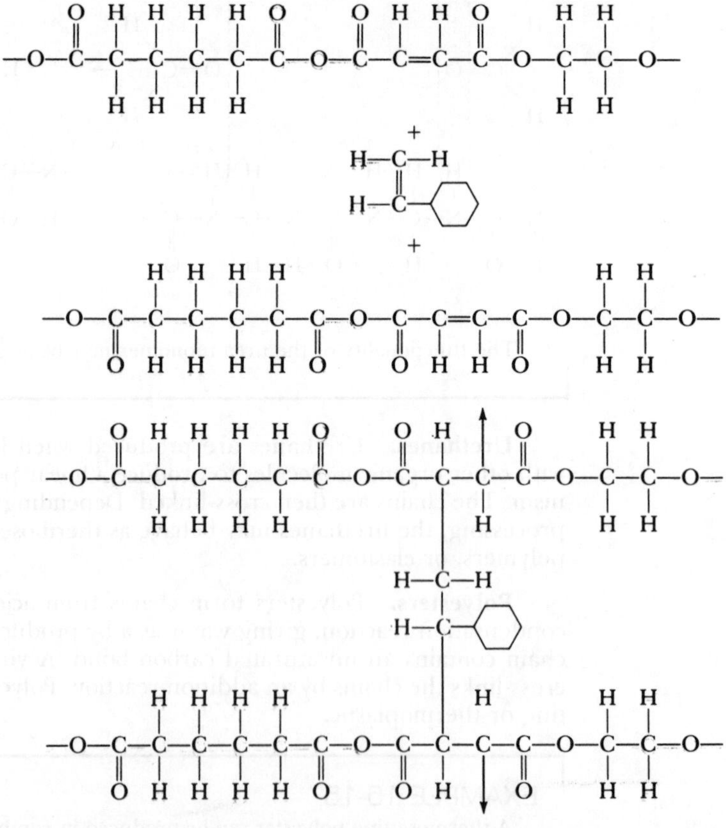

$$
\text{—O—C—C—C—C—C—C—O—C—C=C—C—O—C—C—O—}
$$

$$+$$

$$
\begin{array}{c}
\text{H—C—H} \\
\text{H—C—(hexagon)}
\end{array}
$$

$$+$$

$$
\text{—O—C—C—C—C—C—C—O—C—C=C—C—O—C—C—O—}
$$

$$
\text{—O—C—C—C—C—C—C—O—C—C—C—C—O—C—C—O—}
$$

$$
\begin{array}{c}
\text{H—C—H} \\
\text{H—C—(hexagon)}
\end{array}
$$

$$
\text{—O—C—C—C—C—C—C—O—C—C—C—C—O—C—C—O—}
$$

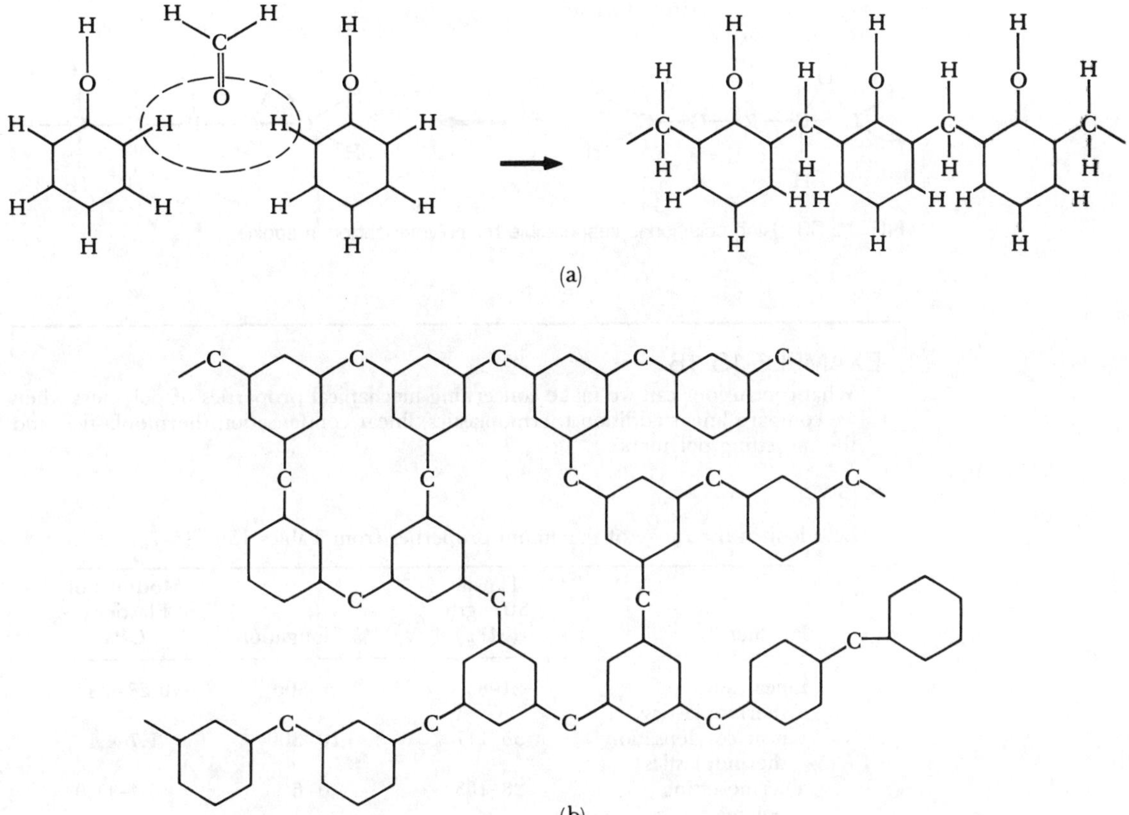

(a)

(b)

FIG. 15-29 Structure of a resole phenolic. In (a), two phenol rings are joined by a condensation reaction through a formaldehyde molecule. Eventually, a linear chain forms. In (b), excess formaldehyde serves as the cross-linking agent, producing a network, thermosetting polymer.

Epoxies. Complex monomers that contain a tight C–O–C ring are polymerized by an addition mechanism into linear chains which then join into framework structures. During polymerization, the C–O–C rings are broken (called *ring scission*) and the bonds are rearranged to join the molecules (Figure 15-30).

Deformation of thermosetting polymers. Because of the rigid three-dimensional framework structure, groups of atoms cannot readily move in thermosetting polymers. In a tensile test, thermosetting polymers should have the same behavior as a brittle metal or ceramic.

Most of the thermosetting polymers have high strength, low ductility, a high modulus of elasticity, and poor impact properties compared to other polymers. As the temperature increases, the properties decrease due to greater stretching of bonds and degradation of the polymer.

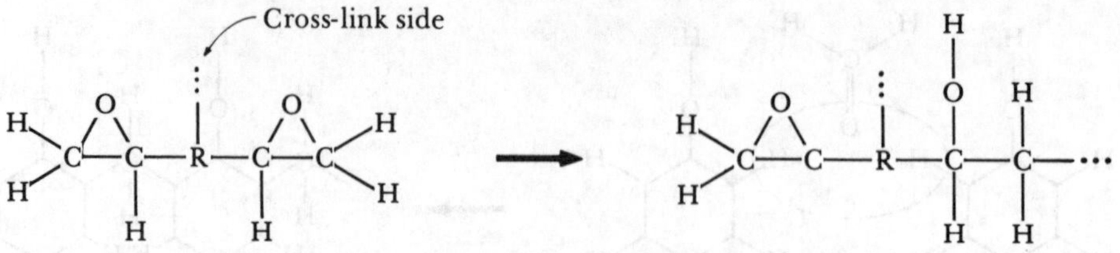

FIG. 15-30   Ring scission is responsible for polymerization in epoxies.

EXAMPLE 15-16

What conclusions can we make concerning mechanical properties of polymers when we compare linear addition thermoplastics, linear condensation thermoplastics, and thermosetting polymers?

Answer:

Let's look at the range of maximum properties from Tables 15-6, 15-7, and 15-9.

Polymer	Tensile Strength (MPa)	% Elongation	Modulus of Elasticity (GPa)
Linear addition thermoplastics	21–83	5–800	0.28–4.1
Linear condensation thermoplastics	55–117	10–300	1.7–4.1
Thermosetting polymers	28–103	0–6	3.4–11.0

The linear addition polymers have the lowest strength and stiffness but the highest ductility. The thermosets have the highest strength and stiffness but are brittle. Most linear condensation thermoplastics have intermediate properties; their molecular structure is normally more complex than that of the addition polymers but they are not cross-linked as are the thermosets.

15-12 Additives to Polymers

Most polymers contain additives which impart special characteristics to the material.

Pigments.   Pigments are used to produce colors in plastics and paints. The pigment must withstand the temperatures and pressures during processing of the polymer, must be compatible with the polymer, and must be stable.

Stabilizers.   Stabilizers prevent deterioration of the polymer due to environmental effects. Antioxidants are added to ABS, polyethylene, and polystyrene. Heat stabilizers are required in processing polyvinyl chloride. Stabilizers also prevent deterioration due to ultraviolet radiation.

Antistatic agents. Most polymers, because they are poor conductors, build up a charge of static electricity. Antistatic agents attract moisture from the air to the polymer surface, improving the surface conductivity of the polymer and reducing the likelihood of a spark or discharge.

Flame retardants. Most polymers, because they are organic materials, are flammable. Additives that contain chlorine, bromine, phosphorous, or metallic salts reduce the likelihood that combustion will occur or spread.

Lubricants. Lubricants, such as wax or calcium stearate, reduce the viscosity of the molten plastic and improve forming characteristics.

Plasticizers. Plasticizers are low molecular weight molecules or chains which, by reducing the glass transition temperature, improve the properties and forming characteristics of the polymer. Plasticizers are particularly important for polyvinyl chloride, which has a glass transition temperature well above room temperature.

Fillers. Fillers are added for a variety of purposes. Perhaps the best known example is the addition of carbon black to rubber, which improves strength and wear resistance of tires. Some fillers, such as short fibers or flakes of inorganic materials, improve the mechanical properties of the polymer. Others, called *extenders,* permit a large volume of a polymer material to be produced with relatively little actual resin. Calcium carbonate, silica, and clay are frequently used extenders.

Blowing agents. Some polymers, including urethane and polystyrene, can be expanded into hollow, cellular foams. The polymer is first produced as small solid beads containing the blowing agent. When the beads are heated, the polymer becomes plastic, the blowing agent decomposes to form a gas within the bead, and the walls of the bead expand. When the preexpanded beads are introduced into a heated mold, the beads are joined together to produce a shape. The expanded foams are excellent insulating materials with an exceptionally low density.

Reinforcement. The strength and rigidity of polymers are improved by introducing glass, polymer, or graphite filaments. For example, fiberglass consists of short filaments of glass in a polymer matrix.

Coupling agents. Coupling agents are added to improve the bonding of the polymer to inorganic filler materials, such as glass-reinforcing fibers. A variety of silanes and titanates are used for this purpose.

15-13 Forming of Polymers

The techniques used to form polymers into useful shapes depend to a large extent on the nature of the polymer—in particular, whether it is thermoplastic or thermosetting. Typical processes are shown in Figures 15-31 and 15-32.

The greatest variety of techniques is used to form the thermoplastic polymers. The polymer is heated to a temperature near or above the melting temperature so that the polymer is plastic or liquid. The polymer is then cast or injected into a mold or forced into or through a die to produce the required shape.

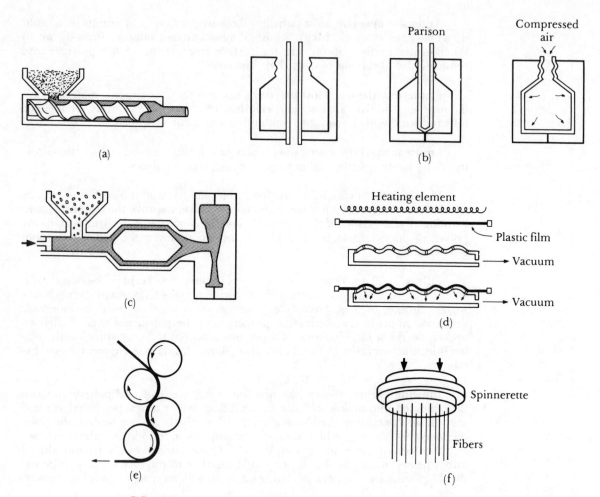

FIG. 15-31   Typical forming processes for thermoplastic polymers. (a) Extrusion, (b) blow molding, (c) injection molding, (d) vacuum forming, (e) calendering, and (f) spinning.

Fewer forming techniques are used for the thermosetting polymers because once polymerization has occurred and the framework structure is established, the thermosetting polymers are no longer capable of being formed. After vulcanization, elastomers also can be formed no further.

Extrusion.   A screw mechanism forces the heated thermoplastic through a die opening to produce solid shapes, films, sheets, tubes, pipes, and even plastic bags. Extrusion can also be used to coat wires and cables.

Blow molding.   A hot glob of plastic polymer, called a *parison*, is introduced into a die and, by gas pressure, expanded against the walls of the die. This process is used to produce plastic bottles, containers, and many other hollow shapes.

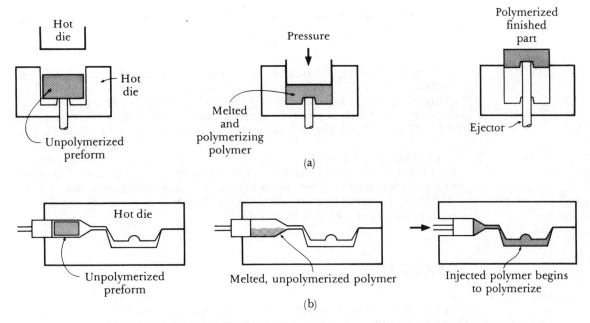

FIG. 15-32 Typical forming processes for thermosetting polymers. (a) Compression molding and (b) transfer molding.

Injection molding. Thermoplastics heated above the melting temperature can be forced into a closed die to produce a molding. This process is similar to the die casting of molten metals. A plunger or a special screw mechanism applies the pressure to force the hot polymer into the die.

Vacuum forming. Thermoplastic sheets heated into the plastic region are draped over a die or pattern connected to a vacuum system. Because of small vents in the die or pattern, the vacuum pulls the hot plastic sheet into conformation with the pattern. A unique application of this technique is the V process for making molds in the metals casting industry.

Calendering. In a *calender,* molten plastic is poured into a set of rolls with a small opening. The rolls squeeze out a thin sheet or film of the polymer. A large amount of polyvinyl chloride sheet is made in this manner.

Drawing and rolling. These processes produce fibers or change the shape of extrusions. In addition to producing the final dimensions, these processes cause crystallization and a preferred orientation of the chains in thermoplastic polymers.

Spinning. Filaments, fibers, and yarns may be produced by spinning, which is really an extrusion process. The molten thermoplastic polymer is forced through a die containing many tiny holes. The die, called a *spinnerette,* can rotate and produce a yarn.

Compression molding.   Thermosetting polymers can be formed by placing the solid material into a heated die. Application of a high pressure and temperature causes the polymer to melt, fill the die, and immediately begin to harden.

Transfer molding.   A double chamber is used in transfer molding of thermosetting polymers. The polymer is heated under pressure in one chamber; after melting, the polymer is injected into the adjoining die cavity. This process combines elements of both compression and injection molding and permits some of the advantages of injection molding to be used for thermosetting polymers.

Casting.   Many polymers can be cast into molds and permitted to solidify. The molds may be plate glass for producing individual thick plastic sheets or moving stainless steel belts for continuous casting of thinner sheets. Hollow rubber balls can be made by a centrifugal type of casting; as the mold rotates, the molten plastic is spun against the mold wall.

SUMMARY

Polymers fit into three categories—thermoplastics, thermosets, and elastomers—depending on the structure and properties that are achieved. The properties of thermoplastics, which include good ductility but low strength, are controlled by preventing the linear chains from sliding, much like our control of slip in metals. The properties of elastomers, which include exceptional elastic elongations, are controlled by cross-linking. The structure of the thermosets is typified by a network that provides high strengths but poor ductility. The importance of structure and processing on the properties is particularly evident in polymer materials.

GLOSSARY

Addition polymers Polymer chains built up by adding monomers together without creating a by-product.

Atactic The structure formed in a polymer composed of nonsymmetrical mers when the nonsymmetrical group is randomly arranged along the chain.

Branching When a separate polymer chain is attached to another chain.

Cis A monomer form in which the unsaturated bonds are located on the same side of the molecule.

Condensation polymer Polymer chains built up by a chemical reaction between two or more molecules, producing a by-product.

Copolymer An addition polymer produced by joining more than one type of monomer.

Cross-linking Attaching chains of polymers together to produce a three-dimensional network polymer.

Degradation temperature The temperature above which a polymer burns, chars, or decomposes.

Degree of polymerization The number of monomers in a polymer.

Elastomers Polymers possessing a highly kinked and partly cross-linked chain structure permitting the polymer to have exceptional elastic deformation.

Extenders Additives or fillers to polymers that provide bulk at a low cost.

Functionality The number of sites on a monomer at which polymerization can occur.

Glass transition temperature The temperature below which the amorphous polymer assumes a rigid glassy structure.

Isomer A molecule that has the same composition as but a structure different from a second molecule.

Isotactic Polymers in which nonsymmetrical groups in the monomer are all located on the same side of the polymer.

Mer The molecule from which a polymer is produced after the double covalent bond has been broken.

Monomer The molecule from which a polymer is produced.

Parison A hot glob of soft or molten polymer that is blown or formed into a useful shape.

Plasticizer An additive that reduces the glass transition temperature, thus improving the formability of a polymer.

Reinforcement Additives to polymers designed to provide significant improvement in strength. Fibers are typical reinforcements.

Spinnerette An extrusion die containing many small openings through which the hot or molten polymer is forced to produce filaments. Rotation of the spinnerette twists the filaments into a yarn.

Stress relaxation A reduction of the stress acting on a material over a period of time at a constant strain due to viscoelastic deformation.

Syndiotactic Polymers in which nonsymmetrical groups in the monomer alternate from one side of the chain to the other.

Thermoplastic polymers Polymers that can be reheated and remelted numerous times, since no by-product forms during processing.

Thermosetting polymers Polymers that polymerize at high temperatures, releasing a by-product and thus restricting their recyclability.

Trans A monomer form in which the unsaturated bonds are located on opposite sides of the molecule.

Unsaturated bond The double or even triple covalent bond joining two atoms together in an organic molecule. When a single covalent bond replaces the unsaturated bond, polymerization can occur.

Viscoelasticity The deformation of a polymer by viscous flow of the chains or segments of the chains when a stress is applied.

Vulcanization Cross-linking elastomer chains by introducing sulfur atoms at elevated temperatures and pressures.

PRACTICE PROBLEMS

1 Draw the structure of ethane, C_2H_6, and explain why ethane does not polymerize.

2 The formula for formaldehyde is HCHO. Draw the structure of the formaldehyde molecule and mer and show how an acetal polymer is produced from the mer by an addition reaction. What is the backbone of the linear chain?

3 Is it possible to combine formaldehyde molecules by a condensation reaction to produce a linear chain based only on carbon atoms, with H_2O being driven off as a reaction by-product? Explain or show.

4 Determine the change in energy each time another formaldehyde molecule is joined to an acetal polymer by the linear addition method, Problem 2.

5 Sketch the structure of the acrylic polymer produced from methyl methacrylate. Explain why crystallization is difficult.

6 Suppose the distance between the centers of two carbon atoms in polyethylene is 1.5 Å. How long is the polymer chain if the degree of polymerization is 550?

7 If the distance between the centers of two carbon atoms is 1.5 Å, determine the number of vinyl chloride mers in a chain that is 1000 Å long.

8 Suppose 1 kg of ethylene is to be polymerized. (a) How many g · moles of ethylene are required? (b) How much energy is released during polymerization? (c) If the specific heat of polyethylene is 2300 J $(kg · K)^{-1}$ and if all of the energy produced during polymerization remains in the polymer, what will be the temperature rise during polymerization? Does this illustrate the importance for cooling during polymerization?

9 Suppose a total of 45.4 kg of 6,6-nylon is produced. (See Example 15-4.) (a) How many individual condensation steps are required? (b) How much water is given off? (c) The specific heat of nylon is 1700 J (kg · K)$^{-1}$. Calculate the temperature change during polymerization if no heat is lost or gained to the surroundings.

10 Determine the change in energy when 2000 g of teflon are produced.

11 How much hydrogen peroxide initiator must be added to 10,000 g of polypropylene to produce an average degree of polymerization of 750?

12 Suppose 1 g of hydrogen peroxide is added to 10,000 g of acrylonitrile. Determine the degree of polymerization and the average molecular weight of the polymer. (See Figure 15-21.)

13 Suppose 5 g of hydrogen peroxide are added to 10,000 g of ethylene. Determine the average molecular weight of the polymer if all of the hydrogen peroxide is consumed during initiation and termination.

14 Determine the degree of polymerization if polyvinyl chloride has a molecular weight of 220,000 g (g · mole)$^{-1}$.

15 Determine the degree of polymerization if dacron, or polyethylene terephthalate, has a molecular weight of 200,000 g (g · mole)$^{-1}$. (See Figure 15-9.)

16 Determine the degree of polymerization when a copolymer containing equal parts of vinyl chloride and vinylidene chloride has a molecular weight of 125,000 g (g · mole)$^{-1}$.

17 Determine the ratio between vinyl chloride and vinyl acetate monomers in a copolymer having a degree of polymerization of 800 and a molecular weight of 60,000 g (g · mole)$^{-1}$. (See Figure 15-20.)

18 From the data below, determine the weight average molecular weight and the number average molecular weight.

Molecular Weight Range [g (g · mole)$^{-1}$]	f_i	x_i
0–5,000	0.03	0.02
5,000–10,000	0.10	0.05
10,000–15,000	0.19	0.13
15,000–20,000	0.22	0.24
20,000–25,000	0.24	0.23
25,000–30,000	0.12	0.17
30,000–35,000	0.06	0.10
35,000–40,000	0.04	0.06

19 From the data below, determine the weight average molecular weight and the number average molecular weight.

Molecular Weight Range [g (g · mole)$^{-1}$]	f_i	x_i
0–2,500	0.01	0.03
2,500–5,000	0.08	0.10
5,000–7,500	0.19	0.22
7,500–10,000	0.27	0.36
10,000–12,500	0.23	0.19
12,500–15,000	0.11	0.07
15,000–17,500	0.06	0.02
17,500–20,000	0.05	0.01

20 How much hydrogen peroxide must be added to 1000 g of polymethyl methacrylate to produce an average molecular weight of 250,000 g (g · mole)$^{-1}$ in each chain?

21 Calculate the degree of polymerization in 6,6-nylon having a molecular weight of 250,000 g (g · mole)$^{-1}$.

22 The triclinic unit cell of crystalline 6,6-nylon has lattice parameters of $a_0 = 4.9$ Å, $b_0 = 5.4$ Å, $c_0 = 17.2$ Å, $\alpha = 48.5°$, $\beta = 77°$, and $\gamma = 63.5°$. The density of nylon is about 1.14 Mg m^{-3}. (a) Calculate the number of monomers of hexamethylene diamine and adipic acid in each unit cell. (b) Calculate the number of carbon, nitrogen, hydrogen, and oxygen atoms in each unit cell. Remember that water is evolved during polymerization.
$$V_{\text{triclinic}} = a_0 b_0 c_0 \sqrt{1 - \cos^2 \alpha - \cos^2 \beta - \cos^2 \gamma + 2 \cos \alpha \cos \beta \cos \gamma}$$

23 Suppose we make BS rubber by combining equal numbers of styrene and butadiene monomers. Determine the wt% S required to cross-link half of the sites.

24 How much sulfur must be added to completely cross-link 100 kg of butadiene rubber?

25 Suppose 2.27 kg of sulfur are added to 172.4 kg of polyisoprene. What fraction of the possible cross-linking sites are used?

26 Suppose a butadiene-acrylonitrile rubber is produced by adding one acrylonitrile mer to nine butadiene mers. How much sulfur would be needed to completely cross-link 45.4 kg of the rubber?

27 An amino polymer can be produced by reacting formaldehyde with melamine. (a) What is the functionality of melamine? (b) Show how polymerization occurs. (c) Calculate the change in energy during the condensation reaction. (d) Calculate the weight of water evolved when 100 g of melamine are completely polymerized with formaldehyde.

28 The monomer for furan—furfuryl alcohol—is shown in Table 15-9. What is the functionality of the monomer?

29 (a) What percent formaldehyde must be added to phenol to produce novalac chains with no cross-linking? (b) How much additional formaldehyde is required to cross-link each phenol monomer to an adjacent chain?

30 Suppose 11.34 kg of formaldehyde are added to 45.4 kg of phenol. (a) Does this ratio produce a resole or novalac chain? Explain. (b) What fraction of the possible cross-link sites are satisfied?

31 22.7 kg of phenol are to be completely polymerized into a framework structure using formaldehyde. (a) How much formaldehyde is required? (b) How much energy is released in each condensation step?

32 The viscosity of polymers is typically 10^{10} N s $\cdot$ m^{-2} at the glass transition temperature and 10^3 N s $\cdot$ m^{-2} at a temperature 35°C above the glass transition temperature. Estimate the activation energy for viscous flow in low density polyethylene.

33 (a) How many kg of formaldehyde are required to produce linear condensation chains in 45.4 kg of urea? (b) After the condensation by-product is formed, what is the total weight remaining? (c) How much formaldehyde must be added to cross-link the amino chains at a quarter of the sites available after chain formation? (d) Calculate the total weight of the amino polymer after the condensation by-product is evolved.

34 A thermosetting polyester produced from adipic acid, maleic acid, and ethylene glycol first produces a linear chain. The polyester contains one ethylene glycol monomer for two adipic acid monomers and two maleic acid monomers. If 500 ethylene glycol monomers are present in each chain, calculate the degree of polymerization and the molecular weight of the chains.

35 In Problem 34, how much water is produced if 22.7 kg of ethylene glycol are used?

36 If 22.7 kg of ethylene glycol are used in Problem 34, how much styrene must be used to completely cross-link the linear chains at all available unsaturated carbon bonds?

37 Suppose a thermosetting polyester is obtained by mixing equal numbers of monomers of adipic acid, maleic acid, and ethylene glycol to obtain a degree of polymerization of 350. The amount of maleic acid added is 100 g. (a) How many carbon atoms are in each chain before cross-linking? (b) How much styrene must be added to provide cross-links at one of every ten unsaturated carbon bonds on each chain?

38 Explain why a thermosetting network polyester cannot be produced using only adipic acid and ethylene glycol. (Compare to the polyester produced from adipic acid, ethylene glycol, and maleic acid.)

39 Suppose 4 kg of sulfur are added to vulcanize 800 kg of polychloroprene. What fraction of the possible cross-linking sites are used?

40 Suppose 178 g of sulfur must be added to 500 g of butadiene-styrene rubber to provide complete cross-linking of the elastomer. How many grams of styrene must be present in the original rubber for this to occur?

41 A 3.45 MPa stress is applied to an elastomer. After 10 weeks at 27°C, the stress is reduced to 3.31 MPa by stress relaxation; after 10 weeks at 50°C, the stress drops to 2.21 MPa. Calculate the activation energy E_η for viscous flow in the elastomer.

42 An amino polymer is produced by reacting urea and formaldehyde. (a) Calculate the change in energy during the condensation reaction each time two urea mers are joined by a formaldehyde molecule. (b) Calculate the weight of water evolved when 4.54 kg of urea are completely cross-linked with the formaldehyde. (c) Calculate the final weight of the amino polymer.

43 Suppose a thermoplastic polymer can be produced in sheet form either by rolling or by continuous casting. In which case would you expect to obtain the higher strength? Explain.

44 You want to produce a complex component from an elastomer. Should you vulcanize the rubber before or after the forming operation? Explain.

45 Compare the maximum tensile strength-to-weight ratios of the following materials, using the tables in Chapters 11, 12, and 15. What conclusions can you reach concerning the use of polymers rather than metals for high-strength applications?

high-density polyethylene
polyvinylidene chloride
polycarbonate
amine
epoxy
> 99% aluminum
pure annealed copper
age-hardened copper–2% beryllium
0.4% C steel (slowly cooled)

46 Compare the maximum modulus of elasticity-to-weight ratios of the following materials, using Table 6-4 and the tables in this chapter. What can you conclude concerning the suitability of polymers for applications requiring high stiffness yet low weight?

low-density polyethylene
polypropylene

Teflon
polyamide (nylon)
polyester
silicone
magnesium alloys
aluminum alloys
copper alloys
steels

CHAPTER 16

Composite Materials

16-1 Introduction

Composites are produced when two materials are joined to give a combination of properties that cannot be attained in the original materials. Composite materials may be selected to give unusual combinations of stiffness, strength, weight, high-temperature performance, corrosion resistance, hardness, or conductivity. Composites can be metal-metal, metal-ceramic, metal-polymer, ceramic-polymer, ceramic-ceramic, or polymer-polymer. Metal-ceramic composites, for example, include cemented carbide cutting tools, silicon carbide fiber-reinforced titanium, and enameled steel.

Composites can be placed into three categories—particulate, fiber, and laminar—based on the shapes of the materials (Figure 16-1). Concrete, a mixture of cement and gravel, is a particulate composite; fiberglass, containing glass fibers embedded in a polymer, is a fiber-reinforced composite; and plywood, having alternating layers of wood veneer, is a laminar composite. If the reinforcing particles are uniformly distributed, particulate composites have isotropic properties; fiber composites may be either isotropic or anisotropic; laminar composites always have anisotropic behavior.

16-2 Particulate-Reinforced Composite Materials

In particulate-reinforced composites, discrete uniformly dispersed particles of a hard brittle material are surrounded by a softer more ductile matrix. In fact, the structure resembles that of many of the two-phase dispersion-strengthened metal alloys. However, in composites, a phase transformation is not used to introduce the dispersed particles.

We can subdivide the particulate composite materials into two general categories based on the size of the particles and the nature by which the particles influence the properties of the composite. These two categories include (a) dispersion-strengthened composites and (b) true particulate composites.

16-3 Dispersion-Strengthened Composites

The particle size in dispersion-strengthened composites is very small, with diameters of 100 Å to 2500 Å. Because the small particles block the movement of

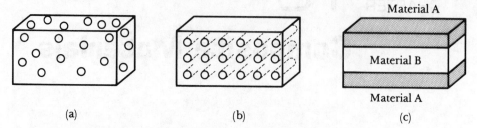

FIG. 16-1   Comparison of the three types of composite materials. (a) Particulate composite, (b) fiber-reinforced composite, and (c) laminar composite.

dislocations, they produce a pronounced strengthening effect. Only small amounts of the dispersed material are required.

At normal temperatures, the dispersion-strengthened composites are no stronger than two-phase metal alloys. However, because the dispersion-strengthened composites do not catastrophically soften by overaging, overtempering, grain growth, or coarsening of the dispersed phase, the strength of the composite decreases gradually as the temperature increases (Figure 16-2). Furthermore, creep resistance is superior to that of metals and alloys (Figure 16-3).

Considerations in selecting the dispersant.   The properties of the dispersion-strengthened composites are optimized if we consider the following guidelines.

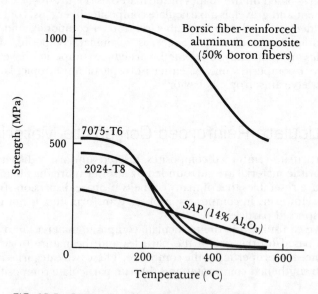

FIG. 16-2   Comparison of the yield strength of dispersion-strengthened sintered aluminum powder (SAP) composite with two conventional two-phase high-strength aluminum alloys. The composite has benefits above about 300°C. A fiber-reinforced aluminum composite is shown for comparison.

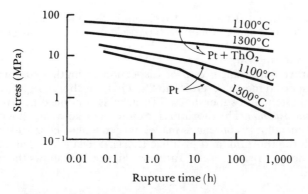

FIG. 16-3 Dispersion-strengthened platinum, containing 12.5% ThO$_2$, has much higher creep resistance than pure platinum.

1. The dispersed phase, typically a hard stable oxide, must be an effective obstacle to slip.

2. The dispersed material must have an optimum size, shape, distribution, and amount.

3. The dispersed material must have a low solubility in the matrix material. Furthermore, no chemical reactions should occur between the dispersant and the matrix. Alumina does not readily dissolve in aluminum; thus alumina is an effective dispersant for aluminum alloys. However, copper oxide will dissolve in copper at high temperatures; the Cu-Cu$_2$O system would not be effective.

4. Good bonding must be achieved between the dispersed material and the matrix. A small solubility of the dispersant in the matrix may help produce a good firm bond.

Examples of dispersion-strengthened composites. Table 16-1 lists some materials of interest. Perhaps the classic example is the sintered aluminum powder, or SAP, composite. The SAP material has an aluminum matrix strengthened by up to 14% Al$_2$O$_3$. The composite can be formed by a powder metallurgy process; aluminum and alumina powders are blended, compacted at high pres-

TABLE 16-1 Examples and applications of selected dispersion-strengthened composites

System	Applications
Ag-CdO	Electrical contact materials
Al-Al$_2$O$_3$	Possible use in nuclear reactors
Be-BeO	Aerospace and nuclear reactors
Co-ThO$_2$, Y$_2$O$_3$	Possible creep-resistant magnetic materials
Ni-20% Cr-ThO$_2$	Turbine engine components
Pb-PbO	Battery grids
Pt-ThO$_2$	Filaments, electrical components
W-ThO$_2$, ZrO$_2$	Filaments, heaters

sures, and sintered. In a second technique, the aluminum powder is treated to give a continuous oxide film on each particle. When the powder is compacted, the oxide film fractures into tiny particles which are surrounded by the aluminum metal during sintering.

Another important group of dispersion-strengthened composites includes thoria-dispersed metals. Even 1% to 2% ThO_2, or thoria, significantly strengthens nickel, tungsten, and superalloys. TD-nickel is produced in several ways, including *internal oxidation*. The thorium is present in nickel as an alloying element; after a powder metallurgy compact is made, oxygen is allowed to diffuse into the metal, react with the thorium, and produce thoria. Yttria (Y_2O_3), alumina (Al_2O_3), and other oxides perform the same function. Figure 16-4 shows the microstructure of TD-nickel.

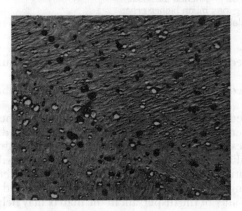

FIG. 16-4 Electron micrograph of TD-nickel. The dispersed ThO_2 particles have a diameter of 3000 Å or less. From *Oxide Dispersion Strengthening*, p. 714, Gordon and Breach, 1968. © AIME.

EXAMPLE 16-1

Suppose 2 wt% ThO_2 is added to nickel. Each ThO_2 particle has a diameter of 1000 Å. How many particles are present in each cubic centimeter?

Answer:

The densities of ThO_2 and nickel are 9.69 Mg m^{-3} and 8.9 Mg m^{-3}, respectively. The volume fraction is

$$f_{ThO_2} = \frac{\dfrac{2}{9.69}}{\dfrac{2}{9.69} + \dfrac{98}{8.9}} = 0.0184 \text{ cm}^3 \text{ ThO}_2 \text{ per cm}^3$$

The volume of each ThO_2 sphere is

$$V_{ThO_2} = \frac{4}{3}\pi r^3 = \frac{4}{3}\pi(0.5 \times 10^{-5} \text{ cm})^3 = 0.52 \times 10^{-15} \text{ cm}^3$$

$$\text{Number of } ThO_2 = \frac{0.0184}{0.52 \times 10^{-15}} = 35.4 \times 10^{12} \text{ particles/cm}^3$$

16-4   True Particulate Composites

The true particulate composites contain large amounts of coarse particles that do not effectively block slip. The particulate composites are designed to produce unusual combinations of properties rather than to improve strength.

Rule of mixtures.   Certain properties of a particulate composite depend only on the relative amounts and properties of the individual constituents. The *rule of mixtures* can accurately predict these properties. The density of a particulate composite, for example, is

$$\rho_c = \Sigma f_i \rho_i = f_1 \rho_1 + f_2 \rho_2 + \cdots + f_n \rho_n \tag{16-1}$$

where ρ_c is the density of the composite, $\rho_1, \rho_2, \ldots, \rho_n$ are the densities of each constituent in the composite, and $f_1, f_2, \ldots, f_n$ are the volume fractions of each constituent. Unfortunately, properties such as hardness and strength cannot be predicted by the rule of mixtures.

EXAMPLE 16-2

A cemented carbide cutting tool used for machining contains 75 wt% WC, 15 wt% TiC, 5 wt% TaC, and 5 wt% Co. Estimate the density of the composite.

Answer:

First, we must convert the weight percentages to volume fraction. The densities of the components of the composite are

$$\rho_{WC} = 15.77 \text{ Mg m}^{-3} \quad \rho_{TiC} = 4.94 \text{ Mg m}^{-3}$$

$$\rho_{TaC} = 14.5 \text{ Mg m}^{-3} \quad \rho_{Co} = 8.90 \text{ Mg m}^{-3}$$

$$f_{WC} = \frac{\dfrac{75}{15.77}}{\dfrac{75}{15.77} + \dfrac{15}{4.94} + \dfrac{5}{14.5} + \dfrac{5}{8.9}} = \frac{4.76}{8.70} = 0.547$$

$$f_{TiC} = \frac{\dfrac{15}{4.94}}{8.70} = 0.349$$

$$f_{TaC} = \frac{\dfrac{5}{14.5}}{8.70} = 0.040$$

$$f_{Co} = \frac{\dfrac{5}{8.90}}{8.70} = 0.064$$

From the rule of mixtures, the density of the composite is

$$\rho_c = \Sigma f_i \rho_i = (0.547)(15.77) + (0.349)(4.94) + (0.040)(14.5)$$

$$+ (0.064)(8.9)$$

$$= 11.50 \text{ Mg m}^{-3}$$

16-5 Applications for Particulate Composites

Particulate composite materials include many combinations of metals, ceramics, and polymers. A number of these systems will be discussed to illustrate the wide range of materials, manufacturing processes, and applications that are used.

Cemented carbides.   Cemented carbides contain hard ceramic particles dispersed in a metallic matrix. Tungsten carbide inserts used for cutting tools in machining operations are typical of this group. Tungsten carbide, WC, is extremely hard (Table 16-2) and can cut quenched and tempered steels. The carbide is also very stiff, so close tolerances can be held during machining, and has a very high melting temperature, so that high temperatures generated during rapid machining can be tolerated. Unfortunately, tools constructed from tungsten carbide are extremely brittle.

To improve toughness, tungsten carbide particles are combined with cobalt powder and pressed into powder compacts. The compacts are heated above the melting temperature of the cobalt. The liquid cobalt surrounds each of the solid tungsten carbide particles (Figure 16-5). After solidification, the cobalt serves as the binder for tungsten carbide and provides good impact resistance. As the tungsten carbide particles at the cutting surface become dull, they either fracture

TABLE 16-2 Hardness of selected materials

Material	Formula	Mohs Hardness	Knoop Hardness
Talc	$3MgO \cdot 4SiO_2 \cdot H_2O$	1	
Gypsum	$CaSO_4 \cdot 2H_2O$	2	32
Silver	Ag		60
Calcite	$CaCO_3$	3	135
Ferrite	α-Fe		154
Fluorite	CaF_2	4	163
Copper	Cu		163
Apatite	$CaF_2 \cdot 3Ca_3(PO_4)_2$	5	430
Pearlite	α-$Fe + Fe_3C$		453
Nickel	Ni		557
Feldspar	$K_2O \cdot Al_2O_3 \cdot 6SiO_2$	6	560
Martensite (high carbon steel)			600
Quartz	SiO_2	7	820
Topaz	$(AlF)_2SiO_4$	8	1340
Cemented carbide	WC-Co		1600
Cementite	Fe_3C		1720
Titanium nitride	TiN	9	1800
Tungsten carbide	WC		1880
Alumina	Al_2O_3	9	2100
Titanium carbide	TiC		2470
Silicon carbide	SiC	9.2	2480
Boron carbide	B_4C	9.3	2750
Diamond	C	10	7000

From *Handbook of Chemistry and Physics, 56th Ed.*, CRC Press, 1975.

FIG. 16-5 Microstructure of tungsten carbide—20% cobalt-cemented carbide. From *Metals Handbook*, Vol. 7, 8th Ed., American Society for Metals, 1972.

or pull out of the cobalt matrix and expose new, sharp-edged particles that continue to provide good cutting. For finish machining, the amount of cobalt binder is intentionally reduced so the particles pull out easily and the tool remains sharp. For rough grinding, more cobalt is added to improve toughness. Table 16-3 shows several classes of cemented carbides.

TABLE 16-3 Grades and compositions for typical cemented carbide cutting tools

Grade	% WC	% Co	% TaC	% TiC	Applications
C-1	90	6–12	0–3		Stainless steels, cast irons, nonferrous alloys
C-2	93	5–6	0–2		
C-3	93	4–6	0–2		
C-4	97	3–4			
C-5	73	8–10	10–12	6–8	Steels
C-6	74	5–9	10–12	7–12	
C-7	76	4–6	5–10	10–14	
C-8	75	4–6	4–8	10–20	

Abrasives. Grinding and cutting wheels are formed from alumina (Al_2O_3), silicon carbide (SiC), cubic boron nitride (BN), and diamond. To provide toughness, the abrasive particles are bonded by a glass or polymer matrix. As the hard particles wear, they fracture or pull out of the matrix, exposing new cutting surfaces.

The typical designation system used for abrasives is described in Table 16-4. Five features of the abrasive are specified. The type and grit refer to the composition and size of the abrasive material. The grade relates to the amount of binder and the ease with which the particles pull out of the matrix. The structure describes the amount of porosity remaining after the composite is formed; porous abrasives are used for rough grinding in order to avoid overheating. The bond refers to the matrix that holds the particles.

TABLE 16-4   Designation of abrasive grinding and cutting wheels

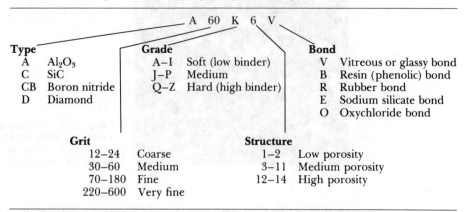

Type	Grade	Bond
A Al$_2$O$_3$	A–I Soft (low binder)	V Vitreous or glassy bond
C SiC	J–P Medium	B Resin (phenolic) bond
CB Boron nitride	Q–Z Hard (high binder)	R Rubber bond
D Diamond		E Sodium silicate bond
		O Oxychloride bond

Grit		Structure	
12–24	Coarse	1–2	Low porosity
30–60	Medium	3–11	Medium porosity
70–180	Fine	12–14	High porosity
220–600	Very fine		

Grit is the number of openings per inch in a screen through which the abrasive particles pass.

EXAMPLE 16-3

Designate an alumina-based abrasive that might be employed for a grinding process on a hard steel requiring an excellent final surface finish.

Answer:

We would select an abrasive that has a very fine particle size. A grit of 600 might be desired.

In order for the abrasive to remain sharp, the particles should pull out easily; a grade with a low amount of binder might be advantageous.

Low porosity will be helpful; for finish grinding, overheating will not be a problem, since we will remove only a little steel at a time.

A glassy bond, which is more rigid than the others, permits us to obtain more accurate dimensions.

Although there are several choices, one possibility is

A   600   B   2   V

Electrical contacts.   Materials used for electrical contacts in switches and relays must have a good combination of wear resistance and electrical conductivity. Otherwise, the contacts erode, causing poor contact and arcing. Particulate composites, such as tungsten-reinforced silver, provide materials having the proper combination of hardness and conductivity.

A tungsten powder compact is made using conventional powder metallurgical processes (Figure 16-6), producing high interconnected porosity. Liquid silver is then vacuum infiltrated to fill the interconnected voids. Both the silver and the tungsten are continuous. Thus, the pure silver efficiently conducts current while the hard tungsten provides wear resistance.

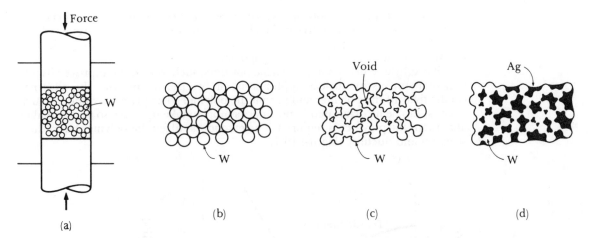

FIG. 16-6 The steps in producing a silver-tungsten electrical composite. (a) Tungsten powders are pressed, (b) a low-density compact is produced, (c) sintering joins the tungsten powders, and (d) liquid silver is infiltrated into the pores between the particles.

EXAMPLE 16-4

A silver-tungsten composite for an electrical contact is produced by first making a porous tungsten powder metallurgy compact, then infiltrating pure silver into the pores. The density of the tungsten compact before infiltration is 14.5 g/cm³. Calculate the volume fraction of porosity and the final weight percent of silver in the compact after infiltration.

Answer:

The densities of pure tungsten and pure silver are 19.3 Mg m^{-3} and 10.49 Mg m^{-3}. We can assume that the density of a pore is zero, so from the rule of mixtures

$$\rho_c = f_W\rho_W + f_{pore}\rho_{pore}$$
$$14.5 = f_W(19.3) + f_{pore}(0)$$
$$f_W = 0.75$$
$$f_{pore} = 1 - 0.75 = 0.25$$

After infiltration, the volume fraction of silver will equal the volume fraction of pores.

$$f_{Ag} = f_{pore} = 0.25$$
$$wt\% \ Ag = \frac{(0.25)(10.49)}{(0.25)(10.49) + (0.75)(19.3)} \times 100 = 15.3\%$$

Polymers. Many engineering polymers which contain fillers and extenders are particulate composites. A classic example is carbon black in vulcanized rubber. Carbon black consists of tiny carbon spheroids only 50 Å to 5000 Å in diameter. The carbon black improves the strength, stiffness, hardness, wear resistance, and heat resistance of the rubber.

Another composite polymer is ABS, in which a copolymer, acrylonitrile and styrene (SAN), forms a matrix in which an elastomer, butadiene-styrene (BS), is embedded as a round particulate. The rubber improves the impact properties and molding characteristics of the SAN copolymer.

Polyethylene may contain metallic powders, such as lead, to improve absorption of neutrons in nuclear applications, or bronze, to improve the electrical conductivity of the polymer and permit the polymer to be plated. Extenders, such as clays, can be incorporated in polymers simply to take up space so that a smaller amount of the polymer is required. The extenders stiffen the polymer but reduce strength and ductility (Figure 16-7).

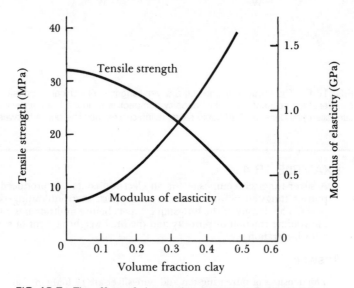

FIG. 16-7 The effect of clay on the properties of polyethylene.

Foundry molds and cores.   Molds and cores used to make metal castings frequently consist of silica sand grains bonded by a matrix of either an organic or an inorganic resin. The sand grains are refractory, insulating materials that do not react with the molten metal. Common binders include phenolic resins, urethane resins, furan resins, and sodium silicate. The resins coat the individual sand grains and provide bridging (Figure 16-8). The voids between the sand grains permit gases to escape from the mold rather than being trapped in the metal casting.

Compocasting.   One unusual technique for producing particulate-reinforced castings is based on the thixotropic behavior of partly liquid-partly solid melts. A liquid alloy is allowed to cool until about 40% solids have formed; during solidification the solid-liquid mixture is vigorously stirred to break up the dendritic structure (Figure 16-9). The resulting solid-liquid slurry has *thixotropic* behavior—the slurry behaves as a solid when no stress is applied, but flows like a liquid when pressure is exerted.

If a particulate material is introduced to the molten metal during cooling and stirring, a uniform dispersion is produced. The thixotropic slurry containing the particulate is injected into a die under pressure; this process is termed *compocast-*

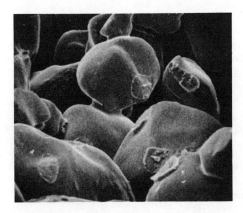

FIG. 16-8 A scanning microscope view of the bridging of a thermosetting polymer that binds sand grains in a shell mold or core for the foundry industry. Note some fractured bonds.

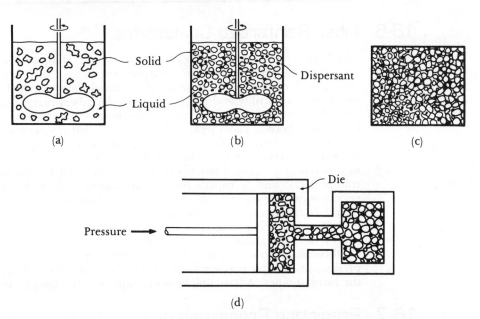

FIG. 16-9 In compocasting, (a) a solidifying alloy is stirred to break up the dendritic network, (b) a reinforcement is introduced into the slurry, (c) when no force is applied, the solid-liquid does not flow, and (d) high pressures cause the solid-liquid mixture to flow into a die.

ing. A variety of materials, including Al_2O_3, SiC, TiC, and glass beads, have been incorporated into aluminum and magnesium alloys by this technique.

EXAMPLE 16-5

We wish to compocast an aluminum alloy containing 9% Cu and 25% Al_2O_3. We find that compocasting is best accomplished when the alloy is die cast at a temperature that gives 60% liquid and 40% solid. From the phase diagram for aluminum-copper (Figure 11-1), estimate the appropriate casting temperature.

Answer:

To determine the temperature at which 60% liquid is present in the $\alpha + L$ region, let's perform the lever law calculation at several temperatures. The liquidus temperature is 630°C and the eutectic temperature is 548°C.

At 600°C: $L = \dfrac{9 - 3}{21 - 3} \times 100 = 33\%$

At 620°C: $L = \dfrac{9 - 2}{14 - 2} \times 100 = 58\%$

At 625°C: $L = \dfrac{9 - 1.5}{11 - 1.5} \times 100 = 79\%$

We obtain 60% L near 620°C.

16-6 Fiber-Reinforced Composites

The fiber-reinforced composites improve strength, fatigue resistance, stiffness, and strength-to-weight ratio by incorporating strong, stiff, brittle fibers into a softer, more ductile matrix. The matrix material transmits the force to the fibers and provides ductility and toughness, while the fibers carry most of the applied force. Unlike dispersion-strengthened composites, the strength of the composite is increased both at room temperature and elevated temperatures (Figure 16-2).

A tremendous variety of reinforcing materials are employed. For centuries, straw has been used to strengthen mud bricks. Steel reinforcing bar is introduced to concrete structures. Glass fibers in a polymer matrix produce fiberglass for transportation and aerospace applications. Fibers made of boron, graphite, and polymers provide exceptional reinforcement. Even tiny single crystals of ceramic materials, called *whiskers,* are being developed.

The reinforcing materials also are arranged in a variety of orientations (Figure 16-10). Short, randomly oriented glass fibers are usually present in fiberglass. Unidirectional arrangements of continuous fibers may be used to deliberately produce anisotropic properties. Fibers can be woven into fabrics or produced in the form of tapes. Alternating layers of tapes can be changed in orientation.

16-7 Predicting Properties of Fiber-Reinforced Composites

The rule of mixtures always predicts the density of fiber-reinforced composites.

$$\rho_c = f_m \rho_m + f_f \rho_f \tag{16-2}$$

where the subscripts m and f refer to the matrix and the fiber. In addition, the rule of mixtures accurately predicts the electrical and thermal conductivity of fiber-reinforced materials along the fiber direction if the fibers are continuous and unidirectional.

$$k_c = f_m k_m + f_f k_f \tag{16-3}$$

$$\sigma_c = f_m \sigma_m + f_f \sigma_f \tag{16-4}$$

where k is the thermal conductivity and σ is the electrical conductivity.

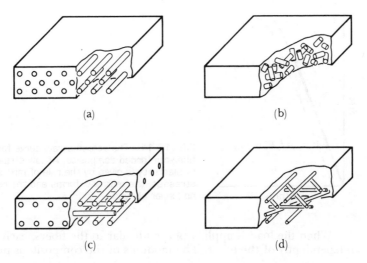

FIG. 16-10 Various morphologies of fiber-reinforced composites.
(a) Continuous unidirectional fibers, (b) randomly oriented
discontinuous fibers, (c) orthogonal fibers, and (d) multiple-ply fibers.

EXAMPLE 16-6

The density of a composite made from boron fibers in an epoxy matrix is 1.8 Mg m^{-3}.
The density of boron is 2.36 Mg m^{-3} and that of epoxy is 1.38 Mg m^{-3}. Calculate the
volume fraction of boron fibers in the composite.

Answer:

If f_B is the volume fraction of boron, then $1 - f_B$ is the volume fraction of epoxy.

$$\rho_c = f_B\rho_B + f_E\rho_E = f_B\rho_B + (1 - f_B)\rho_E$$
$$1.8 = f_B(2.36) + (1 - f_B)(1.38)$$
$$1.8 = 2.36f_B + 1.38 - 1.38f_B$$
$$f_B = \frac{1.80 - 1.38}{2.36 - 1.38} = 0.429$$

Modulus of elasticity. When a load is applied parallel to continuous, unidirectional fibers, the rule of mixtures accurately predicts the modulus of elasticity.

$$E_c = f_m E_m + f_f E_f \tag{16-5}$$

However, when the applied stress is very large, the matrix begins to deform and
the stress-strain curve is no longer linear (Figure 16-11). Since the matrix now
contributes little to the stiffness of the composite, the modulus can be approximated by

$$E_c = f_f E_f \tag{16-6}$$

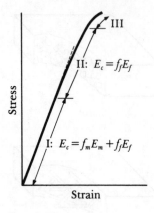

FIG. 16-11 The stress-strain curve for a fiber-reinforced composite. At low stresses, the modulus of elasticity is given by the rule of mixtures. At higher stresses, the matrix deforms and the rule of mixtures is no longer obeyed.

When the load is applied perpendicular to the fibers, each component acts independently of the other. The modulus of the composite is now

$$\frac{1}{E_c} = \frac{f_m}{E_m} + \frac{f_f}{E_f} \tag{16-7}$$

EXAMPLE 16-7

Derive the rule of mixtures for the modulus of elasticity of a fiber-reinforced composite when a stress is applied along the axis of the fibers.

Answer:

The total force acting on the composite is the sum of the forces carried by each constituent.

$$F_c = F_m + F_f$$

Since $F = \sigma A$

$$\sigma_c A_c = \sigma_m A_m + \sigma_f A_f$$

$$\sigma_c = \sigma_m \left(\frac{A_m}{A_c}\right) + \sigma_f \left(\frac{A_f}{A_c}\right)$$

If the fibers have a uniform cross section, the area fraction equals the volume fraction f.

$$\sigma_c = \sigma_m f_m + \sigma_f f_f$$

From Hooke's law, $\sigma = \varepsilon E$. Therefore

$$E_c \varepsilon_c = E_m \varepsilon_m f_m + E_f \varepsilon_f f_f$$

If the fibers are rigidly bonded to the matrix, both the fibers and the matrix must stretch equal amounts, so

$$\varepsilon_c = \varepsilon_m = \varepsilon_f$$
$$E_c = f_m E_m + f_f E_f$$

EXAMPLE 16-8

Derive the equation for the modulus of elasticity of a fiber-reinforced composite when a stress is applied perpendicular to the axis of the fiber.

Answer:

In this example, the strains are no longer equal; instead, the weighted sum of the strains in each component equals the total strain in the composite, while the stresses in each component are equal.

$$\varepsilon_c = f_m \varepsilon_m + f_f \varepsilon_f$$

$$\frac{\sigma_c}{E_c} = f_m \left(\frac{\sigma_m}{E_m}\right) + f_f \left(\frac{\sigma_f}{E_f}\right)$$

Since $\sigma_c = \sigma_m = \sigma_f$

$$\frac{1}{E_c} = \frac{f_m}{E_m} + \frac{f_f}{E_f}$$

Strength.   The strength of a composite depends on the bonding between the fibers and the matrix and is limited by deformation of the matrix. Consequently, the strength is almost always less than that predicted by the rule of mixtures.

Other properties, such as ductility, impact properties, fatigue properties, and creep properties are even more difficult to predict than the tensile properties.

EXAMPLE 16-9

Borsic-reinforced aluminum containing 40 vol% fibers is an important high-temperature, lightweight composite material. Estimate the density, modulus of elasticity, and strength parallel to the fiber axis. Also estimate the modulus of elasticity perpendicular to the fibers.

Answer:

The properties of the individual components are shown below.

Material	Density $(Mg\ m^{-3})$	Modulus of Elasticity (GPa)	Tensile Strength (MPa)
Fibers	2.36	379	2760
Aluminum	2.70	69	35

From the rule of mixtures

$$\rho_c = (0.6)(2.7) + (0.4)(2.36) = 2.56\ \text{Mg m}^{-3}$$

$$E_c = (0.6)(69 \times 10^9) + (0.4)(379 \times 10^9) = 193\ \text{GPa}$$

$$TS_c = (0.6)(35 \times 10^9) + (0.4)(2.76 \times 10^9) = 1125\ \text{MPa}$$

Perpendicular to the fibers

$$\frac{1}{E_c} = \frac{0.6}{69 \times 10^9} + \frac{0.4}{379 \times 10^9} = 9.75 \times 10^{-12}$$

$$E_c = 102.5 \text{ GPa}$$

The actual modulus and strength parallel to the fibers are shown in Figure 16-12. The calculated modulus of elasticity, 193 GPa, is exactly the same as the measured modulus. However, the estimated strength, 1125 MPa, is substantially higher than the actual strength, about 897MPa. We also note that the modulus of elasticity is very anisotropic.

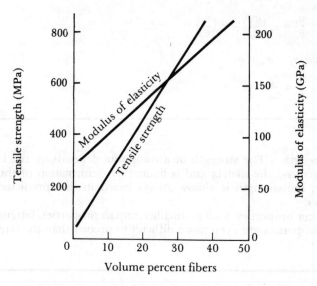

FIG. 16-12   The influence of volume percent Borsic fibers on the properties of Borsic-reinforced aluminum.

EXAMPLE 16-10

Glass fibers in nylon provide reinforcement. If the nylon contains 30 vol% E glass, what fraction of the applied stress is carried by the glass fibers?

Answer:

The modulus of elasticity for each component of the composite is

$$E_{\text{glass}} = 72.4 \text{ GPa} \qquad E_{\text{nylon}} = 2.75 \text{ GPa}$$

Both the nylon and the glass fibers have equal strain if bonding is good, so

$$\varepsilon_c = \varepsilon_m = \varepsilon_f$$

$$\varepsilon_m = \frac{\sigma_m}{E_m} = \varepsilon_f = \frac{\sigma_f}{E_f}$$

$$\frac{\sigma_f}{\sigma_m} = \frac{E_f}{E_m} = \frac{72.4 \times 10^9}{2.75 \times 10^9} = 26.32$$

$$\text{Fraction} = \frac{\sigma_f}{\sigma_f + \sigma_m} = \frac{1}{1 + \frac{\sigma_m}{\sigma_f}} = \frac{1}{1 + \frac{1}{26.32}} = 0.96$$

Almost all of the load is carried by the glass fibers.

Discontinuous fibers. The properties of the composite are more difficult to predict when the fibers are discontinuous. Because the ends of each fiber carry less of the load than the remainder of the fiber, the strength of the composite is lower than that predicted by the rule of mixtures. The error is reduced when the actual length l of the fibers is greater than a critical fiber length l_c, or more precisely, when the length-to-diameter ratio of the fibers l/d exceeds a critical value. This ratio, called the *aspect ratio*, significantly affects the properties of the composite. For example, nylon reinforced with carbon fibers with an aspect ratio of 30 has a tensile strength of 110 MPa; longer fibers with an aspect ratio of 800 produce a strength of 240 MPa.

16-8 Characteristics of Fiber-Reinforced Composites

A large number of factors must be considered when designing a fiber-reinforced composite.

Aspect ratio. Continuous fibers, which give the best properties, are often difficult to produce and introduce into the matrix material. However, discontinuous fibers with a large aspect ratio are more easily incorporated into a matrix yet still produce high stiffness and strength.

Volume fraction of fibers. A greater volume fraction of fibers increases the strength and stiffness of the composite (Figure 16-12). The upper limit, about 80%, is determined by our ability to surround the fibers with the matrix material.

Orientation of fibers. Unidirectional fibers have optimum stiffness and strength when the applied load is parallel to the fibers (Figure 16-13). However, the properties are very anisotropic, as we found in Example 16-9. We may instead use fibers laid out in an orthogonal or cross-ply manner; we sacrifice maximum strength but gain more uniform properties in the composite.

Properties of the fiber. The fiber material should be strong, stiff, lightweight, and have a high melting temperature. The *specific strength* and *specific modulus* of a material are defined by

$$\text{Specific strength} = \frac{\sigma}{\rho} \tag{16-8}$$

$$\text{Specific modulus} = \frac{E}{\rho} \tag{16-9}$$

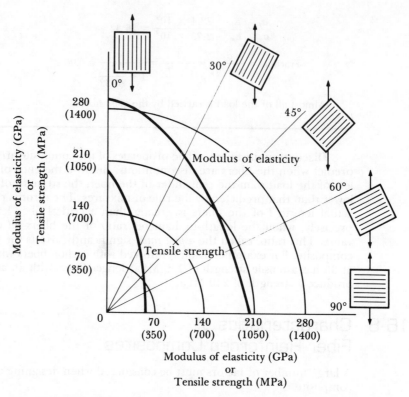

FIG. 16-13   The effect of fiber orientation with respect to the applied stress for fiber-reinforced composites for a boron-reinforced titanium alloy.

where E is the modulus of elasticity, σ is the yield strength, and ρ is the density. Materials with a high specific modulus or strength are preferred for use as fibers. Some typical data for fibers are shown in Table 16-5 and Figure 16-14. The highest specific modulus is found in materials having a low atomic number and covalent bonding, such as graphite and boron. These two elements also have a high strength and melting temperature. They must be used in a composite material, however, because they are too brittle and reactive to be of service alone.

Kevlar, the trade name for an aromatic polyamide polymer strengthened by a backbone containing benzene rings, has excellent mechanical properties, although its melting temperature is low.

Alumina and glass also are lightweight and have a high strength and specific modulus. Beryllium and tungsten, although metallically bonded, have a high modulus that makes them attractive for certain applications. Whiskers have the highest strength and stiffness; however, they are discontinuous and are difficult and expensive to produce.

EXAMPLE 16-11

Compare the strength of Kevlar fibers to graphite fibers at 30°C and 300°C.

Answer:

From Figure 16-15, we find the specific strength of each fiber at the two temperatures of interest.

Material	Specific Strength ($\times 10^6$ m^2 s^{-1})	Density (Mg m^{-3})	Strength (MPa)
Kevlar at 30°C	2.4	1.44	3407
Kevlar at 300°C	1.5	1.44	2152
Graphite at 30°C	1.6	1.49	2420
Graphite at 300°C	1.6	1.49	2420

The graphite fibers, although weaker at low temperatures, maintain their strength to high temperatures.

TABLE 16-5   Properties of selected fiber-reinforcing materials

Material	Density (Mg m^{-3})	Tensile Strength (MPa)	Modulus of Elasticity ($\times 10^6$ GPa)	Melting Temperature (°C)	Specific Modulus ($\times 10^3$ m^2 s^{-1})	Specific Strength (m^2 s^{-1})
E glass	2.55	3448	72.4	<1725	28	1.35
S glass	2.50	4483	86.9	<1725	35	1.79
SiO$_2$	2.19	5862	72.4	1728	33	2.68
Al$_2$O$_3$	3.15	2068	172	2015	55	0.66
ZrO$_2$	4.84	2068	345	2677	71	0.43
Graphite HS (high strength)	1.50	2759	276	3700	184	1.84
Graphite HM (high modulus)	1.50	1862	531	3700	354	1.24
BN	1.90	1380	90	2730	47	0.73
Boron	2.36	3448	379	2030	161	1.46
B$_4$C	2.36	2276	482	2450	204	0.96
SiC	4.09	2068	482	2700	118	0.51
TiB$_2$	4.48	103	510	2980	114	0.002
Be	1.83	1276	303	1277	166	0.70
W	19.4	400	407	3410	21	0.021
Mo	10.2	2207	359	2610	35	0.022
Kevlar	1.44	3620	124		86	2.51
Al$_2$O$_3$ whiskers	3.96	20690	428	1982	108	5.22
BeO whiskers	2.85	13103	345	2550	121	4.60
B$_4$C whiskers	2.52	13793	482	2450	191	5.47
SiC whiskers	3.18	20690	482	2700	151	6.51
Si$_3$N$_4$ whiskers	3.18	13793	379		119	4.34
Graphite whiskers	1.66	20690	703	3700	423	12.46
Cr whiskers	7.2	3021	241	1890	33	0.42
Cu whiskers	8.92	2945	124	1083	14	0.33

Adapted from L. J. Broutman, "Mechanical Properties of Fiber Reinforced Plastics," *Composite Engineering Laminates,* ed. G. H. Dietz, The M.I.T. Press, 1969.

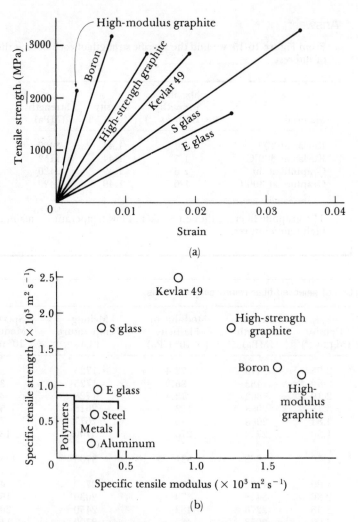

FIG. 16-14 Properties of fibers. (a) The stress-strain curve for individual fibers and (b) comparison of the specific strength and specific modulus of fibers versus metals.

EXAMPLE 16-12

Compare the specific modulus and specific strength of a 1040 annealed steel and a 2024-T4 aluminum alloy to that of boron.

Answer:

The yield strengths for the aluminum alloy and steel are found in Tables 11-2 and 12-4.

Material	Yield Strength (MPa)	Modulus of Elasticity ($\times 10^6$ GPa)	Density (Mg m^{-3})
2024-T4 aluminum	324	69	2.68
1040 steel	353	207	7.86
Boron	2759	379	2.35

From this data, we can calculate the specific strength and specific modulus for each material.

$$\text{Specific strength (2024-T4 aluminum)} = \frac{324 \times 10^6}{2.68 \times 10^3} = 1.2 \times 10^5 \text{ m}^2 \text{ s}^{-1}$$

$$\text{Specific strength (1040 steel)} = \frac{353 \times 10^6}{7.86 \times 10^3} = 4.5 \times 10^4 \text{ m}^2 \text{ s}^{-1}$$

$$\text{Specific strength (boron fiber)} = \frac{2759 \times 10^6}{2.35 \times 10^3} = 1.2 \times 10^6 \text{ m}^2 \text{ s}^{-1}$$

$$\text{Specific modulus (2024-T4 aluminum)} = \frac{69 \times 10^6}{2.68 \times 10^3} = 2.6 \times 10^4 \text{ m}^2 \text{ s}^{-1}$$

$$\text{Specific modulus (1040 steel)} = \frac{207 \times 10^6}{7.86 \times 10^3} = 2.6 \times 10^4 \text{ m}^2 \text{ s}^{-1}$$

$$\text{Specific modulus (boron fiber)} = \frac{379 \times 10^6}{2.35 \times 10^3} = 1.6 \times 10^5 \text{ m}^2 \text{ s}^{-1}$$

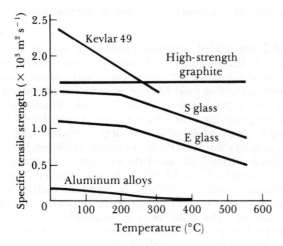

FIG. 16-15 The effect of temperature on the strength of several fibers used for fiber-reinforced composites.

Matrix properties. Matrix materials are usually tough and ductile, transmitting the load to the fibers and preventing cracks in broken fibers from propagating throughout the entire composite. The matrix should also be strong so that it contributes to the overall strength of the composite. Finally, the melting temperature influences the suitability of the matrix. Polymers can be used from a maximum of 80°C for polyesters to 315°C for polyimide resins. Metallic matrices permit higher service temperatures.

16-9 Manufacturing Fibers and Composites

Fibers. Coarse fibers, such as steel-reinforcing bar, are produced by rolling. Finer fibers, such as wire, are made by wire drawing, provided the materials possess sufficient ductility and strain-hardening characteristics. Tungsten, beryllium, stainless steel, and nylon can all be drawn to a small diameter.

Boron and graphite are too brittle and reactive to be made by conventional drawing processes. Boron is produced by vapor decomposition [Figure 16-16(a)].

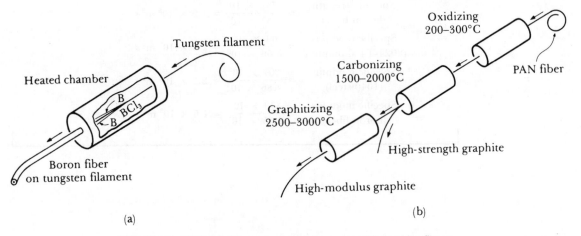

FIG. 16-16   Methods for producing (a) boron and (b) graphite fibers.

A very fine, 12.7 μm heated tungsten filament is used as a substance, passing through a seal into a heated chamber. Vaporized boron compounds, such as BCl_3, are introduced into the chamber, decompose, and permit boron to precipitate onto the tungsten wire. The final fiber may be 25 μm to 200 μm in diameter.

Graphite fibers approximately 7.6 μm in diameter are made by *carbonizing*, or *pyrolizing*, an organic filament which is more easily drawn or spun into thin continuous lengths [Figure 16-16(b)]. The organic filament, known as a *precursor*, is often nylon, polyacrylonitrile (PAN), or pitch. High temperatures decompose the organic polymer, driving off all of the elements but carbon. As the carbonizing temperature increases from 1000°C to 3000°C, the tensile strength decreases while the modulus of elasticity increases. Drawing the carbon filaments at critical times during carbonizing may produce desirable preferred orientations in the final graphite filament. The filaments can be loosely woven into a yarn, or *tow*, which may contain hundreds or thousands of filaments (Figure 16-17).

Whiskers, which are single crystals of exceptional fineness, are discontinuous,

FIG. 16-17   A scanning electron micrograph of a graphite tow, containing many individual graphite filaments.

with aspect ratios of 20 to 1000. Because the whiskers contain no mobile dislocations, slip cannot occur and the whiskers have exceptionally high strengths. The technology for producing whiskers is very complex.

Composites.   The fibers must be embedded in the matrix with the proper alignment and spacing to produce optimum properties. Discontinuous fibers can be mixed with the matrix material to produce either a random or preferred orientation. Continuous fibers are normally unidirectionally aligned as tapes, woven into a fabric in an orthogonal arrangement, or wound around a mandrel. There are several techniques used to surround the fibers with the matrix.

1. Casting: Casting processes force liquid around the fibers. Pouring concrete around steel-reinforcing rods that have been properly assembled is a rough example. In the filament-reinforced composites, the liquid is introduced around the fibers by capillary action, vacuum infiltration, or pressure casting (Figure 16-18). Special coatings on the fibers may be necessary to assure proper wetting of the fibers by the liquid matrix.

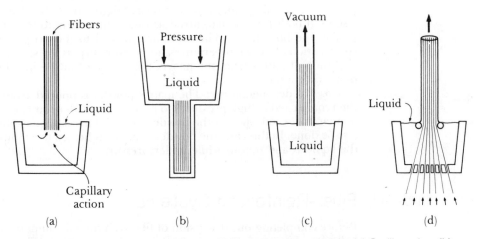

FIG. 16-18   Casting techniques for producing composite materials. (a) Capillary rise, (b) pressure casting, (c) vacuum infiltration, and (d) continuous casting.

2. Prepregs: When the fibers are woven into a fabric, a polymer matrix is infiltrated into each fabric layer. The infiltration is carried out under conditions such that the resin does not polymerize (Figure 16-19). Later, these prepregs are stacked in layers, then heated under pressure so that the resins melt and polymerize to form the solid composite. The orientation of fabric layers can be arranged to produce various cross-plies of fibers.

FIG. 16-19   Unidirectionally aligned boron fibers in a tape prepegged with epoxy.

3. Tapes: Individual fibers can be unwound onto a mandrel, which determines the spacing of the individual fibers, and prepegged with polymer resin, Figure 16-20(a). The tapes, up to 1.2 m wide, are joined to produce wider material or stacked to produce thicker material. Heat and pressure complete the polymerization process.

4. Precoating: A metal matrix may be applied to a fiber using a molten metal bath, plasma spraying, vapor deposition, or electrodeposition. The precoated fibers, often in the form of tapes, are then assembled and bonded by other techniques.

5. Deformation and diffusion bonding: Deformation processes, such as hot pressing and rolling, join layers of tapes [Figure 16-20(b)]. Diffusion bonding is also used both for original introduction of the matrix to the fiber as well as for joining the fiber tapes. The tapes are stacked to the proper thickness, then a combination of high temperature and pressure brings the surfaces close together. Diffusion of atoms from the matrix fills the voids at the interface to produce a dense composite.

6. Powder metallurgy: The matrix powder is poured around the fibers and then compacted at high pressures to produce a powder compact. Sintering at high temperatures consolidates the powder to a solid mass. Liquid-phase sintering can also be done. In this case, the powder compact is heated to a temperature between the liquidus and solidus while under pressure.

16-10   Fiber-Reinforced Systems

Before completing our discussion of fiber-reinforced composites, let's look at the behavior of some of the most common of these materials. Figure 16-21 compares the specific modulus and specific strength of several composites versus metals.

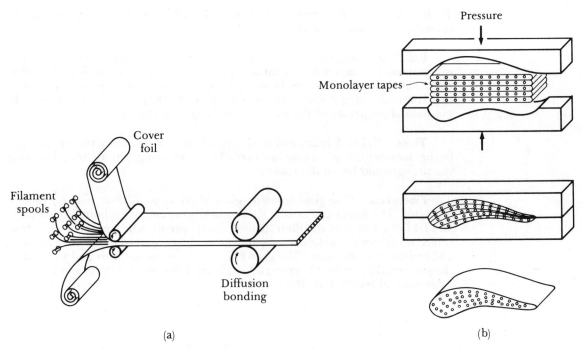

FIG. 16-20   Solid-state techniques for producing composites include (a) rolling and diffusion bonding of tapes and (b) pressing and simultaneous bonding of tapes into shapes.

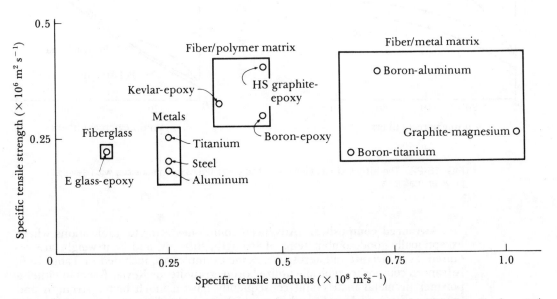

FIG. 16-21   A comparison of the specific modulus and specific strength of several composite materials and metals.

Their values are lower than in Figure 16-14 since now we are looking at the composite, not just the fiber.

Reinforced concrete.   Reinforced concrete is a double composite; concrete is a particulate composite (cement and gravel) that in turn is reinforced by steel rods. The steel rods improve the strength of the concrete and prevent the structure from collapsing if the concrete fails. By prestressing the reinforcing bar, the mechanical properties of the concrete are further enhanced.

Tires.   Nylon, Kevlar, and steel wire are used to reinforce the rubber used in the manufacture of automobile tires. The reinforcing wire or fiber improves the strength and life of the rubber.

Fiberglass.   Fiberglass contains glass fibers in a polymer matrix, such as polyester. The fibers are usually short and discontinuous, although glass yarns and fabrics are also used. Often silane "keying" agents, which coat the glass fiber (called *sizing*) with a molecular organic material, are applied to improve bonding and resistance to moisture. The glass fiber improves the stiffness and strength of the polymer (Figure 16-22), giving a specific modulus and specific strength comparable to good metals and alloys.

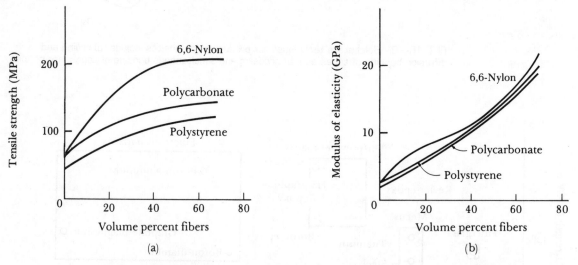

FIG. 16-22   The effect of randomly oriented glass fibers in composites with several polymer matrices.

Advanced composites.   Advanced composites refer to applications where exceptionally good combinations of strength, stiffness, and light weight are required, as in aircraft and aerospace. Some examples are included in Table 16-6. Advanced composites typically utilize boron, graphite, or Kevlar fibers in either a polymer or metal matrix and consequently have a much better strength and fatigue resistance than fiberglass or high-strength alloys (Figures 16-21 and 16-23).

TABLE 16-6 Examples of fiber-reinforced materials and applications

Material	Applications
Borsic aluminum	Fan blades in engines, other aircraft and aerospace applications
Kevlar-epoxy	Aircraft, aerospace (including Space Shuttle), boat hulls,
Kevlar-polyester	sporting goods (including tennis rackets, golf club shafts, fishing rods), flak jackets
Graphite-polymer	Aerospace, automotive, sporting goods
Glass-polymer	Lightweight automotive applications, water and marine applications, corrosion-resistant applications, sporting goods equipment, aircraft and aerospace components

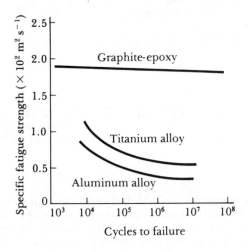

FIG. 16-23 A comparison of the specific fatigue strength for composites and metals.

Advanced composites are used extensively in both structural and skin applications in modern aircraft to take advantage of the strength-to-weight ratio. These composites are effective provided that temperatures remain relatively low. Composites with a metal matrix, such as aluminum, titanium, or nickel, which are strengthened by graphite, boron, or silicon carbide, are used when higher temperatures are encountered (Figure 16-24).

Borsic-reinforced aluminum is an important fiber-reinforced composite that utilizes strong stiff boron fibers. To achieve good bonding between the boron and aluminum, the fiber is first coated with a thin layer of silicon carbide or boron nitride, which are compatible with both boron and aluminum. Fiber-reinforced aluminum has better strength and temperature resistance than the dispersion-strengthened SAP composites (Figure 16-2).

Tungsten fibers are considered for reinforcing cobalt and nickel superalloys. The higher temperature servicability may significantly increase the efficiency of turbine engines, which contain large numbers of the superalloy blades and vanes.

Superconductors. Large quantities of superconducting wire will be required for fusion reactors. Unfortunately, superconducting materials are often hard brittle intermetallic compounds that are difficult to form. This problem can be avoided by using composite materials.

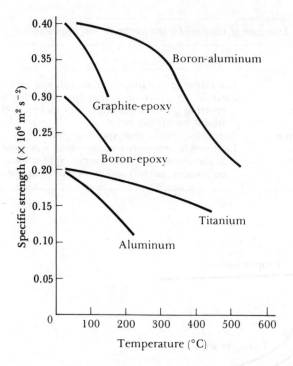

FIG. 16-24 The specific strength versus temperature for several composites and metals.

The intermetallic compound Nb_3Sn has excellent superconducting properties. To produce Nb_3Sn, pure niobium wire is surrounded by copper as they are formed into a wire composite (Figure 16-25). The niobium-copper composite wire is then coated with tin. The tin diffuses through the copper and reacts with the niobium to produce the intermetallic compound.

16-11 Laminar Composite Materials

Laminar composites include very thin coatings, thicker protective surfaces, claddings, bimetallics, laminates, and a host of others. Many laminar composites are designed to improve corrosion resistance while retaining low cost, high strength, or light weight. Other important applications include superior wear or abrasion resistance, improved appearance, and unusual thermal expansion characteristics.

Rule of mixtures. Some properties of the laminar composite materials along the lamellae are estimated from the rule of mixtures. Density, electrical and thermal conductivity, and modulus of elasticity can be calculated with little error.

$$\text{Density} = \rho_c = \Sigma f_i \rho_i$$
$$\text{Electrical conductivity} = \sigma_c = \Sigma f_i \sigma_i$$
$$\text{Thermal conductivity} = k_c = \Sigma f_i k_i$$
$$\text{Modulus of elasticity} = E_c = \Sigma f_i E_i$$

(16-10)

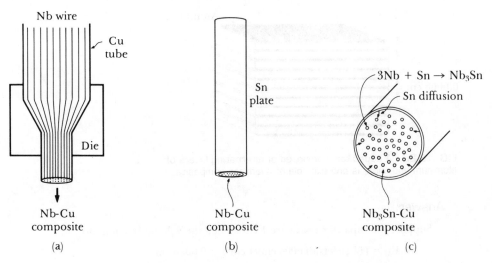

FIG. 16-25 The manufacture of composite superconductor wires. (a) Niobium wire is surrounded with copper during forming. (b) Tin is plated onto Nb-Cu composite wire. (c) Tin diffuses to niobium to produce the Nb₃Sn-Cu composite.

The laminar composites are very anisotropic. The properties perpendicular to the lamellae are

$$\text{Electrical conductivity} = \frac{1}{\sigma_c} = \Sigma \frac{f_i}{\sigma_i}$$

$$\text{Thermal conductivity} = \frac{1}{k_c} = \Sigma \frac{f_i}{k_i} \tag{16-11}$$

$$\text{Modulus of elasticity} = \frac{1}{E_c} = \Sigma \frac{f_i}{E_i}$$

However, many of the really important properties, such as corrosion or wear resistance, depend primarily on only one of the components of the composite, so the rule of mixtures is inappropriate.

EXAMPLE 16-13

Capacitors used to store electrical charge are essentially laminar composites built up from alternating layers of a conductor and an insulator (Figure 16-26). Suppose we construct a capacitor from 10 sheets of mica, each 0.01 cm thick, and 11 sheets of aluminum, each 0.0006 cm thick. The electrical conductivity of aluminum is 3.8×10^5 (A V^{-1} cm^{-1}) and the conductivity of mica is 10^{-13} (A V^{-1} cm^{-1}). Determine the electrical conductivity of the capacitor parallel and perpendicular to the sheets.

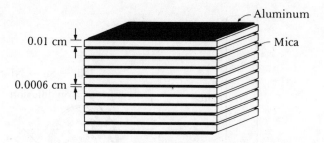

FIG. 16-26 A capacitor, composed of alternating layers of aluminum and mica, is one example of a laminar composite.

Answer:

Suppose the capacitor plates are 1 cm². Then the volume fractions are

$$V_{Al} = (11 \text{ sheets})(0.0006 \text{ cm})(1 \text{ cm}^2) = 0.0066 \text{ cm}^3$$

$$V_{mica} = (10 \text{ sheets})(0.01 \text{ cm})(1 \text{ cm}^2) = 0.1 \text{ cm}^3$$

$$f_{Al} = \frac{0.0066}{0.0066 + 0.1} = 0.062 \qquad f_{mica} = \frac{0.1}{0.0066 + 0.1} = 0.938$$

Parallel

$$\sigma = (0.062)(3.8 \times 10^5) + (0.938)(10^{-13}) = 0.24 \times 10^5 \big| (\text{A V}^{-1} \text{ cm}^{-1})$$

Perpendicular

$$\frac{1}{\sigma} = \frac{0.062}{3.8 \times 10^5} + \frac{0.938}{10^{-13}} = 0.938 \times 10^{13}$$

$$\sigma = \frac{1}{0.938 \times 10^{13}} = 1.07 \times 10^{-13} \big| (\text{A V}^{-1} \text{ cm}^{-1})$$

The composite, or capacitor, has high conductivity parallel to the plates, but acts as an insulator perpendicular to the plates.

16-12 Examples and Applications of Laminar Composites

The number of laminar composites is so varied and their applications and intentions are so numerous that we cannot make generalizations concerning their behavior. Instead we will examine the characteristics of a few commonly used examples.

Laminates. Laminates are layers of materials joined by an organic adhesive. A familiar laminate is plywood, in which an odd number of wood veneer plies are stacked so that the grain is at right angles in each alternating ply. The plies are glued by an adhesive such as a phenolic or amino resin. Plywood permits wood products to be available in large sizes yet be inexpensive and resistant to splitting and warping.

Safety glass is a laminate in which a plastic adhesive, such as polyvinyl butyral, joins two pieces of glass; the adhesive prevents fragments of glass from flying about when the glass is broken. Laminates are used for insulation in motors, for making gears, for printed circuit boards, and for decorative items such as Formica countertops and furniture.

The adhesive laminates combine unusual characteristics including light weight, flame retardance, impact strength, corrosion resistance, easy forming and machining, good ablative ability, and good insulation characteristics.

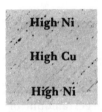

FIG. 16-27
Photomicrograph of the cross section of U.S. silver coinage.

Hard surfacing.   Hard wear-resistant surfaces can be deposited on softer, more ductile materials by fusion welding techniques known as *hard surfacing*. Hard-surfacing alloys include hardenable grades of steel, irons and steels that produce hard carbides, cobalt-base alloys, and certain nonferrous alloys. Composite tungsten carbide rods can also be used to provide tungsten carbide at the wear surface. Similar welding procedures can improve corrosion resistance or heat resistance at surfaces.

Clad metals.   Clad materials are metal-metal composites. A common example is United States silver coinage (Figure 16-27). A Cu-80% Ni alloy is bonded to both sides of a Cu-20% Ni alloy. The ratio of thicknesses is about $\frac{1}{6}:\frac{2}{3}:\frac{1}{6}$. The high-nickel alloy gives a silver color, while the predominantly copper core provides low cost.

Clad materials provide a combination of good corrosion resistance with high strength. *Alclad* is a clad composite in which commercially pure aluminum is bonded to higher strength aluminum alloys. The pure aluminum protects the higher strength alloy from corrosion. The thickness of the pure aluminum layer is about 1% to 15% of the total thickness (Figure 16-28). Alclad is used in aircraft construction, heat exchangers, building construction, and storage tanks where combinations of corrosion resistance, strength, and light weight are desired.

FIG. 16-28
Photomicrograph of Alclad.

Bimetallics.   Temperature indicators and controllers take advantage of the different coefficients of thermal expansion of the two metals in the laminar composite. If two pieces of metal are heated, the metal with the higher coefficient of thermal expansion becomes longer (Figure 16-29). If the two pieces of metal are rigidly bonded together, the difference in their coefficients causes the strip to bend and produce a curved surface. If one end of the strip is fixed, the free end

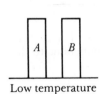

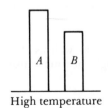

Low temperature

High temperature

(a)

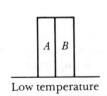

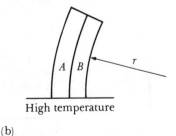

Low temperature

High temperature

(b)

FIG. 16-29 The effect of thermal expansion coefficient on the behavior of materials. (a) Increasing the temperature increases the length of one metal more than the other. (b) If the two metals are joined, the difference in expansion causes a radius of curvature to be produced.

moves. The amount of movement depends on the temperature; by measuring the curvature or deflection of the strip, we can determine the temperature. Likewise, if the free end of the strip activates a relay, the strip can turn on or off a furnace or air conditioner to regulate temperature.

The amount that the bimetallic strip deforms is given by the radius of curvature, r, that is produced. If the two strips are of equal thickness, then

$$\frac{1}{r} = \frac{24(\alpha_2 - \alpha_1)\Delta T}{t\left(14 + \dfrac{E_1}{E_2} + \dfrac{E_2}{E_1}\right)} \tag{16-12}$$

where α_1, α_2 are the coefficients of thermal expansion (in/in. °F), E_1, E_2 are the moduli of elasticity (psi), ΔT is the temperature change (°F), and t is the thickness of the strip (in.). As the difference in expansion coefficient or temperature increases, the radius of curvature decreases, which indicates a larger amount of deflection.

Metals selected for bimetallics must have (a) very different coefficients of thermal expansion, (b) expansion characteristics that are reversible and repeatable, and (c) a high modulus of elasticity, so the bimetallic device can do work. Often the low-expansion strip is made from Invar, an iron-nickel alloy, while the high-expansion strip may be brass, Monel, manganese-nickel-copper, nickel-chromium-iron, or pure nickel. Characteristics of typical bimetallic components are given in Table 16-7.

TABLE 16-7 Properties of some materials that might be components of bimetallics

Material	Coefficient of Thermal Expansion ($\times 10^{-6}$ cm/cm · °C)	Modulus of Elasticity	GPa
Cu	17	18	124
Al	24	10	69
Ni	13	30	207
Steel	12	30	207
Cu-30% Zn brass	16	22	152
Monel	14	26	179
Fused quartz	5.5	10	69
Invar (Fe-36% Ni)	1	21.4	148

FIG. 16-30 Schematic showing the deflection obtained from the bimetallic in Example 16-14.

Bimetallics can act as circuit breakers as well as thermostats; if a current passing through the strip becomes too high, heating causes the bimetallic to deflect and break the circuit.

EXAMPLE 16-14

A bimetallic is formed by roll bonding a 1.27 mm thick strip of copper to a 0.05 in. thick strip of Invar. Determine the radius of curvature produced when the temperature increases from 23°C to 100°C. Estimate the deflection of a 20 mm long bimetallic strip if one end is fixed.

Answer:

$$\alpha_1 = \alpha_{Invar} = 1 \times 10^{-6} \text{ cm/cm} \cdot {}^\circ\text{C}$$

$$\alpha_2 = \alpha_{Cu} = 17 \times 10^{-6} \text{ cm/cm} \cdot {}^\circ\text{C}$$

$$E_1 = E_{Invar} = 148 \text{ GPa} \quad \text{psi}$$

$$E_2 = E_{Cu} = 124 \text{ GPa} \quad \text{si}$$

$$\Delta T = 100 - 23 = 77{}^\circ\text{C}$$

$$t = 2(1.27) = 2.54$$

$$\frac{1}{r} = \frac{24(17 \times 10^{-6} - 1 \times 10^{-6})(77)}{2.54 \times 10^{-3}\left(14 + \dfrac{148 \times 10^9}{124 \times 10^9} + \dfrac{124 \times 10^9}{148 \times 10^9}\right)} = 0.73$$

$$r = 1.37 \text{ m}$$

The deflection is shown schematically in Figure 16-30. The circumference of the circle $= 2\pi r = 2\pi(1.37) = 8.61$ m. Thus angle θ is

$$\frac{0.02 \text{ m}}{3.61 \text{ m}} = \frac{\theta}{360{}^\circ} \quad \text{or} \quad \theta = 0.84{}^\circ$$

Since the bimetallic is short, the angle of deflection is also about 2.12°. The deflection is

$$d = (2 \text{ cm}) \tan\theta = 2 \tan(0.84) = 2(0.015) = 0.03 \text{ cm}$$

Because so little deflection is obtained in a straight bimetallic, thermostats are usually constructed in spiral shapes to accentuate the deflection (Figure 16-31).

FIG. 16-31
The spiral shape of a typical bimetallic thermostat composite.

16-13 Manufacturing Laminar Composites

Several methods are used to produce laminar composites, including a variety of deformation and joining techniques (Figure 16-32).

Rolling. Most of the metallic laminar composites, such as claddings and bimetallics, are produced by hot or cold roll bonding. If the percent deformation

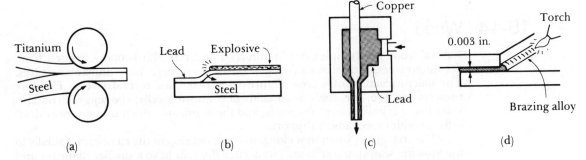

FIG. 16-32 Techniques for producing laminar composites. (a) Roll bonding, (b) explosive bonding, (c) coextrusion, and (d) brazing.

is great enough, the pressure exerted by the rolls breaks up the oxides at the surface, brings the surfaces into atom-to-atom contact, and permits the two surfaces to be welded.

Explosive bonding.   An explosive charge can provide the pressure required to join metals, as described in Chapter 9 and Figure 16-33. This process is particularly well suited for joining very large plate that will not fit into a rolling mill.

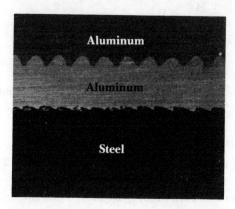

FIG. 16-33   A laminar composite of two aluminum plates and a steel plate formed by explosive bonding.

Coextrusion.   Very simple laminar composites, such as coaxial cable, are produced by coextruding two metals through a die in such a way that the soft material surrounds the harder material. Similarly, a thermoplastic polymer could be coextruded around a metal conductor wire.

Pressing.   For small components, high pressures at elevated temperatures provide welding. Hot pressing is frequently used to cure the adhesive in laminates.

Brazing.   Brazing can join composite plates. The metallic sheets are separated by a very small clearance, perferably about 76 μm, and heated above the melting temperature of the brazing alloy. The molten brazing alloy is drawn into the thin joint by capillary action.

16-14 Wood

Natural wood is a complex fiber-reinforced material. Hardwood, obtained from deciduous trees, and softwood, obtained from evergreens, have similar structures. The macrostructure of a tree (Figure 16-34) contains several layers. The *bark* protects the tree; the *cambium* contains new growing cells; the *sapwood* contains some living cells that store nutrients; and the *heartwood,* which contains only dead cells, provides mechanical support.

The tree grows when new elongated cells develop in the cambium. Initially in the growing season, the cells are large; later the cells have a smaller diameter and higher density. This difference between the early, or *spring,* wood and the late, or *summer,* wood permits us to observe annual growth rings.

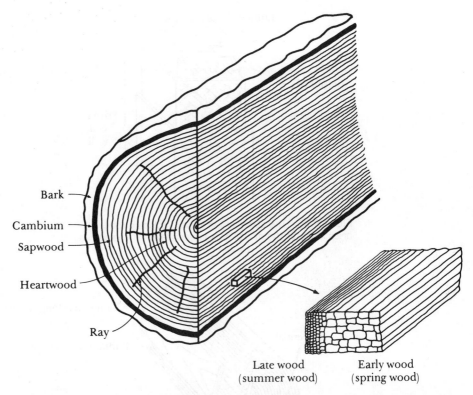

FIG. 16-34 The macrostructure of wood, showing the different layers and the cell size within an annual growth ring.

The cells grow as fibers, or *tracheids,* often having an aspect ratio of 100 or more, and constitute about 95% of the solid material in wood. The fibers are normally longer in softwoods than in hardwoods. The cells, which initially are hollow with a thin flexible *primary wall,* expand during growth. Later the wall thickens as a multilayered *secondary wall* grows into the hollow center, or *lumen,* of the cell (Figure 16-35).

About half of the cell wall is composed of cross-linked cellulose chains with a degree of polymerization of as much as 30,000. These chains, called *microfibrils,* are oriented differently in each layer of the wall. Less cross-linked polymer chains, called *hemicellulose,* are also present. The microfibrils are bonded by an organic cement called *lignin.* The lignin also bonds the cell or fiber to adjoining cells. Finally, a large percentage of water, sometimes more than the weight of the solid material, is present in the cell walls, lumen, and imperfections of the wood.

Several factors influence the behavior of the wood. First, the strength increases as the moisture is reduced. During drying, water is first driven from the lumens; later water is removed from the cell boundaries. During drying, the wood shrinks and distortion or cracking may occur.

Wood is highly anisotropic. Figure 16-36 shows a schematic of a log, describing the three important directions. Most lumber is cut in a tangential-longitudinal or radial-longitudinal manner. Because of the orientation of the fibers, the

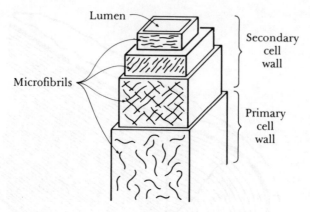

FIG. 16-35   The structure of a cell, including the
orientation of the microfibrils in each layer of the cell wall.

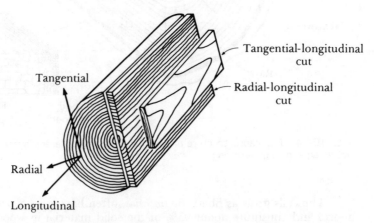

FIG. 16-36   The different directions in a log; because of differences in
cell orientation and the grain, wood has anisotropic behavior.

strength in the longitudinal direction may be 25 to 50 times greater than the
strength in the radial or tangential direction.

The fibers, a composite of microfibrils and lignin, have a longitudinal tensile
strength of about 690 MPa. Clear wood, a composite of fibers and lignin free of
imperfections such as knots, has a longitudinal tensile strength of 69 MPa to 138
MPa. Construction lumber, which contains many imperfections, may have a tensile
strength below 35 MPa. Strength in tension is superior to either compression or
shear. In compression, the fibers buckle; in shear the brittle lignin cannot prevent
the fibers from sliding past one another.

Wood has good toughness because of the slight misorientation of the microfibrils in the middle layer of the secondary wall. Under load, the microfibrils
straighten, permitting some ductility and energy absorption.

Wood has a good specific strength and specific modulus. After drying, the
density of typical hardwoods is 0.3 Mg m^{-3} to 0.8 Mg m^{-3} whereas the density of

typical softwoods is 0.3 Mg m^{-3} to 0.5 Mg m^{-3}. Table 16-8 compares the specific strength and specific modulus of clear wood to other common construction materials.

TABLE 16-8 Comparison of the specific strength and specific modulus of wood to other common construction materials

Material	Specific Strength (GPa)	Specific Modulus (GPa)
Clear wood	4.8	655
Aluminum	3.4	724
1020 steel	1.4	724
Copper	1.0	379
Concrete	0.4	241

After F. F. Wangaard, "Wood: Its Structure and Properties," *J. Educ. Models for Mat. Sci. and Engr.*, Vol. 3, No. 3, 1979.

16-15 Concrete and Asphalt

Concrete and asphalt, common construction materials, are particulate composites in which an aggregate, usually gravel and sand, is bonded in a matrix of Portland cement or bitumen.

Concrete. Concrete is a composite of aggregate, cement, and water. The aggregate, composed of gravel and sand, is bonded by a cementation reaction between the minerals in Portland cement and water. Several factors determine the properties and performance of the concrete.

The aggregate must be clean, strong and durable, must have the appropriate size and distribution of sizes, and must produce a high packing factor. The size distribution is critical in minimizing the amount of open porosity in the finished concrete—poor packing permits water to penetrate the concrete. When the water freezes, the resulting expansion can cause the concrete to disintegrate. An angular rather than round aggregate provides better strength due to mechanical interlocking and a greater surface area for bonding.

The binder, which is usually very fine Portland cement, is composed of various ratios of $3CaO \cdot Al_2O_3$, $2CaO \cdot SiO_2$, and $3CaO \cdot SiO_2$. When water is added to the cement, a hydration reaction occurs in which the water is rigidly attached to the minerals to produce a solid gel. Heat is evolved during this reaction. Each of the minerals behaves differently during hydration: $3CaO \cdot Al_2O_3$ and $3CaO \cdot SiO_2$ produce rapid setting but low strengths. The $2CaO \cdot SiO_2$ reacts more slowly but produces higher strengths (Figure 16-37). By controlling the relative amounts of the minerals in the cement, both the rate of setting and the final strength can be controlled.

The water-cement ratio determines the workability and the final strength of the concrete. *Workability* refers to how easily the concrete slurry can fill all of the space in the form—air pockets trapped due to poor workability reduce the

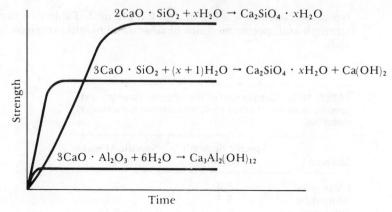

$$2CaO \cdot SiO_2 + xH_2O \rightarrow Ca_2SiO_4 \cdot xH_2O$$

$$3CaO \cdot SiO_2 + (x+1)H_2O \rightarrow Ca_2SiO_4 \cdot xH_2O + Ca(OH)_2$$

$$3CaO \cdot Al_2O_3 + 6H_2O \rightarrow Ca_3Al_2(OH)_{12}$$

FIG. 16-37   The rate of hydration of the minerals in Portland cement. $3CaO \cdot Al_2O_3$ and $3CaO \cdot SiO_2$ set rapidly but produce low strength; $2CaO \cdot SiO_2$ sets slowly but produces high strength.

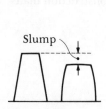

FIG. 16-38
The slump test, in which deformation of a concrete shape under its own weight is measured, is used to describe the workability of the concrete mix.

(a)

(b)

FIG. 16-40
The ideal structure of asphalt (a) compared to the undesirable structure (b) in which round grains, a narrow distribution of grains, and excess binder all reduce the strength of the final material.

strength of the concrete structure. Workability can be measured by the *slump test*. A wet concrete shape is produced (Figure 16-38) and is permitted to stand under its own weight. After some period of time, it deforms. The reduction in height of the form is the slump, which increases as the water-cement ratio increases.

The strength of concrete also depends on the water-cement ratio (Figure 16-39). Normally, a water-cement ratio of 0.45 to 0.55 is typical. When excess water is used, voids are left as the water evaporates.

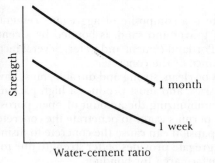

FIG. 16-39   The water-cement ratio and the setting time are important factors in determining the strength of concrete.

Asphalt. Asphalt is a composite of aggregate and bitumen, which is a thermoplastic polymer. The properties of the asphalt are determined by the characteristics of the aggregate, the binder, and their relative amounts.

The aggregate, as in concrete, should be clean, angular, and have a distribution of grain sizes to provide a high packing factor and good mechanical interlocking between the aggregate grains (Figure 16-40).

The binder, composed of thermoplastic polymer chains, bonds the aggregate particles. The binder has a relatively narrow useful temperature range, being brittle at sub-zero temperatures and beginning to melt at relatively low temperatures. Additives, such as gasoline or kerosene, can be used to modify the binder, permitting it to liquefy more easily during mixing while causing the asphalt to cure more rapidly after application.

The ratio of binder to aggregate is important; just enough binder should be added so that the aggregate particles touch but voids are minimized. Excess binder permits viscous deformation of the asphalt under load.

16-16 Sandwich Structures

Sandwich materials have thin layers of a facing material joined to a lightweight filler material, such as a polymer foam. Neither the filler nor the facing material are strong or rigid, but the composite possesses both properties. A familiar example is corrugated cardboard. A corrugated core of paper is bonded on either side to flat thick paper. Neither the corrugated core nor the facing paper are rigid but their combination is.

Another important example is the honeycomb structure used in aircraft applications. A *honeycomb* is produced by gluing thin aluminum strip at selected locations. The honeycomb material is then expanded to produce a very low density cellular panel that by itself is unstable (Figure 16-41). When an aluminum facing sheet is adhesively bonded to either side of the honeycomb, however, a very stiff, rigid, strong, and exceptionally lightweight sandwich is obtained.

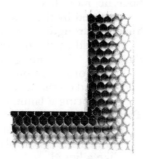

FIG. 16-41 A honeycomb structure. Thin aluminum foil is glued at selected locations, then expanded into a cellular panel. Thicker aluminum facing sheets produce a strong, rigid structure.

SUMMARY

The strengthening mechanisms discussed earlier are difficult to apply to composite materials. Instead composites are designed to provide unusual combinations of properties that cannot be obtained by the typical techniques used to control microstructure and mechanical properties. This is especially true for the laminar and particulate composites, which are almost always designed to satisfy special service requirements other than strength. Fiber-reinforced composites are normally designed to produce unusual combinations of strength, stiffness, temperature resistance, and light weight. The rule of mixtures is suitable for determining the behavior of many simple fiber-reinforced materials.

GLOSSARY

Aspect ratio The length of a fiber divided by its diameter.

Bimetallic A laminar composite material produced by joining two strips of metal with different thermal expansion coefficients, making the material sensitive to temperature changes.

Cambium The layer of growing cells in wood.

Carbonizing Driving off the noncarbon atoms from a polymer fiber, leaving behind a graphite fiber of high strength. Also known as pyrolizing.

Cemented carbides Particulate composites containing hard ceramic particles bonded with a soft metallic matrix. The composite combines high hardness and cutting ability yet still has good shock resistance.

Cladding The good corrosion-resistant or high-hardness layer of a laminar composite formed onto a less expensive or higher strength backing.

Compocasting Injection of a thixotropic mixture of an alloy and a filler material into a die at high pressures to form a composite.

Honeycomb A lightweight but stiff assembly of aluminum strip joined and expanded to form the core of aircraft components.

Lignin Polymer chains in wood that bond the microfibrils in the cell wall and bond the individual fibers to one another.

Lumen The hollow center of a fibrous cell in wood.

Microfibrils Cellulose chains that comprise the cell walls in wood fibers.

Precursor The polymer fiber that is carbonized to produce graphite fibers.

Prepregs Layers of fibers in unpolymerized resins. After the prepregs are stacked to form a desired structure, polymerization joins the layers together.

Rule of mixtures The statement that the properties of a composite material are a function of the volume fraction of each material in the composite.

Sandwich A composite material constructed of a lightweight low-density material surrounded by dense solid layers. The sandwich combines overall light weight with excellent stiffness.

Sizing Coating glass fibers with an organic material to improve bonding and moisture resistance in fiberglass.

Slump test A test to measure the workability of a concrete mix.

Specific modulus The modulus of elasticity divided by the density.

Specific strength The strength of a material divided by the density.

Surfacing A welding technique by which a hard or corrosion-resistant material is deposited on the surface of a second material, producing a laminar composite.

Thixotropic The ability of a partly liquid-partly solid material to maintain its shape until a stress is applied, when it then flows like a liquid.

Tow A bundle of hundreds or thousands of filaments.

Tracheids Hollow fibrous tubes that comprise the structure of wood.

Whiskers Very fine fibers grown in a manner that produces single crystals with no dislocations, thus giving nearly theoretical strengths.

Workability The ease with which a concrete slurry can fill all of the space in a form.

PRACTICE PROBLEMS

1 Suppose spherical aluminum powder 203 μm in diameter has an Al_2O_3 oxide layer 2.54 μm thick.

During the powder metallurgy processing, the Al_2O_3 is broken up to provide dispersion strengthening. What vol% Al_2O_3 is present in the SAP alloy?

2 Nickel is alloyed with 1 wt% Th, converted to a powder, compacted into a desirable shape, and oxidized during sintering. What vol% ThO_2 is produced in the TD-nickel? (The density of ThO_2 is 9.86 Mg m^{-3}.)

3 Calculate the density of a type C-2 cemented carbide cutting tool containing 93 wt% WC, 6 wt% Co, and 1 wt% TaC. (See Example 16-2 for densities.)

4 Designate a SiC cutting wheel that might be appropriate for rough high-speed cutting of steel castings.

5 A silver-tungsten contact material has a density of 16.1 Mg m^{-3}. Calculate (a) the volume fraction of silver in the composite, (b) the volume fraction of pores in the tungsten compact before the silver is infiltrated, and (c) the original density of the tungsten compact before the silver is infiltrated.

6 Suppose we want to fill polystyrene with 40 vol% SiO_2. (a) How much SiO_2 must we add to 1 kg of polystyrene? (b) What will be the density of the composite! ($\rho_{SiO_2} = 2.66$ Mg m^{-3}; $\rho_{polystyrene} = 1.06$ Mg m^{-3}.)

7 How much clay, having a density of 2.4 Mg m^{-3}, must we add to 1 kg of nylon, density 1.14 Mg m^{-3}, to produce a composite having a density of 2.1 Mg m^{-3}?

8 Suppose a phenolic resin is used to coat round silica sand grains 0.04 cm in diameter with a resin layer 0.001 cm thick. How much resin must be added per 100 kg of sand? (The density of sand is 2.2 Mg m^{-3} and that of the resin is 1.28 Mg m^{-3}.)

9 An abrasive material contains 70 vol% Al_2O_3, 10 vol% binder, and 20 vol% pores. Calculate and compare the density of the abrasive composite if the binder is (a) phenolic and (b) silica glass. (The density of phenolic is 1.28 Mg m^{-3}, of glass is 2.2 Mg m^{-3} and of Al_2O_3 is 3.965 Mg m^{-3}.)

10 An abrasive cutting wheel which has a ratio of one part BN to two parts glass binder by volume is produced. The density of the finished wheel is 1.95 Mg m^{-3}. Calculate the vol% porosity in the wheel. (The density of BN is 1.9 Mg m^{-3} and that of the glass binder is 2.2 Mg m^{-3}.)

11 We can consider styrofoam a composite material of gas and thin-walled polystyrene having an overall density of about 0.016 Mg m^{-3}. The density of polystyrene is 1.06 Mg m^{-3}. (a) Calculate the volume fraction of each material in the composite. (b) Assuming for simplicity that the styrofoam beads are cubes 0.1 cm on each edge, estimate the wall thickness of the beads. (c) The thermal conductivity of air is about 0.026 W m^{-1} K^{-1} and that for polystyrene is about 0.125 W m^{-1} K^{-1}. Estimate the thermal conductivity of the styrofoam.

12 Suppose we wanted to produce a compocast Cu-10% Sn alloy containing 35 vol% glass beads (Figure 12-7). (a) Recommend a temperature for accomplishing the compocasting process. (b) Calculate the density after the composite is made. (The density of the copper-tin alloy is 8.77 Mg m^{-3} and that of the glass is 2.2 Mg m^{-3}.)

13 A foundry sand which has an apparent density of 1.52 Mg m^{-3} is composed of sand grains, each 381 μm in diameter. (a) Determine the percent porosity in the sand. (b) The sand grains are uniformly coated with a phenolic resin. The wt% resin added to the sand is 2.5%. Calculate the thickness of the resin layer on each sand grain. (The density of the sand is 2.2 Mg m^{-3} and that of the resin is 1.28 Mg m^{-3}.)

14 A Kevlar-epoxy composite containing 30 vol% Kevlar fibers is prepared. Calculate (a) the wt% Kevlar in the composite, (b) the density of the composite, and (c) the modulus of elasticity parallel to the fibers in the composite. (Assume $\rho_{epoxy} = 1.25$ Mg m^{-3} and $E_{epoxy} = 3.1$ GPa).

15 Fiberglass containing unidirectional S glass fibers in an epoxy matrix has a modulus of elasticity of 18.2 GPa. Determine (a) the volume fraction of glass fibers and (b) the density of the composite. (The density of epoxy is 1.3 Mg m^{-3} and its modulus is about 2.75 GPa.)

16 When we look at a boron-aluminum composite through a microscope, we find that the fibers have a diameter of 0.05 mm and the fibers are 0.15 mm apart (Figure 16-42). Determine the volume fraction

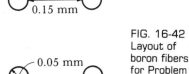

FIG. 16-42 Layout of boron fibers for Problem 16.

of boron fibers and the modulus of elasticity of the composite. (The modulus of elasticity for aluminum is 69 GPa.)

17 The modulus of elasticity of a boron-aluminum fiber composite perpendicular to the fibers is 206.9 GPa. Determine the vol% boron fibers present.

18 Determine the modulus of elasticity for a HS graphite fiber-aluminum matrix composite containing 50 vol% graphite (a) parallel to the fibers and (b) perpendicular to the fibers.

19 (a) Estimate the modulus of elasticity of a 55 vol% HM graphite reinforced polyester composite. The polyester has a modulus of elasticity of 3.45 GPa.

(b) Estimate the modulus after the matrix begins to deform.

20   Calculate the specific modulus for an epoxy-40 vol% Kevlar composite parallel to the fibers. (Assume the modulus of epoxy is 2.76 GPa and the density of epoxy is 1.3 Mg m^{-3}.)

21   Determine the thermal conductivity along the graphite fibers in a phenolic-30 vol% graphite brake pad. (The thermal conductivity of the phenolic is 0.147 W m^{-1} K^{-1} and that of graphite is 0.021 W m^{-1} K^{-1}.)

22   Determine the specific modulus of a composite containing a beryllium matrix and 30 vol% SiC whiskers.

23   Compare the specific modulus of a 50 vol% HM graphite-epoxy fishing pole to that of a 2024-T6 aluminum fishing pole.

24   A lightweight leaf spring for an automobile can be produced from a HM graphite-epoxy composite. What vol% and wt% of the fibers must be present for the spring to have the same modulus of elasticity as steel, 207 GPa?

25   Suppose we produce a cable containing 30 vol% Al fibers in an epoxy matrix. Calculate the electrical conductivity of the cable. (The electrical conductivity of aluminium is 3.8 × 10^3 A V^{-1} m^{-1} and that for epoxy is 10^{-11} A V^{-1} m^{-1}.)

26   Determine the modulus of elasticity of a compos-ite formed from 12.5 mm steel plate and 3 mm titanium plate (a) parallel and (b) perpendicular to the plates. (E_{steel} = 207 GPa; $E_{titanium}$ = 124 GPa.)

27   Suppose we produce a laminated heat shield from a 1-cm copper plate and a 2-cm-thick fused quartz sheet. Determine the thermal conductivity parallel and perpendicular to the heat shield. (The thermal conductivity of copper is 401 W m^{-1} K^{-1} and that for fused silica is 1.32 W m^{-1} K^{-1}.)

28   Estimate the total wt% Ni in a U.S. silver coin. (The density of Ni-20% Cu is 8.91 Mg m^{-3} and of Cu-20% Ni is 8.95 Mg m^{-3}.)

29   A 0.305 m outside diameter spherical tank used to contain a corrosive material must be made from 1 in. thick copper for corrosion resistance. However, an alternate design permits the use of a 38.1 mm composite of 6.35 mm, copper bonded to 31.75 mm aluminum. Compare (a) the modulus of elasticity of the copper versus the composite both parallel and perpendicular to the plate and (b) the weight of the tank made of copper to that of the composite.

30   A 0.2-cm bimetallic strip composed of equal thicknesses of copper and Invar is heated to 100°C. Calculate the radius of curvature produced.

31   A 0.2-cm bimetallic strip composed of equal thicknesses of fused quartz glass and aluminum produces a radius of curvature of 20 cm when heated from 22°C. What is the temperature?

PART **IV**

PHYSICAL PROPERTIES OF ENGINEERING MATERIALS

The physical behavior of materials is described by a variety of electrical, magnetic, optical, thermal, and elastic properties. Most of these properties are determined by the atomic structure, atomic arrangement, and crystal structure of the material. In Chapter 17, we will find that the atomic structure, in particular the energy gap between the electrons in the valence and conduction bands, helps us to divide materials into conductors, semiconductors, and insulators. The atomic structure is responsible for the ferromagnetic behavior discussed in Chapter 18 and explains many of the optical properties such as emission and transparency which are discussed in Chapter 19.

The physical properties can be altered to a significant degree by changing the short- and long-range order of the atoms as well as by introducing and controlling imperfections in the atomic arrangement. We will find that strengthening mechanisms and metal processing techniques, for example, have a significant effect on electrical conductivity of metals. Improved magnets are obtained by introducing lattice defects or by controlling grain size. In this section, we will again demonstrate the importance of the structure-property-processing relationship.

17

Electrical Conductivity

17-1 Introduction

In many applications, the electrical behavior of the material is more critical than the mechanical behavior. Metal wire used to transfer current over long distances must have a high electrical conductivity so that little power is lost by heating of the wire. Ceramic insulators must prevent arcing between conductors. Semiconductor devices used to convert solar energy to electrical power must be as efficient as possible to make solar cells a practical alternative energy source.

To select and use materials for electrical and electronic applications, we must understand how properties such as electrical conductivity are produced and controlled. We must also realize that electrical behavior is influenced by the structure of the material, the processing of the material, and the environment to which the material is exposed. To accomplish these goals, we must examine in greater detail the electronic structure of groups of atoms.

17-2 Relationship Between Ohm's Law and Electrical Conductivity

Most of us are familiar with the common form of Ohm's law

$$V = IR \tag{17-1}$$

where V is the voltage (volts, V), I is the current (amps, A), and R is the resistance (ohms, Ω) to the current flow. The resistance R is a characteristic of the materials that compose the circuit.

$$R = \rho \frac{l}{A} = \frac{l}{\sigma A} \tag{17-2}$$

where l is the length (m) of the conductor, A is the cross-sectional area (m^2) of the conductor, ρ is the electrical *resistivity* (ohm $\cdot$ m), and ϱ, which is the reciprocal of ρ, is the electrical *conductivity* (ohm^{-1} m^{-1}). We can use this equation to design resistors, since we can vary the length or the cross-sectional area of the device.

EXAMPLE 17-1

Calculate the loss of power in a copper transmission line 1500 m long when a current of 50 A is flowing. The copper wire has a diameter of 1 mm and the electrical resistivity is $1.67 \times 10^{-8}\ \Omega \cdot$m.

$$R = \frac{\rho l}{A} = (1.67 \times 10^{-8})\ \frac{(1500\ \text{m})}{\pi/4\ (10^{-3}\ \text{m})^2} = 31.9\ \Omega$$

The power loss in the form of heating of the wire is

$$\text{Power} = VI = I^2 R = (50)^2(31.9) = 7.98 \times 10^6\ \text{W}$$

A second form of Ohm's law is obtained if we combine Equations (17-1) and (17-2).

$$V = IR = \frac{Il}{\sigma A}$$

$$\frac{I}{A} = \sigma\ \frac{V}{l}$$

If we define I/A as the *current density* J (Am^{-2}) and V/l as the *electric field* ξ (V m^{-1}), then

$$J = \sigma\xi \tag{17-3}$$

We can also determine that the current density J is

$$J = nq\bar{v}$$

where n is the number of charge carriers (carriers m^{-3}), q is the charge on each carrier $(1.6 \times 10^{-19}\ \text{C})$, and v is the *average drift velocity* (m s^{-1}) at which the charge carriers move (Figure 17-1). Thus

$$\sigma\xi = nq\bar{v} \quad \text{or} \quad \sigma = nq\ \frac{\bar{v}}{\xi}$$

The term $\bar{v}/\xi$ is called the *mobility* μ $[\text{m}^2\ (\text{V} \cdot \text{s})^{-1}]$

$$\mu = \frac{\bar{v}}{\xi}$$

Finally,

$$\sigma = nq\mu \tag{17-4}$$

The charge q is a constant; from inspection of the equation, we find that we can control the electrical conductivity of a material by controlling the number of charge carriers in the material or by controlling the mobility, or ease of movement, of the charge carriers.

Electrons are the charge carriers in conductors (such as metals), semiconductors, and insulators whereas ions carry the charge in most ionic compounds (Figure 17-2). The mobility depends on atomic bonding, lattice imperfections, microstructure, and, in ionic compounds, diffusion rates.

Table 17-1 includes some useful units and relationships.

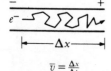

$$\bar{v} = \frac{\Delta x}{\Delta t}$$

FIG. 17-1
Charge carriers, such as electrons, are deflected by atoms or lattice defects and take an irregular path through a conductor. The average rate at which the carriers move is the drift velocity $\bar{v}$.

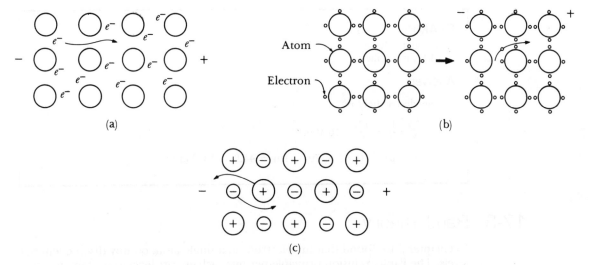

(a)   (b)

(c)

FIG. 17-2 Charge carriers in different materials. (a) Valence electrons in the metallic bond move easily, (b) covalent bonds must be broken in semiconductors and insulators for an electron to be able to move, and (c) entire ions must diffuse to carry charge in many ionically bonded materials.

TABLE 17-1 Some useful relationships and units

Electron volt = 1 eV = 1.6×10^{-19} J
1 amp = 1 coulomb/second = 1 C s^{-1}
1 volt = 1 amp·ohm = 1 A·Ω
k = Boltzmann's constant = 8.63×10^{-5} eV/K = 1.381×10^{-23} J K^{-1}
kT at room temperature = 0.025 eV = 0.04×10^{-19} J

EXAMPLE 17-2

Calculate the mobility of an electron in copper, assuming that all of the valence electrons contribute to the current flow.

Answer:

The valence of copper is one; therefore the number of electrons equals the number of copper atoms in the material. The lattice parameter of copper is 3.615×10^{-10} m and, since copper is FCC, there are 4 atoms/unit cell.

$$n = \frac{(4 \text{ atoms/cell})(1 \text{ electron/atom})}{(3.6151 \times 10^{-10} \text{ m})^3} = 8.467 \times 10^{28} \text{ electrons} \cdot \text{m}^{-3}$$

$$q = 1.6 \times 10^{-19} \text{ C} = 1.6 \times 10^{-19} \text{ A} \cdot \text{s}$$

$$\mu = \frac{\sigma}{nq} = \frac{1}{\rho nq} = \frac{1}{(1.67 \times 10^{-8})(8.467 \times 10^{28})(1.6 \times 10^{-19})}$$

$$= 4.42 \times 10^{-3} \text{ m}^2 \, (\Omega \cdot \text{C})^{-1} = 4.42 \times 10^{-3} \text{ m}^2 \, (\text{V} \cdot \text{s})^{-1}$$

EXAMPLE 17-3

Calculate the average drift velocity for the electrons in a 1 m copper wire when 10 V are applied.

Answer:

The electric field is

$$\xi = \frac{V}{l} = \frac{10}{1} = 10 \text{ V m}^{-1}$$

$$\bar{v} = \mu\xi = (4.42 \times 10^{-3})(10) = 44.2 \times 10^{-3} \text{ m} \cdot \text{s}^{-1}$$

17-3 Band Theory

In Chapter 2 we found that the electrons in a single atom occupy discrete energy levels. The Pauli exclusion principle permits each energy level to contain only two electrons. For example, the $2s$ level of a single atom contains one energy level and two electrons. The $2p$ level contains three energy levels and a total of six electrons.

When N atoms come together to produce a solid, the Pauli principle still requires that only two electrons in the entire solid have the same energy. Each energy level broadens into a band (Figure 17-3). Consequently, the $2s$ band in a solid contains N discrete energy levels and $2N$ electrons, two in each energy level. Each of the $2p$ levels contains N energy levels and $2N$ electrons. Since the three $2p$ bands actually overlap, we could alternately describe a single broad $2p$ band containing $3N$ energy levels and $6N$ electrons.

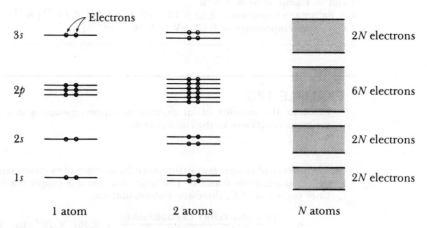

FIG. 17-3 The energy levels broaden into bands as the number of electrons grouped together increases.

EXAMPLE 17-4

How many energy levels are present in the $2s$ band of a pure aluminum crystal 10 mm × 10 mm × 10 mm in size?

Answer:

Aluminum is FCC, has 4 atoms/unit cell, and has a lattice parameter of 4.0496×10^{-10} m.

$$N = \text{number of atoms} = \frac{4 \text{ atoms/cell}}{(4.0496 \times 10^{-10})^3} = 6 \times 10^{28} \text{ atoms} \cdot \text{m}^{-3}$$

$2N$ = energy levels in the $2s$ band

$2N = 12 \times 10^{28}$ energy levels in the $2s$ band $\cdot$ cm^{-3}

17-4   Band Structure of Alkali Metals

The alkali metals in Column IA of the periodic table have only one electron in the outermost s level of their electronic structure. Figure 17-4 shows the development of the band structure in sodium, which has an electronic structure of $1s^2 2s^2 2p^6 3s^1$. The vertical line in Figure 17-4 represents the equilibrium interatomic spacing of the atoms in solid sodium.

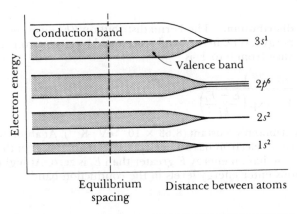

FIG. 17-4 The band structure for sodium. The energy levels broaden into bands. The 3s band, which is only half-filled with electrons, is responsible for conduction in sodium.

The shaded areas in Figure 17-4 indicate the portion of the band in which the energy levels are completely occupied by electrons. The $3s$ valence band of sodium is only half-filled and only the lowest possible energy levels within the $3s$ band are occupied.

An electron moves through the metal and conducts electrical charge when it gains sufficient energy to occupy a higher level in the valence band (Figure 17-5). The electron moves at a velocity determined by the mobility and the strength of the electric field and is accelerated toward the positive terminal of the circuit.

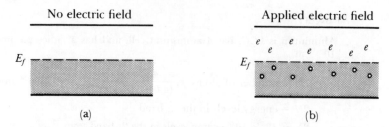

FIG. 17-5 (a) At equilibrium, all of the electrons in the outer energy level have the lowest possible energy. (b) However, when a voltage, or electric field, is applied, some electrons are excited into unfilled levels and conduction occurs.

Fermi energy.   The *Fermi energy* E_f is the energy at which half of the possible energy levels in the band are actually occupied by electrons. At absolute zero, only the energy levels in the bottom half of the $3s$ band of sodium are occupied (Figure 17-5), and the Fermi energy is the energy at the center of the $3s$ band.

When the temperature of the metal is increased, electrons near the Fermi energy are excited into the unoccupied levels. But this creates an equal number of vacated energy levels below the Fermi energy. Consequently, the Fermi energy is unchanged.

Fermi distribution.   The Fermi distribution $f(E)$ gives the probability that a particular energy level E in the band is occupied by an electron. The function $f(E)$, which may vary from 0 to 1, is

$$f(E) = \frac{1}{1 + \exp\left(\dfrac{E - E_f}{kT}\right)} \tag{17-5}$$

where k is Boltzmann's constant $(8.63 \times 10^{-5}\,\text{eV} \cdot \text{K}^{-1})$. At absolute zero, the probability $f(E)$ that an electron has an energy E less than E_f is one; the probability $f(E)$ that an electron has an energy E greater than E_f is zero. At higher temperatures, some electrons enter energy levels in the conduction band.

EXAMPLE 17-5

Calculate the probability that energy levels at E_f and at $E_f + 0.05$ eV, 0.10 eV, 0.50 eV, and 1.00 eV will contain an electron at 27°C.

Answer:

At 27°C, the value $kT = (8.63 \times 10^{-5})(27 + 273) = 0.025$ eV. The probability $f(E)$ is

$$f(E_f) = \frac{1}{1 + \exp\left(\dfrac{E_f - E_f}{0.025}\right)} = \frac{1}{1 + \exp(0)} = 0.50$$

$$f(E_f + 0.05) = \frac{1}{1 + \exp\left(\dfrac{0.05}{0.025}\right)} = \frac{1}{1 + \exp(2)} = 0.12$$

$$f(E_f + 0.10) = \frac{1}{1 + \exp\left(\dfrac{0.10}{0.025}\right)} = \frac{1}{1 + \exp(4)} = 0.02$$

$$f(E_f + 0.50) = \frac{1}{1 + \exp\left(\dfrac{0.50}{0.025}\right)} = \frac{1}{1 + \exp(20)} = 2 \times 10^{-9}$$

$$f(E_f + 1.00) = \frac{1}{1 + \exp\left(\dfrac{1.00}{0.025}\right)} = \frac{1}{1 + \exp(40)} = 4 \times 10^{-18}$$

These probabilities are plotted in Figure 17-6 and compared to the Fermi distribution at 0 K and 1000 K. Since the Fermi distribution is symmetrical, the probabilities of energy levels below the Fermi energy being vacant are also plotted.

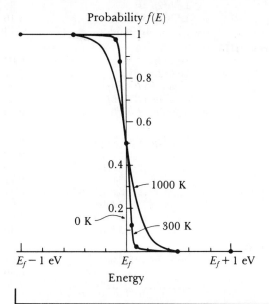

FIG. 17-6 The Fermi distribution at 0 K, 300 K (27°C), and 1000 K.

The Fermi distribution shows that some electrons in a metal possess sufficient energy to exceed the Fermi level and conduct a charge. At the same time, vacant energy levels within the valence band are created. When an electrical field is applied, the electrons in the upper half are accelerated towards the positive terminal of the circuit, while electrons in the lower half decelerate towards the negative terminal by filling vacant sites in the valence band. Both of these events cause a net flow of current.

Conductivity of alkali metals. Sodium and other alkali metals have good electrical conductivity because of the half-filled outermost s band. Table 17-2 compares the electronic structures and conductivities of the alkali metals to other metals.

TABLE 17-2   Electronic structure and electrical conductivity of several groups of metals at 25°C

Metal	Electronic Structure	Electrical Conductivity $(ohm^{-1} \cdot m^{-1})$
Alkali metals		
Li	$1s^2 2s^1$	1.07×10^7
Na	$1s^2 2s^2 2p^6 3s^1$	2.13×10^7
K	$1s^2 2s^2 2p^6 3s^2 3p^6 4s^1$	1.64×10^7
Rb	$........ 4s^2 4p^6 5s^1$	0.86×10^7
Cs	$........ 5s^2 5p^6 6s^1$	0.50×10^7
Alkali earths		
Be	$1s^2 2s^2$	2.50×10^7
Mg	$1s^2 2s^2 2p^6 3s^2$	2.25×10^7
Ca	$1s^2 2s^2 2p^6 3s^2 3p^6 4s^2$	3.16×10^7
Sr	$........ 4s^2 4p^6 5s^2$	0.43×10^7
Aluminum and Group IIIA		
B	$1s^2 2s^2 2p^1$	0.03×10^7
Al	$1s^2 2s^2 2p^6 3s^2 3p^1$	3.77×10^7
Ga	$...3s^2 3p^6 3d^{10} 4s^2 4p^1$	0.66×10^7
In	$...4s^2 4p^6 4d^{10} 5s^2 5p^1$	1.25×10^7
Tl	$...5s^2 5p^6 5d^{10} 6s^2 6p^1$	0.56×10^7
Transition metals		
Sc	$1s^2 2s^2 2p^6 3s^2 3p^6 3d^1 4s^2$	0.77×10^7
Ti	$............ 3d^2 4s^2$	0.24×10^7
V	$............ 3d^3 4s^2$	0.40×10^7
Cr	$............ 3d^5 4s^1$	0.77×10^7
Mn	$............ 3d^5 4s^2$	0.11×10^7
Fe	$............ 3d^6 4s^2$	1.00×10^7
Co	$............ 3d^7 4s^2$	1.90×10^7
Ni	$............ 3d^8 4s^2$	1.46×10^7
Copper and Group IB		
Cu	$1s^2 2s^2 2p^6 3s^2 3p^6 3d^{10} 4s^1$	5.98×10^7
Ag	$.......... 4p^6 4d^{10} 5s^1$	6.80×10^7
Au	$.......... 5p^6 5d^{10} 6s^1$	4.26×10^7

From *ASM Metals Handbook*, Vol. 2, 9th Ed., 1979.

17-5   Band Structure of Other Metals

The band structure and electrical behavior vary for different groups of metals. The band structures for four groups of elements are shown in Figure 17-7 and conductivities are listed in Table 17-2.

The alkali earth metals, Column IIA of the periodic table, have two electrons in their outermost s band. We might expect these metals to have poor electrical conductivity, since there appear to be no unoccupied energy levels into which the electrons can be excited for conduction. Yet these metals have a higher conductivity than the alkali metals because the p band overlaps the s band [Figure 17-7(a)]. Consequently, there are a large number of unoccupied energy levels in the combined $3s$ and $3p$ band to which a magnesium electron can be excited.

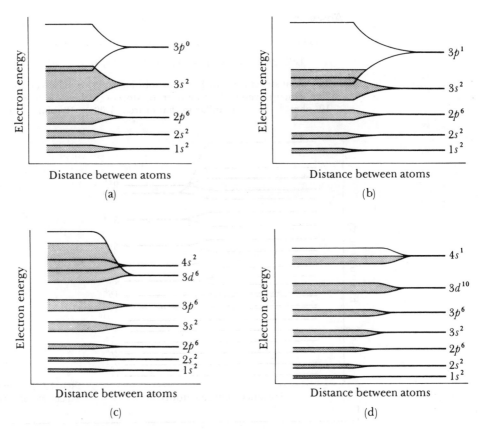

FIG. 17-7 Simplified band structures for selected metals. (a) Magnesium, (b) aluminum, (c) iron, and (d) copper.

Aluminum, which follows sodium and magnesium in atomic number, has a partly filled $3p$ band, which serves as the conduction band [Figure 17-7(b)]. Electrons easily enter unoccupied levels in the $3p$ band.

The transition metals, including scandium through nickel, contain one or two electrons in their outermost s band but possess an unfilled d band. In iron [Figure 17-7(c)], the $4s$ band overlaps the $3d$ band, which contains only six electrons. Electrons may enter the conduction band in the upper half of the overlapping bands. However, complex interactions between the bands prevent the conductivity from being as high as in some of the better conductors.

Group IB metals—copper, silver, and gold—contain one electron in their outermost s band and immediately follow the transition metal groups in atomic number. The inner d band electrons are tightly held by the atom core and do not interact with the electrons in the s band [Figure 17-7(d)]. Copper, silver, and gold have very high conductivities.

EXAMPLE 17-6

From the electronic structure of tungsten, sketch the expected band structure and compare the conductivity, 1.77×10^7 ohm$^{-1} \cdot$ m^{-1}, to metals with a similar band structure.

Answer:

The electronic structure of tungsten, which has an atomic number of 74, is

$$1s^2 2s^2 2p^6 3s^2 3p^6 3d^{10} 4s^2 4p^6 4d^{10} 4f^{14} 5s^2 5p^6 5d^4 6s^2$$

The band structure, which is shown in Figure 17-8, has overlapping $5d$ and $6s$ levels. The structure resembles that of iron. Moreover, the conductivity of tungsten, 1.77×10^7 ohm$^{-1} \cdot$ m^{-1}, is similar to that of iron, 1.00×10^7 ohm$^{-1} \cdot$ m^{-1}.

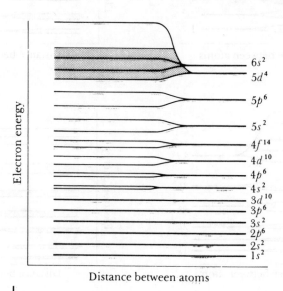

FIG. 17-8 Simplified band structure of tungsten.

Distance between atoms

17-6 Controlling the Conductivity of Metals

The conductivity of a pure, defect-free metal is determined by the electronic structure of the atoms. But we can significantly affect the conductivity by influencing the mobility μ of the carriers. The mobility is proportional to the drift velocity $\bar{v}$, which is low if the electrons collide with imperfections in the lattice. The *mean free path* is the average distance between collisions; a long mean free path permits high mobilities and high conductivities.

If the metal contains no lattice defects and performs at absolute zero degrees, the mean free path is infinite and the electrical resistivity is zero. However, no metals are perfect nor do they operate at absolute zero.

Temperature effect. When the temperature of a metal increases, thermal energy causes the atoms to vibrate (Figure 17-9). At any instant, the atom may not be in its equilibrium position and therefore interacts with and scatters electrons. The mean free path decreases, the mobility of electrons is reduced, and the resistivity increases. The change in resistivity with temperature can be estimated from the equation

$$\rho = \rho_r(1 + a\Delta T) \tag{17-6}$$

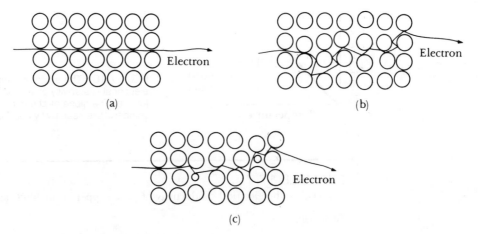

FIG. 17-9 Movement of an electron through (a) a perfect crystal, (b) a crystal heated to a high temperature, and (c) a crystal containing lattice defects. Scattering of the electrons reduces the mobility and conductivity.

where ρ_r is the resistivity at room temperature (25°C), ΔT is the temperature difference between the temperature of interest and room temperature, and a is the *temperature resistivity coefficient*. The relationship between resistivity and temperature is linear over a wide temperature range (Figure 17-10). Examples of the temperature resistivity coefficient are given in Table 17-3. The resistivity of the metal due only to thermal vibration of the atoms is ρ_T.

TABLE 17-3 The temperature resistivity coefficient for selected metals

Metal	Room Temperature Resistivity ($\times 10^{-8} \Omega \cdot$ m)	Temperature Resistivity Coefficient (°C^{-1})
Be	4.0	0.0250
Mg	4.45	0.0165
Ca	3.91	0.0042
Al	2.65	0.0043
Cr	12.90	0.0030
Fe	9.71	0.0065
Co	6.24	0.0060
Ni	6.84	0.0069
Cu	1.67	0.0068
Ag	1.59	0.0041
Au	2.35	0.0040

From *ASM Metals Handbook*, Vol. 2, 9th Ed., 1979.

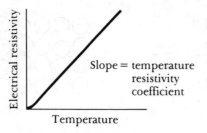

FIG. 17-10 The effect of temperature on the electrical resistivity of a metal with a perfect lattice. The slope of the curve is the temperature resistivity coefficient.

EXAMPLE 17-7

Calculate the electrical conductivity of pure copper at (a) 400°C and (b) −100°C.

Answer:

The resistivity of copper at room temperature is $1.67 \times 10^{-8} \, \Omega \cdot$ m and the temperature resistivity coefficient is $0.0068°\text{C}^{-1}$.

(a) At 400°C

$$\rho = \rho_r(1 + a\Delta T) = (1.67 \times 10^{-8})[1 + 0.0068(400 - 20)]$$

$$\rho = 5.985 \times 10^{-8} \, \Omega \cdot \text{m}$$

$$\sigma = 1/\rho = 1.67 \times 10^7 \, \Omega^{-1} \, \text{m}^{-1}$$

(b) At −100°C

$$\rho = (1.67 \times 10^{-5})[1 + 0.0068(-100 - 20)] = 0.307 \times 10^{-8} \, \Omega \cdot \text{m}$$

$$\sigma = 32.5 \times 10^7 \, \Omega^{-1} \, \text{m}^{-1}$$

Effect of lattice defects. Lattice imperfections scatter electrons and thus reduce the mobility and conductivity of the metal [Figure 17-9(c)]. Greater numbers of defects reduce the mean free path and have a pronounced effect on the conductivity. For example, the increase in the resistivity due to solid solution atoms is

$$\rho_d = b(1 - x)x \tag{17-7}$$

where ρ_d is the increase in resistivity due to the defects, x is the atomic fraction of the impurity or solid solution atoms present, and b is the *defect resistivity coefficient*. In a similar manner, vacancies, dislocations, and surface defects, such as grain boundaries, also reduce the conductivity of the metal. Each defect contributes to an increase in the resistivity of the metal. Thus, the overall resistivity is

$$\rho = \rho_T + \rho_d \tag{17-8}$$

where ρ_d equals the contributions from solid solution atoms, interstitial atoms, vacancies, grain boundaries, and other imperfections. The effect of the defects is independent of temperature (Figure 17-11).

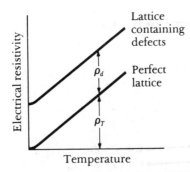

FIG. 17-11   The electrical resistivity of a metal is composed of a constant defect contribution ρ_d and a variable temperature contribution ρ_T.

Effect of processing and strengthening.   Strengthening mechanisms and metal processing techniques affect the electrical properties of a metal in different ways.

Solid solution strengthening is a poor way to obtain high strength in metals intended to have high conductivities. The mean free paths are very short due to the random distribution of the interstitial or substitutional atoms. Figure 17-12

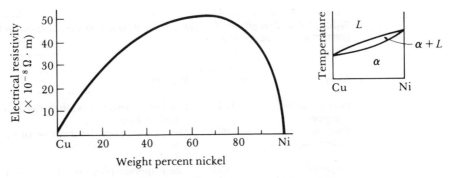

FIG. 17-12   Electrical resistivity in the copper-nickel system. Solid solution strengthening of either pure copper or nickel increases the electrical resistivity.

shows the influence of alloying element in the isomorphous copper-nickel system. A maximum in the resistivity occurs near 60% Ni. The change in resistivity is very large—from about $1.6 \times 10^{-8}\ \Omega \cdot$ m to more than $50 \times 10^{-5}\ \Omega \cdot$ m—as the composition changes from pure copper to Cu-60% Ni. Figure 17-13 shows the effect of zinc on the conductivity of copper; as the amount of zinc increases, the conductivity decreases substantially.

Age hardening and dispersion strengthening reduce the conductivity less than solid solution strengthening. Table 17-4 shows the effect of several strengthening mechanisms on the conductivity of copper and its alloys.

In the dispersion-strengthened two-phase alloys, such as eutectoid and eutectic alloys, the electrical conductivity or resistivity can be estimated from the rule of mixtures. Figure 17-14 shows the electrical resistivity in lead-tin alloys. The change in resistivity initially is very steep since the major effect is solid solution strengthening. However, the change is slower in the two-phase region, which suggests that the rule of mixtures might apply.

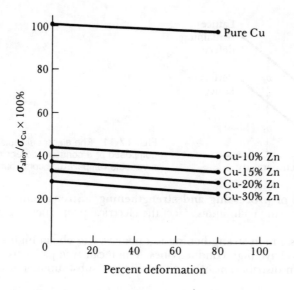

FIG. 17-13 The effect of solid solution strengthening and cold working on the electrical conductivity of copper.

TABLE 17-4   The effect of alloying, strengthening, and processing on the electrical conductivity of copper and its alloys

Alloy	$\dfrac{\sigma_{alloy}}{\sigma_{Cu}} \times 100$	Remarks
Pure annealed copper	101	Few lattice defects to scatter electrons; the mean free path is long.
Pure copper deformed 80%	98	Many dislocations, but because of the tangled nature of the dislocation networks, the mean free path is still long.
Dispersion-strengthened Cu-0.7% Al_2O_3	85	The dispersed phase is not as closely spaced as solid solution atoms nor is it coherent, as in age hardening. Thus, the effect on conductivity is small.
Solution-treated Cu-2% Be	18	The alloy is single phase; however, the small amount of solid solution strengthening from the supersaturated beryllium greatly decreases conductivity.
Aged Cu-2% Be	23	During aging, the beryllium leaves the copper lattice to produce a coherent precipitate. The precipitate does not interfere with conductivity as much as the solid solution atoms.
Cu-35% Zn	28	This alloy is solid solution strengthened by zinc. The conductivity is low but not as low as when beryllium, which has an atomic radius much different from copper, is present.

Strain hardening and grain size control have less effect on conductivity. Since dislocations and grain boundaries are further apart than solid solution atoms, there are large volumes of metal that have a long mean free path. Consequently, cold working is an effective way to increase the strength of a metallic conductor without seriously impairing the electrical properties of that material. In addition,

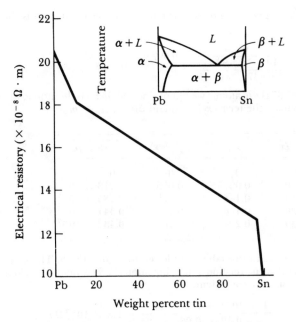

FIG. 17-14 The electrical resistivity in the lead-tin eutectic system. The conductivity is linear in the two-phase region.

the effects of cold working on conductivity can be eliminated by the low tempera-ture recovery heat treatment, in which good conductivity is restored while the strength is retained. Both Figure 17-13 and Table 17-4 illustrate the effect of cold working on conductivity; compared to solid solution strengthening, the effect of cold working is nearly negligible.

EXAMPLE 17-8

Is Equation (17-7) valid for the copper-zinc system? If so, calculate the defect resistiv-ity coefficient for zinc in copper.

Answer:

Let's look at the data in Figure 17-13 at 0% cold work. The conductivity of copper is $5.98 \times 10^7 \ \Omega^{-1} \ m^{-1}$.

% Zn	$\dfrac{\sigma_{alloy}}{\sigma_{Cu}} \times 100$	$\sigma_{alloy} \ (\Omega^{-1} \ m^{-1})$	$\rho_{alloy} \ (\Omega \cdot m)$
0	101	6.00×10^7	0.167×10^{-7}
10	44	2.63×10^7	0.380×10^{-7}
15	37	2.21×10^7	0.452×10^{-7}
20	33	1.97×10^7	0.508×10^{-7}
30	28	1.67×10^7	0.599×10^{-7}

We can convert to atomic fraction x; a sample calculation is shown.

$$x_{Zn} = \frac{\dfrac{10}{65.37}}{\dfrac{10\text{ wt\%}}{65.37} + \dfrac{90\text{ wt\%}}{63.55}} = 0.0975$$

where 65.37 is the atomic weight of zinc and 63.55 is the atomic weight of copper. Now let's calculate the terms $x(1-x)$ and $\Delta\rho = \rho_{alloy} - \rho_{Cu}$.

% Zn	x_{Zn}	$x(1-x)$	$\Delta\rho = \rho_d$
0	0	0	0
10	0.0975	0.088	0.213×10^{-7}
15	0.146	0.125	0.285×10^{-7}
20	0.196	0.158	0.341×10^{-7}
30	0.294	0.208	0.432×10^{-7}

The results in this table are plotted in Figure 17-15. The straight line that results indicates that the relationship in Equation (17-7) is valid. The slope of the graph is the defect resistivity coefficient.

$$b = \frac{0.4 \times 10^{-7} - 0.2 \times 10^{-7}}{0.19 - 0.08} = 1.8 \times 10^{-7}\,\Omega \cdot m$$

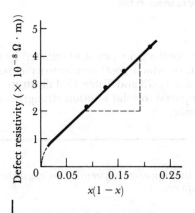

FIG. 17-15   The resistivity due to lattice defects, or zinc substitution atoms, versus the parameter $x(1-x)$. The slope of the graph is the defect resistivity coefficient for zinc in copper. (See Example 17-8.)

Anisotropic electrical behavior.   The electrical conductivity or resistivity in single crystals or in metals with a texture or preferred orientation may vary with direction, since the mean free path may vary. Typical examples of anisotropic behavior are given in Table 17-5.

17-7   Thermocouples

The Fermi distribution in a conductor changes with temperature. Although the Fermi energy is constant, more electrons occupy energy levels above E_f at elevated temperatures. If we heat only one end of the conductor, the Fermi distribution varies from the hot end to the cold end (Figure 17-16). The large number of

TABLE 17-5 Anisotropic electrical resistivity of selected metals

Metal	Crystal Structure	Electrical Resistivity ($\times 10^{-8} \Omega \cdot m$)			
		a_0	b_0	c_0	Polycrystalline
Mg	HCP	4.48		3.74	4.45
Ga	Orthorhombic	17.4	8.1	54.3	15.05

From *ASM Metals Handbook*, Vol. 2, 9th Ed., 1979.

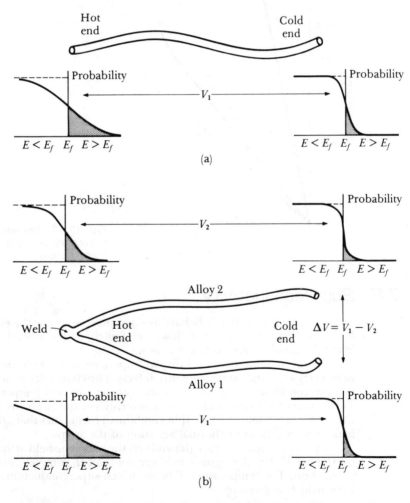

FIG. 17-16 (a) Temperature changes the Fermi distribution along a conductor heated at one end. (b) When two different conductors are joined, their different responses to the temperature create a voltage that is directly related to the temperature.

excited electrons at the hot end move toward the cold end. This produces a voltage between the ends of the conductor.

We cannot measure this voltage when only a single conductor is used. However, suppose we join two wires of different materials. The flow of electrons from the hot junction to a voltmeter is different in each material. A potential, which increases as the temperature increases, develops between the two conductors (Figure 17-17). We can measure the potential and use the device as a *thermocouple* to measure or control temperature.

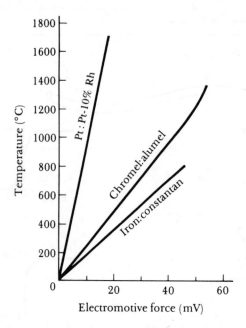

FIG. 17-17   The relationship between voltage and temperature for several thermocouples.

17-8   Superconductivity

Some crystals cooled to 0 K behave as *superconductors,* since the electrical resistivity becomes zero, and a current flows indefinitely in the material. Unfortunately, obtaining absolute zero is not practical.

However, some materials display superconductive behavior above absolute zero, even when the crystal contains defects. The change from normal conduction to superconduction, which occurs abruptly at a critical temperature T_c (Figure 17-18), is related to the magnetic characteristics of the electrons. Electrons having the same energy but opposite spin combine to form pairs that are not affected by lattice imperfections or thermal agitation of the atoms.

The critical temperature depends on the magnetic field acting on the conductor (Figure 17-19). A magnetic field greater than H_c completely suppresses superconduction. The temperature T below which superconduction occurs in a magnetic field H is given by

$$H_c = H_0 \left[1 - \left(\frac{T}{T_c} \right)^2 \right]$$

(17-9)

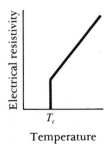

FIG. 17-18
The electrical
resistivity of a
superconductor
becomes zero
below some
critical
temperature T_c.

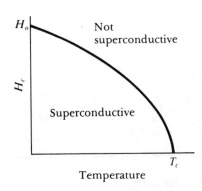

FIG. 17-19   The effect of a magnetic field on
the temperature below which superconductivity
occurs. If the field exceeds H_c at any
temperature below T_c, the metal will not
superconduct.

Although many materials display superconductivity (Table 17-6), all require
that the temperature be extremely low. Superconductive behavior is necessary for
the operation of fusion reactors for power generation and many more applications could be found if materials with a much higher critical temperature were
available.

TABLE 17-6   The critical temperature and magnetic field for
superconductivity of selected metals and compounds

Metal	Critical Temperature (K)	Critical Magnetic Field H_o mA m^{-1}
W	0.015	0.092
Ti	0.39	4.46–7.96
Al	1.18	8.36
Sn	3.72	24.27
Hg	4.15	32.71
Pb	7.23	63.90
Nb	9.25	156.77
La_3Se_4	8.6	
$SnTa_3$	8.35	
Nb_3Sn	18.05	
GaV_3	16.8	
$AlNb_3$	18.0	
$(Al_{0.8}Ge_{0.2})Nb_3$	20.7	

From *Handbook of Chemistry and Physics*, 56th Ed., CRC Press, 1975.

EXAMPLE 17-9

Niobium is to be used as a superconductor in a magnetic field of 39.79 m A $\cdot$ m^{-1}.
What temperature must be obtained for niobium to be superconductive?

Answer:

From Table 17-6, $T_c = 9.25$ K and $H_o = 156.77$ mA m^{-1}.

$$H_c = H_o \left[1 - \left(\frac{T}{T_c} \right)^2 \right]$$

$$39.79 = 156.77 \left[1 - \left(\frac{T}{9.25} \right)^2 \right] = 156.77 \left(1 - \frac{T^2}{85.56} \right)$$

$$T = \sqrt{(0.746)(85.56)} = 7.99 \text{ K}$$

17-9 Energy Gaps—Insulators and Semiconductors

The elements in Group IVA of the periodic table contain two electrons in their outer p shell and have a valence of four. We expect the Group IVA elements to have a high conductivity due to the unfilled p band. However, this behavior is not observed.

These elements are covalently bonded; consequently, the electrons in the outer s and p bands are rigidly bound to the atoms. The restrictions caused by covalent bonding produce a complex change in the band structure, producing *hybridization*. The $2s$ and $2p$ levels of the carbon atoms in diamond can contain up to eight electrons, but there are only four valence electrons available. When N carbon atoms are brought together to form solid diamond, the $2s$ and $2p$ levels interact and produce two nonoverlapping bands (Figure 17-20). Each of the hy-

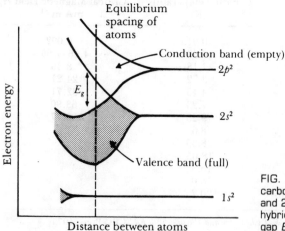

FIG. 17-20   The band structure of carbon in the diamond form. The $2s$ and $2p$ levels combine to form two hybrid bands separated by an energy gap E_g.

brid bands can contain $4N$ electrons. Since there are only $4N$ electrons available, the lower, or valence, band is completely filled while the upper, or conduction, band is empty. A large *energy gap E_g* separates the electrons from the conduction band. Few electrons possess sufficient energy to jump the forbidden zone to the conduction band. Consequently, diamond is an excellent insulator (Table 17-7).

Almost all of the covalently and ionically bonded materials have a band structure containing an energy gap between the valence and conduction bands. Consequently, these materials behave as *electrical insulators*.

TABLE 17-7 Electronic structure and electrical conductivity of the Group IVA elements at 25°C

Metal	Electronic Structure	Electrical Conductivity $(\Omega^{-1}\,m^{-1})$
C (diamond)	$1s^2 2s^2 2p^2$	$< 10^{-16}$
Si	$1s^2 2s^2 2p^6 3s^2 3p^2$	5×10^{-4}
Ge	$\ldots\ldots 4s^2 4p^2$	2
Sn	$\ldots\ldots 5s^2 5p^2$	0.9×10^7

Although germanium, silicon, and tin have the same crystal and band structure as diamond, the energy gap is smaller. In fact, the energy gap in tin is so small that tin behaves as a metal. The energy gap is somewhat larger in silicon and germanium—these elements behave as *semiconductors*. Table 17-8 gives the energy gap for these four elements.

TABLE 17-8 Energy gaps and mobilities for semiconducting metals

Metal	Energy Gap (Eg) (eV)	Electron Mobility (μ_e) $[m^2\,(V \cdot s)^{-1}]$	Hole Mobility (μ_h) $[m^2\,(V \cdot s)^{-1}]$
C (diamond)	5.4	0.18	0.14
Si	1.107	0.19	0.05
Ge	0.67	0.38	0.182
Sn	0.08	0.25	0.24

17-10 Intrinsic Semiconductors

Because the energy gap E_g is small in silicon and germanium, a few electrons possess enough thermal energy to be excited into the conduction band. The excited electrons leave behind unoccupied energy levels, or *holes,* in the valence band. When a voltage is applied to the material, electrons in the conduction band accelerate towards the positive terminal, while holes move towards the negative terminal (Figure 17-21). Current is therefore conducted by the movement of both the electrons and the holes, which act as positively charged electrons. The material behaves as an *intrinsic semiconductor.*

The conductivity is determined by the number of electron-hole pairs.

$$\sigma = n_e q \mu_e + n_h q \mu_h \qquad (17\text{-}10)$$

where n_e is the number of electrons in the conduction band, n_h is the number of holes in the valence band, and μ_e and μ_h are the mobilities of the electrons and holes (Table 17-8). In intrinsic semiconductors,

$$n = n_e = n_h$$

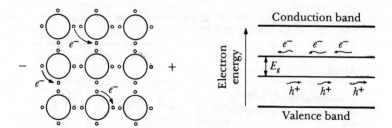

FIG. 17-21   When a voltage is applied to a semiconductor, the electrons move through the conduction band while the electron holes move through the valence band in the opposite direction.

Therefore, the conductivity is

$$\sigma = nq(\mu_e + \mu_h) \tag{17-11}$$

We can use the Fermi distribution to estimate the number of charge carriers and to determine the influence of temperature on the conductivity of semiconductors. In semiconductors, the Fermi energy lies in the center of the forbidden zone between the valence and conduction bands. At 0 K, all of the electrons are in the valence band, where $f(E) = 1$, while all of the levels in the conduction band are unoccupied, $f(E) = 0$ (Figure 17-22).

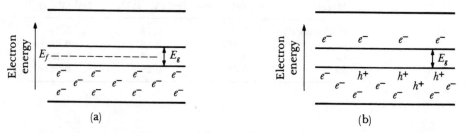

FIG. 17-22   The distribution of electrons and holes in the valence and conduction bands at (a) absolute zero and (b) at an elevated temperature.

As the temperature increases, the Fermi distribution changes and there is some probability that an energy level in the conduction band is occupied (and an equal probability that a level in the valence band is unoccupied, or that a hole is present). The number of electrons in the conduction band, or the number of holes in the valence band, is

$$n = n_e = n_h = n_o \exp\left(\frac{-E_g}{2kT}\right) \tag{17-12}$$

where n_o can be considered a constant, although it actually depends on temperature. Higher temperatures permit more electrons to cross the forbidden zone and hence the conductivity increases.

$$\sigma = n_o q(\mu_e + \mu_n) \exp\left(\frac{-E_g}{2kT}\right) \tag{17-13}$$

The behavior of the semiconductor is opposite to that of metals (Figure 17-23). As the temperature increases, the conductivity of a semiconductor increases because more charge carriers are present, whereas the conductivity of a metal decreases due to lower mobility of the charge carriers.

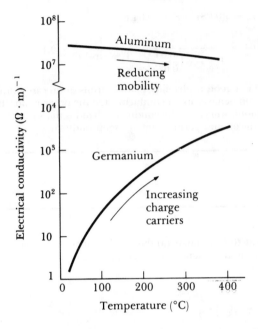

FIG. 17-23 The electrical conductivity versus temperature for semiconductors compared to metals.

EXAMPLE 17-10

Estimate the probability that an electron can gain sufficient energy at 27°C to enter the conduction band in (a) diamond, (b) silicon, (c) germanium, and (d) tin.

Answer:

The Fermi energy E_f for each material is located halfway between the valence and conduction bands. Thus, the energy that the electron must gain to just reach the conduction band is $E_f + \frac{1}{2}E_g$.

(a) In diamond, $E = E_f + \frac{1}{2}(5.4 \text{ eV}) = E_f + 2.7 \text{ eV}$.

$$f(E_f + 2.7) = \frac{1}{1 + \exp\dfrac{(E_f + 2.7 - E_f)}{0.025}} = \frac{1}{1 + \exp(108)} = 1.2 \times 10^{-47}$$

(b) In silicon, $E = E_f + \frac{1}{2}(1.107) = E_f + 0.5535 \text{ eV}$.

$$f(E_f + 0.5535) = \frac{1}{1 + \exp\left(\dfrac{0.5535}{0.025}\right)} = \frac{1}{1 + \exp(22.1)} = 2.5 \times 10^{-10}$$

(c) In germanium, $E = E_f + \frac{1}{2}(0.67) = E_f + 0.335$ eV.

$$f(E_f + 0.335) = \frac{1}{1 + \exp\left(\dfrac{0.335}{0.025}\right)} = \frac{1}{1 + \exp(13.4)} = 1.5 \times 10^{-6}$$

(d) In tin, $E = E_f + \frac{1}{2}(0.08) = E_f + 0.04$ eV.

$$f(E_f + 0.04) = \frac{1}{1 + \exp\left(\dfrac{0.04}{0.025}\right)} = \frac{1}{1 + \exp(1.6)} = 0.17$$

Because there is a good probability that electrons in tin can jump the small forbidden energy gap, tin behaves like a conductor. On the other hand, the probability that electrons in diamond will reach the conduction band is nil, so diamond is an excellent insulator. Both silicon and germanium are semiconductors.

EXAMPLE 17-11

For germanium at 20°C, estimate (a) the number of charge carriers, (b) the fraction of the total electrons in the valence band that are excited into the conduction band, and (c) the constant n_o.

Answer:

From Tables 17-7 and 17-8

$$\sigma = 2 \ (\Omega \cdot m)^{-1} \qquad E_g = 0.67 \text{ eV}$$

$$\mu_e = 0.38 \ m^2 \ (V \cdot s)^{-1} \qquad \mu_h = 0.182 \ m^2 \ (V \cdot s)^{-1}$$

$$2kT = 2(0.025) = 0.05 \text{ eV at } T = 20°C$$

(a) From Equation (17-11)

$$n = \frac{\sigma}{q(\mu_e + \mu_h)} = \frac{2}{(1.6 \times 10^{-19})(0.38 + 0.182)} = 2.22 \times 10^{19}$$

There are 2.22×10^{19} electrons m^{-3} and 2.22×10^{19} holes m^{-3} helping to conduct charge in germanium at room temperature.

(b) The lattice parameter of diamond cubic germanium is 5.6575×10^{-10} m. The total number of electrons in the valence band of germanium is

$$\text{Total electrons} = \frac{(8 \text{ atoms/cell})(4 \text{ electrons/atom})}{(5.6575 \times 10^{-10} \text{ m})^3} = 0.175 \times 10^{30}$$

$$\text{Fraction} = \frac{2.22 \times 10^{19}}{0.175 \times 10^{30}} = 1.24 \times 10^{-10} \text{ of the available electrons}$$

(c) From Equation (17-12)

$$n_o = \frac{n}{\exp\left(\dfrac{-E_g}{2kT}\right)} = \frac{2.22 \times 10^{19}}{\exp\left(\dfrac{-0.67}{0.05}\right)} = 1.45 \times 10^{25} \text{ carriers } m^{-3}$$

17-11 Extrinsic Semiconductors

We cannot accurately control the behavior of an intrinsic semiconductor since slight variations in temperature change the conductivity. However, by intentionally adding a small number of impurity atoms to the material, we can produce an *extrinsic semiconductor*. The conductivity of the extrinsic semiconductor depends primarily on the number of impurity, or *dopant*, atoms and in a certain temperature range may even be independent of temperature. Conductivity is therefore controllable and stable.

n-type semiconductors. Suppose we add an impurity atom such as antimony, which has a valence of five, to silicon or germanium. Four of the electrons from the antimony atom participate in the covalent bonding process, while the extra electron enters an energy level in a donor state just below the conduction band (Figure 17-24). Since the extra electron is not tightly bound to the atoms,

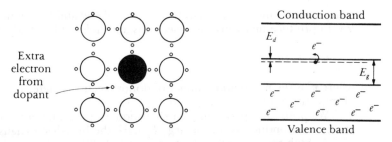

FIG. 17-24 When a dopant atom with a valence greater than four is added to silicon, an extra electron is introduced and a donor energy state is created. Now electrons are more easily excited into the conduction band.

only a small energy E_d is required for the electron to enter the conduction band. The energy gap controlling conductivity is now E_d rather than E_g. No corresponding holes are created when the donor electrons enter the conduction band.

Some intrinsic semiconduction still occurs, with a few electrons gaining enough energy to jump the large E_g gap. The number of charge carriers is

$$n_{total} = n_e(\text{dopant}) + n_e(\text{intrinsic}) + n_h(\text{intrinsic})$$

or

$$n_o = n_{od} \exp\left(\frac{-E_d}{kT}\right) + 2n_o \exp\left(\frac{-E_g}{2kT}\right) \tag{17-14}$$

where n_{od}, n_o are approximately constant. At low temperatures, few intrinsic electrons and holes are produced and the number of electrons is about

$$n_{total} = n_{od} \exp\left(\frac{-E_d}{kT}\right) \tag{17-15}$$

As the temperature increases, more of the donor electrons jump the E_d gap until, eventually, all of the donor electrons enter the conduction band. At this point, we have reached *donor exhaustion* (Figure 17-25). The conductivity is virtually con-

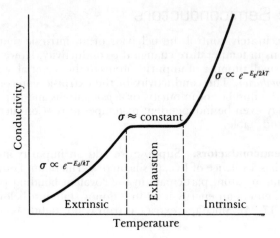

FIG. 17-25 The effect of temperature on the conductivity of an extrinsic semiconductor. At low temperatures, the conductivity increases as more donor electrons enter the conduction band. At moderate temperatures, donor exhaustion has occurred. At high temperatures, intrinsic semiconduction becomes important.

stant, since no more donor electrons are available and the temperature is still too low to produce many intrinsic electrons and holes. The conductivity is

$$\sigma = n_d q \mu_e \tag{17-16}$$

where n_d is the maximum number of donor electrons, determined by the number of impurity atoms that are added.

We prefer to operate in the exhaustion, or plateau, temperature range. Generally, semiconductors with a large E_g have the broadest exhaustion plateau.

At high temperatures, the term $\exp(-E_g/2kT)$ becomes significant and the conductivity increases again according to

$$\sigma = q n_d \mu_e + q(\mu_e + \mu_h) n_o \exp\left(\frac{-E_g}{2kT}\right) \tag{17-17}$$

EXAMPLE 17-12

Calculate the number of extrinsic charge carriers per cubic meter in an n-type semiconductor when one out of every 1,000,000 atoms in silicon is replaced by an antimony atom. Estimate the conductivity of the semiconductor in the exhaustion zone.

Answer:

The lattice parameter of diamond cubic silicon is 5.4307×10^{-10} m.

$$n_d = \frac{\left(1 \frac{\text{electron}}{\text{Sb atom}}\right)\left(10^{-6} \frac{\text{Sb atom}}{\text{Si atom}}\right)\left(8 \frac{\text{Si atoms}}{\text{unit cell}}\right)}{(5.4307 \times 10^{-10} \text{ m})^3}$$

$$n_d = 5 \times 10^{22} \text{ electrons} \cdot \text{m}^{-3}$$

$$\sigma = n_d q \mu_e = (5 \times 10^{22})(1.6 \times 10^{-19})(0.19) = 1.52 \times 10^3 \, \Omega^{-1} \cdot \text{m}^{-1}$$

EXAMPLE 17-13

How many carriers are required to give a conductivity of $10 \times 10^3 \, \Omega^{-1} \cdot m^{-1}$ in the exhaustion region of silicon? How many antimony atoms would have to be added to silicon?

Answer:

$$\sigma = n_d q \mu_e = n_d(0.19)(1.6 \times 10^{-19}) = 10 \times 10^3$$

$$n_d = \frac{10 \times 10^3}{(0.19)(1.6 \times 10^{-19})} = 3.29 \times 10^{23} \text{ electrons m}^{-3}$$

$$3.29 \times 10^{23} = \frac{\left(1 \frac{\text{electron}}{\text{Sb atom}}\right)\left(x \frac{\text{Sb atom}}{\text{Si atom}}\right)\left(8 \frac{\text{Si atoms}}{\text{unit cell}}\right)}{(5.4307 \times 10^{-10} \text{ m})^3}$$

$$x = 6.59 \times 10^{-6} \text{ Sb atoms/Si atom, or 6.59 Sb atoms/}10^6 \text{ Si atoms}$$

p-type semiconductors. When we add an impurity such as gallium, which has a valence of three, to a semiconductor, there are not enough electrons to complete the covalent bonding process. An electron hole is created in the valence band which can be filled by electrons from other locations in the band (Figure 17-26). The holes act as acceptors of electrons. These hole sites have a somewhat

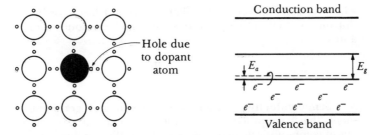

FIG. 17-26 When a dopant atom with a valence of less than four is substituted into the silicon lattice, an electron hole is created in the structure and an acceptor energy level is created just above the valence band. Little energy is required to excite the electron holes into motion.

higher than normal energy and create an acceptor level of possible electron energies just above the valence band (Table 17-9). An electron must gain an energy of only E_a in order to create a hole in the valence band. The hole then moves and carries the charge. Now we have a *p-type semiconductor*.

Again some intrinsic semiconduction may occur. The number of charge carriers is

$$n_t = n_h(\text{acceptor}) + n_e(\text{intrinsic}) + n_h(\text{intrinsic})$$

$$n_t = n_{oa} \exp\left(\frac{-E_a}{kT}\right) + 2n_o \exp\left(\frac{-E_g}{2kT}\right) \tag{17-18}$$

TABLE 17-9   The donor and acceptor energy gaps in electron volts when silicon and germanium semiconductors are doped

Dopant	Silicon		Germanium	
	E_d	E_a	E_d	E_a
P	0.045		0.0120	
As	0.049		0.0127	
Sb	0.039		0.0096	
B		0.045		0.0104
Al		0.057		0.0102
Ga		0.065		0.0108
In		0.160		0.0112

At low temperatures, the acceptor levels predominate.

$$n_t = n_{oa} \exp\left(\frac{-E_a}{kT}\right) \tag{17-19}$$

Eventually, the temperature is high enough to cause *acceptor saturation* and

$$\sigma = n_a q \mu_h \tag{17-20}$$

where n_a is the maximum number of acceptor levels, or holes, introduced by the dopant. At higher temperatures, intrinsic semiconduction becomes important and

$$\sigma = n_a q \mu_h + q(\mu_e + \mu_h)n_o \exp\left(\frac{-E_g}{2kT}\right) \tag{17-21}$$

Semiconducting compounds.   Silicon and germanium are the only elements that have practical application as semiconductors. However, a large number of compounds display the same effect. We can divide the compounds into two classes—stoichiometric semiconductors and nonstoichiometric, or defect, semiconductors.

The *stoichiometric semiconductors,* usually intermetallic compounds, have crystal structures and band structures similar to DC silicon and germanium. Examples are given in Table 17-10. Elements from Group III and Group V of the periodic table are classic examples. Gallium from Group III and arsenic from Group V combine to form a compound GaAs, with an average of four valence electrons per atom. The $4s^2 4p^1$ levels of gallium and the $4s^2 4p^3$ levels of arsenic produce two hybrid bands, each capable of containing $4N$ electrons. An energy gap of 1.35 eV separates the valence and conduction bands. The GaAs compound can be doped to produce either an *n*-type or *p*-type semiconductor. The large energy gap E_g leads to a broad exhaustion plateau and high mobilities of charge carriers in the compound lead to high conductivities.

The *nonstoichiometric,* or *defect, semiconductors* are ionic compounds containing an excess of either anions (producing a *p*-type semiconductor) or cations (producing an *n*-type semiconductor). A number of oxides and sulfides have this behavior. For example, if an extra zinc atom is added to ZnO, the zinc atom enters the

TABLE 17-10   Energy gaps and mobilities for semiconducting compounds

Compound	Energy Gap (E_g) (eV)	Electron Mobility (μ_e) [m^2 (V·s)$^{-1}$]	Hole Mobility (μ_h) [m^2 (V·s)$^{-1}$]
Zns	3.54	0.018	0.0005
ZnTe	2.26	0.034	0.01
CdTe	1.44	0.12	0.005
GaP	2.24	0.03	0.01
GaAs	1.35	0.88	0.04
GaSb	0.67	0.40	0.14
InSb	0.165	7.80	0.075
InAs	0.36	3.30	0.046
ZnO	3.2	0.018	
CdS	2.42	0.04	
CdSe	1.74	0.065	
PbS	0.37	0.06	0.06
PbTe	0.25	0.16	0.06
CdSnAs$_2$	0.26	2.20	0.025

From *Handbook of Chemistry and Physics, 56th Ed.,* CRC Press, 1975.

structure as an ion, Zn^{2+}, giving up two electrons which contribute to the number of charge carriers. These electrons can be activated by a small increase in energy to carry current (Figure 17-27).

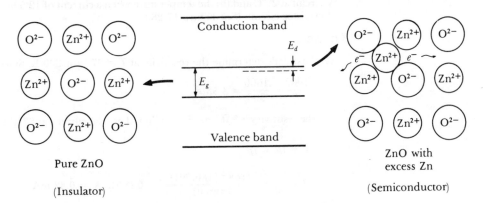

FIG. 17-27   Extra interstitial zinc atoms can ionize and introduce extra electrons, creating an *n*-type defect semiconductor in ZnO.

17-12 Applications of Semiconductors to Electrical Devices

Many electronic devices have been developed using the characteristics of semiconduction. A few of these are described here.

Thermistors.   The conductivity of semiconductors increases with temperature (Figure 17-28). By knowing the relationship between conductivity and temperature, we can use the semiconductor to measure temperature. Thermistors also have other uses, including use as a fire alarm. When the thermistor heats, it passes a larger current through a circuit and activates the alarm.

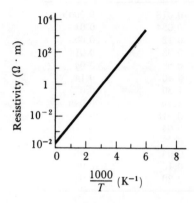

FIG. 17-28   The electrical resistivity of an $Fe_3O_4 \cdot MgCr_2O_4$ thermistor versus temperature.

EXAMPLE 17-14

A thermistor made from an $Fe_3O_4 \cdot MgCr_2O_4$ defect semiconductor is 1 mm in diameter and 10 mm long. Calculate (a) the current produced in the thermistor in a 16-V circuit at 27°C and (b) the temperature when a current of 12.5 mA flows in the 16-V thermistor circuit. (See Figure 17-28.)

Answer:

(a) First, let's determine the resistivity at $T = 27°C$ or 300 K. Since

$$\frac{1000}{T} = \frac{1000}{300} = 3.33$$

the resistivity is $3 \ \Omega \cdot m$. From Ohm's law, the current at 27°C is

$$V = IR = I\rho \frac{l}{A}$$

$$I = \frac{VA}{\rho l} = \frac{(16)(\pi/4)(0.001)^2}{(3)(0.01)} = 0.00042 \ A = 0.42 \ mA$$

(b) When the current is 12.5 mA (0.0125 A)

$$\rho = \frac{VA}{Il} = \frac{(16)(\pi/4)(0.001)^2}{(0.0125)(0.01)} = 0.10 \ \Omega \cdot m$$

From Figure 17-28, $0.10 \ \Omega \cdot m$ corresponds to $1000/T = 2.2$

$$T = \frac{1000}{2.2} = 455 \ K = 181°C$$

Pressure transducers. The band structure and the energy gap are a function of the spacing between the atoms in the material (Figure 17-29). When pressure is applied to the semiconductor, atoms are forced closer together, the energy gap decreases, and the conductivity increases. If we measure the conductivity, we can in turn calculate the pressure acting on the material.

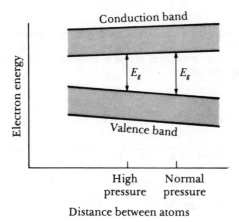

FIG. 17-29 Pressure squeezes the atoms closer together in a semiconductor, reducing the energy gap and increasing the electrical conductivity.

Magnetometers. Semiconductors measure the strength of a magnetic field by using the *Hall effect*. A charge carrier moving through a magnetic field is deflected to one side of the material through which the charge is moving (Figure 17-30). Electrons are deflected in one direction, holes in the other. This creates a

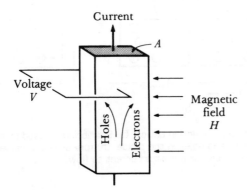

FIG. 17-30 The Hall effect. Electrons and holes are deflected by the magnetic field. By measuring the voltage, the strength of the magnetic field can be calculated.

voltage drop across the material which is related to the current and the magnetic field by

$$V_H = HJR_H \tag{17-22}$$

where V_H is the Hall voltage, J is the current density, H is the magnetic field, and R_H is the Hall coefficient, a constant given by

$$R_H = \frac{1}{n_h q} = -\frac{1}{n_e q} \tag{17-23}$$

If we apply a given current through a known cross section of the material and measure the voltage, we can calculate the strength of the magnetic field. The Hall effect also permits us to determine whether we have a p-type or n-type semiconductor, since the voltage drop will have opposite signs.

EXAMPLE 17-15

A silicon rod 1.25 mm in diameter is doped with one antimony atom per million silicon atoms. A current of 10 A is passed through the rod when the rod is in a magnetic field. If 6 V is measured across the silicon semiconductor, calculate the strength of the magnetic field.

Answer:

From Example 17-12, $n_e = 5 \times 10^{22}$ electrons $\cdot$ m^{-3}

$$R_H = -\frac{1}{n_e q} = \frac{-1}{(5 \times 10^{22})(-1.6 \times 10^{-19})} = 0.125 \times 10^{-3}\ \text{m}^3 \cdot \text{C}^{-1}$$

$$H = \frac{V}{J R_H} = \frac{6\ \text{V}}{\left[\dfrac{10\ \text{A}}{\pi/4 (1.25 \times 10^{-3})^2}\right](0.125 \times 10^{-3})} = 5.89 \times 10^{-3}\ \text{V} \cdot \text{C} (\text{A} \cdot \text{m})^{-1}$$

Rectifiers (p-n junction devices). Rectifiers convert alternating current to direct current. The rectifiers are produced by joining an n-type to a p-type semiconductor, forming a p-n junction (Figure 17-31). Electrons move towards the n-

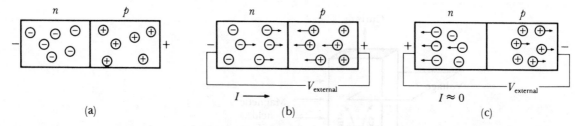

(a) (b) (c)

FIG. 17-31 Behavior of a p-n junction device. (a) Equilibrium caused by electrons concentrating in the n-side and holes in the p-side, (b) forward bias causes a current to flow, and (c) reverse bias does not permit a current to flow.

type junction while holes move towards the p-type junction. The resulting electrical imbalance creates a voltage, or *contact potential*, across the junction.

If we place an external voltage on the p-n junction so that the negative terminal is at the n-type side, a net current is produced. This is called *forward bias* [Figure 17-31(b)].

However, if the applied voltage is reversed, creating a *reverse bias*, both the holes and electrons move away from the junction. With no charge carriers at the junction, the junction behaves as an insulator and virtually no current flows [Figure 17-31(c)]. Because the p-n junction permits current to flow only in one direction, it passes only half of an alternating current.

When the reverse bias becomes too large, however, any carriers that do leak through the insulating barrier of the junction are highly accelerated, excite other charge carriers, and cause a high current in the reverse direction (Figure 17-32).

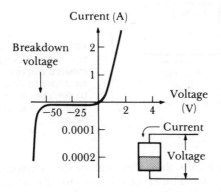

FIG. 17-32 A typical current-voltage characteristic for a *p-n* junction. Note the different scales in the first and third quadrants. At a reverse bias that exceeds the breakdown voltage, the junction no longer acts as a rectifier but may act as a Zener diode.

We can use this phenomenon to design voltage-limiting devices. By proper doping and construction of the *p-n* junction, the breakdown or *avalanche voltage* can be preselected. When the voltage in the circuit exceeds the breakdown voltage, a high current flows through the junction and is diverted from the rest of the circuit. These devices are called *Zener diodes* and are used to protect circuitry from accidental high voltages.

Transistors. A *transistor,* which is used to amplify electrical signals, is a sandwich of either *n-p-n* or *p-n-p* semiconductor materials (Figure 17-33). There

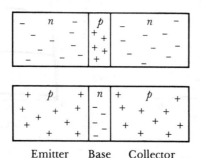

Emitter   Base   Collector   FIG. 17-33 The *n-p-n* and *p-n-p* transistors.

are three zones in the transistor—the *emitter,* the *base,* and the *collector.* As in the *p-n* junction, electrons are initially concentrated in the *n*-type material and holes are concentrated in the *p*-type material.

Figure 17-34 shows a *n-p-n* transistor and its electrical circuit. The electrical signal that is to be amplified is connected between the base and the emitter of the transistor. The output from the transistor, or the amplified signal, is connected between the emitter and collector ends of the transistor. The circuit is connected so that a forward bias is produced between the emitter and the base, while a reverse bias is produced between the base and the collector. The forward bias causes electrons to leave the emitter and enter the *p*-type base.

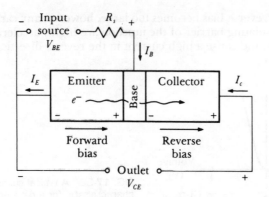

FIG. 17-34   A circuit for an *n-p-n* transistor. The input creates a forward and reverse bias that causes electrons to move from the emitter through the base, and into the collector, creating an amplified output.

If the base is exceptionally thin, almost all of the electrons pass through the base and enter the collector. The reverse bias causes the electrons to accelerate through the collector, the circuit is complete, and an output signal is produced.

The base current, which is the signal that is eventually amplified, creates the forward and reverse biases that move the electrons through the transistor and produce the output collector current. By controlling the base current, we in turn control the collector current that is produced (Figure 17-35).

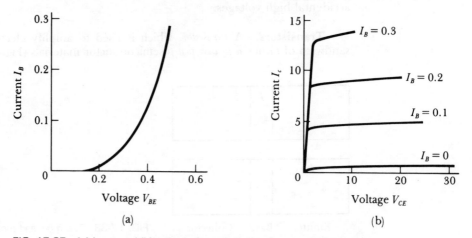

FIG. 17-35   (a) Input and (b) output characteristics for a typical transistor.

The amplification β of the transistor is the ratio of the collector current to the input current.

$$\beta = \frac{I_C}{I_B} = \frac{\text{output current}}{\text{input current}} \qquad (17\text{-}24)$$

Typical amplifications are 10 to 300.

17-13 Manufacture and Fabrication of Semiconductor Devices

Preparation of semiconductors and semiconductor devices requires special technologies. Semiconductor materials are single crystals, very pure except for the controlled amounts of dopants, and normally are very small in size.

Producing pure single crystals. *Zone refining* is one technique by which semiconducting materials such as silicon and germanium are purified. A rod of the material is moved at a very slow speed through a furnace with a very narrow high temperature zone [Figure 17-36(a)]. The rod is molten over only a short

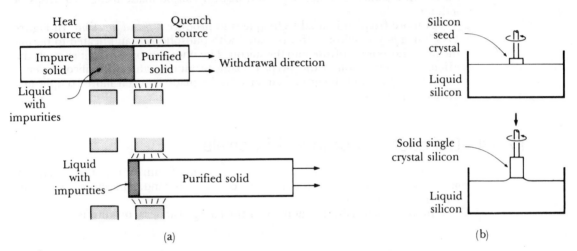

(a)　　　　　　　　　　　　　　　　　　(b)

FIG. 17-36 Producing pure single crystal semiconductor materials by (a) zone refining, where impurities segregate into a narrow molten zone and are carried to the end of the rod, and (b) growth of single crystals by slowly withdrawing a seed crystal from a molten bath of metal.

length at any particular time. The impurities, which have a low solubility in the solid metal, collect in the molten zone. As the molten zone moves down the rod, the impurities are gradually carried to the end of the rod. After several passes, the impurities will have collected at the end, which is then removed from the rod.

In a second technique, a single crystal seed is dipped into the properly doped liquid metal and then slowly withdrawn. Surface tension causes the molten metal to adhere to and then solidify onto the seed crystal as it is withdrawn [Figure 17-36(b)].

Fabrication of junctions. A single crystal of silicon or germanium can be used as a substrate onto which the dopant is deposited. Two techniques are illustrated in Figure 17-37.

An *alloyed junction* can be produced by placing a drop of indium on a disc of germanium, then heating the combination above the melting point of indium. Germanium diffuses into the indium. On cooling, the molten material solidifies as a single crystal. If the original germanium is an *n*-type semiconductor, then the

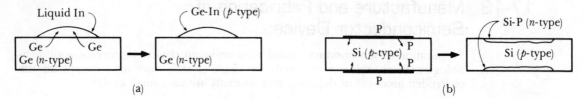

FIG. 17-37 Fabrication of *p-n* and *n-p-n* junction devices by (a) alloying and (b) diffusion.

germanium-indium alloy is a *p*-type semiconductor and a *p-n* junction is produced. In a similar manner, a *p-n-p* transistor can be made using two drops of indium.

A more frequently used technique is to produce *diffused junctions*. In Figure 17-37(b), a *p*-type silicon wafer is coated with phosphorous and heated to permit the phosphorous to diffuse into the silicon. The phosphorous content exceeds the original concentration of the *p*-type dopant, producing an *n*-type silicon layer.

Transistors and integrated circuits can be made by even more complicated diffusion processes.

17-14 Conductivity of Ionic Materials

Conduction in ionic materials occurs by movement of entire ions, since the energy gap is too large for electrons to gain the conduction band. Therefore, most ionic materials behave as insulators.

In ionic materials, the mobility of the charge carriers, or ions, is

$$\mu = \frac{ZqD}{kT} \tag{17-25}$$

where D is the diffusion coefficient, k is Boltzmann's constant, T is the absolute temperature, q is the charge, and Z is the valence of the ion. The mobility is many orders of magnitude lower than the mobility of electrons, hence the conductivity is very small.

$$\sigma = nZq\mu \tag{17-26}$$

Impurities and vacancies increase conductivity; vacancies are necessary for diffusion in substitutional types of crystal structures, and impurities can also diffuse and help carry the current. High temperatures also increase conductivity because the rate of diffusion increases. The conductivity in molten ionic materials may be much higher than normal, due to the greater mobility of the ions in the liquid.

EXAMPLE 17-16

Suppose that the electrical conductivity of MgO is determined primarily by the diffusion of the Mg^{2+} ions. Estimate the mobility of the Mg^{2+} ions and calculate the electrical conductivity of MgO at 1800°C.

Answer:

From Figure 5-6, the diffusion coefficient of Mg^{2+} ions in MgO at 1800°C is 10^{-14} $m^2 \cdot s^{-1}$. For MgO, $Z = 2$/ion, $q = 1.6 \times 10^{-19}$ C, $k = 1.38 \times 10^{-23}$ J K^{-1}, and $T = 2073$ K.

$$\mu = \frac{ZqD}{kT} = \frac{(2)(1.6 \times 10^{-19})(10^{-14})}{(1.38 \times 10^{-23})(2073)} = 1.12 \times 10^{-13} \text{ C} \cdot \text{m}^2 \text{ (J} \cdot \text{s)}^{-1}$$

Since $C = A \cdot s$ and $J = A \cdot V \cdot s$

$$\mu = 1.12 \times 10^{-13} \text{ m}^2 \text{ (V} \cdot \text{s)}^{-1}$$

MgO has the NaCl structure, with four magnesium ions per unit cell. The lattice parameter is 3.96×10^{-10} m, so the number of Mg^{2+} ions per cubic meter is

$$n = \frac{4 \text{ Mg}^{2+} \text{ ions/cell}}{(3.96 \times 10^{-10} \text{ m})^3} = 6.4 \times 10^{28} \text{ ions m}^{-3}$$

$$\sigma = nZq\mu = (6.4 \times 10^{28})(2)(1.6 \times 10^{-19})(1.12 \times 10^{-13})$$

$$= 22.94 \times 10^{-4} \text{ C} \cdot \text{m}^2 \text{ (m}^3 \cdot \text{V} \cdot \text{s)}^{-1}$$

Since $C = A \cdot s$ and $V = A \cdot \Omega$

$$\sigma = 2.294 \times 10^{-3} \text{ (}\Omega \cdot \text{m)}^{-1}$$

SUMMARY

Electrical conductivity is particularly sensitive to atomic bonding, atomic structure, and processing of the material. Consequently, metals have a high conductivity which is reduced when the temperature increases or lattice imperfections are introduced by alloying or processing. Semiconductors and insulators have lower conductivities due to the nature of the ionic or covalent bonds. However, their conductivity can be increased by higher temperatures or by the introduction of the right type of lattice imperfections, or doping. By taking advantage of the influence of temperature and structure on conductivity, a large variety of special electronic devices can be constructed and utilized.

GLOSSARY

Acceptor saturation When all of the extrinsic acceptor levels in a p-type semiconductor are filled.

Alloyed junction A junction device obtained by melting one material onto a second material.

Avalanche voltage The reverse bias voltage that causes a large current flow in a p-n junction.

Conduction band The unfilled energy levels into which electrons can be excited to provide conductivity.

Current density The current flowing through a given cross-sectional area.

Defect resistivity coefficient Relates the effect of lattice imperfections to the conductivity.

Diffused junction Junction devices obtained

by diffusing the dopant into the base material.

Donor exhaustion When all of the extrinsic donor levels in an *n*-type semiconductor are filled.

Doping Addition of controlled amounts of impurities to increase the number of charge carriers in a semiconductor.

Drift velocity The average rate at which electrons or other charge carriers move through the material.

Electric field The voltage gradient, or volts per unit length.

Energy gap The energy between the top of the valence band and the bottom of the conduction band that a charge carrier must obtain before it can transfer a charge.

Extrinsic semiconductor A semiconductor prepared by adding impurities or dopants which determine the number of charge carriers.

Fermi energy The energy midway between the valence and conduction bands.

Forward bias Connecting a junction device so that holes and electrons flow toward the junction to produce a net current flow.

Hall effect A charge carrier moving through a magnetic field is deflected. By measuring the voltage, the strength of the field can be measured.

Holes Unfilled energy levels in the valence band. Because electrons move to fill these holes, the holes move and a current is carried.

Hybridization The valence and conduction bands are separated by an energy gap. This leads to the semiconductive behavior of silicon and germanium.

Intrinsic semiconductor A semiconductor in which the temperature determines the conductivity.

Mobility The ease with which a charge carrier moves through a material.

Rectifiers *p-n* junction devices that permit current to flow in only one direction in a circuit.

Reverse bias Connecting a junction device so that holes and electrons flow away from the junction, preventing a net current flow.

Superconductivity Flow of current through a material that has no resistance to that flow.

Temperature resistivity coefficient Relates the effect of temperature to the conductivity.

Thermistor A semiconductor device that is particularly sensitive to changes in temperature, permitting it to serve as an accurate measure of temperature.

Thermocouple A pair of conductor wires. The voltage produced between the wires when the temperature is changed can be used to measure temperature.

Transistor A semiconductor device that can be used to amplify electrical signals.

Valence band The energy levels filled by electrons in their lowest energy states.

Zener diode A *p-n* junction device which, with a very large reverse bias, causes a current to flow.

Zone refining A solidification technique used to purify materials for semiconductor devices.

PRACTICE PROBLEMS

1 Compare the power loss in a 3000 m transmission line 2 mm in diameter through which a current of 25 A is flowing if the wire is made of (a) aluminum, (b) iron, and (c) silicon.

2 We would like a 0.01 mm diameter chromium wire to offer a resistance of 500 Ω. How long should the wire be?

3 What voltage must be applied to a 1 mm diameter chromium wire 1 m long in order to produce a current of 10 A?

4 A 50 mm long nickel rod is intended to offer a resistance of 0.25 Ω. Calculate the required diameter of the wire.

5 A copper wire 300 mm long offers a resistance of 0.0001 Ω when a voltage of 1 mV is applied. (a) Determine the current. (b) Determine the diameter of the wire. (c) Suppose the wire heats to 500°C. Calculate the resistivity at 500°C and determine the new resistance and current.

6 The current density in a silver wire 100 m in

length is 70×10^6 A m^{-2}. Calculate the voltage and electric field.

7 A current of 400 A is passed through an arc in an arc welding process. The diameter of the arc is about 10 mm. The arc extends 3 mm between the tip of the electrode and the metal being welded. A voltage of 40 V is applied across the arc. Calculate (a) the current density in the arc, (b) the electric field across the arc, and (c) the electrical conductivity of the gases in the arc during welding.

8 Calculate the mobility of an electron in calcium if all of the valence electrons contribute to the current flow.

9 Calculate the fraction of the valence electrons that contribute to a current flow in pure gold if the mobility of the electrons is found to be 0.056 m^2 (V·s)$^{-1}$.

10 Determine the average drift velocity of an electron in aluminum when 12 V are applied to a 300 mm long wire. Assume that all of the valence electrons participate in carrying the current.

11 Describe the band structure you expect for zinc. How would you expect its conductivity to compare to that of copper and magnesium? Explain.

12 Calculate the probability that energy levels at $E_f + 0.1$, $E_f + 0.2$, and $E_f + 0.3$ will be occupied at (a) 50 K, (b) 300 K, and (c) 700 K.

13 Calculate the electrical resistivity of gold at $-200°C$ and at $500°C$.

14 The measured electrical conductivity of a magnesium wire inside a furnace is 3.8×10^6 ohm^{-1}·m^{-1}. Calculate the temperature of the furnace.

15 Suppose the electrical conductivity of iron containing 5 at % impurity at $500°C$ is 0.02×10^8 ohm^{-1}·m^{-1}. Determine the contribution to resistivity due to temperature and impurities by (a) calculating the expected resistivity of pure iron at $500°C$, (b) calculating the resistivity due to impurities, and (c) calculating the defect resistivity coefficient.

16 Based on the data in Figure 17-12, estimate the defect resistivity coefficient for nickel in copper.

17 The electrical resistivity of a Cu-5% Al alloy is 9.8×10^{-4} Ω·m. Calculate (a) the at% Al and (b) the defect resistivity coefficient.

18 The electrical resistivity of an Al-1.2% Mg alloy is 3.4×10^{-4} Ω·m. Calculate the defect resistivity coefficient.

19 The defect resistivity coefficient for zinc in copper was calculated in Example 17-8 and for aluminum in copper was calculated in Problem 17. The resistivity of a Cu-20% Ni alloy is 28×10^{-4} Ω·m and for Cu-10% Sn is 16×10^{-4} Ω·m. Calculate the coefficient for the copper-nickel and copper-tin alloys

and compare for all of the alloys. Is the defect resistivity coefficient dependent on the atom size difference between the impurity and copper atoms?

20 A cooling curve for a solidifying metal is obtained by immersing a chromel-alumel thermocouple in the melt. A thermal arrest is indicated at 35 mV. What is the melting point of the metal?

21 Figure 17-38 shows the output from an iron-constantan thermocouple immersed in a lead-tin alloy. Assuming the alloy is hypoeutectic, estimate (a) the liquidus temperature, (b) the eutectic temperature, and (c) the composition of the alloy.

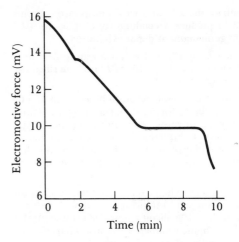

FIG. 17-38   The cooling curve in terms of the voltage output of an iron-constantan thermocouple for a hypoeutectic lead-tin alloy. (See Problem 22.)

22 Will lead be superconductive at 5 K when a magnetic field of 31.8×10^3 A m^{-1} is present? Explain.

23 Estimate the number of charge carriers and the fraction of the total electrons in the valence band that are excited into the conduction band in pure silicon at room temperature. Calculate the constant n_o.

24 Using the value for the constant n_o in Example 17-11, and assuming that the mobilities are constant with temperature, plot the electrical conductivity of germanium from $-50°C$ to $+500°C$.

25 Estimate the probability that an electron will enter the conduction band in diamond, silicon, and germanium at $500°C$ and $1000°C$.

26 (a) How many charge carriers are required to produce a conductivity of 150×10^3 (Ω·m)$^{-1}$ in germanium in the exhaustion range? (b) What atom fraction of arsenic must be added to the germanium to produce this conductivity?

27 (a) Calculate the number of extrinsic charge carriers in an *n*-type semiconductor when one out of every 10^7 atoms in germanium is replaced with a phosphorous atom. (b) Estimate the conductivity of the doped germanium in the exhaustion range.

28 (a) Calculate the number of extrinsic charge carriers in a *p*-type semiconductor when one out of every 5×10^6 atoms in silicon is replaced with a gallium atom. (b) What is the wt% Ga in silicon? (c) Estimate the conductivity of the doped silicon in the saturation zone.

29 Estimate the at% and wt% phosphorous atoms required to produce a conductivity of 2×10^3 $(\Omega \cdot m)^{-1}$ in germanium at donor exhaustion.

30 Estimate the at% and wt% Ga required to produce a conductivity of $3 \times 10^{-2} (\Omega \cdot m)^{-1}$ in silicon at acceptor exhaustion.

31 Assume that donor exhaustion occurs at 150°C. (a) Based on the E_d for phosphorous in germanium, and neglecting intrinsic semiconduction, determine the value of n_{od} in Equation (17-15) when one out of 10^7 atoms are phosphorous. (b) Estimate the number of extrinsic charge carriers and the conductivity at −100°C.

32 Assume that exhaustion occurs at 200°C in an extrinsic silicon semiconductor doped with five indium atoms per 10^7 silicon atoms. Determine (a) the value of n_{oa}, (b) the number of extrinsic charge carriers at 0°C, and (c) the conductivity at 0°C.

33 Suppose GaAs is prepared so that the composition is 49.99 at% Ga and 50.01 at% As. (a) Is this a *p*-type or an *n*-type semiconductor? (b) Estimate the

conductivity when the extrinsic charge carriers are all active. The lattice parameter is 5.68 Å and four gallium and four arsenic atoms are in each unit cell.

34 Suppose InSb contains one extra antimony atom for every 10^4 antimony atoms. Estimate the electrical conductivity due to the extrinsic semiconduction. The lattice parameter is 2.86 Å and there are four antimony atoms in each unit cell.

35 Compare the percent change in the electrical conductivity of intrinsic silicon, ZnS, and InSb when the temperature changes from −100°C to +100°C. Which material would be more suitable for a thermistor?

36 Compare the percent change in conductivity, assuming conduction is due to extrinsic charge carriers, for silicon doped with (a) antimony, (b) gallium, and (c) indium when the temperature increases from 0°C to 50°C. (Assume that exhaustion is not yet reached.)

37 A germanium crystal 10 mm long and 1 mm in diameter is doped with antimony atoms. A current of 5 A is passed through the rod at a 5-V potential, producing a magnetic field of 300 V · C $(A \cdot m)^{-1}$. Calculate the number of antimony atoms per 10^6 germanium atoms in the semiconductor.

38 Calculate the temperature when a 25 mA current flows in a 24-V $Fe_3O_4 \cdot MgCr_2O_4$ thermistor circuit when the thermistor is 100 mm long and 1.5 mm in diameter.

39 The electrical conductivity of CaO is expected to be due primarily to diffusion of Ca^{2+} ions. Estimate the mobility of the Ca^{2+} ions and the electrical conductivity at 1000°C and 1500°C, using Figure 5-6. The lattice parameter of CaO is 4.62 Å and there are 4 Ca^{2+} ions per cell.

18

Dielectric and Magnetic Properties

18-1 Introduction

The response of a material to an electric field can be used to advantage even when no charge is transferred. These effects are described by the dielectric properties of the material. Dielectric materials possess a large energy gap between the valence and conduction bands; thus the materials have a high electrical resistivity. Two important applications for dielectric materials include electrical insulators and capacitors. Insulators, used to prevent the transfer of charge in an electric circuit, include the plastic covering on electrical wires and the ceramic "bells" used in high-voltage power lines. Capacitors are used to store electric charge. Other characteristics of dielectrics include electrostriction, piezoelectricity, and ferroelectricity.

The effects of a magnetic field on a material are equally profound. Some magnetic materials possess a permanent magnetization for applications ranging from toys to computer storage; other magnetic materials are used in electric motors and transformers.

Although the responses of a material to an electric or magnetic field are based on very different phenomena, a similar approach is used to describe the two effects. For this reason, dielectric and magnetic behavior will be discussed together in this chapter and some of the similarities will become apparent in our discussion.

18-2 Dipoles

In both electrical and magnetic materials, the application of a field causes the formation and movement of dipoles. *Dipoles* are atoms or groups of atoms that have an unbalanced charge. In an imposed field, the dipoles become aligned in the material. Alignment of the dipoles causes *polarization* in an electric field and *magnetization* in a magnetic field. The ease with which polarization and magnetization occur determines the behavior of the dielectric or magnetic material.

18-3 Polarization in an Electric Field

When an electric field is applied to a material, dipoles are induced within the atomic or molecular structure and become aligned with the direction of the field.

In addition, any permanent dipoles already present in the material are aligned with the field. The material is polarized. The polarization P (C/m²) is

$$P = Zqd \qquad (18\text{-}1)$$

where Z is the number of charge centers that are displaced per cubic meter, q is the electronic charge, and d is the displacement between the positive and negative ends of the dipole. Four mechanisms cause polarization—electronic polarization, ionic polarization, molecular polarization, and space charges (Figure 18-1).

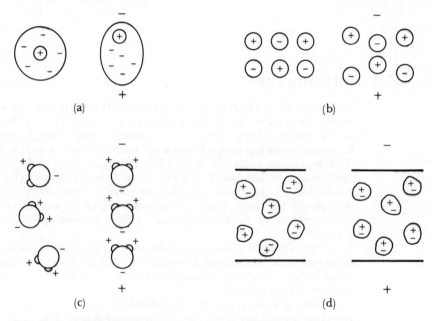

FIG. 18-1   Polarization mechanisms in materials. (a) Electronic polarization, (b) ionic polarization, (c) molecular polarization, and (d) space charges.

Electronic polarization.   When an electric field is applied to an atom, the electronic arrangement is distorted, with electrons concentrating on the side of the nucleus near the positive end of the field. The atom acts as a temporary, induced dipole. This effect is small and temporary.

EXAMPLE 18-1

Suppose that the average displacement of the electrons relative to the nucleus in a copper atom is 1×10^{-8} Å when an electric field is imposed on a copper plate. Calculate the polarization.

Answer:

The atomic number of copper is 29, so there are 29 electrons in each copper atom. The lattice parameter of copper is 3.615 Å. Thus

$$Z = \frac{(4 \text{ atoms/cell})(29 \text{ electrons/atom})}{(3.615 \times 10^{-10} \text{ m})^3} = 2.46 \times 10^{30} \text{ electrons/m}^3$$

$$P = Zqd = \left(2.46 \times 10^{30} \frac{\text{electrons}}{\text{m}^3}\right)\left(1.6 \times 10^{-19} \frac{\text{C}}{\text{electron}}\right)(10^{-8}\text{Å})(10^{-10} \text{ m/Å})$$

$$= 3.94 \times 10^{-7} \text{ C/m}^2$$

Ionic polarization. When an ionically bonded material is placed in an electric field, the bonds between the ions are elastically deformed. Consequently, the charge is minutely redistributed within the material. Depending on the direction of the field, cations and anions either move closer together or further apart. These temporarily induced dipoles provide polarization and may also change the overall dimensions of the material.

EXAMPLE 18-2

The ionic polarization observed in a NaCl crystal is 4.3×10^{-8} C/m^2. Calculate the displacement between the Na$^+$ and Cl$^-$ ions.

Answer:

In this example, there is one electric charge on each Na$^+$ ion. In the NaCl unit cell, which has a lattice parameter of 5.5 Å, there are four Na$^+$ ions.

$$Z = \frac{(4 \text{ Na}^+ \text{ ions/cell})(1 \text{ charge/Na}^+ \text{ ion})}{(5.5 \times 10^{-10} \text{ m})^3}$$

$$= 2.4 \times 10^{28} \text{ charges/m}^3$$

$$d = \frac{P}{Zq} = \frac{4.3 \times 10^{-8}}{(2.4 \times 10^{28})(1.6 \times 10^{-19})} = 11.2 \times 10^{-18} \text{ m} = 11.2 \times 10^{-8} \text{ Å}$$

Molecular polarization. Some materials contain natural dipoles. When a field is applied, the dipoles rotate to line up with the imposed field. When the field is removed, the dipoles may remain in alignment, causing permanent polarization.

The permanent dipoles are present in asymmetrical molecules such as water and organic polymers, which have asymmetrical mers. Certain ceramic crystal structures lack a center of symmetry and also behave as dipoles.

EXAMPLE 18-3

Compare the tendency of the following organic molecules to act as dipoles and produce molecular polarization—CH_4, CH_3Cl, CH_2Cl_2, $CHCl_3$, and CCl_4.

Answer:

The structure of each molecule is shown schematically below. The chlorine atoms are strong centers of negative charge and hydrogen atoms are weak centers of positive charge.

CH$_4$

$$\begin{array}{c} H \\ | \\ H-C-H \\ | \\ H \end{array}$$

Because the molecule is symmetrical, no orientation polarization is expected.

CH$_3$Cl

$$\begin{array}{c} H \\ \backslash \\ H-C-Cl \\ | \\ H \end{array}$$

A large dipole effect is created, since the positive hydrogen atoms are displaced from the chlorine atom.

CH$_2$Cl$_2$

$$\begin{array}{c} H \qquad Cl \\ \backslash \ / \\ C \\ / \ \backslash \\ H \qquad Cl \end{array}$$

The dipole effect is smaller, since the large chlorine atoms become predominant in the structure.

CHCl$_3$

$$\begin{array}{c} Cl \\ | \\ H-C-Cl \\ | \\ Cl \end{array}$$

CCl$_4$

$$\begin{array}{c} Cl \\ | \\ Cl-C-Cl \\ | \\ Cl \end{array}$$

Again the molecule is symmetrical and no orientation polarization is expected.

The measured dipole moments for each of the molecules, given in C · m are: CH$_4$ 0; CH$_3$Cl 6.2 × 10^{-30}; CH$_2$Cl$_2$ 5.0 × 10^{-30}; CHCl$_3$ 3.0 × 10^{-30}; CCl$_4$ 0.

Space charges. A charge may develop at interfaces between phases within a material, normally as a result of the presence of impurities. The charge moves on the surface when the material is placed in an electric field. This type of polarization is not an important factor in most common dielectrics.

18-4   Dielectric Properties and Capacitors

A *capacitor* is an electrical device used to store charge received from a circuit. The capacitor may smooth out fluctuations in the signal, accumulate charge to prevent damage to the rest of the circuit, store charge for later distribution, or even change the frequency of the electric signal. Capacitors are designed so that the charge is stored in a polarized material between two conductors (Figure 18-2). The material between the conductors must easily polarize yet have a high electrical resistivity to prevent the charge from passing from one plate to the next. Dielectric materials satisfy both of these requirements.

Dielectric constant. The charge Q (coulombs, C) that can be stored by a capacitor is

$$Q = CV \tag{18-2}$$

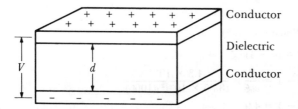

FIG. 18-2 A simple plate capacitor, in which a dielectric of thickness d stores a charge that is proportional to the voltage applied between the conductor plates.

where V is the voltage across the plates of the capacitor and C is the *capacitance*. The units for capacitance are C/V, or farads (F).

The capacitance depends on both the material between the plates and the design of the device. For a simple parallel plate capacitor with only two plates

$$C = \varepsilon \frac{A}{d} \tag{18-3}$$

where A is the area of each plate and d is the distance between the plates. The *permittivity* ε is the ability of the material to polarize and store a charge. The *relative permittivity* or *dielectric constant* κ is the ratio between the permittivity of the material and the permittivity of a vacuum, ε_0 (Figure 18-3).

$$\kappa = \frac{\varepsilon}{\varepsilon_0} \tag{18-4}$$

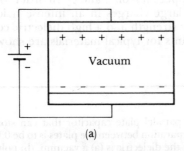

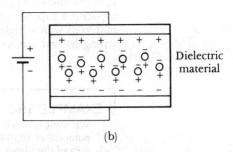

FIG. 18-3 A charge can be stored at the conductor plates in a vacuum. However, when a dielectric is placed between the plates, the dielectric polarizes and additional charge is stored. (a) Shows a total of 12 units of charge and (b) shows a total of 22 units.

The permittivity ε_0 of a vacuum is 8.85×10^{-12} F/m or 8.85×10^{-14} F/cm. The dielectric constant κ, which depends on the material, temperature, and frequency, is related to polarization.

$$P = (\kappa - 1)\varepsilon_0 \xi \tag{18-5}$$

where ξ is the strength of the electric field (V/m).

EXAMPLE 18-4

Suppose sodium chloride has a polarization of 4.3×10^{-8} C/m² in an electric field of 1000 V/m. Calculate the dielectric constant for sodium chloride.

Answer:

From Equation (18-5)

$$P = (\kappa - 1)\varepsilon_0 \xi$$

$$\kappa - 1 = \frac{P}{\varepsilon_0 \xi} = \frac{4.3 \times 10^{-8}}{(8.85 \times 10^{-12})(1000)}$$

$$\kappa - 1 = 4.9$$

$$\kappa = 5.9$$

For a capacitor containing n parallel conductor plates, the capacitance is

$$C = \varepsilon_0 \kappa \frac{A}{d} (n - 1) \tag{18-6}$$

Many large plates, a small separation between the plates, and a high dielectric constant improve the ability of the capacitor to store a charge.

Dielectric strength. Small separations and high voltages cause the capacitor to break down and discharge. The breakdown voltage, or dielectric strength, must be considered in addition to the dielectric constant. The *dielectric strength* is the maximum electric field ξ that can be maintained between the plates. The dielectric strength therefore places an upper limit on C and Q. In order to construct smaller capacitors capable of storing large charges in an intense field, we must select materials with a high dielectric strength and a high dielectric constant. Dielectric strengths and dielectric constants for typical materials are shown in Table 18-1.

EXAMPLE 18-5

We want to make a simple parallel plate capacitor that can store 4×10^{-5} C at a potential of 10,000 V. The separation between the plates is to be 0.02 cm. Calculate the area of the plates required if the dielectric is (a) a vacuum, (b) polyethylene, (c) water, and (d) barium titanate.

Answer:

$$C = \frac{Q}{V} = \frac{4 \times 10^{-5}}{10,000} = 4 \times 10^{-9}\ \text{F}$$

$$A = \frac{Cd}{\kappa \varepsilon_0} = \frac{(4 \times 10^{-9})(0.02)}{(8.85 \times 10^{-14})\kappa} = \frac{900}{\kappa}$$

(a) For vacuum, $\kappa = 1$, $A = 900$ cm^2

(b) For polyethylene, $\kappa = 2.26$, $A = 398$ cm^2

(c) For water, $\kappa = 78.3$, $A = 11.5$ cm^2

(d) For barium titanate, $\kappa = 3000$, $A = 0.3$ cm^2

The benefit of a high dielectric constant on the size of capacitors is obvious.

TABLE 18-1 Properties of selected dielectric materials

Material	Dielectric constant			Resistivity $(VA^{-1} m)$	Dielectric strength $(Vm^{-1}) \times 10^6$
	60 cycle s^{-1}	10^6 cycle s^{-1}	10^8 cycle s^{-1}		
Phenol-formaldehyde	7.5	4.7	4.3	10^{10}	12
Polyethylene	2.3	2.3	2.3	10^{13}–10^{16}	20
Teflon	2.1	2.1	2.1		
Polystyrene	2.5	2.5	2.5	10^{16}	20
Polyvinyl chloride (amorphous)	7	3.4		10^{14}	40
Polyvinyl chloride (glass)	3.4	3.4			
6,6-Nylon		3.3	3.2		
Rubber	4	3.2	3.1		20
Epoxy		3.6	3.3		
Paraffin wax		2.3	2.3	10^{13}–10^{17}	10
Fused silica	3.8	3.8	3.8	10^9–10^{10}	10
Fused quartz		3.9			
Soda-lime glass	7	7		10^{13}	10
Pyrex glass	4.3	4		10^{14}	14
Alumina	9	6.5		10^9–10^{12}	6
Barium titanate		3000		10^6–10^{13}	12
TiO$_2$		14–110		10^{11}–10^{16}	8
Mica		7		10^{11}	40
Water		78.3		10^{12}	
Gases		1.0006–1.02		10^{11}	
Vacuum		1			

From *Handbook of Chemistry and Physics, 56th Ed.*, CRC Press, 1975, and other sources.

EXAMPLE 18-6

A mica capacitor 258 mm^2 and 2.54 μm thick is to have a capacitance of 0.0252 μF. (a) How many plates are needed? (b) What is the maximum allowable voltage?

Answer:

From Table 18-1, typical properties for mica are $\kappa = 7$ and a dielectric strength of 40×10^6 V/m.

(a) $C = \varepsilon_0 \kappa \dfrac{A}{d}(n-1) = 0.0252 \times 10^{-6}$ F

$$n - 1 = \frac{Cd}{\varepsilon_0 \kappa A} = \frac{(0.0252 \times 10^{-6})(2.54 \times 10^{-6})}{(8.85 \times 10^{-12})(7)(0.258 \times 10^{-6})}$$

$$n - 1 = 4$$

$$n = 5$$

To be able to store the required charge, we need five conductor plates, with four layers of dielectric.

(b) $(40 \times 10^6 \text{ V/m}) = \dfrac{V}{d}$

$V = (40 \times 10^6)(2.54 \times 10)$

18-5 Controlling Dielectric Properties

Dielectrics used for capacitors should exhibit a large degree of polarization over a wide range of temperatures and frequencies. Several important factors influence dielectric behavior.

Electrical resistivity.   All effective dielectrics have a high electrical resistivity to prevent leakage or discharge of the stored energy. Table 18-1 shows that common dielectrics have resistivities of 10^{11} VA^{-1} m or greater.

Structure.   Most of the dielectric materials for capacitors fall into one of three groups—liquids composed of polar molecules, polymers, and certain ceramics (Table 18-2). All possess permanent dipoles that move easily in an electric field yet still produce high dielectric constants.

TABLE 18-2   Capacitor construction and materials

Types of Capacitors	Capacitor Shape and Characteristics
Mica (2.5 μm) Lead foil (5 μm)	Plate capacitors with good temperature stability, good for radio frequencies, bulky
Mica—sprayed silver coating	Plate capacitors, same capacitance as other mica capacitors
Glass-metal foil	Plate capacitors, moisture resistance, same capacitance as mica capacitors
Kraft paper (with wax or oil)—foil or metallized with tin	Rolled tubes, usually with several layers of paper between each set of conductors
Plastic (such as polyester)—foil	Rolled tubes
Ceramic—sprayed silver	Plate or tube type
Electrolytic Al-Al_2O_3—liquid Ta-TaO—liquid	

Water, which has a high dielectric constant, is corrosive, relatively conductive, and difficult to use in constructing capacitor devices. Organic oils or waxes are more effective. These materials contain relatively long chainlike molecules that serve as dipoles yet are easily aligned. Often they are impregnated into paper, which itself is a dielectric.

In amorphous polymers, segments of the chains possess sufficient mobility to cause polarization. Capacitors frequently use polyester (such as Mylar), polystyrene, polycarbonate, and cellulose (paper) as dielectrics. Glass, an amorphous ceramic, behaves in much the same way. Glassy polymers and crystalline materials have lower dielectric constants and dielectric strengths than their amorphous counterparts.

Polymers with asymmetrical chains have a higher dielectric constant, even though the chains may not easily align, because the strength of each molecular dipole is greater. Thus, polyvinyl chloride and polystyrene have dielectric strengths greater than polyethylene.

Barium titanate, $BaTiO_3$, a crystalline ceramic, also has an asymmetrical structure at room temperature (Figure 18-4). The titanium ion is displaced

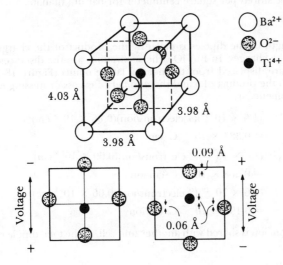

FIG. 18-4 The crystal structure of barium titanate, $BaTiO_3$. Because of the displacement of O^{2-} and Ti^{4+} ions, the unit cell is a permanent dipole and produces excellent polarization.

slightly from the center of the unit cell, causing the crystal to be tetragonal and permanently polarized. In an alternating field, the titanium ion moves back and forth between its two allowable positions to assure that polarization is aligned with the field. The unique crystal structure and its rapid response to the applied field cause barium titanate and similar materials to have an extraordinarily high dielectric constant. However, since polarization is highly anisotropic, the crystal must be properly aligned with respect to the applied field.

EXAMPLE 18-7

Rank the expected dielectric constant for polyethylene, polystyrene, rubber, and epoxy based on their molecular structure and ability to be polarized. Compare your ranking to the measured values at 10^6 cycles s^{-1} in Table 18-1.

Answer:

Polyethylene contains a simple carbon backbone with only small hydrogen atoms attached. Consequently, the chains are easily rearranged, giving a low residual polarization and low dielectric constant of about 2.3.

Polystyrene is an asymmetrical chain with a benzene ring attached to the chain. The mobility of the chains is thus reduced and a slightly higher dielectric constant, 2.5, is observed.

Rubber is cross-linked to further reduce the mobility of the polymer chains, causing still higher dielectric constants, about 3.2.

Epoxy is a rigid three-dimensional network polymer. Permanent dipoles are locked into place, giving high dielectric constants, about 3.6.

EXAMPLE 18-8

Calculate the maximum polarization per cubic centimeter and the maximum charge that can be stored per square centimeter for barium titanate.

Answer:

The strength of the dipoles is given by the product of the charge and the distance between the charges. In $BaTiO_3$, the separations are the distances that the Ti^{4+} and O^{2-} ions are displaced from the normal lattice points (Figure 18-4). The charge on each ion is the product of q and the number of excess or missing electrons. Thus the dipole moments are

$$Ti^{4+}: \quad (1.6 \times 10^{-19})(4 \text{ electrons/ion})(0.06 \times 10^{-8} \text{ cm})$$
$$= 0.384 \times 10^{-27} \text{ C} \cdot \text{cm/ion}$$

$$O^{2-}_{(top)}: \quad (1.6 \times 10^{-19})(2 \text{ electrons/ion})(0.09 \times 10^{-8} \text{ cm})$$
$$= 0.288 \times 10^{-27} \text{ C} \cdot \text{cm/ion}$$

$$O^{2-}_{(side)}: \quad (1.6 \times 10^{-19})(2 \text{ electrons/ion})(0.06 \times 10^{-8} \text{ cm})$$
$$= 0.192 \times 10^{-27} \text{ C} \cdot \text{cm/ion}$$

Each oxygen ion is shared with another unit cell, so the total dipole moment in the unit cell is

$$\begin{aligned}
\text{Dipole} \atop \text{moment} &= (1 \text{ Ti}^{4+}/\text{cell})(0.384 \times 10^{-27}) + \\
&\quad (1 \text{ O}^{2-} \text{ at top/cell})(0.288 \times 10^{-27}) + \\
&\quad (2 \text{ O}^{2-} \text{ at sides/cell})(0.192 \times 10^{-27}) \\
&= 1.056 \times 10^{-27} \text{ C} \cdot \text{cm/cell}
\end{aligned}$$

The polarization per cubic centimeter is

$$P = \frac{1.056 \times 10^{-27} \text{ C} \cdot \text{cm/cell}}{(3.98 \times 10^{-8} \text{ cm})^2(4.03 \times 10^{-8} \text{ cm})}$$
$$= 1.65 \times 10^{-5} \text{ C/cm}^2$$

The total charge on a $BaTiO_3$ crystal 1 cm × 1 cm is

$$Q = PA = (1.65 \times 10^{-5})(1)^2$$
$$= 1.65 \times 10^{-5} \text{ C}$$

Frequency. Dielectric materials are often used in alternating-current circuits. The dipoles must therefore switch directions, often at a high frequency, in order for the electronic device to perform satisfactorily.

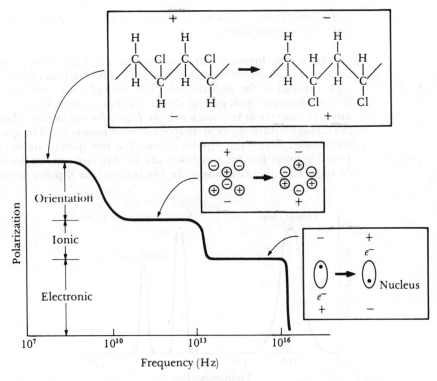

FIG. 18-5 The response of the material to frequency depends on the polarization mechanism. Permanent dipoles do not respond as rapidly as electronic and ionic dipoles.

Electronic polarization occurs easily even at frequencies as high as 10^{16} Hz, since no rearrangement of atoms is necessary (Figure 18-5). Ionic polarization also occurs readily up to 10^{13} Hz; only a simple elastic distortion of the bonds between the ions is required. However, materials that rely on molecular polarization are very sensitive to frequency, since entire atoms or groups of atoms must be rearranged. Consequently, the response of the dielectric material to an alternating field is reduced and complete polarization may not occur.

The structure also influences the frequency effect. Gases and liquids polarize at higher frequencies than solids. Amorphous polymers and ceramics polarize at higher frequencies than their crystalline counterparts. Polymers with bulky asymmetrical groups attached to the chain polarize only at low frequencies.

Voltage. Increasing the voltage of the applied field forces permanent dipoles into alignment more easily and completely. Eventually, all of the dipoles are aligned, a saturation polarization is achieved, and further increases in voltage have little effect on polarization. However, high voltages may cause breakdown of the dielectric.

Temperature. When the temperature increases, permanent dipoles have a greater mobility, polarize more easily, and give a higher dielectric constant. However, the higher temperatures again permit the dielectric to break down and may cause the crystal structure to change to a less polar condition, which greatly reduces the polarization.

Dielectric losses.   Some energy is lost as heat when a dielectric material is polarized in an alternating electrical field. The fraction of the energy lost during each reversal is the *dielectric loss*. The energy losses are due primarily to two factors—current leakage and dipole friction. Losses due to current leakage are low if the electrical resistivity is high. *Dipole friction* occurs when reorientation of the dipoles is difficult, as in complex organic molecules. The greatest loss occurs at frequencies where the dipoles almost, but not quite, can be reoriented (Figure 18-6). At lower frequencies, losses are low because the dipoles have time to move. At higher frequencies, losses are low because the dipoles do not move at all.

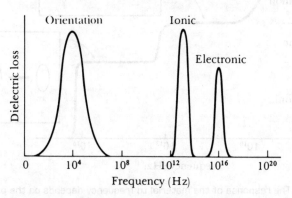

FIG. 18-6   The influence of frequency on dielectric loss. Losses are greatest at frequencies where one of the contributions to polarization is lost.

We can intentionally select a frequency so that materials with permanent dipoles have a high dielectric loss and materials that polarize only by electronic or ionic contributions have a low dielectric loss. Consequently, the permanent dipole materials heat, but the other materials remain cool. Microwave ovens are used to cure many polymer adhesives; the materials to be joined, including metals, have a low loss factor, while the adhesive has a high loss factor. The heat produced in the adhesive due to dielectric losses initiates the thermosetting reaction.

EXAMPLE 18-9

Explain why microwave ovens can be used to heat food but do not heat the container.

Answer:

The organic food substances are highly complex, polar molecules with a high dielectric loss, while the cooking ware is composed of material with a lower loss.

18-6 Dielectric Properties and Electrical Insulators

Materials used to insulate an electric field from the surroundings must also be dielectric. Electrical insulators must possess a high electrical resistivity, a high dielectric strength, and a low loss factor; however, high dielectric constants are not helpful.

The high electrical resistivity, which results from the large energy gap between the valence and conduction bands, prevents current leakage.

A high dielectric strength prevents catastrophic breakdown of the insulator at high voltages. *Internal failure* of the insulator occurs if impurities provide donor or acceptor levels that permit electrons to be excited into the conduction band. *External failure* is caused by arcing along the surface of the insulator or through interconnected porosity within the insulator body. In particular, adsorbed moisture on the surface of ceramic insulators presents a problem. Glazes on ceramic insulators seal off porosity and reduce the effect of surface contaminants.

The small dielectric constant prevents polarization, so charge is not stored locally at the insulator. Low dielectric constants are desirable for insulators, but high constants are required for capacitors.

18-7 Piezoelectricity and Electrostriction

Polarization changes the dimensions of the material, an effect called *electrostriction*. This might occur as a result of atoms acting as egg-shaped particles rather than spheres, or the bonds between ions changing in length, or by distortions due to the orientation of the permanent dipoles in the material.

However, certain dielectric materials display a further property. When a dimensional change is imposed on the dielectric, polarization occurs and a voltage or field is created (Figure 18-7). Dielectric materials that display this reversible

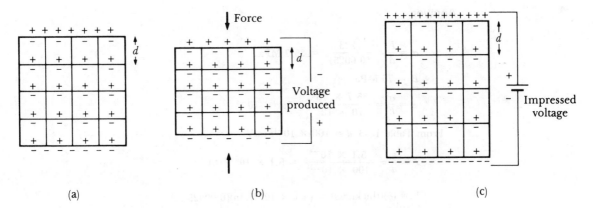

FIG. 18-7 The piezoelectric effect. (a) Piezoelectric crystals have a charge difference due to permanent dipoles. (b) A compressive force reduces the distance between charge centers, changes the polarization, and introduces a voltage. (c) A voltage changes the distance between charge centers, causing a change in dimensions.

behavior are *piezoelectric*. We can write the two reactions that occur in piezoelectrics as

$$\text{Field produced by stress} = \xi = g\sigma \qquad (18\text{-}7)$$

$$\text{Strain produced by field} = \varepsilon = d\xi \qquad (18\text{-}8)$$

where ξ is the electric field (V/m), σ is the applied stress (Pa), ε is the strain, and g and d are constants. Typical values for d are given in Table 18-3. The constant g is related to d through the modulus of elasticity E.

$$E = \frac{1}{gd} \qquad (18\text{-}9)$$

TABLE 18-3 The piezoelectric constant d for selected materials

Material	Piezoelectric Constant d $(C/Pa \cdot m^2 = m/V)$
Quartz	2.3×10^{-12}
BaTiO$_3$	100×10^{-12}
PbZrTiO$_6$	250×10^{-12}
PbNb$_2$O$_6$	80×10^{-12}

EXAMPLE 18-10

A 223 N force is applied to a 2.5 × 2.5 mm wafer of barium titanate that is 0.25 mm thick. Calculate the strain produced by the force and the voltage that is created. The modulus of elasticity is 70 GPa.

Answer:

$$\sigma = \frac{F}{A} = \frac{223}{(0.0025)^2} = 35.7 \text{ MPa}$$

$$E = 70 \text{ MPa}$$

$$\varepsilon = \frac{\sigma}{E} = \frac{35.7 \times 10^6}{70 \times 10^9} = 5.1 \times 10^{-4}$$

From Table 18-3, $d = 100 \times 10^{-12}$ m/V

$$\xi = \frac{\varepsilon}{d} = \frac{5.1 \times 10^{-4}}{100 \times 10^{-12}} = 5.1 \times 10^6 \text{ V/m}$$

$$V = (\xi)(\text{thickness}) = (5.1 \times 10^6 \text{ V/m})(0.00025 \text{ m})$$

$$= 1275 \text{ V}$$

Materials which are permanently polarized display this effect. For example, we can measure a potential difference across a barium titanate crystal because the titanium ion is always shifted slightly from the body-centered position in the unit cell. The most common of these materials include quartz, barium titanate, lead titanate, lead zirconate, CdS, and ZnO.

The piezoelectric effect is used in *transducers,* which convert acoustical waves (sound) into electric fields or electric fields into acoustical waves (Figure 18-8).

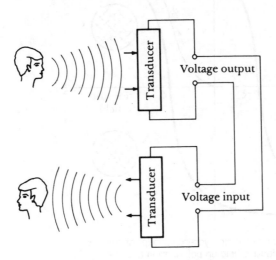

FIG. 18-8 Schematic diagram showing the use of piezoelectric transducers for the telephone.

Sound of a particular frequency produces a strain in a piezoelectric material. The dimensional changes polarize the crystal, creating an electric field. In turn, the electric field is transmitted to a second piezoelectric crystal; the electric field produces dimensional changes in the second crystal which produce an acoustical wave that is amplified. This description depicts the telephone. Similar electromechanical transducers are used for stereo record players and other audio devices. The piezoelectric effect is also employed in tuners for radios; by imposing the correct strain in the crystal, only the desired frequency is picked up and amplified.

18-8 Ferroelectricity

The presence of polarization in a material after the electric field is removed can be explained in terms of a residual alignment of permanent dipoles. Barium titanate is again an excellent example. Materials that retain a net polarization when the field is removed are called *ferroelectric.*

Many materials do not display this behavior, even though they have permanent dipoles. The dipoles become randomly arranged when the field is removed, the polarization of each dipole is canceled by neighboring dipoles, and no net polarization results. However, in ferroelectric materials, the orientation of one dipole influences the surrounding dipoles to have an identical alignment. We can examine this behavior by describing the effect of an electric field on polarization (Figure 18-9).

Let's begin with a crystal whose dipoles are randomly oriented so there is no net polarization. When a field is applied, the dipoles begin to line up with the field,

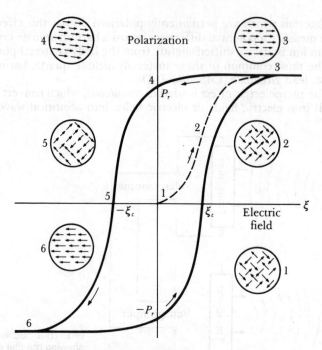

FIG. 18-9 The ferroelectric hysteresis loop, showing the
influence of the electric field on polarization and the
alignment of the dipoles.

points 1 to 3 in Figure 18-9. Eventually the field aligns all of the dipoles and the maximum, or *saturation,* polarization P_s is obtained, point 3. When the field is subsequently removed, a *remanent* polarization P_r remains, point 4, due to the coupling between the dipoles. The material is permanently polarized. The ability to retain polarization permits the ferroelectric material to retain information, making the material useful in computer circuitry.

When a field is applied in the opposite direction, the dipoles must be reversed. A *coercive field* ξ_c must be applied to remove the polarization and randomize the dipoles, point 5. If the reverse field is increased further, saturation occurs with the opposite polarization, point 6. As the field continues to alternate, a *hysteresis loop* is described showing how the polarization of the ferroelectric varies with the field. The area contained within the hysteresis loop is related to the energy required to cause polarization to switch from one direction to the other.

The ferroelectric behavior depends on temperature. Above the critical *Curie temperature* dielectric and consequently ferroelectric behavior are lost (Figure 18-10). In some materials, such as barium titanate, the Curie temperature corresponds to a change in crystal structure, so that the permanent dipoles in each unit cell no longer exist. Curie temperatures of typical ferroelectrics are shown in Table 18-4.

Ferroelectric materials are always dielectric, piezoelectric, and have a high dielectric constant which make them particularly suitable for use in capacitors and piezoelectric transducers.

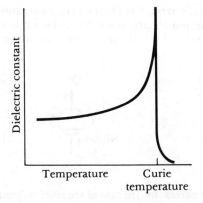

FIG. 18-10 The effect of temperature on the dielectric constant of barium titanate. Above the Curie temperature, the molecular polarization is lost due to a change in crystal structure and barium titanate is no longer ferroelectric.

TABLE 18-4 The Curie temperature for selected ferroelectrics

Material	Curie Temperature (°C)
$SrTiO_3$	−200
$Cd_2Nb_2O_7$	−88
Rochelle salt	24
$BaTiO_3$	120
$PbZrO_3$	233
$PbTa_2O_6$	260
$KNbO_3$	435
$PbTiO_3$	490
$PbNb_2O_6$	570
$NaNbO_3$	640

18-9 Magnetization versus Polarization

In dielectric materials, polarization occurs when induced or permanent electric dipoles are oriented by an interaction between the material and the electrical field. In an analagous manner, *magnetization* occurs when induced or permanent magnetic dipoles are oriented by an interaction between the magnetic material and a magnetic field. Magnetization enhances the influence of the magnetic field, permitting larger magnetic energies to be stored than if the material were absent. This energy can be stored permanently or temporarily and can be used to do work.

18-10 Magnetic Dipoles and Magnetic Moments

Each electron in an atom has two magnetic moments. A *magnetic moment* is simply the strength of the magnetic field associated with the electron. This moment, called the *Bohr magneton*, is equal to

$$\text{Bohr magneton} = \frac{qh}{4\pi m_e} = 9.27 \times 10^{-24} \text{ A} \cdot \text{m}^2 \qquad (18\text{-}10)$$

where q is the charge on the electron, h is Planck's constant, and m_e is the mass of the electron. The magnetic moments are due to the orbital motion of the electron around the nucleus and the spin of the electron about its own axis (Figure 18-11).

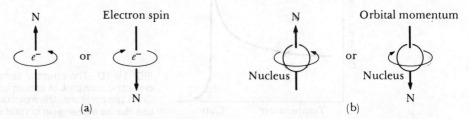

FIG. 18-11 Origin of magnetic dipoles. (a) The spin of the electron produces a magnetic field with a direction dependent on the quantum number m_s. (b) Electrons orbiting about the nucleus create a magnetic field about the atom.

EXAMPLE 18-11

Magnetization is given in units of Am^{-1}, or oersted where 1 Oe $= (10^3/4\pi)$ Am^{-1}. We will find later that each iron atom contains four electrons that act as magnetic dipoles. Calculate thew maximum magnetization that we might expect in iron. The lattice parameter of BCC iron is 2.866 Å. (Note that this example is similar to Example 18-1, when we calculated polarization.)

Answer:

$$M = \left(\frac{2 \text{ atoms/cell}}{(2.866 \times 10^{-10} \text{ m})^3}\right) \left(4 \frac{\text{Bohr magnetons}}{\text{atom}}\right) \left(9.27 \times 10^{-24} \frac{\text{A} \cdot \text{m}^2}{\text{magnetons}}\right)$$

$$= (0.085 \times 10^{30})(4)(9.27 \times 10^{-24}) = 3.15 \times 10^6 \text{ A/m}$$

When we discussed the electronic structure and quantum numbers in Chapter 2, we pointed out that each discrete energy level could contain two electrons, each having an opposite spin. The magnetic moments of each electron pair in an energy level are opposed and, consequently, whenever an energy level is completely full, there is no *net* magnetic moment.

Based on this reasoning, we expect any atom of an element with an odd atomic number to have a net magnetic moment from the unpaired electron. However, this is not the case. In most of these elements, the unpaired electron is a valence electron. Because the valence electrons from each atom interact, the magnetic moments, on average, cancel and no net magnetic moment is associated with the material.

However, certain elements, such as the transition metals, have an inner energy level that is not completely filled. The elements scandium through copper, whose electronic structures are shown in Table 18-5, are typical. Except for chromium and copper, the valence electrons in the $4s$ level are paired; the unpaired

TABLE 18-5   The electron spins in the 3d energy level
in transition metals, with arrows indicating the direction of
the spin

Metal	3d					4s
Sc	↑					↑↓
Ti	↑	↑				↑↓
V	↑	↑	↑			↑↓
Cr	↑	↑	↑	↑	↑	↑
Mn	↑	↑	↑	↑	↑	↑↓
Fe	↑↓	↑	↑	↑	↑	↑↓
Co	↑↓	↑↓	↑	↑	↑	↑↓
Ni	↑↓	↑↓	↑↓	↑	↑	↑↓
Cu	↑↓	↑↓	↑↓	↑↓	↑↓	↑

electrons in chromium and copper are canceled by interactions with other atoms. Copper also has a completely filled 3d shell and thus does not display a net moment.

The electrons in the 3d level of the remaining transition elements do not enter the shell in pairs. Instead, as in manganese, the first five electrons have the same spin. Only after half of the 3d level is filled do pairs with opposing spins form. Therefore, each atom in a transition metal has a permanent magnetic moment, equal in strength to the number of unpaired electrons. Each atom behaves as a magnetic dipole.

The response of the atom to an applied magnetic field depends on how the magnetic dipoles represented by each atom react to the field. Most of the transition elements react in such a way that the magnetic moments cancel one another. However, in nickel, iron, and cobalt, a magnetic field causes all of the dipoles to line up with the magnetic field and produces a desirable amplification of the effect of the field.

18-11 Magnetization, Permeability, and the Magnetic Field

Let's look at the relationship between the magnetic field and magnetization. When a magnetic field is applied in a vacuum, lines of magnetic flux are produced (Figure 18-12). A greater number of lines of flux increases the work that the magnetic field can accomplish. The flux density, or *inductance*, is related to the applied field by

$$B = \mu_0 H \qquad (18\text{-}11)$$

where B is the inductance (tesla or Wb m^{-2}), H is the magnetic field (Am^{-1}), and μ_0 is the *magnetic permeability* of a vacuum. The permeability in a vacuum is a constant, $4\pi \times 10^{-7}$ Wb A^{-1} m^{-1}.

When we place a material within the magnetic field, the magnetic inductance is determined by the manner in which induced and permanent magnetic dipoles interact with the field. The magnetic inductance now is

$$B = \mu H \qquad (18\text{-}12)$$

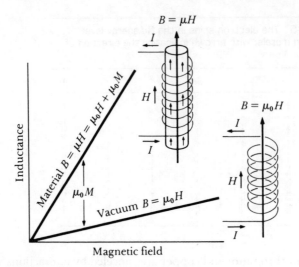

FIG. 18-12   A current passing through a coil sets up
a magnetic field H with a flux density B. The flux
density is higher when a magnetic core is placed within
the coil.

where μ is the magnetic permeability of the material in the field. If the magnetic moments reinforce the applied field, $\mu > \mu_0$; but if the magnetic moments oppose the field, $\mu < \mu_0$.

We can describe the influence of the magnetic material by the *relative permeability* μ_r, where

$$\mu_r = \frac{\mu}{\mu_0} \qquad (18\text{-}13)$$

A large relative permeability means that the material has amplified the effect of the magnetic field. Thus, the relative permeability has the same importance that the dielectric constant, or relative permittivity, has in dielectrics.

The *magnetization M* represents the increase in the magnetic inductance due to the core material, so we can rewrite the equation for inductance as

$$B = \mu_0 H + \mu_0 M \approx \mu_0 M \qquad (18\text{-}14)$$

The *magnetic susceptibility* χ, which is the ratio between magnetization and the applied field, gives the amplification produced by the material.

$$\chi = \frac{M}{H} \qquad (18\text{-}15)$$

Because the term $\mu_0 M$ is often much greater than $\mu_0 H$, we can frequently equate $B = \mu_0 M$. We sometimes interchangeably refer to either inductance or magnetization. Normally, we are interested in producing a high inductance B or magnetization M. This is accomplished by selecting materials that have a high relative permeability or magnetic susceptibility.

EXAMPLE 18-12

By combining Equations (18-12) and (18-14), derive the relationship between magnetic susceptibility and relative permeability.

Answer:

$$B = \mu H = \mu_0 H + \mu_0 M$$

$$\frac{\mu H}{\mu_0} = H + M$$

$$\mu_r H = H + M$$

$$\mu_r H - H = M$$

$$(\mu_r - 1)H = M$$

$$\mu_r - 1 = \frac{M}{H}$$

$$\mu_r = 1 + \chi$$

EXAMPLE 18-13

A magnetic field of 2387 A m^{-1} is applied to a material with a relative permeability of 5000. Calculate the magnetization and inductance.

Answer:

$$\mu_r = 1 + \chi = 1 + \frac{M}{H} = 5000$$

$$\frac{M}{H} = 5000 - 1 = 4999$$

$$M = 4999(2387) = 1.193 \times 10^7 \text{ A m}^{-1}$$

$$\mu = \mu_r \mu_0 = (5000)(4\pi \times 10^{-7}) = 6.28 \times 10^{-3}$$

$$B = \mu H = (6.28 \times 10^{-3})(2387) = 14.99 \text{ tesla}$$

18-12 Interactions Between Magnetic Dipoles and the Magnetic Field

When a magnetic field is applied to a collection of atoms, several types of behavior may be observed (Figure 18-13).

Diamagnetic behavior. A magnetic field acting on any atom induces a magnetic dipole for the entire atom by influencing the magnetic moment due to the orbiting electrons. These dipoles oppose the magnetic field, causing the magnetization to be less than zero. This behavior, called *diamagnetism*, gives a relative

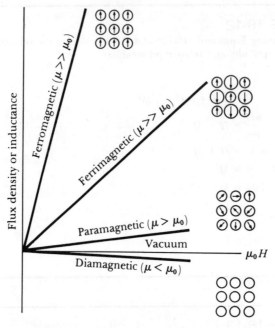

FIG. 18-13 The effect of the core material on the flux density. No magnetic moment is produced in diamagnetic materials. Progressively stronger moments are present in paramagnetic, ferrimagnetic, and ferromagnetic materials for the same applied field.

permeability of about 0.99995. Diamagnetic behavior has no important applications for magnetic materials or devices.

Paramagnetism.   When materials have unpaired electrons, a net magnetic moment due to electron spin is associated with each atom. When a magnetic field is applied, the dipoles line up with the field, causing a positive magnetization. However, because the dipoles do not interact, extremely large magnetic fields are required to align all of the dipoles. This effect, called *paramagnetism*, is only significant at high temperatures. The relative permeability is less than about 1.01.

Ferromagnetism.   *Ferromagnetic* behavior is due to unfilled energy levels in the $3d$ level (for iron, nickel, and cobalt) or the $4f$ level (for gadolinium). In ferromagnetic materials, the permanent unpaired dipoles line up with the imposed magnetic field. The dipoles are easily aligned in the magnetic field due to mutual reinforcement of the dipoles. A very large amplification of the imposed field is produced even for small magnetic fields, giving high relative permeabilities.

Antiferromagnetism.   In some materials, the magnetic moments produced in neighboring dipoles line up in opposition to one another in the magnetic field, even though the strength of each dipole is very high (Figure 18-14). These materials, including manganese and MnO, are *antiferromagnetic* and have zero magnetization. The difference between ferromagnetism and antiferromagnetism is in the interactions between neighboring dipoles—whether they reinforce or oppose one another.

Ferrimagnetism.   In ceramic materials, different ions have different magnetic moments. In a magnetic field, the dipoles of ion A may line up with the field

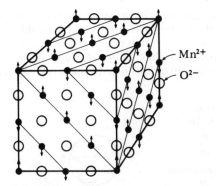

FIG. 18-14 The crystal structure of MnO consists of alternating layers of {111} type planes of oxygen and manganese ions. The magnetic moments of the manganese ions in every other (111) plane are oppositely aligned. Consequently, MnO is antiferromagnetic.

while dipoles of ion *B* oppose the field. But because the strengths of the dipoles are not equal, a net magnetization results. The *ferrimagnetic* materials can provide good amplification of the imposed field.

18-13 Domain Structure

Ferromagnetic materials have their powerful influence on magnetization because of the positive interaction between the dipoles of neighboring atoms. Within the grain structure of a ferromagnetic material, a substructure composed of magnetic domains is produced, even in the absence of an external field. *Domains* are regions in the material in which all of the dipoles are aligned. In a material that has never been exposed to a magnetic field, the individual domains have a random orientation. The net magnetization in the material as a whole is zero.

Boundaries, called *Bloch walls,* separate the individual domains, much like grain boundaries. The Bloch walls are narrow zones in which the direction of the magnetic moment gradually and continuously changes from that of one domain to that of the next (Figure 18-15). These domains are typically very small, about 0.005 cm or less, while the Bloch walls are about 1000 Å thick.

Orientation of domains in a magnetic field. When a magnetic field is imposed on the material, domains that are already lined up with the field grow at the expense of unaligned domains. In order for the domains to grow, the Bloch walls

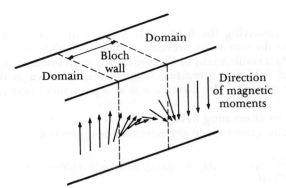

FIG. 18-15 The magnetic moments in adjoining atoms change direction continuously across the boundary between domains.

must move. The imposed magnetic field provides the force required for the walls to migrate. As the strength of the field increases, favorably oriented domains continue to grow and a greater net magnetization occurs (Figure 18-16). The *saturation magnetization*[1], produced when all of the domains are properly oriented, is the greatest amount of magnetization that the material can obtain.

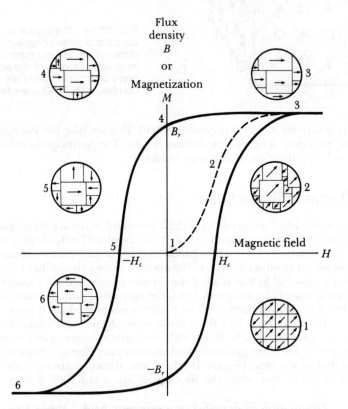

FIG. 18-16   The ferromagnetic hysteresis loop, showing the effect of the magnetic field on inductance or magnetization. The dipole alignment leads to saturation magnetization, point 3, a remanence, point 4, and a coercive field, point 5.

Effect of removing the field.   When the field is removed, the resistance offered by the domain walls prevents regrowth of the domains into random orientations. As a result, many of the domains remain oriented near the direction of the original field and a residual magnetization, known as the *remanence,* is present in the material. The material acts as a permanent magnet.

Effect of an alternating field.   If we now apply a field in the reverse direction, the domains grow with an alignment in the opposite direction. A *coercive field*

[1] Note that we are interchangeably using magnetization and inductance, since in ferromagnetic materials, $\mu_0 M \gg \mu_0 H$.

H_c is required to force the domains to be randomly oriented and cancel one another's effect. Further increases in the strength of the field eventually align the domains to saturation in the opposite direction.

As the field continually alternates, the magnetization versus field relationship traces out a *hysteresis loop*. The hysteresis loop describes the strength and direction of the magnetization in an alternating magnetic field.

18-14 Application of the Magnetization-Field Curve

The behavior of a material in a magnetic field is related to the size and shape of the hysteresis loop (Figure 18-17). Let's look at three applications for magnetic materials.

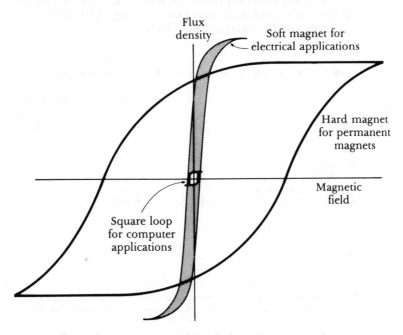

FIG. 18-17 Comparison of the hysteresis loops for three applications of ferromagnetic materials—electrical applications, computer applications, and permanent magnets.

Magnetic materials for electrical applications. Ferromagnetic materials are used to enhance the magnetic field produced when an electric current is passed through the material. The magnetic field is then expected to do work. Applications include cores for electromagnets, electric motors, transformers, generators, and other electrical equipment. These devices utilize an alternating field, so that the core material is continually cycled through the hysteresis loop.

Electrical magnetic materials, often called *soft* magnets, should have the following characteristics.

1. A high saturation magnetization permits the material to do the most work.

We obtain the highest values of magnetization if we select a material that has a high magnetic permeability.

2. The domains can be reoriented with the smallest imposed field if the coercive field is very small.

3. A small hysteresis loop gives small energy losses when a material is cycled through an alternating field.

4. A small remanence is desired. If the remanence is large in an electromagnet used to transfer steel scrap from a freight car to a steel mill for recycling, the scrap is not released when the imposed field is removed.

5. The dipoles must be easily realigned at the frequency at which the material operates. If the frequency is so high that the domains cannot be realigned each cycle, then the device may heat, just as in dielectric materials, due to *dipole friction*.

6. If the electrical resistivity is too low, current may leak from the field as eddy currents and again may cause heating of the device. Typical soft magnetic materials are described in Table 18-6.

TABLE 18-6   Properties of selected soft or electrical magnetic materials

Material	Maximum Relative Permeability	Saturation Magnetization (tesla)	Coercive Field (A m^{-1})
99.91% pure iron	5,000	2.15	79.58
99.95% pure iron	180,000	2.15	3.98
Fe-3% Si (oriented)	30,000	2.00	11.93
Fe-4% Si (not oriented)	7,000	1.97	39.79
4S Permalloy (54.7% Fe, 45% Ni, 0.3% Mn)	25,000	1.60	23.87
Supermalloy (79% Ni, 15.7% Fe, 5% Mo, 0.3% Mn)	800,000	0.80	159.16
A6 Ferroxcube (Mn,Zn)Fe$_2$O$_4$		0.40	
B2 Ferroxcube (Ni,Zn)Fe$_2$O$_4$		0.30	

From *Handbook of Chemistry and Physics, 56th Ed.*, CRC Press, 1975, and other sources.

Magnetic materials for computer memories.   Ferromagnetic materials are used to store bits of information in computers. Memory is stored by magnetizing the material in a certain direction. For example, if the north pole is up, the bit of information stored is 1. If the north pole is down, then a 0 is stored.

For this application, materials with a square hysteresis loop, a low remanence, a low saturation magnetization, and a low coercive field are preferable. The square loop assures that a bit of information placed in the material by a field remains stored; a steep and abrupt change in magnetization is required to remove the information from storage in the ferromagnet. Furthermore, the magnetization is produced by small external fields, so the coercive field, saturation magnetization, and remanence should be low.

Magnetic materials for permanent magnets.   Finally, ferromagnetic materials are used to make strong permanent magnets (Table 18-7). Permanent magnets require high remanence, high permeability, high coercive fields, and high power.

TABLE 18-7   Selected properties of hard or permanent magnetic materials

Material	Remanence (tesla)	Coercive field (A m^{-1})	$(BH)_{max}$ (Wb A^{-1} m^{-1})
Steel (0.9% C, 1.0% Mn)	1.000	3,979	1,592
Alnico I (21% Ni, 12% Al, 5% Co, bal Fe)	0.710	35,015	11,141
Alnico V (24% Co, 14% Ni, 8% Al, 3% Cu, bal Fe)	1.310	50,931	47,748
Alnico XII (35% Co, 18% Ni, 8% Ti, 6% Al, bal Fe)	0.580	75,601	12,328
Cunico (50% Cu, 29% Co, 21% Ni)	0.340	52,522	63,66
Cunife (60% Cu, 20% Fe, 20% Ni)	0.540	43,769	11.937
Silmanal (86.6% Ag, 8.8% Mn, 4.4% Al)	0.055	477,480	597
Co$_5$Sm	0.950	756,010	
BaO · 6Fe$_2$O$_3$	0.400	190,992	
SrO · 6Fe$_2$O$_3$	0.340	262,614	

From *Handbook of Chemistry and Physics, 56th Ed.*, CRC Press, 1975, and other sources.

The *power* of the magnet is related to the size of the hysteresis loop, or the maximum product of B and H. The area of the largest rectangle that can be drawn in the second or fourth quadrants of the B-H curve is related to the energy required to demagnetize the magnet (Figure 18-18). For the product to be large, both the remanence and the coercive field should be large.

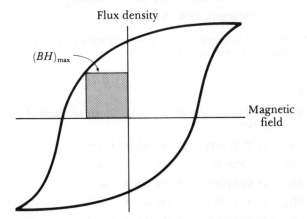

FIG. 18-18   The largest rectangle drawn in the second or fourth quadrant of the B-H curve gives the maximum BH product. (BH)$_{max}$ is related to the power, or energy required to demagnetize the permanent magnet.

Development of strong permanent magnets, often said to be magnetically *hard,* is aimed at improving both the magnetic permeability and the stability of the domains. We will see how this is achieved in the following sections.

EXAMPLE 18-14

Determine the power, or BH product, for the magnetic material whose properties are shown in Figure 18-19.

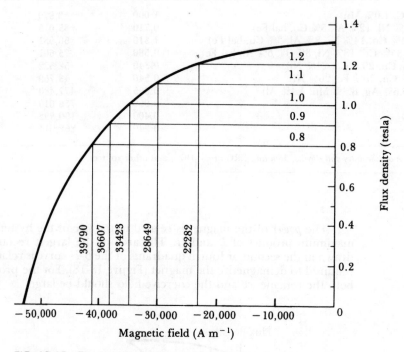

FIG. 18-19   The fourth quadrant of the B-H curve for a permanent magnetic material. (See Example 18-14.)

Answer:

Several rectangles have been drawn in the fourth quadrant of the B-H curve. The BH product in each is

$$BH_1 = (1.2)(22282) = 26,739 \text{ Wb A}^{-1} \text{ m}^{-1}$$

$$BH_2 = (1.1)(28649) = 31,514 \text{ Wb A}^{-1} \text{ m}^{-1}$$

$$BH_3 = (1.0)(33424) = 33,424 \text{ Wb A}^{-1} \text{ m}^{-1}$$

$$BH_4 = (0.9)(36607) = 32,946 \text{ Wb A}^{-1} \text{ m}^{-1}$$

$$BH_5 = (0.8)(39790) = 31,832 \text{ Wb A}^{-1} \text{ m}^{-1}$$

Thus, the power is about 33424 Wb A^{-1} m^{-1}.

18-15   Temperature Effects

When the temperature of a ferromagnetic material is increased, the added thermal energy reduces the magnetic permeability or magnetization and permits the domains to become randomly oriented. Consequently, magnetization, remanence, and the coercive field are all reduced at high temperatures [Figure 18-20(a)]. If the temperature exceeds the critical *Curie temperature*, ferromagnetic behavior is no longer observed [Figure 18-20(b)]. The Curie temperature (Table 18-8) depends on the type of magnetic material and can be changed by alloying elements.

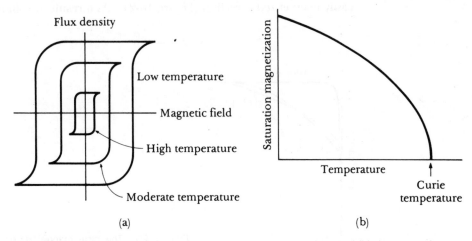

FIG. 18-20   The effect of temperature on (a) the hysteresis loop and (b) the saturation remanence. Ferromagnetic behavior disappears above the Curie temperature.

TABLE 18-8   Curie temperatures for selected magnetic materials

Material	Curie Temperature (°C)
Gd	16
Ni	358
Fe	770
Co	1131

From *Handbook of Chemistry and Physics, 56th Ed.*, CRC Press, 1975.

18-16 Magnetic Materials

Let's look at typical metallic alloys and ceramic materials used in magnetic applications and discuss how their properties and behavior can be enhanced.

Magnetic metals.   Iron, nickel, and cobalt are readily magnetized to serve as permanent magnets. However, the domains are easily reoriented in these pure metals and both the relative permeability and the *BH* product are small compared to more complex alloys. Some improvement in the magnetic properties is gained by introducing defects into the microstructure.

Silicon iron.   Introduction of 3% to 5% Si to iron produces an alloy that can, through proper processing, have excellent magnetization with a very small hysteresis loop. The silicon iron is frequently used in motors and generators.

We take advantage of the anisotropic magnetic behavior of silicon iron in the following way. Unusually small hysteresis loops and coercive fields are obtained when the crystal structure of the silicon iron is lined up with the field in the most easily magnetized direction (Figure 18-21). As a result of rolling and subsequent

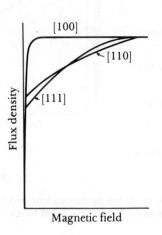

FIG. 18-21   The initial magnetization curve for silicon iron is highly anisotropic; magnetization is easiest when the ⟨100⟩ directions are aligned with the field.

annealing, a sheet texture is formed in which the ⟨100⟩ directions in each grain are aligned. Because the silicon iron is most easily magnetized in the ⟨100⟩ directions, the hysteresis loop and energy losses are small. Similar behavior is obtained in an iron-nickel alloy called Permalloy.

Metallic glasses.   Amorphous metallic glasses, produced by extraordinarily high cooling rates during solidification, also have excellent magnetic properties. The metallic glasses can be produced in the form of thin tapes which can be stacked together to produce larger cores.

Magnetic tapes.   Magnetic materials for information storage must have a square loop and a low coercive force, permitting very rapid transmission of information. Magnetic tape, produced by evaporating, sputtering, or plating a magnetic material onto a plastic tape, helps satisfy these requirements. Use of the Fe-81.5% Ni alloy or certain ceramic oxides, such as Fe_3O_4, provides the proper hysteresis loop. Because the magnetic material is very thin, domains can rotate and thus respond rapidly.

Complex metallic alloys for permanent magnets.   Improved permanent magnets are produced by making the grain size so small that only one domain is present in each grain. Now the boundaries between domains are grain boundaries rather than Bloch walls; the domains can only change their orientation by rotating, which requires greater energy than domain growth.

Two techniques are used to produce these magnetic materials—phase transformations and powder metallurgy. Alnico, one of the most common of the complex metallic alloys, has a single-phase BCC structure at high temperatures. But when Alnico slowly cools below 800°C, a second BCC phase rich in iron and cobalt precipitates. This second phase is so fine that each precipitate particle is a single domain, producing a very high remanence, coercive field, and power. Often the alloys are permitted to cool and transform when in a magnetic field to align the domains as they form.

A second technique—powder metallurgy—is used for another group of metallic alloys, including cobalt-samarium. A composition giving Co_5Sm, an intermetallic compound, has excellent magnetization (Figure 18-22). The brittle interme-

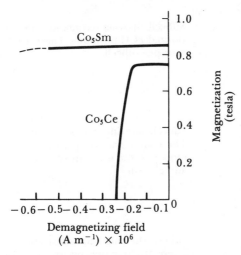

FIG. 18-22 Demagnetizing curves for Co_5Sm and Co_5Ce, representing a portion of the hysteresis loop.

tallic is crushed and ground to produce a fine powder in which each particle is a domain. The powder is then compacted when in an imposed magnetic field to align the powder domains. Careful sintering to avoid growth of the particles produces a solid powder metallurgy magnet.

Ferrimagnetic ceramic materials. Common magnetic ceramics are the ferrites, which have a spinel crystal structure (Figure 18-23). Each metallic ion in the crystal structure behaves as a dipole. Although the dipole moments of each type of ion may oppose one another, the strengths of the dipoles are different, a net magnetization develops, and ferrimagnetic behavior is observed.

We can understand the behavior of these ceramic magnets by looking at magnetite, Fe_3O_4. Magnetite contains two different iron ions, Fe^{2+} and Fe^{3+}, so we could rewrite the formula for magnetite as $Fe^{2+}Fe_2^{3+}O_4^{2-}$. The magnetite, or spinel, crystal structure is based on a FCC arrangement of oxygen ions, with iron ions occupying selected interstitial sites. Although the spinel unit cell actually contains eight of the FCC arrangements, we need only examine just one of the FCC subcells.

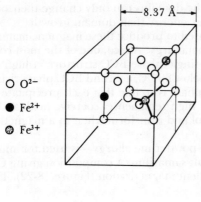

O O^{2-}
● Fe^{2+}
◉ Fe^{3+}

Complete unit cell
for magnetite, Fe_3O_4.
For simplicity, only
one of the eight
identical FCC subcells
is shown in detail.

(a)

Subcell showing oxygen ions in
normal FCC positions. There are four
ions per subcell.

(b)

Subcell showing tetrahedral sites.
There are eight sites per subcell.
Only one site is occupied with
an Fe^{3+} ion.

(c)

Subcell showing octahedral sites.
There are four sites per subcell. One
is occupied by an Fe^{2+} ion and
a second contains an Fe^{3+} ion.

(d)

FIG. 18-23 (a) The structure of magnetite, Fe_3O_4. The details of the
complex arrangement of the Fe^{2+}, Fe^{3+}, and O^{2-} ions are shown in (b),
(c), and (d).

1. The oxygen ions are in the FCC positions of the subcell; thus there are a
total of four oxygen ions.

2. Octahedral sites, which are surrounded by six oxygen ions, are present at
each edge and the center of the subcell. The sites at the 12 edges are each shared
by four subcells; thus the number of octahedral sites belonging uniquely to each
subcell is 12 edges/4 subcells plus one center site giving a total of four sites. One
Fe^{2+} and one Fe^{3+} ion occupy octahedral sites.

3. Tetrahedral sites, surrounded by four oxygen ions, have indices in the
subcell such as $\frac{1}{4},\frac{1}{4},\frac{1}{4}$ and represent the center of oxygen tetrahedra. There are

eight tetrahedral sites in the subcell. One Fe^{3+} ion occupies one of the tetrahedral sites.

4. When Fe^{2+} ions form, the two $4s$ electrons of iron are removed but all of the $3d$ electrons remain. Because there are four unpaired electrons in the $3d$ level of iron, the magnetic strength of the Fe^{2+} dipole is four Bohr magnetons. However, when Fe^{3+} forms, both $4s$ electrons and one of the $3d$ electrons are removed. The Fe^{3+} ion has five unpaired electrons in the $3d$ level and thus has a strength of five Bohr magnetons.

5. The ions in the tetrahedral sites of the magnetite line up so that their magnetic moments oppose the applied magnetic field, but the ions in the octahedral sites reinforce the field (Figure 18-24). Consequently, the Fe^{3+} ion in the

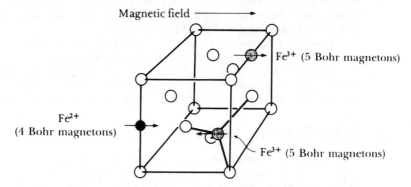

FIG. 18-24 The subcell of magnetite. The ions in the octahedral sites line up with the magnetic field, but ions in tetrahedral sites oppose the field. A net magnetic moment is produced by this ionic arrangement.

tetrahedral site neutralizes the Fe^{3+} ion in the octahedral site—the Fe^{3+} ions have antiferromagnetic behavior. However, the Fe^{2+} ion in the octahedral site is not opposed by any other ion and therefore reinforces the magnetic field. In a reversing magnetic field, magnetite displays a hysteresis loop, just as in ferromagnetic materials.

EXAMPLE 18-15

Calculate the total magnetic moment per cubic centimeter in magnetite.

Answer:

In the subcell (Figure 18-24), the total magnetic moment is four Bohr magnetons, obtained from the Fe^{2+} ion, since the magnetic moments from the two Fe^{3+} ions are canceled by each other.

In the unit cell overall, there are eight subcells, so the total magnetic moment is 32 Bohr magnetons per cell.

The size of the unit cell, with a lattice parameter of 8.37×10^{-8} cm, is

$$V_{cell} = (8.37 \times 10^{-8})^3 = 5.86 \times 10^{-22} \text{ cm}^3$$

The magnetic moment per cubic centimeter is

$$\frac{\text{Total}}{\text{moment}} = \frac{32 \text{ Bohr magnetons/cell}}{5.86 \times 10^{-22} \text{ cm}^3/\text{cell}} = 5.46 \times 10^{22} \text{ magnetons/cm}^3$$

$$= (5.46 \times 10^{22})(9.27 \times 10^{-21} \text{ A} \cdot \text{m}^2/\text{magneton})$$

$$= 0.51 \text{ A} \cdot \text{m}^2/\text{cm}^3 = 5.1 \times 10^5 \text{ A/m}$$

This represents the magnetization M at saturation.

Soft, electrical magnets are obtained when the Fe^{2+} ion is replaced by various mixtures of manganese, zinc, nickel, and copper. The nickel and manganese ions have magnetic moments that partly cancel the effect of the two iron ions, but a net ferrimagnetic behavior, with a small hysteresis loop, is obtained. Common ferrites include $NiFe_2O_4$, $MnFe_2O_4$, and $(Zn,Mn)Fe_2O_4$.

Computer ferrites are made by additions of manganese and magnesium or cobalt to produce a square loop hysteresis behavior.

Hard ceramic magnets normally are not obtained from the ferrites; other crystal structures typically give a stronger ferrimagnetic effect. Typical ceramics for permanent magnets include $BaO \cdot 6Fe_2O_3$, $PbO \cdot 6Fe_2O_3$, and $CoO \cdot Fe_2O_3$.

18-17 Eddy Current Losses

The ceramic materials have a great advantage over metallic magnets for electrical applications. Because of the high electrical conductivity of the metallic magnets, *eddy currents* are produced, causing power losses and heating. The ceramic materials, however, have a high electrical resistivity and their use in electrical devices keeps the eddy current losses at a minimum.

In metallic magnetic materials, such as silicon iron, composite material techniques have been used to solve the problem of eddy current losses. Thin sheets of silicon iron are laminated with a layer of an insulating material. The laminated layers are then built up to the desired overall thickness. The laminant increases the resistivity of the composite magnets and reduces the eddy current losses. Because eddy current losses are also reduced by using thin sheet, the laminated materials are again beneficial. The laminates are successful at low and intermediate frequencies.

At very high frequencies, eddy current losses are even more significant because the domains are unable to realign. In this case, a composite material containing domain-sized magnetic particles in a polymer matrix may be used. The particles, or domains, rotate easily in the soft plastic and eddy current losses are again minimized.

SUMMARY

Dielectric and magnetic properties are related to the ability of a material to store an electric charge or to retain magnetization. In either case, the behavior of the material is enhanced when it contains permanent dipoles that can be controlled. The permanent dipoles can be influenced by temperature, microstructure, and processing to produce a wide range of beneficial properties.

GLOSSARY

Antiferromagnetism Opposition of adjacent magnetic dipoles, causing zero net magnetization.

Bloch walls The boundaries between magnetic domains.

Bohr magneton The strength of a magnetic moment.

Capacitor An electrical device constructed from alternating layers of a dielectric and conductor, which is capable of storing a charge.

Coercive field The strength of the electric or magnetic field required to eliminate polarization or magnetization from a material.

Curie temperature The temperature above which ferroelectric or ferromagnetic behavior is lost.

Diamagnetism The effect caused by the magnetic moment due to the orbiting electrons, which produces a slight opposition to the imposed magnetic field.

Dielectric constant The ratio of the permittivity of a material to the permittivity of a vacuum, thus describing the relative ability of a material to polarize and store a charge.

Dielectric loss The fraction of energy lost each time an electric field in a material is reversed.

Dielectric strength The maximum electric field that can be maintained between two conductor plates.

Dipole friction Loss in energy when a cyclic electric field is applied to a dielectric material due to the inability of the dipoles to realign rapidly.

Dipoles Atoms or groups of atoms having an unbalanced charge or moment.

Domains Small regions within a material in which all of the dipoles are aligned.

Eddy currents Currents induced into a material by the imposition of an electric field.

Electronic polarization Polarization of an atom when the electrons are displaced to one side of the atom.

Electrostriction The dimensional change that occurs in a material when an electric field is acting on it.

Ferrimagnetism Magnetic behavior obtained when two types of dipoles, having different strengths, oppose one another but a net magnetization remains.

Ferroelectricity Alignment of domains so that a net polarization remains after the electric field is removed.

Ferromagnetism Alignment of domains so that a net magnetization remains after the magnetic field is removed.

Hard magnet Ferromagnetic material that has a large hysteresis loop and remanence.

Hysteresis loop The loop traced out by the polarization or magnetization as the electric or magnetic field is cycled.

Inductance The flux density produced by a magnetic field.

Ionic polarization Polarization of an ionic material by the relative displacement of the anions and cations.

Magnetic moment The strength of the magnetic field associated with an electron.

Magnetic permeability The ratio between the magnetic field and the inductance or magnetization.

Magnetic susceptibility The ratio between magnetization and the applied field.

Magnetization The sum of all of the magnetic moments per unit volume.

Molecular polarization Polarization caused by the asymmetrical nature of certain molecules or crystal structures.

Paramagnetism The net magnetic moment caused by alignment of the electron spins when a magnetic field is applied.

Permittivity The ability of the material to polarize and store a charge within the material.

Piezoelectricity The ability in some materials for a change in electric field to change the dimensions of the material, while a change in dimensions produces an electric field.

Polarization Alignment of dipoles so that a charge can be permanently stored.

Power The strength of a permanent magnet as expressed by the maximum product of the inductance and magnetic field.

Remanence The polarization or magnetization that remains in a material after it has been removed from the field due to permanent alignment of the dipoles.

Saturation　When all of the dipoles have been aligned by the field, producing the maximum polarization or magnetization.

Soft magnet　Ferromagnetic material that has a small hysteresis loop and little energy loss in an alternating field.

Space charge　An electrical charge, which can move in the presence of an electric field, that develops on surfaces or at interfaces within a material, thus contributing to polarization.

PRACTICE PROBLEMS

1　Calculate the polarization that occurs when the electrons in BCC iron are displaced 1×10^{-9} Å by an electric field.

2　Calculate the polarization that occurs when the electrons in FCC lead are displaced by 9×10^{-10} Å by an electric field.

3　Calculate the displacement of the electrons in a silicon crystal when an electric field causes a polarization of 8×10^{-8} C/m^2.

4　Calculate the displacement of the electrons in aluminum when an electric field causes a polarization of 5×10^{-8} C/m^2.

5　Calculate the ionic polarization expected for FeO if an electric field causes a displacement of 5×10^{-8} Å between the ions. (FeO has the sodium chloride structure with a lattice parameter of 4.12 Å; the valence of the iron ions is 2.)

6　Calculate the ionic polarization expected for CsCl if an electric field causes a displacement of 9×10^{-8} Å between the ions. The lattice parameter is 4.018 Å.

7　Calculate the displacement between the ions expected in MgO if a polarization of 5×10^{-8} C/m^2 is obtained. The lattice parameter is 3.96 Å.

8　Suppose magnesium oxide has a dielectric constant of 4.2. Determine the displacement between the magnesium and oxygen ions when 5000 V is applied across a 1 mm thick magnesium oxide crystal.

9　The dielectric constant measured for a sodium chloride crystal, which has a lattice parameter of 5.5 Å, is 6.12. Calculate the strength of the electric field that causes a displacement of 8×10^{-9} Å between the ions.

10　Calcium fluoride has the fluorite crystal structure (see Chapter 14) with a lattice parameter of 5.43 Å and a dielectric constant of 7.36. If an electric field of 800 V/m is applied, calculate the displacement between the ions.

11　(a) Determine the polarization obtained in fused silica when an electric field of 1000 V/m is applied. (b) How thick must the fused silica be to prevent breakdown when a voltage of 10,000 V is used?

12　A parallel plate capacitor containing a 0.001-cm film of mica must not break down. What is the maximum voltage that can be applied?

13　Calculate the capacitance of a parallel plate capacitor containing 10 layers of teflon 1 cm^2 and 0.001 cm thick.

14　A barium titanate capacitor 2 cm^2 and 0.001 cm thick is to have a capacitance of 1.06 μF. (a) How many plates of barium titanate are required? (b) What is the maximum voltage that can be applied? (c) How many layers would be required if the dielectric material is polyethylene rather than barium titanate?

15　Calculate the capacitance of a parallel plate capacitor containing 14 layers of mica 3 cm^2 and 0.002 cm thick.

16　A parallel plate capacitor is produced by sandwiching layers of 1 cm × 1 cm × 0.002 cm polystyrene sheet between conductors. (a) How many layers of polystyrene and how many conductor plates are required to produce a capacitance of 0.22 μF? (b) If a voltage of 20 V is applied across the entire capacitor, calculate the charge that can potentially be stored. (c) Calculate the electrical field in V/mm across each of the sheets of polystyrene, assuming an equal voltage drop across each sheet. Will the capacitor break down?

17　Compare the electric fields produced when a 100-kg force is applied to a 0.5 cm × 0.5 cm piezoelectric crystal of (a) barium titanate and (b) quartz. (The modulus of elasticity of barium titanate is 68.96 GPa and of quartz is 71.72 GPa.)

18　A strain of 1×10^{-5} cm/cm is produced in piezoelectric quartz having a cross-sectional area of 0.5 cm^2, producing a voltage of 100 V. Calculate (a) the thickness of the quartz crystal and (b) the force required to produce the voltage.

19　Calculate the stress and strain produced in a 1 cm × 1 cm × 0.002 cm barium titanate crystal when 75 V are applied across the device.

20　A voltage of 10 V must be applied across a crystal of ferroelectric A in Figure 18-25 to reduce polarization to zero. Calculate the thickness of the crystal.

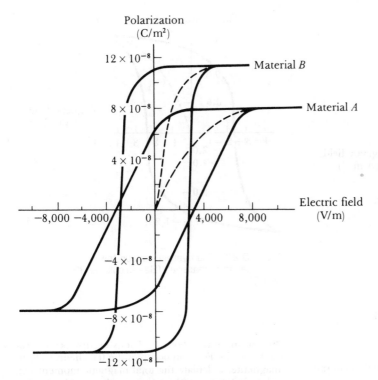

FIG. 18-25 The hysteresis loop for two ferroelectric materials. (See Problems 20 to 24.)

21 What voltage must be applied across a 0.1-mm crystal of ferroelectric B in Figure 18-25 to eliminate polarization and randomize the dipoles?

22 Ferroelectric A in Figure 18-25, which originally is not polarized, is placed in an electric field which causes polarization of 6×10^{-8} C/m^2. Determine the field required and calculate the dielectric constant at that point.

23 An electric field of 2000 V/m is applied to ferroelectric B (Figure 18-25), producing polarization. Determine the polarization and calculate the dielectric constant at that point.

24 Plot a graph showing how the dielectric constant of originally unpolarized ferroelectric A in Figure 18-25 changes as the field increases.

25 Calculate and compare the maximum magnetization we would expect in iron, nickel, and cobalt.

26 Calculate the maximum magnetization we would expect in gadolinium. What is the number of Bohr magnetons for each atom? There are seven electrons in the 4f level of gadolinium.

27 Suppose a BCC Fe-20 wt% Co alloy, which has a

lattice parameter of 2.88 Å, is prepared. Calculate (a) the atomic % Co and (b) the maximum magnetization of the alloy, assuming no interactions between the cobalt and iron atoms.

28 Suppose a BCC iron-molybdenum alloy has a maximum magnetization of 2.86×10^6 A m^{-1}. The lattice parameter of the solid solution is 2.91 Å. Calculate the at% and wt% Mo in the alloy, assuming no interactions between the molybdenum and iron atoms.

29 Compare the magnetization and inductance obtained when a magnetic field of 159.16 A m^{-1} is applied to the (a) 4S Permalloy and (b) Supermalloy materials.

30 Determine the magnetization and inductance obtained when a field of 63.66 A m^{-1} is applied to (a) an oriented Fe-3% Si alloy and (b) 99.95% pure iron.

31 A ferromagnetic material has a saturation magnetization of 1.5 tesla and a relative permeability of 8000. What magnetic field must be applied to give saturation?

32 Determine the power of the magnetic material shown in Figure 18-26.

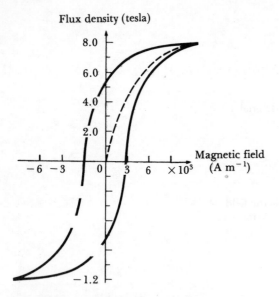

FIG. 18-26   The hysteresis loop for a ferromagnetic material. (See Problems 32 and 34.)

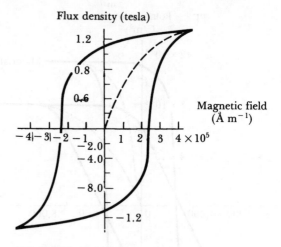

FIG. 18-27   The hysteresis loop for a ferromagnetic material. (See Problems 33 and 35.)

33 Determine the power of the magnetic material shown in Figure 18-27.

34 In Figure 18-26, a magnetic field of 40 oersted is applied to an originally nonmagnetized material. Determine (a) the magnetization produced and (b) the relative permeability at that field.

35 In Figure 18-27, a magnetic field produces a magnetization of 1.0 tesla in an originally nonmagnetized material. Determine (a) the field required to produce this magnetization and (b) the relative permeability at that field.

36 Suppose that both Fe^{3+} ions occupy octahedral sites and the Fe^{2+} ion occupies the tetrahedral site in magnetite. Calculate the total magnetic moment per cubic centimeter and compare to the actual value for normal magnetite.

37 Suppose that the Fe^{3+} ions in the octahedral sites of magnetite are replaced by neutral ions. Calculate the expected total magnetization per cubic centimeter.

38 Calculate the magnetic moment per cubic centimeter you would expect if the Fe^{2+} ion in the octahedral site of magnetite is replaced by (a) Co^{2+} and (b) Ni^{2+}.

CHAPTER **19**

Optical, Thermal, and Elastic Properties

19-1 Introduction

Optical, thermal, and elastic properties are related to the interaction of a material with radiation in the form of waves or particles of energy. The frequency, wavelength, and energy of the radiation are determined by the source. For example, gamma rays are produced by changes in the structure of the nucleus of the atom; X rays, ultraviolet radiation, and the visible spectrum are produced by changes in the electronic structure of the atom. Infrared radiation, microwaves, and radio waves are low-energy, long-wavelength radiation caused by vibration of the atoms or crystal structure. When radiation interacts with a material, a variety of effects are produced, including absorption, colors, fluorescence, conduction of heat, and elastic behavior.

OPTICAL PROPERTIES

19-2 Emission of Continuous and Characteristic Radiation

Energy, or radiation in the form of waves or particles called *photons*, can be emitted from a material. The important characteristics of the photons—their energy E, wavelength λ, and frequency ν—are related by the equation

$$E = h\nu = \frac{hc}{\lambda} \tag{19-1}$$

where c is the speed of light ($3 \times 10^8 \, \text{m} \cdot \text{s}^{-1}$) and h is Planck's constant ($6.62 \times 10^{-34} \, \text{J} \cdot \text{s}$). This equation permits us to either consider the photon as a particle of energy E or as a wave with a characteristic wavelength and frequency.

Continuous spectrum. A stimulus such as an accelerated electron is decelerated when it strikes a material. As the electron decelerates, energy is given up and emitted as photons. Each time the electron strikes an atom, more of its energy is given up. Each interaction, however, may be more or less severe, so the electron gives up a different fraction of its energy each time, producing photons of different wavelengths (Figure 19-1). A *continuous spectrum,* or *white radiation,* is produced (Figure 19-2).

If the electron were to lose all of its energy in one impact, the minimum wavelength of the emitted photons would be equivalent to the original energy of the stimulus. Thus the continuous spectrum has a *short wavelength limit* λ_{SWL}. When

Incoming electron

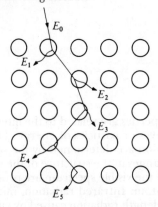

$E_1 + E_2 + E_3 + E_4 + E_5 = E_0$

FIG. 19-1   When an accelerated electron strikes and interacts with a material, its energy may be reduced in a series of steps. In the process, several photons of different energies E_1 to E_5 are emitted, each with a unique wavelength.

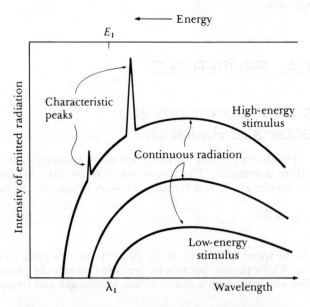

FIG. 19-2   The continuous and characteristic spectra of radiation emitted from a material. Low-energy stimuli produce a continuous spectrum of low-energy, long-wavelength photons. A more intense, higher energy spectrum is emitted when the stimulus is more powerful until eventually characteristic radiation is observed.

the energy of the stimulus increases, the short wavelength limit decreases and the number and energy of the emitted photons increase, giving a more intense continuous spectrum.

EXAMPLE 19-1

A voltage of 10,000 V is applied to a heated tungsten filament, which then emits electrons. Calculate the short wavelength limit that the electron can stimulate if it strikes a target material.

Answer:

First, let's calculate the energy of the emitted electrons.

$$E = (10,000 \text{ V})(1.6 \times 10^{-19} \text{ J} \cdot \text{V}^{-1}) = 1.6 \times 10^{-15} \text{ J}$$

$$hc = (6.62 \times 10^{-34} \text{ J} \cdot \text{s})(3 \times 10^{8} \text{ m} \cdot \text{s}^{-1}) = 19.86 \times 10^{-26} \text{ J} \cdot \text{m}$$

$$\lambda_{SWL} = \frac{hc}{E} = \frac{19.86 \times 10^{-26}}{1.6 \times 10^{-15}} = 1.24 \times 10^{-10} \text{ m} = 1.24 \text{ Å}$$

Characteristic spectrum.   If the incoming stimulus has a sufficient energy, an electron from an inner energy level is excited into an outer energy level. To restore equilibrium, the empty inner level is filled by electrons from a higher level.

There are discrete energy differences between any two energy levels. When an electron drops from one level to a second level, a photon having that particular energy and wavelength is emitted. Photons with this energy and wavelength comprise the *characteristic spectrum* and are X rays. The characteristic spectrum appears as a series of peaks superimposed on the continuous spectrum (Figure 19-2).

We can show this effect in Figure 19-3. If an electron is excited from the K

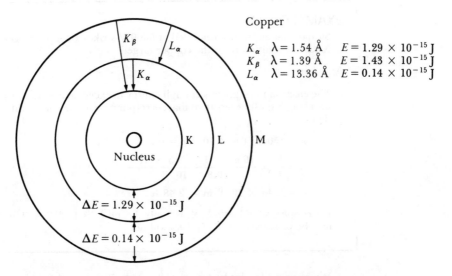

FIG. 19-3 Characteristic X rays are produced when electrons change from one energy level to a lower energy level. The energy and wavelength of the X rays are fixed by the energy differences between the energy levels.

shell, electrons may fill that vacancy from any outer shell. Normally, electrons in the next closest shell fill the vacancies. Thus, photons with energies $\Delta E = E_K - E_L$ (K_α X rays) or $\Delta E = E_K - E_M$ (K_β X rays) are emitted. If an electron from the L shell fills the K shell, then an electron from the M shell may fill the L shell, giving a photon with energy $\Delta E = E_L - E_M$ (L_α X rays) which has a longer wavelength or lower energy. Note that we need a more energetic stimulus to produce K_α X rays compared to L_α X rays. Examples of a portion of the characteristic spectra for several elements are included in Table 19-1.

TABLE 19-1   Characteristic emission lines and absorption edges for selected elements

Metal	K_α (Å)	K_β (Å)	L_α (Å)	Absorption Edge (Å)
Al	8.337	7.981		7.951
Si	7.125	6.768		6.745
S	5.372	5.032		5.018
Cr	2.291	2.084		2.070
Mn	2.104	1.910		1.896
Fe	1.937	1.757		1.743
Co	1.790	1.621		1.608
Ni	1.660	1.500		1.488
Cu	1.542	1.392	13.357	1.38
Mo	0.711	0.632	5.724	0.620
W	0.211	0.184	1.476	0.178

From B. Cullity, *Elements of X-ray Diffraction, 2nd Ed.*, Addison-Wesley, 1978.

EXAMPLE 19-2

Suppose an electron accelerated at 5000 V strikes a copper target. Will K_α, K_β, or L_α X rays be emitted from the copper target?

Answer:

The electron must possess enough energy to excite an electron to a higher level, or its wavelength must be less than that corresponding to the energy difference between the shells.

$$E = (5000)(1.6 \times 10^{-19}) = 0.8 \times 10^{-15} \, J$$

$$\lambda = \frac{hc}{E} = \frac{(6.62 \times 10^{-34})(3 \times 10^8)}{0.8 \times 10^{-15}}$$

$$= 2.48 \times 10^{-10} \, m = 2.48 \, \overset{\circ}{A}$$

For copper, K_α is 1.542 Å, K_β is 1.392 Å, and L_α is 13.357 Å. Therefore, the L_α peak may be produced but K_α and K_β will not.

We have so far looked primarily at the emission of rather short wavelength radiation or photons that ordinarily are classified as X rays. Depending on the source of the photons, we could have different wavelengths of radiation. The entire spectrum of electromagnetic radiation is shown in Figure 19-4. The continuous spectrum produced when a stimulus strikes a material may contain all radiation having a longer wavelength than that of the stimulus. Thus, a stimulus producing X rays could also produce ultraviolet radiation. However, the longer wavelength radiation is often absorbed by the material.

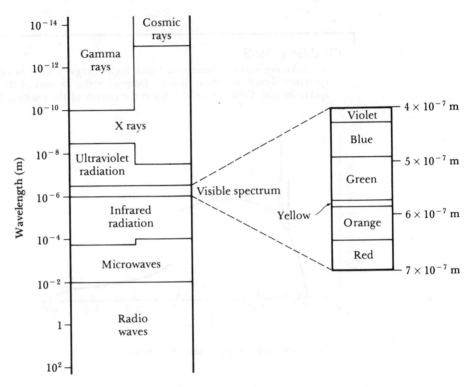

FIG. 19-4   The electromagnetic spectrum of radiation.

19-3   Examples of Emission Phenomena

Let's look at some particular examples of emission phenomena which are of some familiarity and importance.

X rays.   As we have already noted, X rays are produced when the inner shell electrons are stimulated, normally by an incident electron. The most useful of the X rays are the most energetic, shortest wavelength photons produced by filling the K and L levels. These X rays can be used to determine the composition of the material.

If we bombard an unknown material with very high energy photons, the material emits both the characteristic and the continuous spectra. We can then measure the wavelength or energy of the characteristic peaks. If we match the emitted characteristic wavelengths with those expected for various materials, the identity of the material can be determined. We can also measure the intensity of the characteristic peaks. By comparing measured intensities to standard intensities, we can estimate the amount of each emitting atom and determine the composition of the material. We can perform this test on large samples of the material using X-ray fluorescent analysis or on a microscopic scale using the electron microprobe or the scanning electron microscope, permitting us to identify individual phases or even inclusions in the microstructure.

EXAMPLE 19-3

An unknown metal is bombarded with high-energy X rays, producing the emission spectrum shown in Figure 19-5. (a) Determine the identity of the material from the spectrum and Table 19-1. (b) What is the energy of the exciting X rays?

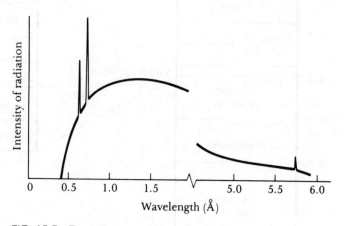

FIG. 19-5   The emission spectrum for Example 19-3.

Answer:

(a) From Figure 19-5, we find three characteristic peaks. The wavelengths are about 0.62 Å, 0.71 Å, and 5.75 Å. From Table 19-1, we find that molybdenum has $K_\beta = 0.632$ Å, $K_\alpha = 0.711$ Å, and $L_\alpha = 5.724$ Å, providing a good match. The material is most likely molybdenum.

(b) The energy of the stimulus corresponds to the energy giving the short wavelength limit of the spectrum. The $\lambda_{SWL} = 0.4$ Å.

$$E = \frac{hc}{\lambda_{SWL}} = \frac{(6.62 \times 10^{-34})(3 \times 10^8)}{0.4 \times 10^{-10}}$$

$$= 4.965 \times 10^{-15} \, J$$

The X rays are produced by a voltage of

$$E = \frac{4.965 \times 10^{-15}}{1.6 \times 10^{-19}} = 31,000 \, V$$

EXAMPLE 19-4

From the energy spectra of the phases in Figure 19-6, determine the probable identity of each phase.

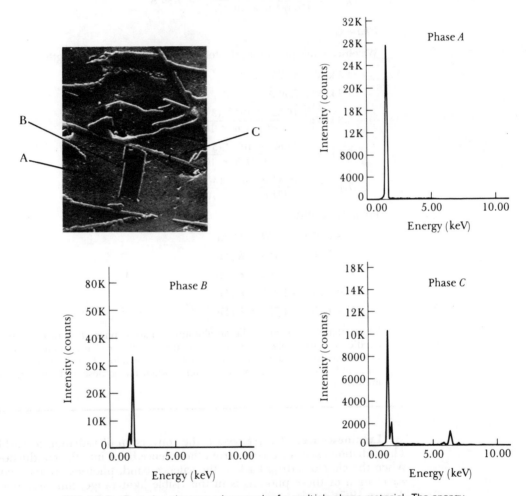

FIG. 19-6 Scanning electron micrograph of a multiple-phase material. The energy distribution of the emitted radiation from the three phases marked A, B, and C are shown. The identity of each phase is determined in Example 19-4.

Answer:

Phase A

The energy of the single peak is about 1.5 keV = 1500 eV

$$\lambda = \frac{(6.62 \times 10^{-34} \text{ J} \cdot \text{s})(3 \times 10^{8} \text{ m} \cdot \text{s}^{-1})}{(1500 \text{ eV})(1.6 \times 10^{-19} \text{ J} \cdot \text{eV}^{-1})} = 8.275 \text{ Å}$$

Phase B

The energies of the two peaks are 1.5 keV = 1500 eV and 1.7 keV = 1700 eV. The wavelength of the second, large peak is

$$\lambda = \frac{(6.62 \times 10^{-34})(3 \times 10^8)}{(1700)(1.6 \times 10^{-19})} = 7.30 \text{ Å}$$

Phase C

We observe five peaks with the following energies and wavelengths.

1.5 keV: 8.275 Å

1.7 keV: 7.30 Å

5.8 keV: $\dfrac{(6.62 \times 10^{-34})(3 \times 10^8)}{(5800)(1.6 \times 10^{-19})} = 2.14$ Å

6.4 keV: $\dfrac{(6.62 \times 10^{-34})(3 \times 10^8)}{(6400)(1.6 \times 10^{-19})} = 1.94$ Å

7.1 keV: $\dfrac{(6.62 \times 10^{-34})(3 \times 10^8)}{(7100)(1.6 \times 10^{-19})} = 1.75$ Å

From Table 19-1

$\lambda = 8.275 \simeq 8.337 = K_\alpha$ Al

$\lambda = 7.30 \simeq 7.125 = K_\alpha$ Si

$\lambda = 2.14 \simeq 2.104 = K_\alpha$ Mn

$\lambda = 1.94 \simeq 1.937 = K_\alpha$ Fe

$\lambda = 1.75 \simeq 1.757 = K_\beta$ Fe

Thus, phase *A* appears to be an aluminum matrix, phase *B* appears to be a silicon needle, perhaps containing some aluminum, and phase *C* appears to be an Al-Si-Mn-Fe compound. Actually, this is an aluminum-silicon alloy. The stable phases are aluminum and silicon, with inclusions forming when manganese and iron are present as impurities.

Luminescence.   *Luminescence* is the conversion of radiation to visible light. The radiation excites electrons from the valence band into the conduction band. When the electrons drop back to the valence band, photons are emitted. If the wavelength of these photons is in the visible light range, luminescence occurs (Figure 19-7).

In metals, the energy of the emitted photon is very small, since the valence and conduction bands overlap, and the wavelength is longer than the visible light spectrum. Therefore, luminescence does not occur. However, in certain ceramic materials, the energy gap between the valence and conduction band is such that an electron dropping through this gap produces a photon in the visible range. One wavelength, corresponding to the energy gap E_g, predominates.

Two different effects may be observed in luminescent materials—fluorescence and phosphorescence. In *fluorescence*, luminescence ceases when the stimulus is removed. All excited electrons drop back to the valence band and the photons are emitted within about 10^{-8} s.

However, *phosphorescent* materials have impurities which introduce a donor

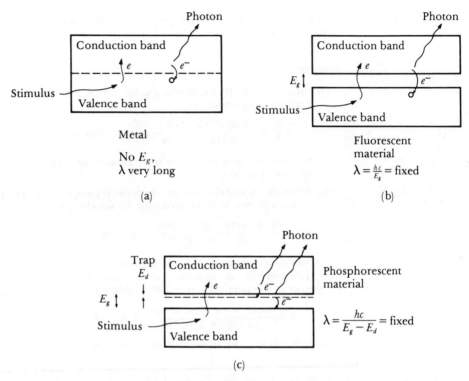

FIG. 19-7 Luminescence occurs when photons have a wavelength in the visible spectrum. (a) In metals, there is no energy gap so luminescence does not occur. (b) Fluorescence occurs when there is an energy gap. (c) Phosphorescence occurs when the photons are emitted over a period of time due to donor traps in the energy gap.

level within the energy gap. The stimulated electrons first drop into the donor level and are trapped. The electrons must then escape the trap before returning to the valence band. There is a delay of more than 10^{-8} s before the photons are emitted. When the source is removed, electrons in the traps gradually escape and emit light over some additional period of time. The intensity of the luminescence is given by

$$\ln\left(\frac{I}{I_0}\right) = -\frac{t}{\tau} \tag{19-2}$$

where τ is the *relaxation time,* a constant for the material. After time t following removal of the source, the intensity of the luminescence is reduced from I_0 to I. Phosphorescent materials are very important in the operation of television screens. In this case, the relaxation time must not be too long or the images begin to overlap.

EXAMPLE 19-5

The wavelength of the luminescent radiation is determined by impurities in the material, which in turn determine the energy gap. (a) Determine the wavelength of a photon required to excite an electron in ZnS. (b) An impurity in ZnS gives an energy trap 1.38 eV below the conduction band. Calculate the wavelength and determine the type of radiation produced during luminescence.

Answer:

(a) From Table 17-10, the energy gap for pure ZnS is 3.54 eV. A photon must have a maximum wavelength of

$$\lambda = \frac{hc}{E} = \frac{(6.62 \times 10^{-34})(3 \times 10^8)}{(3.54)(1.6 \times 10^{-19})} = 3506 \text{ Å}$$

to excite an electron into the conduction band. This wavelength corresponds to ultraviolet radiation.

(b) When the excited electron returns to the valence band, it first enters the trap. This causes a photon to be emitted with the wavelength

$$\lambda = \frac{(6.62 \times 10^{-34})(3 \times 10^8)}{(1.38)(1.6 \times 10^{-19})} = 8995 \text{ Å}$$

This wavelength, which is in the infrared spectrum, is not visible.

However, the electron then drops from the trap to the valence band. The energy of the photon, $3.54 - 1.38 = 2.16$ eV, corresponds to a wavelength of

$$\lambda = \frac{(6.62 \times 10^{-34})(3 \times 10^8)}{(2.16)(1.6 \times 10^{-19})} = 5747 \text{ Å}$$

This photon is in the visible spectrum and gives yellow light.

EXAMPLE 19-6

In a television screen, the intensity of the phosphorescent material decreases by a factor of 10 within 10^{-3} s. What is the relaxation time?

Answer:

If $I_0 = 10$, then $I = 1$.

$$\ln \left(\frac{I}{I_0} \right) = \ln \frac{1}{10} = \frac{-t}{\tau} = \frac{-10^{-3}}{\tau}$$

$$-2.3 = \frac{-10^{-3}}{\tau}$$

$$\tau = 4.3 \times 10^{-4} \text{ s}$$

Lasers.   In certain materials, electrons excited by the stimulus produce photons which in turn excite additional photons of identical wavelength (Figure 19-8). Consequently, a large amplification of the photons emitted in the material occurs. By proper choice of stimulant and material, the wavelength of the photons can be in the visible range. The output of the laser is a beam of photons that are parallel, of the same wavelength, and coherent. In a *coherent* beam, the wavelike nature of the photons is in phase so that destructive interference does not occur. Lasers are

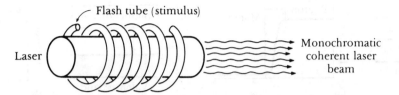

FIG. 19-8 The laser converts a stimulus into a beam of coherent photons.

useful in heat treating and melting of metals, welding, surgery, mapping, and a variety of other applications.

Thermal emission.   When a material is heated, electrons are thermally excited to higher energy levels, particularly in the outer energy levels where the electrons are less tightly bound to the nucleus. The electrons immediately drop back to their normal levels and release photons with low energy and long wavelength.

As the temperature increases, the thermal agitation increases, and the maximum energy of the emitted photons increases. A continuous spectrum of radiation is emitted, with a minimum wavelength and an intensity distribution dependent on the temperature. The photons may include wavelengths in the visible spectrum; consequently the color of the material changes with temperature. At low temperatures, the wavelength of the radiation is too long to be visible. As the temperature increases, emitted photons have shorter wavelengths. At high temperatures, all wavelengths in the visible range are produced and the emitted spectrum is white light. We can, by measuring the intensity of a narrow band of the emitted wavelengths with a pyrometer, estimate the temperature of the material.

19-4 Interaction of Photons with a Material

Photons, either characteristic or continuous, cause a number of optical phenomena when they interact with the electronic or crystal structure of a material.

Absorption.   If the incoming photons interact with the material, they give up their energy and are absorbed. Most metals absorb photons with wavelengths shorter than infrared radiation, including visible light. Because the valence band is not filled in metals, almost any photon has a sufficient energy to excite an electron to a higher energy level in the conduction band [Figure 19-9(a)]. Therefore metals, unless they are exceptionally thin, absorb visible light and are opaque.

Absorption is given by

$$\ln \left(\frac{I}{I_0} \right) = -\mu x \tag{19-3}$$

where x is the path through which the photons move and μ is the *linear absorption coefficient* of the material for the photons. The absorption coefficient is related to the density of the material, the wavelength of the radiation, and the energy gap between the conduction and valence bands.

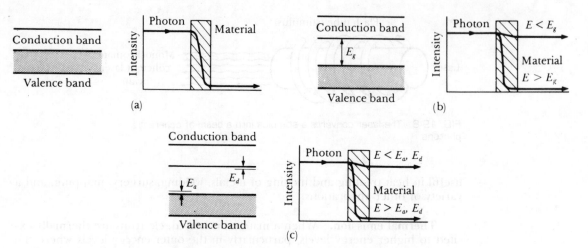

FIG. 19-9 Relationship between absorption and the energy gap. (a) Metals, (b) insulators and intrinsic semiconductors, and (c) extrinsic semiconductors.

EXAMPLE 19-7

Suppose an X-ray source is shielded from the surroundings by aluminum foil. If 95% of the energy of the X-ray beam is to penetrate the foil, determine the maximum thickness of the aluminum. The linear absorption coefficient is 42 m^{-1}.

Answer:

The final intensity of the beam must be at least $0.95I_0$. Thus

$$\ln \frac{(0.95I_0)}{I_0} = -(42)(x)$$

$$\ln (0.95) = -0.051 = -42x$$

$$x = \frac{-0.051}{-42} = 1.2 \times 10^{-3} \text{ m} = 1.2 \text{ mm}$$

Transmission. If the photons do not possess enough energy to excite an electron to a higher energy level, the photons may be transmitted rather than absorbed and the material is transparent [Figure 19-9(b)]. Whether a photon is absorbed or transmitted depends on the relationship between the energy of the photon and the energy gap between the conduction and valence bands. In metals, there is no gap and almost all photons are absorbed unless the metal is exceptionally thin. In simple, high-purity ceramics and polymers, the energy gap is large and the material is transparent to even high-energy photons, including visible light. In semiconductors, however, electrons can be excited into the acceptor levels or out of the donor levels. The photons are absorbed unless their energies are smaller than the donor or acceptor energy gaps [Figure 19-9(c)]. Semiconductors are therefore opaque to short wavelength radiation but transparent to very long wavelength photons.

The degree of transmission is also related to the overall atomic arrangement. Amorphous materials, such as glass and simple polymers, may be transparent. However, when the material crystallizes, photons can interact with the crystal structure and are at least partially absorbed.

EXAMPLE 19-8

Determine the shortest wavelength you would expect to be transmitted in the following materials: intrinsic silicon, phosphorus-doped silicon, diamond, tin.

Answer:

From the energy gaps in Tables 17-8 and 17-9

Intrinsic silicon: $E_g = 1.107$ eV

$$\lambda = \frac{(6.62 \times 10^{-34})(3 \times 10^8)}{(1.107)(1.6 \times 10^{-19})} = 11,200 \text{ Å}$$

Phosphorus-doped silicon: $E_d = 0.045$ eV

$$\lambda = \frac{(6.62 \times 10^{-34})(3 \times 10^8)}{(0.045)(1.6 \times 10^{-19})} = 276,000 \text{ Å}$$

Diamond: $E_g = 5.4$ eV

$$\lambda = \frac{(6.62 \times 10^{-34})(3 \times 10^8)}{(5.4)(1.6 \times 10^{-19})} = 2,300 \text{ Å}$$

Tin: $E_g = 0.08$ eV

$$\lambda = \frac{(6.62 \times 10^{-19})(3 \times 10^8)}{(0.08)(1.6 \times 10^{-19})} = 155,000 \text{ Å}$$

Of these four materials, only diamond transmits visible light.

Refraction. Even when photons are transmitted by the material, the photon loses some of its energy and therefore has a slightly longer wavelength. The photon then behaves as though the speed of light in the material has been reduced and the beam of photons changes directions (Figure 19-10). If α and β are respectively the angles that the incident and diffracted beams make with the normal to the surface of the material, then

$$n = \frac{c}{v} = \frac{\lambda_{\text{vacuum}}}{\lambda} = \frac{\sin \alpha}{\sin \beta} \tag{19-4}$$

The ratio n is the *index of refraction, c* is the speed of light in a vacuum and v is the speed of light in the material.

The photons interact with, and are refracted more, when the electrons in the material are more easily polarized. Consequently, we expect to find a relationship between the index of refraction and the dielectric constant of the material.

$$n = \sqrt{\kappa} \tag{19-5}$$

Materials that polarize easily, such as dielectric materials, have a high index of

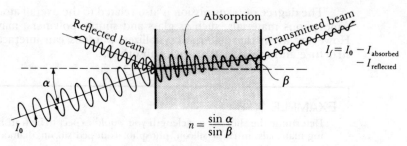

FIG. 19-10   A beam of photons changes direction as a result of interactions as it passes through a material. The change in direction is given by the index of refraction n.

refraction. The index of refraction is also larger for denser materials. Examples of the index of refraction for several materials are given in Table 19-2.

TABLE 19-2   Index of refraction of selected materials for photons of wavelength 5890 Å

Material	Index of Refraction
Ice	1.309
NaCl	1.544
Quartz	1.544
Diamond	2.417
TiO_2	2.7
Water	1.333
Plastics	1.5

From *Handbook of Chemistry and Physics*, *56th Ed.*, CRC Press, 1975.

EXAMPLE 19-9

Suppose a beam of photons in a vacuum strikes a sheet of polyethylene at an angle of 10° to the normal to the surface of the polymer. Calculate the index of refraction of polyethylene and find the angle between the incident beam and the beam as it passes through the polymer.

Answer:

The index of refraction is related to the dielectric constant. From Table 18-1, $\kappa = 2.3$.

$$n = \sqrt{\kappa} = \sqrt{2.3} = 1.52$$

The angle β is

$$n = \frac{\sin \alpha}{\sin \beta}$$

$$\sin \beta = \frac{\sin \alpha}{n} = \frac{\sin 10°}{1.52} = \frac{0.174}{1.52} = 0.114$$

$$\beta = 6.55°$$

Reflection. If the surface of the material is smooth and if the incoming photons have a low energy, a portion of the incoming photons are reflected from the material's surface. The *reflectivity* R is related to the index of refraction. In a vacuum,

$$R = \text{percent reflection} = \left(\frac{n - 1}{n + 1}\right)^2 \times 100 \qquad (19\text{-}6)$$

Materials with a high index of refraction have a higher reflectivity than materials with a low index—light is transmitted rather than reflected in these materials.

EXAMPLE 19-10

What fraction of the photons striking polyethylene are reflected?

Answer:

The index of refraction for polyethylene, as calculated in Example 19-9, is 1.52. Thus, the reflectivity is

$$R = \left(\frac{n - 1}{n + 1}\right)^2 \times 100 = \left(\frac{1.52 - 1}{1.52 + 1}\right)^2 \times 100 = \left(\frac{0.52}{2.52}\right)^2 \times 100$$

$$= 4.3\%$$

EXAMPLE 19-11

A beam of photons passes through a 20 mm thick glass, with 60% of the original beam intensity being transmitted. If the dielectric constant of the glass is 3.2, calculate the linear absorption coefficient of the photons in the glass.

Answer:

A portion of the beam is reflected. The index of refraction is

$$n = \sqrt{\kappa} = \sqrt{3.2} = 1.79$$

$$R = \left(\frac{n - 1}{n + 1}\right)^2 \times 100 = \left(\frac{1.79 - 1}{1.79 + 1}\right)^2 \times 100 = 28.3\%$$

Thus, the intensity of the beam entering the glass is $100 - 28.3 = 71.7\%$. We will let $I_0 = 71.7\%$. The final beam has an intensity of 60%. The rest of the beam is absorbed.

$$\ln \left(\frac{I}{I_0} \right) = -\mu x$$

$$\ln \left(\frac{60}{71.7} \right) = -20\mu \quad \ln (0.837) = -20\mu$$

$$\mu = \frac{0.178}{20} = 8.9 \text{ m}^{-1}$$

Selective absorption, transmission, or reflection. Unusual optical behavior is observed when photons are selectively absorbed, transmitted, or reflected. We have already found that semiconductors transmit long-wavelength photons but absorb short-wavelength radiation. There are a variety of other cases where similar selectivity produces unusual optical properties.

1. Color: Copper and gold reflect only a certain range of wavelengths and absorb the remaining photons. Copper absorbs the shorter wavelength photons on the blue or violet end of the visible spectrum but reflects photons with longer wavelengths at the red end of the spectrum. Since we see only the reflected light, copper appears red.

2. Color center: Impurities, which create donor or acceptor energy levels in certain ceramics and polymers, cause absorption of particular wavelength photons. A photon of the correct wavelength causes an electron to enter a higher energy level and the photon is absorbed. The remainder of the light is transmitted. This selective absorption and transmission of visible light photons is responsible for the rich and varied color of natural and synthetic minerals, including glass. Coloring agents in ceramics often are transition or rare earth ions, which have $3d$ or $4f$ energy levels that are not completely filled. The photons can easily react with these ions and be absorbed.

EXAMPLE 19-12

Polychromatic glass, used for sunglasses, contains silver atoms. The glass darkens in sunlight but becomes transparent in darkness. Explain this phenomenon.

Answer:

In bright light, the silver ions in the glass gain an electron by excitation by the photons and are reduced from Ag^+ to metallic silver atoms. Thus, absorption of photons occurs. When the incoming light diminishes in intensity, the silver reverses to silver ions and no absorption occurs.

---

3. X rays: A material may selectively absorb X rays that interact with the inner orbital electrons (Figure 19-11). In X-ray diffraction, filters are used to absorb almost all of the continuous spectrum, permitting only a single desired

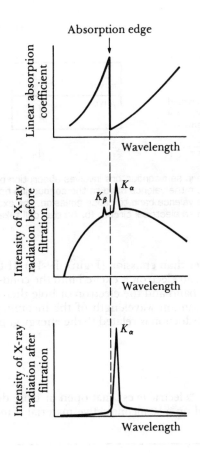

FIG. 19-11 Elements have a selective lack of absorption of certain wavelengths. If a filter is selected with an absorption edge between the K_α and K_β peaks of an X-ray spectrum, all X rays except K_α are absorbed.

characteristic wavelength to pass. At wavelengths just greater than the *absorption edge*, practically no absorption occurs in the filter. This characteristic wavelength is then used in X-ray diffraction experiments. Note that the K_β peak is eliminated along with most of the continuous spectrum if the filter has an absorption edge lying between K_α and K_β. Absorption edges for several elements are included in Table 19-1.

EXAMPLE 19-13

What filter material should you use to isolate the K_α peak of nickel?

Answer:

From Table 19-1, $K_\alpha = 1.660$ Å and $K_\beta = 1.500$ Å for nickel. We need a filter with an absorption edge between these characteristic peaks. Cobalt, with an absorption edge of 1.608 Å, will work.

---

4. Photoconduction: *Photoconduction* occurs in semiconducting materials if the semiconductor is part of an electric circuit. In this case, the stimulated elec-

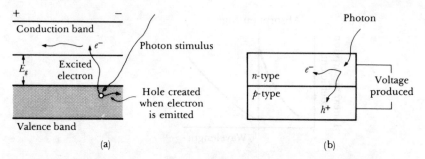

FIG. 19-12   (a) Photoconduction in semiconductors involves absorption of a stimulus by exciting electrons from the valence band to the conduction band. Rather than dropping back to the valence band to cause emission, the excited electrons carry a charge through an electrical circuit. (b) A solar cell takes advantage of this effect.

trons produce a current rather than emission (Figure 19-12). If the energy of an incoming photon is sufficient, an electron is excited into the conduction band or a hole is created in the valence band and the electron or hole then carries a charge through the circuit. The maximum wavelength of the incoming photon that is required to produce photoconduction is related to the energy gap in the semiconductive material.

$$\lambda_{max} = \frac{hc}{E_g} \qquad\qquad (19\text{-}7)$$

We can use this principle for "electric eyes" that open or close doors or switches when a beam of light focused on a semiconductive material is interrupted.

EXAMPLE 19-14

What is the maximum wavelength of light energy that should be used in a photoconductive electric eye circuit if intrinsic silicon is used as the semiconductor at room temperature?

Answer:

Photons must have an energy greater than the energy gap in silicon, 1.107 eV.

$$E = 1.107 = \frac{hc}{\lambda} = \frac{(6.62 \times 10^{-34} \, J \cdot s)(3 \times 10^8 \, m \cdot s^{-1})}{\lambda}$$

$$\lambda = \frac{(6.62 \times 10^{-34} \, J \cdot s)(3 \times 10^8 \, m \cdot s^{-1})}{(1.107 \, eV)(1.6 \times 10^{-19} \, J \cdot eV^{-1})}$$

$$= 1.12 \times 10^{-6} \, m$$

Solar cells are *p-n* junctions designed so that photons excite electrons into the conduction band. The electrons move to the *n*-side of the junction, while holes move through the valence band to the *p*-side of the junction. This produces a contact voltage due to the charge imbalance. If the junction device is connected to an electric circuit, the junction acts as a battery to power the circuit.

Diffraction. In X-ray diffraction, the wavelike nature of X rays is used to determine information about the crystal structure and atom location within the material. The wavelike X-ray radiation interacts with the electronic dipoles and is scattered in all directions. The radiation scattered from one atom interacts destructively with radiation from other atoms except in certain directions (Figure 19-13). In these directions, the scattered radiation is reinforced rather than destroyed.

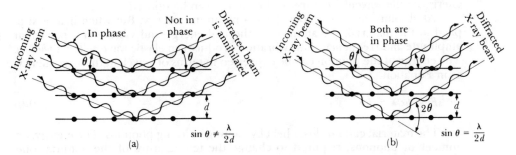

FIG. 19-13 (a) Destructive and (b) reinforcing interactions between X rays and the crystal structure of a material. Reinforcement occurs at angles that satisfy Bragg's law.

The radiation, which is of the identical wavelength as the original X-ray beam, is reinforced in the angles given by Bragg's law.

$$\sin \theta = \frac{\lambda}{2d_{hkl}} \tag{19-8}$$

where the angle θ is half of the angle between the diffracted beam and the original beam direction, λ is the wavelength, and d_{hkl} is the interplanar spacing that causes constructive reinforcement of the beam. As we discussed in Chapter 3, analysis of the angle θ permits us to identify d_{hkl} and eventually the identity, lattice parameter, and other information concerning the crystal.

EXAMPLE 19-15

Determine the angle 2θ between the transmitted and diffracted K_α X-ray beam of nickel when the (111) planes in copper ($a_0 = 3.615$ Å) are responsible for diffraction.

Answer:

From the interplanar spacing equation, Equation (3-6)

$$d_{111} = \frac{a_0}{\sqrt{h^2 + k^2 + l^2}} = \frac{3.615}{\sqrt{1^2 + 1^2 + 1^2}} = 2.087 \text{ Å}$$

From Table 19-1, the wavelength of K_α X rays of nickel is 1.660 Å.

$$\sin \theta = \frac{\lambda}{2d_{111}} = \frac{1.660}{2(2.087)} = 0.398$$

$$\theta = 23.45°$$

$$2\theta = 46.9°$$

THERMAL PROPERTIES

19-5   Heat Capacity

Thermal properties, including heat capacity, thermal conductivity, and thermal expansion, are influenced by atomic vibration and, in the case of thermal conductivity, transfer of energy by electrons. The vibration may be expressed as an energy or the wavelike nature of the energy can be utilized.

At absolute zero, the atoms have a minimum energy. But when heat is supplied to the material, the atoms gain thermal energy and vibrate at a particular amplitude and frequency. This vibration produces an elastic wave called a *phonon*. The energy of the phonon can be expressed in terms of wavelength or frequency, as in Equation (19-1),

$$E = hc/\lambda = h\nu \tag{19-9}$$

The material gains or loses heat by gaining or losing phonons. The energy, or number of phonons, required to change the temperature of the material one degree is of interest. We can express this energy as the heat capacity or the specific heat.

The *heat capacity* is the energy required to raise the temperature of one mole of a material one degree. The heat capacity can be expressed either at constant pressure, C_p, or at a constant volume, C_v. At high temperatures, the heat capacity for a given volume of material approaches.

$$C_p = 3R = 25 \text{ J (mol} \cdot \text{K)}^{-1} \tag{19-10}$$

where R is the gas constant $[8.314 \text{ J (mol} \cdot \text{K)}^{-1}]$. However, the heat capacity is not a constant, as shown in Figure 19-14.

The *specific heat* is the energy required to raise the temperature of a particular weight or mass of a material one degree. The relationship between specific heat and heat capacity is

$$\text{Specific heat} = c = \frac{\text{heat capacity}}{\text{atomic weight}} \tag{19-11}$$

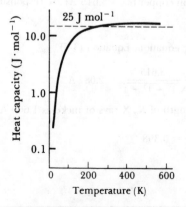

FIG. 19-14   The heat capacity as a function of temperature for metals.

In most engineering calculations, the specific heat is more conveniently used. The specific heat of typical materials is given in Table 19-3.

TABLE 19-3   The specific heat of selected materials at 27°C

Material	Specific Heat $kJ (kg \cdot K)^{-1}$
Al	0.90
Cu	0.39
B	1.03
Fe	0.44
Pb	0.16
Mg	1.02
Ni	0.44
Si	0.70
Ti	0.52
W	0.13
Zn	0.39
Water	4.19
He	5.20
N	1.04
Polymers	0.84–1.47
Diamond	0.52

EXAMPLE 19-16

Calculate the temperature after 20.95 kJ are introduced to 250 grams of tungsten originally at 25°C.

Answer:

The specific heat of tungsten is 0.13 kJ $(kg \cdot K)^{-1}$

$$\Delta T = \frac{20.95 \text{ kJ}}{(0.25 \text{ kg}) [0.13 \text{ kJ } (kg \cdot K)^{-1}]} = 645 \text{ K}$$

$$T_{final} = 25 + 645 = 670°C$$

EXAMPLE 19-17

Suppose the temperature of 50 g of niobium increases 75°C when heated for a period of time. Estimate the specific heat and determine the heat in calories required.

Answer:

The atomic weight of niobium is 92.91 g/g · mole. We can use Equation 19-11 to estimate the heat required to raise the temperature of one gram one °C.

$$c \approx \frac{25}{92.91} = 0.269 \text{ J} \, (\text{g} \cdot \text{K})^{-1}$$

Thus the total heat required is

$$\text{heat} = [0.269 \text{ J} \, (\text{g} \cdot \text{K})^{-1}] \, (50 \text{ g})(75 \text{ K})$$

$$= 1009 \text{ J}$$

EXAMPLE 19-18

Calculate the expected specific heat of aluminum, copper, iron, and silicon based on the classical heat capacity of $25 \text{ J} \cdot \text{mol}^{-1}$ and compare to the measured values in Table 19-3.

Answer:

The weight in grams of one mole of each metal is the atomic weight.

Metal	Atomic Weight (g/mol)	$\dfrac{25 \text{ J mol}^{-1}}{\text{Atomic Weight}}$ (J g^{-1})	Measured Specific Heat $\text{kJ} \, (\text{kg} \cdot \text{K})^{-1}$
Al	26.982	0.93	0.90
Cu	63.546	0.39	0.39
Fe	55.847	0.45	0.44
Si	32.064	0.78	0.70

The results, although not precise, do closely correspond to the measured values for specific heat.

19-6 Thermal Expansion

An atom that gains thermal energy and begins to vibrate behaves as though it has a larger atomic radius. The average distance between the atoms and the overall dimensions of the material increase. The change in the dimensions of the material Δl per unit length is given by the *linear coefficient of thermal expansion* α.

$$\alpha = \frac{1}{l} \frac{\Delta T}{\Delta l} \tag{19-12}$$

where ΔT is the increase in temperature and l is the initial length.

The linear coefficient of thermal expansion is related to the strength of the atomic bonds. In order for the atoms to move from their equilibrium separation, energy must be introduced to the material. If a very deep energy trough caused by strong atomic bonding is characteristic of the material, the atoms separate to a lesser degree and have a low linear coefficient of thermal expansion (Figure 19-15). This relationship also indicates that materials having a high melting temperature, also due to strong atomic attractions, have low linear coefficients of

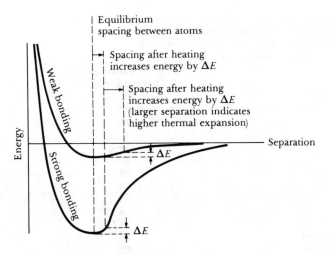

FIG. 19-15   The energy-separation curve for two atoms.
Materials that display a steep curve with a deep trough have
low linear coefficients of thermal expansion.

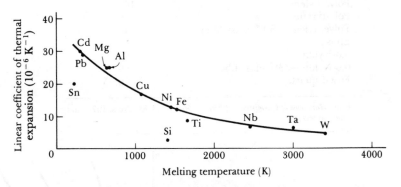

FIG. 19-16   The relationship between the linear coefficient of thermal
expansion and the melting temperature in metals at 25°C. Higher melting
point metals expand to a lesser degree.

thermal expansion (Figure 19-16). Materials like lead and polyethylene expand a
great deal compared to diamond, tungsten, or ceramics (Table 19-4).

Several precautions must be taken when calculating dimensional changes in
materials.

1. The expansion characteristics of some materials, particularly single crystals or materials having a preferred orientation, may be anisotropic.

2. Allotropic materials have abrupt changes in their dimensions when the
phase transformation occurs (Figure 19-17). These abrupt changes contribute to
cracking of refractories on heating or cooling and quench cracks in steels.

3. The linear coefficient of expansion is not constant at all temperatures.
Normally, α is listed in handbooks either as a complicated temperature-dependent function or is given as a constant for only a particular temperature range.

TABLE 19-4   The linear coefficient of thermal expansion at room temperature for selected materials

Material	Linear Coefficient of Thermal Expansion ($\times 10^{-6}$ K^{-1})
Al	25
Cu	16.6
Fe	12
Pb	29
Mg	25
Ni	13
Si	3
Ti	8.5
W	4.5
1020 steel	12
Gray iron	12
Stainless steel	17.3
3003 aluminum alloy	23.2
Yellow brass	18.9
Invar (Fe-36% Ni)	1.54
Polyethylene	100
Polystyrene	70
Polyethylene—30% glass fiber	48
Epoxy	55
6,6-Nylon	80
6,6-Nylon—33% glass fiber	20
Fused quartz	0.55

From *Handbook of Chemistry and Physics, 56th Ed.*, CRC Press, 1975, and other sources.

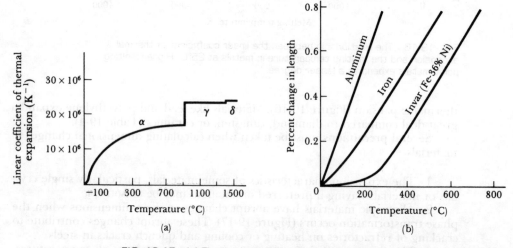

FIG. 19-17   (a) The linear coefficient of thermal expansion of iron changes abruptly at temperatures where an allotropic transformation occurs. (b) The expansion of Invar is very low due to the magnetic properties of the material at low temperatures.

4. Interaction of the material with electric or magnetic fields produced by magnetic domains may prevent normal expansion until temperatures above the Curie temperature are reached. This is the case for Invar, an Fe-36% Ni alloy, which undergoes practically no dimensional changes at temperatures below the Curie temperature, about 200°C. This makes Invar attractive as a material for bimetallics (Figure 19-17).

EXAMPLE 19-19

Explain why, in Figure 19-16, the linear coefficient of thermal expansion for silicon and tin do not fall on the curve. How would you expect germanium to fit into this figure?

Answer:

Both silicon and tin are covalently bonded. The strong covalent bonds are more difficult to stretch than the metallic bonds (a deeper trough in the energy-separation curve) so these elements have a lower coefficient. Since germanium also is covalently bonded, its thermal expansion should be less than that predicted by Figure 19-16.

EXAMPLE 19-20

An aluminum casting solidifies at 660°C. At that point, the casting is 250 mm long. What is the length after the casting cools to room temperature?

Answer:

The linear coefficient of thermal expansion for aluminum is 25×10^{-6} K^{-1}. The temperature change from solidification to room temperature is $660 - 27 = 633°C$.

$$\Delta l = \alpha l \Delta T = (25 \times 10^{-6})(250)(633) = 3.96 \text{ mm}$$

$$l_f = l - \Delta l = 250 - 3.96 = 246.04 \text{ mm}$$

The contraction in castings is normally compensated for by making the original pattern oversized by an amount calculated from the linear coefficient of thermal expansion. In this case, the original pattern should be $250 + 250(0.0158) = 254$ mm long to permit a final aluminum casting exactly 250 mm long.

19-7   Thermal Conductivity

The *thermal conductivity* K is a measure of the rate at which heat is transferred through a material. The conductivity relates the heat Q transferred across a given plane of area A per second when a temperature gradient $\Delta T / \Delta x$ exists (Figure 19-18).

$$\frac{Q}{A} = K \frac{\Delta T}{\Delta x} \tag{19-13}$$

Note that the thermal conductivity K plays the same role in heat transfer that the diffusion coefficient D does in mass transfer.

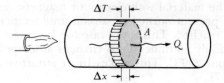

FIG. 19-18 When one end of a bar is heated, a heat flux Q/A flows toward the cold end at a rate determined by the temperature gradient produced in the bar.

If valence electrons are easily excited into the conduction band, thermal energy can be transferred by the electrons. The amount of energy transferred depends on the number of excited electrons and their mobility; thus we expect a relationship between thermal and electrical conductivity.

$$\frac{K}{\sigma T} = L = 2.3 \times 10^{-8} \text{ W} \cdot \Omega \cdot \text{K}^{-2} \tag{19-14}$$

where L is the Lorentz constant.

Thermally induced vibrations of the atoms cause the emission of phonons which also transfer energy through the material. We expect higher temperatures to increase the rate of heat transfer due to higher energy phonons.

EXAMPLE 19-21

A window glass 10 mm thick and 1.2 m $\times$ 1.2 m square separates a room at 25°C from the outside, at 40°C. Calculate the amount of heat entering the room through the window each day.

Answer:

From Table 19-5, the thermal conductivity of a soda-lime glass, typical of windows, is 0.96 W $(\text{m} \cdot \text{K})^{-1}$.

$$\frac{\Delta T}{\Delta x} = \frac{40 - 25}{10} = 1.5 \text{ K mm}^{-1} = 1.5 \times 10^3 \text{ K m}^{-1}$$

$$\frac{Q}{A} = K \frac{\Delta T}{\Delta x} = (0.96)(1.5 \times 10^3) = 1.44 \times 10^3 \text{ W m}^{-2}$$

$$A = (1.2 \text{ m} \times 1.2 \text{ m}) = 1.44 \text{ m}^2$$

$$t = (1 \text{ day})(24 \text{ h/day})(3600 \text{ s/h}) = 8.64 \times 10^4 \text{ s}$$

$$\frac{\text{Heat}}{\text{Day}} = \left(\frac{Q}{A}\right)(t)(A) = (1.44 \times 10^3)(8.64 \times 10^4)(1.44)$$

$$= 179 \times 10^6 \text{ J} \cdot \text{day}^{-1}$$

Thermal conductivity in metals. The electronic contributions are the dominant factor in the conduction of thermal energy in metals and alloys, since the valence band is not completely filled. But the thermal conductivity also depends on lattice defects, microstructure, and processing of the metal. Thus cold-worked metals, solid solution-strengthened metals, and two-

TABLE 19-5 Thermal conductivity of selected materials at 27°C

Material	Thermal Conductivity W (m · K)$^{-1}$	Material	Thermal Conductivity W (m · K)$^{-1}$
Al	238	Gray iron	80
Cu	400	aluminum manganese alloy	280
Fe	80	Yellow brass	222
Mg	100	Cu-30% Ni	50
Pb	35	Ar	0.018
Si	150	Carbon (graphite)	335
Ti	22	Carbon (diamond)	2320
W	170	Soda-lime glass	0.96
Zn	117	Vitreous silica	1.34
Zr	23	Vycor glass	1.25
low carbon steel	100	Fireclay	0.27
Ferrite	75	Silicon carbide	88
Cementite	50	6,6-Nylon	121
304,S15 stainless steel	30	Polyethylene	188

phase alloys might display lower conductivities compared to their defect-free counterparts.

We expect higher temperatures to reduce the mobility and the thermal conductivity of metals. However, higher temperatures also increase the energy of the electrons and permit heat to be transferred by lattice vibration. In metals, the thermal conductivity often initially decreases with temperature, becomes nearly constant, and then increases slightly (Figure 19-19).

Thermal conductivity in ceramics. Lattice vibrations, or phonons, are responsible for transfer of heat in ceramic or insulating materials. In these materials, the energy gap is too large for many electrons to be excited into the conduction band except at very high temperatures. Normally, ceramics have higher thermal conductivities at higher temperatures due to higher energy phonons and some electronic contributions.

Other factors influence the thermal conductivity of ceramics. Materials with a close-packed structure, low density, and high modulus of elasticity produce high-energy phonons that encourage high thermal conductivities. Crystalline solids have higher conductivities than their amorphous or glassy counterparts because less scattering of phonons occurs. Porosity reduces thermal conductivity. The best insulating brick, for example, contains a large fraction of porosity. We can consider this material as a composite material, where porosity is one of the materials.

EXAMPLE 19-22

Suppose we are able to introduce 30 vol% porosity into a soda-lime glass. If we assume that the thermal conductivity of the porosity is zero, how much heat do we save in Example 19-21 per day?

Answer:

Let's use the rule of mixtures to estimate the thermal conductivity of our glass-porosity composite.

$$K_c = K_g f_g + K_p f_p = (0.96)(0.7) + (0)(0.3) = 0.672 \text{ W (m} \cdot \text{K)}^{-1}$$

$$\frac{\text{Total}}{\text{heat}} = \left(\frac{Q}{A}\right)(t)(A) = (K)\left(\frac{\Delta T}{\Delta x}\right)(t)(A)$$

$$= (0.672)(1.5 \times 10^3)(8.64 \times 10^4)(1.44) = 12.5 \times 10^7 \text{ J} \cdot \text{day}^{-1}$$

$$\frac{\text{Heat}}{\text{savings}} = 17.9 \times 10^7 - 12.5 \times 10^7 = 5.4 \times 10^7 \text{ J} \cdot \text{day}^{-1}$$

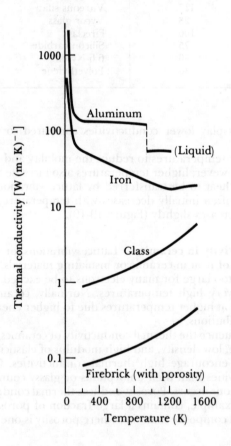

FIG. 19-19 The effect of temperature on the thermal conductivity of selected materials.

Thermal conductivity in semiconductors. Heat is conducted in semiconductors by both phonons and electrons. At low temperatures, phonons are the principal carriers of energy, but at higher temperatures, electrons are excited through the small energy gap into the conduction band and the thermal conductivity increases significantly.

ELASTIC PROPERTIES

19-8 Elastic Behavior

In Chapter 6 we defined the modulus of elasticity, or Young's modulus, as a measure of the stiffness of a material. In this section we will relate elastic behavior to atomic bonding, note that elastic waves, or phonons, can be produced, and discover that some unusual physical properties of a material can be described and explained.

Atomic bonding. The modulus of elasticity is related to the force required to move the atoms away from their normal equilibrium spacing (Figure 19-20).

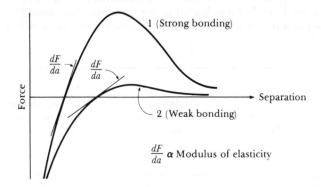

FIG. 19-20 The force-separation curve for two materials, showing the relationship between atomic bonding and the modulus of elasticity.

The modulus of elasticity is the slope of the force-separation curve at the equilibrium spacing. Tightly bound atoms display a deep energy trough, a steep force-separation curve, and a high slope or modulus of elasticity. High melting point materials, which have strong bonding, have a high modulus.

Elastic constants. The modulus of elasticity is actually one of several elastic properties (Figure 19-21). This modulus describes the elastic deformation when a material is uniaxially stretched or compressed (note that the same slope in the force-separation curve is obtained for either small tensile or compressive forces). The modulus E is defined by Hooke's law

$$E = \frac{\sigma}{\varepsilon} \tag{19-15}$$

where σ is the stress and ε is the resulting strain.

Poisson's ratio relates the longitudinal elastic deformation produced by a simple tensile or compressive stress to the lateral deformation that must simultaneously occur.

$$\mu = - \frac{\varepsilon_{lateral}}{\varepsilon_{longitudinal}} \tag{19-16}$$

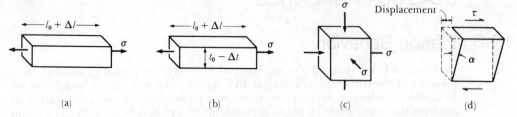

FIG. 19-21   Description of the four elastic constants and how they are obtained. (a) Modulus of elasticity, (b) Poisson's ratio, (c) compressibility, and (d) shear modulus.

For ideal materials, we can show that Poisson's ratio $\mu = 0.5$. However, in real materials we find that less lateral strain develops than would be predicted based on conservation of volume; typically Poisson's ratios of $\mu = 0.3$ are measured.

EXAMPLE 19-23

Suppose a 10 mm $\times$ 10 mm $\times$ 100 mm bar is pulled in tension until the bar is 101 mm long. If all of the deformation is elastic, determine the other dimensions of the bar. What should be the Poisson's ratio?

Answer:

If we assume that the volume is conserved and that the bar is isotropic in behavior, then

$$V_0 = 10 \times 10 \times 100 = 10 \times 10^3 \text{ mm}^3$$

$$V_f = (101)(w)(w) = 10 \times 10^3 \text{ mm}^3$$

$$w = \sqrt{\frac{10^4}{101}} = 9.95 \text{ mm}$$

$$\varepsilon_{\text{longitudinal}} = \frac{101 - 100}{100} = 0.01$$

$$\varepsilon_{\text{lateral}} = \frac{9.95 - 10}{10} = -0.005$$

$$\mu = -\frac{-0.005}{0.010} = 0.5$$

However, this result is not usually observed.

The *modulus of compressibility*, or the *bulk modulus*, B describes the decrease in the volume of a material subjected to a pressure exerted from all sides.

$$B = \frac{\sigma}{\Delta V/V} = \frac{1}{\beta} \tag{19-17}$$

The *compressibility* β is the reciprocal of the bulk modulus.

The *modulus of rigidity*, or the *shear modulus*, G describes the angle through which a material is distorted by a shear stress τ.

$$G = \frac{\tau}{\gamma} = \frac{\tau}{\tan \alpha} \tag{19-18}$$

If any two of these four elastic properties are known, the others can be calculated using the following relationships.

$$B = \frac{E}{3(1 - 2\mu)}$$

$$G = \frac{E}{2(1 + \mu)} \tag{19-19}$$

$$\frac{1}{E} = \frac{1}{9B} + \frac{1}{3G}$$

In anisotropic materials, we may need as many as three moduli of elasticity, Poisson's ratios, and shear moduli to completely describe the elastic behavior of the material.

EXAMPLE 19-24

Aluminum has a modulus of elasticity of 69 GPa and a Poisson's ratio of 0.3. If a shear stress τ of 34 MPa is placed on a 25 mm aluminum cube, calculate the displacement of the top surface relative to the bottom surface.

Answer:

We need to find the shear modulus G to make our calculation. From Equation (19-19)

$$G = \frac{E}{2(1 + \mu)} = \frac{69}{2(1 + 0.3)} = 26.5 \text{ GPa}$$

From Equation (19-18)

$$\gamma = \tan \alpha = \frac{\tau}{G} = \frac{34}{26.5 \times 10^3} = 0.0013$$

The displacement is

$$\tan \alpha = \frac{\text{displacement}}{\text{height}} = 0.0013$$

$$\text{Displacement} = (0.0013)(25) = 0.032 \text{ mm}$$

The elastic constants depend on temperature. As the temperature increases, the bonding energy decreases, the energy trough becomes shallower, and the slope of the force-separation curve is lower.

The elastic constants may also depend on the applied stress. In gray cast iron or concrete, the modulus of elasticity continually decreases as the stress increases. Consequently, we either specify the range of stresses for which the modulus applies or we define an average modulus. The former is the *tangent modulus,* the latter is the *secant modulus* (Figure 19-22).

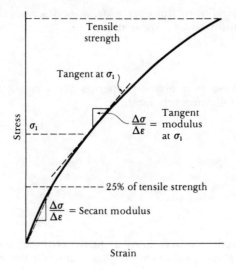

FIG. 19-22 The tangent or secant moduli of elasticity are used to describe the elastic behavior of materials that do not have a linear elastic region in the stress-strain curve.

On the other hand, the modulus of rubber increases as the stress increases (Figure 19-23). The low initial modulus represents the uncoiling of the chains, causing large strains for low stresses. However, after the chains are uncoiled, further elastic deformation requires stretching of the bonds in the carbon backbone, which requires a much higher stress.

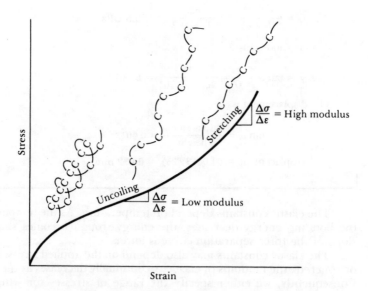

FIG. 19-23 The elastic behavior of a rubber is not linear due to uncoiling, then stretching of the chains.

19-9   Anelastic and Thermoelastic Behavior

Normally, we consider that an elastic strain is instantly recovered when the load is removed, so that recovery follows exactly the same path taken during loading. Because this is not always observed, some unusual elastic behavior may result.

Thermoelasticity is an effect in which a rapid elastic strain causes the material to lose thermal energy and cool. This effect is associated with an increase in volume because Poisson's ratio is less than the ideal. However, when the material warms back to ambient temperature, a further expansion occurs (Figure 19-24). If the stress is suddenly removed, the temperature increases slightly; after cooling the material contracts back to its original dimensions. Eventually, the normal total strain is produced when the stress is applied and the strain is completely recovered when the stress is removed. However, a hysteresis loop path is produced compared to loading at normal rates.

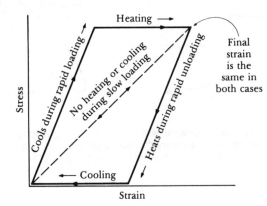

FIG. 19-24 Thermoelastic stress-strain behavior.

Anelasticity, of which thermoelasticity is a special case, is a time-dependent elastic behavior. A portion of the elastic strain is produced instantly, but the remaining strain occurs over a period of time. This time is required to permit thermal equilibrium, as in thermoelasticity, or to permit rearrangement of point defects, dislocations, or other discontinuities in the material. For example, carbon atoms in BCC iron may shift from $\frac{1}{2},0,0$ positions to $0,\frac{1}{2},0$ or $0,0,\frac{1}{2}$ positions when the stress is applied and reversed (Figure 19-25). A certain time is required for diffusion to occur in order to permit the rearrangement of the carbon atoms.

Anelastic behavior becomes of engineering importance when an alternating elastic strain is imposed on the material (Figure 19-26). The hysteresis loop produced by an alternating stress represents lost energy which may be converted to heat. This could, for example, cause overheating of an automobile tire operating at high speeds or in an underinflated condition.

Anelastic behavior is most pronounced at a particular frequency (Figure 19-27), which is related to the rate at which the atoms move to their final positions. At a low frequency, there is ample time for atoms to become rearranged, but at high frequencies no atoms are rearranged. At the critical frequency, a large anelastic effect occurs.

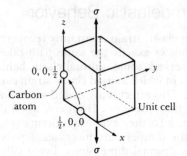

FIG. 19-25 Movement of interstitial atoms may explain the anelastic behavior of iron.

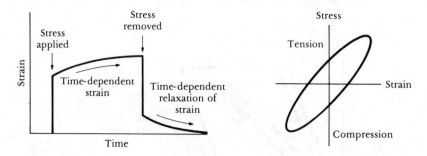

FIG. 19-26 Because of the time-dependent strain, a hysteresis loop is created when an anelastic material undergoes an alternating stress.

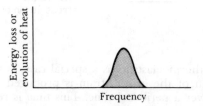

FIG. 19-27 The effect of frequency on the anelastic behavior. At critical frequencies, large losses of energy or evolution of heat are observed.

SUMMARY

The optical, thermal, and elastic properties of a material depend on interactions between radiation or energy and the atomic structure and arrangement. As a consequence, we can influence some of these properties by altering the atomic arrangement through changes in the composition and processing of the material. Thus, many absorption characteristics can be altered to provide changes in color or transparency of a material. However, since the atomic structure is not easily altered, some properties of a material are fixed, including many emission phenomena, thermal expansion coefficients, and heat capacities.

GLOSSARY

Absorption edge The wavelength at which the absorption characteristics of a material abruptly change.

Anelasticity Time-dependent elastic behavior.

Bulk modulus The modulus of compressibility, an elastic property describing the decrease in volume of a material subjected to a compressive load.

Characteristic spectrum The spectrum of radiation emitted from a material which occurs at fixed wavelengths corresponding to particular energy level differences within the atomic structure of the material.

Compressibility The reciprocal of the bulk modulus

Continuous spectrum Radiation emitted from a material having all wavelengths longer than a critical short wavelength limit.

Fluorescence Emission of radiation from a material only when the material is actually being stimulated.

Heat capacity The energy required to raise the temperature of one mole of a material one degree.

Index of refraction Relates the change in velocity and direction of radiation as it passes through a transparent material.

Laser A beam of monochromatic coherent radiation produced by the controlled emission of photons.

Linear absorption coefficient Describes the ability of a material to absorb radiation.

Linear coefficient of thermal expansion Describes the amount by which each unit length of a material changes when the temperature of the material changes by one degree.

Lorentz constant Relates thermal and electrical conductivity.

Luminescence Conversion of radiation to visible light.

Phonon An elastic wave that transfers energy through a material.

Phosphorescence Emission of radiation from a material after the material is stimulated.

Photoconduction Production of a current due to stimulation of electrons into the conduction band by light radiation.

Photons Energy or radiation produced from atomic, electronic, or nuclear sources that can be treated as particles or waves.

Poisson's ratio The ratio between lateral and longitudinal elastic deformation of a material.

Reflectivity The percent of incident radiation that is reflected.

Relaxation time The time required for $1/e$ of the electrons to drop from the conduction to the valence band in luminescence.

Secant modulus The approximate modulus of elasticity of a material determined at 25% of the tensile strength.

Shear modulus The modulus of rigidity, which describes the angle through which a material is elastically distorted by an applied shear stress.

Short wavelength limit The shortest wavelength or highest energy radiation emitted from a material under particular conditions.

Specific heat The energy required to raise the temperature of one gram of a material one degree.

Tangent modulus The modulus of elasticity of a material as measured by the local slope of the stress-strain curve.

Thermal conductivity Measures the rate at which heat is transferred through a material.

Thermal emission Emission of photons from a material due to excitation of the material by heat.

Thermoelasticity An anelastic behavior in which a sudden change in the strain in a material causes the material to heat or cool.

X rays Electromagnetic radiation produced by changes in the electronic structure of atoms.

PRACTICE PROBLEMS

1 The voltage acting on a heated tungsten filament is 100,000 V. Calculate the wavelength of the shortest X ray that is emitted.

2 The wavelength of the shortest X ray emitted from a tungsten filament is 0.93 Å. Calculate the voltage acting on the filament.

3 What is the minimum accelerating voltage required to produce the K_α characteristic peak in (a) tungsten, (b) iron, and (c) aluminum?

4 A voltage of 11,900 V is required to produce the K_α characteristic X rays from bromine. Calculate the wavelength of the K_α peak.

5 Estimate the difference in energy between electrons in (a) the K and L shells, (b) the K and M shells, and (c) the L and M shells of molybdenum.

6 A tungsten filament attached to a 10-V source produces radiation. What portion of the electromagnetic spectrum does the most intense radiation from the filament belong to?

7 Electrons produced by an accelerating voltage of 15,000 V strike a sample of molybdenum. Will K_α, K_β, or L_α X rays be produced?

8 The results of an X-ray fluorescent analysis test are shown in Figure 19-28. Determine (a) the accelerating voltage and (b) the identity of the elements in the sample.

9 The identity of an inclusion in a metal can be determined by obtaining the spectrum of emitted radiation, shown in Figure 19-29. Determine the probable identity of the inclusion.

10 The results of an energy dispersive analysis of X rays emitted from a sample are shown in Figure 19-30. Convert the energies to wavelengths and determine the identity of the elements in the specimen.

11 Sketch the intensity versus the energy of X rays emitted from a chromium-manganese-tungsten alloy.

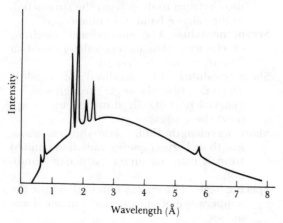

FIG. 19-28 The spectrum from an X-ray fluorescent analysis. (See Problem 8.)

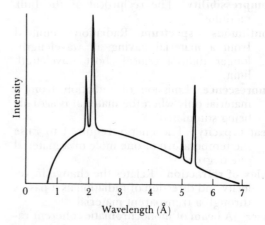

FIG. 19-29 The spectrum emitted from an inclusion. (See Problem 9.)

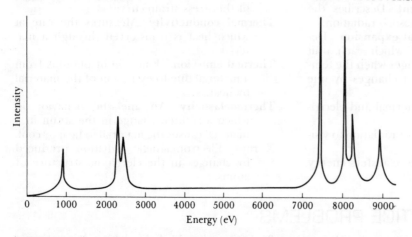

FIG. 19-30 The energy spectrum for emitted X rays. (See Problem 10.)

You will not be able to assign actual intensities to the peaks.

12 Determine the maximum wavelength and position in the electromagnetic spectrum of photons required to excite an electron into the conduction band in CdS, CdSe, and PbTe. (See Table 17-10.) Is the radiation emitted from these materials in the visible portion of the spectrum?

13 We wish to introduce impurities into ZnO that will produce visible light of 6000 Å during phosphorescence. Determine the position of the electron trap within the gap between the conduction and valence bands that is required.

14 The intensity of a phosphorescent source decreases by 50% in 10^{-3} s. Determine (a) the relaxation time and (b) the relative intensity after 10^{-2} s.

15 Suppose that the relaxation time for a phosphorescent source is 1.5×10^{-2} s. Calculate the time required for the intensity of the source to decrease to 10% of the original value.

16 The intensity of an X-ray beam is reduced by 50% when passing through a 1 mm thick copper sheet. Calculate the linear absorption coefficient in copper.

17 The linear absorption coefficient of lead for X rays is 3969 m^{-1}. Estimate the thickness of a lead sheet required to reduce the X-ray intensity to a value of 1% of the original intensity.

18 The intensity of the continuous radiation from an X-ray source and the linear absorption coefficient for X rays in iron are listed below for several wavelengths. Plot the initial intensity I_i and the final intensity I_f versus wavelength after the X rays pass through a 2 mm thick iron filter.

Wavelength (Å)	Initial Intensity	Linear Absorption Coefficient (m^{-1})
2.291	10	1461
2.103	50	1155
1.937	90	925
1.790	100	756
1.659	92	5044
1.542	80	4117
1.436	60	3431
0.711	10	489

19 Will ZnS, ZnO, CdSe, and PbS transmit visible light? Show.

20 A light source with a wavelength of 8×10^{-9} m is directed at a semiconductor device. Will photoconduction occur in (a) germanium, (b) silicon, (c) CdS, and (d) InSb?

21 Compare the wavelength of the photon required to produce photoconduction in intrinsic germanium versus indium-doped germanium.

22 From Table 18-1, calculate and compare the index of refraction and the reflectivity of polyethylene, Teflon, and soda-lime glass.

23 A beam of photons striking a dielectric material at an angle of 28° from the normal changes direction to 15° from the normal during transmission. Calculate the index of refraction and estimate the fraction of the beam that is reflected.

24 Suppose that 40% of the intensity of a beam of photons entering a material is transmitted through a 5 mm thick material with a dielectric constant of 1.544. (a) Determine the fraction reflected, the fraction absorbed, and the fraction transmitted. (b) Calculate the linear absorption coefficient of the photons in the material.

25 The linear absorption coefficient of a 20 mm thick material is 130 m^{-1}. The intensity of an original beam of photons is 100 and that of the transmitted beam is 5. Calculate the reflectivity, index of refraction, and dielectric constant of the material.

26 What filter would you use to isolate the K_α peak of copper characteristic X rays?

27 What filter would you use to isolate the K_α peak of iron characteristic X rays?

28 Calculate the angle θ of the K_α X-ray beam of copper diffracted from the (110), (200), and (220) planes of BCC iron, which has a lattice parameter of 2.866 Å.

29 A K_α X-ray beam of iron is diffracted from a (222) plane of a cubic metal powder, giving a Bragg angle θ of 36°. Determine the lattice parameter of the material.

30 Suppose K_α radiation of aluminum is used to obtain a diffraction pattern from the (111) plane of copper. Is this possible? Prove.

31 A 2θ angle of 134.32° is observed for the X ray beam diffracted from the (220) planes of nickel. What type of K_α X-ray radiation was used?

32 Calculate the temperature after 10.475 kJ are introduced to 1 kg of zinc at 20°C. Will the zinc melt?

33 The temperature of a 30 mm nickel cube increases by 100°C. How much heat was introduced?

34 Suppose 25.14 kJ are introduced to magnesium at 20°C. What is the maximum amount of magnesium that can be melted?

35 Does the classical heat capacity of 25 $J \cdot mol^{-1}$ predict the specific heat of (a) diamond, (b) lead, and (c) nitrogen? Show.

36 Determine the size of the pattern for making a 0.2 m × 0.2 m × 0.02 m magnesium sand casting.

37 Lead is to be poured into an iron permanent mold heated to 150°C. Calculate the size of the cavity machined into the iron block at 27°C if the finished lead casting is to be 0.25 m × 0.25 m × 0.03 m. (Assume that the mold remains at the same temperature during pouring and solidification.)

38 Suppose a fused quartz enamel is bonded to a gray iron casting 250 mm long. Compare the length to which each will try to expand when the temperature increases from 27°C to 500°C. What will probably happen to the enamel coating?

39 Compare the rate at which heat is transferred through a 0.2 m × 0.2 m × 0.02 m plate of (a) copper, (b) lead, and (c) vitreous silica when the temperature difference between the top and bottom of the plate is 125°C.

40 One end of a copper rod 30 mm in diameter and 200 mm long is heated to 600°C. The other end is immersed in 1000 mL of water at 27°C. (a) Assuming a constant temperature gradient, calculate the heat Q transferred to the cold end each second. (b) Estimate the time required to warm the water to 30°C if no heat is lost to the surroundings.

41 A heat shield is made by laminating five layers of 10 mm thick copper to four layers of vitreous silica each 1 mm thick. (a) Determine the heat transferred across each square centimeter of the shield in 10 min when the temperature difference across the composite is 800°C. (b) Compare the heat transferred in part (a) to the case where a solid 50 mm thick copper plate is used.

42 Review the differences between ductile and gray cast irons (Chapter 13). Then explain why gray iron has a higher thermal conductivity than ductile iron.

43 Determine the change in diameter of an aluminum bar 75 mm in diameter when an elastic stress of 13.8 MPa is applied along the bar. Poisson's ratio is 0.3.

44 The Poisson's ratio for steel is 0.33 and the modulus of elasticity is 207 GPa. Calculate the bulk modulus, compressibility, and shear modulus for steel.

45 Calculate the displacement that occurs when a shear stress of 69 MPa acts on a 75 mm cube of steel. The shear modulus is 77.9 GPa.

46 A stress-strain curve is shown in Figure 19-31 for gray cast iron. Determine (a) the secant modulus of elasticity and (b) the tangent modulus of elasticity at 138 MPa and 207 MPa.

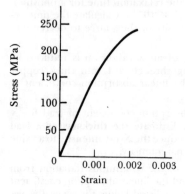

FIG. 19-31 The stress-strain curve for gray cast iron. (See Problem 46.)

PROTECTION AGAINST DETERIORATION AND FAILURE OF MATERIALS

In this section, the failure of materials by corrosion, wear, and fracture is discussed. We will again find that deterioration or failure is related to the structure, properties, and processing of the materials. In Chapter 20, corrosion and wear will be examined; electrochemical corrosion will be found to be particularly important. In addition to determining the mechanism for corrosion, we will look at techniques for controlling and preventing damage to the material by these processes.

In Chapter 21, we will review the mechanical failure of materials and pick up some hints on how to identify the cause for fracture. We will also examine a number of techniques by which we can nondestructively test a material to determine if it is subject to fracture. Finally, we will briefly discuss fracture mechanics, in which we combine the principles of fracture analysis and nondestructive testing to help assure that catastrophic failures do not occur.

PART V

CHAPTER 20

Corrosion and Wear

20-1 Introduction

The composition and physical integrity of a solid material are altered in a corrosive environment. In chemical corrosion, the material is dissolved by a corrosive liquid. In electrochemical corrosion, metal atoms are removed from the solid material due to an electric circuit that is produced. Metals and certain ceramics react with a gaseous environment, usually at elevated temperatures, and the material may be destroyed by formation of oxides or other compounds. Polymers undergo cross-linking or degradation when exposed to oxygen at elevated temperatures. Materials may also be altered when exposed to radiation or even bacteria. Finally, a variety of wear and wear-corrosion mechanisms alter the shape of materials. Billions of dollars are required to repair the damage done by corrosion each year.

20-2 Chemical Corrosion

In *chemical corrosion,* or direct solution, the material dissolves in a corrosive liquid medium. The material continues to dissolve in the liquid until either the material is consumed or the liquid is saturated. A simple example is salt dissolving in water.

Liquid metal attack. Liquid metals first attack a solid at high energy locations, such as grain boundaries. If these regions continue to be attacked preferentially, cracks eventually grow (Figure 20-1). Often this form of corrosion is complicated by wetting, formation of compounds or fluxes that accelerate the attack, or electrochemical corrosion.

Selective leaching. One particular element in an alloy may be selectively dissolved, or leached, from the solid. *Dezincification* occurs in brass containing more than 15% Zn. Both copper and zinc are dissolved by aqueous solutions at high temperatures; the zinc ions remain in solution while the copper ions are replated onto the brass (Figure 20-2). Eventually, the brass becomes porous and weak.

Graphitic corrosion of gray cast iron occurs when iron is selectively dissolved in water or soil, leaving behind interconnected graphite flakes and a corrosion product. Localized graphitic corrosion often causes leakage or failure of buried gray iron gas lines, leading to explosions.

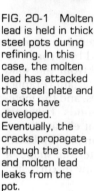

FIG. 20-1 Molten lead is held in thick steel pots during refining. In this case, the molten lead has attacked the steel plate and cracks have developed. Eventually, the cracks propagate through the steel and molten lead leaks from the pot.

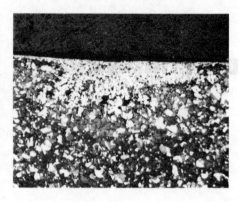

FIG. 20-2   Photomicrograph of a copper deposit in brass, showing the effect of dezincification.

Fluxing of ceramics.   Ceramic refractories used to contain molten metal during melting or refining may be dissolved by the slags that are produced on the metal surface. Frequent replacement of the refractories is thus required.

Solvents for polymers.   Polymers dissolve in liquid solvents having similar structures. Polyethylene, which has a straight chainlike structure, dissolves readily in organic solvents whose molecules resemble the ethylene molecule. Polystyrene dissolves more easily in organic solvents such as benzene that have a similar molecular structure.

General principles of chemical corrosion.   Several common features are observed in chemical corrosion.

1. Small ions or molecules dissolve faster than more complicated structures. A wax dissolves more rapidly in a liquid organic solvent than does polyethylene. Ionic salts dissolve faster in a flux than do complex silicate ions.

2. Solution occurs more rapidly when the solid and the dissolving liquid have similar structures.

3. Solution is accelerated at higher temperatures due to more rapid dissociation rates and higher solubilities.

Prevention of chemical corrosion is relatively difficult. Low temperatures and protective coatings may be helpful. However, the best way to prevent chemical corrosion is to avoid contact between the solid material and the liquid solvent. Thus, a refractory material selected to contain molten metal should not react with the slag that is produced on the molten metal.

EXAMPLE 20-1

Suppose a slag that is produced during melting and refining of steel at 1600°C contains 50% SiO_2 and 50% FeO. The refractory containing the steel could be made from Al_2O_3 or MgO. From the ternary phase diagrams (Figure 20-3), which refractory would appear to work best for these conditions?

(a)

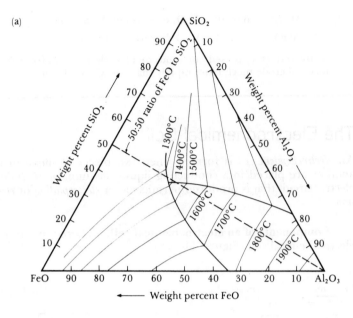

(b)

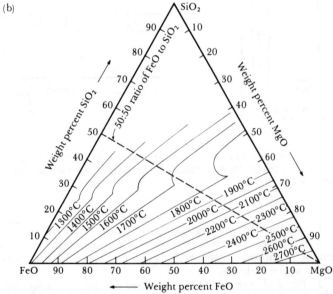

FIG. 20-3   The liquidus plots for (a) SiO₂-FeO-Al₂O₃ and (b)
SiO₂-FeO-MgO.

Answer:

In each curve, we can draw a line that runs from the 50% SiO₂-50% FeO point to pure
Al₂O₃ or MgO. When we do this, we find that, at 1600°C, the solubility of Al₂O₃ or
MgO in the slag is

Al_2O_3:   Up to 40% Al_2O_3 dissolves in the SiO_2-FeO liquid

MgO: Up to 24% MgO dissolves in the SiO_2-FeO liquid

As the refractory is used numerous times, each time with fresh slag, the Al_2O_3 refractory will erode much more rapidly than the MgO refractory.

20-3   The Electrochemical Cell

An *electrochemical cell* is formed when two pieces of metal in contact with one another are placed in a conducting liquid medium, or *electrolyte*. The complete electric circuit that is produced permits either *electroplating* or *electrochemical corrosion*.

Components of an electrochemical cell.   There are four components in an electrochemical cell (Figure 20-4).

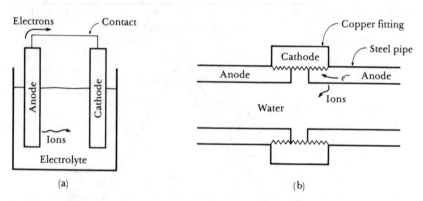

FIG. 20-4   The components in an electrochemical cell. (a) A possible electroplating set-up. (b) A corrosion cell between a steel water pipe and a copper fitting.

1. Anode: The *anode* gives up electrons to the circuit and corrodes.
2. Cathode: The *cathode* receives electrons from the circuit by means of a chemical, or cathode, reaction. Ions that combine with the electrons produce a by-product at the cathode.
3. Physical contact: The anode and cathode must be electrically connected, usually by physical contact, to permit the electrons to flow from the anode to the cathode.
4. Electrolyte: A liquid electrolyte must be in contact with both the anode and cathode. The electrolyte is conductive, thus completing the circuit. The electrolyte provides the means by which metallic ions leave the anode surface and assures that ions move to the cathode to accept the electrons.

This description of an electrochemical cell defines either electrochemical corrosion or electroplating. The two processes differ depending on the purpose of the cell, the source of the electric potential by which a current is caused to flow in the circuit, and the type of reaction that occurs at the cathode.

Anode reaction. The anode, which is a metal, undergoes an *oxidation reaction* by which metal atoms are ionized. The metal ions enter the electrolytic solution while the electrons leave the anode through the electrical connection.

$$M \rightarrow M^{n+} + ne^- \tag{20-1}$$

Because metal atoms leave the anode, the anode corrodes.

Cathode reaction in electroplating. In electroplating, a *reduction reaction*, which is the reverse of the anode reaction, occurs at the cathode.

$$M^{n+} + ne^- \rightarrow M \tag{20-2}$$

The metal ions, either intentionally added to the electrolyte or formed by the anode reaction, combine with electrons at the cathode. The metal then plates out and covers the cathode surface.

Cathode reactions in corrosion. Except in unusual conditions, plating of a metal does not occur during electrochemical corrosion. Instead, the reduction reaction forms a gas, solid, or liquid by-product at the cathode (Figure 20-5).

1. The hydrogen electrode: In oxygen-free liquids, such as hydrochloric acid (HCl) or stagnant water, the most common cathode reaction is the evolution of hydrogen.

$$2H^+ + 2e^- \rightarrow H_2 \uparrow \tag{20-3}$$

This is the *hydrogen electrode*.
2. The oxygen electrode: In aereated water, oxygen is available to the cathode and hydroxyl, or OH^-, ions are formed.

$$O_2 + 2H_2O + 4e^- \rightarrow 4OH^- \tag{20-4}$$

The *oxygen electrode* enriches the electrolyte in OH^- ions. These ions react with positively charged metallic ions, such as Fe^{2+}, finally producing a solid product, such as $Fe(OH)_2$, or rust.
3. The water electrode: In oxidizing acids, the cathode reaction produces water as a by-product.

$$O_2 + 4H^+ + 4e^- \rightarrow 2H_2O \tag{20-5}$$

If a continuous supply of both oxygen and hydrogen is available, the *water electrode* produces neither a buildup of solid rust nor a high concentration or dilution of ions at the cathode.

20-4 The Electrode Potential in Electrochemical Cells

In electroplating, an imposed voltage is required to cause a current to flow in the cell. But in corrosion a potential naturally develops when a material is placed in a solution. Let's see how the potential required to drive the corrosion reaction develops.

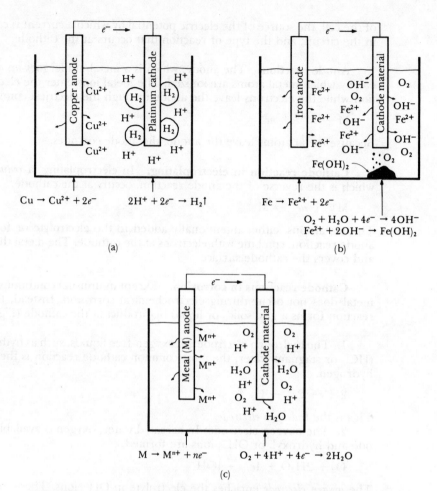

FIG. 20-5 The anode and cathode reactions in typical electrolytic corrosion cells. (a) The hydrogen electrode, (b) the oxygen electrode, and (c) the water electrode.

Electrode potential. When a perfect ideal metal is placed in an electrolyte, an *electrode potential* is developed which is related to the tendency of the material to give up its electrons. However, the driving force for the oxidation reaction is offset by an equal but opposite driving force for the reduction reaction. No net corrosion occurs. Consequently, we cannot measure the electrode potential for a single electrode material.

Electromotive force series. To determine the tendency of a metal to give up its electrons, we measure the *potential difference* between the metal and a standard electrode using a half-cell (Figure 20-6). The metal electrode to be tested is placed in a 1-M solution of its ions. A reference standard electrode is also placed in a 1-M solution of its ions. The two electrolytes are in electrical contact but are not permitted to mix with one another. Each electrode establishes its own elec-

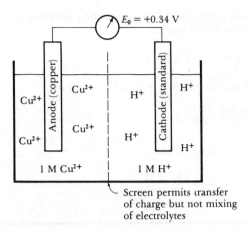

FIG. 20-6   The half-cell used to measure the electrode potential of copper under standard conditions. The electrode potential of copper is the potential difference between it and the standard hydrogen electrode in an open circuit. Since E_0 is greater than zero, copper is cathodic compared to the hydrogen electrode.

trode potential. By measuring the voltage between the two electrodes when the circuit is open, we obtain the potential difference. The potential of the hydrogen electrode, which is taken as our standard reference electrode, is arbitrarily set equal to zero volts. If the metal has a greater tendency to give up electrons than hydrogen, then the potential of the metal is negative—the metal is anodic with respect to the hydrogen electrode.

The *electromotive force,* or emf, *series* shown in Table 20-1 compares the electrode potential E_0 for each metal to the hydrogen electrode under standard conditions of 25°C and 1-M solution of ions in the electrolyte. Note that the measurement of the potential is made when the electric circuit is open. We shall see that the voltage difference begins to change as soon as the circuit is closed.

Effect of concentration on the electrode potential.   The electrode potential depends on the concentration of the electrolyte. The *Nernst equation* permits us to estimate the electrode potential in nonstandard solutions.

$$E = E_0 + \frac{0.0592}{n} \log (C_{ion}) \tag{20-6}$$

where E is the electrode potential in a solution containing a concentration C_{ion} of the metal in molar units, n is the valence of the metallic ion, and E_0 is the standard electrode potential in a 1-M solution. Note that when $C_{ion} = 1$, $E = E_0$.

EXAMPLE 20-2

Suppose 1 g of copper as Cu^{2+} is dissolved in 1000 g of water to produce an electrolyte. Calculate the electrode potential of the copper half-cell in this electrolyte.

Answer:

From chemistry, we know that a standard 1-M solution of Cu^{2+} is obtained when we add 1 mol of Cu^{2+} (an amount equal to the atomic mass of copper) to 1000 g of water.

The atomic mass of copper is 63.546 g/mol. The concentration of the solution when only 1 g of copper is added must be

$$C_{ion} = \frac{1}{63.546} = 0.0157 \text{ M}$$

From the Nernst equation, with $n = 2$ and $E_0 = +0.34$ V,

$$E = E_0 + \frac{0.0592}{n} \log (C_{ion})$$

$$E = 0.34 + \frac{0.0592}{2} \log (0.0157)$$

$$= 0.34 + (0.0296)(-1.8)$$

$$= 0.29 \text{ V}$$

TABLE 20-1 The electromotive force (emf) series for selected elements

Metal	Electrode Potential (V)
Anodic $Li \rightarrow Li^+ + e^-$	-3.05
$Mg \rightarrow Mg^{2+} + 2e^-$	-2.37
$Al \rightarrow Al^{3+} + 3e^-$	-1.66
$Ti \rightarrow Ti^{2+} + 2e^-$	-1.63
$Mn \rightarrow Mn^{2+} + 2e^-$	-1.63
$Zn \rightarrow Zn^{2+} + 2e^-$	-0.76
$Cr \rightarrow Cr^{3+} + 3e^-$	-0.74
$Fe \rightarrow Fe^{2+} + 2e^-$	-0.44
$Ni \rightarrow Ni^{2+} + 2e^-$	-0.25
$Sn \rightarrow Sn^{2+} + 2e^-$	-0.14
$Pb \rightarrow Pb^{2+} + 2e^-$	-0.13
$H_2 \rightarrow 2H^+ + 2e^-$	0.00
$Cu \rightarrow Cu^{2+} + 2e^-$	$+0.34$
$4(OH)^- \rightarrow O_2 + 2H_2O + 4e^-$	$+0.40$
$Ag \rightarrow Ag^+ + e^-$	$+0.80$
$Pt \rightarrow Pt^{4+} + 4e^-$	$+1.20$
$2H_2O \rightarrow O_2 + 4H^+ + 4e^-$	$+1.23$
Cathodic $Au \rightarrow Au^{3+} + 3e^-$	$+1.5$

Rate of corrosion or plating. The amount of metal plated on the cathode in electroplating or removed from the metal by corrosion can be determined from *Faraday's equation.*

$$w = \frac{ItM}{nF} \tag{20-7}$$

where w is the weight plated or corroded (g/s), I is the current (A), M is the atomic mass of the metal, n is the valence of the metal ion, t is the time (s), and F is Faraday's constant, 96,500 C.

To determine the rate of electroplating, we only need to fix the current in the cell—current is a controllable variable. However, the rate of corrosion is more difficult to control and measure. To protect metals from destruction by corrosion, we want the current to be as small as possible, but to increase the rate of electroplating, the current should be as high as possible.

EXAMPLE 20-3

Copper is electroplated onto one side of a 1 cm × 1 cm cathode using a current of 10 A. Calculate (a) the weight of copper plated per hour and (b) the time required to make a copper plate 0.1 cm thick.

Answer:

(a) From Faraday's equation, using $n = 2$ and $M = 63.546$ g/mol,

$$w = \frac{ItM}{nF} = \frac{(10)(3600 \text{ s/h})(63.546)}{(2)(96,500)}$$

$$= 11.85 \text{ g/h}$$

(b) The time required to produce a 0.1 cm thick layer on a 1 cm² cathode is

$$\rho_{Cu} = 8.96 \text{ Mg m}^{-3} = 8.96 \text{ g/cm}^3 \qquad A = 1 \text{ cm}^2$$

$$\text{Volume/h} = \frac{11.85}{8.96} = 1.323 \text{ cm}^3/\text{h}$$

$$\text{Volume required} = \text{thickness} \times \text{area} = 0.1 \text{ cm} \times 1 \text{ cm}^2 = 0.1 \text{ cm}^3$$

$$\text{Time} = \frac{0.1}{1.323} = 0.076 \text{ h} = 4.6 \text{ min}$$

20-5 The Corrosion Current in the Electrochemical Cell

Let's place a piece of copper in an electrolytic solution (Figure 20-7) without producing an external voltage or a complete circuit. Although no net change in the surface of the copper occurs, the oxidation and reduction reactions begin and

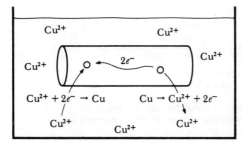

FIG. 20-7   When copper is placed in an electrolyte, the anode and cathode reactions occur but no net corrosion or plating takes place. There is an exchange of electrons from one location to another, however, giving an exchange current density.

continue in equilibrium as a particular number of electrons dissolve at the anode region and deposit at the cathode region. The *exchange current density* i_0 (A/cm²) represents the number of electrons involved in this equilibrium process.

Suppose we now apply a voltage to the copper (Figure 20-8). This voltage is

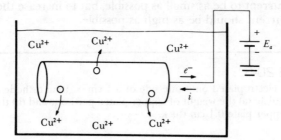

FIG. 20-8 When an overpotential E_a is applied to the copper, making it the anode, the anode and cathode reactions at the copper are no longer in equilibrium. A net current density i results as electrons leave the copper and the copper goes into solution.

the *overvoltage* E_a. If the overvoltage drives the reaction in the anodic direction, then $Cu \rightarrow Cu^{2+} + 2e^-$ becomes the favored reaction and a net current flows in the circuit. The *current density* i that is produced is related to the overvoltage that we apply.

$$i = i_0 \exp\left(\frac{E_a}{B_a}\right) \tag{20-8}$$

where i_0 is the exchange current density, E_a is the overpotential applied to the anode, and B_a is the *Tafel constant*, or slope, at the anode.

When we apply an overpotential to produce a net current density, we polarize our circuit and upset equilibrium. The graph of Equation (20-8) is the *anode polarization curve* (Figure 20-9). The Tafel slope, given in Table 20-2, determines

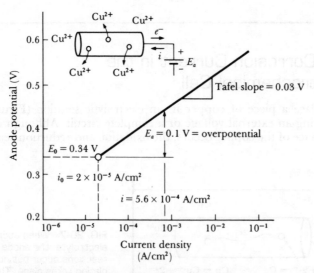

FIG. 20-9 Anode polarization curve for copper. When a voltage E_a is applied to make the copper anodic, a current will flow, causing the copper to corrode.

TABLE 20-2 Tafel slopes and exchange current densities for selected electrodes

Electrode	Tafel Slope (V)	Exchange Current Density (A/cm²)
Cathode reactions		
$2H^+ + 2e^- \rightarrow H_2$ (on Pt substrate)	0.03	1×10^{-3}
$O_2 + 4H^+ + 4e^- \rightarrow 2H_2O$ (on Pt substrate)	0.05	4×10^{-13}
$Fe^{2+} + 2e^- \rightarrow Fe$	0.12	1×10^{-8}
$Cu^{2+} + 2e^- \rightarrow Cu$	0.12	2×10^{-5}
$Zn^{2+} + 2e^- \rightarrow Zn$	0.12	2×10^{-5}
$Ni^{2+} + 2e^- \rightarrow Ni$	0.12	2×10^{-9}
Anode reactions		
$Fe \rightarrow Fe^{2+} + 2e^-$	0.02	1×10^{-8}
$Cu \rightarrow Cu^{2+} + 2e^-$	0.03	2×10^{-5}
$Zn \rightarrow Zn^{2+} + 2e^-$	0.02	2×10^{-5}

Because of differences in electrolytes, these values may change dramatically. Adapted from H. Uhlig, *Corrosion and Corrosion Control*, John Wiley & Sons, 1963.

the amount of polarization. The overpotential makes the anode less anodic, or the total anode potential E_{anode} becomes more positive.

$$E_{anode} = E_0 + E_a \tag{20-9}$$

By applying the overpotential, we have a net current flow from the electrode. The magnitude of the current density determines the rate of corrosion.

If we apply a potential in the opposite direction, any copper ions in solution plate out as the reaction $Cu^{2+} + 2e^- \rightarrow Cu$ becomes favored. There is a net flow of electrons to the electrode and the current density is

$$i = i_0 \exp \left(\frac{E_c}{B_c} \right) \tag{20-10}$$

where E_c is the cathode overpotential and B_c is the Tafel slope for the cathode. The equation gives the *cathode polarization curve* (Figure 20-10). The current density is related to the rate of electroplating. The total cathode potential is

$$E_{cathode} = E_0 - E_c \tag{20-11}$$

The overvoltage causes the cathode to become more anodic, or more negative.

The rate at which metal is corroded or electroplated depends on the current density and therefore on the overpotential that is applied or produced. In Examples 20-4 to 20-7, we will examine how the combined anode and cathode polarization curves determine the final cell potential and current density.

Electroplating. If we let the anode and cathode be the same material, then the exchange current density i_0 and the electrode potential E_0 are identical for both anode and cathode (Figure 20-11). During electroplating, polarization be-

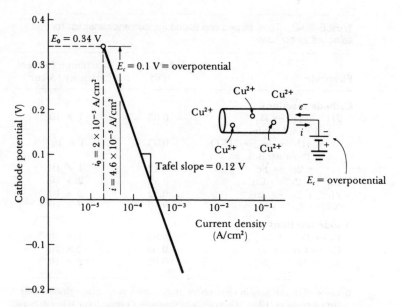

FIG. 20-10 Cathode polarization curve for copper. When the applied overpotential E_c makes the copper cathodic, the current density in the circuit causes copper to plate out of the electrolyte onto the cathode.

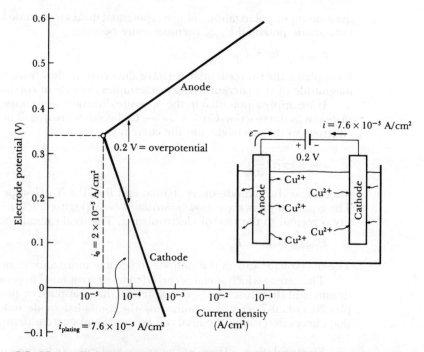

FIG. 20-11 Electroplating of copper. When an external 0.2 V is applied to the circuit, the polarization curves show that a current density of 7.6×10^{-5} A/cm^2 is produced. The copper anode corrodes and enters the electrolyte, while copper ions plate out on the copper cathode.

gins when the circuit is closed and the overpotential is applied to the circuit. Part of the applied overpotential causes anodic polarization while the remainder causes cathodic polarization. The current density is obtained at the point where the difference between the anode and cathode polarization curves equals the overpotential. With this current density we can calculate the rate at which copper is electroplated onto the cathode.

EXAMPLE 20-4

Suppose 0.2 V is applied to a copper electroplating cell. Determine the rate of plating if the cathode has a surface area of 1000 cm².

Answer:

We find from Figure 20-11 that a current density of 7.6×10^{-5} A/cm² is produced by the applied voltage source. Thus, the current is

$$I = (7.6 \times 10^{-5})(1000) = 0.076 \text{ A}$$

$$w = \frac{ItM}{nF} = \frac{(0.076)(3600)(64.546)}{(2)(96,500)} = 0.09 \text{ g/h}$$

The thickness of the copper plate, after one hour, will only be

$$\text{Thickness} = \frac{0.09}{(8.96)(1000)} = 1 \times 10^{-5} \text{ cm}$$

In electroplating, we must use overpotentials greater than 0.2 V to be practical.

EXAMPLE 20-5

Suppose 2 V is applied to the copper electroplating cell in Example 20-4. Calculate the current density and rate of plating.

Answer:

An overvoltage of 2 V is too large for the graph in Figure 20-11. However, the overpotential of 2 V is the sum of the overpotentials on the anode and cathode.

$$\Delta E = E_a + E_c = 2.0 \text{ V}$$

By rearranging Equations (20-8) and (20-10) and substituting for E_a and E_c

$$B_a \ln \left(\frac{i}{i_0}\right) + B_c \ln \left(\frac{i}{i_0}\right) = 2.0$$

$$(B_a + B_c) \ln \left(\frac{i}{i_0}\right) = 2.0$$

From Table 20-2

$$(0.03 + 0.12) \ln \left(\frac{i}{2 \times 10^{-5}} \right) = 2$$

$$\ln \left(\frac{i}{2 \times 10^{-5}} \right) = \frac{2}{0.15} = 13.3$$

$$\frac{i}{2 \times 10^{-5}} = \exp(13.3) = 5.97 \times 10^5$$

$$i = 11.94 \ \text{A/cm}^2 \qquad I = (11.94)(1000) = 11,940 \ \text{A}$$

$$w = \frac{(11,940)(3600)(63.546)}{(2)(96,500)} = 14,150 \ \text{g/h}$$

Metal anode-hydrogen electrode corrosion cell.   Let's connect a zinc anode to a platinum electrode at which hydrogen in solution in the electrolyte can combine to give the cathode reaction (Figure 20-12). When the circuit is completed,

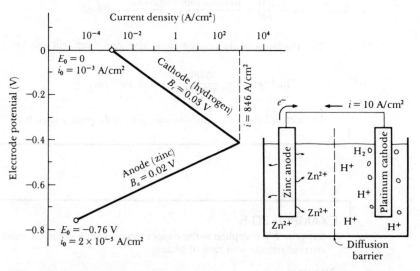

FIG. 20-12   When an electrolytic cell is formed between zinc and the hydrogen electrode, zinc corrodes and hydrogen is evolved. The electrodes polarize until the potential difference is zero. Corrosion occurs in this case at a current density of 846 A/cm².

the zinc ionizes while electrons flow to the cathode, combine with H^+ ions, and produce hydrogen gas. Eventually, equilibrium is established when the number of electrons released by the anode equals the number of electrons accepted at the cathode. Consequently, the current at the anode equals the current at the cathode (assuming equal-sized anode and cathode areas). Equilibrium is obtained when the anode and cathode polarization curves meet, giving a corrosion current density $i_{corrosion}$ that dictates the rate at which corrosion occurs. The final potential of the electrodes under these conditions is somewhere between the original individual electrode potentials of the anode and cathode and is the same for both.

EXAMPLE 20-6

Calculate the current density in an iron anode that is connected to a hydrogen electrode.

Answer:

In this case, equilibrium is reached and the current density is established when the potential difference between the two electrodes is zero.

$$\Delta E = E_{anode} - E_{cathode} = 0$$

$$E_0^{Fe} + E_a - (E_0^H - E_c) = 0$$

$$E_a + E_c = E_0^H - E_0^{Fe}$$

$$B_a^{Fe} \ln \left(\frac{i}{i_0(Fe)} \right) + B_c^H \ln \left(\frac{i}{i_0(H)} \right) = E_0^H - E_0^{Fe}$$

From Tables 20-1 and 20-2

$$(0.02) \ln \frac{i}{1 \times 10^{-8}} + (0.03) \ln \frac{i}{1 \times 10^{-3}} = 0 - (-0.44) = 0.44$$

$$(0.02) \ln(i) - (0.02) \ln(10^{-8}) + (0.03) \ln(i) - (0.03) \ln(10^{-3}) = 0.44$$

$$(0.05) \ln(i) - (0.02)(-18.42) - (0.03)(-6.91) = 0.44$$

$$(0.05) \ln(i) = 0.44 - 0.368 - 0.207 = -0.135$$

$$\ln(i) = -2.7$$

$$i = 0.067 \text{ A/cm}^2$$

Metal anode-metal cathode corrosion cell. A final idealized example is a corrosion cell between a zinc electrode in a 1-M solution of zinc ions connected to a copper cathode immersed in a 1-M solution of copper ions (Figure 20-13). The two solutions are electrically connected but mixing cannot occur between the two solutions. Since we have standard conditions, we can use the exchange current densities and the standard electrode potentials for each electrode.

When the circuit is closed, the anode reaction begins at the zinc electrode and the zinc corrodes. At the cathode, copper ions plate out in order to accept the electrons. The final current density and electrode potential are obtained from the intersection of the polarization curves for the zinc anode and the copper cathode. We can use this current density to calculate the rate at which the zinc corrodes or the copper plates out.

EXAMPLE 20-7

Calculate the amount of zinc loss and copper plated per hour for the electrolytic cell shown in Figure 20-13.

Answer:

We can use Faraday's equation for both calculations, noting that the current density must be 0.051 A/cm².

$$w(\text{zinc loss}) = \frac{(i_{\text{corrosion}})tM}{nF}$$

$$= \frac{(0.051)(3600)(65.37)}{(2)(96,500)}$$

$$= 0.062 \text{ g/cm}^2 \cdot \text{h}$$

$$w(\text{copper plate}) = \frac{(i_{\text{plating}})tM}{nF}$$

$$= \frac{(0.051)(3600)(63.546)}{(2)(96,500)}$$

$$= 0.060 \text{ g/cm}^2 \cdot \text{h}$$

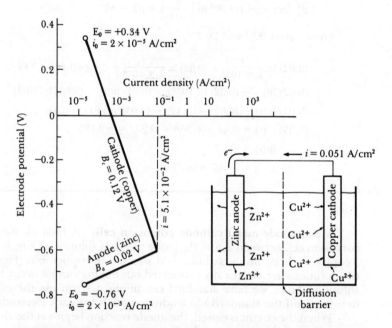

FIG. 20-13   When zinc and copper are in an electrolytic cell, the zinc corrodes and copper is plated out of solution. The current density, 0.051 A/cm², determines the rate of corrosion and plating.

20-6   Sources of Polarization

A change in the potential of an anode or cathode in an electrochemical cell is called *polarization*. Polarization occurs by three mechanisms.

Activation polarization.   An activation energy is required to force the solution of ions from the anode or the combination of electrons with ions at the cathode. We supply this energy by applying an overpotential. This *activation polarization* occurs in every electrochemical cell and leads to the Tafel slope and the polarization curves that we have just discussed. A steep Tafel slope indicates more polarization. If we increase the degree of polarization, or produce steeper Tafel slopes, the current density decreases. Polarization minimizes the rate of corrosion but reduces the rate of electroplating.

Concentration polarization.   After corrosion begins, the concentration of ions at the anode or cathode surface may change. For example, a higher concentration of metal ions may be produced at the anode if the ions are unable to diffuse rapidly into the electrolyte. Hydrogen ions may be depleted at the cathode in a hydrogen electrode, or a high OH^- concentration may develop at the cathode in an oxygen electrode. When this occurs, either the anode or cathode reaction is stifled because fewer electrons are released at the anode or accepted at the cathode.

In any of these examples, the current density decreases due to *concentration polarization*. Figure 20-14 shows the effect of depleting the hydrogen ions at the

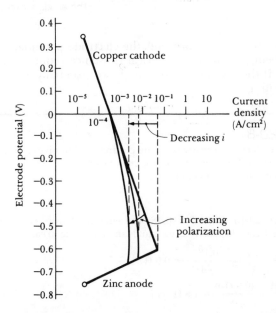

FIG. 20-14 Concentration polarization at the cathode reduces the current density in the cell. In this example, polarization may be due to depletion of copper ions in the electrolyte.

cathode in a hydrogen electrode. The cathode polarization curve bends at high current densities, giving a lower final corrosion current density. Figure 20-15 shows a similar effect when anode polarization occurs.

Normally, polarization is less pronounced when the electrolyte is highly concentrated, the temperature is increased, or the electrolyte is vigorously agitated. Each of these factors increases the current density, speeding electroplating but also encouraging electrochemical corrosion.

Resistance polarization.   There is some resistance to the flow of a current

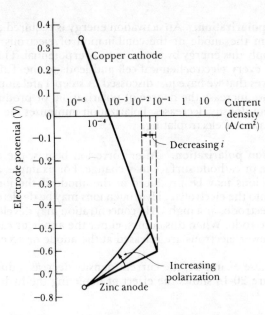

FIG. 20-15 Concentration polarization at the anode reduces the current density in the cell. In this case, polarization might be due to a buildup of excess zinc ions at the anode surface.

through the electrolyte. This prevents the electrode potentials at the anode and cathode from being equal. The electrical resistance reduces the current density (Figure 20-16). If the resistance is very high, we essentially have no circuit and no corrosion or plating.

EXAMPLE 20-8

Calculate the electrical resistivity of the electrolyte that gives a resistance polarization equivalent to 0.18 V in the copper-zinc cell. Assume that the electrodes have a surface area of 10 cm² and are 5 cm apart.

Answer:

From Figure 20-16, for $E_R = 0.18$, the current density is $i = 0.02$ A/cm². Thus assuming that the electrolyte area equals the electrode area.

$$R = \frac{E_R}{I} = \frac{E_R}{iA} = \frac{0.18}{(0.02)(10)} = 0.9 \ \Omega$$

$$\rho = \frac{RA}{l} = \frac{(0.9)(10)}{5} = 1.8 \ \Omega \cdot cm$$

20-7   Types of Electrochemical Corrosion

In this section we will look at some of the more common forms that electrochemical corrosion takes.

Uniform attack.   When a metal is placed in an electrolyte, some regions are anodic to other regions. However, the location of these regions moves and even

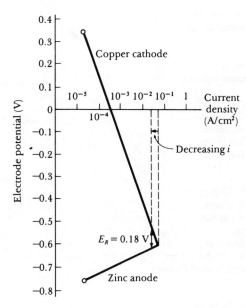

FIG. 20-16 Resistance polarization reflects the electrical resistivity of the electrolyte. A voltage E_R is required to cause conduction of the ions through the electrolyte.

reverses from time to time. Since the anode and cathode regions continually shift, the metal corrodes uniformly even without contact with a second material.

Galvanic attack. Galvanic attack occurs when certain areas always act as anodes, while other areas always act as cathodes. These electrochemical cells are called *galvanic cells* and can be separated into three types—composition cells, stress cells, and concentration cells.

Composition cells. Composition cells, or *dissimilar metal* corrosion, develop when two metals or alloys, such as copper and iron, form an electrolytic cell. Because of the effect of alloying elements and electrolyte concentration on polarization, the emf series may not tell us which regions corrode and which are protected. Instead, we use a *galvanic series*, in which the different alloys are ranked according to their anodic or cathodic tendencies in a particular environment (Table 20-3). We may find a different galvanic series for seawater, freshwater, and industrial atmospheres.

EXAMPLE 20-9

A brass fitting used in a marine application is joined by soldering with lead-tin solder. Will the brass or the solder corrode?

Answer:

From the galvanic series, we find that all of the copper-base alloys are more cathodic than a 50% Pb-50% Sn solder. Thus, the solder is the anode and corrodes.

TABLE 20-3 The galvanic series in seawater

Anodic	Magnesium
	Magnesium alloys
	Zinc
	Galvanized steel
	5052 aluminum
	3003 aluminum
	1100 aluminum
	6053 aluminum
	Alclad
	Cadmium
	2017 aluminum
	2024 aluminum
	Low-carbon steel
	Cast iron
	410 stainless steel (active)
	50% Pb-50% Sn solder
	316 stainless steel (active)
	Lead
	Tin
	Cu-40% Zn brass
	Manganese bronze
	Nickel-base alloys (active)
	Cu-35% Zn brass
	Aluminum bronze
	Copper
	Cu-30% Ni alloy
	Nickel-base alloys (passive)
	Stainless steels (passive)
	Silver
	Titanium
	Graphite
	Gold
Cathodic	Platinum

After *ASM Metals Handbook*, Vol. 10, 8th Ed., 1975.

Composition cells develop in two-phase alloys, where one phase is more anodic than the other. Since ferrite is anodic to cementite in steel, small microcells cause steel to galvanically corrode (Figure 20-17). Almost always, a two-phase alloy has less resistance to corrosion than a single-phase alloy of a similar composition.

Intergranular corrosion occurs when precipitation of a second phase or segregation at grain boundaries produces a galvanic cell. In zinc alloys, for example, impurities such as cadmium, tin, and lead segregate at the grain boundaries during solidification. The grain boundaries are anodic compared to the remainder of the grains and corrosion of the grain boundary metal occurs (Figure 20-18). In austenitic stainless steels, chromium carbides can precipitate at grain boundaries [Figure 20-17(b)]. The formation of the carbides removes chromium from the austenite adjacent to the boundaries. The low-chromium austenite at the

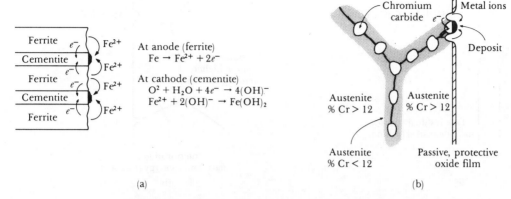

At anode (ferrite)
$$Fe \rightarrow Fe^{2+} + 2e^-$$

At cathode (cementite)
$$O^2 + H_2O + 4e^- \rightarrow 4(OH)^-$$
$$Fe^{2+} + 2(OH)^- \rightarrow Fe(OH)_2$$

(a)   (b)

FIG. 20-17   Example of microgalvanic cells in two-phase alloys. (a) In steel, ferrite is anodic to cementite. (b) In austenitic stainless steel, precipitation of chromium carbide makes the austenite in the grain boundaries anodic.

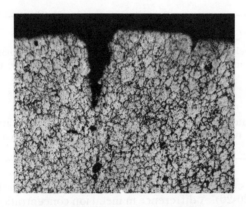

FIG. 20-18   Photomicrograph of intergranular corrosion in a zinc die casting. Segregation of impurities to the grain boundaries produces microgalvanic corrosion cells.

grain boundaries is anodic to the remainder of the grain and corrosion occurs at the grain boundaries.

Stress cells.   Stress cells develop when a metal contains regions with different local stresses. The most highly stressed, or high-energy, regions act as anodes to the less stressed cathodic areas (Figure 20-19). Regions with a finer grain size, or a higher density of grain boundaries, are anodic to coarse grain regions of the same material. Highly cold-worked areas are anodic to less cold-worked areas. High applied stresses accelerate corrosion.

Stress corrosion occurs by galvanic action but other mechanisms, such as adsorption of impurities at the tip of an existing crack, may also occur. Failure occurs as a result of corrosion and the applied stress. Higher applied stresses reduce the time required for failure.

Fatigue failures are also initiated or accelerated when corrosion occurs. *Corrosion fatigue* can reduce fatigue properties by initiating cracks, perhaps by producing pits or crevices, and by increasing the rate at which the cracks propagate.

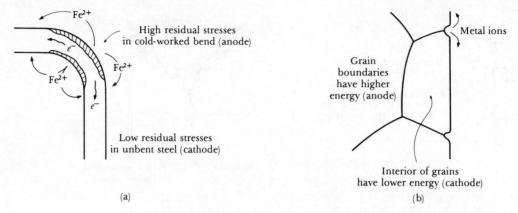

FIG. 20-19 Examples of stress cells. (a) Cold work required to bend a steel bar introduces high residual stresses at the bend, which then is anodic and corrodes. (b) Because grain boundaries have a high energy, they are anodic and corrode.

EXAMPLE 20-10

A cold-drawn steel wire is formed into a nail by additional deformation, producing the point at one end and the head at the other. Where will the most severe corrosion of the nail occur?

Answer:

Since the head and point have been cold worked an additional amount compared to the shank of the nail, the head and point serve as anodes and corrode most rapidly.

Concentration cells.   Concentration cells develop due to differences in the electrolyte (Figure 20-20). A difference in metal ion concentration causes a difference in electrode potential, according to the Nernst equation. The metal in con-

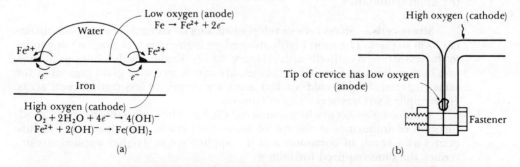

FIG. 20-20 Concentration cells. (a) Corrosion occurs beneath a water droplet on a steel plate due to low oxygen concentration in the water. (b) Corrosion occurs at the tip of a crevice due to limited access of oxygen.

tact with the most concentrated solution is the cathode; the metal in contact with the dilute solution is the anode.

The *oxygen concentration* cell (often referred to as oxygen starvation) occurs when the cathode reaction is the oxygen electrode, $2H_2O + O_2 + 4e^- \rightarrow 4OH^-$. Electrons flow from the low-oxygen region, which serves as the anode, to the high-oxygen region, which serves as the cathode.

Deposits, such as rust or water droplets, shield the underlying metal from oxygen. Consequently, the metal *under* the deposit is the anode and corrodes. This causes one form of *pitting corrosion. Waterline corrosion* is similar—metal above the waterline is exposed to oxygen, while metal beneath the waterline is deprived of oxygen. Hence, the metal under the water corrodes. Normally, the metal far below the surface corrodes more slowly than metal just below the waterline due to differences in the distance that electrons must travel. Because cracks and crevices have a lower oxygen concentration than the surrounding base metal, the tip of a crack or crevice is the anode, causing *crevice corrosion.*

Pipe buried in soil may corrode because of differences in the composition of the soil. Velocity differences may cause concentration differences—stagnant water contains low oxygen concentrations, while fast-moving aereated water contains higher oxygen concentrations; metal near the stagnant water is anodic and corrodes.

EXAMPLE 20-11

Two pieces of steel are joined mechanically by crimping the edges. Why would this be a bad idea if the steel is then exposed to water? If the water contains salt, would corrosion be affected?

Answer:

By crimping the steel edges, we produce a crevice. The region in the crevice is exposed to less air and moisture, so it behaves as the anode in a concentration cell. The steel in the crevice corrodes.

Salt in the water increases the conductivity of the water, permitting electrons and ions to move at a more rapid rate. This causes a higher current density and thus faster corrosion due to less resistance polarization.

20-8 Protection Against Electrochemical Corrosion

The problem of corrosion of metals is serious but not hopeless. A number of techniques combat corrosion, including design, coatings, inhibitors, cathodic protection, passivation, and materials selection.

Design. By properly designing metal structures, corrosion can be slowed or even avoided. Some of the factors that should be considered are as follows.

1. Prevent the formation of galvanic cells. For example, steel pipe is frequently connected to brass plumbing fixtures, producing a galvanic cell that causes the steel to corrode. By using intermediate plastic fittings to electrically insulate the steel and brass, this problem can be minimized.

2. Make the anode area much larger than the cathode area. For example, copper rivets can be used to fasten steel sheet (Figure 20-21). Because of the small area of the copper rivets, a limited cathode reaction occurs. The copper accepts few electrons and the steel anode reaction proceeds slowly. If, on the other hand, steel rivets are used for joining copper sheet, the small steel anode area gives up many electrons which are accepted by the large copper cathode area. Corrosion of the steel rivets is then very rapid.

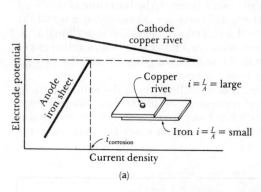

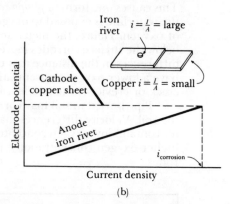

(a)   (b)

FIG. 20-21 The effect of anode area on corrosion rate. In each case iron is the anode and the current through the copper and iron must be equal. (a) If the anode area is large, the anode current density is small and corrosion occurs slowly. (b) If the anode area is small, the anode current density is large and corrosion occurs rapidly.

EXAMPLE 20-12

In Example 20-7, we assumed that the copper and zinc electrodes had the same area. Calculate the weight loss of zinc if (a) the copper cathode area is 100 cm² and the zinc anode area is 1 cm², (b) the copper cathode area is 1 cm² and the zinc anode area is 100 cm².

Answer:

The potential difference between the anode and cathode should be zero in the absence of resistance polarization.

$$\Delta E = E_a - E_c = 0$$

$$E_0^{Zn} + E_a^{Zn} - (E_0^{Cu} - E_c^{Cu}) = 0$$

$$E_a^{Zn} + E_c^{Cu} = E_0^{Cu} - E_0^{Zn}$$

$$B_a^{Zn} \ln\left(\frac{i_{Zn}}{i_0(Zn)}\right) + B_c^{Cu} \ln\left(\frac{i_{Cu}}{i_0(Cu)}\right) = E_0^{Cu} - E_0^{Zn}$$

From Tables 20-1 and 20-2

$$(0.02) \ln\left(\frac{i_{Zn}}{2 \times 10^{-5}}\right) + (0.12) \ln\left(\frac{i_{Cu}}{2 \times 10^{-5}}\right) = 0.34 + 0.76 = 1.1$$

$$(0.02) \ln (i_{Zn}) - (0.02) \ln (2 \times 10^{-5}) + (0.12) \ln (i_{Cu})$$
$$- (0.12) \ln (2 \times 10^{-5}) = 1.1$$

$$(0.02) \ln (i_{Zn}) - (0.02)(-10.82) + (0.12) \ln (i_{Cu})$$
$$- (0.12)(-10.82) = 1.1$$
$$(0.02) \ln (i_{Zn}) + (0.12) \ln (i_{Cu}) = 1.1 - 0.216 - 1.30 = -0.416$$

The current flowing in the circuit must be the same in both the copper and the zinc. Thus

$$I_{Cu} = I_{Zn}$$
$$i_{Cu} A_{Cu} = I_{Zn} A_{Zn}$$
$$i_{Cu} = (i_{Zn}) \frac{A_{Zn}}{Z_{Cu}}$$

(a) If $A_{Cu} = 100$ cm^2 and $A_{Zn} = 1$ cm^2, then $i_{Cu} = i_{Zn}/100$.

$$(0.02) \ln (i_{Zn}) + (0.12) \ln \left(\frac{i_{Zn}}{100}\right) = -0.416$$

$$(0.02) \ln (i_{Zn}) + (0.12) \ln (i_{Zn}) - (0.12) \ln (100) = -0.416$$

$$\ln (i_{Zn}) = \frac{-0.416 + (0.12)(4.605)}{0.14} = \frac{0.137}{0.14} = 0.979$$

$$i_{Zn} = 2.66 \text{ A/cm}^2$$

$$I_{Zn} = (2.66)(1 \text{ cm}^2) = 2.66 \text{ A}$$

$$w = \frac{ItM}{nF} = \frac{(2.66)(3600)(65.37)}{(2)(96,500)} = 3.24 \text{ g/h}$$

(b) If $A_{Cu} = 1$ cm^2 and $A_{Zn} = 100$ cm^2, then $i_{Cu} = 100 i_{Zn}$.

$$(0.02) \ln (i_{Zn}) + (0.12) \ln (100 i_{Zn}) = -0.416$$

$$(0.02) \ln (i_{Zn}) + (0.12) \ln (i_{Zn}) + (0.12) \ln (100) = -0.416$$

$$(0.14) \ln (i_{Zn}) = -0.416 - (0.12)(4.605) = -0.969$$

$$\ln (i_{Zn}) = \frac{-0.969}{0.14} = -6.92$$

$$i_{Zn} = 9.9 \times 10^{-4} \text{ A/cm}^2$$

$$I_{Zn} = (9.9 \times 10^{-4})(100 \text{ cm}^2) = 0.099 \text{ A}$$

$$w = \frac{ItM}{nF} = \frac{(0.099)(3600)(65.37)}{(2)(96,500)} = 0.12 \text{ g/h}$$

When the zinc anode area is much larger than the copper cathode area, the corrosion rate of the zinc is greatly reduced.

3. Design components so that fluid systems are closed rather than open and so that stagnant pools of liquid do not collect. Partly filled tanks undergo waterline corrosion. Open systems continuously dissolve gas, providing ions that participate in the cathode reaction and encourage concentration cells.

4. Avoid crevices between assembled or joined materials (Figure 20-22). Welding may be a better joining technique than brazing, soldering, or mechanical fasteners. Galvanic cells develop in brazing or soldering, since the filler metals have a different composition than the metal being joined. Mechanical fasteners produce crevices that lead to concentration cells. However, if the filler metal is closely matched to the base metal, welding may prevent these cells from developing.

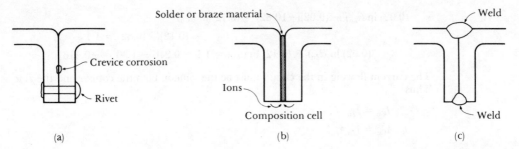

FIG. 20-22   Alternative methods for joining two pieces of steel. (a) Fasteners may produce a concentration cell, (b) brazing or soldering may produce a composition cell, and (c) welding with a filler metal that matches the base metal may avoid galvanic cells.

Coatings.   Coatings are used to isolate the anode and cathode regions. Temporary coatings, such as grease or oil, provide some protection but are easily disrupted. Organic coatings, such as paint, or ceramic coatings, such as enamel or glass, provide better protection. However, if the coating is disrupted, a small anodic site is exposed that undergoes rapid, localized corrosion.

Metallic coatings include tin-plated and galvanized (zinc-plated) steel (Figure 20-23). A continuous coating of either metal isolates the steel from the electrolyte. However, when the coating is scratched, exposing the underlying steel, the two coatings behave differently. The zinc continues to be effective, since zinc is anodic

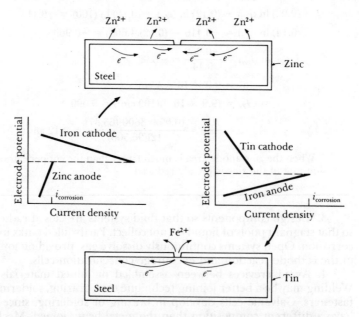

FIG. 20-23   Tin-plated steel and zinc-plated steel are protected differently. Zinc protects steel even when the coating is scratched, since zinc is anodic to steel. Tin does not protect steel when the coating is disrupted, since steel is anodic with respect to tin.

to steel. Since the area of the exposed steel cathode is small, the zinc coating corrodes at a very slow rate and the steel remains protected. However, steel is anodic to tin, so a tiny steel anode is created when tinplate is scratched. Rapid corrosion of the steel subsequently occurs. Composite materials, such as Alclad, are used to improve corrosion resistance of high-strength two-phase aluminum alloys.

Chemical conversion coatings are produced by a chemical reaction with the surface. Liquids such as zinc acid orthophosphate solutions form an adherent phosphate layer on the metal surface. The phosphate layer is, however, rather porous and is more often used to improve paint adherence. Stable adherent nonporous nonconducting oxide layers form on the surface of aluminum, chromium, and stainless steel. These oxides exclude the electrolyte and prevent galvanic cells.

Inhibitors. Some chemicals, added to the electrolyte solution, migrate preferentially to the anode or cathode surface and produce concentration or resistance polarization (Figure 20-24). Chromate salts perform this function in automobile radiators. However, if insufficient inhibitor is added, local small anode areas are unprotected and corrosion is accelerated.

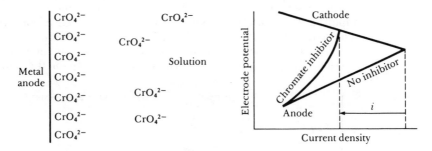

FIG. 20-24 Inhibitors may concentrate at the anode, causing severe concentration polarization, and significantly reduce the rate of corrosion of the anode.

Cathodic protection. We can protect against corrosion by supplying the metal with electrons and forcing the metal to be a cathode (Figure 20-25). Cathodic protection can be produced by using a sacrificial anode or an impressed voltage.

A *sacrificial anode* is attached to the material to be protected, forming an electrochemical circuit. The sacrificial anode corrodes, supplies electrons to the metal, and thereby prevents an anode reaction at the metal. The sacrificial anode, typically zinc or magnesium, is consumed and must eventually be replaced. Applications include preventing corrosion of buried pipelines, ships, off-shore drilling platforms, and water heaters.

An *impressed voltage* is obtained from a direct current source connected between an auxiliary anode and the metal to be protected. Essentially, we have a battery connected so that electrons flow to the metal, causing the metal to be the cathode. The auxiliary anode, such as scrap iron, corrodes.

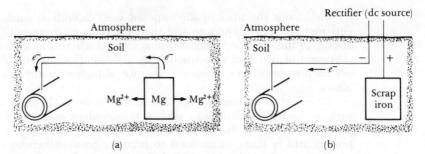

FIG. 20-25   Cathodic protection of a buried steel pipeline. (a) A sacrificial magnesium anode assures that the galvanic cell makes the pipeline the cathode. (b) An impressed voltage between a scrap iron auxiliary anode and the pipeline assures that the pipeline is the cathode.

Passivation or anodic protection.   Metals near the anodic end of the galvanic series are *active* and serve as anodes in most electrolytic cells. However, if these metals are made *passive,* or more cathodic, they corrode at slower rates than normal. *Passivation* is accomplished by producing strong anodic polarization, preventing the normal anode reaction. Thus the term *anodic protection.*

We can produce passivation by exposing the metal to highly concentrated oxidizing solutions. If iron is dipped in very concentrated nitric acid, the iron rapidly and uniformly corrodes to form a thin protective iron hydroxide coating. The coating protects the iron from subsequent corrosion in nitric acid.

We can also cause passivation by increasing the potential on the anode above a critical level (Figure 20-26). An overvoltage E_1 on the anode causes rapid

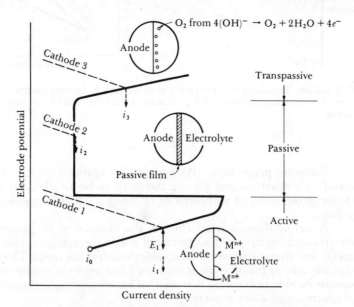

FIG. 20-26   Anodic protection, or passivation, is achieved when a large overvoltage causes a passive film to coat the anode surface. Thus, the current density is reduced. If the overvoltage is too high, evolution of oxygen upsets the passive film and the current density again increases.

corrosion. But when the applied potential is increased, as with cathode 2, a passive film forms on the metal surface, causing strong anode polarization, and the current decreases to a very low level, i_2. The metal is now in a passive state.

Passivation of aluminum is called *anodizing* and a thick oxide coating is produced. This oxide layer can be dyed to produce attractive colors. The coating remains passive even when the impressed voltage is removed. However, if the potential of the anode is too high, the metal enters the *transpassive* region; if the passive film breaks down, a high corrosion current causes rapid pitting.

Materials selection and treatment.   Corrosion can be prevented or minimized by selecting appropriate materials and heat treatments. In castings, for example, segregation causes tiny localized galvanic cells which accelerate corrosion. We can improve corrosion resistance with an homogenization heat treatment. When metals are formed into finished shapes by bending, differences in the amount of cold work and residual stresses cause local stress cells. These may be minimized by a stress relief anneal or a full recrystallization anneal.

The heat treatment is particularly important in austenitic stainless steels (Figure 20-27). When the steel cools slowly from 870°C to 425°C, chromium carbides precipitate at the grain boundaries. The austenite at the grain boundaries may contain less than 12% chromium, which is the minimum required to produce a passive oxide layer. The steel is sensitized. Because the grain boundary regions

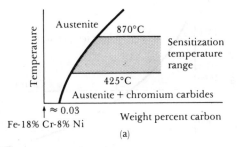

(a)

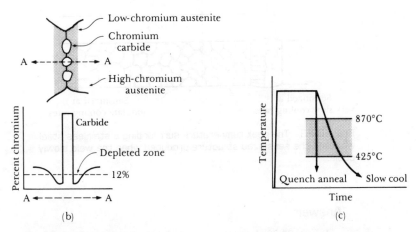

(b)

(c)

FIG. 20-27   (a) Intergranular corrosion in austenitic stainless steel. (b) Slow cooling permits chromium carbides to precipitate at grain boundaries. (c) A quench anneal to dissolve the carbides may prevent intergranular corrosion.

are small and highly anodic, rapid corrosion of the austenite at the grain boundaries occurs. There are several techniques by which we can minimize the problem.

1. If the steel contains less than 0.03% C, the chromium carbides do not form.

2. If the percent chromium is very high, the austenite may not be depleted to below 12% Cr, even if the chromium carbides form.

3. Addition of titanium or niobium ties up the carbon as TiC or NbC, preventing the formation of chromium carbide. The steel is said to be stabilized.

4. The sensitization temperature range—425°C to 870°C—should be avoided during manufacture or service.

5. In a quench anneal heat treatment, the stainless steel is heated above 800°C, causing the chromium carbides to dissolve. The structure, now containing 100% austenite, is rapidly quenched to prevent formation of carbides.

EXAMPLE 20-13

The peak temperatures obtained near a weld on a 304 stainless steel are shown in Figure 20-28. Where will corrosion most likely occur as the cooling rate after welding increases?

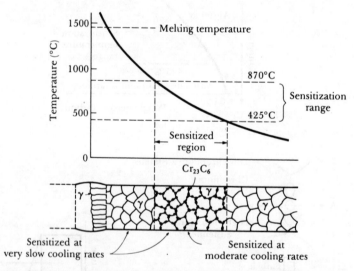

FIG. 20-28   The peak temperature surrounding a stainless steel weld and the sensitized structure produced when the weld slowly cools.

Answer:

Figure 20-28 shows the region of the weld that will heat above the sensitization range and the region that will be held in the sensitization range. The region that is held in the sensitization range may eventually contain precipitated carbides. Carbides will precipitate in the other region only if the weld cools slowly. Thus, for slow cooling, the entire heat-affected area may be sensitized and will corrode. For fast cooling, only the region that heated into the sensitization range will corrode.

20-9 Oxidation and Other Gas Reactions

Materials of all types may react with oxygen and other gases. These reactions can, like corrosion, alter the composition, properties, or integrity of the material.

Oxidation of metals. Metals may react with oxygen to produce an oxide at the surface. We are interested in three aspects of this reaction—the ease with which the metal oxidizes, the nature of the oxide film that forms, and the rate at which oxidation occurs.

The ease with which oxidation occurs is given by the *free energy of formation* for the oxide (Figure 20-29). There is a large driving force for the oxidation of magnesium and aluminum, but there is little tendency for the oxidation of nickel or copper.

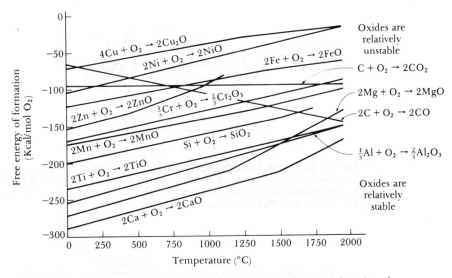

FIG. 20-29   The free energy of formation of selected oxides as a function of temperature. A large negative free energy indicates a more stable oxide.

EXAMPLE 20-14

Explain why we should not add alloying elements such as chromium to pig iron before the pig iron is converted to steel in a basic oxygen furnace at 1700°C.

Answer:

In a basic oxygen furnace, we lower the carbon content of the metal from about 4% to much less than 1% by blowing pure oxygen through the molten metal. If chromium were already present before the steel making began, Figure 20-29 tells us that chromium would oxidize before the carbon, since chromium oxide has a lower free energy of formation, or is more stable, than carbon monoxide (CO). All of the expensive chromium that we add is lost before the carbon is removed from the pig iron.

The type of oxide film determines the rate at which oxidation occurs and whether the oxide causes the metal to be passive. Three types of behavior are observed, depending on the relative volumes of the oxide and the metal (Figure 20-30). We can determine this ratio from the Pilling-Bedworth equation for the following oxidation reaction.

$$nM + mO_2 \rightarrow M_nO_{2m} \tag{20-12}$$

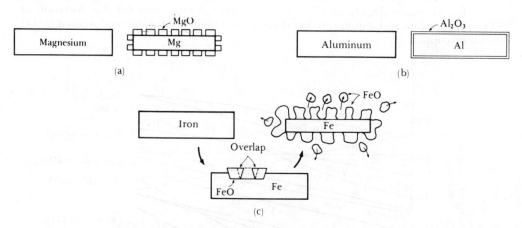

(a)

(b)

(c)

FIG. 20-30   Three types of oxides may form, depending on the volume ratio between the metal and the oxide. (a) Magnesium produces a porous oxide film, (b) aluminum forms a protective adherent nonporous oxide film, and (c) iron forms an oxide film that spalls off the surface and provides poor protection.

The *Pilling-Bedworth ratio* is

$$\text{P-B ratio} = \frac{\text{oxide volume per metal atom}}{\text{metal volume per atom}}$$

$$= \frac{(M_{\text{oxide}})(\rho_{\text{metal}})}{n(M_{\text{metal}})(\rho_{\text{oxide}})} \tag{20-13}$$

where M is the atomic or molecular mass, ρ is the density, V is the volume, and n is the number of metal atoms in the oxide, as defined in Equation 20-12. Examples of the Pilling-Bedworth ratio for several metal-oxide combinations are shown in Table 20-4.

If the Pilling-Bedworth ratio is less than one, the oxide occupies a smaller

TABLE 20-4   The Pilling-Bedworth ratio for selected
metal-metal oxide systems

Metal and Oxide	Density of Oxide ($Mg\ m^{-3}$)	Pilling-Bedworth Ratio
Mg-MgO	3.6	0.8
Al-Al_2O_3	4.0	1.3
Ti-TiO_2	5.1	1.5
Zr-ZrO_2	5.6	1.5
Fe-Fe_2O_3	5.3	2.1
Cr-Cr_2O_3	5.1	2.1
Cu-Cu_2O	6.2	1.6
Ni-NiO	6.9	1.6
Si-SiO_2	2.7	1.9
U-UO_2	11.1	1.9
W-WO_3	7.3	3.3

From J. West, *Basic Corrosion and Oxidation*, 1980, Ellis Horwood.

volume than the metal from which it formed. The oxide film is porous, typical of
metals such as magnesium.

If the ratio is equal to one, the volumes of the oxide and metal are nearly
equal and an adherent, nonporous and protective film forms, typical of aluminum
and chromium.

If the ratio is greater than one, the oxide is greater than that of the metal and,
initially, the oxide forms a protective layer. However, as the thickness of the film
increases, high tensile stresses develop in the oxide. The oxide may flake from the
surface, exposing fresh metal which continues to oxidize. This nonadherent oxide
layer is typical of iron.

EXAMPLE 20-15

The density of aluminum is 2.7 g/cm^3 and that of Al_2O_3 is about 4 g/cm^3. Describe
the characteristics of the aluminum oxide film. Compare to the oxide film that
forms on tungsten; the density of tungsten is 19.254 g/cm^3 and that of WO_3 is 7.3
g/cm^3.

Answer:

For 2Al + $\frac{3}{2}O_2 \rightarrow Al_2O_3$, the molecular weight of Al_2O_3 is 101.96 and that of
aluminum is 26.981.

$$\text{P-B} = \frac{M_{Al_2O_3}\rho_{Al}}{nM_{Al}\rho_{Al_2O_3}} = \frac{(101.96)(2.7)}{(2)(26.981)(4)} = 1.28$$

For tungsten, W + $\frac{3}{2}O_2 \rightarrow WO_3$, the molecular weight of WO_3 is 231.85 and that
of tungsten is 183.85.

$$\text{P-B} = \frac{M_{WO_3}\rho_W}{nM_W\rho_{WO_3}} = \frac{(231.85)(19.254)}{(1)(183.85)(7.3)} = 3.33$$

Since P-B $\simeq$ 1 for aluminum, the Al_2O_3 film is nonporous and adherent, providing protection to the underlying aluminum. However, P-B > 1 for tungsten, so the WO_3 is nonadherent and nonprotective.

The rate at which oxidation occurs depends on the access of oxygen to the metal atoms. A *linear* rate of oxidation occurs when the oxide is porous (as in magnesium) and oxygen has continued access to the metal surface.

$$y = kt \tag{20-14}$$

where y is the thickness of the oxide, t is the time, and k is a constant that depends on the metal and temperature.

A *parabolic* relationship is observed when diffusion of ions or electrons through a nonporous oxide layer is the controlling factor. This relationship is observed in iron, copper, and nickel.

$$y = \sqrt{kt} \tag{20-15}$$

Finally, a *logarithmic* relationship is observed for the growth of thin oxide films that are particularly protective, as for aluminum and possibly chromium.

$$y = k \ln (ct + a) \tag{20-16}$$

where k, a, and c are constants.

EXAMPLE 20-16

At 1000°C, pure nickel follows a parabolic oxidation curve given by the constant $k = 3.9 \times 10^{-12}$ cm^2/s in an oxygen atmosphere. If this relationship is not affected by the thickness of the oxide film, calculate the time required for a 0.1-cm nickel sheet to completely oxidize.

Answer:

Assuming the sheet oxidizes from both sides

$$y = \sqrt{kt} = \sqrt{(3.9 \times 10^{-12})(t)} = \frac{0.1 \text{ cm}}{2 \text{ sides}} = 0.05 \text{ cm}$$

$$t = \frac{(0.05)^2}{3.9 \times 10^{-12}} = 6.4 \times 10^8 \text{ s} = 20.5 \text{ years}$$

Selective oxidation.   The curves in Figure 20-29 show that different metals have different tendencies to oxidize. The metal having the largest negative free energy of formation oxidizes first when more than one metal is present. This behavior is used to refine metals. In steel making, oxygen blown into liquid pig iron reacts first with carbon, rather than reacting with iron. The carbon content of pig iron is lowered until the desired steel is produced. We utilize the same principle in designing oxidation-resistant alloys. When chromium is added to steel, the chromium oxidizes first, produces a protective chromium oxide film, and protects the underlying metal.

Another example of selective oxidation is the *decarburization* of steel during hot working or heat treatment. Oxygen reacts with the carbon at the surface of the steel, leaving behind a low-carbon surface layer that reduces the mechanical properties of the steel at the surface.

Oxidation of ceramics.   Most oxide ceramics are not significantly affected by oxygen, even at high temperatures. However, the high-temperature usefulness of silicon carbide, boron nitride, and graphite is limited by their oxidation.

Oxidation of polymers.   Exposure of polymers to oxygen, usually at elevated temperatures, alters the polymerized structure. In rubber, for example, oxygen provides additional cross-linking, causing the rubber to become harder. Oxygen may also cause depolymerization, or chain scission, permitting small molecules to escape as a gas, or cause charring or even burning of the polymer at high temperatures. Polymers based on silicon rather than carbon backbones are more resistant to oxidation and can be used at higher temperatures.

20-10   Radiation Damage

A variety of problems occur in materials exposed to radiation, causing changes in the structure and properties of the material.

Metals.   High-energy radiation, such as neutrons, may knock an atom out of its normal lattice site, creating interstitials and vacancies. These point defects reduce electrical conductivity and cause ductile materials to become harder and more brittle. Annealing may help eliminate radiation damage.

Ceramics.   Radiation also creates point defects in ceramic materials. Normally, little effect on mechanical properties is observed, since ceramics are brittle, but physical properties, such as thermal conductivity and optical properties, may be impaired.

Polymers.   Even low-energy radiation can alter the structure in polymers. Chains can be broken, reducing the degree of polymerization or causing branching to occur, thus reducing the strength of the polymer.

20-11   Wear and Erosion

Wear and erosion remove material from a component by mechanical attack of solids or liquids. Corrosion and mechanical failure also contribute to this type of attack.

Adhesive wear.   Adhesive wear, also known as scoring, galling, or seizing, occurs when two solid surfaces slide over one another under pressure. Surface projections, or *asperities*, are plastically deformed and eventually welded together by the high local pressures (Figure 20-31). As sliding continues, these bonds are broken, producing cavities on one surface, projections on the second surface, and frequently tiny abrasive particles, all of which contribute to further wear of the surfaces.

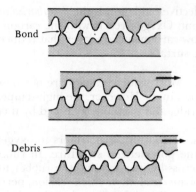

FIG. 20-31   The asperities on two rough surfaces may initially be bonded. A sufficient force breaks the bonds and the surfaces slide. As they slide, asperities may be fractured, wearing away the surfaces and producing debris.

Several factors help minimize the rate of adhesive wear.

1. Low loads reduce the rate of wear.

2. If both surfaces have high hardnesses that are approximately the same, the wear rate is low.

3. Smooth surfaces reduce the likelihood that asperities will bond, giving slower wear.

4. Prevention of adhesions will minimize wear. Some materials adsorb gases or form oxides at the surface which prevent adhesion, particularly at low loads. Gray cast iron contains graphite flakes which provide excellent self-lubrication at the surfaces, particularly when the matrix is pearlite rather than soft ferrite. Finally, adhesive wear is less likely when a lubricant is used.

Abrasive wear.   Abrasive wear occurs when material is removed from the surface by contact with hard particles, which may either be present at the surface of a second material or may be present as loose particles between two surfaces (Figure 20-32). Unlike adhesive wear, no bonding occurs. This type of wear is common in machinery such as plows, scraper blades, crushers, and grinders used to handle abrasive materials and may also occur when hard particles are uninten-

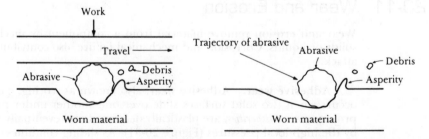

FIG. 20-32   Abrasive wear, caused either by trapped or free-flying abrasives, produces troughs in the material, piling up asperities which may fracture into debris.

tionally introduced into moving parts of machinery. Abrasive wear is also used for grinding operations to intentionally remove material.

Materials with a high hardness and high hot strength are most resistant to abrasive wear. Typical materials used for abrasive wear applications include quenched and tempered steels, carburized or surface-hardened steels, manganese steels which strengthen by work hardening during use, cobalt alloys such as Stellite, composite materials, white cast irons, and hard surfaces produced by welding.

Liquid erosion. The integrity of a material may be destroyed by erosion caused by high pressures associated with a moving liquid. The liquid causes strain hardening of the metal surface, leading to localized deformation, cracking, and loss of material. Two types of liquid erosion deserve mention.

Cavitation occurs when a liquid containing a dissolved gas enters a low-pressure region. Gas bubbles, which precipitate and grow in the liquid, collapse when the pressure subsequently increases (Figure 20-33). The high pressure, local

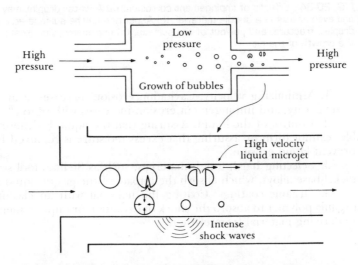

FIG. 20-33 Cavitation occurs when gas bubbles grow from the liquid in a low-pressure region, then collapse on reentering a high-pressure region. Collapse due to the implosion of the gas bubbles creates high intensity shock waves or high velocity microjets of liquid which erode the material surface.

shock wave that is produced may exert a pressure of thousands of atmospheres against the surrounding material. Cavitation is frequently encountered in propellors, dams and spillways, and hydraulic pumps.

Liquid impingement occurs when liquid droplets carried in a rapidly moving gas strike a metal surface (Figure 20-34). High localized pressures develop because of the initial impact and the rapid lateral movement of the droplets from the impact point along the metal surface. Water droplets carried by steam may erode turbine blades in steam generators and nuclear power plants.

Liquid erosion can be minimized by proper materials selection and design, including the following.

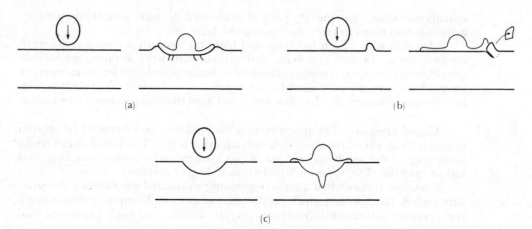

FIG. 20-34   Effects of impingement corrosion. (a) A water droplet may create a crater and even cracks in a ductile material. (b) Asperities on the surface arrest the spreading droplet, fracture, and pull out of the surface. (c) Impingement on existing pits accelerates the growth of the pit.

1. Minimizing velocity. Cavitation erosion increases at an exponential rate with velocity, and impingement erosion increases with v^5 to v^6.

2. Control of the liquid. Assuring that the liquid is deaereated so that bubbles cannot form or assuring that excess moisture is removed from steam helps prevent liquid erosion.

3. Selecting hard, tough materials, such as Stellite, tool steels, titanium, or nickel-base alloys, which absorb the impact of the gas or liquid droplets.

4. Organic coatings. Coating the material with an elastomer permits the organic polymer to absorb the shock of cavitation or impingement and protect the underlying material from erosion.

SUMMARY

Deterioration of a material by corrosion is primarily a chemical process which is determined by the mutual solubility of materials or the tendency of a material to give up its electrons in an electrochemical cell. To a certain extent, we can control this deterioration by excluding the corrosive environment, including the use of coatings. However, the structure of the material also plays an important role in many corrosive situations; producing more homogeneous structures free of stresses and high-energy locations by controlling the composition, processing, and heat treatment will improve corrosion resistance.

Deterioration by wear is reduced by improved control over the structure and properties of the material. In particular, alloy selection and materials processing which provide high hardness lead to lower rates of wear.

GLOSSARY

Abrasive wear   Removal of material from surfaces by the cutting action of hard particles.

Adhesive wear   Removal of material from surfaces of moving equipment by momentary local bonding, then bond fracture, at the surfaces.

Anode   The location at which corrosion occurs as electrons and ions are given up.

Anodizing   An anodic protection technique in which a thick oxide layer is deliberately produced on a metal surface.

Asperities   Small projections on the surface of moving parts that participate in adhesive wear.

Cathode   The location at which electrons are accepted and a by-product is produced during corrosion.

Cavitation   Erosion of a material surface by the pressures produced when a gas bubble collapses within a moving liquid.

Chemical corrosion   Removal of atoms from a material by virtue of the solubility or chemical reaction between the material and the surrounding liquid.

Composition cells   Electrochemical corrosion cells produced between two materials having a different composition. Also known as a galvanic cell.

Concentration cells   Electrochemical corrosion cells produced between two locations on a material at which the composition of the electrolyte is different.

Corrosion fatigue   Accelerated failure of a material by combined corrosion and a cyclic load.

Crevice corrosion   A special concentration cell in which corrosion occurs in crevices because of the low concentration of oxygen.

Decarburization   Selective oxidation of carbon from the surface of a steel.

Dezincification   A special chemical corrosion process by which both zinc and copper atoms are removed from brass, but the copper is replated back onto the metal.

Electrochemical cell   A cell in which electrons and ions can flow by separate paths between two materials, producing a current which in turn leads to corrosion or plating.

Electrochemical corrosion   Corrosion produced by the development of a current in an electrochemical cell which removes ions from the material.

Electrode potential   Related to the tendency of a material to corrode. The potential is the voltage produced between the material and a standard electrode.

Electrolyte   The conductive medium through which ions move to carry current in an electrochemical cell.

Electroplating   The precipitation of ions on the cathode in an electrochemical cell.

Emf series   The arrangement of elements according to their electrode potential, or their tendency to corrode.

Exchange current density   The local flow of current or electrons on the surface of a material when the circuit is not completed.

Galvanic series   The arrangement of alloys according to their tendency to corrode in a particular environment.

Graphitic corrosion   A special chemical corrosion process by which iron is leached from cast iron, leaving behind a weak spongy mass of graphite.

Hydrogen electrode   The cathode reaction at which electrons and hydrogen ions from the corrosion circuit combine to produce hydrogen gas.

Impingement   Erosion of a material surface due to collisions with a rapidly moving liquid.

Impressed voltage   A cathodic protection technique by which a direct current is introduced into the material to be protected, thus preventing the anode reaction.

Inhibitors   Additions to the electrolyte that preferentially migrate to the anode or cathode, cause polarization, and reduce the rate of corrosion.

Intergranular corrosion   Corrosion at grain boundaries because grain boundary segregation or precipitation produce local galvanic cells.

Overvoltage   The voltage introduced externally or which develops in a corrosion cell which causes the current to flow and produce corrosion or plating.

Oxidation Reaction of a metal with oxygen to produce a metallic oxide. This normally occurs most rapidly at high temperatures.

Oxidation reaction The anode reaction, by which electrons are given up to the electrochemical cell.

Oxygen electrode The cathode reaction by which electrons and ions combine to produce OH^- ion groups in the electrolyte.

Oxygen starvation The concentration cell in which low-oxygen regions of the electrolyte cause the underlying material to behave as the anode and corrode.

Passivation Producing strong anodic polarization by causing a protective coating to form on the anode surface and interrupt the electric circuit.

Pilling-Bedworth ratio Describes the type of oxide film that forms on a metal surface during oxidation.

Polarization Changing the voltage between the anode and cathode to reduce the rate of corrosion. Activation polarization is related to the energy required to cause the anode or cathode reaction; concentration polarization is related to changes in the composition of the electrolyte; and resistance polarization is related to the electrical resistivity of the electrolyte.

Reduction reaction The cathode reaction, by which electrons are accepted from the electrochemical cell.

Sacrificial anode Cathodic protection by which a more anodic material is connected electrically to the material to be protected. The anode corrodes to protect the desired material.

Stress cells Electrochemical corrosion cells produced by differences in imposed or residual stresses at different locations in the material.

Stress corrosion Deterioration of a material in which an applied stress accelerates the rate of corrosion.

Tafel constant Describes the activation polarization when an electrochemical cell is completed.

Water electrode The cathode reaction by which electrons and ions combine to produce water in the electrolyte.

PRACTICE PROBLEMS

1 Suppose 3 g of Ag^+ ions are dissolved in 1000 g of water to produce an electrolyte. Calculate the electrode potential of the silver half-cell.

2 Suppose 25 g of Ni^{2+} ions are dissolved in 1000 g of water to produce an electrolyte. Calculate the electrode potential of the nickel half-cell.

3 A half-cell produced by dissolving zinc in water produces an electrode potential of -0.77 V. Calculate the amount of zinc that must be added to 1000 g of water to produce this potential.

4 A half-cell produced by dissolving aluminum in water produces an electrode potential of -1.69 V. Calculate the amount of aluminum that must be added to 1000 g of water to produce this potential.

5 An electrode potential in a gold half-cell is found to be 1.48 V. Determine the concentration of Au^{3+} ions in the electrolyte.

6 (a) Determine the rate at which silver is electroplated onto a 20 cm^2 cathode when a current of 25 A is produced in the circuit. (b) Determine the thickness of the silver plate after 1 h.

7 What current is required to produce a 0.001-cm thick gold layer on a 4-cm^2 cathode in 30 min?

8 Zinc is to be plated onto a cathode using a current of 15 A. In 3 h, a thickness of 0.08 cm is produced. Calculate the area of the cathode that was used.

9 A 0.05-cm thick layer of nickel is to be electroplated onto an automobile bumper having an area of 5000 cm^2 using a current of 7.5 A. How long will this take?

10 A current density of 5×10^{-3} A/cm^2 is applied to a cathode during electroplating of copper. What overpotential is required?

11 A current density of 9×10^{-2} A/cm^2 is applied to a cathode during electroplating of zinc. What overpotential is required?

12 Calculate (a) the current density and (b) the rate of plating of copper when a potential of 0.5 V is applied to a 10,000-cm^2 cathode.

13 Calculate (a) the current density and (b) the rate of plating of copper when a potential of 1.2 V is applied to a 5000-cm^2 cathode.

14 Calculate (a) the current density and (b) the rate of plating of zinc from a zinc anode to a 500-cm² zinc cathode if a 0.3-V potential is introduced.

15 Suppose copper and iron compose an electrochemical cell. Determine the amount of iron lost and copper plated per hour if the anode and cathode are both 100 cm².

16 Suppose copper and iron compose an electrochemical cell. Determine the amount of iron lost and copper plated per hour if the iron anode is 1 cm² and the copper cathode is 100 cm².

17 Suppose copper and iron compose an electrochemical cell. Determine the amount of iron lost and copper plated per hour if the iron anode is 100 cm² and the copper cathode is 1 cm².

18 Calculate the current and the rate of corrosion of a 150-cm² iron anode connected to a standard hydrogen electrode if the electrolyte at the iron anode contains 8 g of iron per 1000 g of water.

19 A scratch having a 2-cm² area is made in the galvanized coating on a 400-cm² iron sheet, exposing the underlying iron. Estimate the current produced when the zinc coating begins to corrode.

20 When a 75-cm² zinc anode is connected to a hydrogen electrode, a contact potential of 0.2 V is observed. If the anode and cathode are 4 cm apart, determine (a) the current density and (b) the electrical resistivity of the electrolyte.

21 Suppose the electrolyte in a copper-zinc electrolytic cell has a resistivity of 1.5 Ω · cm. If the anode and cathode surface areas are each 100 cm² and are 5 cm apart, calculate the current produced with and without the resistance polarization.

22 A zinc plate is attached to an iron plate by an aluminum bolt, as shown in Figure 20-35. Indicate the locations at which corrosion will occur and explain why corrosion occurs at these locations.

23 Annealed copper sheets are joined by a very slow welding process. Corrosion occurs in the base metal about 1 cm away from the edge of the fusion zone. Describe the changes in structure in the heat-affected zone during welding and explain why corrosion occurs where it does.

24 Cast iron pipe is coated on the inside with tar, which provides a protective coating. Acetone in a chemical laboratory is routinely drained through the pipe. After some period of time the pipe begins to corrode. Explain why.

25 Explain why you would expect the point and head of a nail to corrode more rapidly than the shank of the nail. Why will the shank corrode?

26 A 1060 steel is exposed to the environment except for a few spatters of paint. After a period of

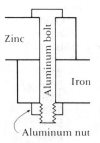

FIG. 20-35
Joining of zinc and iron plate with an aluminum bolt produces a corrosion cell. (See Problem 22.)

time, the exposed portion of the steel is slightly rusted and corrosion pits are observed under the paint spatters. Explain why both areas corrode.

27 Suppose a plumber designs a piping system in which a short length of pipe is a dead-end. Where will corrosion probably occur? Explain.

28 Tubing made from 70% Cu-30% Zn brass is used to provide cooling water in a recirculating cooling system. After several years, the tubing begins to leak. The outside of the tubing *appears* to be solid material. Explain the probable cause for the leakage.

29 An aluminum nut is very securely tightened onto a bolt used in a marine environment. After several months, the nut contains numerous cracks. Explain why cracking might have occurred.

30 A weld in an austenitic stainless steel is found to corrode in the heat-affected zone a short distance from the fusion zone. Explain why this occurs, why the weld and base metal do not corrode, and how the problem could be corrected.

31 An anodized aluminum sheet is connected to a lead sheet and, contrary to the emf and galvanic series, the lead sheet corrodes. Explain why this might be reasonable.

32 Suppose steel is plated with nickel. Describe the conditions under which the steel will be protected. Compare to the case when steel is coated with a thin layer of aluminum.

33 What effect will an increase in the electrical resistivity of the soil have on the corrosion rate of a buried steel pipeline? Explain.

34 Determine the Pilling-Bedworth ratio for the following metals and predict the behavior of the oxide that forms on the surface.

Metal	Metal Density ($Mg\ m^{-3}$)	Oxide	Oxide Density ($Mg\ m^{-3}$)
Ca	1.55	CaO	2.62
K	0.86	K_2O	2.32
Ti	4.54	TiO_2	4.26
Mn	7.44	MnO	5.43
Cr	7.19	Cr_2O_3	5.21
Li	0.534	Li_2O	2.013
Zn	7.133	ZnO	5.61
Sb	6.62	Sb_2O_3	5.2

35 A steel held at 700°C for 3000 h is oxidized to a depth of 1.16 mm below the surface. If steel oxidizes at a parabolic rate, determine (a) the constant k and (b) the weight loss in grams from a 30 cm × 20 cm × 10 cm steel block after 1 year. The density of the steel is 7.8 $Mg\ m^{-3}$.

CHAPTER 21

Failure—Origin, Detection, and Prevention

21-1 Introduction

In spite of our understanding of the behavior of materials, failures frequently occur. The sources of these failures include improper design, materials selection, materials processing, and abuse. The engineer must anticipate potential failures and consequently exercise good design, materials and processing selection, quality control, and testing to prevent failures. When failures do occur, the engineer must determine the cause so that failures can be prevented in the future.

Although the topic of failure analysis is far too complex to cover in one chapter, we will discuss a few general principles. First, we will identify the fracture mechanism by which failure occurs. Then we will discuss some considerations that may help us prevent failures, including nondestructive testing techniques. Finally, we will examine fracture mechanics.

21-2 Determining the Fracture Mechanism in Metal Failures

Failure analysis requires a combination of technical understanding, careful observation, detective work, and common sense. Knowledge of the history of the failed component, including the applied stress, the environment, the temperature, the intended structure and properties, and unusual changes in any of these factors, help make identification of the cause of failure much easier.

An understanding of fracture mechanisms may also lead to the cause of failure. In this section we will concentrate on identifying the mechanism by which a metal fails when subjected to a stress. We will consider five common fracture mechanisms—ductile, brittle, fatigue, creep, and stress corrosion failures.

Ductile fracture. Ductile fracture normally occurs in a *transgranular* manner (through the grains) in metals that have good ductility and toughness. Often a considerable amount of deformation, including necking, is observed in the failed component. The deformation occurs before the final fracture. Ductile fractures are usually due to simple overloads, or applying too high a stress to the material.

Ductile fracture in a simple tensile test begins by the nucleation, growth, and coalescence of microvoids at the center of the test bar (Figure 21-1). *Microvoids* form when a high stress causes separation of the metal at grain boundaries or interfaces between the metal and inclusions. As the local stress continues to in-

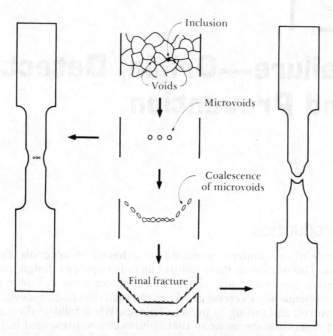

FIG. 21-1   When a ductile material is pulled in a tensile test, necking begins and voids form, starting near the center of the bar, by nucleation at grain boundaries or inclusions. As deformation continues, a 45° shear lip may form, producing a final cup and cone fracture.

crease, the microvoids grow, connect, and produce larger cavities. Eventually, the metal-to-metal contact area is too small to support the load and final fracture occurs.

Deformation by slip also contributes to the ductile fracture of a metal. We know that slip occurs when the resolved shear stress reaches the critical resolved shear stress and that the resolved shear stresses are highest at a 45° angle to the applied tensile stress (Schmid's law).

These two aspects of ductile fracture give the failed surface characteristic features that, like clues, help us to determine when a metal fails by ductile fracture. In thick metal sections, we expect to find evidence of necking, with a significant portion of the fracture surface having a flat face where microvoids first nucleated and coalesced, and a small *shear lip*, where the fracture surface is at a 45° angle to the applied stress. The shear lip, indicating that slip occurred, gives the fracture a cup and cone appearance (Figure 21-2). Simple macroscopic observation of this fracture may be sufficient to identify the mode of failure as ductile fracture.

Examination of the fracture surface at a high magnification, perhaps using a scanning electron microscope, reveals a dimpled surface (Figure 21-3). The dimples are traces of the microvoids produced during fracture. Normally, these microvoids are round, or equiaxed, when a normal tensile stress produces the failure [Figure 21-4(a)]. However, on the shear lip, the dimples are oval shaped, or

FIG. 21-2   The cup and cone fracture observed when a ductile material, in this case an annealed low carbon steel, fractures in a tensile test.

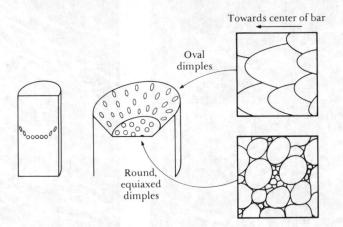

FIG. 21-3   Dimples form during ductile fracture. Equiaxed dimples form in the center where microvoids grow. Elongated dimples, pointing toward the origin of failure, form on the shear lip.

elongated, with the ovals pointing toward the center of the tensile bar where fracture began [Figure 21-4(b)].

In thin plate, less necking is observed and the entire fracture surface is a shear face (Figure 21-5). Microscopic examination of the fracture surface shows elongated dimples rather than equiaxed dimples, indicating a greater proportion of 45° slip than in the thicker metals.

EXAMPLE 21-1

A chain used to hoist heavy loads fails. Examination of the failed link indicates considerable deformation and necking prior to failure. List some of the possible reasons for failure.

Answer:

This description suggests that the chain failed in a ductile manner by a simple tensile overload. Two factors could be responsible for this failure.

1. The load exceeded the hoisting capacity of the chain. Thus, the stress due to the load exceeded the yield strength of the chain, permitting failure. Comparison of the load to the manufacturer's specifications will indicate that the chain was not intended for such a heavy load. This is the fault of the user!

2. The chain was of the wrong composition or was improperly heat treated. Consequently, the yield strength was lower than intended by the manufacturer and could not support the load. This is the fault of the manufacturer!

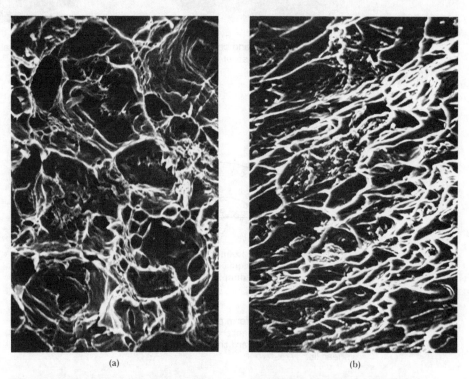

(a)   (b)

FIG. 21-4 Scanning electron micrographs of an annealed low carbon steel exhibiting ductile fracture in a tensile test. (a) Equiaxed dimples at the flat center of the cup and cone, and (b) elongated dimples at the shear lip.

FIG. 21-5 Ductile fracture of a thin annealed eutectoid steel plate in a tensile test. Necking is observed, but the entire fracture is a shear lip rather than a cup and cone.

Brittle fracture.   Brittle fracture occurs in high-strength metals or metals with poor ductility and toughness. Furthermore, metals fail in a brittle manner at low temperatures, in thick sections, at high strain rates (such as impact), or when flaws play an important role. Brittle fractures are frequently observed when impact rather than overload causes failure.

In brittle fracture, little or no plastic deformation is required. Initiation of the crack normally occurs at small flaws which cause a concentration of stress. The crack may move at a rate approaching the velocity of sound in the metal. Normally, the crack propagates most easily along specific crystallographic planes, often the {100} planes, by *cleavage*. In some cases, however, the crack may take an *intergranular* (along the grain boundaries) path, particularly when segregation or inclusions weaken the grain boundaries.

Brittle fracture can be identified by observing the features on the failed surface. Normally, the fracture surface is flat and perpendicular to the applied stress in a tensile test (Figure 21-6). If failure occurred by cleavage, each fractured

FIG. 21-6   Brittle fracture of a quenched eutectoid steel tensile bar. Because the microstructure is entirely martensite, a flat brittle fracture surface is obtained.

grain is flat, differently oriented, and gives a crystalline, or "rock candy" appearance to the fracture surface (Figure 21-7). Often the layman claims that the metal failed because it crystallized. Of course, we know that the metal was crystalline to begin with and the surface appearance is due to the cleavage faces.

Another common fracture feature is the *chevron pattern* (Figure 21-8), produced by separate crack fronts propagating at different levels in the material. A radiating pattern of surface markings, or ridges, fan away from the origin of the crack (Figure 21-9). The chevron pattern is visible with the naked eye or a magnifying glass and helps us identify both the brittle nature of the failure process as well as the origin of the failure.

EXAMPLE 21-2

An engineer investigating the cause of an automobile accident finds that the right rear wheel has broken off at the axle. The axle is bent. The fracture surface reveals a chevron pattern pointing towards the surface of the axle. Suggest a possible cause for the fracture.

Answer:

The evidence suggests that the axle did not break prior to the accident. The deformed axle means that the wheel was still attached when the load was applied. The chevron pattern indicates that the wheel was subjected to an intense impact blow, which was transmitted to the axle, causing failure. The preliminary evidence suggests that the driver lost control, crashed, and the force of the crash caused the axle to break. Further examination of the fracture surface, microstructure, composition, and properties may verify that the axle was properly manufactured.

FIG. 21-7 Scanning electron micrograph of a brittle fracture surface on a quenched 1080 steel.

FIG. 21-8 The chevron pattern in a quenched alloy steel. The steel failed in a brittle manner by an impact blow.

Fatigue fracture.   A metal fails by fatigue when an alternating stress greater than the endurance limit is applied. Fracture occurs by a three-step process involving (a) nucleation of a crack, (b) slow cyclic propagation of the crack, and (c) catastrophic failure of the metal. Cracks nucleate at locations of highest stress and lowest local strength. Normally, nucleation sites are at or near the surface, where the stress is at a maximum, and include surface defects such as scratches or pits, sharp corners due to poor design or manufacture, inclusions, grain boundaries, or dislocation concentrations.

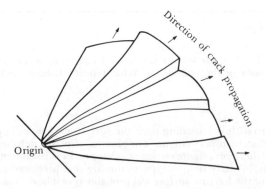

FIG. 21-9 The chevron pattern forms as the crack propagates from the origin at different levels. The pattern points back to the origin.

Once nucleated, the crack grows towards lower stress regions. Because of the stress concentration at the tip, the crack propagates a little bit further during each cycle until the load carrying capacity of the remaining metal is approached. The crack then grows spontaneously, usually in a brittle manner.

Fatigue failures are often easy to identify. The fracture surface, particularly near the origin, is typically smooth. The surface becomes rougher as the original crack increases in size and finally may be fibrous during final crack propagation.

Microscopic and macroscopic examination reveal a fracture surface including a beach mark pattern and striations (Figure 21-10). *Beach marks* are normally formed when the load is changed during service or when the loading is intermittent, perhaps permitting time for oxidation inside the crack. *Striations*, which are on a much finer scale, may show the position of the crack tip after each cycle. Observation of beach marks always suggests a fatigue failure; unfortunately the absence of beach marks does not rule out fatigue failure.

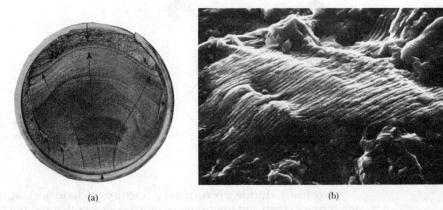

(a)   (b)

FIG. 21-10 Fatigue fracture surface. (a) At low magnifications, the beach mark pattern indicates fatigue as the fracture mechanism and points to the origin of the failure. (b) At very high magnifications, closely spaced striations formed during fatigue are observed. (a) From C. A. Cottell, "Fatigue Failures with Special Reference to Fracture Characteristics," *Failure Analysis: The British Engine Technical Reports*, American Society for Metals, 1981, p. 318.

EXAMPLE 21-3

A crankshaft in a diesel engine fails. Examination of the crankshaft reveals no plastic deformation. The fracture surface is smooth. In addition, several other cracks appear at other locations in the crankshaft. What type of failure mechanism would you expect?

Answer:

Since the crankshaft is a rotating part, the surface experiences cyclical loading. We should immediately suspect fatigue. The absence of plastic deformation supports our suspicion. Furthermore, the presence of other cracks is consistent with fatigue—the other cracks didn't have time to grow to the size that produced catastrophic failure. Examination of the fracture surface will probably reveal beach marks or fatigue striations.

Creep and stress rupture.   At elevated temperatures, a metal undergoes thermally induced plastic deformation even though the applied stress is below the nominal yield strength. The typical creep curve shows three regions—primary creep, as dislocations and other lattice imperfections are rearranged to cause rapid plastic deformation; secondary or steady-state creep, when dislocation climb and cross-slip cause a steady, continuous plastic deformation; and tertiary creep, where necking, void nucleation and coalescence, or grain boundary sliding cause rapid deformation and failure (Figure 21-11). *Creep failures* are defined as excessive deformation or distortion of the metal part, even if fracture has not occurred. *Stress-rupture failures* are defined as the actual fracture of the metal part.

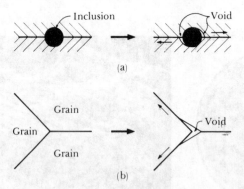

FIG. 21-11   Grain boundary sliding during creep causes (a) the creation of voids at an inclusion trapped at the grain boundary and (b) the creation of a void at a triple point where three grains are in contact.

Normally, ductile stress-rupture fractures include necking of the metal during tertiary creep and the presence of many cracks that did not have an opportunity to produce final fracture. Furthermore, grains near the fracture surface tend to be elongated. Ductile stress-rupture failures are generally transgranular and occur at high creep rates, short rupture times, and relatively low exposure temperatures.

Brittle stress-rupture failures are usually intergranular, show little necking, and occur more often at slow creep rates and high temperatures. Equiaxed grains

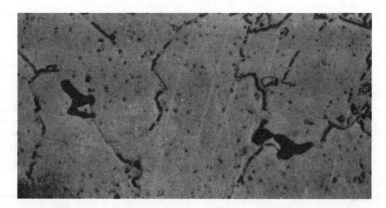

FIG. 21-12   Creep cavities formed at grain boundaries in an austenitic stainless steel. From *Metals Handbook*, vol. 7, 8th Ed., American Society for Metals, 1972.

are observed near the fracture surface. Brittle failure typically occurs by formation of voids at the intersection of three grain boundaries and precipitation of additional voids along grain boundaries by diffusion processes (Figure 21-12).

Stress-corrosion fractures. Stress-corrosion fractures occur at stresses well below the yield strength of the metal due to attack by a corrosive medium. Deep fine corrosion cracks are produced even though the metal as a whole shows little uniform attack. The stresses can be either externally applied or can be stored residual stresses. Stress-corrosion failures are often identified by microstructural examination of the nearby metal. Ordinarily, extensive branching of the cracks along grain boundaries is observed (Figure 21-13). The location at which cracks initiated may be identified by the presence of a corrosion product.

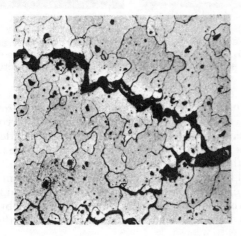

FIG. 21-13 Photomicrograph of a metal near a stress-corrosion fracture, showing the many intergranular cracks formed as a result of the corrosion process. From *Metals Handbook*, vol. 7, 8th Ed., American Society for Metals, 1972.

EXAMPLE 21-4

A titanium pipe used to transport a corrosive material at 400°C is found to fail after several months. How would you determine the cause for the failure?

Answer:

Since a period of time at a high temperature was required before failure occurred, we might first suspect a creep or stress-corrosion mechanism for failure. Microscopic examination of the material near the fracture surface would be advisable. If many tiny branched cracks leading away from the surface are noted, then stress-corrosion is a strong possibility. However, if the grains near the fracture surface are elongated, with many voids between the grains, creep is a more likely culprit.

21-3   Fracture in Nonmetallic Materials

In ceramic materials, the ionic or covalent bonds permit little or no slip. Consequently, failure is a result of brittle rather than ductile fracture. Most crystalline ceramics fail by cleavage along widely spaced, close-packed planes. The fracture surface is typically smooth and frequently no characteristic surface features point to the origin of the fracture [Figure 21-14(a)].

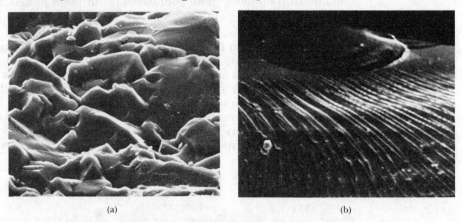

(a)   (b)

FIG. 21-14   Scanning electron micrographs of fracture surfaces in ceramics. (a) The fracture surface of Al_2O_3, showing the cleavage faces. (b) The fracture surface of glass, showing the mirror zone and tear lines characteristic of conchoidal fracture.

Glasses also fracture in a brittle manner. Frequently, a *conchoidal* fracture surface is observed. This surface contains a very smooth mirror zone near the origin of the fracture, with tear lines comprising the remainder of the surface [Figure 21-14(b)]. The tear lines point back to the mirror zone and the origin of the crack, much like the chevron pattern in metals.

Polymers can fail by either a ductile or a brittle mechanism. Below the glass transition temperature, thermoplastic polymers fail in a brittle manner, much like a ceramic glass. Likewise the hard thermosetting polymers fail by a brittle mechanism. Thermoplastics, however, fail in a ductile manner above the glass transition temperature, giving evidence of extensive deformation and even necking prior to failure. The ductile behavior is a result of sliding of the polymer chains, which is not possible in glassy or thermosetting polymers.

Fracture in fiber-reinforced composite materials is more complex. Typically, these composites contain strong brittle fibers surrounded by a soft ductile matrix, as in boron-reinforced aluminum. When a tensile stress is applied along the fibers, the soft aluminum deforms in a ductile manner, with void formation and coalescence eventually producing a dimpled fracture surface. As the aluminum deforms, the load is no longer transmitted effectively to the fibers; the fibers break in a brittle manner until there are too few fibers left intact to support the final load.

Fracture becomes easier if bonding between the fibers is poor. Voids can form between the fibers and the matrix, causing *pull-out*. Voids can also form between layers of the matrix if composite tapes or sheets are not properly bonded (Figure 21-15).

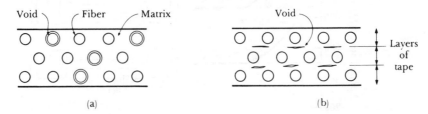

FIG. 21-15 Fiber-reinforced composites can fail by several mechanisms. (a) Due to weak bonding between the matrix and fibers, fibers can pull out of the matrix, creating voids. (b) If the individual layers of the matrix are poorly bonded, the matrix may debond, creating voids.

EXAMPLE 21-5

Describe the difference in fracture mechanism between a boron-reinforced aluminum composite and a glass fiber-reinforced epoxy composite.

Answer:

In the boron-aluminum composite, the aluminum matrix is soft and ductile; thus we expect the matrix to fail in a ductile manner, while the boron fibers fail in a brittle manner.

Both glass fibers and epoxy are brittle; thus the composite as a whole should display little evidence of ductile fracture.

21-4 Source and Prevention of Failures in Metals

We can prevent metal failures by several approaches—design of components, selection of appropriate materials and processing techniques, and consideration of the service conditions.

Design.   Components must be designed to (a) permit the material to withstand the maximum stress that is expected to be applied during service, (b) avoid stress raisers that cause the metal to fail at lower than expected loads, and (c) assure that deterioration of the material during service does not cause failure at lower than expected loads.

Creep, fatigue, or stress-corrosion failures occur at stresses well below the yield strength. The design of the component must be based on the appropriate creep, fatigue, or stress-corrosion data, not yield strength. Designers must not introduce galvanic cells when components are fabricated, particularly from different materials.

Stress raisers produced by designed-in notches such as sharp fillets or keyways must be avoided. These sharp corners concentrate stresses so that fatigue or corrosion cracks can more easily nucleate (Figure 21-16).

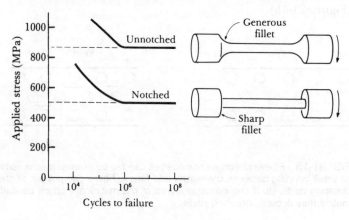

FIG. 21-16   Notches, introduced by poor design, significantly reduce fatigue failure, as shown here for heat-treated 3.5% nickel alloy steel.

Materials selection.   A tremendous variety of materials is available to the engineer for any application, with many being capable of withstanding high applied stresses (Figure 21-17). Selection of a material is based both on the ability of the material to serve and on the cost of the material and its processing.

The engineer must consider the condition of the material. For example, age-hardened, cold-worked, or quenched and tempered alloys lose their strength at high temperatures. Figure 21-18 shows the temperature ranges over which a number of groups of alloys can operate as a function of the applied stress.

Materials processing.   All finished components are at one time passed through some type of processing—casting, forming, machining, joining, or heat treatment—producing the appropriate shape, size, and properties. However, a variety of flaws can be introduced. The engineer must design to compensate for the flaws or the engineer must detect their presence and either reject the material or correct the flaw. Detection of flaws will be discussed in the next section. Figure 21-19 illustrates some of the typical flaws that might be introduced.

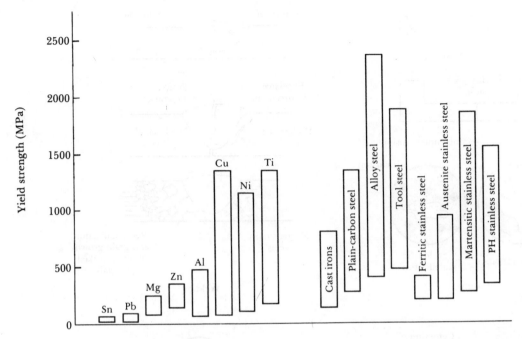

FIG. 21-17 Comparison of the range of properties available for many important metals and alloys. A wide range of properties is possible for each alloy system, depending on composition and treatment.

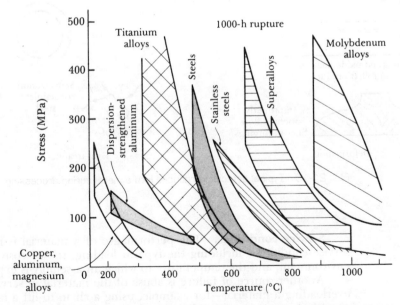

FIG. 21-18 Stress-rupture plots showing the range of suitable temperatures for several alloys.

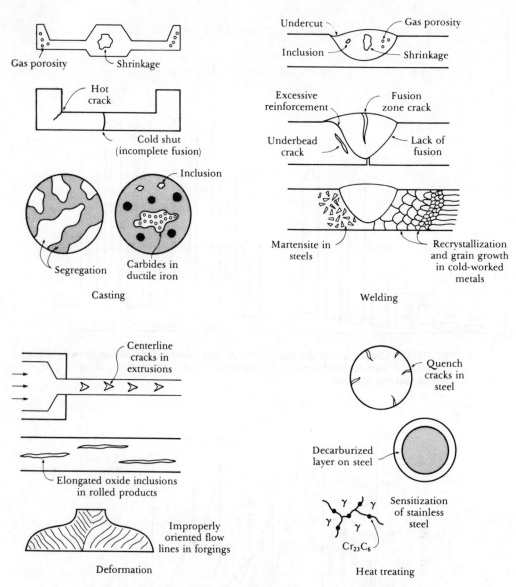

FIG. 21-19   Typical defects introduced into a metal during processing.

Service conditions.   The performance of a material is influenced by the service conditions, including the type of loading, the corrosive or atmospheric environment, and the temperature to which an assembly is exposed.

Another source of failure is abuse of the material in service. This includes overloading a material—for example, using a chain to lift a tank whose weight exceeds the capacity of the chain. An ordinary carpenter's hammer should not be used as a crow bar by striking the hammer with another metallic instrument.

Flakes of metal may spall off of the face of the hammer, which is heat treated to a high hardness, causing injury.

Improper maintenance, such as inadequate lubrication of moving parts, can lead to adhesive wear, overheating, and oxidation. If overheated, the microstructure changes and decreases the strength or ductility of the metal.

EXAMPLE 21-6

An alloy steel, which is welded using an electrode that produces a high hydrogen atmosphere, is found to fail in the heat-affected zone near the weld. What factors may have contributed to the failure?

Answer:

If we assume that the joint is designed so that the maximum stresses can be withstood, failure probably occurred as a result of flaws introduced during welding. Since the steel failed in the heat-affected zone rather than the fusion zone, possible problems include the following.

(a) Severe grain growth may have weakened the metal near the fusion zone.

(b) Since an alloy steel has good hardenability, martensite may have formed during cooling. A large grain size also increases the likelihood of martensite.

(c) Hydrogen, dissolved in the fusion zone, may have diffused to the heat-affected zone and caused hydrogen embrittlement, or underbead cracking.

(d) Because of temperature differences in the steel during welding, residual stresses may build up in the weld. Tensile residual stresses reduce the local load-bearing ability of the weld.

(e) Elongated inclusions in the steel could act as stress raisers and initiate cracks.

Additional information obtained concerning the initial state of the steel, the welding parameters, and the microstructure would permit us to pin down the cause for failure.

EXAMPLE 21-7

A steel rope, composed of many tiny strands of steel, passes over a 50 mm diameter pulley. After several months, the steel rope fails, dropping its heavy load. Suggest a possible reason for failure.

Answer:

The long period of time required before failure occurred suggests that fatigue might be the culprit. Each time the steel rope passes over the pulley, the outer strands of the steel experience a high stress. This stress may exceed the endurance limit. After passing over the pulley enough times, the strands begin to fail by fatigue. This increases the stress on the remaining strands, accelerating their failure, until the rope is overloaded and breaks. A larger diameter pulley reduces the stress on the fibers to below the endurance limit, so the rope will not fail.

EXAMPLE 21-8

After a helicopter crash, the teeth on a gear in the transmission are found to have worn away. The gear, which is a carburized alloy steel, is intended to have a surface hardness of R_c60. However, the hardness measured on a portion of one tooth that is still intact reveals a hardness of R_c30. Suggest possible causes for the failure.

Answer:

We know that the rate of wear increases when the hardness decreases. Thus, we confidently blame the failure on the soft gear. However, we still need to determine why the gear was soft. One explanation would be that the gear was not carburized or heat treated. Microscopic examination would reveal this. Suppose that examination shows the presence of a proper case depth and an overtempered martensitic structure. This would suggest that the gear, originally properly manufactured, overheated during use, softened, and then began to wear. In this case, failure may have been due to loss of oil from the gear box, permitting the gear to overheat.

21-5　Detection of Potentially Defective Materials

Care in design, materials selection, materials processing, and service conditions help prevent failure of metals. However, how do we determine whether our engineering process has been successful? To assure that our final part performs successfully, we may perform two types of tests—destructive tests and nondestructive tests.

Destructive testing.　Special tests are used to determine the properties of the finished product. For example, chain manufacturers routinely pull their product to failure to determine if the chain can withstand the rated load. By so doing, the effectiveness of the manufacturing process can be monitored. We use statistical methods to show that, if a certain sample is good, then all other parts will also be good.

Proof test.　In many instances a *proof test* can be designed. We load a part to its rated capacity and see if the part remains intact. If the part is never abused or used to support loads higher than the proof test, we can usually be confident that the part will perform properly.

Hardness test.　In some situations, a hardness test can be used to assure proper heat treatment. We may be able to determine whether an aluminum alloy has been aged to the required properties, or a steel has been adequately tempered, or a gray cast iron has been fully annealed. However, the hardness test does not tell us the microstructure, or if a crack is present at the surface, or if a large shrinkage cavity lies within a casting.

Radiography.　In radiographic testing, the transmission and absorption characteristics of a material are utilized to produce a visual image of any flaws in a material. There are several components in a radiographic technique.

1. A source of a penetrating radiation is required. X rays emitted from a

tungsten target are most common, but occasionally gamma rays or neutrons are required.

2. A detection system is needed. Normally, special film is used to detect the amount of radiation that is transmitted through the material. Other detectors include fluorescent screens, Geiger counters, Xerographs, or television screens.

3. The flaw, or discontinuity, in the material must have a different absorption characteristic than the material.

In *X-ray radiography* (Figure 21-20), an X-ray tube provides the source of the

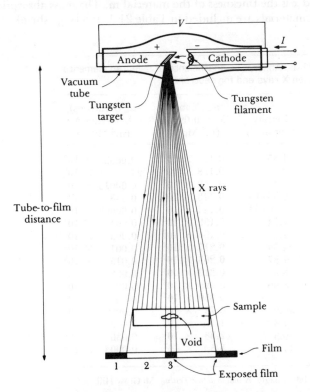

FIG. 21-20 The elements in an X-ray radiographic nondestructive testing set-up.

radiation. Electrons are emitted from a tungsten filament cathode and accelerated at a high voltage onto an anode material, also tungsten. The beam excites electrons in the inner shells of the tungsten target and a continuous spectrum of X rays is emitted when the electrons fall back to their equilibrium states. The emitted X rays are directed to the material to be inspected. A small fraction of the X rays is transmitted through the material to expose the film.

The intensity I_0 of the incoming X-ray radiation depends on three factors.

1. High tube voltages produce higher energy X rays.

2. Longer tube-to-film distances permit the X rays to spread out. The intensity decreases with the square of the tube-to-film distance.

3. Increasing the tube current increases the intensity of the emitted X-ray beam.

The intensity I of the transmitted X-ray beam depends on the absorption coefficient and the thickness of the material.

$$I = I_0 \exp(-\mu x) = I_0 \exp(-\mu_m \rho x) \qquad (21\text{-}1)$$

where I_0 is the intensity of the incoming beam, μ is the linear absorption coefficient (m^{-1}), μ_m is the mass absorption coefficient ($m^2\ Mg^{-1}$), $\dot{\rho}$ is the density Mg m^{-3}, and x is the thickness of the material m. The mass absorption coefficients for selected materials are included in Table 21-1. If a large shrinkage cavity is present

TABLE 21-1   Absorption coefficients of selected elements for tungsten X rays and for neutrons

Element	Density ($Mg\ m^{-3}$)	μ_m (X rays), $\lambda = 0.098$ Å ($m^2\ Mg^{-1}$)	μ_m (neutrons) $\lambda = 1.08$ Å ($m^2\ Mg^{-1}$)
Be	1.85	0.131×10^2	0.0003×10^2
B	2.3	0.138×10^2	24×10^2
C	2.2	0.142×10^2	0.00015×10^2
N	0.00116	0.143×10^2	0.048×10^2
O	0.00133	0.144×10^2	0.00002×10^2
Mg	1.74	0.152×10^2	0.001×10^2
Al	2.7	0.156×10^2	0.003×10^2
Ti	4.54	0.217×10^2	0.001×10^2
Fe	7.87	0.265×10^2	0.015×10^2
Ni	8.9	0.310×10^2	0.028×10^2
Cu	8.96	0.325×10^2	0.021×10^2
Zn	7.133	0.350×10^2	0.0055×10^2
Mo	10.2	0.79×10^2	0.009×10^2
Sn	7.3	1.17×10^2	0.002×10^2
W	19.3	2.88×10^2	0.036×10^2
Pb	11.34	3.5×10^2	0.0003×10^2

After W. McGonnagle, *Nondestructive Testing*, McGraw-Hill, 1961.

inside a casting, the absorption of the X rays by the cavity is lower than in the solid metal and consequently the intensity of the transmitted beam is greater. The film is exposed to more radiation and, after developing, is much darker (Figure 21-21).

EXAMPLE 21-9

In Figure 21-20, let's assume that the material being inspected is a copper plate 25 mm thick and that a discontinuity 6.25 mm thick is present at point 3. The discontinuity contains air. Estimate the intensity I at locations 1, 2, and 3 if the copper plate is placed on the film 762 mm from the source. The average wavelength of the tungsten X rays is 0.098 Å.

Answer:

Point 1

The beam must pass through air only. If we assume that air is 80% N_2 and 20% O_2, then

$$\mu_m, \text{air} = f_{O_2}\mu_{m,O_2} + f_{N_2}\mu_{m,N_2}$$
$$= (0.2)(0.144 \times 10^2) + (0.8)(0.143 \times 10^2)$$
$$= 0.143 \times 10^2 \text{ m}^2 \text{ Mg}^{-1}$$

$$\rho_{\text{air}} = f_{O_2}\rho_{O_2} + f_{N_2}\rho_{N_2}$$
$$= (0.2)(1.33 \times 10^{-3}) + (0.8)(1.16 \times 10^{-3})$$
$$= 1.19 \times 10^{-3} \text{ Mg m}^{-3}$$

$$\frac{I}{I_0} = \exp(-\mu_m\rho x) = \exp(-14.3)(1.19 \times 10^{-3})(0.762)$$

$$= \exp(-0.013) = 0.987$$

Virtually no absorption occurs in air.

Point 2

Since virtually no absorption occurs in air, let's neglect absorption of the beam before it reaches the copper. Thus, $x = 1$ in. $= 2.54$ cm. From Table 21-1, $\mu_{m,Cu} = 0.325 \times 10^2$ m^2 Mg^{-1}.

$$\frac{I}{I_0} = \exp(-\mu_m\rho x) = \exp(-32.5)(8.96)(0.025) = \exp(-7.28)$$

$$= 6.89 \times 10^{-4}$$

Practically none of the beam is transmitted.

Point 3

Since the 6.25 mm cavity absorbs on X rays compared to the copper, we can ignore it and consider the thickness to be 18.75 mm.

$$\frac{I}{I_0} = \exp(-\mu_m\rho x) = \exp(32.5)(8.96)(0.01875) = \exp(-5.46)$$

$$= 4.25 \times 10^{-3}$$

The X-ray beam intensity is about one order of magnitude greater when it passes through the discontinuity. Thus, the film will be darker beneath the discontinuity.

Obtaining a radiograph that sensitively detects the presence of a discontinuity in the material requires optimization of the film, accelerating voltage, exposure, and various geometric factors (Table 21-2). The *sensitivity* is defined as

$$\text{Percent sensitivity} = \frac{\Delta x}{x} \times 100 \tag{21-2}$$

where x is the total thickness of the metal and Δx is the smallest change in the thickness of the metal that is visible on the film. The sensitivity tells us the minimum size of a flaw that can be detected. If the sensitivity is very small, we can detect the presence of very small discontinuities.

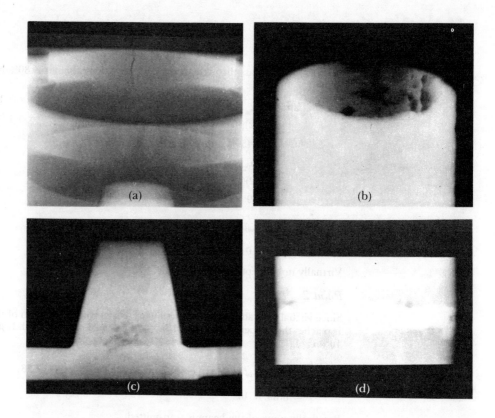

FIG. 21-21   Radiographs of (a) magnesium casting containing crack and hydrogen porosity; (b) graphitic corrosion of a cast iron pipe; (c) solidification shrinkage in an aluminum casting; (d) incomplete penetration of a steel weld.

TABLE 21-2   Factors producing good sensitivity and definition in X-ray radiography

---

Fine-grained film emulsion
Low X-ray energy or long wavelength
Low accelerating voltage
Long tube-to-film distance
Thin metal samples
Large difference in relative absorption coefficients
   of material and discontinuity
Orienting part so discontinuities are near the film

EXAMPLE 21-10

The sensitivity of a particular X-ray radiographic set-up is 3%. What is the thickness of the smallest discontinuity that we can detect in a casting 25 mm thick? . . . in a casting 75 mm thick?

Answer:

$$\text{Percent sensitivity} = 3 = \frac{\Delta x}{x} \times 100$$

If $x = 25$ mm, then $\Delta x = (3\%)(25)/100 = 0.75$ mm
If $x = 75$ mm, then $\Delta x = (3\%)(75)/100 = 2.25$ mm

The geometry and physical properties of the discontinuity also affect how easily it can be detected (Figure 21-22).

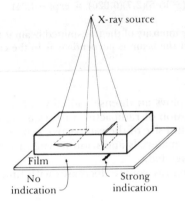

FIG. 21-22 The importance of orientation on the detection of discontinuities in a material by radiography.

1. Shrinkage voids or gas porosity in castings can be detected if their size falls within the sensitivity of the radiographic technique. The voids or pores are normally round so the direction of the X rays compared to the orientation of the casting is not critical. The voids or pores are filled with either a gas or a vacuum which have very different absorption coefficients than the metal.

2. Nonmetallic inclusions in rolled products may be flattened during deformation processing. If the X rays are perpendicular to the inclusion, the thin inclusions may not cause a noticeable change in intensity. Either the geometry between the X-ray source and the flaw must be changed or some other nondestructive testing technique may be required.

3. Detection of cracks caused by casting, welding, incipient fatigue, or stress-corrosion is dependent on the crack geometry. If the crack is perpendicular to the incident radiation, it may not be detected, but the crack may be easily observed if it is parallel to the radiation.

EXAMPLE 21-11

A crack 30 mm long and 0.1 mm thick is present in a 50 mm × 50 mm × 50 mm block of aluminum. Determine the relative intensity of the transmitted tungsten X-ray beam if the X rays are parallel to the crack. Repeat for the case where the X rays are perpendicular to the crack.

Answer:

Let's assume that the crack does not absorb X rays. From Table 21-1, $\mu_{m,\text{Al}} = 0.156 \times 10^2$ m^2 Mg^{-1} and $\rho_{\text{Al}} = 2.7$ Mg m^{-3}.

No crack: $x = 50$ mm

$$\frac{I}{I_0} = \exp(-\mu_m\rho x) = \exp(-15.6)(2.7)(0.05) = \exp(-2.1) = 0.12$$

Perpendicular to crack: $x = 50 - 0.1 = 49.9$ mm

$$\frac{I}{I_0} = \exp(-\mu_m\rho x) = \exp(-15.6)(2.7)(0.0499) = \exp(-2.1) = 0.12$$

Parallel to crack: $x = 50 - 30 = 20$ mm

$$\frac{I}{I_0} = \exp(-\mu_m\rho x) = \exp(-15.6)(2.7)(0.020) = \exp(-0.84) = 0.43$$

There is a large change in the intensity of the transmitted beam if the beam is parallel to the crack, but no change if the beam is perpendicular to the crack.

Gamma ray radiography employs an intense radiation of a single wavelength produced by nuclear disintegration of radioactive materials. The intensity of the radiation depends on the type and size of the radioactive source. Normally, cobalt 60 is used for thick, absorptive materials and iridium 192 or cesium 137 is used when less intense radiation is needed.

The intensity of the gamma ray source decreases with time.

$$I = I_0 \exp(-\lambda t) \tag{21-3}$$

where λ is the decay constant for the material and t is the time. The *half-life* is the time required for I to decrease to $0.5I_0$. The half-life of cobalt 60 is about 5.27 years and that for iridium 192 is 74 days. Usually, the source is no longer suitable after about two half-lives have elapsed.

EXAMPLE 21-12

Determine the decay constant for cobalt, which has a half-life of 5.27 years.

Answer:

In Equation (21-3), $I = 0.5I_0$ when $t = 5.27$ years.

$$0.5I_0 = I_0 \exp(-5.27\lambda)$$
$$0.5 = \exp(-5.27\lambda)$$
$$\ln(0.5) = -5.27\lambda$$
$$\lambda = \frac{-0.693}{-5.27} = 0.131 \text{ years}^{-1} = 4.15 \times 10^{-9} \text{ s}^{-1}$$

Neutron radiography is occasionally used because the neutrons are absorbed by nuclear rather than electronic interactions. Materials that have similar absorption coefficients for X rays or gamma rays may be easily distinguished by neutron radiography (Table 21-1). However, the source for the neutrons is a nuclear reactor, which is not widely available.

Ultrasonic testing.   A material may both transmit and reflect elastic waves. An ultrasonic transducer utilizing the piezoelectric effect introduces a series of elastic pulses into the material at a high frequency, usually greater than 100,000 Hz. The elastic wave, or phonon, is transmitted through the material at a rate that depends on the modulus of elasticity and the density of the material. For a thin rod

$$\nu \simeq \sqrt{\frac{E}{\rho}} \frac{(N \cdot m^{-2})}{(kg \cdot m^{-3})} \tag{21-4}$$

where E is the modulus of elasticity, and ρ is the density of the material. Examples of the ultrasonic velocity ν for several materials are given in Table 21-3. More complicated expressions are required for pulses propagating in bulkier materials.

TABLE 21-3   Ultrasonic velocities for selected materials

Material	Velocity $(m\ s^{-1} \times 10^3)$	Modulus of Elasticity (GPa)	Density $(Mg\ m^{-3})$
Al	5.055	69	2.7
Cu	3.720	124	8.96
Pb	1.111	14	11.34
Mg	4.584	41	1.74
Ni	4.903	214	8.9
60% Ni-40% Cu	4.484	179	8.9
Ag	2.692	76	10.49
Stainless steel	4.991	197	7.91
Sn	2.745	55	7.3
W	4.592	406	19.25
Air	0.330		1.3×10^{-3}
Glass	5.571	72	2.32
Lucite	1.697	3.4	1.18
Polyethylene (HD)	0.882	0.7	0.96
Quartz	5.103	69	2.65
Water	1.499		1.00

From W. McGonnagle, *Nondestructive Testing*, McGraw-Hill, 1961.

Three common techniques are used to ultrasonically inspect a material.

1. The pulse-echo or reflection method: An ultrasonic pulse is generated and transmitted through the material (Figure 21-23). When the elastic wave strikes an interface, a portion is reflected and returns to the transducer. Both the initial and the reflected pulse can be displayed on an oscilloscope.

From the oscilloscope display, we measure the time required for the pulse to travel from the transducer to the reflecting interface back to the transducer. If we know the velocity at which the pulse travels in the material, we can determine how far the elastic wave traveled and can calculate the distance below the surface at which the reflecting interface is located. If there are no flaws in the metal, the beam reflects from the opposite side of the metal and our measured distance corresponds to twice the wall thickness.

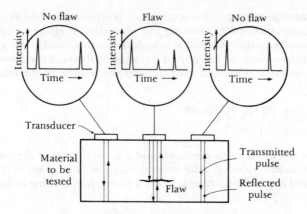

FIG. 21-23   The pulse-echo ultrasonic test. The time required for a pulse to travel through the metal, reflect off a discontinuity on the opposite side, and return to the transducer is measured with an oscilloscope.

If a discontinuity is present and properly oriented beneath the transducer, at least a portion of the pulse reflects from the discontinuity and registers at the transducer in a shorter period of time. Now our calculations show that a discontinuity lies within the material and even tell us the depth of the discontinuity below the surface. By moving the transducer we can estimate the size of the discontinuity. Automatic scanning techniques can move the transducer over the surface and display the results of the entire scan, showing the exact location of flaws. If we also combine holography with ultrasonics, we can obtain a three-dimensional picture of the discontinuities.

EXAMPLE 21-13

In a pulse-echo ultrasonic inspection of an aluminum rod, the oscilloscope shows three peaks. The first, at time zero, is the initial transmitted pulse; the second, at time 1.63×10^{-5} s, is the reflection from an internal discontinuity; and the third peak, at 2.44×10^{-5} s, is the reflection from the opposite surface of the material. Calculate the thickness of the material and the depth below the surface of the flaw.

Answer:

From Table 21-3, we expect the ultrasonic velocity in aluminum to be 5.055×10^3 ms^{-1}. Thus, the total distance that the pulse traveled in each case is

Discontinuity: Distance $= (5.055 \times 10^3)(1.63 \times 10^{-5})$

$= 82$ mm

Back surface: Distance $= (5.055 \times 10^3)(2.44 \times 10^{-5})$

$= 123$ mm

Since the total distance includes the return path, the actual depth of the discontinuity is $82/2 = 41$ mm and the actual thickness of the rod is $123/2 = 61.5$ mm.

2. Through-transmission method: In this method an ultrasonic pulse is generated at one transducer and detected at the opposite surface by a second transducer (Figure 21-24). The initial and transmitted pulses are displayed on an oscilloscope. The loss of energy from the initial to the transmitted pulse depends on whether or not a discontinuity is present in the metal.

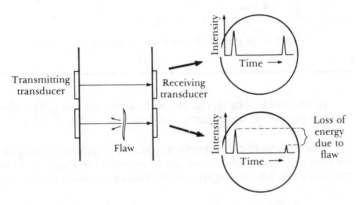

FIG. 21-24   The through-transmission ultrasonic test. The presence of a discontinuity reflects a portion of the transmitted beam, thus reducing the intensity of the pulse at the receiving transducer.

3. Resonance method: In both the reflection and transmission methods, the phonon, or elastic wave, is treated as an energy particle. In the resonance method, we utilize the wavelike nature of the phonon. A continuum of pulses is generated and travels as an elastic wave through the material (Figure 21-25). By selecting a wavelength or frequency so that the thickness of the metal is a whole number of half-wavelengths, a stationary elastic wave is produced and reinforced in the metal. A discontinuity in the material prevents resonance from occurring. However, this technique is used more frequently to accurately determine the thickness of the material.

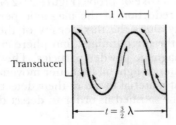

FIG. 21-25   In the resonance method of ultrasonic testing, a stationary elastic wave is produced by changing the wavelength or frequency of the ultrasonic pulse until a whole number of half-wavelengths matches the thickness of the material.

EXAMPLE 21-14

The thickness of copper is to be determined by the resonance method. A frequency of 1,213,000 Hz is required to produce 12 half-wavelengths at resonance. Determine the thickness of the copper.

Answer:

From Table 21-3, the ultrasonic velocity in copper is 3.720×10^3 ms^{-1}.

$$\lambda = \frac{v}{\nu} = \frac{3.720 \times 10^3}{1,213,000} = 3.0 \text{ mm}$$

$$\frac{\lambda}{2} = 1.5 \text{ mm}$$

$$\frac{\lambda}{2}(12) = 18 \text{ mm} = \text{thickness of the copper}$$

In ultrasonic testing, the discontinuity should be perpendicular to the beam. A crack parallel to the beam is not detected.

Magnetic particle inspection.   Magnetic particle testing detects discontinuities near the surface of ferromagnetic materials. A magnetic field is induced in the material to be tested (Figure 21-26), producing lines of flux. The magnetic field

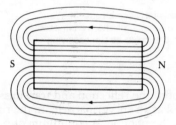 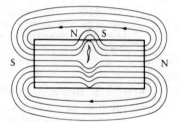

FIG. 21-26   A flaw in a ferromagnetic material causes a disruption of the normal lines of magnetic flux. If the flaw is at or near the surface, lines of flux leak from the surface. Magnetic particles are attracted to the flux leakage and indicate the location of the flaw.

can be introduced by placing the metal in a coil, passing a current through the metal, or touching the metal with a yoke or probe (Figure 21-27). If a discontinuity is present in the metal, the reduction in the magnetic permeability of the material due to the discontinuity alters the flux density of the magnetic field. Leakage of the lines of flux into the surrounding atmosphere creates local north and south poles which attract magnetic powder particles. The particles may be added dry or in a fluid such as water or light oil for better movement; they may be dyed or coated with a fluorescent material to aid in their detection.

Several requirements must be satisfied in order to detect discontinuities by magnetic particle inspection.

1. The discontinuity must be perpendicular to the lines of flux. Thus, different methods of introducing the magnetic field detect differently oriented discontinuities.

2. The discontinuity must be near the surface, or the flux lines merely crowd together rather than leaking from the material. Magnetic particle testing

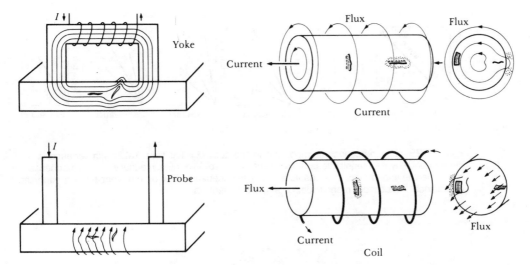

FIG. 21-27 Magnetic particle testing techniques and the relationship between the flux pattern and the orientation of the flaw.

is well suited for locating quench cracks, fatigue cracks, or cracks induced by grinding, all of which occur at the surface.

3. The discontinuity must have a lower magnetic permeability than the metal.

4. Only ferromagnetic materials can be tested.

EXAMPLE 21-15

Two cylindrical bars of a steel are joined by friction welding. Describe a magnetic particle test that will determine if the bars are properly joined.

Answer:

The potential weld defect is perpendicular to the cylindrical bars. Thus, we want to produce longitudinal magnetization to detect any flaw. Either a yoke or a coil technique produces the proper orientation of the magnetic flux lines with the weld.

Eddy current testing. An alternating current flowing in a coil produces an impedance, which is a vector quantity defined by the reactance and the resistance of the coil (Figure 21-28). When the coil is placed in close proximity to a conductive material, an electric field produces eddy currents. Eddy currents follow a fixed path in the metal, determined by the primary field, electrical properties of the material, and electromagnetic fields set up in the metal. The eddy currents in turn interact with the electric field and properties of the coil, changing the normal impedance. The through-coil method and the probe method are typical eddy current tests (Figure 21-29).

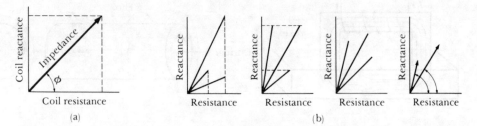

FIG. 21-28   The impedance is important in the eddy current test. (a) The impedance is defined both by its magnitude and its direction, ϕ. (b) We must measure two components to define the impedance; it is possible that the impedance may be different even though the resistance, reactance, magnitude, or angle are identical.

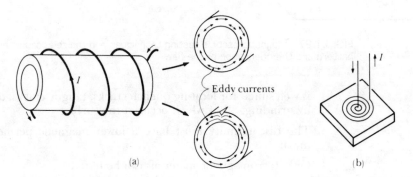

FIG. 21-29   (a) The through-coil method and (b) the probe method for eddy current inspection.

By measuring the effect of the material on the current and impedance of the coil, we can deduce the characteristics of the material. Factors that affect the coil impedance include (a) the electrical conductivity of the material (thus the composition or heat treatment may be detected)(Figure 21-30), (b) discontinuities in the

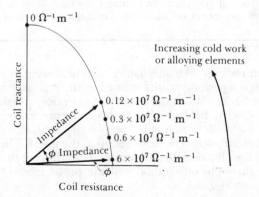

FIG. 21-30   Eddy current test of copper with different conductivities. High conductivity gives a lower impedance magnitude, smaller angle, higher coil resistance, and lower coil reactance.

material, particularly those near the surface, and (c) the magnetic permeability of the material. Interpretation of the eddy current testing results is very difficult since the impedance is a vector quantity. We can measure reactance or resistance components, the direction or phase, or the magnitude of the impedance, but more than one must be measured to completely define the impedance (Figure 21-28).

Furthermore, many factors besides the material's properties and discontinuities are important, including the size of the material, its shape, the location of the material within the coil or the closeness of the probe to the material, coil design, frequency of the current, and the magnitude of the current. Generally, eddy current testing is used as a go or no-go test that is standardized on good parts. If the impedance or other measured coil characteristics are the same when other parts are tested, the parts may be assumed to be good.

EXAMPLE 21-16

Figure 21-30 shows the effect of electrical conductivity of copper and its alloys on the impedance of a coil. Which of the four components of the impedance would provide us with the most sensitive measurement of conductivity?

Answer:

The magnitude of the impedance and the coil resistance change very little from $0.12 \times 10^7 \ \Omega^{-1} \ m^{-1}$ to $6.0 \times 10^7 \ \Omega^{-1} \ m^{-1}$; neither of these components of the impedance will be very sensitive. However, both the angle ϕ and the coil reactance change significantly and may be sensitive measures of the conductivity of the copper.

Liquid penetrant inspection.   Discontinuities such as cracks that penetrate to the surface can be detected by the *dye penetrant* technique. A liquid dye is drawn by capillary action into a thin crack that might otherwise be invisible. A four-step process is involved (Figure 21-31). The surface is first thoroughly cleaned. A liquid dye is sprayed onto the surface and permitted to stand for a period of time, during which the dye is drawn into any surface discontinuities. Excess dye is then cleaned from the metal surface. Finally, a developing solution is sprayed onto the surface. The developer reacts with any dye that remains, drawing the dye from the cracks. The dye then can be observed, either because it changes the color of the developer or because it fluoresces under ultraviolet lamps.

Electromagnetic testing.   This process is often used for sorting materials but also may be used for analysis of microstructure and heat treatment. Electromagnetic testing can analyze the hysteresis loop characteristics of ferromagnetic materials to sort steels based on composition, heat treatment, hardness, residual stress, or case depth or can be used to determine amounts of ferrite present in austenitic stainless steels (Figure 21-32).

21-6   Fracture Mechanics

Fracture mechanics is the discipline that studies the behavior of structures containing cracks or other small flaws. All materials contain some flaws, which may or

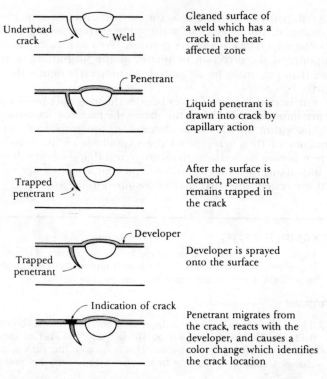

Underbead crack — Weld: Cleaned surface of a weld which has a crack in the heat-affected zone

Penetrant: Liquid penetrant is drawn into crack by capillary action

Trapped penetrant: After the surface is cleaned, penetrant remains trapped in the crack

Developer: Developer is sprayed onto the surface
Trapped penetrant

Indication of crack: Penetrant migrates from the crack, reacts with the developer, and causes a color change which identifies the crack location

FIG. 21-31 The elements of the dye penetrant test, used to detect cracks that penetrate to the surface.

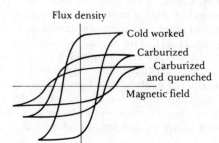

Flux density
Cold worked
Carburized
Carburized and quenched
Magnetic field

FIG. 21-32 Hysteresis loop analysis is an electromagnetic testing method used to evaluate the condition of steels.

may not be detectable using typical nondestructive testing techniques. We wish to know the maximum stress that a material can withstand if it contains flaws of a certain size and geometry.

A typical fracture mechanics test may be performed by applying a tensile stress to a thick specimen prepared with a flaw of known size and geometry (Figure 21-33). If the specimen is thick enough, a "plain strain" condition is produced which gives the worst behavior of the material. The stress required to propagate a crack from the prepared flaw can be measured. The *stress intensity factor K* then can be calculated. In simple tests, the stress intensity factor is

$$K = f\sigma\sqrt{a\pi} \tag{21-5}$$

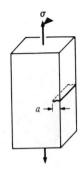

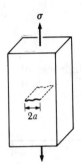

FIG. 21-33
Schematic
drawing of
fracture
toughness
specimens with
edge and internal
flaws.

where f is a geometry factor for the specimen and flaw, σ is the applied stress, and a is the flaw size as defined in Figure 21-33. For thick plate, $f = 1$. If K is greater or equal to a critical value K_{I_c}, the flaw grows and the material fails. The *critical fracture toughness* K_{I_c} is a property of the material (Table 21-4). A similar approach can be used to determine the ease with which a flaw grows in torsion, impact, fatigue, or other loading conditions.

TABLE 21-4   The critical fracture toughness for selected materials

Material	Critical Fracture Toughness $MPa \cdot m^{1/2}$	Yield Strength (MPa)
Al-Cu alloy	24	496
Ti-6% Al-4% V	115	910
	55	1034
817M40 steel	99	862
	60	1517
H-11 tool steel	38	1793
Al-Si-Mg alloy	27	276
Maraging steel	104	1607

The fracture mechanics approach allows us to design and select materials while taking into account the inevitable presence of flaws. There are three variables we need to consider—the property of the material (K_{I_c}), the stress σ that the material must withstand, and the size of the flaw a.

We may be able to set an upper limit on the size of any flaw that is present by nondestructive testing. For example, ultrasonic testing or X-ray radiography may detect any flaw longer than 0.1 mm. This fixes the largest size a and gives us the worst condition that the material will face. If we know the magnitude of the applied stress, we can select a material that has a fracture toughness K_{I_c} large enough to prevent the flaw from growing. Or, if the material has already been specified, we can calculate the maximum permitted stress that can act on the material. Finally, if we know both the applied stress and the fracture toughness of the material, we can determine if our nondestructive testing capability is adequate.

The ability of the material to resist the growth of a crack depends on a large variety of factors, including the following.

1. Larger flaws reduce the permitted stress. Special manufacturing techniques have been devised to improve the cleanliness of metals in order to improve fracture toughness. For example, liquid aluminum is passed though a ceramic filter to remove impurity particles, whereas the AOD process (argon-oxygen decarburization) has been developed to produce steels containing fewer oxide inclusions.

2. Increasing the strength of a given metal tends to reduce fracture toughness (Table 21-4).

3. Thicker materials have a lower stress intensity factor K than thin materials [Figure 21-34(a)].

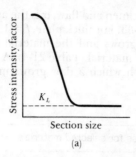
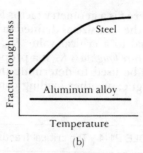

FIG. 21-34   The effect of (a) section size, (b) temperature, and crystal structure on the stress intensity factor and fracture toughness of materials.

4. Increasing the temperature normally increases the fracture toughness of BCC and HCP metals. However, the fracture toughness of FCC metals is relatively unaffected by temperature [Figure 21-34(b)].

EXAMPLE 21-17

For a large plate, the geometry factor f is one. Suppose a steel casting alloy has a critical fracture toughness of 90 MPa $\cdot$ m$^{1/2}$. The steel will be exposed to a stress of 310 MPa during service. Calculate the minimum size of a crack at the surface that will grow. Repeat for an internal crack.

Answer:

$$K_{I_c} = f\sigma\sqrt{a\pi}$$
$$90 = (1)(310)\sqrt{a\pi}$$

For a surface crack: $a = \dfrac{1}{\pi}\left(\dfrac{90}{310}\right)^2 = 0.027$ m (27 mm)

For an internal crack: $2a = 54$ mm

SUMMARY

We often are able to determine the manner in which a material fails by examining the characteristic features on the fracture surface or the microstructure adjacent to the fracture. This analysis helps us to improve our materials selection or engineering. By combining nondestructive testing and fracture toughness techniques, we may be able to prevent failures. The nondestructive testing techniques, which rely on the physical properties of the material, are consequently very sensitive to the material's structure, including incipient cracks or other discontinuities.

GLOSSARY

Beach marks Marks on the surface of a fatigue fracture which represent the position of the crack front at various times during failure.

Brittle fracture Fracture of a material with little or no deformation.

Chevron pattern Markings caused by the merging of crack fronts in brittle fracture. The markings form arrows which point back toward the origin of the brittle fracture.

Cleavage Brittle fracture along particular crystallographic planes in the grains of the material.

Conchoidal fracture A characteristic fracture surface in glass, with a mirror zone near the origin of the fracture and tear lines pointing toward the origin.

Creep failure Excessive deformation or distortion of a material at high temperatures, without fracture occurring.

Critical fracture toughness The stress concentration required to cause a flaw to grow in a material.

Ductile fracture Fracture of a material with significant deformation required.

Eddy current testing A nondestructive testing technique to detect flaws or evaluate structure and properties by determining the reaction between the material and an electric field.

Electromagnetic testing Evaluating the structure or treatment of a ferromagnetic material by determining its response to a magnetic field.

Fatigue fracture Fracture of a material due to a cyclical application of a load.

Intergranular fracture Fracture of a material along the grain boundaries.

Liquid penetrant inspection A nondestructive testing technique in which a liquid, drawn into a surface imperfection by capillary action, is exposed by a dye or ultraviolet light.

Magnetic particle inspection A nondestructive testing technique that relies on the interruption of lines of magnetic flux by imperfections near the surface.

Mass absorption coefficient Related to the absorption of X rays or other radiation by a material.

Microvoids Tiny voids at the fracture surface formed by separation of the material at grain boundaries or other interfaces during ductile failure.

Proof test Loading a material to its designed capacity to determine if it is capable of proper service.

Radiography A nondestructive testing technique that relies on a difference between the absorption of radiation by the material and flaws in the material.

Sensitivity A measure of the minimum size of a flaw that can be detected in a material by a particular radiographic set-up.

Shear lip The surface formed by ductile fracture that is at a 45° angle to the direction of the applied stress.

Stress-corrosion fracture Fracture caused by a combination of corrosion and a stress below the yield strength.

Stress intensity factor The concentration of stress produced by a flaw in a material.

Stress-rupture fracture Fracture of a material due to prolonged exposure at a high temperature.

Striations Microscopic traces of the location of a fatigue crack.

Transgranular fracture Fracture of a material through the grains rather than along the grain boundaries.

PRACTICE PROBLEMS

1 A 75 mm diameter shaft on a water pump fails at the root of a thread after several months of use. The shaft is intended to have a quenched and tempered martensitic microstructure, giving a hardness of R_c35. (a) What fracture mechanism would we expect is responsible for the failure? How would we verify the fracture mechanism? (b) The reason that the fracture occurred is due to misuse, improper materials selection, poor manufacturing process, or improper heat treatment. How would we go about determining the cause of the failure?

2 A 50 mm diameter 0.8% carbon steel shaft is water quenched during heat treatment. Shortly after the shaft is put into service under a static load, failure occurs. The load is found to be well below the nominal yield strength of the material. Can you suggest a fracture mechanism and a cause for the failure? What additional information would you need to verify your theory?

3 A roller-bearing race, which was not properly hardened during heat treatment, is found to develop cracks and eventually pits during service. What fracture mechanism is probably responsible? Explain.

4 An X-ray beam passes through a zinc casting. The casting, which is 20 mm thick, contains a void 7 mm in diameter. Compare the intensity of the beam passing through the solid casting to that of the beam passing through the casting containing the void. Assume the void contains a vacuum.

5 Repeat Problem 4 for a neutron radiographic system.

6 An X-ray beam passes through a welded joint in a 75 mm thick aluminum plate. The weld contains a 3.2 mm diameter tungsten inclusion that melted off of the welding electrode. Compare the intensity of the beam passing through the inclusion to that passing only through aluminum.

7 X-ray radiography is used to detect shrinkage voids in an iron casting. The film adjacent to the casting is 65 times more dense than the film beneath the solid casting and 30 times more dense than the film beneath a portion of the casting containing a void. Estimate (a) the thickness of the casting and (b) the thickness of the void if the film-to-camera distance is 1 m.

8 An X-ray radiographer wishes to determine the thickness of a 25 mm thick roll-bonded composite material composed of a plate of copper bonded to a plate of titanium. The relative intensity of the transmitted beam to original beam is 0.0016. Calculate the thickness of the titanium plate.

9 Suppose a copper-zinc alloy has severe segregation, with portions of a 20 mm thick casting containing 75% Zn and other portions containing only 20% Zn. Calculate the relative intensities at the two locations using (a) X-ray radiography and (b) neutron radiography. With which process will we be more likely to detect the segregation?

10 Suppose we can detect a difference in intensity of an X-ray beam of 5%. Can we see the section size change from 120 mm to 117.5 mm in X-ray radiography of beryllium?

11 Calculate the decay constant for radioactive Ir^{192}.

12 The decay constant for Ce^{137} is 6.6×10^{-8} s^{-1}. Calculate the half-life of the isotope.

13 An ultrasonic pulse in a 75 mm copper plate returns to the transducer after 1.8×10^{-5} s. Is there a flaw in the copper? Explain.

14 An ultrasonic pulse in a 150 mm long silver rod returns to the transducer after 4.19×10^{-5} s. Is there a flaw in the silver? Explain.

15 The density of a phenolic polymer is 1.28 Mg m^{-3} and the modulus of elasticity is 3.79 GPa. Estimate the velocity of sound in the polymer.

16 An ultrasonic pulse is introduced into an 200 mm long rod of an unknown metal. After 7.048×10^{-5} s, a pulse returns to the transducer. Determine (a) the velocity of the ultrasonic pulse and (b) the probable identity of the material.

17 An ultrasonic pulse in a 0.3 mm cylinder of an aluminum-copper alloy returns to the transducer after 1.09×10^{-4} s. Calculate the ultrasonic velocity in the alloy.

18 Determine the time required for an ultrasonic pulse to pass through a 50 mm copper plate sandwiched to a 150 mm lead plate.

19 A laminar composite material 150 mm thick is prepared by weld depositing a layer of lead onto nickel. In a through-transmission ultrasonic test, the time required for the pulse to pass through the composite is 2.97×10^{-5} s. How thick is the layer of lead?

20 In a through-transmission ultrasonic test, a pulse requires 6.49×10^{-5} s to pass through a 125 mm thick rod of a material. Determine (a) the ultrasonic velocity in the material and (b) the probable identity of the material.

21 A frequency of 2.7×10^6 Hz produces 15 half-wavelengths in a resonance ultrasonic test in a sheet of lucite plastic. What is the thickness of the sheet?

22 Suppose we want to determine the thickness of a polyethylene film by the ultrasonic resonance technique. What frequency must we use if we want to produce 6 half-wavelengths in a 0.01 mm thick film?

23 Suppose an aluminum casting contains microscopic gas porosity. Explain how ultrasonic testing might be used to detect the presence and severity of porosity.

24 Review the difference in microstructure between ductile cast iron and gray cast iron. Describe how the through-transmission ultrasonic testing technique might be used to sort the two types of iron.

25 Describe techniques by which you could continuously monitor the thickness of aluminum foil as it is being rolled.

26 Which, if any, of the nondestructive testing techniques might be useful in determining the following.

(a) Measuring the case depth of a carburized steel

(b) Determining the amount of ferrite in an austenitic stainless steel

(c) Detecting debonding of the matrix in fiber-reinforced composites

(d) Detection of quench cracks in steels

27 A thin-walled tubing made by a resistance welding process is shown in Figure 21-35. Describe the nondestructive testing techniques that might be used to determine if the weld is a good joint.

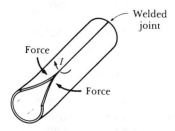

FIG. 21-35 Resistance welding of thin-walled tubing. (See Problem 27.)

28 Figure 21-36 shows three types of cracks in welds. Which nondestructive testing process would be satisfactory and which would not be appropriate for detecting each type of crack?

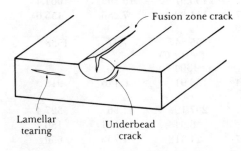

FIG. 21-36 Three types of cracks in a fusion weld. (See Problem 28.)

29 A heat-treated steel, which has a fracture toughness of 60 MPa $m^{1/2}$, is designed to carry a stress of 1345 MPa. Calculate the largest flaw length (a) at the surface and (b) at the center that will not propagate.

30 Estimate the size of the largest internal and largest surface flaw that will propagate when an Al-Si-Mg alloy (Table 21-4) is loaded to the yield point.

31 After X-ray radiography we are confident that there are no surface flaws in a Ti-6% Al-4% V alloy larger than 0.2 mm in length. The fracture toughness of the alloy is 55 MPa $\cdot$ $m^{1/2}$. What is the maximum stress that we can apply to the alloy without the possible flaws becoming a problem?

32 We have an ultrasonic inspection technique that is capable of detecting any internal flaws up to 0.2 mm in length in a steel with a fracture toughness of 100 MPa $\cdot$ $m^{1/2}$. What is the maximum stress we can apply without the flaw enlarging to form a crack?

33 A manufacturing process introduces flaws of up to 0.4 mm in length at the surface. If the finished component must withstand a stress of 1206 MPa, what must be the fracture toughness of the alloy that we select?

34 A machine component that must withstand a stress of 310 MPa is manufactured by a machining process that introduces flaws at the surface as deep as 0.2 mm. What should be the fracture toughness of the material that is selected for the component?

35 A lap joint between two sheets of copper is to be made by brazing (Figure 21-37). Describe the nondestructive testing technique that might be used to determine if the joint is completely filled.

FIG. 21-37 A brazed lap joint. (See Problem 35.)

APPENDIX A   Selected physical properties of metals

Metal		Atomic Number	Crystal Structure	Lattice Parameter (Å)	Atomic Mass (g · g mole)	Density (Mg · m⁻³)	Melting Temperature (°C)
aluminum	Al	13	FCC	4.04958	26.981	2.699	660.4
antimony	Sb	51	hex	$a = 4.307$ $c = 11.273$	121.75	6.697	630.7
arsenic	As	33	hex	$a = 3.760$ $c = 10.548$	74.9216	5.778	816
barium	Ba	56	BCC	5.025	137.3	3.5	729
beryllium	Be	4	hex	$a = 2.2858$ $c = 3.5842$	9.01	1.848	1290
bismuth	Bi	83	hex	$a = 4.546$ $c = 11.86$	208.98	9.808	271.4
boron	B	5	rhomb	$a = 10.12$ $\alpha = 65.5°$	10.81	2.3	2300
cadmium	Cd	48	HCP	$a = 2.9793$ $c = 5.6181$	112.4	8.642	321.1
calcium	Ca	20	FCC	5.588	40.08	1.55	839
cerium	Ce	58	HCP	$a = 3.681$ $c = 11.857$	140.12	6.6893	798
cesium	Cs	55	BCC	6.13	132.91	1.892	28.6
chromium	Cr	24	BCC	2.8844	51.996	7.19	1875
cobalt	Co	27	HCP	$a = 2.5071$ $c = 4.0686$	58.93	8.832	1495
copper	Cu	29	FCC	3.6151	63.54	8.93	1084.9
gadolinium	Gd	64	HCP	$a = 3.6336$ $c = 5.7810$	157.25	7.901	1313
gallium	Ga	31	ortho	$a = 4.5258$ $b = 4.5186$ $c = 7.6570$	69.72	5.904	29.8
germanium	Ge	32	FCC	5.6575	72.59	5.324	937.4
gold	Au	79	FCC	4.0786	196.97	19.302	1064.4
indium	In	49	tetra	$a = 3.2517$ $c = 4.9459$	114.82	7.286	156.6
iron	Fe	26	BCC FCC BCC	2.866 3.589	55.847 (>912°C) (>1394°C)	7.87	1538
lanthanum	La	57	HCP	$a = 3.774$ $c = 12.17$	138.91	6.146	918
lead	Pb	82	FCC	4.9489	207.19	11.36	327.4
lithium	Li	3	BCC	3.5089	6.94	0.534	180.7
magnesium	Mg	12	HCP	$a = 3.2087$ $c = 5.209$	24.312	1.738	650
manganese	Mn	25	cubic	8.931	54.938	7.47	1244
mercury	Hg	80	rhomb		200.59	13.546	−38.9
molybdenum	Mo	42	BCC	3.1468	95.94	10.22	2610
nickel	Ni	28	FCC	3.5167	58.71	8.902	1453
niobium	Nb	41	BCC	3.294	92.91	8.57	2468

Metal		Atomic Number	Crystal Structure	Lattice Parameter (Å)	Atomic Mass (g · g mole)	Density (Mg · m⁻³)	Melting Temperature (°C)
palladium	Pd	46	FCC	3.8902	106.4	12.02	1552
platinum	Pt	78	FCC	3.9231	195.09	21.45	1769
potassium	K	19	BCC	5.344	39.09	0.855	63.2
rhenium	Re	75	HCP	$a = 2.760$ $c = 4.458$	186.21	21.04	3180
rhodium	Rh	45	FCC	3.796	102.99	12.41	1963
rubidium	Rb	37	BCC	5.7	85.467	1.532	38.9
selenium	Se	34	hex	$a = 4.3640$ $c = 4.9594$	78.96	4.809	217
silicon	Si	14	FCC	5.4307	28.08	2.33	1410
silver	Ag	47	FCC	4.0862	107.868	10.49	961.9
sodium	Na	11	BCC	4.2906	22.99	0.967	97.8
strontium	Sr	38	FCC BCC	6.0849 4.84	87.62 (>557°C)	2.6	768
tantalum	Ta	73	BCC	3.3026	180.95	16.6	2996
tellurium	Te	52	hex	$a = 4.4565$ $c = 5.9268$	127.6	6.24	449.5
thorium	Th	90	FCC	5.086	232	11.72	1755
tin	Sn	50	FCC	6.4912	118.69	5.765	231.9
titanium	Ti	22	HCP BCC	$a = 2.9503$ $c = 4.6831$ 3.32	47.9 (>882°C)	4.507	1668
tungsten	W	74	BCC	3.1652	183.85	19.254	3410
uranium	U	92	ortho	$a = 2.854$ $b = 5.869$ $c = 4.955$	238.03	19.05	1133
vanadium	V	23	BCC	3.0278	50.941	6.1	1900
yttrium	Y	39	HCP	$a = 3.648$ $c = 5.732$	88.91	4.469	1522
zinc	Zn	30	HCP	$a = 2.6648$ $c = 4.9470$	65.38	7.133	420
zirconium	Zr	40	HCP BCC	$a = 3.2312$ $c = 5.1477$ 3.6090	91.22 (>862°C)	6.505	1852

Element	Atomic Radius (Å)	Valence	Ionic Radius (Å)
aluminum	1.432	+3	0.51
antimony		+5	0.62
arsenic		+5	2.22
barium	2.176	+2	1.34
beryllium	1.143	+2	0.35
bismuth		+5	0.74
boron	0.46	+3	0.23
bromine	1.19	−1	1.96
cadmium	1.49	+2	0.97
calcium	1.976	+2	0.99
carbon	0.77	+4	0.16
cerium	1.84	+3	1.034
cesium	2.65	+1	1.67
chlorine	0.905	−1	1.81
chromium	1.249	+3	0.63
cobalt	1.253	+2	0.72
copper	1.278	+1	0.96
fluorine	0.6	−1	1.33
gallium	1.218	+3	0.62
germanium	1.225	+4	0.53
gold	1.442	+1	1.37
hafnium		+4	0.78
hydrogen	0.46	+1	1.54
indium	1.570	+3	0.81
iodine	1.35	−1	2.20
iron	1.241 (BCC)	+2	0.74
	1.269 (FCC)	+3	0.64
lanthanum	1.887	+3	1.016
lead	1.75	+4	0.84
lithium	1.519	+1	0.68
magnesium	1.604	+2	0.66
manganese	1.12	+2	0.80
		+3	0.66
mercury	1.55	+2	1.10
molybdenum	1.363	+4	0.70
nickel	1.243	+2	0.69
niobium	1.426	+4	0.74
oxygen	0.60	−2	1.32
palladium	1.375	+4	0.65
phosphorous	1.10	+5	0.35
platinum	1.387	+2	0.80
potassium	2.314	+1	1.33
rubidium	2.468	+1	0.70
selenium		−2	1.91
silicon	1.176	+4	0.42
silver	1.445	+1	1.26
sodium	1.858	+1	0.97
strontium	2.151	+2	1.12
sulfur	1.06	−2	1.84
tantalum	1.43	+5	0.68
tellurium		−2	2.11
thorium	1.798	+4	1.02
tin	1.405	+4	0.71
titanium	1.475	+4	0.68
tungsten	1.371	+4	0.70
uranium	1.38	+4	0.97
vanadium	1.311	+3	0.74
yttrium	1.824	+3	0.89
zinc	1.332	+2	0.74
zirconium	1.616	+4	0.79

APPENDIX C

SI prefixes:

10^{-12}	pico	p
10^{-9}	nano	n
10^{-6}	micro	μ
10^{-3}	milli	m
10^3	kilo	k
10^6	mega	M
10^9	giga	G
10^{12}	tera	T

SI base units:
length—meter (m)
mass—kilogram (kg)
time—second (s)
electric current—ampere (A)
thermodynamic temperature—kelvin (K)
amount of substance—mole (mol)
luminous intensity—candela (cd)

A list of certain physical constants expressed in SI units:

Atomic mass unit	amu	1.6606×10^{-27} kg
Avagadro constant	N	6.0225×10^{23} mol^{-1}
Boltzmann constant	k	1.3805×10^{-23} J K^{-1}
Electron charge	q	1.602×10^{-14} C
Electron mass	m_e	9.11×10^{-31} kg
Electron volt	eV	1.602×10^{-14} J
Faraday constant	F	96500 C mol^{-1}
Gas constant	R	8.314 J mol^{-1} K^{-1}
Neutron mass	m_n	1.6748×10^{-27} kg
Planck constant	h	6.626×10^{-34} Js
Proton mass	m_p	1.6725×10^{-27} kg
Speed of light	c	2.9979×10^8 m s^{-1}

Conversions

Quantity	*SI unit*	*Conversion factor*
Length	meter (m)	1 m = 39.37 in
		= 3.28 ft
		= 10^{10} Å
Area	square millimeter (mm^2)	1 mm^2 = 1.55×10^{-3} in^2
	square metre (m^2)	1 m^2 = 1550 in^2
		= 10.76 ft^2
Volume	cubic millimeter (mm^3)	1 mm^3 = 6.12×10^{-5} in^3
	cubic meter (m^3)	1 m^3 = 35.29 ft^3
Mass	kilogram (kg)	1 kg = 2.20 lbm
	tonne (t)	1 t = 0.98 ton
Density	kg m^{-3}	1 kg m^{-3} = 0.062 lbm ft^{-3}
	Mg m^{-3}	1 Mg m^{-3} = 62.4 lbm ft^{-3}
Force	newton (N)	1 N = 0.2248 lbf
Stress, pressure	pascal (Pa) or N m^{-2}	1 Pa = 145×10^{-6} l bf in^{-2}
	MPa	1 MPa = 0.065 tonf in^{-2}
	bar = 10^5 Pa	
Work, energy	joule (J) = Nm = Ws	1 J = 0.7376 ft lbf
		= 0.239 calorie
		= 0.947×10^{-3} BTU
		= 6.24×10^{18} eV
Temperature	kelvin (K)	°C + 273.15
Magnetic flux density	tesla (T) = Wb m^{-2}	1 T = 10^4 gauss
Magnetizing field	A m^{-1}	1 A m^{-1} = 0.0126 oersted
Magnetic flux	weber (Wb)	1 Wb = 10^8 maxwell
Electric charge	coulomb (C) = As	

Answers to Selected Odd-Numbered Practice Problems

CHAPTER 1

1. Because Al_2O_3 is a ceramic material, it is very brittle.
3. Metals: brass, lead-tin solder, magnesium alloys; ceramics: sodium chloride, silicon carbide, graphite; polymers: epoxy, rubber; composites: concrete, reinforced concrete, fiberglass, asphalt.
5. Engine block: casting, machining; brick: extrusion, sintering; paper clip: bending; wrench: forging, casting, machining, heat treatment; plywood: adhesive bonding; plastic toys: injection molding; plastic water pipe: extrusion; steel transmission gear: forging, machining, heat treatment.

CHAPTER 2

1. 24.3247 g/g · mole.   3. 64% Ga^{69} − 36% Ga^{71}.
5. 32 electrons; at. no. 60.   7. 3.   9. inert.
11. electropositive.   13. gold.   15. (a) 5.8×10^{-15}.
19. 109.47°.   21. MgS.   23. MgF_2.   25. Na^+.
27. metallic, B; ionic, C; Van der Waals, A.

CHAPTER 3

1. (a) 3.516 Å; (b) 8.97 Mg m^{-3}.   3. 1.36 Å.
5. BCC.   7. (a) 0.386; (b) 5.91 Mg m^{-3}.
9. (a) 2 atoms/cell; (b) 0.69.
11. (a) 5.658 Å; (b) 5.326 Mg m^{-3}.
13. (a) 3.302 Å; (b) 1.43 Å.
15. (a) 4 atoms/cell; (b) 0.531.
17. 2 atoms/lattice point.
19. (a) 1.27%; (b) expansion.
21. −0.6% contraction.   23. 30.42 Å^3.
25. A: [110]; C: [12̄1]; E: [1̄02]; G: [616].
27. A: (001); C: (023); E: (030).

CHAPTER 4

29. A: [210]; C: (011̄0); E: [1̄00]; G: (101̄1).
41. (a) yes; (b) no.   43. no.   45. (111).
47. [1̄31̄0].
49. (100) = 0.59; (200) = 0.59; (110) = 0.83.
51. (100) = 0.302; (001) = 0.785; (a) not equivalent.
53. 1.476 Å.   55. (22̄0).   57. 1.982 Å.
59. 3.1477 Å; Mo.
61. (a) ⟨100⟩; (b) 0.627; (c) 3.615 Mg m^{-3}.
63. (a) ⟨111⟩; (b) 2; (c) 8; (d) PF = 0.684; density = 4.311 Mg m^{-3}.
65. 2.928 Å.   67. 1.172 Å.   69. 4.
71. 35 C atoms; 70 H atoms.

CHAPTER 4

1. [110]; 1.237 Å.   3. [111]; 1.01 Å.
9. [100]; [1̄00].   11. (a) 28.2 MPa; (b) 0 MPa.
13. 0.57 MPa.
15. (a) 81 MPa; (b) slip will not occur.
19. $f_{Au} = 0.25$.   21. 0.000875.
23. 7.1887 Mg m^{-3}.
25. (a) P.F. = 0.685; (b) 7.9318 Mg m^{-3}
27. 0.04 Schottky defects/cell.   29. 8.   31. 7.
33. 0.0142°.

CHAPTER 5

1. $D_0 = 0.785 \times 10^{-4}$ m^2 s^{-1}; $Q = 184025$ J mol^{-1}
3. $D_0 = 0.0032 \times 10^{-4}$ m^2 s^{-1}; $Q = 299585$ J mol^{-1}
5. (a) 1×10^{-6} at % As, 1×10^{-3} at% As; (b) −9.99 at % m^{-1}; (c) -4.396×10^{27} As atoms (m$^3 \cdot$ m)$^{-1}$
7. (a) −25 wt% Cu mm^{-1}, −25.1 at % Cu mm^{-1}, -0.22×10^{32} Cu atoms (m$^3 \cdot$ m)$^{-1}$; (b) 1.31×10^{-9} Cu atoms (m$^2 \cdot$ s)$^{-1}$; (c) 5.835 $\times 10^9$ Cu atoms (m$^2 \cdot$ s)$^{-1}$.

9. (a) -4.7×10^{30} H atoms $(m^3 \cdot m)^{-1}$; (b) 1.22×10^{23} H atoms $(m^2 \cdot s)^{-1}$; (c) 3.73×10^{22} H atoms $(m^2 \cdot s)^{-1}$.

11. -25.7×10^{17} N atoms $(m^3 \cdot m)^{-1}$ 13. 738°C.

15. $Q_{Mg} = 138270$ J mol^{-1}; $Q_{Zn} = 113130$ J mol^{-1}; $Q_{Mo} = 439950$ J mol^{-1}.

17. 0.28% C.   19. 950°C.   21. 0.202% N.

23. 3 h.   25. 0.192% C in BCC iron.

27. 0.0144% N.   29. 8.76×10^9 s.   31. 756°C.

CHAPTER 6

1. yes.   3. 460.715 m
5. (a) 3.1 kN; (b) 0.0006.   7. yes.
9. (a) 37.6% elongation; (b) 67% reduction in
11. area.
(a) 2.24 GPa; (b) 1.29 GPa; (c) 414 MPa;
13. (d) 13.9%; (e) 21.9%.
15. (a) 195 GPa; (b) 1.29 GPa; (c) 1.59 GPa.
(a) 1276 MPa; (b) 1000 MPa; (c) 172.4 GPa
(d) 24.8%.
17. (a) no; (b) yes.   19. 61 kg.
21. 2×10^6 cycles. 25. (a) -125°C, -50°C, 25°C.
27. 37 mm.   29. 17.4 mm^2.   31. 127 mm.
33. 552°C.   35. (a) 475 BHN.
37. 1379 MPa.   39. 19.

CHAPTER 7

1. (a) 345°C; (b) 1108°C.   3. 11.75 Å
5. 305 atoms.   7. 1.255×10^7 atoms.
9. 5965 atoms.   13. 0.068 mm.   15. planar.
17. (a) 0.067 dendritic; (b) 0.267 dendritic; (c) 0.533 dendritic.
19. (a) $k = 0.004$; (b) $n = 0.33$.   21. 30.6 min.
23. $H = D > 105$ mm.
25. (a) 0.77; (b) 0.96; (c) 0.84; (d) sphere.
27. $D = H = 133$ mm.   31. 0.6 mm; 0.06 mm.
33. $n = 2$ and $B = 0.02$ min · mm^{-2}.
35. (a) 1600°C; (b) 1450°C; (c) 150°C; (d) 8 min; (e) 10 min; (f) nickel.
37. 233 mm.   39. 1.67% expansion.
41. 0.05 mm^2 H$_2$ per 100 g

CHAPTER 8

1. one.   3. Mo–W, Ge–Si.   5. (a) Al; (b) none.
7. (a) 100% NiO; (b) 65% NiO; (c) 40% NiO.
11. 1100°C: α, 20% Ni in α, 100% α; 1180°C: $\alpha + L$, 25% Ni in α, 17% Ni in L, 37.5% α, 62.5% L; 1300°C: L, 20% Ni in L, 100% L.
15. 2400°C: S, 70% MgO in S, 100% S; 2600°C: $S + L$, 82% MgO in S, 64% MgO in L, 33% S, 67% L; 2800°C: L, 70% MgO in L, 100% L.
17. (a) 30% Ni in α, 21% Ni in L; (b) 24.6% Ni.
19. (a) 86% Sb in α, 48% Sb in L; (b) 55.6% Ni.
21. (a) 430°C; (b) 72% Sb; (c) 71% α, 29% L.
23. (a) 435°C; (b) 74% Sb in α, 27% Sb in L.

25. (a) 85% Ni in α, 77% Ni in L, 37.5% α, 62.5% L; (b) 86.0 at% Ni in α, 78.4 at% Ni in L, 37.3 at% α, 62.7 at% L.
27. (a) 13 mol% S, 87 mol% L; (b) 11.7% S, 88.3% L.
29. (a) 22.5% Sb; (b) 66% Sb in α, 20% Sb in L, 5.4% α, 94.6% L.
31. (a) 83% Sb; (b) 10% Sb.
39. (a) 19.1 h; (b) 56,240 h.   41. 7.17×10^{12} s.

CHAPTER 9

1. A: $n = 0.36$, $K = 655$ MPa; B: $n = 1$, $K = 130316$ MPa; material B.
3. (a) 65.7%; (b) 207 MPa tensile, 186 MPa yield, 3% elongation.
5. (a) 27.1%; (b) 469 MPa tensile, 393 MPa yield, 18%
7. yes.
9. (a) 84%; (b) 400 MPa tensile, 393 MPa yield, 1% elongation.
11. (a) 40%; (b) 26 mm.   13. (a) 30%; (b) 45 mm.
15. 85.8%.   17. 69–79 mm.   19. 50.2–53.7 mm.
23. 49.96 mm if cold worked; 51.27 mm if hot worked.
25. (a) 384 N; (b) no.
27. (a) 0.24 mm, 92%; (b) 0.48 mm, 84%; (c) 0.99 mm, 67%.
31. hot worked.   37. 2840 N mm^{-1}; 3291 N mm^{-1}

CHAPTER 10

1. (a) θ is nonstoichiometric; (b) α, γ, β, η; (c) 1100°C: peritectic, 900°C: monotectic, 680°C: eutectic, 600°C: peritectoid, 300°C: eutectoid.
3. (a) β, γ, ζ, δ, ϵ, η are nonstoichiometric; (b) peritectics at 800°C, 750°C, 410°C; eutectoids at 580°C, 570°C, 520°C, 350°C; peritectoids at 630°C, 580°C; eutectic at 220°C; reaction at 630°C is not defined.
5. (a) θ, ζ, ϵ are stoichiometric, η is nonstoichiometric; (b) 1250°C: peritectic, 1230°C: peritectic, 1200°C: eutectic, 1180°C: eutectic, 1170°C: peritectoid, 950°C: peritectoid, 780°C: eutectoid, 760°C: eutectoid, 550°C: eutectoid.
7. 3.1%; 10.2%; the Pb-12% Sn alloy.
9. (a) 30% Zn in Cu, 38% Zn in Cu, 18% Zn in Cu; (b) 0% Cu in Zn, 2% Cu in Zn, 3% Cu in Zn.
11. (a) 7% Sn; (b) 12% Sn in α, 32% Sn in L, 85% α, 15% L; (c) liquidus = 290°C, solidus = 230°C, solvus = 170°C; (d) 15% Sn in α, 100% α; (e) 2% Sn in α, 100% Sn in β, 87% α, 13% β.
13. 41% α; 59% θ; θ, brittle.
15. (a) hypereutectic; (b) 98% Sn in β; (c) 61.9% Sn in L, 97.5% Sn in β, 49.2% L, 50.8% β; (d) 19% Sn in α, 97.5% Sn in β, 22.3% α, 77.7% β; (e) 61.9% Sn in eutectic, 97.5% Sn in primary β, 49.2% eutectic, 50.8% primary β; (f) 2% Sn in α, 100% Sn in β, 20% α, 80% β.
17. (a) hypoeutectic; (b) 1% Si in α; (c) 1.65% Si in α, 12.6% Si in L, 69.4% α, 30.6% L; (d) 1.65% Si in

α, 99.83% Si in β, 96.6% α, 3.4% β; (e) 1.65% Si in primary α, 12.6% Si in eutectic, 69.4% primary α, 30.6% eutectic; (f) 0% Si in α, 100% Si in β, 95% α, 5% β.

19. 70.8% Sn.   **21.** 75.5% Sn.   **23.** 7.67% Si.

27. (a) 8.62 Mg m^{-3}; (b) 71.4% Sn.

29. (a) $L_{(4.3\%C)} = \gamma_{(2.11\%C)} + Fe_3C_{(6.67\%C)}$; (b) brittle; (c) 67.3% γ, 32.7% Fe$_3$C, 32% primary γ, 68% eutectic.

31. (a) 1200°C; (b) 300°C; (c) 900°C; (d) 650°C; (e) 250°C; (f) 550 s; (g) 470 s.

33. $\alpha/\beta = 0.59$.   **35.** 1×10^{-1} cm.

37. 100% L; 20% δ, 80% L; 80% γ, 20% L; 100% γ.

39. 63.9%.   **41.** 0.13% C.   **43.** 0%; 5.4%; 5.1%.

45. (a) 1000°C, Mg$_2$Si; (b) 1020°C, β; (c) 610°C, α.

47. 20% Al-20% Mg-60% Si.

49. (a) Ni-3% Mo-40% Cr in α, Ni-7% Mo-74% Cr in β, 44% α, 56% β; (b) Ni-60% Mo-10% Cr in δ, 100% δ; (c) Ni-4% Mo-39% Cr in α, Ni-10% Mo-65% Cr in β, Ni-14% Mo-55% Cr in σ, 25% α, 33% β, 42% σ.

CHAPTER 11

1. 5.6%; 94.4%.   **3.** (a) 3.2% Cu; (b) 3.9 vol%.

7. b, d, e.

11. 95.8% α, 4.2% Fe$_3$C, 62.8% primary α, 37.2% pearlite.

13. 0.13.   **15.** 0.35% C; hypoeutectoid.

17. 1.48% C; hypereutectoid.   **19.** 7.846 Mg m^{-3}.

21. 1.35% C.   **23.** 51.6% α, 48.4% γ_2; ductile.

25. 73% α, 27% γ_2, 44% primary α, 56% eutectoid.

27. 700°C; all phases are brittle.

29. 0.29 Å versus 0.53 Å.

31. 2×10^{-4} mm; 379 MPa.

35. (a) γ; (b) $P + \gamma$; (c) P; (d) $B + \gamma$; (e) B; (f) M; (g) M.

37. (a) 100% M with 0.3% C in M; (b) 65% M with 0.43% C in M; (c) 43% M with 0.66% C in M; (d) no martensite forms.

39. 760°C; 0.38% C.   **41.** 0.5% contraction.

45. 450–600°C.

CHAPTER 12

1. (a) 40.6 mm, 2.99 kg m^{-1}; (c) 10.7 mm, 0.04 kg m^{-1}; (e) 23.4 mm, 3.86 kg m^{-1}; (g) 12.7 mm, 0.91 kg m^{-1}; (i) 6.9 mm, 0.16 kg m^{-1}.

7. Al$_6$Mn; 4.7%.

11. (a) 86% α, 6.4% β, 7.6% θ; (b) 71% α, 19% β, 10% θ.

13. 413.   **15.** EZ32.   **17.** Mg$_2$Si.

19. 27000-OS025.

21. 12.4% Be; 0% CuBe in TB00; 8.2% CuBe in O70.

23. 4 cm.   **27.** 190.

29. (b) about 10% β and α' (martensite); (c) 89% α, 11% β.

33. Ti (4.4×10^9); Al (3.1×10^9); Mg (2.5×10^9); Cu (2.2×10^9); Monel (1.4×10^9); W (1.3×10^9).

35. yes.

CHAPTER 13

1. 4840.

3. 91.3% α; 8.7% Fe$_3$C; 22.7% primary α; 77.3% pearlite.

5. 1020.

9. (a) $\alpha + P$; (b) $\alpha + B$; (c) $\alpha + B$; (d) $\alpha + M$; (e) B; (f) $B + M$.

13. 1050 = 200 s; 1080 = 500 s; 10110 = 700 s; 4340 = 2000 s.

15. (a) $\alpha + P$; (b) B; (c) $\alpha + P + M$; (d) $B + M$; (e) M; (f) tempered M.

17. 1020: 0.2% C, R_c49; 1040: 0.4% C, R_c60; 4340: 0.4% C, R_c60; 5160: 0.6% C, R_c65.

19. 200°C.

21. (a) $\alpha + P + B$; (b) P; (c) $B + M$.

23. (a) $A_1 = 850$°C, 0.25% C; (b) $A_1 = 690$°C, 0.50% C; (c) $A_1 = 820$°C, 0.45% C.

25. (a) surface = R_c43, center = R_c41; (b) surface = R_c59, center = R_c54.

27. 8640, 4340.   **29.** (a) 0.7; (b) 8°C s^{-1}

33. A, 43; B, 42; C, 42; D, 41; E, 34.

35. 1020: $F + P + B$; 1080: P; 4340: M.

37. surface = $P + M$; center = P.

41. 0.6%.

45. 4.1% CE; austenite; hypoeutectic; austenite.

49. 0.21% Mn.   **51.** 10–15 mm.   **53.** 4.55% C.

CHAPTER 14

1. 2.009 Å.

3. $\frac{1}{2}$,0,0; 0,$\frac{1}{2}$,0; 0,0,$\frac{1}{2}$; $\frac{1}{2}$,$\frac{1}{2}$,$\frac{1}{2}$.

5. 4; zincblende.   **7.** no; yes.

9. 5.4 Å; 0.60; 11.1 Mg m^{-3}

11. 5.96 Å; 0.53.   **13.** 0.99 Å.

15. metasilicate.   **17.** orthosilicate.

19. 7.1×10^{20} vacancies/cm^3; 50.3 at% O; 22.48 wt% O.

21. 2.11; yes.   **23.** 46 g.

25. (a) 2.55×10^5 J mol^{-1}; (b) 3095°C.

27. (a) 1600°C: 0% S, 1400°C: 23% S, 1200°C: 37.5% S, 1000°C: 50% S, 800°C: 58% S; (b) less.

29. 91% Al$_2$O$_3$.   **31.** 67% Mg$_2$SiO$_4$; 33% L.

33. 235 g CaO per kg ore.

35. (a) 1760°C; (b) first solid = $3Al_2O_3 \cdot 2SiO_2$ (or 28% SiO$_2$-72% Al$_2$O$_3$); (d) 68% SiO$_2$, 11% CaO, 21% Al$_2$O$_3$; (e) 55% L, 45% of $3Al_2O_3 \cdot 2SiO_2$.

37. 34.9% CaO $-$ 38.5% Al$_2$O$_3$ $-$ 26.6% SiO$_2$; 1500°C; L = 57%, $2CaO \cdot Al_2O_3 \cdot SiO_2$ = 43%.

39. (a) 6.4 kg cement; (b) 1.8 kg water.

CHAPTER 15

1. no unsaturated double bonds.
3. (a) no.   **7.** 408 mers.
9. (a) 2.4×10^{26}; (b) 7.2 kg; (c) 88.5°C.
11. 10.8 g.   **13.** 68,000.   **15.** 2083.
17. $f_{\text{vinyl chloride}} = 0.509$.
19. wt. avg. = 9875; nu. avg. = 8550.
21. 2212.   **23.** 9.2%.   **25.** 0.028.
27. (a) 3; (c) 155 kJ $(g \cdot mol)^{-1}$ released; (d) 9.7 kg
29. (a) 24%; (b) 8.2%.
31. (a) 10.9 kg; (b) 331 kJ $(g \cdot mol)^{-1}$ released.
33. (a) 22.7 kg; (b) 54.4 kg; (c) 5.7 kg; (d) 56.7 kg.
35. 32.9 kg water.
37. (a) 1400; (b) 8.965 kg
39. 0.0138   **41.** 83729 J mole^{-1}.

CHAPTER 16

1. 7.32 vol%.   **3.** 15.057 Mg m^{-3}.
5. (a) 0.36; (b) 0.36; (c) 12.35 Mg m^{-3}.
7. 6.7 kg.   **9.** (a) 2.9 Mg m^{-3}; (b) 2.9955 Mgm^{-3}.
11. (a) $f_{\text{polystyrene}} = 0.015$; (b) 0.0005 cm; (c) 6.66×0.028 J s^{-1} cm^{-1} °C^{-1}.
13. (a) 30.8%; (b) 1×10^{-4} cm.
15. (a) 0.184; (b) 1.52 Mg m^{-3}.   **17.** 0.813.
19. (a) 294 GPa; (b) 292 GPa.
21. 0.628 J s^{-1} cm^{-1} °C^{-1}.
23. 1.95×10^7 m for composite; 2.62×10^6 m for Al.
25. 1.14×10^5 $(\Omega \cdot cm)^{-1}$.
27. 1.33 W/cm · K; 0.02 W/cm · K.
29. (a) 78 GPa parallel, 74 GPa perpendicular; (b) copper: 9408 kg composite: 5870 kg.
31. 382°C.

CHAPTER 17

1. (a) 1.58×10^4 W; (b) 5.97×10^4 W; (c) 1.19×10^{15} W.
3. 1.65 V.   **5.** (a) 10 A; (b) 7.99 mm; (c) 2.35 A.
7. (a) 5.09 A mm^{-2}; (b) 13.3 V mm^{-1}; (c) 3.83×10^2 (ohm · m)$^{-1}$.
9. 0.081.
13. (a) 2.35×10^{-9} ohm · m; (b) 6.82×10^{-8} ohm · m.
15. (a) 39.7×10^{-8} ohm · m; (b) 10.3×10^{-8} ohm · m; (c) 217×10^{-6}.
17. (a) 11 at% Al; (b) 83×10^{-6}.
21. (a) 280°C; (b) 180°C; (c) Pb-20% Sn.
23. (a) 1.3×10^{10}; (b) 6.51×10^{-14}; (c) 5.36×10^{19}.
25. (a) 2.645×10^{-18} for diamond, 0.00025 for Si, 0.0067 for Ge; (b) 2.12×10^{-11} for diamond, 0.0065 for Si, 0.047 for Ge.
27. (a) 4.42×10^{15}; (b) 2.68×10^2 (ohm · m)$^{-1}$.
29. 7.44×10^{-5} at% P; 3.17×10^{-5} wt% P.
31. (a) 6.14×10^{15}; (b) 2.75×10^{15}; 1.67×10^2 (ohm·m)$^{-1}$.
33. (a) n-type; (b) 12250×10^2 (ohm · m)$^{-1}$.
35. (a) 4.3×10^{10}%; (b) 3.5×10^{29}%; (c) 1837%.
37. 54,100 Sb atoms/10^6 Ge atoms.
39. 4.68×10^{-8} (ohm · m)$^{-1}$; 8.49×10^{-5} (ohm · m)$^{-1}$.

CHAPTER 18

1. 3.53×10^{-8} C/m^2.   **3.** 7.15×10^{-9} Å.
5. 9.15×10^{-8} C/m^2.   **7.** 2.43×10^{-8} Å.
9. 68 V/m.   **11.** (a) 2.478×10^{-8} C/m^2; (b) 1 mm.
13. 1.86×10^{-9} F.   **15.** 1.3×10^{-8} F.
17. (a) 5.68×10^6 V/m; (b) 2.38×10^8 V/m.
19. 3.75×10^{-4}; 25.86 MPa.   **21.** 0.3 V.
23. (a) 9×10^{-8} C/m^2; (b) 6.08.
25. Fe, 3.15×10^6 A/m; Ni, 1.71×10^6 A/m; Co, 2.51×10^6 A/m.
27. (a) 19.1 at% Co; (b) 2.96×10^6 A/m.
29. (a) tesla, 49,998 tesla; (b) 160 tesla, 159.998 tesla
31. 149.2 A m^{-1}.   **33.** 17,500,000.
35. (a) 2.39×10^5 A m^{-1}; (b) 4.3.   **37.** 0.127 A · m^2/cm^3.

CHAPTER 19

1. 0.124 Å.
3. (a) 58,825 V; (b) 6400 V; (c) 1489 V.
5. (a) 17,460 V; (b) 19,640 V; (c) 2,168 V.
7. L_α X rays are produced.   **9.** MnS.
13. 2.07 eV above the conduction band.
15. 0.0345 s.   **17.** 1.16 mm.
19. ZnS and ZnO do not.
21. intrinsic, 1.85×10^{-3} mm; extrinsic, 0.11 mm.
23. 1.81; 8.3%.   **25.** 32.4; 3.67; 13.5.
27. manganese.   **29.** 5.708 Å.
31. chromium.
33. 109527 J.
35. (a) 2.1 J g^{-1}; (b) 0.12 J g^{-1}; (c) 1.8 J g^{-1}.
37. 251.8 mm × 251.8 mm × 30.2 mm
39. (a) 100560 J · s^{-1}; (b) 8799 J · s^{-1}; (c) 335 J · s^{-1}.
41. (a) 155 J · mm^{-2}; (b) 3862 J · mm^{-2}.
43. -4.57×10^{-3} mm.   **45.** 0.067 mm.

CHAPTER 20

1. 0.708 V.   **3.** 30.   **5.** 19 g.
7. 0.063 A.   **9.** 271 h.   **11.** 1.01 V.
13. (a) 0.0596 A/cm^2; (b) 353 g/h.
15. 0.186 g Fe/h; 0.21 g Cu/h.
17. 0.0036 g Fe/h; 0.0041 g Cu/h.
19. 1.248×10^{-6} A.
21. no resistance, 5.17 A; with resistance, 1.9 A.
29. stress corrosion cracking.
35. (a) 1.24×10^{-9} cm^2/s; (b) 3415 g/yr.

CHAPTER 21

1. (a) fatigue.   **3.** fatigue.
5. 0.925 vs 0.950.   **7.** (a) 20.1 mm; (b) 3.7 mm.
9. (a) 0.0054 at 75% Zn, 0.0035 at 20% Zn; (b) 0.866 at 75% Zn, 0.734 at 20% Zn.
11. 1.08×10^{-7} s^{-1}.   **13.** flaw is present.
15. 1722 ms^{-1}.   **17.** 5588 ms^{-1}.
19. 12.7 mm.   **21.** 7.4 mm.   **27.** eddy current.
29. surface, 0.64 mm; internal, 1.27 mm.
31. 2193 MPa.   **33.** 268 MPa.

INDEX

A_1 temperature 352
A_3 temperature 352
Abrasive 513, 514
Abrasive wear 704
ABS 482, 516
Absorption 639, 728
Absorption edge 632, 645
Acceptor saturation 578
Acetal 480
A_{cm} temperature 352
Acrylic 478
Activation energy
 for diffusion 95, 96
 and melting point 103
 in stress relaxation 491
 and structure 101
 for viscous flow 428, 472
Activation polarization 685
Active (in corrosion) 696
Addition polymer 455
Adhesive wear 703
Adipic acid 465
Advanced composites 532
Age hardening 280–288
 in Al alloys 281, 316
 in Cu alloys 332
 in Mg alloys 324
 in Monel 334
 in PH stainless steels 390
 requirements for 284
 in superalloys 337
 in titanium alloys 342
AISI-SAE steels 353
Alclad 12, 537, 695
Alkali metals 555, 558
Allotropic transformation 60
 in cobalt 300
 in iron 60, 288, 301, 652

Allotropic transformation
 (continued)
 in silica 421
 in titanium 339
 in zirconium 345
Alloy 1
Alloyed junction 585
Aluminum alloys 313–321
 designation system 315
 properties 318, 320
 temperature effect 124, 325
 temper designation 315, 317
Aluminum bronze 328
Amine 493, 494
Amorphous polymers 474
Amplification 584
Anelasticity 661
Angstrom 43
Anion 34
Anisotropy 58
 after annealing 224
 in cold-worked alloys 213
 in dielectrics 599
 and electrical conductivity 567
 after hot working 227
 modulus of elasticity 58
 in selected alloys 215
 in solidification 163
Annealing
 of cast iron 396
 elimination of cold work 219–
  225
 of glass 429, 445
 of steel 355
 of titanium 340, 342
Annealing texture 224
Anode 672
Anode polarization curve 678

Anode potential 679
Anode reaction 673, 679
Anodic protection 696
Anodizing 697
Antiferromagnetism 612
Antistatic agent 499
Apparent porosity 444
Artificial aging 284
Aspect ratio 523
 in whiskers 529
 in wood 541
Asperities 703
Asphalt 544
ASTM grain size number 86
Atactic 484
Athermal reaction 300
Atomic bonding 31–36, 412
Atomic mass 19, 746–747
Atomic number 19, 746–747
Atomization 112
Atoms per cell 46
 diamond cubic 61
Ausforming 371
Austempering
 in cast iron 397
 in steel 358
Austenite 290, 351
 austenite grain size 296
Austenitizing 356
Avalanche voltage 583
Avogadro's number 19

Bainite 299, 351
 lower bainite 300
 properties 301
 upper bainite 300

Band structure   554–560
  in semiconductors   570
  in sodium   555
Barium titanate   599, 604
Basal plane   57
Beach marks   717
Benzene ring   458
  in phenol   460
  in styrene   458
Beryllium   324
  effect of temperature   324
  modulus of elasticity   324
Bimetallic   537
Binding energy   37, 38
Bitumen   544
Bloch wall   613
Blowing agents   499
Blow molding   500
Body-centered cubic   44
  characteristics   49
  close-packed directions   47
  sketch   45
Bohr magnetron   607
Boltzmann's constant   553, 648
Bond energy in polymers   461
Boron   521
Borosilicate glass   429, 450
Borsic aluminum   521, 533
Bragg's law   59, 647
Branching   479, 481
Bravais lattice   43, 44
Brazing   539, 693
Brinell hardness   142
  and tensile strength   145
Brittle fracture   714–716
Bronze   328
BS rubber   482, 488
Bulk density   444
Bulk modulus   658
Burger's vector   72
Butadiene   484, 488

Calendering   500, 501
Cambium   540
Capacitance   595
Capacitors   535, 594
  types   598
Carbon black   499, 515
Carbon equivalent   391
Carbonitriding   379
Carbonizing   528
Carburizing   378
  gas carburizing   379
  liquid carburizing   379
  pack carburizing   379
Case depth
  carburizing   378, 381
  surface hardening   378
Castability   320

Cast iron   390–404
  chilled iron   394
  compacted graphite iron   404
  ductile iron   401–403
  gray iron   391, 397–398
  malleable iron   399–401
  mottled iron   394
  white iron   399
Cathode   672
Cathode polarization   679
Cathode potential   679
Cathode reaction   673, 679
Cathodic protection   695
Cation   34
Cavitation   705
Cellulose   480
  wood   541
Cement   445, 543
Cementation   445
  reactions   446, 543
Cemented carbides   511–513
Cementite   291, 351
Ceramic bond   442
Cesium chloride   63, 416, 417
Chain scission   703
Characteristic spectrum   631
  table   632
Charge carriers   552
Charpy test   132
Chemical conversion coating   695
Chemical corrosion   669–672
Chemical reduction   112
Chill depth   394, 395
Chilled iron   394
Chill zone   163
China   447
Chvorinov's rule   158, 166
Cis   484, 487
Clad metals   537
Clay   420, 423, 425
Cleavage   715
Climb of dislocations   139
Close-packed directions   46, 56
  diamond cubic   61
Close-packed planes   56
Cobalt   300, 334
  properties   335
Coefficient of thermal expansion
    538, 650, 652
Coercive field
  magnetization   614, 617
  polarization   606
Coextrusion   539
Coherent light   638
Coherent precipitate   279
Cold indentation welding   229
Cold shut   724
Cold work
  in aluminum   235
  in brass   235

Cold work (continued)
  in copper   211
  and corrosion   689
  equation   209
  in OF copper   236
  and properties   217
  vs. solid solution strengthening
      218
  in steel   236
  See also Strain hardening
Collector   583
Colloids   425
Color center   644
Columbium. See Niobium
Columnar zone   163
  in directional solidification   171
Combined carbon   400
Compacted graphite cast iron   390,
    404
  microstructure   404
  properties   404
  specifications   404
  treatment   404
Compocasting   516
Composites   4, 507–545
  asphalt   544
  concrete   543
  fiber   518
  laminar   534
  particulate   507
  sandwich   545
  wood   540
Composition cell   687
Compressibility   658
Compression molding   501, 502
Concentration cell   690
Concentration gradient   97
Concentration polarization   685
Conchoidal fracture   720
Concrete   4, 543, 507
  cementation reactions   544
  structure   4
Condensation polymer   455, 465
Conditioning   389
Conduction band   555
Contact potential   582
Continuous casting
  of fiber composites   529
  of plastic   502
Continuous cooling transformation
    diagram   364
  alloying elements   367
  in 1020 steel   367
  in 1080 steel   366
  in 4340 steel   369
Continuous spectrum   630
Cooling curve
  eutectic   248
  hypoeutectic   252
  and microstructure   279

Cooling curve (*continued*)
    pure metal   162
    solid solution alloy   195
Cooling rate
    effect on aluminum alloys   279,
        321
    effect on eutectoid   296
    effect of process   321
    solidification   161
Coordination number   47, 49
    vs. radius ratio   62, 414
Copolymer   481
    forms   482
    styrene-acrylonitrile   482
    styrene-butadiene   482
    vinyl chloride-vinyl acetate   482
Copper alloys   326
    designation system   327
    properties   326
    temper designation   327
Coring   197
Corrosion   669–699
    chemical corrosion   669–672
    corrosion rate   676
    electrochemical corrosion   672–
        699
    pitting in aluminum   12
    protection   691–699
Coupling agent   499
Covalent bond   31, 38
    in ceramics   412
Creep failure   691
Creep properties   138
    creep curve   140
    creep temperature   139
    of polymers   472
    representation of data   141
    stress effect   140
    temperature effect   140
Creep rate   140
Creep resistance
    in platinum   509
    in polyethylene   472
    in superalloys   334
Cristobalite   63, 413
Critical radius   152
Critical resolved shear stress   78
    table   79
Cross-linking   3
    in elastomers   487
    in thermosets   492–496
Cross-slip   81
Crystalline polymer   475
    effect of deformation   475, 586
    properties   477, 484
Crystal structure   42
Cup and cone fracture   713
Curie temperature
    in ferroelectrics   606
    in magnets   619, 653

Current density   552, 678
Cyaniding   379

Dacron   465, 466, 480
Debye-Scherrer X-ray diffraction
    59, 647
Decarburization   703
Decay constant   732
Defect resistivity coefficient   562
    for zinc in copper   565
Defects (during processing)   724
Defect semiconductor   578
Deformation
    of ceramics   428
    of elastomers   487, 660
    of metals   78–82, 121–143,
        207–219, 225–233
    of thermoplastic polymers   470
    of thermosets   497
Deformation bonding processes
    228–229
Deformation processes   210
Degradation temperature   472
Degree of polymerization   467
    in addition   467
    in condensation   468
    examples   468
    number average   469
    and properties   477
    weight average   469
    in wood   541
Dendrite   157, 178
    dendrite fraction   158
    dendrite size   159
Dendritic growth   157
Density   48, 63
    of elements   746–747
    of oxides   701, 710
    of polymers   478, 480, 488,
        493
Deoxidation   170, 296
Desulfurization   402
Devitrification   445
    in Pyroceram   450
Dezincification   669
Diamagnetism   611
Diamond cubic   61
Die casting   172
Dielectric constant   594
    and index of refraction   641
    and polarization   595
    of typical materials   597
Dielectric loss   602
Dielectrics   594–607
    controlling properties   598
    frequency effect   600
    structure effect   598
    temperature effect   601
    voltage effect   601

Dielectric strength   596
    of typical materials   597
Diffraction   59, 647
Diffused junction   585
Diffusion   93–115
    and carburizing   380
    comparison of types   108, 109
    in conductivity   586
    in eutectic   248
    in eutectoid   292
    grain boundary   108
    in grain growth   110
    and homogenization   198
    interstitial   94, 101
    in ionic crystals   113
    in peritectic   247
    in polymers   113
    in sintering   112
    surface   108
    vacancy   94, 101
    volume   108
Diffusion bonding   111
    in fiber composites   530
Diffusion coefficient   97
    effect of diffusion type   108,
        109
    effect of temperature   99, 100
Dihedral angle   277
Dimethyl terephthalate   465, 466
Dipole friction
    in dielectrics   602
    in magnets   616
Dipoles   591
    in dielectrics   592
    in magnets   608
Directions of a form   51
Dislocation   72
    in ceramics   424
    edge   72
    screw   72
Dispersion-strengthened compos-
    ites   507
    creep resistance   509
    effect of temperature   508
    examples   509
    requirements   509
Dispersion strengthening   239
    in composites   507
    in eutectic   252–257
    in eutectoid   294–300
    exceeding solubility   247, 255
    principles   239
Displacive transformation   421
Dissimilar metal corrosion   687
Domain   613
Donor exhaustion   575
Dopant   575
Drawing of cast iron   400
Drift velocity   552
Ductile fracture   711–714

Ductile iron   390, 401–404
    Larson-Miller parameter   142
    microstructure   403
    production   402
    properties   403
    specification   403
Ductility   131
Dye penetrant   739

Earing   224
Earthenware   447
Eddy current   616, 624
Eddy current test   737
Elastic constants   657
    effect of temperature   659
Elastomer   456, 487–492
    properties   488
    structure   488
Electrical conductivity   551
    anisotropic behavior   567
    and atomic bonding   31, 33
    comparison for materials   3
    in Cu-Ni alloys   563
    in Cu-Zn alloys   564
    in dielectrics   598, 602
    in ionic materials   586
    and lattice defects   562
    in Pb-Sn alloys   565
    vs. processing method   563
    of selected conductors   558
    of semiconductors   571
    vs. strengthening mechanism
        563
    and temperature   558
Electrical resistivity   551
    of dielectrics   597
Electric eye   646
Electric field   552, 591
Electrochemical cell   672
Electrochemical corrosion   672–
    699
Electrode potential   674–676
Electrolyte   672
Electromagnetic spectrum   633
Electromagnetic testing   739
Electromotive force (emf) series
    675, 676
Electron beam welding   174
Electronegativity   26
Electroplating   673, 679
Electrostriction   603
Elongation, %   131
Embeddability   334
Embryo   151
Emitter   583
Enamel   450
Endurance limit   137
Endurance ratio   138

Energy gap   570
    in absorption and transmission
        640
    in dielectrics   603
    in luminescence   637
    in photoconduction   646
    vs. pressure   581
    for selected materials   571, 578,
        579
Epitaxial growth   173
Epoxy   493, 497
Equiaxed zone   164
Error function   102
Ethane   460
Ethylene   460
Ethylene glycol   465, 466
Eutectic   242
    in cast irons   393
    cooling curve   248
    general properties   252
    microstructure   248, 254
    nonequilibrium   258
    in Pb-Sn alloys   245
    properties of Al-Si   257
    properties of Pb-Sn   255
    solidification   247
    use   244
Eutectic cells   397
Eutectic grains   252, 256
Eutectoid   242, 288
    control of properties   294
    in Cu-Al   332
    in Fe-C   291, 352
    in Ti   340
    use   244
Excessive reinforcement   724
Exchange current density   678,
    679
Explosive bonding   144, 229
    in laminar composites   539
Extenders   499, 516
External failure   603
Extrinsic semiconductor   575
Extrusion   210
    of ceramics   438, 440
    of polymers   500

Face-centered cubic   44
    characteristics   49
    close-packed directions   47
    sketch   45
Fading   403
Faraday's constant   677
Faraday's equation   676
Fatigue failure   716
Fatigue life   136
Fatigue properties   136
    in composites   533

Fatigue properties (continued)
    and corrosion   689
    notch sensitivity   138, 722
Fatigue strength   137
Feldspar   447
Fermi distribution   526
    in semiconductors   572
    temperature effect   557, 567
    in thermocouples   567
Fermi energy   556, 572
Ferrimagnetism   612
Ferrite   290, 351, 396
Ferroelectricity   604
Ferromagnetism   612
Fiber
    aspect ratio   523
    characteristics   523
    manufacture   528
    properties   525, 526
    temperature effect   527
Fiberglass   4, 499, 532
Fiber-reinforced composites   4,
        518–534
    Borsic-reinforced aluminum
        522
    estimating properties   518
    fatigue   533
    manufacture   529, 439
    orientation   519, 524
    properties   531, 532
Fiber texture   213
Fick's first law   97
Fick's second law   102
    in carburizing   380
    segregation   199
Fictive temperature   425
Figure of merit   230
Firing   441
First-stage graphitization   400
Flame retardant   499
Fluidity   320
Fluorescence   636
Fluorite   416, 418
Flux, in ceramics   432, 447, 451,
        670
Flux, in diffusion   97
Flux, magnetic   609, 736
Folded chain model   477
Formaldehyde   492, 494
Forsterite   419
Forward bias   582
Foundry molds   516
Fracture   711–721
Fracture mechanisms   739
Fracture toughness   741
Frank-Read source   209
Free energy
    formation of oxides   699
    strain energy   275

Free energy (*continued*)
  surface free energy 151, 275
  volume free energy 151, 275
Freezing range 188
  in aluminum alloys 321
  and shrinkage 200
Frenkel defect 82, 424
Friction welding 229
Fringed micelle model 477
Functionality 459
  in dacron 465
  in ethylene 459
  in phenol 460
  in urea 495
Furan 494
Fusion welding 174
Fusion zone 173, 175

Galvanic cell 687, 691
Galvanic series 687, 688
Galvanized steel 694
Gamma loop 383
Gamma ray radiography 732
Gas constant 99
Gas-metal arc welding 174
Gas-tungsten arc welding 174
  of titanium 344
Gels 425
Gibb's phase rule 189
  in NiO-MgO system 202
  in three-phase reactions 244
Glasses
  composition 450
  metallic 156, 620
  oxygen-silicon ratio 427
  in polymers 474
  short-range order 41
  silicates 426
  and soda 427
Glass formers 426
Glass forming 438, 439
Glass transition temperature 474,
    475
  and melting temperature 476
  and plasticizers 499
  and structure 475
Glaze 450
Grain 85
Grain boundary 85
  and corrosion 689
  energies 89
  liquid metal attack 669
Grain growth 110, 221
  temperature 223
Grain refinement 156
  in aluminum 320
Grain size strengthening 86,
    156

Graphite 419, 449
  in cast iron 391
  fibers 524, 528
Graphitic corrosion 669
Gravity segregation 254
Gray cast iron 390, 397
  and cooling rate 399
  microstructure 398
  properties 398
  specifications 398
Green sand mold 172
Grit 514
Grog 448
Grossman chart 375
Guinier-Preston zone 282, 316
Gutta-percha 484

Half-cell 674
Half-life 732
Hall effect 581
Hall-Petch equation 86
Hardenability 367
Hardenability curve 372–376
  for selected steels 373, 410
Hardening (of Al-Si alloys) 254,
    320
  microstructure 257
Hard magnet 617
Hardness
  Brinell 143
  comparison of materials 512
  Knoop 144
  microhardness 144
  Moh's 512
  Rockwell 144
  and strength 145
  Vickers 144
Hard surfacing 537
H coefficient 365
Heartwood 540
Heat-affected zone 175
  in age hardenable alloys 287
  cold-worked metals 225
  in stainless steel 698
  in steel 382
Heat capacity 649
Hemicellulose 541
Heterogeneous nucleation 155
  effect on grain size 156
Hexagonal close-packed 49
  characteristics 49
Hexagonal unit cell 44
Hexamethylene diamine 465
Holes 571
Homogeneous nucleation 152
Homogenization heat treatment
    198, 697
Honeycomb 545

Hooke's law 130, 657
Hot crack 724
Hot shortness 198
  in age hardenable alloys 281
  in eutectic alloys 259
Hot working 225
  characteristics 226
HSLA steels 377
Hume-Rothery rules 183
Hybridization 570
Hydrogen electrode 673, 675
Hydrogen embrittlement
  of copper 12
  of steel 725
Hydrogen peroxide 461–463
Hypereutectic alloy
  microstructure 250
Hypereutectoid alloy 292
Hypoeutectic alloy
  cooling curve 252
  microstructure 250
  solidification 249
Hypoeutectoid alloy 292
Hysteresis loop
  in anelastic materials 661, 662
  in magnetism 615
  in polarization 606

Impact energy 133
Impact properties 132
  and crystal structure 134
  notch sensitivity 134
  and temperature 133
  and true stress, strain 135
Impact test 132
Impressed voltage 695
Index of refraction 641, 642
  and dielectric constant 641
Inductance 609
Induction welding 229
Inertia welding 229
Ingot 162
Ingot structure 163
Inhibitors 695
Injection molding 500, 501
Inoculation 156
  in cast iron 397, 403
  eutectic grains 252
Insulator 570
Interatomic spacing 37
Interfacial energy 277
Intergranular corrosion 387, 688,
    697
Intergranular fracture 715
Interlamellar spacing
  eutectic 252
  eutectoid 296
  growth rate effect 253

Interlamellar spacing (*continued*)
  and properties of pearlite 297
  and strength 252
  and transformation tempera-
    ture 298
Intermediate oxide 426
Intermetallic compounds
  nonstoichiometric 240
  stoichiometric 240
Internal failure 603
Internal oxidation 510
Interplanar spacing 58
Interstitial 82
  in ceramics 415, 422
  sites in unit cell 415
Intrinsic semiconductor 571
Invar 538, 652
Investment casting 172
Ionic bond 33, 38
  in ceramic compounds 412
Isocyanate 495
Isomorphous phase diagram 187
  strength relationship 194, 347
Isopleth 266, 385, 389
Isoprene 484, 488
Isostatic pressing 440
Isotactic 484
Isothermal anneal 358
Isothermal heat treatment 357
  interrupted 360
Isothermal plot 263, 264, 435
Isothermal transformation 297
Isothermal transformation diagram.
  See TTT diagram
Isotope 19
Isotropy 58

Jiggering 440
Jominy distance 372
  and bar diameter 375, 409
  and cooling rate 374
  and plate thickness 376
Jominy test 372

Kaolinite 420, 442, 449
Kevlar 524, 532
Keying agent 532
Killed steel 170
Kirkendall effect 107
Kraft paper 598

Lack of fusion defect 724
Lamellar structure
  eutectic 247
  eutectoid 291
Laminants 536

Laminar composite 534–540
  bimetallics 537
  capacitors 535
  cladding 537
  hard surface 537
  laminates 536
  manufacture 539
  rule of mixtures 534
Larson-Miller parameter 142
Laser 638
Laser welding 174
Latent heat of fusion 153, 154
Lattice 42
Lattice parameter 43
  of metals 746–747
Lattice points 42
Leaching 669
Lever law 192, 266
Lignin 541
Liquid impingement 705
Liquid metal attack 669
Liquidus 187
Liquidus plot 262, 317, 435, 437
Linear absorption coefficient 639
Linear polymer 455
Long-range order 42, 416
Lorentz constant 654
Lumen 541
Luminescence 636

Magnesite 448
Magnesium alloys 322–324
  designation system 324
  properties 325
  and temperature 325
Magnetic field 581, 609
Magnetic moment 607
Magnetic particle test 736
Magnetic susceptibility 610
Magnetite 621
Magnetization 607–624
  application 615
  control 619
  and core material 612
  factors affecting 615
  and magnetic field 614
  properties 616, 617
  and temperature 618
Magnetometer 581
Malleable cast iron 390, 399
  heat treatment 400
  microstructure 401
  properties 401
  specifications 401
Manganese bronze 328
Maraging steel 377
Marquench 364, 365
Martemper 364, 365

Martensite 300
  in cobalt 300
  in Cu-Al alloys 333
  hardness versus carbon 303
  properties 302
  in steel 301
  tempering 304
  in titanium 304, 342
Martensite start and finish 301
  and alloying 371
  and carbon 363
Mass absorption coefficient 728
Materials processing defects 724
Materials processing methods 9
  casting 172
  of ceramics 438–446
  of composites 509–518, 528–
    530, 539–540
  forming 210
  joining 174, 229, 539
  of polymers 499–502
  powder metallurgy 112
  of semiconductors 585–586
Matrix 239
Mean free path 560
Melamine 493, 494
Melting temperature
  metals 746–747
  polymers 473
Memory alloy 304
Mer 456
Metallic bond 31, 38
Metallic glass 156, 620
Metasilicate 420
Methane 460
Mica 597
Microalloyed steel 377
Microconstituent 239
  calculation of 251
  eutectic 247
  eutectoid 291
  primary 251
  proeutectic 251
Microfibrils 541
Microvoids 711
Miller-Bravais indices 55
Miller indices
  for directions 50
  for planes 51
Mirror zone 720
Miscibility gap 260
Mobility 552
  in compounds 579
  in semiconductors 571
Modification 254, 320
  microstructure 256
  properties 255
Modifiers 426
Modulus of elasticity
  and atomic bonding 130, 657

Modulus of elasticity (*continued*)
  and melting temperature 130
  in piezoelectrics 604
  of polymers 478, 480, 493
  of selected metals 130, 538, 733
  in stress-strain diagram 124
Modulus of rigidity 659
Moh's hardness 512
Molar solution 675
Mold constant 158
Molybdenum 345
Monel 193, 334, 538
Monoclinic unit cell 44
Monomer 456
  for addition thermoplastics 478
  for condensation thermoplastics 480
Monotectic 242, 260
Montmorillonite 420, 423
Mottled cast iron 393, 394
Mullite 432, 448
Mylar 599

Natural aging 284
Necking 128, 486, 712
Neoprene 488
Nernst equation 675
Network polymer 455
Neutron radiography 732
Nickel alloys 334
  creep 142
Nickel silver 328
Niobium 345
  in stainless steel 388, 698
Nitriding 379
Nodulizing 402
Nondestructive testing 726–739
Nonequilibrium solidification
  eutectic alloys 258
  solid solution alloys 194
Nonstoichiometric semiconductor 578
Normalizing
  in cast iron 396
  in steel 355
Novalac 494
n-type semiconductor 575
Nucleation
  in cast iron 397, 400, 403
  heterogeneous 155
  homogeneous 152
  in solidification 151
  solid state 275
Nucleus 152
Nylon 465, 480, 522, 532

Octahedral site 415, 622
Offset yield strength 127

Ohm's law 551
Olivine 419, 423
  refractory 448
Orthorhombic unit cell 44
Orthosilicate 419, 422
Overage 282
  during use 285
  during welding 287
Overvoltage 678
Oxidation 699–703
  of refractory metals 345
Oxidation reaction 673, 700
Oxyacetylene welding 174
Oxygen concentration cell 691
Oxygen electrode 673
Oxygen starvation 691

Packing factor 47, 49
  diamond cubic 62
Paramagnetism 612
Parison 500
Particulate composite 4, 507–518
  dispersion strengthening 507–510
  effect of temperature 508
  true 511–518
Passivation 696
Passive (in corrosion) 696
Pauli exclusion principle 21, 554
Pearlite 291, 351
  in cast iron 396
  colonies 296
  growth 292
  and properties 296
  and transformation temperature 297
Periclase 448
Periodic table 27, 28
Peritectic 242, 259
  in Fe-C 259
  segregation 244, 260
  solidification 259
Peritectoid 242
  segregation 244
Permanent mold casting 172
Permeability 609
Permittivity 595
Perovskite structure 418
Phase 181
Phase diagram
  binary 187
  isomorphous 187
  ternary 261
Phase diagrams
  Al-Cu 276
  Al-Cu-Si 349
  Al-Mg 316, 323
  Al-Mg-Si 262, 317

Phase diagrams (*continued*)
  Al-Mn 316
  Al-Sb 241
  Al-Si 256
  Al-Zn 338
  Bi-Sb 204
  $CaO-Al_2O_3$ 433
  $CaO-SiO_2-Al_2O_3$ 435
  $CaO-SiO_2-Na_2O$ 437
  $CaO-ZrO_2$ 433
  Cu-Al 330, 332
  Cu-Be 330
  Cu-Ni 187
  Cu-Pb 260
  Cu-Si 330
  Cu-Sn 329
  Cu-Zn 329
  Fe-C 259, 289, 352
  Fe-Cr 383
  Fe-Cr-C 267, 385
  Fe-Cr-Ni-C 389
  $FeO-Al_2O_3-SiO_2$ 671
  $FeO-MgO-SiO_2$ 671
  Fe-Si-C 392
  Mg-Mn 323
  Mg-Si 323
  Mg-Sn 323
  Mo-Rh 241
  Mo-W 347
  Ni-Mo-Cr 264
  NiO-MgO 201
  Pb-Sn 245
  $SiO_2-Al_2O_3$ 431
  SiO_2-CaO 431
  SiO_2-MgO 431
  SiO_2-Na_2O 431
  Ti-Al 339
  Ti-Mn 270, 339
  Ti-Mo 339
  Ti-Sn 339
  Ti-V 350
Phenol 460
Phenolic 492, 493
Phonon 648
Phosphor bronze 328
Phosphorescence 636
Photoconduction 645
Photon 629
Piezoelectricity 604
Pigment 498
Pilling-Bedworth equation 700, 701
Pipe 164
Pitting corrosion 12, 691
Planar density 54, 79
Planar growth 156
Planck's constant 620
Planes of a form 54
Plasticizer 499
Plexiglass 478

Plywood 4, 536
p-n junction 582
Point defects 82
Poisson's ratio 657
Polarization 591
   in corrosion 678, 679, 684
   dimensional changes 593, 604
   and frequency 601
   mechanisms 592
   and temperature 601
   and voltage 601
Polyacrylonitrile 474, 528
Polyamide 480
Polybutadiene 488
Polybutylene 488
Polycarbonate 480, 532
Polychloroprene 488
Polychlorotrifluoroethylene 479
Polyester 465, 480, 493, 495
Polyether 480, 493
Polyethylene 456
   and clay 516
   corrosion 670
   crystalline 64
   growth rate 461
   initiation 461, 462
   stress rupture 472
   structure 457
   termination 461, 463
Polygonization 220
Polyimide 481
Polyisoprene 488
Polymer 456
Polymerization 2, 455–498
Polymethyl methacrylate 478
Polymorphism. See Allotropy
Polypropylene 457, 478
Polystyrene 478, 532, 670
Polyvinyl chloride 36, 478, 499
Polyvinylidene chloride 479
Porcelain 447
Porosity 168, 444
Portland cement 543
Potential difference 674
Pouring temperature 161
Powder metallurgy 112
   in cemented carbides 512
   in composites 509, 530
   in electrical contacts 514
   in fiber composites 530
   in magnets 621
Power (of magnet) 617
Precipitate 239
Precipitation hardening. See Age
   hardening
Precursor 528
Preferred orientation 213
Prepreg 530
Pressure transducer 581
Process anneal 353
Proof test 726

p-type semiconductor 577
Pull-out 721
Pulverization 112
Purple plague 108
Pyrex 428, 450
Pyroceram 450
Pyrolizing 528
Pyrometer 639
Pyrosilicate 419

Quantum numbers 20, 22
Quartz 421
Quench and temper
   in aluminum bronze 332
   in stainless steel 389
   in steel 362
   in titanium 342
Quench anneal 387, 697
Quench rate 281, 364

Radiation damage 703
Radiography 726–732
Recalescence 162
Reconstructive transformation 421
Recovery 220
Recrystallization 221
   grain size 223
Recrystallization temperature 221
   factors affecting 222
   for selected metals 223
Rectifiers 582
Reduction in area, % 132
Reduction reaction 673
Reflection 643
Reflectivity 643
   and index of refraction 643
Refraction 641
Refractories 446, 447
   compositions 448
Refractory metals 345
   coatings 345
   oxidation 345
   properties 345, 347
   types 345
Reinforced concrete 532
Relative permeability 610
Relative permittivity 595
Relaxation time
   in luminescence 637
   in stress relaxation 491
Remanence
   magnetism 614, 617
   polarization 606
Residual stress
   in aging 287
   in corrosion 690
   in deformation 216
   in quenched steel 364
Resistance polarization 685

Resistance welding 174
Resistivity 551
Resole 492
Resonance test 730
Retained austenite 363
Reverse bias 582
Rhombohedral unit cell 44
Rimmed steel 170
Ring scission 497
Riser 164
Rock candy fracture 715
Rockwell hardness 143
Roll bonding
   laminar composites 539
Rule of mixtures
   fiber composites 518
   laminar composites 534
   thermal conductivity 655
   true particulate 511
Rupture time (in creep) 140

Sacrificial anode 695
Safety glass 537
SAN copolymer 482
Sandwich 545
SAP 508, 509
Sapwood 540
Saturation magnetization 614
Saturation polarization 606
SBR rubber 488
Schmid's law 76
Schottky defect 82, 424
Secant modulus 659
Secondary dendrite arm spacing
   159
   in eutectic alloys 254
   and homogenization 199
   and properties 160
   and solidification time 159
Secondary hardening peak
   in stainless steel 387
   in tool steel 377
Second-stage graphitization 400
Segregation 197
   in age hardening 281
   in corrosion 688, 697
   interdendritic 197
   macrosegregation 200
   microsegregation 197
   in peritectic 260
Self-diffusion 93, 96
Semiconductors 570–586
Sensitivity 729
Sensitization 387, 697
Shear lip 712
Shear modulus 659
Sheet texture 213, 620
Shielded-metal arc welding 174
Short-range order 41, 412
Short wavelength limit 630

Shot peening 217
Shrinkage 164
  and freezing range 200
  interdendritic 168
  for selected metals 165
Sievert's law 169
Silica (SiO$_2$) 33, 63, 412, 421
Silicates 419
Silicon bronze 328
Silicon carbide 448, 513, 517
Silicone 488, 490, 493
Simple cubic 44
  characteristics 49
  close-packed directions 47
  sketch 45
Sintering 112, 441, 509
Sizing 532
Slag 451
Slip 74
Slip casting 440
Slip direction 74, 75, 80
Slip plane 74, 75, 80
Slip system 74
Slump test 544
Small angle grain boundaries 87
Soda-lime glass 429
Sodium chloride 34, 62, 416, 417
Soft magnet 615
Solar cell 6, 645
Solidification 151
  cast irons 390
  ceramics 433
  eutectic alloys 247
  hypereutectics 249
  monotectic 260
  nonequilibrium 194, 258
  peritectic 259
  pure metals 151
  solid-solution alloys 193, 246
Solidification time
  casting process 171
  and Chvorinov's rule 158
  and dendrite size 159, 160
  and interlamellar spacing 253
  local solidification time 161
Solid solution 182
Solid solution strengthening 83,
    184, 245
  and conductivity 187
  and creep 187
  and ductility 187
  in eutectic alloys 252
  factors affecting 184
Solidus 187
Solubility 181
Solubility limit 245
Solution treatment 281
Solvus 245, 247
Space charge 594
Specific heat 156, 648
  of selected metals 649

Specific modulus 523
  in fiber composites 531
  for fibers 525, 526
  of wood 543
Specific strength 523
  in fiber composites 531
  for fibers 525, 526
  and temperature 527, 534
  of wood 543
Spheroidite 356
Spinel structure 621
Spinnerette 500, 501
Spinning 500, 501
Splat cooling 156
Spring wood 540
Stabilization 388, 698
Stabilizer 498
Stacking faults 88
  energies 89
Stacking sequence 57
Stainless steel 383–390
  austenitic 386
  composition 384
  and corrosion 387, 688, 697
  ferritic 384
  martensitic 386
  precipitation hardening 389
  properties 384
Steel 288, 351–390
  and alloying elements 367, 368
  compositions 353
  isothermal heat treatment 357
  microstructure 353
  properties 295, 357, 372
  quench and tempering 362
  simple heat treatment 355
  stainless steel 383
  tool steel 377
Stellite 705, 706
Stiffness 130
Stoichiometric semiconductors
    578
Stoneware 447
Strain
  engineering strain 122
  true strain 128
Strain energy 275
Strain hardening 207–219
Strain hardening coefficient 208
Strength-to-weight ratio 313
  various materials 13
Stress 5
  engineering stress 122
  true stress 128
Stress cell 689
Stress corrosion 689, 719
Stress intensity factor 740
Stress relaxation 490
Stress relief anneal 220, 697
Stress-rupture curve 141, 336,
    509, 723

Stress-rupture failure 718
Stress-strain diagram
  for aluminum alloy 124
  and ductility 129
  engineering vs. true 129
  fiber composites 520, 526
  polymers 473
  and temperature 132, 133
  for titanium 149
Striations 717
Stringer 215
Submerged arc welding 174
Substitutional defect 82
Summer wood 540
Superalloy 334
  microstructure 337
  properties 335
  and temperature 336
Superconductors 568
  in composites 533
  in selected materials 569
Superheat 161
Superplasticity 231
  in titanium alloys 345
  in zinc-aluminum 338
Supersaturated solid solution 281
Surface defects 85
Surface energy 154
Surface hardening 378–380
Syndiotactic 484

Tafel constant 678, 679
Tangent modulus 659
Tantalum 345
  creep 142
Tapes 530
TD nickel 510
Teflon 479
Temperature resistivity coeffi-
    cient 561
  for selected metals 561
Temper designation 315
  for aluminum 317
  for copper 327
  for magnesium 322
Tempered glass 445
Tempered martensite 305, 362,
    371, 387
Tempering 304
  and alloy content 371
  and properties 305, 362
  in stainless steel 387
Tensile strength 128
Tensile test 121
Ternary phase diagram 261
Tetrafluoroethylene. See Teflon
Tetragonal unit cell 44, 55, 301
Tetrahedral site 415, 622
Tetrahedron 31, 32
  carbon 458

Tetrahedron (*continued*)
 diamond   61
 in silica   33, 413
Thermal arrest   161
Thermal conductivity   653
 in ceramics   655
 in metals   654
 in semiconductors   656
 table   655
 and temperature   656
Thermal emission   639
Thermal expansion   650
 and bond strength   651
 in hot working   227
 and melting temperature   651
Thermistor   580
Thermocouple   566
Thermoelasticity   661
Thermoplastics   4, 455–487
 forming   500
 monomers   478, 480
 properties   478, 480
 property vs. structure   487
Thermosets   4, 455, 492–498
 forming   501
 properties   493
 structure   493
Thermostats   538
Thixotropic   516
Thoria   510
Three-phase reaction
 identification   243
 phase rule   244
Through-transmission test   734
Tie line   189
 in eutectics   249
 finding martensite   302
 in ternary diagram   265, 436
Tie triangle   265, 435
Tilt boundaries   87
Tin bronze   328
Tin-plated steel   694
Titanium alloys   338
 processing   344
 properties   340
 and temperature   325, 341, 344
Tool steel   377
 tempering   372
 typical compositions   377
Toughness   133
Tow   528
Tracheids   541
Trans   484
Transducer   605
Transfer molding   501, 502
Transgranular fracture   711
Transistor   583
Transition metals   27, 609
Transition temperature   134, 346
Transmission   640

Transpassive region   697
Triclinic unit cell   44
Tridymite   421
True particulate composite   507–518
 cemented carbides   512
 compocasting   516
 electrical contacts   514
 polymers   515
 rule of mixtures   511
True porosity   444
TTT diagram   297
 and alloying element   367
 bay   371
 for eutectoid steel   298
 for 1050 steel   359
 for 10110 steel   359
 for 4340 steel   369
Tungsten   345, 533
Twin boundaries   88
 energies   89
Twist boundaries   87

Ultrasonic testing   733
Ultrasonic velocity   733
Ultrasonic welding   229
Underbead crack   724
Undercooling   152, 157
 in ceramics   425
 maximum during solidification   154
 in polymers   475
Undercut   724
Uniform attack   686
Unit cell   43
Unsaturated bond   458
 in cross-linking   487
Urea   493, 494
Urethane   493, 495
U.S. coinage   537

Vacancy   82
Vacuum forming   500, 501
Valence   23
Valence band   555
Van der Waals bond   35, 38
 in clay   420
 in polymers   472
 in PVC   36
Vapor decomposition   528
Vermicular graphite   404
Vinyl compounds   478
Vinylidene compounds   478
Viscoelasticity   470
Viscosity
 in glasses   428
 in polymers   470
 in stress relaxation   491
 temperature   429, 472

Viscous flow   428
 in elastomers   487
 in glasses   428
 in polymers   470, 472
 in stress relaxation   491
 temperature   472
 time   472
Visible spectrum   633
Vitrification   442
Vulcanization   487
Vycor   450

Wear   703–706
 abrasive   704
 adhesive   703
 erosion   705
Weldability   381
Welding
 aging   286
 cold-worked metal   225
 stainless steel   698
 steel   381
 titanium   344
Whiskers   518
 manufacture   529
 properties   525
White cast iron   390, 399
 microstructure   8, 399
White radiation   630
Widmanstatten structure   277, 342
Wire drawing   218
Wood   540
 properties   543
 structure   541
Workability   543
Working range   429
Wrought alloys   314
Wurtzite structure   416, 418

X-ray radiography   727
X rays   59, 533, 631, 647
 absorption   644

Yield strength   126
 double yield point   127
 offset yield strength   127
Young's modulus. *See* Modulus of
 elasticity

Zener diode   583
Zinc alloys   337
 intergranular corrosion   688
 properties   337
Zincblende structure   416, 418
Zircon   448
Zirconia   432, 448
Zirconium   345
Zone refining   585